DIFFERENTIAL EQUATIONS

A FIRST COURSE

Third Edition

DIFFERENTIAL EQUATIONS

A FIRST COURSE
Third Edition

Martin M. Guterman
Tufts University

Zbigniew H. Nitecki
Tufts University

SAUNDERS COLLEGE PUBLISHING
A Harcourt Brace Jovanovich College Publisher

Fort Worth   Philadelphia   San Diego   New York   Orlando   Austin
San Antonio   Toronto   Montreal   London   Sydney   Tokyo

Text Typeface: Times Roman
Compositor: Maryland Composition
Acquisitions Editors: Jay Ricci, Robert Stern
Managing Editor: Carol Field
Senior Project Manager: Marc Sherman
Manager of Art and Design: Carol Bleistine
Associate Art Director: Doris Bruey
Art and Design Coordinator: Caroline McGowan
Cover Designer: Lawrence R. Didona
New Text Artwork: Grafacon
Text Artwork: J&R Technical Services, Inc.
Production Manager: Bob Butler
Director of EDP: Tim Frelick
Product Manager: Monica Wilson

Cover Credit: © 1991 Bishop/Phototake, Inc., NYC

Printed in the United States of America

DIFFERENTIAL EQUATIONS: A FIRST COURSE, 3rd edition

ISBN 0-03-072878-9

Library of Congress Catalog Card Number: 91-053093

1234 016 987654321

THIS BOOK IS PRINTED ON **ACID-FREE, RECYCLED** PAPER

This book is an introduction to standard topics in differential equations for the average sophomore engineering or science student. The fundamental first two chapters treat first-order equations and linear constant-coefficient equations, including a brief view of 2×2 systems via elimination. The chapters that follow give mutually independent treatments of systems (linear and nonlinear), the Laplace transform, power series solutions, numerical methods, and Fourier series methods for partial differential equations. With an appropriate choice of topics, this book can serve as the basis for various courses to follow a two- or three-semester calculus sequence; the possibilities range from a basic survey of general methods for solving a single differential equation to an integrated introduction to linear differential equations and systems that incorporates the rudiments of linear algebra (without assuming previous exposure to determinants, matrices or vectors).

Our exposition is aimed at the beginning user of differential equations; it guides the reader through the underlying ideas of the subject while maintaining a hands-on experience of specific problems. The discussion proceeds from the concrete to the abstract by means of many worked-out examples and observation of general patterns. Boxed summaries reiterate the main points to remember, including explicit problem-solving procedures. These serve as handy reference points for the reader. The notes that follow the summaries discuss technical fine points and specific shortcuts or difficulties that arise in practice. Exercises at the end of each section are arranged roughly in order of increasing difficulty and abstraction; especially involved problems are starred. Review problems at the end of each chapter provide an opportunity for the reader to check his or her understanding of the chapter as a whole. The answers to odd-numbered exercises appear at the end of the book.

The introductory section at the beginning of each chapter considers a single class of physical models (*e.g.*, populations, springs, electrical circuits, or heat flow) as a practical motivation for the mathematical discussion that follows. This can be treated as independent reading when no class time is available for modeling or applications.

We have organized the book with an eye toward flexibility; possible arrangements of topics are sketched in the diagram on page viii. The chapter-by-chapter contents of the mathematical discussion are summarized as follows:

First-Order Equations (Chapter 1), motivated by population models, are handled primarily by separation of variables and variation of parameters. Optional sections, which can be skipped or deferred until later, discuss further applications, substitution methods, the Existence and Uniqueness theorem, graphing solutions, stability theory, and exact equations.

Linear Differential Equations (Chapter 2) are motivated by models involving damped springs and treated by the method of characteristic roots. Our treatment of the general theory (the Wronskian and linear independence of solutions) proceeds from concrete, hands-on examples to general principles, highlighting the relevance of the Existence and Uniqueness theorem to the solution of specific o.d.e.'s. The methods of undetermined coefficients and variation of parameters are treated as natural extensions of earlier techniques. Some further consideration of physical models and a brief treatment of 2×2 differential systems by elimination can be covered or skipped at the user's discretion. Appendix A discusses the calculation of determinants (of arbitrary size) and Cramer's rule from the ground up, for the benefit of readers who have not previously encountered this material.

Our discussion of **Linear Systems of Differential Equations** (Chapter 3) is motivated by electrical circuit models and exploits the analogies between the solution of a single differential equation and that of a system. This material occurs earlier than usual because we feel it forms a natural sequel to the discussion in Chapter 2. Basic tools from linear algebra are developed as the need for them arises. A fundamental computational tool in this chapter is row reduction of matrices (3.6), which in itself is one of the most useful techniques for a student of engineering or science to learn at this stage. The depth and generality of the treatment of systems can be controlled by the extent to which the material on complex eigenvalues (3.8) and generalized eigenvectors (3.9 and 3.10) is covered.

In a chapter new to this edition (Chapter 4), the **Qualitative Theory of Systems of O.D.E.'s** is motivated by a discussion of interacting population models which extends to 2×2 systems our treatment of stability for first-order equations from Section 1.8. The subsequent treatment of the qualitative theory of linear and nonlinear 2×2 systems does not depend on Section 4.1, but builds on the second-order case of linear systems from Sections 3.7 to 3.9. Our approach is geometric. The relation between characteristic roots and phase portraits of linear systems, together with the Hartman-Grobman linearization theorem (seldom found in treatments at this level), yield stability criteria for equilibria. Constants of motion and Lyapunov functions introduce a global perspective on the behavior of solutions. A final section treats limit cycles (van der Pol equation and Poincaré-Bendixson theorem) and points to the possibility of chaos in higher-order systems (Lorenz equations).

The motivation for Chapter 5, **The Laplace Transform,** comes from problems with discontinuous forcing terms. The chapter develops an operational calculus for handling initial-value problems via the Laplace transform, including the adaptation to systems.

The motivation for Chapter 6, **Linear Equations with Variable Coefficients: Power Series,** comes from the Cauchy-Euler, Legendre, and Bessel equations as they arise in temperature distribution problems. After a review of power series, we develop power series solution of ordinary differential equations with polynomial coefficients, including the Frobenius method.

In Chapter 7, **Numerical Approximation of Solutions,** we discuss the Euler, average-slope, and Runge-Kutta methods. We believe that it is both desirable and practicable to give students a hands-on experience of calculating numerical schemes on electronic devices. Our treatment is geared toward the use of computers. We discuss the ideas of each scheme, then formulate it in terms of a general algorithm, and finally implement this algorithm in a BASIC program. The deliberate simplicity of our programs insures their portability and accessibility to students with no previous experience in programming computers. An important section, which should be included in any treatment of numerical methods, is Section 7.6, treating examples of the limitations of numerical methods. The ideas of Chapter 7 are applied in Appendix B in a sketch of the proof of the Existence and Uniqueness theorem for ordinary differential equations.

Chapter 8, **Partial Differential Equations: Fourier Series,** introduces partial differential equations via the heat equation. Separation of variables and Fourier series (including sine series and cosine series) are used to solve various boundary-value problems. These techniques are then also applied to the one-dimensional wave equation and the two-dimensional Dirichlet problem. In the final section we show how expansion in other orthogonal families of functions can be needed to solve certain higher-dimensional problems.

Readers familiar with earlier editions of this book will note many changes throughout the text and exercises. There has been a net increase of 21 worked examples and 86 exercises over the previous edition. In particular, new examples have been added to theoretical discussions to further promote the reader's concrete understanding. At the same time, several esoteric notes have been cut to keep the discussion focused and accessible.

Among the other changes in this edition are the following:

1. The treatment of general theory for nth order linear o.d.e.'s (Sections 2.2 to 2.4) has been streamlined, and the treatment of determinants has been moved to Appendix A.
2. Coverage of the qualitative theory of systems has been expanded to a new chapter (Chapter 4).
3. A new section has been added to Chapter 1 (Section 1.5) demonstrating the usefulness of substitutions in the solution of o.d.e.'s.
4. The treatment of chemical reactions in Section 1.4 has been rewritten to apply to more general reactions.

Many people have contributed to this book directly or indirectly by their advice, general interest, and moral and material support. We remain indebted to the many individuals who contributed to the first and second editions of this book and to *Differential Equations with Linear Algebra*. We have benefited greatly from the input of our colleagues and students, especially Bill Reynolds, George Leger, Fulton Gonzalez, and Andrew Hyman. In preparing the present edition, we were greatly aided by the thoughtful comments of Frank Chole-winski, Lowell Hansen, Stephen Merrill, Mary Beth Ruskai, and Rolland Trapp. We are particularly grateful to our colleague Bill Schlesinger for the new computer-generated figures which grace this volume. Finally, we thank the editorial and production staff at Saunders College Publishing, including Bob Stern, Marc Sherman, Doris Bruey, Linda Davoli, Monica Wilson, and Jay Ricci.

We dedicate this book to
Sonia, Lila, Beth,
Alicia, and Lily
with love and gratitude.

M.G.
Z.N.

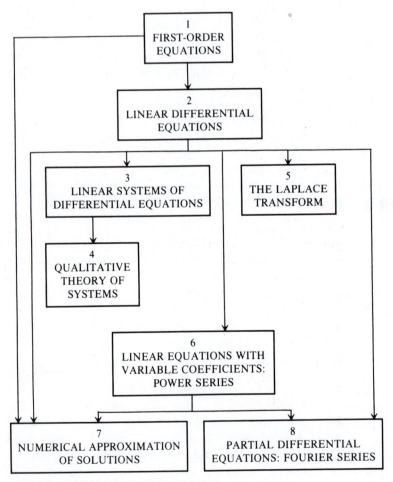

POSSIBLE ARRANGEMENT OF TOPICS.

C O N T E N T S

C H A P T E R

1

FIRST-ORDER EQUATIONS

1.1 INTRODUCTION

In the late seventeenth century, Isaac Newton in England (1665, 1687) and Gottfried Wilhelm Leibniz in Germany (1673) synthesized several centuries of mathematical thought to create a language and method for describing and predicting the motion of bodies in various physical situations. The invention of calculus was immediately followed by a period of intense mathematical activity, and the effect of these ideas on the development of mathematics, science, and technology makes this event surely one of the most important in the history of Western thought. During the development of calculus, differential equations and their solutions played the central role. They arose as mathematical formulations of physical problems, and attempts at their solution motivated much of the mathematical development of calculus.

The role of differential equations in the modeling of physical phenomena is well illustrated by **Newton's second law** of motion, familiar to all physics students in the mnemonic form

$$F = ma.$$

In the situation that most interested Newton (gravity), the force F is the weight of the body, the constant m is its mass, and a is its acceleration. Although we are really interested in the position of the body, the equation tells us about neither the position nor its rate of change, but rather about the rate of change of the rate of change of the position. In the language of calculus, if $x = x(t)$ represents the position of the body at time t, then the velocity is the derivative of x, $v = dx/dt$, and the acceleration is the derivative of the velocity, $a = dv/dt = d^2x/dt^2$. Thus, Newton's second law

$$F = m\frac{d^2x}{dt^2}$$

is an equation involving a derivative of the interesting variable—that is, it is a **differential equation.**

It was Newton's brilliant observation that in many physical situations the relation between rates of change of observable quantities is simpler than the relation between the quantities themselves. This is at the same time the source of the power of differential equations and the central problem in using them to predict physical phenomena. For if Newton's second law is to lead to useful physical predictions, we must translate this statement about the second derivative of position into a prediction of the position of the body at some time in the future; that is, we must express x as a function of time:

$$x = \phi(t).$$

In general, the highest order of differentiation that appears in a differential equation is called the **order** of the equation. Thus, Newton's law is a second-order differential equation. A **solution** of an nth-order differential equation is a function $x = \phi(t)$ with derivatives at least up to order n, that when substituted into the equation yields an identity on the domain of definition of $\phi(t)$. The problem of obtaining solutions to a given differential equation is a purely mathematical one and forms the subject of this book.

Example 1.1.1

Determine which of the functions

$$x_1 = 16t^2, \quad x_2 = 16t^2 + t - 1, \quad x_3 = t^2 + t - 1$$

are solutions of the second-order differential equation

$$\frac{d^2x}{dt^2} = 32.$$

Note that this equation is a special case of Newton's second law. If distance is measured in feet and time in seconds, then the equation models the motion of a ball acted on by gravity and no other forces.

The derivatives of x_1, x_2, and x_3 are

$$\frac{dx_1}{dt} = 32t, \quad \frac{dx_2}{dt} = 32t + 1, \quad \frac{dx_3}{dt} = 2t.$$

The second derivatives are

$$\frac{d^2x_1}{dt^2} = 32, \quad \frac{d^2x_2}{dt^2} = 32, \quad \frac{d^2x_3}{dt^2} = 2.$$

Thus, x_1 and x_2 are solutions of our differential equation, whereas x_3 is not.

Example 1.1.2

Find all values of k for which

$$x = e^{kt}$$

is a solution of the third-order differential equation

$$\frac{d^3x}{dt^3} - \frac{dx}{dt} = 0.$$

The first three derivatives of $x = e^{kt}$ are

$$\frac{dx}{dt} = ke^{kt}, \quad \frac{d^2x}{dt^2} = k^2e^{kt}, \quad \frac{d^3x}{dt^3} = k^3e^{kt}.$$

Substitution into the differential equation gives

$$k^3e^{kt} - ke^{kt} = 0.$$

In order for this to be an identity, we need

$$k^3 - k = 0.$$

This equation has three roots:

$$k = 0, \quad k = 1, \quad k = -1.$$

Thus, the desired solutions of the differential equation are

$$x = e^{0t} = 1, \quad x = e^t, \quad x = e^{-t}.$$

Although guesswork and substitution play a role in solving differential equations, we will spend most of our time discussing more systematic approaches. We will be especially interested in methods for finding a formula that describes *all* the solutions of a given differential equation. We refer to such a formula as the **general solution** of the equation. If a differential equation has the special form

$$\frac{d^nx}{dt^n} = E(t),$$

then we can obtain the general solution by integrating n times.

▮ Example 1.1.3

Find the general solution of the differential equation

$$\frac{d^2x}{dt^2} = 32.$$

Integrating both sides of the differential equation with respect to t

$$\int \frac{d^2x}{dt^2}\, dt = \int 32\, dt$$

gives

$$\frac{dx}{dt} = 32t + c_1.$$

Now we integrate again to obtain the general solution:

$$\int \frac{d^2x}{dt^2}\, dt = \int (32t + c_1)\, dt$$

$$x = 16t^2 + c_1 t + c_2.$$

The general solution obtained in Example 1.1.3 involves two "arbitrary constants," c_1 and c_2, which resulted from taking two indefinite integrals. The physical significance of these constants becomes clearer when we think of this equation as representing the motion of a falling ball under the force of gravity. Whereas the differential equation takes into account the gravitational force on the ball, this is hardly enough to predict the ball's position. We need to know where it started from and whether it was dropped or thrown. Without such information, we can make only a general prediction, vague enough to apply to all possible circumstances of the ball. To make a **specific** prediction without ambiguity, we need to know the initial position (the value of x when $t = 0$) and the initial velocity (the value of $x' = dx/dt$ when $t = 0$). Note that the initial position is

$$x(0) = 16(0)^2 + c_1(0) + c_2 = c_2,$$

and the initial velocity is

$$x'(0) = 32(0) + c_1 = c_1.$$

We see that in this case the values of the two arbitrary constants in the general solution are numerically equal to the **initial conditions** that determine a specific solution. We shall consider the role of initial conditions in determining specific solutions as we study the various kinds of differential equations.

Of course, the process of finding the general solution in Example 1.1.3 was very easy. Before we turn our attention to more complex examples we consider some population models, with an eye toward understanding the different kinds of differential equations that can arise in modeling various phenomena. We will solve the equations of these examples later in the text. Other physical models leading to similar differential equations are considered in the exercises at the end of this section and in Section 1.4.

Example 1.1.4

The simplest model of population growth was first formulated by the English economist Thomas R. Malthus in 1798. He assumed that a population changes at a rate proportional to the size of the population. If $x = x(t)$ stands for the number of individuals in the population after t years, then this means that

$$\frac{dx}{dt} = \lambda x,$$

where λ is a constant. We refer to

$$\lambda = \frac{1}{x}\frac{dx}{dt}$$

as the **per capita growth rate** of the population. For example, a population that grows at the annual per capita rate of 5% would satisfy the first-order differential equation

$$\frac{dx}{dt} = \frac{5}{100}x.$$

Example 1.1.5

The rate of change of the population in the previous example was a multiple of the population. In some cases there may be an additional component (immigration) that depends only on time.

Suppose a disease-causing organism reproduces in its host with a daily per capita growth rate of 100%. Suppose also that its presence causes the host's resistance to deteriorate, so that on the tth day after the initial infection, t

thousand organisms are able to enter the host from the surrounding environment. Let $x = x(t)$ be the number of organisms in the host, measured in thousands. Then the rate of change of x has two components: reproduction contributes x to dx/dt, and new organisms entering from the surrounding environment contribute t. The total rate of change is

$$\frac{dx}{dt} = x + t.$$

This equation, like the previous one, is of first order. Note that here the variable t appears explicitly on the right side.

Example 1.1.6

Another model of population growth was formulated in 1837 by the Belgian mathematical biologist P. Verhulst to deal with the effects of a limited environment. This model can be obtained from the one in Example 1.1.4 by modifying our assumptions about the per capita growth rate $(1/x)(dx/dt)$.

Suppose a population consisting of $x = x(t)$ thousand organisms would, in an unlimited environment, have a per capita growth rate of 5% per year. Assume the environment is limited and can support at most a population of 10 thousand. Then as the population approaches 10 thousand, we would expect the per capita growth rate to decline. The simplest way to take account of this fact is to multiply the unlimited per capita growth rate by a factor that approaches zero as x approaches 10; the simplest such factors are constant multiples of $10 - x$. Thus, we expect our population to satisfy

$$\frac{1}{x}\frac{dx}{dt} = 0.05\alpha(10 - x)$$

where α is a constant. Since a small population experiences little competition, the limited environment per capita growth rate should approach the unlimited per capita growth rate as x approaches zero:

$$\lim_{x \to 0} 0.05\alpha(10 - x) = 0.05.$$

It follows that $\alpha = 1/10$. Our model has the equation

$$\frac{1}{x}\frac{dx}{dt} = 0.005(10 - x)$$

or

$$\frac{dx}{dt} = 0.005(10 - x)x.$$

This equation, like the ones in Examples 1.1.4 and 1.1.5, is of first order. Unlike the others, the expression for dx/dt involves an x^2 term.

Example 1.1.7

Suppose two neighboring countries, with populations $x_1(t)$ and $x_2(t)$, have natural per capita growth rates (birth rate minus death rate) of 15% and 10%, respectively. Suppose that 4% of the first population moves to the second country each year, whereas 3% of the second population moves to the first country each year. Then the rate of change of each of the populations is made up of three components: the natural growth rate, emigration, and immigration. Thus

$$\frac{dx_1}{dt} = 0.15x_1 - 0.04x_1 + 0.03x_2 = 0.11x_1 + 0.03x_2$$

$$\frac{dx_2}{dt} = 0.10x_2 - 0.03x_2 + 0.04x_1 = 0.04x_1 + 0.07x_2.$$

In this case we have a **system** of two differential equations, each involving both variables x_1 and x_2 in an unavoidable way.

Each example so far has involved only ordinary derivatives (as opposed to partial derivatives). We refer to such equations as **ordinary differential equations** (abbreviated **o.d.e.'s**). An equation like

$$\frac{\partial u}{\partial t} = c \frac{\partial^2 u}{\partial x^2},$$

which involves partial derivatives, is called a **partial differential equation (p.d.e.)**.

In this book we concentrate primarily on a special class of o.d.e.'s and systems (linear ones) for which a systematic solution procedure can be formulated and which are used in a broad variety of physical models. The precise delineation of this class will occur piecemeal as we study various specific instances.

In this chapter we look at first-order equations, like the ones in Examples 1.1.4 through 1.1.6. These are the simplest from the point of view of calculus, since they involve only first derivatives. Yet a large variety of phenomena can be described by using such equations.

We close this section with a summary of the basic definitions that play a large role throughout the book.

SOME BASIC DEFINITIONS

An **ordinary differential equation** (abbreviated **o.d.e.**) is an equation whose unknown x is a function of one independent variable t. The equation relates values of x and its derivatives to values of t.

The **order** of an o.d.e. is the highest order of differentiation of x appearing in the equation.

A **solution** of an nth-order o.d.e. is a function $x = \phi(t)$, with derivatives at least up to order n, that when substituted into the o.d.e. yields an identity on the domain of definition of $\phi(t)$.

The **general solution** of an o.d.e. of order n is a formula (usually involving n arbitrary constants) that describes all solutions of the equation. A **specific solution** (or, equivalently, the value of each of the constants in the general solution) is determined by certain **initial conditions,** such as the starting point and the starting velocity.

EXERCISES

1. Determine the order of each of the following o.d.e.'s.

   a. $t^4 \dfrac{d^3x}{dt^3} + t \dfrac{dx}{dt} - x = t^7$

   b. $\left(\dfrac{dx}{dt}\right)^5 + \dfrac{d^4x}{dt^4} - t^3x^7 + t^7 = 0$

   c. $x^8 \dfrac{dx}{dt} + \dfrac{d^7x}{dt^7} = x + t^9$

   d. $(x')^2x''' = x^4x'' + t^5x'$

In Exercises 2 through 6, check to see whether the given function $x = \phi(t)$ is a solution of the given o.d.e.

2. $x = t^5$; $\dfrac{d^2x}{dt^2} - \dfrac{20x}{t^2} = t^3$

3. $x = e^{3t}$; $\dfrac{d^3x}{dt^3} - 9\dfrac{d^2x}{dt^2} = 0$

4. $x = te^{3t}$; $\dfrac{d^2x}{dt^2} - 9\dfrac{dx}{dt} = 6e^{3t}$

5. $x = \ln(-t)$, $t < 0$; $tx' = 1$

6. $x = \begin{cases} t^2, & t > 0 \\ 3t^3, & t < 0 \end{cases}$; $tx' - 2x = 0$

In Exercises 7 through 12, find all values of the constant k for which the given function $x = \phi(t)$ is a solution of the given o.d.e.

7. $x = t^k$, $t > 0$; $t^2 x'' - 6x = 0$ 8. $x = e^{kt}$; $\dfrac{d^2 x}{dt^2} - x = 0$

9. $x = k$; $\dfrac{d^7 x}{dt^7} + \dfrac{dx}{dt} - x = 7$ 10. $x = t^k$, $t > 0$; $16t^2 xx'' + 3x^2 = 0$

11. $x = ke^t$; $\dfrac{d^2 x}{dt^2} + 5\dfrac{dx}{dt} = 3e^t$ 12. $x = kte^{3t}$; $x'' - 3x' = e^{3t}$

In Exercises 13 through 17, find (a) the general solution of the o.d.e. and (b) the specific solution satisfying the given initial condition.

13. $\dfrac{d^2 x}{dt^2} = 3t + 1$; $x(0) = 2$, $x'(0) = 3$

14. $\dfrac{d^2 x}{dt^2} = \dfrac{-1}{(t + 1)^2}$, $t > -1$; $x(0) = 2$, $x'(0) = 3$

15. $x''' = 6$; $x(1) = x'(1) = x''(1) = 0$

16. $\dfrac{d^3 x}{dt^3} = e^{-t}$; $x(0) = x'(0) = x''(0) = 1$

17. $\dfrac{d^2 x}{dt^2} = te^t$; $x(0) = x'(0) = 0$

Exercises 18 through 27 describe some more models leading to first-order o.d.e.'s.

18. *Compound Interest:* When interest is compounded continuously, the rate of change of the principal is proportional to the principal. The constant of proportionality is called the *interest rate.* Set up a differential equation to model the principal $x = x(t)$ in an account accruing interest at 8% per year, compounded continuously.

19. *Savings:* A savings account pays 8% interest per year, compounded continuously. In addition, the income from another investment is credited to the account continuously, at the rate of $400 per year. Set up a differential equation to model this account.

20. *Radioactive Decay:* The atoms of a radioactive substance tend to decompose into atoms of a more stable substance at a rate proportional to the number $x = x(t)$ of unstable atoms present. Set up a differential equation for $x = x(t)$.

21. *Drug Absorption:* A drug is absorbed by the body at a rate proportional to the amount $x(t)$ present in the bloodstream. Set up an o.d.e. for the amount of the drug in the bloodstream of a patient if the drug is absorbed at the rate of $0.5x(t)$ per hour, and if the patient is simultaneously receiving the drug intravenously at the constant rate of 15 milligrams per hour.

22. *Varying Dosage:* Suppose the doctor adjusts the dosage of the drug in the preceding exercise so that the rate at which it is added to the patient's bloodstream declines at a constant rate from 15 milligrams per hour at the outset to zero 12 hours later. Set up an o.d.e. for the amount of drug in the bloodstream during this 12-hour period.

23. *Chemical Solutions:* Fifty pounds of a chemical are added to a tank of water and gradually dissolve. Assume the water is constantly stirred so that the concentration of the chemical is uniform throughout the tank.
    a. If the tank is very large, then the chemical enters into the solution at a rate proportional to the amount $x(t)$ of the chemical not yet dissolved. Set up an o.d.e. for $x(t)$ if the chemical dissolves at the rate of $0.25x(t)$ per hour.
    b. If the tank is small, then, as more and more of the chemical dissolves, the solution approaches saturation (the point at which no more of the chemical can enter into solution). As this happens, the rate at which the chemical dissolves tends to zero. Assume that at saturation the solution would contain 25 lb of the chemical. Modify the o.d.e. from part (a) to account for the small size of the tank by multiplying the right side with a constant multiple of the difference between this number and the amount $a(t)$ of the chemical already in solution. To determine the value of the constant, use the fact that when $a(t)$ is close to zero, the chemical should behave almost as if it were in a large tank.

24. *A Mixing Problem:* A tanker carrying 100,000 gallons of oil runs aground off Nantucket. Water pours into the tanker at one end at 1000 gallons per hour while the polluted water–oil mixture pours out at the other end, also at 1000 gallons per hour. Set up a differential equation to predict the amount $x = x(t)$ of oil in the tanker. (*Hint:* What percentage of the polluted mixture is oil?)

25. *Memorization:* Empirical studies suggest that each person has a certain maximum number of symbols that he or she can remember at any one time; the rate at which new symbols can be memorized is proportional to the difference between this maximum and the number of symbols already memorized. Set up a differential equation to model the rate at which a person can memorize new symbols.

26. *Prices:* Assume that the rise in the price $p = p(t)$ of a product is proportional to the difference between the demand $w(t)$ and the supply $s(t)$ (known as the excess demand) and that the demand depends on the price as a first-degree polynomial. Set up a differential equation for the price.

27. *Water Skiing:* A water skier W is pulled by a 20-ft rope attached to the end of a boat B (see Figure 1.1). The boat moves along the x-axis at a constant speed, and the rope is always taut. Initially, the boat is at the origin O and the skier is at the point $(0, 20)$.

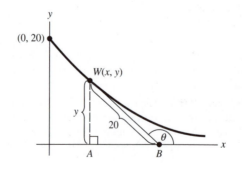

FIGURE 1.1

   a.   Find the length of *AB* in terms of *y*.
   b.   Find tan θ in terms of *y*.
   c.   Use the fact that the rope is tangent to the curve traveled by the water skier to obtain an expression for *dy/dx* in terms of *y*.

The curve traveled by the water skier is called a **tractrix.**

1.2 SEPARATION OF VARIABLES

As we saw in Section 1.1, certain differential equations can be solved by the simple device of "integrating both sides." In this section we explore an extension of this technique to a large class of first-order equations.

The method we will discuss is most easily described with the use of the notational convention of **differentials.** If we are given a function $x = x(t)$, then we introduce a new variable *dt*, and we define the differential *dx* by the rule

$$dx = \frac{dx}{dt}\, dt.$$

For example, if $x = \sin t$, then $dx/dt = \cos t$ so that $dx = \cos t\, dt$. Our manipulation of differential expressions in this section may seem cavalier to those of you who did not use differentials in calculus. The notes at the end of the section indicate some dangers as well as some justification of these manipulations.

Let's begin our discussion with an example.

Example 1.2.1

Solve the equation

$$\frac{dx}{dt} = x^2.$$

We multiply both sides by *dt* to get

$$dx = x^2\, dt.$$

If we try to integrate both sides, we run into a problem. Since *x* is a function of *t*, we can't integrate the right side unless we have an expression for x^2 in terms of *t*. Fortunately, we can play a little with the differential form before rushing in with an integral sign. If we divide both sides of the last equation by x^2, we obtain

$$\frac{1}{x^2}\, dx = dt$$

which *is* integrable in a perfectly reasonable way:

$$\int \frac{1}{x^2}\, dx = \int dt$$

$$\frac{-1}{x} = t + c$$

$$x = \frac{-1}{t + c}.$$

When we divided by x^2, we implicitly assumed $x \neq 0$. It is easy to check that

$$x = 0$$

is also a solution of the equation.

The method we used in Example 1.2.1 can be applied to any o.d.e. that can be written in the form

$$\frac{dx}{dt} = f(x)g(t).$$

To guarantee that the integrals that arise exist, and to avoid other technical problems (see Section 1.6), we restrict our attention to intervals of t-values on which $g(t)$ is continuous, and we look for solutions whose values lie in intervals of x-values on which $f(x)$ and $f'(x)$ are continuous. We begin by multiplying by dt to rewrite the o.d.e. in differential form

$$dx = f(x)g(t)\, dt.$$

We then divide by $f(x)$ to get all terms involving x on one side of the equation and all terms involving t on the other:

$$\frac{1}{f(x)}\, dx = g(t)\, dt.$$

Now that we have separated the variables, we integrate both sides,

$$\int \frac{1}{f(x)}\, dx = \int g(t)\, dt$$

to obtain an expression of the form

$$F(x) = G(t) + c.$$

(Note that a constant of integration need only be introduced on one side.) We can then try to solve this equation to obtain an explicit expression for x as a function of t; in cases where this is difficult or impossible, we content ourselves with the preceding formula, which describes x implicitly.

As in the example, when we divided by $f(x)$ we assumed $f(x) \neq 0$. If there is a number r such that $f(r) = 0$, then $x = r$ is also a solution of our o.d.e.

O.d.e.'s that can be written in the form $dx/dt = f(x)g(t)$ are said to be **separable.** The method we have described for dealing with them is called **separation of variables** and was first formulated by Johann Bernoulli in 1694.

Let's consider some more examples.

Example 1.2.2

Solve

$$\frac{dx}{dt} = \lambda x,$$

where λ is a constant.

We multiply by dt, divide by x, and integrate:

$$dx = \lambda x\, dt$$

$$\frac{1}{x}\, dx = \lambda\, dt$$

$$\int \frac{1}{x}\, dx = \int \lambda\, dt$$

$$\ln|x| = \lambda t + c.$$

We can exponentiate both sides of this equation to get

$$|x| = e^{\lambda t + c} = e^c e^{\lambda t}.$$

Since c represents an arbitrary constant, e^c represents an arbitrary positive constant. We can write

$$|x| = k e^{\lambda t}$$

as long as we remember that k is positive. Solving for x gives us an ambiguity

$$x = \pm k e^{\lambda t}$$

that can be resolved by allowing k to also be negative. Finally, noting that we divided by x, we check that $x = 0$ is a solution of our equation. This solution would be included in the preceding formula if we allowed $k = 0$. Thus, the formula

$$x = ke^{\lambda t},$$

where k represents an arbitrary constant, constitutes a description of all solutions of the equation; it is the general solution of our o.d.e.

Note that when $t = 0$ we have

$$x(0) = ke^0 = k.$$

It might be more suggestive to call this value x_0 and to write the general solution in the form

$$x = x_0 e^{\lambda t}.$$

Example 1.2.3

Solve the equation

$$\frac{dx}{dt} = \frac{t}{t^2 + 1} x.$$

We multiply by dt to obtain the differential form, separate the variables by dividing by x, and integrate:

$$\frac{1}{x} dx = \frac{t}{t^2 + 1} dt$$

$$\int \frac{1}{x} dx = \int \frac{t}{t^2 + 1} dt$$

$$\ln|x| = \frac{1}{2} \ln(t^2 + 1) + c.$$

We exponentiate the resulting equation and solve for x:

$$x = \pm e^c (t^2 + 1)^{1/2} = k(t^2 + 1)^{1/2}.$$

We check that $x = 0$ is a solution of our equation and note that it is included in our formula for x if we allow $k = 0$. When $t = 0$, $x(0) = k$, so we can write the general solution in the form

$$x = x_0(t^2 + 1)^{1/2}.$$

This problem models the growth of a population with varying per capita growth rate $t(t^2 + 1)$. Note that even though the per capita growth *rate* is eventually decreasing in this model, the population *size* continues to increase indefinitely.

Let's summarize our method.

SEPARATION OF VARIABLES

Given a **separable** first-order o.d.e.

$$\frac{dx}{dt} = f(x)g(t):$$

1. Multiply by dt to obtain the differential form

$$dx = f(x)g(t)\, dt.$$

2. Divide by $f(x)$ to separate the variables

$$\frac{1}{f(x)}\, dx = g(t)\, dt.$$

3. Integrate the separated form

$$\int \frac{1}{f(x)}\, dx = \int g(t)\, dt$$

to obtain an equation

$$F(x) = G(t) + c.$$

4. Try to solve the preceding equation for x in terms of t.
5. Check to see if any of the roots of $f(x) = 0$ yield constant solutions $x = r$ that were overlooked in steps 1 through 4.

Notes

1. An explanation

Our formal manipulation of dx and dt can be justified quite rigorously. The basis for this justification is the chain rule: if $y = F(x)$ and $x = x(t)$, then

$$\frac{dy}{dt} = \frac{dy}{dx}\frac{dx}{dt} = F'(x)\frac{dx}{dt}.$$

This says that if $F(x)$ is a function whose derivative is $f(x)$, then substituting $x = x(t)$ will give a function of t whose derivative is $f(x(t))x'(t)$. This translates into the integral statement

$$\int f(x) \frac{dx}{dt} \, dt = \int f(x) \, dx.$$

Thus, at any stage of our manipulations, a formal equation involving dx on one side and dt on the other can be interpreted rigorously by replacing the dx by $(dx/dt) \, dt$ and putting in integral signs. Our individual manipulations are justified by the fact that integrals of equal functions are equal (up to a constant of integration), and our various multiplications and divisions do not change the equality of the integrands on either side.

An advantage of the differential notation is its symmetry. Sometimes our final answer is more naturally interpreted as an expression for t in terms of x, and the prediction $x = x(t)$ is only given implicitly. The differential notation helps us ignore this problem until after we integrate.

2. A warning

Differential notation does not make sense if we have either a $(dx/dt)^2$ or a higher derivative.

Blind formal manipulation of $(dx/dt)^2$ as if it were $(dx)^2/(dt)^2$ leads to nonsensical expressions like $\int A(x)(dx)^2$. Of course, sometimes an equation involving higher powers of dx/dt can be solved for dx/dt and then can be separable. Any such algebra should be performed *before* switching to differential notation.

It is a common mistake, but a serious one, to think that higher derivatives can be interpreted as quotients of differentials or powers of differentials. This also leads to total nonsense. *Separating variables is strictly a first-order method*, and even then it doesn't always work.

EXERCISES

In Exercises 1 through 14, solve the o.d.e. by separation of variables. (You may omit step 4 of the summary if necessary.)

1. $3 \dfrac{dx}{dt} = 2x$

2. $3 \dfrac{dx}{dt} = 2x + 1$

3. $e^{2t}x' + e^t = 1$

4. $\sin t \, dx + \cos t \, dt = 0, \quad 0 < t < \pi$

5. $(x + 1)(t^2 + 1) = t \dfrac{dx}{dt}, \quad t > 0$

6. $(t^2 - 1)x' + 2x = 0, \quad t > 1$

7. $(10x^4 + 6)x' = x^5 + 3x + 2$

8. $x' = 2tx^2 + x^2 + 8xt + 4x + 8t + 4$

9. $\cos^2 2t \dfrac{dx}{dt} - 4x^2 = 0, \quad -\dfrac{\pi}{2} < t < \dfrac{\pi}{2}$

10. $x^2 \cos^4 3t \dfrac{dx}{dt} - \sin 3t = 0, \quad -\dfrac{\pi}{6} < t < \dfrac{\pi}{6}$

11. $(t^2 - t - 2)x' - x = 0, \quad t > 2$ 12. $(e^t + 1)x' = xe^t$

13. $x \dfrac{dx}{dt} = 4te^{2t+x}$ 14. $\dfrac{dx}{dt} = (x \sin t)^2$

In Exercises 15 through 21, find the specific solution that satisfies the given initial conditions.

15. $3 \dfrac{dx}{dt} + 5x = 0, \quad x(0) = 3$ 16. $(t + 1)x' + tx = 0, \quad x(0) = 2$

17. $t^2 \dfrac{dx}{dt} = x^2, \quad x(1) = 2$ 18. $x^2 \dfrac{dx}{dt} = t^2, \quad x(1) = 2$

19. $t^2 \dfrac{dx}{dt} = x^2 + 1, \quad x(1) = 0$ 20. $e^{-t} \dfrac{dx}{dt} + tx^2 = 0, \quad x(0) = 1$

21. $\dfrac{dx}{dt} = \sqrt{xt}, \quad x(9) = 4$

Exercises 22 through 30 refer to the models in Section 1.1.

22. *A Mixing Problem:* Solve the differential equation of Exercise 24, Section 1.1, modeling a mixing problem. How much oil is left in the tanker after 10 days (240 hours)?

23. *Radioactive Decay:*
    a. Solve the differential equation of Exercise 20, Section 1.1, modeling radioactive decay.
    b. Find an expression for the constant of proportionality in terms of the *half-life* of the substance (the time it takes for half of a given sample to decay).
    c. The half-life of white lead is 22 years. Find a formula predicting the change in size of a sample of white lead.

24. *Compound Interest:*
    a. Solve the differential equation of Exercise 18, Section 1.1, modeling compound interest.
    b. Calculate the percentage increase in such a deposit over one year. (This is called the *effective annual rate*.)

25. *A Population:* How long does it take for the population in Example 1.1.4 to double if the per capita growth rate is 5%?

26. *Cramming for Exams:* Alex Smart memorizes symbols according to the model in Exercise 25, Section 1.1. Based on his experience of last term, he knows he can remember at most 10,000 symbols and, if t is measured in days, the constant of proportionality is $\gamma = 1/100$. He now knows 6500 symbols. One week from now Alex is going to have a math exam requiring 225 new symbols and an organic chemistry exam requiring 275 new symbols. Will he be able to memorize them all in time?

27. *Another Population:*
    a. Solve the differential equation of Example 1.1.6.
    b. Show that any solution of the equation with $x(0) > 0$ satisfies $\lim_{t \to \infty} x(t) = 10$.

28. *Populations:* Suppose a population satisfies the Malthusian model $dx/dt = \lambda x$ (Example 1.1.4).

    a.   Show that if λ is a *positive* constant (births outnumber deaths) and $x(0) \neq 0$, then x grows without bound as $t \to \infty$. Note that *this model predicts its own obsolescence*, since it predicts that the population will ultimately outgrow its (finite) environment. The model can only be realistic as long as the population has a negligible effect on its environment.

    b.   Show that if λ is a *negative* constant (deaths outnumber births), then x approaches 0 as $t \to \infty$. Note that *this model predicts eventual extinction*, since x will ultimately drop below the value corresponding to one individual.

29. *Chemical Solutions:* Solve the o.d.e.'s of Exercise 23(a) and 23(b), Section 1.1, and determine in each case what happens to $x(t)$ as $t \to \infty$.

30. *Water Skiing:* Solve the o.d.e. of Exercise 27, Section 1.1, to determine the equation of the curve traveled by the water skier.

1.3 FIRST-ORDER LINEAR EQUATIONS

In this section we consider a large class of first-order equations that, although not always separable themselves, can be solved by considering a related separable equation. These equations, called linear o.d.e.'s, are important because they and their higher-order cousins encompass many elementary physical models. The basic strategy we use for first-order equations in this section will apply as well to higher-order linear equations in later chapters.

Before formulating a general definition, let's consider the equation that describes the population model in Example 1.1.5:

$$\frac{dx}{dt} = x + t.$$

Note that this equation is *not* separable. Recall that dx/dt was made up of two parts: reproduction contributed x, and immigration contributed t. One way to study such a model is to concentrate first on the isolated behavior of the organism, that is, the growth caused only by reproduction. Once we understand this type of growth, we can try to see how to modify our prediction when immigration is present.

Mathematically, the first step involves solving the equation obtained from the preceding one by dropping the t term:

$$\frac{dx}{dt} = x.$$

This equation is easily separable. Its general solution is

$$x = ke^t,$$

where $k = x(0)$.

Now, how does this prediction change when immigration is present? We note that if the immigration all occurred at one time t_1, then for $t \geq t_1$ the formula for x would be just like the one we got in the isolated case, except that the parameter k would be set higher than the real initial population because of the immigration. By analogy, it might be feasible to use this modified prediction for our actual model as well, provided the parameter k is continuously readjusted to account for the continuing immigration. In mathematical terms we expect the parameter k, which was constant in the isolated model, now to be a function of t. Thus, we expect a solution to the original equation to have the form

$$x = k(t)e^t.$$

The problem, of course, is to find the adjusted parameter $k(t)$. We do this by substituting this form of x back into the original differential equation. We calculate

$$\frac{dx}{dt} = k'(t)e^t + k(t)e^t$$

so the equation $dx/dt = x + t$ reads

$$k'(t)e^t + k(t)e^t = k(t)e^t + t$$

or

$$k'(t)e^t = t.$$

Dividing by e^t and integrating by parts (with $u = t$ and $dv = e^{-t}\, dt$), we have

$$k(t) = \int k'(t)\, dt = \int te^{-t}\, dt = -te^{-t} - e^{-t} + c.$$

Finally, we multiply $k(t)$ by e^t to get

$$x = k(t)e^t = -t - 1 + ce^t.$$

This formula describes all those solutions that have the form $k(t)e^t$. Since any function can be written in this form, $f(t) = [f(t)e^{-t}]e^t$, it actually describes *all* solutions—that is, it is the general solution of our o.d.e.

The method used in this example can be applied to problems in which the effect of immigration is any function of t and in which the terms dx/dt and x in the equation are multiplied by functions of t. (To guarantee that the integrals which arise exist, we have to make some restrictions on the functions.) It is customary to rewrite equations of this type by gathering the terms describing the isolated system on one side of the equation and the extra term, involving the external effects (immigration in our example), on the other side.

Definition: *A first-order o.d.e. is* **linear** *if it can be written in the form*

$$a_1(t)\,\frac{dx}{dt} + a_0(t)x = E(t)$$

where $a_1(t)$, $a_0(t)$, and $E(t)$ are functions of t (possibly constant). The equation is **normal** *on an interval I of the form $\alpha < t < \beta$ provided the functions $a_1(t)$, $a_0(t)$, and $E(t)$ are continuous on I and $a_1(t)$ is never zero on I.*

The first step in the solution of our example was to solve the equation we got by replacing $E(t)$ by zero (this corresponded to ignoring immigration). We give such equations a special name:

Definition: *A linear o.d.e. is* **homogeneous** *if it is of the form*

$$a_1(t)\,\frac{dx}{dt} + a_0(t)x = 0.$$

Otherwise, the o.d.e. is **nonhomogeneous.**

Our method had three basic steps. First, we used separation of variables to solve the homogeneous equation obtained by replacing $E(t)$ with zero. Second, we replaced the parameter k appearing in the solution of the homogeneous equation with a variable $k(t)$, substituted the resulting form for x into the original equation, and obtained a formula for $k'(t)$. Third, we integrated the formula for $k'(t)$ to obtain $k(t)$ and hence the desired solution x. When we solved for $k'(t)$, the terms involving $k(t)$ canceled. If we could verify that this was part of a general pattern, it would shorten the work required by our method.

As we work out the general pattern for a normal first-order linear o.d.e., it is convenient to divide the form in the definition of linearity

$$a_1(t)\,\frac{dx}{dt} + a_0(t)x = E(t)$$

by $a_1(t)$ to obtain what we shall call the **standard form** of the equation:

(N)
$$\frac{dx}{dt} + r(t)x = q(t)$$

(here $r(t) = a_0(t)/a_1(t)$ and $q(t) = E(t)/a_1(t)$).

If we apply our method starting from standard form, we first consider the related homogeneous equation

(H)
$$\frac{dx}{dt} + r(t)x = 0.$$

This equation is separable. Its general solution is of the form

$$x = kh(t)$$

where k is an arbitrary constant and $h(t) = e^{-\int r(t)\,dt}$. Note that $h(t)$ is a solution of (H).

To solve the nonhomogeneous equation, we allow the parameter to vary and look for a solution of the form

$$x = k(t)h(t).$$

We substitute this into the left side of (N), using the product rule to differentiate:

$$\frac{dx}{dt} + r(t)x = k'(t)h(t) + k(t)h'(t) + r(t)k(t)h(t)$$

$$= k'(t)h(t) + k(t)[h'(t) + r(t)h(t)].$$

Since $h(t)$ is a solution of the homogeneous equation, the quantity in brackets is zero. By setting the resulting expression for $dx/dt + r(t)x$ equal to $q(t)$, we obtain the simple formula

(V)
$$k'(t)h(t) = q(t).$$

From here, we proceed as in the example: Divide by $h(t)$, integrate to get $k(t)$, and multiply by $h(t)$ to find the solution x.

Example 1.3.1

Solve

$$t\frac{dx}{dt} + x = t^3, \quad 0 < t < +\infty.$$

Note that the o.d.e. is normal on the given interval. (It is also normal on the interval $-\infty < t < 0$, but is not normal on any interval containing $t = 0$.)

We first divide by t to put the equation in standard form:

(N)
$$\frac{dx}{dt} + \frac{1}{t} x = t^2.$$

Next we solve the related homogeneous equation:

(H)
$$\frac{dx}{dt} + \frac{1}{t} x = 0$$

$$\frac{1}{x} dx = -\frac{1}{t} dt$$

$$\ln|x| = -\ln|t| + c$$

$$= \ln \frac{1}{|t|} + c$$

$$x = \pm \frac{e^c}{t} = k\frac{1}{t}.$$

We now let the parameter vary and try for a solution of (N) in the form

$$x = k(t)\frac{1}{t}.$$

As a result of the analysis preceding this example, we know that substitution of x into (N) yields the equation

(V)
$$k'(t)\frac{1}{t} = t^2.$$

We solve for $k'(t)$ and integrate to find $k(t)$:

$$k'(t) = t^3$$

$$k(t) = \frac{t^4}{4} + c.$$

Finally, we substitute into our formula $x = k(t)/t$ to obtain the general solution of (N):

$$x = \frac{c}{t} + \frac{t^3}{4}.$$

■   **Example 1.3.2**   ·

Find the general solution of

(N)
$$\frac{dx}{dt} - \frac{t}{t^2 + 1} x = e^{-t}(t^2 + 1)^{1/2}.$$

Then find the specific solution that satisfies the initial condition $x(0) = 1$. Note that the o.d.e. is normal on the entire real line, $-\infty < t < \infty$.

This equation is already in standard form, so we begin by solving the related homogeneous equation

(H)
$$\frac{dx}{dt} - \frac{t}{t^2 + 1} x = 0.$$

In Example 1.2.3 we found that the general solution of this equation is

$$x = k(t^2 + 1)^{1/2}.$$

We now seek a solution to (N) in the form

$$x = k(t)(t^2 + 1)^{1/2}.$$

The varying parameter $k(t)$ has to satisfy

(V)
$$k'(t)(t^2 + 1)^{1/2} = e^{-t}(t^2 + 1)^{1/2}.$$

We solve for $k'(t)$ and integrate to find $k(t)$:

$$k'(t) = e^{-t}$$
$$k(t) = \int e^{-t}\, dt = -e^{-t} + c.$$

Finally, we substitute this into our formula for x to obtain the general solution of (N):

$$x = c(t^2 + 1)^{1/2} - e^{-t}(t^2 + 1)^{1/2}.$$

To find the specific solution satisfying $x(0) = 1$, we substitute $t = 0$ into the general solution:

$$1 = x(0) = c - 1.$$

This gives $c = 2$, so the specific solution is

$$x = 2(t^2 + 1)^{1/2} - e^{-t}(t^2 + 1)^{1/2}.$$

The answer we get by means of this method is valid only if we restrict attention to an interval I on which the o.d.e. is normal. This restriction is necessary to guarantee the existence of the integrals that arise and to allow division by $a_1(t)$ at the outset. It turns out [Exercise 31(c)] that the answer always takes the form $x = ch(t) + p(t)$, where $h(t)$ is a solution of the related homogeneous equation. If we are given a value t_0 and a number α, then, as in Example 1.3.2, it is always possible to find exactly one value of c so that the corresponding specific solution satisfies the initial condition $x(t_0) = \alpha$ [Exercise 31(d)].

The general pattern for solving first-order linear o.d.e.'s can be summarized in the following streamlined version of our method, which is called **variation of parameters.** It was first formalized by Johann Bernoulli in 1697.

FIRST-ORDER LINEAR DIFFERENTIAL EQUATIONS

A first-order equation is **linear** if it can be written in the form

$$a_1(t) \frac{dx}{dt} + a_0(t)x = E(t).$$

The equation is **normal** on an interval I provided that $a_1(t)$, $a_0(t)$, and $E(t)$ are continuous on I and $a_1(t)$ is never zero on I.

To solve a normal linear first-order o.d.e.:

0. Divide by $a_1(t)$ to obtain the **standard form**

    (N) $$\frac{dx}{dt} + r(t)x = q(t).$$

1. Separate variables in the **related homogeneous equation**

    (H) $$\frac{dx}{dt} + r(t)x = 0$$

    to obtain the general homogeneous solution

$$x = kh(t).$$

2. We expect solutions of (N) to be of the form

$$x = k(t)h(t).$$

Substituting this into (N) yields the equation

    (V) $$k'(t)h(t) = q(t)$$

which we solve for $k'(t)$.

3. Integrate the formula for $k'(t)$ to obtain $k(t)$, being sure to include the constant of integration:

$$k(t) = \int k'(t) \, dt + c.$$

Multiply $k(t)$ by $h(t)$ to obtain x:

$$x = k(t)h(t).$$

The final solution always has the form

$$x = ch(t) + p(t).$$

If t_0 is in I and if α is any number, then there is exactly one solution that satisfies the initial condition

$$x(t_0) = \alpha.$$

Note

A warning

The division by $a_1(t)$ to obtain standard form is needed if the simple formula $k'(t)h(t) = q(t)$ for the substitution is to hold. If we don't divide, the result of the substitution on the left will be $a_1(t)k'(t)h(t)$ (see Exercise 29).

EXERCISES

1. Which of the following o.d.e.'s are linear? Of the linear ones, which are homogeneous? (When answering this question, take x to be the dependent variable; $x' = dx/dt$.)

   a. $x' + x + t = 0$ b. $x' + xt = 0$

   c. $x't + x = 0$ d. $x'x + t = 0$

   e. $x' + x + t^2 = 0$ f. $(x')^2 + x + t = 0$

   g. $x' + x^2 + t = 0$ h. $x't + xt^2 + 1 = 0$

2. Which of the following linear o.d.e.'s are normal on $0 < t < 2$?

   a. $(t - 1)\dfrac{dx}{dt} - 5x = 3t$ b. $t\dfrac{dx}{dt} - e^t x = \sin t$

   c. $3\dfrac{dx}{dt} - 5x = \csc \pi t$ d. $t \sin t \dfrac{dx}{dt} + 20x = \ln t$

3. Determine the largest intervals on which the given o.d.e. is normal.

   a. $(t^2 + 1)\dfrac{dx}{dt} + x = \sin t$ b. $(t - 1)^2 \dfrac{dx}{dt} - tx = 5$

   c. $5\dfrac{dx}{dt} - tx = \sqrt{t + 1}$ d. $(t^2 - 4)\dfrac{dx}{dt} + 3x = 5t$

In Exercises 4 through 18 find the general solution on the indicated interval. (If no interval is given, take it to be $-\infty < t < \infty$.)

4. $\dfrac{dx}{dt} - x = e^t$

5. $3x' + 2x = 1$

6. $\dfrac{dx}{dt} + x = e^{-t} \sin 2t$

7. $\dfrac{dx}{dt} + 3x = t$

8. $t\dfrac{dx}{dt} - 3x = t^3 + 2t, \quad t > 0$

9. $x' + x \tan t = \cos t, \quad -\dfrac{\pi}{2} < t < \dfrac{\pi}{2}$

10. $x' + x \tan t = \sin t, \quad -\dfrac{\pi}{2} < t < \dfrac{\pi}{2}$

11. $dx + (3x + t^2 + 1)\, dt = 0$

12. $tx' + x = \sin t, \quad t > 0$

13. $(t^2 - 1)\dfrac{dx}{dt} + x = (t - 1)^{1/2}, \quad t > 1$

14. $2\dfrac{dx}{dt} - x = te^t$

15. $\dfrac{dx}{dt} - 2x = \sin t$

16. $(t^2 - 1)(x' - 3x) = te^{3t}, \quad t > 1$

17. $(t^2 - 1)(x' - 3x) = e^{3t}, \quad t > 1$

18. $(t + x)\, dx + dt = 0$ (*Hint:* Solve for t in terms of x.)

In Exercises 19 through 24, solve the initial-value problem.

19. $\dfrac{dx}{dt} - 2x = 8, \quad x(0) = 1$

20. $\dfrac{dx}{dt} + 3x = 8e^t, \quad x(0) = 0$

21. $(t^2 + 1)x' - tx = 0, \quad x(0) = 3$

22. $tx' - x = t^3, \quad x(1) = 0$

23. $\dfrac{dx}{dt} - tx = t, \quad x(0) = \dfrac{1}{2}$

24. $x' - x = e^t \sin t, \quad x(0) = 1$

Exercises 25 through 28 refer to our models.

25. *Savings:* Predict the growth of the savings account in Exercise 19, Section 1.1, starting with an initial deposit of $1000.

26. *Drug Absorption:* Suppose that at the outset the blood of the patient in Exercise 21, Section 1.1, contains none of the drug in question. How much of the drug will be in his bloodstream (a) after 1 hour, (b) after 6 hours, and (c) after 12 hours?

27. *Varying Dosage:* Suppose that at the outset the blood of the patient in Exercise 22, Section 1.1, contains none of the drug in question. How much of the drug will be in his bloodstream (a) after 1 hour, (b) after 6 hours, and (c) after 12 hours?

28. *Prices:* Assume the price of Blue Mountain coffee behaves as described in Exercise 26, Section 1.1. Assume also that (1) the supply is constant at $s = 500,000$ lb per year, (2) if the coffee were free, the demand would be 10 million lb per year, (3) if the price were to hit $40 per pound, no one would buy it, (4) the price rises 5¢ for every 100,000 lb per year of excess demand, and (5) the present price is $4.50 per pound. Predict the price after six months and after one year. (*Caution:* Watch your units.)

Exercises 29 through 36 involve more theoretical considerations.

29. Show that if we perform variation of parameters for the linear first-order o.d.e. $a_1(t)x' + a_0(t)x = E(t)$ without first putting it in standard form, then the substi-

tution $x = k(t)h(t)$, where $h(t)$ is a solution of the related homogeneous equation, leads to the equation $a_1(t)k'(t)h(t) = E(t)$.

30. Try variation of parameters on the nonlinear equation $x' + x^2 = t$ as follows: (a) Find the general solution of $x' + x^2 = 0$ by separating variables; (b) then replace the parameter with a variable and substitute back. *What goes wrong?*

31. Suppose $a_1(t)dx/dt + a_0(t)x = E(t)$ is normal on the interval I.

   a. Use separation of variables to show that the general solution on I of the associated homogeneous equation is $x = kh(t)$, where

$$h(t) = e^{-\int a_0(t)/a_1(t)\, dt}.$$

   b. Why can $h(t)$ never be 0?

   c. Use variation of parameters to show that the general solution on I of the nonhomogeneous o.d.e. is $x = ch(t) + p(t)$, where

$$p(t) = \int \frac{1}{h(t)} \frac{E(t)}{a_1(t)}\, dt.$$

   d. Let t_0 be a value in I, and let α be any number. Show that there is exactly one choice of c

$$c = \frac{\alpha - p(t_0)}{h(t_0)},$$

so that the corresponding specific solution satisfies $x(t_0) = \alpha$.

32. a. Show that the general solution on the interval $0 < t$ or on the interval $t < 0$ of the equation

(*) $\qquad\qquad\qquad tx' - 2x = 0$

is $x = ct^2$.

   b. Show that for any choice of c_1 and c_2,

$$x = \begin{cases} c_1 t^2 & \text{for } t \le 0 \\ c_2 t^2 & \text{for } t > 0 \end{cases}$$

is a continuous solution of (*) on $-\infty < t < \infty$. Note that when $c_1 \ne c_2$, this solution is *not* of the form $x = ct^2$ for any constant c.

   c. Why doesn't this contradict the observation we made just before the summary about the form of the general solution of a first-order linear o.d.e.?

33. a. Show that *every* continuous solution of $tx' - 2x = 0$ must satisfy $x(0) = 0$. (*Hint:* What is the general solution on the interval $t > 0$?)

   b. Why doesn't this contradict the observation we made at the end of the summary about the uniqueness of solutions satisfying a given initial condition?

34. a. Show that *no* solution of $t^3x' - 2t^2x = 1$ can be continuous at $t = 0$. (*Hint:* What is the general solution on the interval $t > 0$?)

   b. Why doesn't this contradict the observation we made just before the summary about the existence of solutions satisfying a given initial condition?

35. Show that all homogeneous linear o.d.e.'s of first order are separable. Which nonhomogeneous linear o.d.e.'s (of first order) are separable?

36. In the model solved at the beginning of this section, the growth *without* immigration is ke^t, and the growth *with* immigration is $ce^t - t - 1$. We expect immigration to increase the population, but here we are subtracting the positive quantity $t + 1$ from the growth without immigration. Explain this apparent contradiction. [*Hint:* Interpret k and c in terms of $x(0)$.]

37. *Another Approach:*

    a. Show that for any linear first-order o.d.e. in standard form

       (N) $$\frac{dx}{dt} + r(t)x = q(t)$$

       multiplication by the function

       $$\rho(t) = e^{\int r(t)\,dt}$$

       leads to an equation that can be rewritten in the form

       $$\frac{d}{dt}(e^{\int r(t)\,dt}x) = e^{\int r(t)\,dt}q(t).$$

    b. Integrate this equation and solve for x to find

       $$x = \left(\int e^{\int r(t)\,dt}q(t)\,dt + c\right)e^{-\int r(t)\,dt}.$$

    c. Check that the integrals in this expression are the same integrals that arise in solving (N) by the method described in the text.

1.4 APPLICATIONS OF FIRST-ORDER DIFFERENTIAL EQUATIONS

In Section 1.1 we saw how first-order differential equations occur in a number of population models. In this section we consider several models in other areas of application where first-order o.d.e.'s play an important role.

To put our discussion of models and of o.d.e.'s throughout this book in perspective, let us step back briefly to consider mathematical models and differential equations in the general context of formulating scientific theories or "explanations" of physical phenomena. At its core all scientific and technological thought is based on observation; any scientific theory begins from experimental data, and in the end its validity is tested by reference to other data. Of course, the very process of observation—and, more to the point, of collecting data—is by its nature selective. One begins by isolating certain quantities as central to understanding or describing a given phenomenon and rejecting others as marginal or irrelevant to this understanding. One then formulates a model by proposing certain precise mathematical relationships between the relatively small number of "relevant" measurable quantities. Often, these quantities are dynamic—they change with time—and it is thus common to see models formulated in terms of differential equations. This poses the serious mathematical problem of "solving" the o.d.e.'s to obtain explicit predictions of behavior from the implicit description given by the modeling o.d.e.'s. Once these solutions have been obtained, it is necessary from the point of view of the scientist or engineer to go back and compare the behavior predicted abstractly by the model with that observed for the actual phenomenon being modeled. When there is tolerable agreement between abstract prediction and observed data, we regard the model as "true," and use it to make further predictions concerning the phenomena at hand or to guide our design of products based on these phenomena.

In this book we shall not explicitly compare the solutions of our o.d.e.'s to precise experimental data. Instead we shall make certain qualitative observations about the predicted behavior and ask ourselves if these qualitative features are reasonable in terms of our experience of the phenomena we are modeling. This kind of "thought experiment" is a first step toward testing an abstract mathematical model in terms of actual physical performance.

Our first example is a mixing problem involving a polluted pond.

Example 1.4.1

Water flows into one end of a pond at the rate of 10 cubic meters per minute (m^3/min) and out of the other end at the same rate. The volume of the pond is 1000 m^3. Initially, the pond contains 100 grams (g) of pollutants. The concentration of pollutants in the water flowing into the pond is 2 g/m^3. Upon entering the pond, the pollutants spread so quickly that the concentration of pollutants is uniform throughout the pond. Find a formula for the number of grams of pollutant in the pond, $x = x(t)$.

The net rate of change in x is the difference between the rate R_{in} at which pollutants are flowing into the pond and the rate R_{out} at which they are flowing out:

$$\frac{dx}{dt} = R_{in} - R_{out}.$$

To find R_{in}, we multiply the rate at which water flows into the pond by the concentration of pollutants in the inflowing water:

$$R_{in} = (10 \text{ m}^3/\text{min})(2 \text{ g/m}^3) = 20 \text{ g/min}.$$

Similarly, R_{out} is the product of the rate at which water flows out with the concentration of pollutants in the outflowing water:

$$R_{out} = (10 \text{ m}^3/\text{min})\left(\frac{x}{1000} \text{ g/m}^3\right) = \frac{x}{100} \text{ g/min}.$$

Combining these equations, we see that

$$\frac{dx}{dt} = 20 - \frac{x}{100}.$$

This o.d.e. can be solved by separation of variables or the method for linear o.d.e.'s described in Section 1.3. The general solution is

$$x = ce^{-t/100} + 2000.$$

Substitution of $t = 0$ yields

$$100 = c + 2000,$$

so that $c = -1900$. Thus,

$$x = -1900e^{-t/100} + 2000.$$

Note that

$$\lim_{t\to\infty} x = \lim_{t\to\infty}(-1900e^{-t/100} + 2000) = 2000,$$

so that the amount of pollutant in the pond tends toward 2000 g. This is the amount we would have if the concentration of pollutants in the pond were the same as that in the water entering the pond.

Our next example deals with a warm body in cooler surroundings. The simplest model for the rate at which such a body loses heat assumes that the rate of change in temperature is proportional to the difference between the temperature of the body x and that of its environment y. Thus,

$$\frac{dx}{dt} = -\gamma(x - y)$$

where γ is a constant. (Since we are assuming the body is warmer than its surroundings, we expect the temperature to decrease. Thus, γ is positive.) This model is called **Newton's law of cooling.**

Example 1.4.2

A glass of water at 50°F is placed in a freezer and loses heat according to Newton's law of cooling. The freezer is kept at the constant temperature of 30°F. After one hour the temperature of the water is 40°F. How long after the water is placed in the freezer will it reach 32°F?

Since the temperature of the surroundings is $y = 30$, the temperature x of the water satisfies

$$\frac{dx}{dt} = -\gamma(x - 30).$$

This can be solved by separation of variables to get

$$x = 30 + ke^{-\gamma t}.$$

Substitution of $t = 0$ yields

$$50 = 30 + k,$$

so that $k = 20$. Thus,

$$x = 30 + 20e^{-\gamma t}.$$

We now substitute $t = 1$,

$$40 = 30 + 20e^{-\gamma}$$

and solve for γ:

$$e^{-\gamma} = \frac{1}{2}$$

$$-\gamma = \ln\frac{1}{2} = -\ln 2$$

$$\gamma = \ln 2.$$

Thus,

$$x = 30 + 20e^{-t\ln 2} = 30 + 20(2^{-t}).$$

To find the time at which the water freezes, we substitute $x = 32$,

$$32 = 30 + 20e^{-t\ln 2}$$

and solve for t:

$$e^{-t\ln 2} = \frac{1}{10}$$

$$-t\ln 2 = -\ln 10$$

$$t = \frac{\ln 10}{\ln 2} \approx 3.32 \text{ hours.}$$

Note that

$$\lim_{t \to \infty} x = \lim_{t \to \infty}(30 + 20e^{-t\ln 2}) = 30$$

so that the temperature of the water approaches the temperature in the freezer.

We next consider a simple circuit (Figure 1.2) containing a resistor, a coil, and a voltage source. (We will consider more complicated circuits in Exercise 3 of Section 2.1 and in Section 3.1.) Experiments show that the voltage drop (measured in *volts*) across a resistor is proportional to the current (measured in *amperes*) through the resistor,

$$V_{res} = RI_{res}$$

and that the voltage drop across a coil is proportional to the rate at which the current through the coil is changing:

$$V_{coil} = L\frac{dI_{coil}}{dt}.$$

The positive constants R and L are called the **resistance** (measured in *ohms*) of the resistor and the **inductance** (measured in *henrys*) of the coil. According to **Kirchhoff's laws,** the current through the resistor in our simple circuit is equal to the current through the coil,

$$I_{res} = I_{coil} = I,$$

and the sum of the voltage drops across the resistor and the coil is equal to the voltage generated by our source:

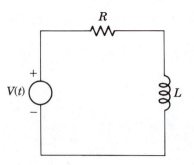

FIGURE 1.2

$$V_{\text{res}} + V_{\text{coil}} = V(t).$$

Thus, we have

$$RI + L\frac{dI}{dt} = V(t).$$

Example 1.4.3

Find a formula for the current in the circuit of Figure 1.2 with $R = 4$, $L = 2$, and $V(t) = \cos 3t$.

The current satisfies the o.d.e.

$$4I + 2\frac{dI}{dt} = \cos 3t.$$

This is a linear o.d.e., which we can solve by the method of Section 1.3. We first rewrite the equation in standard form:

(N) $$\frac{dI}{dt} + 2I = \frac{1}{2}\cos 3t.$$

The related homogeneous equation

(H) $$\frac{dI}{dt} + 2I = 0$$

has general solution

$$I = ke^{-2t}.$$

To find solutions to (N) of the form

$$I = k(t)e^{-2t}$$

we must solve

(V) $$k'(t)e^{-2t} = \frac{1}{2}\cos 3t.$$

The solution of this equation (obtained by integrating by parts twice) is

$$k(t) = \int \frac{1}{2}e^{2t}\cos 3t\, dt = \frac{2}{26}e^{2t}\cos 3t + \frac{3}{26}e^{2t}\sin 3t + c.$$

Thus, the general solution of (N) is

$$I = ce^{-2t} + \left(\frac{2}{26} \cos 3t + \frac{3}{26} \sin 3t \right).$$

Note that the term ce^{-2t} in our solution has the property that

$$\lim_{t \to \infty} ce^{-2t} = 0.$$

Such a term is called a **transient.** When we subtract the transient term, which dies out in the long run, we are left with the **steady-state** part of the solution,

$$\frac{2}{26} \cos 3t + \frac{3}{26} \sin 3t,$$

which is periodic and has the same period as the voltage source.

We next turn to an example involving Newton's second law, $F = ma$.

Example 1.4.4

A parachutist falling toward earth is subject to two forces: her weight $w = 32m$ (where m is her mass) and the drag created by the parachute. Assume that the drag is $8|v|$, that the parachutist starts from rest ($v(0) = 0$), and that she weighs 128 lb. Find formulas for the velocity $v = v(t)$ and for the distance $x = x(t)$ that the parachutist has fallen after t seconds.

Since x increases as the parachutist falls, the downward direction is the positive direction in this problem. The force due to the parachutist's weight acts in the positive x direction, and the force due to the drag acts in the negative x direction. Since the velocity is always positive (so that $|v| = v$), the resultant of the forces is

$$F = 128 - 8v.$$

By Newton's second law,

$$F = ma = 4 \frac{dv}{dt},$$

so the velocity satisfies

$$4 \frac{dv}{dt} = 128 - 8v.$$

The **general** solution of this o.d.e. (obtained using separation of variables or the method of Section 1.3) is

$$v = 16 + ce^{-2t}.$$

Since $v(0) = 0$, $c = -16$. Thus,

$$v = 16(1 - e^{-2t}).$$

Now, $v = dx/dt$, so integration yields a formula for x:

$$x = \int 16(1 - e^{-2t})\, dt = 16t + 8e^{-2t} + k.$$

When $t = 0$, $x = 0$, so $k = -8$. Thus,

$$x = 16t + 8e^{-2t} - 8.$$

Note that

$$\lim_{t \to \infty} v(t) = \lim_{t \to \infty} 16(1 - e^{-2t}) = 16$$

so the velocity approaches a value of 16 ft/sec. After 2 seconds, the velocity is

$$v(2) = 16(1 - e^{-4}) \approx 15.7.$$

This value is already so close to the limiting velocity that, by this time, a casual observer would think that the parachutist was falling at a constant rate.

A basic model for the dynamics of a chemical reaction is given by the **law of mass action,** which says that the rate of a reaction is proportional to the product of the concentrations of the reactants. The constant of proportionality is called the **rate constant** of the reaction. Our next example and Exercises 18 through 23 deal with hypothetical chemical reactions that are governed by this model.

Example 1.4.5

When chemicals A and B react, one molecule of A combines with one molecule of B to form one molecule of substance C:

$$A + B \rightarrow C.$$

Suppose that at the outset the concentration of A is 2 moles per liter, that of B is 3 moles per liter, and that of C is 0 (a mole is Avogadro's number $N_0 = 6.02252 \times 10^{23}$ molecules). After 1 second, the concentration of C is 0.2 moles per liter. Find a formula for the concentration $x = x(t)$ of substance C after t seconds.

Since formation of 1 mole of C requires 1 mole of A, the concentration of A after t seconds will be $2 - x$. Similarly, the concentration of B will be $3 - x$. By the law of mass action,

$$\frac{dx}{dt} = k(2 - x)(3 - x).$$

To solve the o.d.e. we separate variables

$$\frac{1}{(2 - x)(3 - x)} = k \, dt.$$

In order to integrate the left side, we first expand it in partial fractions (see note 2 in Section 4.3). We look for an expression of the form

$$\frac{1}{(2 - x)(3 - x)} = \frac{a}{2 - x} + \frac{b}{3 - x}.$$

Multiplication by $(2 - x)(3 - x)$ gives

$$1 = a(3 - x) + b(2 - x) = (3a + 2b) - (a + b)x.$$

Equating the coefficients of 1 and of x on both sides gives

$$3a + 2b = 1$$
$$a + b = 0$$

which we solve to find

$$a = 1, \quad b = -1.$$

Thus, our separated o.d.e. can be written as

$$\left(\frac{1}{2 - x} + \frac{-1}{3 - x} \right) dx = k \, dt.$$

Integration gives

$$-\ln(2 - x) + \ln(3 - x) = kt + c,$$

so that

$$\ln \left(\frac{3 - x}{2 - x} \right) = kt + c.$$

Thus,

$$\frac{3 - x}{2 - x} = Ce^{kt}.$$

Since $x(0) = 0$, $C = 3/2$. Substitution gives

$$\frac{3 - x}{2 - x} = \frac{3}{2} e^{kt}.$$

Our knowledge of $x(1)$ allows us to determine the rate constant. Since $x(1) = 0.2$, we have $e^k = 28/27$. Thus,

$$k = \ln(28/27) \approx 3.6 \times 10^{-2}.$$

Substitution of this value gives

$$\frac{3 - x}{2 - x} = \frac{3}{2} \left(\frac{28}{27} \right)^t.$$

Finally, we solve this equation for x:

$$x = \frac{6(28/27)^t - 6}{3(28/27)^t - 2}.$$

Note that

$$\lim_{t \to \infty} x = 2$$

so the concentration of substance C tends toward 2 moles per liter. Since the formation of each mole of substance C requires one mole of A, and the concentration of A at the outset was 2 moles per liter, this is the maximum amount of C that can be obtained.

EXERCISES

Mixing Problems

1. A tank contains 100 liters of water in which 20 grams of salt are dissolved. Brine containing 2 grams per liter of salt is poured into the tank at the rate of 4 liters

per minute, and the well-mixed solution flows out at the same rate from a spigot at the bottom of the tank.
a. How much salt is in the tank after t minutes?
b. What happens to the amount of salt in the tank as t gets very large?

2. Repeat the previous exercise under the assumption that the brine flows in (and the solution flows out) at the rate of 8 liters per minute.

3. Suppose that after 10 minutes, the brine flowing into the tank of Exercise 1 is replaced by pure water (still flowing at 4 liters per minute). How much salt will there be in the tank 10 minutes after the switch?

4. Suppose that the spigot in the tank of Exercise 1 is adjustable. Find the amount of salt in the tank after 10 minutes if the solution flows out at the rate of (a) 5 liters per minute and (b) 3 liters per minute. (*Hint:* Be sure to take into account the fact that the volume of the solution in the tank varies.)

5. Suppose that polluted water flows in and out of another pond, just as in Example 1.4.1, except that this pond has a factory alongside that dumps more pollutants into the pond at the rate of 5 grams per minute. As before, the pond contains 100 grams of pollutants at the outset.
a. Find a formula for the amount of pollutant in the pond after t minutes.
b. What happens to this amount as t gets very large?

Newton's Law of Cooling

6. Boiling water (212°F) is poured into a teapot and cools. The temperature in the room is a constant 65°F. After 2 minutes the tea has a temperature of 191°F.
a. What will the temperature of the tea be 16 minutes after the water is added?
b. When tea reaches a temperature below 100°F, it is tepid and tastes bad. How long can we talk to our guests before pouring the tea?

7. When the water in the kettle reached a boil (100°C), John turned off the flame. Some time later his roommate entered the kitchen. The temperature of the water at that time was 80°C, and 10 min later it was 75°C. The temperature in the kitchen was a constant 20°C. How much time passed between the time John turned off the flame and the time his roommate entered the kitchen?

8. A pail of water at 70°F is placed outdoors when the temperature is 40°F. The water loses heat according to Newton's law of cooling, with constant of proportionality $\gamma = 1/10$ when time is measured in hours. Suppose the temperature outdoors decreases steadily at the rate of 3°F per hour.
a. What is the temperature outdoors at time t?
b. What is the temperature of the water at time t?
c. What is the temperature of the water after 5 hours?

Circuits

9. Find a formula for the current in the circuit of Figure 1.2 with $R = 2$, $L = 4$, and $V(t) = 12$, and assuming that the current at time $t = 0$ is $I(0) = 3$. What happens to the current as $t \to \infty$?

10. Find a formula for the current in the circuit of Figure 1.2 with $R = 1$, $L = 1$, and $V(t) = \sin 2t$, and assuming that the current at time $t = 0$ is $I(0) = 4$. **Find the transient part $T(t)$ and the steady-state part $S(t)$ of the current.**

11. Find a formula for the current in the circuit of Figure 1.2 with $R = 2$, $L = 1$, and $V(t) = 10e^{-2t}$, assuming $I(0) = I_0$. What happens to the current as $t \to \infty$?

12. Show that if $V(t) = V$ is a constant, then the current in the circuit of Figure 1.2 will approach a constant value as $t \to \infty$. What is this value?

Newton's Second Law of Motion

13. We assumed in Example 1.4.4 that the parachutist started from rest. In fact, it takes time for a parachute to open, and during that time the parachutist gains speed. Suppose the velocity of the parachutist in Example 1.4.4 is 10 ft/sec when the parachute opens.
    a. Find formulas for the velocity $v(t)$ and the distance $x(t)$ from the point where the parachute opens.
    b. What happens to $v(t)$ as t gets very large?

14. Assume that the drag of the parachute worn by the parachutist in Example 1.4.4 is $2v^2$ (rather than $8|v|$) and that everything else is unchanged. Find a formula for the velocity $v(t)$, and determine what happens to $v(t)$ as t gets very large.

15. An object weighing $W = 20$ lb slides down an inclined plane 20 ft long (see Figure 1.3). Assume that there are no forces resisting the motion of the body.
    a. Find a formula for the distance $x(t)$ from the top of the plane to the object. (*Hint:* What is the component of the weight in the direction parallel to the plane?)
    b. How long does it take for the object to reach the bottom if it starts at rest from the top of the plane?
    c. How long does it take for the object to reach the bottom if it starts at 3 ft/sec from a point 2 ft from the top?

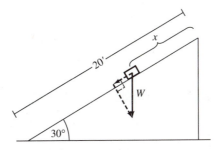

FIGURE 1.3

16. How long will it take the object in Exercise 15 to reach the bottom of the plane if it starts at rest from the top and encounters a resisting force (due to friction) of 1/10 the component of the weight in the direction perpendicular to the plane?

17. Suppose the object in Exercise 15 starts at rest from the top of the inclined plane and slides down, encountering air resistance equal to twice its velocity.
    a. How far down the plane will the object be after 1 second?
    b. How far down the plane will the object be after 2 seconds?
    c. Estimate how long it will take for the object to reach the bottom, using the fact that the exponential term in the solution to the o.d.e. decays very quickly.

Chemical Reactions

18. Suppose, as in Example 1.4.5, that one molecule of A combines with one molecule of B to form one molecule of C. This time, however, assume that experiments have shown that if time is measured in seconds, then the rate constant of the reaction is $k = 5 \times 10^{-2}$. Find a formula for the concentration $x(t)$ of C, and determine what happens to $x(t)$ as $t \to \infty$ if at the outset the concentration of C is 0, the concentration of A is 1 mole per liter, and the concentration of B is
    a. 3 moles per liter   b. 2 moles per liter   c. 1 mole per liter.

19. Suppose that one molecule of A combines with one molecule of B to form one molecule of C. Assume that at the outset the concentrations of A and B are 2 moles per liter, and that of C is 0. After one second, the concentration of C is 0.2 moles per liter. Find a formula for the concentration $x(t)$ of substance C. How long does it take for the concentration of C to reach 1 mole per liter?

20. Suppose that one molecule of A combines with one molecule of B to form one molecule of C. Assume that at the outset the concentrations of A, B, and C are 1 mole per liter, 3 moles per liter, and 0, respectively. After 1 second, the concentration of C is 0.2 moles per liter. What is the concentration of C after 2 seconds?

21. Suppose chemicals A and B react to form C exactly as in Example 1.4.5, with rate constant $k = \ln(28/27)$. This time, however, the initial concentrations of A, B, and C are 2 moles per liter, 3 moles per liter, and 1 mole per liter, respectively. As before, let $x(t)$ be the concentration of substance C. After t seconds, the concentration of C will have increased by $x(t) - 1$. Since formation of one molecule of C requires one molecule of each of A and B, the concentrations of A and B after t seconds will be $(2 - [x - 1])$ and $(3 - [x - 1])$, respectively.
    a. Use the law of mass action to obtain a differential equation for $x(t)$.
    b. Find a formula for the concentration $x(t)$ of substance C.
    c. What happens to $x(t)$ as $t \to \infty$?

22. Suppose that one molecule of A and two molecules of B react to form one molecule of C:
$$A + 2B \to C.$$
    Suppose at the outset the concentrations of A, B, and C are 1 mole per liter, 2 moles per liter, and 0, respectively. Let $x(t)$ be the concentration of C after t seconds. Since the formation of one molecule of C requires one molecule of A and two molecules of B, the concentrations of A and B after t seconds are $(1 - x)$ and $(2 - 2x)$, respectively. To apply the law of mass action in this case, we include the reactant B twice. Thus,
$$\frac{dx}{dt} = k(1 - x)(2 - 2x)^2 = 4k(1 - x)^3.$$
    a. Find a formula for $x(t)$.
    b. What happens to $x(t)$ as $t \to \infty$?

23. Suppose that one molecule of substance A combines with one molecule of substance B to form one molecule of substance C. Suppose that at the outset the concentration of A is a moles per liter, that of B is b moles per liter, and that of C is 0. Find an expression for the concentration $x(t)$ of C and determine what happens to $x(t)$ as $t \to \infty$
    a. if $a > b$ b. if $a = b$.

Any equation that can be written in the form $F(x, y) = c$ represents a family of curves in the xy-plane, with each specific value of c determining a member of the family. Another family of curves $G(x, y) = k$ is said to be a family of **orthogonal trajectories** to the original family, provided that any time two of the curves $F(x, y) = c_1$ and $G(x, y) = k_1$ intersect, the tangent lines to the curves are perpendicular. Starting from a given family $F(x, y) = c$, the orthogonal trajectories can often be found as follows:

   i.   Differentiate the equation $F(x, y) = c$ to obtain an o.d.e. satisfied by each curve in the given family. Write this o.d.e. in the form $dy/dx = f(x, y)$.

  ii.   Since the tangent lines to curves from the two families are perpendicular at points of intersection, we can find an o.d.e. satisfied by each curve in the new family, namely, $dy/dx = -1/f(x, y)$.

 iii.   Solve this o.d.e. to obtain the orthogonal trajectories.

Use this method to find the orthogonal trajectories to the families in Exercises 24 through 28.

24. $y = cx^2$ 25. $x^2 + y^2 = c$

26. $xy = c$ 27. $x^2 - y^2 = c$

28. $y = ce^{-x}$

29. *Telephone Wire:* Assume that a flexible wire of uniform density δ hangs suspended between two points as in Figure 1.4. (Note in particular that we have placed the axes so that the lowest point A on the wire is on the y-axis.) Assume that the wire is in equilibrium and that the only forces on the wire are the horizontal tension $\mathbf{T}_0$ at A, the tension $\mathbf{T}$ tangent to each point P on the curve, and the gravitational force. Since there is no horizontal motion, the horizontal component of $\mathbf{T}$ counterbalances $\mathbf{T}_0$:

$$|\mathbf{T}| \cos \theta = |\mathbf{T}_0|.$$

Since there is no vertical motion, the vertical component of $\mathbf{T}$ counterbalances the weight of the wire from A to P. If we let s represent the length of the wire from A to P, this gives

$$|\mathbf{T}| \sin \theta = \delta s.$$

  a.   Find dy/dx in terms of s. (*Hint:* What is the relation between dy/dx and $\tan \theta$?)

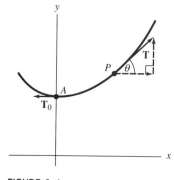

FIGURE 1.4

   b.   Differentiate to get d^2y/dx^2 in terms of ds/dx.
   c.   Use the fact that $ds/dx = \sqrt{(dy/dx)^2 + 1}$ to obtain a second-order o.d.e. for
        y in terms of x.
   d.   Rewrite the o.d.e. from (c) as a first-order o.d.e. for $u = dy/dx$.
   e.   Solve this o.d.e. to find u.
   f.   Integrate to obtain y.
   This curve is known as a **catenary**.

1.5   EXTENSIONS OF OUR METHODS: SUBSTITUTIONS

Sometimes a substitution will transform an o.d.e. into an equation that we know
how to solve. In this section we discuss several examples.

   A first-order o.d.e. is said to have **homogeneous coefficients** if it can be
written in the form

$$\frac{dx}{dt} = G\left(\frac{x}{t}\right).$$

To solve an o.d.e. of this type, we set

$$v = \frac{x}{t}.$$

Then

$$x = vt \qquad \text{so} \qquad \frac{dx}{dt} = v + t\,\frac{dv}{dt}.$$

Substitution into our o.d.e. yields

$$v + t\,\frac{dv}{dt} = G(v).$$

This equation is separable:

$$t\,\frac{dv}{dt} = G(v) - v$$

$$\frac{1}{G(v) - v}\,dv = \frac{1}{t}\,dt.$$

We can integrate to find v. Then $x = vt$ is the solution of our o.d.e.

Example 1.5.1

Find the general solution of

$$t^2 \frac{dx}{dt} = xt + 3x^2, \quad t > 0.$$

We rewrite this as

$$\frac{dx}{dt} = \frac{xt + 3x^2}{t^2} = \frac{x}{t} + 3 \left(\frac{x}{t} \right)^2.$$

If we set

$$v = \frac{x}{t},$$

then

$$x = vt \qquad \text{so} \qquad \frac{dx}{dt} = v + t \frac{dv}{dt}.$$

Substitution into the o.d.e. yields

$$v + t \frac{dv}{dt} = v + 3v^2.$$

We solve this o.d.e. by separation of variables:

$$t \frac{dv}{dt} = 3v^2$$

$$v^{-2} \, dv = \frac{3}{t} \, dt$$

$$-v^{-1} = 3 \ln t + c$$

$$v = \frac{-1}{3 \ln t + c}.$$

Since $x = vt$,

$$x = \frac{-t}{3 \ln t + c}.$$

The o.d.e. in Example 1.5.1 is an example of a general class of o.d.e.'s, each of which has homogeneous coefficients. Recall that a polynomial in the two variables x and t is a sum of monomials of the form $ax^i t^j$. The degree of the monomial $ax^i t^j$ is $i + j$. The o.d.e. in Example 1.5.1 is of the form

$$A(x, t) \frac{dx}{dt} = B(x, t),$$

where $A(x, t) = t^2$ and $B(x, t) = xt + 3x^2$ are polynomials in t and x, all of whose monomial terms have the same degree. This property ensures that the equation has homogeneous coefficients (see Exercise 21).

Fact: *If $A(x, t)$ and $B(x, t)$ are sums of monomials all of which have the same degree, then the o.d.e. $A(x, t)\, dx/dt = B(x, t)$ has homogeneous coefficients.*

Example 1.5.2

Find the general solution of

$$x^2 t \frac{dx}{dt} = x^3 - t^3, \quad t > 0.$$

The functions $A(x, t) = x^2 t$ and $B(x, t) = x^3 - t^3$ are sums of monomials all of which have degree 3. Thus, the equation has homogeneous coefficients. If we set

$$v = \frac{x}{t},$$

then

$$x = vt \qquad \text{so} \qquad \frac{dx}{dt} = v + t \frac{dv}{dt}.$$

Substitution into the o.d.e. yields

$$v^2 t^3 \left(v + t \frac{dv}{dt} \right) = v^3 t^3 - t^3.$$

We divide by $v^2 t^3$ and separate the variables:

$$v + t\frac{dv}{dt} = v - v^{-2}$$

$$t\frac{dv}{dt} = -v^{-2}$$

$$v^2\, dv = -\frac{1}{t}\, dt$$

$$\frac{1}{3}v^3 = -\ln t + c$$

$$v = (-3\ln t + 3c)^{1/3}.$$

Since $x = vt$,

$$x = t(-3\ln t + 3c)^{1/3}.$$

An equation of the form

$$x' + r(t)x = q(t)x^n$$

is called a **Bernoulli equation.** When $n = 0$ or $n = 1$, the equation is linear. When $n \neq 1$, we set

$$v = x^{1-n}.$$

Then

$$x = v^{1/(1-n)} \qquad \text{so} \qquad \frac{dx}{dt} = \frac{1}{1-n}v^{n/(1-n)}\frac{dv}{dt}.$$

Substitution into our o.d.e. yields

$$\frac{1}{1-n}v^{n/(1-n)}\frac{dv}{dt} + r(t)v^{1/(1-n)} = q(t)v^{n/(1-n)}.$$

Multiplication by $1 - n$ and division by $v^{n/(1-n)}$ yields

$$\frac{dv}{dt} + (1-n)r(t)v = (1-n)q(t).$$

This is a linear o.d.e., which we can solve for v. Then $x = v^{1/(1-n)}$ is the solution of our o.d.e.

Example 1.5.3

Find the general solution of

$$\frac{dx}{dt} + 6x = 30x^{2/3}$$

This is a Bernoulli equation with $n = 2/3$. We set

$$v = x^{1-(2/3)} = x^{1/3}.$$

Then

$$x = v^3 \quad \text{so} \quad \frac{dx}{dt} = 3v^2 \frac{dv}{dt}.$$

Substitution into our o.d.e. yields

$$3v^2 \frac{dv}{dt} + 6v^3 = 30v^2.$$

Division by $3v^2$ yields

$$\frac{dv}{dt} + 2v = 10.$$

We leave it to you to check that the general solution of this linear o.d.e. is

$$v = ce^{-2t} + 5.$$

Since $x = v^3$,

$$x = (ce^{-2t} + 5)^3.$$

A simple substitution will enable us to solve higher-order o.d.e.'s of the form

$$\frac{d^n x}{dt^n} + r(t) \frac{d^{n-1} x}{dt^{n-1}} = q(t).$$

If we set

$$v = \frac{d^{n-1} x}{dt^{n-1}},$$

then

$$\frac{dv}{dt} = \frac{d^n x}{dt^n}.$$

Substitution into our o.d.e. yields

$$\frac{dv}{dt} + r(t)v = q(t).$$

This is a linear first-order equation. We can solve it for $v = d^{n-1}x/dt^{n-1}$, and then integrate $n - 1$ times to find x.

Example 1.5.4

Find the general solution of

$$t\frac{d^3 x}{dt^3} - \frac{d^2 x}{dt^2} = t^2, \quad t > 0.$$

If we set

$$v = \frac{d^2 x}{dt^2},$$

then

$$\frac{dv}{dt} = \frac{d^3 x}{dt^3}.$$

Substitution into the o.d.e. yields

$$t\frac{dv}{dt} - v = t^2, \quad t > 0.$$

This is a linear first-order equation. We leave it to you to check that the solution of this equation is

$$v = c_1 t + t^2.$$

Since $v = d^2 x/dt^2$,

$$\frac{d^2 x}{dt^2} = c_1 t + t^2.$$

We integrate twice to find x:

$$\frac{dx}{dt} = c_1 \frac{t^2}{2} + c_2 + \frac{t^3}{3}$$

$$x = c_1 \frac{t^3}{6} + c_2 t + c_3 + \frac{t^4}{12}.$$

We close with a description of the key features of the substitutions we have discussed.

EXTENSIONS OF OUR METHODS: SUBSTITUTIONS

To solve an o.d.e. with **homogeneous coefficients**

$$\frac{dx}{dt} = G\left(\frac{x}{t}\right)$$

set

$$v = \frac{x}{t}.$$

Then

$$x = vt \qquad \text{so} \qquad \frac{dx}{dt} = v + t\frac{dv}{dt}.$$

Substitution into the o.d.e. yields a separable equation for v. Solve this equation and then find $x = vt$.

To solve a **Bernoulli equation**

$$\frac{dx}{dt} + r(t)x = q(t)x^n$$

set

$$v = x^{1-n}.$$

Then

$$x = v^{1/(1-n)} \qquad \text{so} \qquad \frac{dx}{dt} = \frac{1}{1-n}v^{n/(1-n)}\frac{dv}{dt}.$$

Substitution into the o.d.e. yields a linear first-order equation. Solve this equation for v and then use the fact that $x = v^{1/(1-n)}$ to obtain x.

To solve a higher order o.d.e. of the form

$$\frac{d^n x}{dt^n} + r(t)\frac{d^{n-1}x}{dt^{n-1}} = q(t),$$

set

$$v = \frac{d^{n-1}x}{dt^{n-1}}.$$

Then

$$\frac{dv}{dt} = \frac{d^n x}{dt^n}.$$

Substitution into the o.d.e. yields a linear first-order equation. Solve this equation to find v and then integrate $n - 1$ times to find x.

EXERCISES

In Exercises 1 through 4:

(a) Decide whether the given o.d.e. has homogeneous coefficients.
(b) If it does, use a substitution along the lines of Examples 1.5.1 and 1.5.2 to find the general solution on the interval $t > 0$.

1. $t^2 \dfrac{dx}{dt} - xt = x^2$

2. $(x + t) \dfrac{dx}{dt} = x$

3. $t \dfrac{dx}{dt} + x = t^2$

4. $t^2 \dfrac{dx}{dt} = x^2 + 4xt$

In Exercises 5 through 8:

(a) Decide whether the given o.d.e. is a Bernoulli equation.
(b) If it is, perform a substitution along the lines of Example 1.5.3 to find the general solution on the interval $t > 0$.

5. $\dfrac{dx}{dt} + 3x = x^2$

6. $\dfrac{dx}{dt} + tx^3 = x^4$

7. $x^2 \dfrac{dx}{dt} - tx^3 = 2t$

8. $\dfrac{dx}{dt} + 6x = 5x^{1/2}$

In Exercises 9 through 11, find the general solution along the lines of Example 1.5.4.

9. $\dfrac{d^2x}{dt^2} - \dfrac{dx}{dt} = 1$

10. $\dfrac{d^3x}{dt^3} - 3\dfrac{d^2x}{dt^2} = 6$

11. $t \dfrac{d^4x}{dt^4} - \dfrac{d^3x}{dt^3} = 0, \quad t > 0$

In Exercises 12 through 20:

(a) Find the general solution using one of the methods discussed in this section.
(b) Find the specific solution that satisfies the given initial conditions.

12. $x \dfrac{dx}{dt} - 4x^2 = 1; \qquad x(0) = 1$

13. $t^2 \dfrac{dx}{dt} = x^2 - xt + t^2, \quad t > 0; \qquad x(1) = 4$

14. $(xt - 2t^2) \dfrac{dx}{dt} = 2x^2 - 6xt, \quad t > 0; \qquad x(1) = 1$

15. $t \dfrac{d^2x}{dt^2} + \dfrac{dx}{dt} = 8t^3, \quad t > 0; \qquad x(1) = x'(1) = 0$

16. $t^3 \dfrac{dx}{dt} - xt^2 = x^2, \quad t > 0; \qquad x(1) = 2$

17. $t \dfrac{d^3x}{dt^3} + 3 \dfrac{d^2x}{dt^2} = 4t + 3, \quad t > 0; \qquad x(1) = 3, x'(1) = 2, x''(1) = 1$

18. $t \dfrac{dx}{dt} = x + t^2 \cos^2 \left(\dfrac{x}{t}\right), \quad t > 0; \qquad x(2) = \dfrac{\pi}{2}$

19. $\dfrac{dx}{dt} - 4x^2 - x = 0; \qquad x(0) = 1$

20. $\dfrac{d^2x}{dt^2} - \dfrac{dx}{dt} = e^t; \qquad x(0) = 0, x'(0) = 1$

21. a. Let $P(x, t)$ be a sum of monomials, all of which have the same degree k:

$$P(x, t) = \sum a_i x^i t^{k-i}.$$

Show that the substitution $x = vt$ leads to an expression of the form

$$P(x, t) = \left(\sum a_i v^i\right) t^k.$$

b. Show that if $A(x, t)$ and $B(x, t)$ are sums of monomials, all of which have degree k, then the o.d.e. $A(x, t) \, dx/dt = B(x, t)$ has homogeneous coefficients.

1.6   THE EXISTENCE AND UNIQUENESS THEOREM

We saw in Section 1.3 that if we restrict attention to an interval I on which the o.d.e.

$$a_1(t) \dfrac{dx}{dt} + a_0(t)x = E(t)$$

is normal, then we can list all solutions of this o.d.e. by a formula involving a single constant of integration:

$$x = ch(t) + p(t).$$

If we are given numbers t_0 and α, with t_0 in I, then there will be exactly one value of the constant c for which the resulting solution satisfies the initial condition

$$x(t_0) = \alpha.$$

It turns out that, even in cases where we are not clever enough to display the solutions, it is possible to say something about the theoretical existence of solutions and their relation to initial conditions. This theorem will appear in various guises in this book. The first version was formulated by Augustin Cauchy (1820), and several basic modifications were made by Joseph Liouville (1838), Rudolf Lipschitz (1876), and Emile Picard (1890). We state here a very limited version.

Theorem: Existence and Uniqueness of Solutions: *Suppose that a first-order o.d.e. can be written in the form*

$$\frac{dx}{dt} = f(t, x)$$

with both $f(t, x)$ and $\partial f / \partial x$ continuous in a rectangular region in the tx-plane, $t_1 < t < t_2$, $x_1 < x < x_2$ (where any of the bounds may be infinite). Then for any numbers t_0 and α, with (t_0, α) in the region, there is an open interval containing t_0, $a < t < b$, on which there exists precisely one solution satisfying the initial condition $x(t_0) = \alpha$.

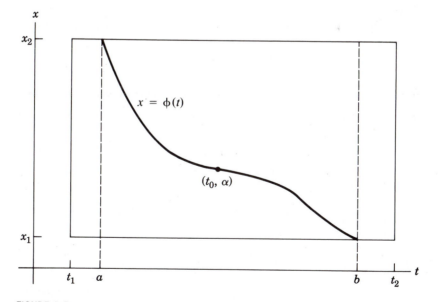

FIGURE 1.5

Note that this theorem guarantees both the existence and the uniqueness of solutions. That is, it tells us

1. There is a function $x = \phi(t)$, defined and continuous for $a < t < b$ that satisfies $x' = f(t, x)$ and $x(t_0) = \alpha$ (see Figure 1.5 on p. 51).

2. If $x = \psi(t)$ is any function defined and continuous for $a < t < b$ that also satisfies $x' = f(t, x)$ and $x(t_0) = \alpha$, then $\phi(t) = \psi(t)$ for $a < t < b$.

Let's look at an example that illustrates the hypotheses of the theorem.

Example 1.6.1

Describe all those points (t_0, α) at which the existence and uniqueness theorem does *not* guarantee a solution of the o.d.e.

$$t \frac{dx}{dt} = 2x$$

that satisfies $x(t_0) = \alpha$. Determine the largest rectangular region of the tx-plane on which the hypotheses of the theorem hold.

This o.d.e. can be rewritten in the form $dx/dt = f(t, x)$ with $f(t, x) = 2x/t$. Both $f(t, x) = 2x/t$ and $\partial f/\partial x = 2/t$ are continuous unless $t = 0$. Thus the existence and uniqueness theorem does not apply at points of the form $(0, \alpha)$ for any α.

In order for the theorem to apply in a rectangular region, the region must not contain any points of the form $(0, \alpha)$. The largest such rectangular regions are the half planes

$$0 < t < \infty, \quad -\infty < x < \infty$$

and

$$-\infty < t < 0, \quad -\infty < x < \infty.$$

Although an o.d.e. that does not satisfy the hypothesis may still have unique solutions satisfying the given initial conditions, the failure of the hypotheses to hold should sound a warning.

Example 1.6.2

Show that the o.d.e.

$$t \frac{dx}{dt} = 2x$$

has no solutions satisfying

$$x(0) = 1$$

and infinitely many solutions satisfying

$$x(0) = 0.$$

Recall from Example 1.6.1 that the existence and uniqueness theorem does not guarantee solutions satisfying these initial conditions.

If $x = x(t)$ is a solution of the o.d.e., then substitution of $t = 0$ into the o.d.e. gives

$$0 = 2x(0).$$

In particular, x cannot satisfy $x(0) = 1$.

Let's begin our investigation of solutions satisfying $x(0) = 0$ by avoiding the troublesome value $t = 0$. To find a solution valid for $t > 0$, we can separate variables by dividing by x and multiplying by dt:

$$\frac{1}{x} dx = \frac{2}{t} dt.$$

Integration gives

$$\ln|x| = 2 \ln|t| + k$$

so that

$$x = \pm e^{2 \ln|t| + k} = \pm e^{\ln t^2} e^k = ct^2 \quad \text{for } t > 0.$$

Since we divided by x, we must check whether $x = 0$ is a solution. It is, but it is of the form $x = ct^2$ with $c = 0$. Similarly, solutions have to satisfy

$$x = kt^2 \quad \text{for } t < 0.$$

We leave it to you to check that any function satisfying

$$x = \begin{cases} kt^2 & \text{for } t < 0 \\ 0 & \text{for } t = 0 \\ ct^2 & \text{for } t > 0 \end{cases}$$

(see Figure 1.6) is continuous and satisfies the o.d.e.

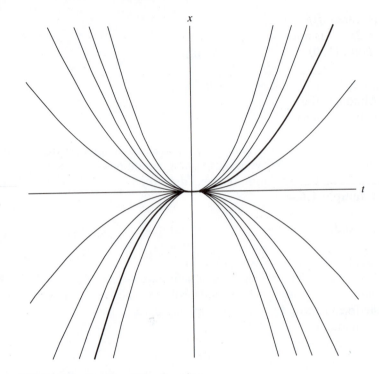

FIGURE 1.6   Graphs of the functions

$$x = \begin{cases} kt^2 \text{ for } t < 0 \\ ct^2 \text{ for } t > 0 \end{cases}$$

Example 1.6.3

The functions

$$x(t) = 0 \quad \text{and} \quad x(t) = (t - 1)^3$$

are both solutions of the o.d.e.

$$\frac{dx}{dt} = 3x^{2/3}$$

and both satisfy the initial condition

$$x(1) = 0.$$

In this case, $f(t, x) = 3x^{2/3}$ is continuous throughout the tx-plane, but $\partial f/\partial x = 2x^{-1/3}$ is not continuous when $x = 0$. Thus, the theorem does not apply at the point $(1, 0)$.

Although the existence and uniqueness theorem guarantees a solution defined on an interval of t-values, it is important to note that this interval may be narrower than the rectangle in which $f(t, x)$ and $\partial f/\partial x$ are continuous.

Example 1.6.4

The o.d.e.

$$\frac{dx}{dt} = x^2$$

is of the form $dx/dt = f(t, x)$, where $f(t, x) = x^2$. Here $f(t, x)$ and $\partial f/\partial x = 2x$ are continuous for $-\infty < t < \infty$ and $-\infty < x < \infty$. Show that the solution of this o.d.e. that satisfies

$$x(0) = 1$$

is not defined at $t = 1$.

We saw in Example 1.2.1 that the solutions of this o.d.e. are $x = 0$ and the functions

$$x = \frac{-1}{t + c}.$$

The only way to satisfy the initial condition $x(0) = 1$ is to take $c = -1$ in the preceding formula:

$$x = \frac{-1}{t - 1}.$$

This function is not defined at $t = 1$.

We summarize with a restatement of the theorem. A sketch of the proof of this theorem is given in Appendix B to Chapter 7, using ideas introduced in that chapter.

EXISTENCE AND UNIQUENESS THEOREM

Suppose that $f(t, x)$ and $\partial f/\partial x$ are continuous in a rectangular region of the tx-plane and that (t_0, α) is a point in that region. Then there is an interval $a < t < b$ containing t_0 on which there exists precisely one solution of the o.d.e.

$$\frac{dx}{dt} = f(t, x)$$

that satisfies the initial condition

$$x(t_0) = \alpha.$$

EXERCISES

In Exercises 1 through 8, describe all those points (t_0, α) at which the existence and uniqueness theorem does *not* guarantee a solution of the given o.d.e. that satisfies $x(t_0) = \alpha$.

1. $t\dfrac{dx}{dt} + 3x = t - 2$

2. $t(x - 1)\dfrac{dx}{dt} = 1$

3. $x(t^2 + 1)\dfrac{dx}{dt} = 1$

4. $(x - t)\dfrac{dx}{dt} = 1$

5. $\dfrac{dx}{dt} = (xt - t)^{1/3}$

6. $x(t^2 - 1)\dfrac{dx}{dt} = 1$

7. $\dfrac{dx}{dt} = |x|$

8. $(x^2 + t^2)\dfrac{dx}{dt} = \sin t$

In Exercises 9 through 16, you are given an o.d.e. and an initial condition $x(t_0) = \alpha$.

(a) Determine the largest rectangular region of the t-x plane that contains the point (t_0, α) and on which the hypotheses of the existence and uniqueness theorem hold for the given o.d.e.

(b) By solving the o.d.e., find the largest interval of t-values on which the solution of the o.d.e. that satisfies $x(t_0) = \alpha$ is defined and has values in the rectangular region found in (a).

9. $\dfrac{dx}{dt} = 5xt^{2/3}, \quad x(0) = 1$

10. $\dfrac{dx}{dt} = 5(xt)^{2/3}, \quad x(0) = 1$

11. $\dfrac{dx}{dt} = x^2, \quad x(0) = 3$

12. $\dfrac{dx}{dt} = x^2t, \quad x(0) = 1$

13. $\dfrac{dx}{dt} = xt^{1/2}, \quad x(1) = 1$

14. $\dfrac{dx}{dt} = (xt)^{1/2}, \quad x(1) = 1$

15. $xt^2 \dfrac{dx}{dt} = 1, \quad x(1) = 2$ 16. $\sin t \dfrac{dx}{dt} = 1, \quad x\left(\dfrac{\pi}{2}\right) = 0$

In Exercises 17 through 22, you are given an o.d.e. and a point (t_0, α) at which the existence and uniqueness theorem does not apply. By investigating the behavior of solutions valid near the given point, determine whether there are no solutions, more than one solution, or a unique solution satisfying $x(t_0) = \alpha$.

17. $t \dfrac{dx}{dt} = x - 2, \quad (t_0, \alpha) = (0, 2)$ 18. $(t - 1) \dfrac{dx}{dt} = -x, \quad (t_0, \alpha) = (1, 2)$

19. $(t - 1) \dfrac{dx}{dt} = -x, \quad (t_0, \alpha) = (1, 0)$ 20. $(t - 1) \dfrac{dx}{dt} = x, \quad (t_0, \alpha) = (1, 0)$

21. $(t - 1) \dfrac{dx}{dt} = x, \quad (t_0, \alpha) = (1, 2)$ 22. $(t^2 - 1) \dfrac{dx}{dt} = 2x, \quad (t_0, \alpha) = (-1, 0)$

23. Show that the solution of $dx/dt = x^2$ that satisfies $x(0) = a$ is $x = a/(1 - at)$ so that the interval on which the solution is defined depends on a.

24. Show that for any nonnegative numbers a and b, the function

$$x = \begin{cases} (t + a)^5 & \text{if } t < -a \\ 0 & \text{if } -a \le t \le b \\ (t - b)^5 & \text{if } t > b \end{cases}$$

is a solution of $dx/dt = 5x^{4/5}$ and satisfies $x(0) = 0$. Why doesn't this contradict the uniqueness part of the theorem?

25. Show that the functions

$$x(t) = t^3 \quad \text{and} \quad x(t) = \begin{cases} 0 & \text{if } t \le 0 \\ t^3 & \text{if } t > 0 \end{cases}$$

are both solutions of the o.d.e. $dx/dt = 3x^{2/3}$ and both satisfy the initial condition $x(1) = 1$. Why doesn't this contradict the uniqueness part of the theorem?

1.7   GRAPHING SOLUTIONS

Our main approach to differential equations will be the one followed in Sections 1.2 through 1.5. We try to express a solution by a formula involving familiar functions (polynomials, exponentials, logarithms, and trigonometric functions). This is known as a "closed-form solution." Unfortunately, many equations cannot be solved in closed form. We are forced to represent solutions of these equations in different ways.

Aside from a formula, there are three important ways of representing a function: a convergent power series, a table of values, and a graph. In this section we learn how to sketch graphs of solutions to a first-order o.d.e., using only the information provided directly by the equation.

Let's see how this is done for the equation

$$\frac{dx}{dt} = x.$$

What does the equation tell us about the graph of x as a function of t? The left side, dx/dt, is the slope of the curve. Thus, at any point on the graph of a solution, the slope of the curve is numerically equal to x, the height of the point above the t-axis. Conversely, any curve that satisfies this condition at all its points is the graph of a solution.

To represent this information geometrically, we choose some points (t, x) of the tx-plane and draw little line segments with slope $m = x$ to represent the tangents to solution curves at these points. In this example, solution curves will all cross a given horizontal line $x = k$ at the same slope, $m = k$. Thus, it is convenient to pick a few values of k and then to draw many segments, all

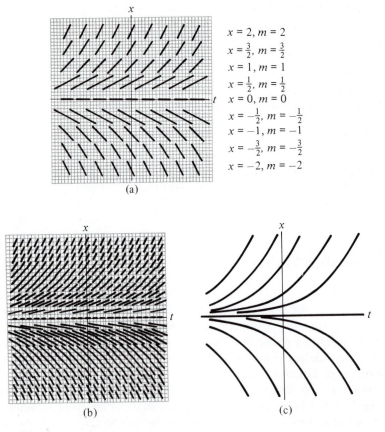

$x = 2, m = 2$
$x = \frac{3}{2}, m = \frac{3}{2}$
$x = 1, m = 1$
$x = \frac{1}{2}, m = \frac{1}{2}$
$x = 0, m = 0$
$x = -\frac{1}{2}, m = -\frac{1}{2}$
$x = -1, m = -1$
$x = -\frac{3}{2}, m = -\frac{3}{2}$
$x = -2, m = -2$

(a)

(b)

(c)

FIGURE 1.7 $x' = x$

crossing $x = k$ with slope $m = k$. In Figure 1.7(a) we have plotted segments for the values $k = -2, -3/2, -1, -1/2, 0, 1/2, 1, 3/2$, and 2.

In Figure 1.7(b) we have plotted segments for values of k in increments of 1/5 rather than 1/2. At this stage, the segments seem to form curves. The curves fitted to these segments [Figure 1.7(c)] give us a family of graphs of solutions to the original equation.

We have succeeded in graphing solutions without first solving the equation. Of course, our graphs are only approximately correct. Their accuracy depends in part on the number of "tangent" segments we have drawn. The more densely we scatter these segments, the more accurate our final graphs are likely to be. Unfortunately, the more segments we draw, the more likely we are to get writer's cramp [Figure 1.7(b), for example, required over 400 segments]. Computers with plotting devices can be very helpful in drawing the segments.

As an aid to drawing segments by hand, it is useful to locate curves along which $m = dx/dt$ is constant. A single slope will serve for all the segments along such a curve. The curves along which dx/dt is constant are called **isoclines** of the equation. In the preceding example, the isoclines were the horizontal lines $x = k$. We will see in Example 1.7.3 that isoclines need not be straight lines.

It is important to remember that even when an isocline is a straight line, its slope is generally different from the slope m of the segments crossing it. In the preceding example, every isocline had slope 0 regardless of the value of m. However, it may happen that a *particular* isocline has slope coinciding with m at all its points (this occurred in the preceding example for $k = 0$). The segments we draw on this isocline, instead of actually crossing it, will lie along it. This isocline will itself be a solution.

The isocline for $k = 0$ is of special interest even when it does not coincide with a solution curve. The segments crossing this isocline will always be horizontal. Any maxima or minima of solutions to the equation will lie on this isocline.

Example 1.7.1

$$\frac{dx}{dt} = t + 1$$

The isoclines in this example are the lines $t + 1 = k$. The segments crossing the isocline $t + 1 = k$ all have slope $m = k$. Since the isoclines are all vertical, no isocline can itself be a solution curve. The line $t = -1$ is the isocline for $k = 0$; segments cross the line horizontally.

In Figure 1.8(a) we have plotted isoclines for values of k in increments of 1/5 and sketched in segments crossing these isoclines. In Figure 1.8(b) we have fitted solution curves to these segments.

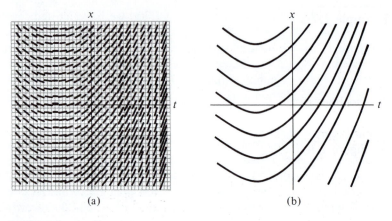

(a)   (b)

FIGURE 1.8 $x' = t + 1$

Example 1.7.2

$$\frac{dx}{dt} = x + t$$

The isoclines are the lines $x + t = k$. The segments crossing $x + t = k$ have slope $m = k$. Since every isocline is a line of slope -1, the particular isocline for $k = -1$ has a slope coinciding with m at every point. Therefore, $x + t = -1$ is a solution of the equation. The isocline for $k = 0$ is the line $x = -t$. We have plotted segments in Figure 1.9(a) and fitted curves to them in Figure 1.9(b).

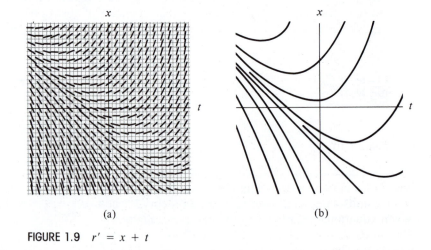

(a)   (b)

FIGURE 1.9 $r' = x + t$

Example 1.7.3

$$\frac{dx}{dt} = x^2 + t$$

The isoclines in this example are the parabolas $x^2 + t = k$. Segments are horizontal alone the isocline $x^2 = -t$. We have sketched some isoclines and segments in Figure 1.10(a and b) and fitted some solution curves to these segments in Figure 1.10(c).

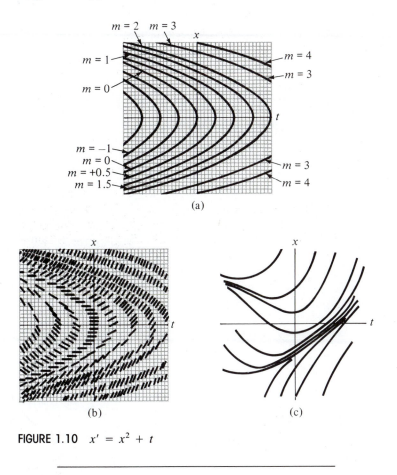

(a)

(b)   (c)

FIGURE 1.10 $x' = x^2 + t$

Our graphs can be used to obtain qualitative information about solutions of a given equation. Let's see, for example, what Figure 1.7(c) tells us about solutions of $dx/dt = x$. Solutions move away from the *t*-axis (i.e., $|x|$ increases) as *t* increases. The *t*-axis ($x = 0$) is itself a solution. No solution changes sign.

We notice that the solution curves fill up the *tx*-plane and do not cross each other. In particular, each point on the *x*-axis ($t = 0$) is crossed by some solution, and no two solutions cross the *x*-axis at the same point. This last observation is a geometric restatement of the existence and uniqueness theorem (Section 1.6) for this equation.

From Figure 1.9(b) we can see that solutions of $dx/dt = x + t$ that lie above the line $x + t = -1$ approach that line as *t* decreases, have a minimum at $x = -t$ (why?), and rise to $+\infty$ as $t \to \infty$. Solutions that lie below the line $x + t = -1$ also approach this line as *t* decreases; as *t* increases, *x* decreases at a progressively faster rate and tends to $-\infty$ as $t \to \infty$.

Let's summarize this technique.

GRAPHING SOLUTIONS OF $\dfrac{dx}{dt} = f(t, x)$

1. Try to sketch **isoclines** $f(t, x) = k$ for a few specific values of *k*. The value $k = 0$ is of special interest.

2. For each specific value of *k*, draw lots of line segments of slope $m = k$ along the corresponding isocline $f(t, x) = k$. These segments will be horizontal for $k = 0$.

3. Find any values of *k* for which the isocline $f(t, x) = k$ is a straight line of slope *k*. Any isocline of this type is itself a solution curve. Draw it!

4. Fit solution curves tangent to the segments you have drawn. Note that any maxima or minima occur along the isocline with $k = 0$.

Note

On approximations

In situations where quantitative information about solutions is needed, graphs obtained by this method can be used to yield approximations. In practical situations the differential equations are based on measurements, which in turn are subject to experimental error. Thus, practically speaking, a reasonably good approximate solution to an o.d.e. should be all we need.

Of course, an approximation is not very useful unless we have a good idea of its accuracy. The analysis of approximation schemes from the point of view of accuracy is a delicate and difficult subject. See Section 7.2 for an example of such an analysis.

EXERCISES

Plot graphs of solutions for the following o.d.e.'s in the region $-2 \leq t \leq 2$, $-2 \leq x \leq 2$. Use at least five isoclines and sketch at least three different solutions.

1. $x' = 3x + 2t$

2. $x' = x + 3t$

3. $\dfrac{dx}{dt} = xt$

4. $\dfrac{dx}{dt} = x + t^2$

5. $\dfrac{dx}{dt} = x - t^2$

6. $\dfrac{dx}{dt} = x^2 + t^2$

7. $tx' = x$

8. $\dfrac{dx}{dt} = x^2 - t$

9. $\dfrac{dx}{dt} = x^2 - t^2$

10. $\dfrac{dx}{dt} = \dfrac{x}{t + 2}$

11. $\dfrac{dx}{dt} = \dfrac{x}{t^2 - 4}$

12. $tx' = 1$ (Note that $x = \ln t$ is by definition the solution of this equation satisfying $x(1) = 0$. Be sure to include the graph of this solution.)

1.8 STABILITY IN FIRST-ORDER EQUATIONS

The graphical methods of the previous section illustrate a *qualitative* approach to differential equations: Instead of solving the o.d.e. in terms of explicit formulas, we try to extract information about the behavior of solutions directly from the equation itself. Our concern in the preceding section was to obtain approximate graphs of solutions. In this section we discuss the long-term behavior of solutions.

Let's begin with an example.

Example 1.8.1 The Logistic Equation

The population model in Example 1.1.6 is a special case of the o.d.e.

$$(1) \qquad\qquad \frac{dx}{dt} = \lambda x(P - x).$$

Let's investigate the long-term behavior of solutions to equation (1) in case λ and P are positive constants.

We start by looking for constant solutions. One such solution is the zero function $x(t) = 0$, whose graph in the tx-plane is the t-axis. Another is the horizontal line $x(t) = P$. Any solution that hits either of these lines must be equal to one of these solutions (Exercise 28).

To analyze the long-term behavior of other solutions $x(t)$, we first look at whether these solutions are increasing or decreasing. This is determined by the sign of the derivative dx/dt. The two stationary positions $x = 0$ and $x = P$ divide the real number line into three intervals, on each of which the sign of dx/dt does not change:

i.   When $x > P$, the right-hand side of equation (1) has one negative factor, $P - x$, and the other factors are all positive. Thus, dx/dt is negative, and any solution $x(t)$ that lies in this interval is a decreasing function.

ii. When $0 < x < P$, all factors on the right side of equation (1) are positive. Here dx/dt is positive, so solutions that lie in this interval are increasing.

iii. When $x < 0$, the factor $P - x$ is positive and λx is negative. Thus, dx/dt is negative, and solutions in this interval are decreasing.

One way to display these observations is to draw what is known as the **phase portrait** of the o.d.e. We represent a solution x by a marker that can slide along the real number line, subject to the o.d.e. The behavior of x depends on the initial position $x(0)$. A marker placed at 0 or P remains stationary. A marker placed to the right of P (so that $x(0) > P$) or to the left of 0 (so that $x(0) < 0$) will move to its left, whereas a marker placed between 0 and P (so that $0 < x(0) < P$) will move to its right. These directions of motion are indicated by arrowheads (Figure 1.11).

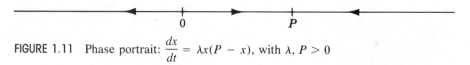

$$\begin{array}{ccc} & | & | \\ & 0 & P \end{array}$$

FIGURE 1.11   Phase portrait: $\dfrac{dx}{dt} = \lambda x(P - x)$, with $\lambda, P > 0$

Can we say anything more about the long-term behavior of these solutions without actually solving the o.d.e.? Since nonconstant solutions are perpetually decreasing or perpetually increasing, we need to know about the long-term behavior of such functions. In general, decreasing functions $x(t)$ come in two varieties (Figure 1.12): *bounded* ones, which never go below some specific lower bound; and *unbounded* ones, which ultimately go below every finite value (so that $\lim_{t \to \infty} x(t) = -\infty$). As $t \to \infty$ a bounded decreasing function $x(t)$ will approach a horizontal asymptote whose height is $L = \lim_{t \to \infty} x(t)$; as this happens, the slope dx/dt will approach 0. Similarly, an increasing function $x(t)$ either increases without bound (so that $\lim_{t \to \infty} x(t) = \infty$) or else (if it is bounded) approaches a horizontal asymptote at $L = \lim_{t \to \infty} x(t)$ and its slope approaches zero. Note that if $x(t)$ is a (decreasing or increasing) bounded solution of equation (1), then as dx/dt approaches zero, the right side of equation (1) will also approach zero, so that $\lambda L(P - L) = 0$. It follows that solutions of equation (1) approach $L = 0$ or $L = P$ (in the bounded case) or $\pm\infty$ (in the unbounded case).

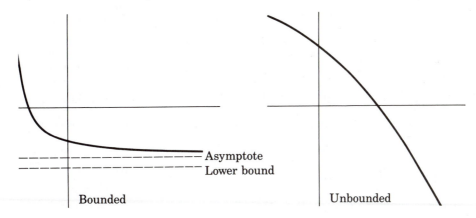

Asymptote

Lower bound

Bounded   Unbounded

FIGURE 1.12   Decreasing functions

Now, if $x(t)$ is a solution with $x(0) < 0$, then $x(t)$ is decreasing. It cannot approach either $L = 0$ or $L = P$, so it cannot be bounded. Thus, *solutions of equation (1) that start from negative initial values decrease without bound.*

If $x(0) > P$, then $x(t)$ is decreasing, and since it cannot cross the line $x = P$, it is bounded below. Since $x(t)$ cannot approach 0 (why not?), it must

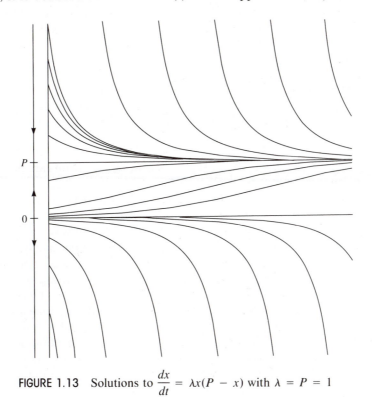

FIGURE 1.13   Solutions to $\dfrac{dx}{dt} = \lambda x(P - x)$ with $\lambda = P = 1$

approach P. Thus, *solutions of* (1) *that start above P decrease, approaching P as* $t \rightarrow \infty$.

A solution of (1) that starts between 0 and P is increasing and bounded above by P. It follows that *every solution of* (1) *with* $0 < x(0) < P$ *is perpetually increasing and approaches P as* $t \rightarrow \infty$.

In Figure 1.13 (page 65), we have aligned the phase portrait of equation (1) with the (vertical) x-axis and sketched some graphs of solutions. Note that the behavior we have predicted is borne out by these graphs.

If we view equation (1) as modeling a population, then $x(0) > 0$. Based on our analysis of the long-term behavior of solutions, we see that no matter what the initial (nonzero) population size is, the population will approach the "saturation" value P.

The procedure and arguments in the preceding example can be applied to a much wider class of first-order o.d.e.'s. In the equations we consider, the value of dx/dt will be determined purely by the value of x, independent of t:

(E)
$$\frac{dx}{dt} = f(x).$$

Equations of this form are called **autonomous.** We will also assume that $f(x)$ and $f'(x)$ are continuous (so that the existence and uniqueness theorem discussed in Section 1.6 applies).

To determine the long-range behavior of solutions to (E), we first find the values of x for which the right side of (E) is zero. These are the locations of the constant solutions. We will refer to a constant solution $x = c$ as an **equilibrium** of (E).

Fact: *The constant function* $x(t) = c$ *is an equilibrium of* $dx/dt = f(x)$ *if and only if* $f(c) = 0$.

We mark these stationary positions on the real number line. They divide the line into intervals, on each of which the sign of $f(x)$ is unchanging. On each of these intervals the direction of motion is determined by the sign of $f(x)$: The motion is toward the right on intervals where $f(x)$ is positive and toward the left where $f(x)$ is negative. We indicate each such direction by an arrowhead, thus obtaining the **phase portrait** of (E).

Now, we note that if a solution $x(t)$ approaches a finite limiting value L, then as it does so the slope dx/dt approaches zero. Thus, the zeros of $f(x)$ are the only possible candidates for the value of L.

Fact: *If x(t) is a solution of the o.d.e. dx/dt = f(x) (with f(x) continuous) and if L = lim$_{t \to \infty}$x(t) is finite, then f(L) = 0.*

To determine the limiting behavior of the solution $x(t)$ with a given initial value $x(0)$, we first locate $x(0)$ on the phase portrait. If $x(0)$ is at a zero of $f(x)$, the solution is constant. If not, note the direction of the arrowhead in the interval containing $x(0)$ and look for the first zero of $f(x)$ encountered when we move in this direction from $x(0)$. If there is none, then $x(t)$ is unbounded. If there is one, it must be the limiting value for the solution starting at $x(0)$.

From this analysis of the limiting behavior of solutions, we see that equilibria of (E) fall into three classes:

1. If $f(c) = 0$ and if the arrowheads on the two intervals of the phase portrait adjacent to c both point toward c, then all solutions of (E) that start near c will approach c as $t \to \infty$. In this case we call the equilibrium $x = c$ an **attractor.**

2. If $f(c) = 0$ and if the arrowheads on the intervals adjacent to c both point away from c, then all solutions of (E) that start near c will move away from c as $t \to \infty$. In this case we call $x = c$ a **repeller.**

3. If $f(c) = 0$ and if the two arrowheads on the intervals adjacent to c point in the same direction, then some solutions of (E) that start near c will move away from c and others will move toward c. In this case the equilibrium $x = c$ is neither a repeller nor an attractor.

The three types of equilibria exhibit different **stability** properties. Generally, a solution to an o.d.e. is **stable** if every solution with initial conditions close to the given one continues to follow closely the given solution forever. Attractors are examples of stable solutions, since any solution starting near an attractor will eventually be indistinguishable from it. On the other hand, a given solution of an o.d.e. is **unstable** if small deviations from its initial conditions can yield solutions that, in the long run, differ substantially from the given one. Equilibria of (E) that fall into the second and third classes are unstable, since in either case some solutions starting near c will move away from c.

Stability properties of solutions are extremely important for the modeling or design of physical systems by differential equations. Imagine a device whose behavior (output) is specified by a solution to an o.d.e., which we control by varying the initial conditions (regarded as input). Then stability of solutions is generally desirable, since it guarantees that we don't need absolute precision in the input to obtain reasonable output. On the other hand, if we are using a similar device to test for some specific condition in the input by observing the output, then the sensitivity to initial conditions implicit in an unstable solution can be exploited to design extremely accurate tests.

Let's consider some more examples.

Example 1.8.2

Sketch the phase portrait of

$$\frac{dx}{dt} = x(1 - x^2).$$

Determine the stability of all equilibria and the long-term behavior of all solutions.

First note that $x = c$ is a solution of the o.d.e. if and only if $c(1 - c^2) = 0$. Thus, there are three equilibria: $x = -1$, $x = 0$, and $x = 1$. Mark these on the number line.

The rest of the line consists of four intervals, on each of which the direction of motion is unchanging:

i.   For $x < -1$, both factors on the right side of the o.d.e. are negative. Thus, dx/dt is positive, and solutions with $x(0) < -1$ are increasing. These solutions approach -1 as $t \to \infty$.

ii. For $-1 < x < 0$, the first factor is negative and the second is positive. Thus, dx/dt is negative, and solutions with $-1 < x(0) < 0$ are decreasing. These solutions also approach -1 as $t \to \infty$.

iii. If $0 < x < 1$, then both factors are positive. Solutions with $0 < x(0) < 1$ increase and approach 1 as $t \to \infty$.

iv. If $x > 1$, then the first factor is positive, but the second is negative. Solutions with $x(0) > 1$ decrease and approach 1 as $t \to \infty$.

The phase portrait is sketched in Figure 1.14. It is easy to see that $x = 0$ is a repeller, and $x = \pm 1$ are attractors.

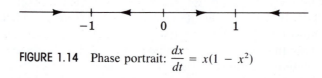

FIGURE 1.14   Phase portrait: $\dfrac{dx}{dt} = x(1 - x^2)$

Example 1.8.3

Analyze the long-term behavior of solutions to

$$\frac{dx}{dt} = x^2(1 - x^2).$$

As in the preceding example, the equilibria of the o.d.e. are $x = -1$, $x = 0$, and $x = 1$. The only way for the right side of the o.d.e. to be negative is for the second factor to be negative. This occurs only if x is smaller than -1 or larger than 1. Thus, the phase portrait is as in Figure 1.15.

From the phase portrait we see that $x = -1$ is a repeller, $x = 1$ is an attractor, and $x = 0$ is neither a repeller nor an attractor. Indeed, solutions with $x(0) < -1$ are decreasing and unbounded. If $-1 < x(0) < 0$, then $x(t)$ is increasing and approaches 0 as $t \to \infty$. Solutions with $0 < x(0) < 1$ are increasing and approach 1; those with $1 < x(0)$ are decreasing and approach 1.

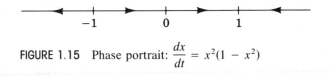

FIGURE 1.15   Phase portrait: $\dfrac{dx}{dt} = x^2(1 - x^2)$

From the phase portrait of an o.d.e. it is always easy to determine the stability of all equilibria. To decide the stability of a single, known equilibrium, the following derivative test for stability is sometimes convenient.

Derivative Test for Stability: *Suppose $f(x)$ and $f'(x)$ are continuous and $f(c) = 0$ so that $x = c$ is an equilibrium of $dx/dt = f(x)$. If $f'(c) \neq 0$, then the stability of $x = c$ is determined by the sign of $f'(c)$:*

1.   *If $f'(c) < 0$, then $x = c$ is an attractor.*
2.   *If $f'(c) > 0$, then $x = c$ is a repeller.*

If $f'(c) = 0$, then the test is inconclusive.

The derivative test is a consequence of our analysis of the implications of the phase portrait. Suppose we know that $f(c) = 0$ and that $f'(c) < 0$. The sign of the derivative tells us that $f(x)$ is decreasing through $x = c$. Thus, $f(x)$ is positive for x slightly to the left of c, and negative for x slightly to the right. This means that the arrowheads on the two intervals of the phase portrait adjacent to c both point toward c, so $x = c$ is an attractor. A similar argument (Exercise 26) gives the case $f'(c) > 0$. To see that the test is inconclusive if $f'(c) = 0$, see Exercise 27.

The usefulness of this test is illustrated by Example 1.8.4.

Example 1.8.4

Determine the stability of the equilibrium $x = 0$ to the o.d.e.

$$\frac{dx}{dt} = x \sin x - \sin 2x.$$

Here

$$f(x) = x \sin x - \sin 2x$$

$$f'(x) = \sin x + x \cos x - 2 \cos 2x.$$

We verify that $f(0) = 0$ and calculate that $f'(0) = -2 < 0$. Hence, the zero solution is an attractor.

The systematic study of stability properties for solutions to differential equations dates back at least to the work of the French mathematician H. Poincaré, the Russian mathematician A. M. Lyapunov, and the English mathematician E. J. Routh in the late nineteenth century. A century of intense study of stability questions has resulted in a very fine-tuned terminology, distinguishing many formulations of stability on subtle, but important, technical grounds. We have concentrated here on one basic formulation of stability, traditionally called *asymptotic stability*, and on an associated variety of instability. We stress that *phase portrait methods, as well as the arguments we used to conclude statements about limiting behavior from the phase portrait, depend heavily on our o.d.e. being autonomous.* In particular, we relied heavily on the fact that the points at which dx/dt was zero and the intervals on which it was positive or negative did not change with time. There are pitfalls in trying to use these methods when dx/dt depends on t as well as on x, even when the phase portrait itself makes some kind of sense.

Let's summarize.

STABILITY IN FIRST-ORDER EQUATIONS

Given an **autonomous** first-order o.d.e.

(E)
$$\frac{dx}{dt} = f(x)$$

with $f(x)$ and $f'(x)$ continuous:

1. Mark off on the real number line the values of c for which $f(c) = 0$. These are the values for which $x = c$ is an **equilibrium** of (E).

2. These marks divide the number line into intervals on each of which the sign of $f(x)$ does not change. On each of these intervals place an arrowhead pointing right if $f(x)$ is positive and left if $f(x)$ is negative.

The resulting diagram is the **phase portrait** of (E).

If $x(t)$ is a solution of (E), then the long-term behavior of $x(t)$ can be determined from the phase portrait as follows. If $x(0)$ is a zero of $f(x)$, then $x(t)$ is constant. Otherwise, locate the interval in which $x(0)$ lies. If the arrow on this interval points right, then $x(t)$ is always increasing. If the arrow points left, then $x(t)$ is decreasing. As $t \to \infty$, $x(t)$ approaches the endpoint of the interval toward which the arrowhead points (finite or infinite).

If every solution with initial condition close to that of $x(t)$ has long-term behavior that closely follows that of $x(t)$, then we say that $x(t)$ is a **stable** solution of (E). Examples of stable solutions are provided by those equilibria $x(t) = c$, that have the property that solutions starting near c all approach c as $t \to \infty$. Such solutions are called **attractors**. If $f(c) = 0$ and if the arrows in the intervals of the phase portrait adjacent to c both point toward c, then $x(t) = c$ is an attractor. Alternatively, if $f(c) = 0$ and $f'(c) < 0$, then $x(t) = c$ is an attractor.

If small deviations from $x(0)$ result in solutions that, in the long run, differ substantially from $x(t)$, then $x(t)$ is **unstable.** There are two ways in which an equilibrium $x(t) = c$ can be unstable. If the arrows in the intervals of the phase portrait adjacent to c point in the same direction, then solutions starting on one side of c approach c, whereas those starting on the other side move away. If the arrows point in opposite directions, then all solutions move away from c; in this case we refer to $x = c$ as a **repeller.** Repellers can also be detected without reference to the phase portrait: If $f(c)$ and $f'(c) > 0$, then $x = c$ is a repeller.

EXERCISES

In Exercises 1 through 14:

(a) Sketch the phase portrait.
(b) Find all equilibria and determine their stability.
(c) Discuss the long-term behavior of all solutions.

1. $\dfrac{dx}{dt} = (x - 1)(x + 2)$

2. $\dfrac{dx}{dt} = 6 - x - x^2$

3. $\dfrac{dx}{dt} = (x - 1)(x + 2)(x + 3)$

4. $\dfrac{dx}{dt} = (x - 1)(x + 2)^2$

5. $\dfrac{dx}{dt} = (x - 1)^2(x + 2)$

6. $\dfrac{dx}{dt} = x^3 - x$

7. $\dfrac{dx}{dt} = \dfrac{(3 - x)(1 + x)}{x^2 - x + 2}$

8. $\dfrac{dx}{dt} = 16 - x^4$

9. $\dfrac{dx}{dt} = x^4 - x^3 - 2x^2$

10. $\dfrac{dx}{dt} = (x + 2)(x + 1)^2(2 - x)^3$

11. $\dfrac{dx}{dt} = (x + 5)(x - 1)^2(x - 3)(x - 4)^2$

12. $\dfrac{dx}{dt} = (x - 1)^{4/3}(x + 4)^{1/3}$

13. $\dfrac{dx}{dt} = (x^2 - 9)(\sin x - 2)$

14. $\dfrac{dx}{dt} = \dfrac{2x - xe^x}{1 + e^x}$

In Exercises 15 through 19, determine whether the equilibrium $x = c$ is an attractor or a repeller.

15. $\dfrac{dx}{dt} = 2x^7 - 3x^6 + x^5 - 2x + 1, \quad c = \dfrac{1}{2}$

16. $\dfrac{dx}{dt} = 24x^3 - 22x^2 + x + 2, \quad c = \dfrac{2}{3}$

17. $\dfrac{dx}{dt} = x + 1 - \sqrt{1 + x^2}, \quad c = 0$

18. $\dfrac{dx}{dt} = x \cos x + x - \dfrac{\pi}{2}, \quad c = \dfrac{\pi}{2}$

19. $\dfrac{dx}{dt} = xe^{-x} + \sin x, \quad c = 0$

20. *Chemical Solutions:* In Exercise 23(b), Section 1.1, we modeled a chemical dissolving in a small tank of water. Use the methods of this section to determine the long-term behavior of the amount $x(t)$ of the chemical that is not yet dissolved.

21. *A Mixing Problem:* Use the methods of this section to determine the long-term behavior of the amount of pollutant in the pond of Exercise 5, Section 1.4.

22. *A Parachutist:* Use the methods of this section to determine the long-term behavior of the velocity of the parachutist in Exercise 14, Section 1.4.

23. *Chemical Reactions:* Use the methods of this section to determine the long-term behavior of the reactions in Exercise 23, Section 1.4.

24. *Growth of an Organism:* For simplicity suppose that a cell is shaped as a sphere of radius r and that its density is constant. The *von Bertalanffy growth model* assumes that two features affect the growth of the cell: intake of nutrients, which occurs uniformly over the surface of the cell, and respiration of cell material proportional to the volume. This leads to an o.d.e. of the form

$$\frac{dr}{dt} = \alpha r^2 - \beta r^3$$

where α and β are positive constants. Discuss the long-term behavior of the cell radius r.

Some more abstract problems:

25. a. Show that any autonomous linear first-order o.d.e. can be written in the form $dx/dt = q - px$, where p and q are constants.

   b. Show that if $p \neq 0$, then this equation has only one equilibrium, and that this equilibrium is an attractor if and only if $p > 0$.

26. Suppose that $f(x)$ and $f'(x)$ are continuous. Show that if $f(c) = 0$ and $f'(c) > 0$, then the equilibrium $x = c$ of the o.d.e. $dx/dt = f(x)$ is a repeller.

27. Find examples to show that if $f(c) = f'(c) = 0$, then the equilibrium $x = c$ of $dx/dt = f(x)$ can be an attractor, a repeller, or neither. (*Hint:* Try taking $f(x) = \pm x^n$ for various values of n.)

28. Suppose that $f(x)$ and $f'(x)$ are continuous and that $x = c$ is an equilibrium of the o.d.e. $dx/dt = f(x)$. Use the existence and uniqueness theorem from Section 1.6 to show that if $x_1(t)$ is a solution of the o.d.e. that satisfies $x_1(t_0) = c$, then $x_1(t) = c$ for all t.

1.9   EXACT DIFFERENTIAL EQUATIONS

The solutions of the o.d.e.'s we dealt with in Sections 1.2 and 1.3 were given explicitly by formulas for x in terms of t or implicitly by formulas of the form $F(x) = G(t) + c$. In this section we deal with equations whose solutions are given by formulas of the form $F(t, x) = c$.

We first note that if x is a function of t, then so is $F(t, x)$. We can use the chain rule for partial derivatives to calculate dF/dt:

$$\frac{dF}{dt} = \frac{\partial F}{\partial t}\frac{dt}{dt} + \frac{\partial F}{\partial x}\frac{dx}{dt} = \frac{\partial F}{\partial t} + \frac{\partial F}{\partial x}\frac{dx}{dt}.$$

Then

$$dF = \frac{\partial F}{\partial t}\,dt + \frac{\partial F}{\partial x}\,dx.$$

Example 1.9.1

If $F(t, x) = t^3 x^2 + e^{-t} - x^5$, then

$$\frac{\partial F}{\partial t} = 3t^2 x^2 - e^{-t} \qquad \text{and} \qquad \frac{\partial F}{\partial x} = 2t^3 x - 5x^4$$

so that

$$dF = (3t^2 x^2 - e^{-t})\,dt + (2t^3 x - 5x^4)\,dx.$$

Now suppose we are given a differential equation that can be written in the form

$$M(t, x)\,dt + N(t, x)\,dx = 0.$$

We will say the equation is **exact** if there is a function $F(t, x)$ so that

$$\frac{\partial F}{\partial t} = M \quad \text{and} \quad \frac{\partial F}{\partial x} = N.$$

If the equation is exact, it can be rewritten as

$$dF = 0.$$

The solution of this equation is

$$F = c.$$

Let's consider an example.

Example 1.9.2

Solve

$$(3t^2 \sin^2 x) \, dt + (2t^3 \sin x \cos x - 2e^{2x}) \, dx = 0.$$

If we can find a function $F(t, x)$ satisfying

$$\frac{\partial F}{\partial t} = 3t^2 \sin^2 x \quad \text{and} \quad \frac{\partial F}{\partial x} = 2t^3 \sin x \cos x - 2e^{2x},$$

then the solution of the o.d.e. will be $F = c$. Let's begin by noting that if $\partial F/\partial t = 3t^2 \sin^2 x$, then we can recover F by integrating this equation with respect to t, holding x constant. Since x is being held constant, we have to allow the "constant of integration" to be a function of x:

$$F = t^3 \sin^2 x + g(x).$$

To fit the second of our conditions, note on the one hand, that partial differentiation with respect to x of our expression for F yields

$$\frac{\partial F}{\partial x} = 2t^3 \sin x \cos x + \frac{dg(x)}{dx}.$$

On the other hand, we want $\partial F/\partial x = 2t^3 \sin x \cos x + 2e^{2x}$. This requires

$$2t^3 \sin x \cos x + \frac{dg(x)}{dx} = 2t^3 \sin x \cos x - 2e^{2x}.$$

Thus, we must have

$$\frac{dg(x)}{dx} = -2e^{2x}.$$

We integrate this equation to find $g(x)$. Since any function $g(x)$ that satisfies this equation will yield a function $F(t, x)$ with the required properties, we needn't include a constant of integration:

$$g(x) = -e^{2x}.$$

We substitute this value into our expression for F:

$$F = t^3 \sin^2 x - e^{2x}.$$

The solution of our o.d.e. is

$$t^3 \sin^2 x - e^{2x} = c.$$

In the next example we see what goes wrong if we attempt to find F for an equation that is not exact.

Example 1.9.3

Given the equation

$$2x \, dt + (2 + x)t \, dx = 0,$$

let's try to find a function $F(t, x)$ so that

$$\frac{\partial F}{\partial t} = 2x \quad \text{and} \quad \frac{\partial F}{\partial x} = (2 + x)t.$$

If we integrate the formula for $\partial F/\partial t$ with respect to t, holding x constant, we get

$$F = 2xt + g(x).$$

If we set the partial derivative with respect to x of this expression for F equal to $(2 + x)t$, we get

$$2t + \frac{dg(x)}{dx} = (2 + x)t.$$

Then

$$\frac{dg(x)}{dx} = xt.$$

This equation is impossible to solve: The left side is a function of x alone, but the right side is a function of x and t.

This last example illustrates the need for a simple test for exactness. If such a test had been available, then we would have known ahead of time that the attempt to find F was doomed to failure.

Our search for a test begins by noting that if $M\,dt + N\,dx = 0$ is exact, then by definition there is a function $F(t, x)$ with $\partial F/dt = M$ and $\partial F/\partial x = N$. It follows that

$$\frac{\partial M}{\partial x} = \frac{\partial}{\partial x}\left(\frac{\partial F}{\partial t}\right) = \frac{\partial^2 F}{\partial x\,\partial t} = \frac{\partial^2 F}{\partial t\,\partial x} = \frac{\partial}{\partial t}\left(\frac{\partial F}{\partial x}\right) = \frac{\partial N}{\partial t}.$$

Suppose conversely that

$$\frac{\partial M}{\partial x} = \frac{\partial N}{\partial t}.$$

Let's look for a function $F(t, x)$ with

$$\frac{\partial F}{\partial t} = M \qquad \text{and} \qquad \frac{\partial F}{\partial x} = N.$$

We can integrate the first of these equations with respect to t, holding x constant (since x is constant, we have to allow the "constant" of integration to be a function of x):

$$F = \int M\,\partial t + g(x).$$

Then

$$\frac{\partial F}{\partial x} = \frac{\partial}{\partial x}\int M\,\partial t + \frac{dg(x)}{dx}.$$

We want this partial derivative to equal N:

$$\frac{\partial}{\partial x}\int M\,\partial t + \frac{dg(x)}{dx} = N.$$

This requires

$$\frac{dg(x)}{dx} = N - \frac{\partial}{\partial x} \int M \, \partial t.$$

This last equation makes sense only if the right side is a function of x alone—that is, only if the partial derivative with respect to t of the right side is 0. This is the case here:

$$\frac{\partial}{\partial t}\left(N - \frac{\partial}{\partial x} \int M \, \partial t \right) = \frac{\partial N}{\partial t} - \frac{\partial^2}{\partial t \, \partial x} \int M \, \partial t$$

$$= \frac{\partial N}{\partial t} - \frac{\partial^2}{\partial x \, \partial t} \int M \, \partial t = \frac{\partial N}{\partial t} - \frac{\partial M}{\partial x} = 0.$$

We can integrate the formula for dg/dx and complete the determination of F by substituting the value for $g(x)$ into our expression for F.

We have obtained a simple test for exactness. *The equation $M \, dt + N \, dx$ is exact if and only if $\partial M/\partial x = \partial N/\partial t$.* If we had applied this test to the equation of Example 1.9.3, where

$$M = 2x \qquad \text{and} \qquad N = (2 + x)t,$$

we would have found that the equation is not exact, since

$$\frac{\partial M}{\partial x} = 2 \qquad \text{and} \qquad \frac{\partial N}{\partial t} = 2 + x \neq 2.$$

On the other hand, if we had applied the test to the equation in Example 1.9.2, where

$$M = 3t^2 \sin^2 x \qquad \text{and} \qquad N = 2t^3 \sin x \cos x - 2e^{2x},$$

we would have found that the equation is exact, since

$$\frac{\partial M}{\partial x} = 6t^2 \sin x \cos x = \frac{\partial N}{\partial t}.$$

In the discussion that led to this test, we used theorems from multidimensional calculus without worrying about whether their hypotheses were satisfied. (For example, we used the "fact" that $\partial^2 F/\partial x \, \partial t = \partial^2 F/\partial t \, \partial x$.) The precise statement of our exactness test involves conditions on the region of the tx-plane in which we are working. Roughly speaking, a region is **simply connected** if it has no holes. More precisely, a region is simply connected if it contains

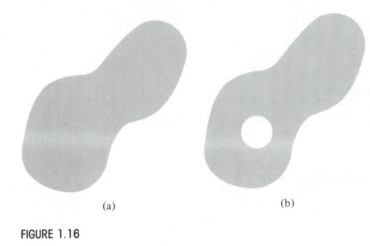

(a)   (b)

FIGURE 1.16

the interior of every simple closed curve lying in it. The region in Figure 1.16(a) is simply connected, whereas the region in Figure 1.16(b) is not. To ensure that our test works, we require that M and N be continuous and have continuous first partial derivatives in a simply connected region.

Fact: *Suppose that M, N, $\partial M/\partial x$, and $\partial N/\partial t$ are continuous in a simply connected region of the tx-plane. Then the equation*

$$M\, dt + N\, dx = 0$$

is exact if and only if

$$\frac{\partial M}{\partial x} = \frac{\partial N}{\partial t}.$$

The derivation of this test used the same method that worked in Example 1.9.2. Let's see how this works in another example.

Example 1.9.4

Find the solution of

$$(2t^3x - 5x^4)\frac{dx}{dt} = -3t^2x^2 - 1$$

that satisfies the initial condition

$$x(1) = 2.$$

We first rewrite the o.d.e. in the form $M \, dt + N \, dx = 0$:

$$(3t^2x^2 + 1) \, dt + (2t^3x - 5x^4) \, dx = 0.$$

This equation is exact, since

$$\frac{\partial}{\partial x}(3t^2x^2 + 1) = 6t^2x = \frac{\partial}{\partial t}(2t^3x - 5x^4).$$

If we find a function $F(t, x)$ with

$$\frac{\partial F}{\partial t} = 3t^2x^2 + 1 \quad \text{and} \quad \frac{\partial F}{\partial x} = 2t^3x - 5x^4,$$

then the general solution will be $F = c$.

We first integrate the equation $\partial F/\partial t = 3t^2x^2 + 1$ with respect to t, holding x constant:

$$F = t^3x^2 + t + g(x).$$

We set the partial derivative with respect to x of the resulting expression for F equal to N:

$$2t^3x + \frac{dg(x)}{dx} = 2t^3x - 5x^4.$$

We solve for dg/dx and integrate to find $g(x)$:

$$\frac{dg(x)}{dx} = -5x^4$$

$$g(x) = -x^5.$$

We substitute this value into our expression for F:

$$F = t^3x^2 + t - x^5.$$

The general solution of our o.d.e. is

$$t^3x^2 + t - x^5 = c.$$

Substituting our initial condition $x(1) = 2$ into this equation, we obtain

$$(1)(2)^2 + 1 - 2^5 = c.$$

That is,

$$c = -27.$$

The required solution is

$$t^3 x^2 + t - x^5 = -27.$$

Let's summarize.

EXACT DIFFERENTIAL EQUATIONS

A differential equation of the form

(E) $$M \, dt + N \, dx = 0$$

is **exact** if there is a function $F(t, x)$ so that

$$\frac{\partial F}{\partial t} = M \quad \text{and} \quad \frac{\partial F}{\partial x} = N.$$

Test for Exactness

The o.d.e. (E) is exact if and only if

$$\frac{\partial M}{\partial x} = \frac{\partial N}{\partial t}.$$

Solving Exact Equations

If the equation (E) is exact, we find the solutions as follows.

1. Integrate the equation $\partial F/\partial t = M$ with respect to t, holding x constant (we must allow a function of x as the "constant" of integration):

    $$F = \int M \, \partial t + g(x).$$

2. Set the partial with respect to x of the resulting expression for F equal to N:

    $$\frac{\partial F}{\partial x} = \frac{\partial}{\partial x} \int M \, \partial t + \frac{dg(x)}{dx} = N.$$

3. Solve for dg/dx:

    $$\frac{dg(x)}{dx} = N - \frac{\partial}{\partial x} \int M \, \partial t.$$

4. Integrate the formula for dg/dx to find g. Since any function satisfying this formula yields a function F with the required properties, we needn't include a constant of integration:

$$g(x) = \int \left(N - \frac{\partial}{\partial x} \int M \, \partial t \right) dx.$$

5. Obtain F by substituting the value for g into the formula for F found in step 1.

6. The solution of the o.d.e. (E) is

$$F = c.$$

Notes

1. On finding F

The integration in step 1 is sometimes hard. In these cases it may be easier first to work with the equation $\partial F/\partial x = N$. We can integrate this equation with respect to x, holding t constant (the "constant" of integration will be a function of t):

$$F = \int N \, \partial x + k(t).$$

Substituting this expression into the equation $\partial F/\partial t = M$ will yield an equation for dk/dt. If the o.d.e. is exact, we can solve for $k(t)$ and thereby find F.

As an example, let's solve the o.d.e.

$$(-x \sin^3 t + 2x \sin t \cos^2 t + 2t) \, dt + \sin^2 t \cos t \, dx = 0.$$

Here

$$M = -x \sin^3 t + 2x \sin t \cos^2 t + 2t \qquad \text{and} \qquad N = \sin^2 t \cos t.$$

We leave it to you to check that $\partial M/\partial x = \partial N/\partial t$, proving that the equation is exact. Integrating $\partial F/\partial t = M$ would require us to calculate $\int (\sin^3 t - 2 \sin t \cos^2 t) \, dt$. Instead, let's first integrate N with respect to x, holding t constant:

$$F = \int N \, \partial x = \int \sin^2 t \cos t \, \partial x = x \sin^2 t \cos t + k(t).$$

We now take the partial derivative with respect to t of this expression and set it equal to M:

$$x(-\sin^3 t + 2 \sin t \cos^2 t) + \frac{dk(t)}{dt} = -x \sin^3 t + 2x \sin t \cos^2 t + 2t.$$

We solve for $dk(t)/dt$ and integrate:

$$\frac{dk(t)}{dt} = 2t$$

$$k(t) = t^2.$$

Thus,

$$F = x \sin^2 t \cos t + t^2.$$

The solution of the o.d.e. is

$$x \sin^2 t \cos t + t^2 = c.$$

2.   Integrating factors
    If an equation $M\,dt + N\,dz = 0$ is not exact, it may be possible to multiply by a function $\rho(t, x)$ so that the resulting equation $\rho M\,dt + \rho N\,dx = 0$ is exact. Such a function ρ is called an **integrating factor,** and we find the solutions of the original equation by solving the new (exact) equation. For example, the equation

$$(3t + 2x)\,dt + (t^2 + t + tx)\,dx = 0$$

is not exact [since $\partial(3t + 2x)/\partial x = 2$, but $\partial(t^2 + t + tx)/\partial t = 2t + 1 + x$]. If we multiply this equation by $\rho = te^x$, we get

$$(3t^2 + 2xt)e^x\,dt + (t^3 + t^2 + t^2x)e^x\,dx = 0,$$

which is exact [since $\partial(3t^2 + 2xt)e^x/\partial x = (3t^2 + 2xt + 2t)e^x = \partial(t^3 + t^2 + t^2x)e^x/\partial t$]. Thus, $\rho = te^x$ is an integrating factor. We can now use the method of this section to find the general solution of our o.d.e.:

$$t^3 e^x + t^2 x e^x = c.$$

Exercises 16 through 25 include other examples of integrating factors.

EXERCISES

Test the equations in Exercises 1 through 11 for exactness and solve those that are exact.

1. $t\,dt + x\,dx = 0$

2. $x\,dt + t\,dx = 0$

3. $x\,dt - t\,dx = 0$

4. $(2x^2 + 2t + 1)\,dt + (4x^3 + 4tx)\,dx = 0$

5. $(2xt^2 + 2x)\,dt + (2tx^2 + 2t)\,dx = 0$

6. $(t^3 + x \sin t)\,dt + (2x - \cos t)\,dx = 0$

7. $xe^{tx}\,dt + e^{tx}\,dx = 0$

8. $xe^{tx}\,dt + te^{tx}\,dx = 0$

9. $x^2 \cos t\,dt + (xt \cos tx + \sin tx)\,dx = 0$

10. $(2x \sin t \cos t + e^x + t)\,dt + (\sin^2 t + te^x)\,dx = 0$

11. $\dfrac{x(1 - t^2)}{(t^2 + 1)^2}\,dt + \dfrac{t}{t^2 + 1}\,dx = 0$

In Exercises 12 through 15, find the solution of the o.d.e. that satisfies the given initial condition.

12. $6tx \, dt + (3t^2 + 4 \sin x \cos x) \, dx = 0, \quad x(0) = \dfrac{\pi}{4}$

13. $\left(x + 1 - \dfrac{2xt}{(t^2 - 1)^2}\right) dt + \left(\dfrac{1}{(t^2 - 1)} + t\right) dx = 0, \quad x(0) = 1$

14. $(2xt + e^t \sin x) + (2x + t^2 + e^t \cos x)\dfrac{dx}{dt} = 0, \quad x(0) = 2$

15. $x(t + 1)e^t + (1 + te^t)\dfrac{dx}{dt} = 0, \quad x(0) = 1$

In Exercises 16 and 17, check that the given function ρ is an integrating factor (see Note 2) and solve the o.d.e.

16. $(1 - x - t) \, dt + (1 + x + t) \, dx = 0, \quad \rho = e^{x-t}$
17. $(x^2 + x + xt) \, dt + (3x + 2t) \, dx = 0, \quad \rho = xe^t$

The o.d.e.'s in Exercises 18 through 22 have integrating factors that are functions of t alone [$\rho = \rho(t)$]; in Exercises 23 through 25, the integrating factors are functions of x alone [$\rho = \rho(x)$].

(a) By applying the exactness test to the exact equation $\rho M \, dt + \rho N \, dx = 0$, find a differential equation for ρ.

(b) Solve the o.d.e. for ρ.

(c) Solve $\rho M \, dt + \rho N \, dt = 0$.

18. $xt^2 \, dt + t^3 \, dx = 0$
19. $(6x + 5t) \, dt + 2t \, dx = 0$
20. $(3x - 3e^{2x}) \, dt + (1 - 2e^{2x})(t + 1) \, dx = 0$
21. $\dfrac{1}{2}x \, dt + (t + t^{1/2} \sin x) \, dx = 0, \quad t > 0$
22. $(x^2 + e^x) \, dt + (2x + e^x)(\tan t) \, dx = 0, \quad -\dfrac{\pi}{2} < t < \dfrac{\pi}{2}$
23. $(x + x^2 \cos t) \, dt + (2t + 3x \sin t) \, dx = 0$
24. $(4t + 2e^t)x \, dt + (3t^2 + 3e^t) \, dx = 0$
25. $\cos t \, dt + (2x + \sin t + 1) \, dx = 0$
26. a. Show that if $M \, dt + N \, dx = 0$ has an integrating factor ρ, then

$$M \frac{\partial \rho}{\partial x} + \rho \frac{\partial M}{\partial x} = N \frac{\partial \rho}{\partial t} + \rho \frac{\partial N}{\partial t}.$$

b. Show that if the integrating factor is a function of t alone [$\rho = \rho(t)$], then

$$\frac{d\rho}{dt} = \frac{\rho}{N}\left(\frac{\partial M}{\partial x} - \frac{\partial N}{\partial t}\right).$$

c. Show that if the integrating factor is a function of x alone [$\rho = \rho(x)$], then

$$\frac{d\rho}{dx} = \frac{\rho}{M}\left(\frac{\partial N}{\partial t} - \frac{\partial M}{\partial x}\right).$$

REVIEW PROBLEMS

The methods we have discussed in this chapter are just a few of the many known methods for solving first-order o.d.e.'s. For Exercises 1 through 19, determine whether the given equation can be solved by any of these methods. If so, (a) find the general solution and (b) find the specific solution satisfying the given initial condition.

1. $(t^2 - 1)\dfrac{dx}{dt} + x = 0, \quad t > 1; \qquad x(2) = 1$

2. $x' = x + t^2; \qquad x(0) = 1$

3. $x' = x + x^2; \qquad x(0) = 1$

4. $x' + 5x = t; \qquad x(0) = 0$

5. $tx' = -x, \quad t > 0; \qquad x(2) = 1$

6. $tx' + 4x = t^3, \quad t > 0; \qquad x(2) = 1$

7. $3t^2 xx' = tx + x + t + 1, \quad t > 0; \qquad x(2) = 1$

8. $\cos t \dfrac{dx}{dt} = x \sin t, \quad -\dfrac{\pi}{2} < t < \dfrac{\pi}{2}; \qquad x(0) = 1$

9. $x' \cos t + x \sin t = \cos t, \quad -\dfrac{\pi}{2} < t < \dfrac{\pi}{2}; \qquad x(0) = 1$

10. $3t\, dx + 4xt\, dt = 0, \quad t > 0; \qquad x(3) = 1$

11. $(t + x)\, dx + (x - t)\, dt = 0; \qquad x(1) = 1$

12. $(e^t - 1)\, dx + e^{2t}\, dt = 0, \quad t > 0; \qquad x(\ln 2) = 1$

13. $t^2 x' - (t + 1)x + t = 0, \quad t > 0; \qquad x(1) = 1$

14. $x' + 5x = tx^3; \qquad x(0) = 1$

15. $(x^2 t + t^2 x)\dfrac{dx}{dt} = (x^3 + t^3), \quad t > 0; \qquad x(2) = 1$

16. $e^{x-t}\dfrac{dx}{dt} - t^2 = 0; \qquad x(0) = \ln 2$

17. $\dfrac{dx}{dt} = x^2 t \sin 2t; \qquad x(0) = 1$

18. $2\dfrac{dx}{dt} - x - t^2 = 0; \qquad x(0) = 1$

19. $2\dfrac{dx}{dt} - x = \sqrt{t}e^t; \qquad t > 0; \quad x(1) = 0$

20. Plot graphs of three or four solutions of $x' = 3x - 2t$ in the region $-2 < t < 2$, $-2 < x < 2$.

21. Plot graphs of three or four solutions of $x' = t^2 - x$ in the region $-2 < t < 2$, $-2 < x < 2$.

22. Discuss the long-term behavior of the solutions of $dx/dt = (x - 1)(x - 2)$ that satisfy
    a. $x(0) = 0$ b. $x(0) = 1$
    c. $x(0) = 1.5$ d. $x(0) = 2.5$

23. Discuss the long-term behavior of the solutions of $dx/dt = (x - 1)^2(x - 2)$ that satisfy
    a. $x(0) = 0$ b. $x(0) = 1$
    c. $x(0) = 1.5$ d. $x(0) = 2.5$

24. *Savings Account:* A savings account is opened with a deposit of $10,000 and ac- crues interest at 7% per year, compounded continuously. Money is withdrawn from the account continuously at the rate of $1000 per year. How long will it take until there is no money left in the account?

25. *An Epidemic (or Rumor):*
    a. When a contagious disease (or a rumor) spreads in a very large population, the rate at which it spreads is proportional to the number $x(t)$ of people who have the disease. Write an o.d.e. for the spread of a disease if each person who has it infects two people per day.
    b. How many people will have the disease of part (a) one week after it is intro- duced into the population by one person?
    c. In a smaller population the rate of spread of a disease decreases as the number of people who don't have it decreases. Assume that the disease in part (a) spreads among the members of a college community numbering 6000. To ac- count for the small population size, multiply the per capita growth rate in (a) by a constant multiple of the number of people who don't have the disease. Determine the constant using the fact that when the number of people who have the disease is small, the per capita growth rate should be close to the rate of spread in a large population.
    d. How many people will have the disease of part (c) one week after it is intro- duced into the population by one person?

26. *Pollution:* A pond containing 1000 cubic meters has two inlets and one outlet. Water containing 2 grams of pollutant per cubic meter flows into one inlet at the rate of 6 cubic meters per minute. The water flowing into the other inlet contains 3 grams of pollutant per cubic meter and flows at the rate of 4 cubic meters per minute. Water flows out through the outlet at the rate of 10 cubic meters per minute.
    a. Find the amount $x(t)$ of pollutant in the pond after t minutes if the water in the pond is pure at the outset.
    b. What happens to $x(t)$ as $t \to \infty$?

27. *A Chemical Reaction:* Suppose that when chemicals A and B react, one molecule of A combines with one molecule of B to form one molecule of C, and that the reaction obeys the law of mass action. At the outset, the concentrations of A, B, and C are 2 moles per liter, 4 moles per liter, and 0, respectively. After 1 second, the concentration of C is 0.4 moles per liter. Find a formula for the concentration $x(t)$ of C after t seconds.

28. *House Heating:* (A calculator may be useful here.) Suppose the heating system of a house could, if the house were perfectly insulated, raise the temperature 10°F per hour. However, the house is not perfectly insulated. It loses heat according to Newton's law of cooling.
    a. If the heat is turned off when the house is at 70°F and the temperature outside is a constant 30°F, then the house temperature reaches 60°F after 2 h. How long does it take, from the time the heat is turned off, for the house to reach 50°F?

b. When the temperature inside reaches 50°F, the heat is turned back on. What will the temperature be 2 hours later?

c. When the temperature inside reaches 70°F, the temperature outside drops steadily at 15°F per hour (from the initial temperature of 30°). If the heating system remains on, what will the temperature of the house be when the temperature outside reaches 0°F?

29. *A Space Launch:* An object of mass m, shot upward so that it travels along a straight line through the center of the Earth, is subject to a gravitational force $F = -mgR^2/x^2$, where g is the gravitational constant, R is the radius of the Earth, and x is the distance of the object from the center of the Earth. Assume that this is the only force acting on the object (we are ignoring air resistance, the gravitational pull of the moon, and so on).

a. Using Newton's second law of motion, set up a differential equation for the distance x of the object from the center of the earth.

b. Using the fact that $a = d^2x/dt^2 = dv/dt = (dx/dt)(dv/dx) = v(dv/dx)$ to obtain a differential equation for velocity v in terms of x.

c. Solve the equation in (b). Express the arbitrary constant in terms of the initial velocity v_0, using $x(0) = R$.

d. If the velocity is 0 at some time, then the object falls back to Earth. If the velocity stays positive—that is, if it is never 0—then the object escapes into space. Show that the object escapes if and only if $v_0^2 \geq 2gR$. (*Hint:* What would x have to be when $v = 0$?)

e. Find a formula for x at time t if the initial velocity is $v_0 = \sqrt{2gR}$.

LINEAR DIFFERENTIAL EQUATIONS

2.1 SOME SPRING MODELS

In this section we consider some elementary models, involving damped spring systems, that lead to "linear" o.d.e.'s of order higher than one. The basic model for a spring is given by **Hooke's law** (Robert Hooke was a contemporary of Newton). This states that *a spring stretched or squeezed from its natural length L to length L + x exerts a restoring force proportional to the displacement x*. When the spring is *stretched* ($x > 0$), the proportionality to x means the force has magnitude kx for some positive constant k. However, since this restoring force acts to *shorten* the spring, it will be negative:

$$F_{\text{spr}} = -kx.$$

You should check that when the spring is *squeezed* ($x < 0$), this same formula gives a force acting to *lengthen* the spring. The constant k is called the **spring constant.**

Example 2.1.1   Undamped Springs

Suppose a mass of m grams moves along a horizontal track but is attached to a wall at one end of the track by a spring with natural length L cm and spring constant k dynes/cm. In modeling the motion, it is most convenient to specify the position of the mass, *not* by the distance from the end of the track, but rather by the deviation x of the length of the spring from equilibrium (see Figure 2.1). Hooke's law tells us that at position x (that is, when the mass is $L + x$ from the wall), the spring exerts a force $F_{\text{spr}} = -kx$. On the other hand, Newton's second law tells us that

$$F = ma = m\,\frac{d^2x}{dx^2}.$$

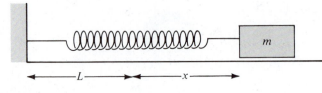

FIGURE 2.1

Equating these two descriptions of the force on the mass leads to the second-order o.d.e.

$$m \frac{d^2x}{dx^2} + kx = 0.$$

Example 2.1.2   Damped Springs

The model in Example 2.1.1 assumed that there were no forces resisting the motion of the mass. In practice, the mass might encounter significant resistance from the air (or other surrounding medium) or might be attached to a device that acts as a shock absorber. For example, the mass might be attached to a dashpot consisting of a piston moving in a stationary container filled with oil (see Figure 2.2). In either case, the mass would be subject to a resisting force that depends on the velocity and opposes the direction. The simplest model for such resisting forces is **viscous damping,** in which the magnitude of the force is proportional to the velocity:

$$F_{\text{damp}} = -b \frac{dx}{dt}.$$

Here, the (positive) constant of proportionality b is called the **damping coefficient.**

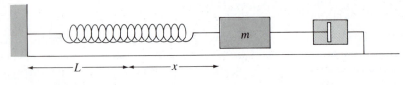

FIGURE 2.2

When we take into account both the spring force and viscous damping, the total force on the mass is

$$F_{\text{tot}} = F_{\text{spr}} + F_{\text{damp}} = -kx - b \frac{dx}{dt}.$$

Newton's second law gives us the second-order o.d.e.

$$m \frac{d^2x}{dt^2} = -kx - b \frac{dx}{dt}$$

or

$$m \frac{d^2x}{dt^2} + b \frac{dx}{dt} + kx = 0.$$

Example 2.1.3 External Forces

The spring and damping forces in the preceding examples were internal in the sense that they depended only on the position and velocity of the mass in the system. When an outside force is imposed on a system of this type, we add an "external" force term

$$F_{ext} = E(t).$$

In the presence of internal (spring and damping) and external forces, the total force is

$$F_{tot} = F_{spr} + F_{damp} + F_{ext}.$$

Newton's second law leads to the second-order equation

$$m \frac{d^2x}{dt^2} + b \frac{dx}{dt} + kx = E(t).$$

Two common forms of the external force term are constants $E(t) = E$ (e.g., gravity acting on a system hung vertically) and oscillating forces, such as the sinusoidal term $E(t) = E \sin \beta t$ with amplitude E and period $2\pi/\beta$.

Example 2.1.4 A Coupled-Spring System

Suppose two masses of m_1 and m_2 grams, respectively, move along a horizontal track and encounter no resisting forces. Assume that they are attached to each other by a spring with natural length L_2 and spring constant k_2, and that the first mass is attached to a wall at the end of the track by a spring with constant k_1 and natural length L_1. A description of the configuration of this system requires two positions, one for each mass. When both springs are at equilibrium, the first mass is L_1 cm from the wall, and the second is

$L_1 + L_2$ cm from the wall. Let's specify the positions in general by x_1 and x_2 when the first mass is $L_1 + x_1$ cm from the wall and the second mass is $L_1 + L_2 + x_2$ cm from the wall (see Figure 2.3).

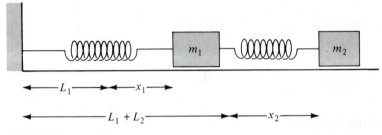

FIGURE 2.3

To apply Newton's second law, we must analyze the forces on each mass. The second mass is acted on only by the connecting spring. The length of this spring is the distance between the two masses,

$$(L_1 + L_2 + x_2) - (L_1 + x_1) = L_2 + (x_2 - x_1),$$

which represents a deviation of $x_2 - x_1$ cm from its natural length. Thus, the force on the second mass is

$$F_2 = -k_2(x_2 - x_1).$$

The situation of the first mass is slightly more complicated. The spring attaching it to the wall is pulled to a length of $L_1 + x_1$ cm, so it exerts a force $F_1 = -k_1x_1$. However, the connecting spring also acts on the first mass, pulling with a force equal in magnitude and opposite in direction to the force it exerts on the second mass, that is, $-F_2$. Thus, the total force on the first mass is

$$F_1 - F_2 = -k_1x_1 + k_2(x_2 - x_1) = -(k_1 + k_2)x_1 + k_2x_2.$$

Applying Newton's law to each mass in turn leads to two equations

$$m_1 \frac{d^2x_1}{dt^2} = -(k_1 + k_2)x_1 + k_2x_2$$

$$m_2 \frac{d^2x_2}{dt^2} = k_2x_1 - k_2x_2.$$

Since each equation involves both x_1 and x_2, we cannot hope to solve either equation separately. Note, however, that we can solve the first equation for x_2 in terms of x_1 and d^2x_1/dt^2:

$$x_2 = \frac{m_1}{k_2} \frac{d^2x_1}{dt^2} + \frac{k_1 + k_2}{k_2} x_1.$$

We can differentiate this equation twice to get

$$\frac{d^2x_2}{dt^2} = \frac{m_1}{k_2}\frac{d^4x_1}{dt^4} + \frac{k_1 + k_2}{k_2}\frac{d^2x_1}{dt^2}.$$

We can now substitute our expressions for x_2 and d^2x_2/dt^2 into the second equation to get

$$\frac{m_2m_1}{k_2}\frac{d^4x_1}{dt^4} + \frac{m_2(k_1 + k_2)}{k_2}\frac{d^2x_1}{dt^2} = k_2x_1 - m_1\frac{d^2x_1}{dt^2} - (k_1 + k_2)x_1.$$

Multiplying by k_2 and collecting terms gives

$$m_2m_1\frac{d^4x_1}{dt^4} + (m_2k_1 + m_2k_2 + m_1k_2)\frac{d^2x_1}{dt^2} + k_2k_1x_1 = 0.$$

This fourth-order o.d.e. can be solved for x_1. We can then obtain x_2 by substituting into our expression for x_2 in terms of x_1 and d^2x_1/dt^2.

EXERCISES

Springs

1. *A Vertical Spring:* An object weighing w lb (so that its mass is $w/32$ slugs) is hung from a vertical spring with spring constant k lb/ft. The forces on the object are the gravitational force (=weight), the restoring force of the spring, and damping with constant b lb/(ft/sec). Write an o.d.e. modeling the motion of the object.

2. *One Mass, Two Springs:* A mass of m grams moves along a horizontal track of length B cm and is attached to walls at each end of the track by springs with natural lengths L_1 and L_2 cm and spring constants k_1 and k_2 dynes/cm, respectively (see Figure 2.4). The mass is subject to viscous damping with constant b dynes/(cm/sec).
   a. Set up a differential equation for the amount $x = x(t)$ by which the first spring is stretched. [*Hint:* If the first spring is stretched by x, then the second spring is stretched by $B - (L_2 + L_1 + x)$.]
   b. Check that we get the same equation if we assume the mass is attached to only one wall, by a spring of natural length $L = L_1$ with spring constant $k = k_1 + k_2$, and subject to an external force $E(t) = k_2(B - L_1 - L_2)$.

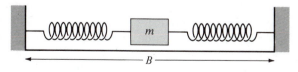

FIGURE 2.4

3. *A Moving Spring:* A 16-lb weight hangs suspended from the bottom of an elevator by a spring with natural length L ft and spring constant k lb/ft. The elevator rises at a constant rate of 2 ft/sec, and the motion of the weight is subject to viscous damping with constant b lb/(ft/sec). Let $z = z(t)$ denote the height of the weight above the ground.
   a. Use the fact that the forces acting on the weight are the gravitational force (= weight), the restoring force of the spring, and the damping force to obtain an expression for d^2z/dt^2 in terms of dz/dt, z, and the distance $y = y(t)$ from the bottom of the elevator to the ground. (*Hint:* When the mass is z ft from the ground, the spring is stretched $y - z - L$ ft.)
   b. Differentiate this expression and substitute $dy/dt = 2$ to obtain a third-order o.d.e. for z.

*4. *Coupled Springs with Friction:* Write an o.d.e. modeling the motion of the first mass (x_1) in Example 2.1.4 when $L_1 = L_2 = L$, $m_1 = m_2 = m$, $k_1 = k_2 = k$, and each mass is subject to damping with constant b.

Other Models

5. *An LRC Circuit:* The current I in the circuit of Figure 2.5 is related to the charge Q on the capacitor by the equations

$$\frac{dQ}{dt} = I$$

$$V(t) - RI - L\frac{dI}{dt} - \frac{1}{C}Q = 0.$$

Here, C, L, and R are constants, and $V(t)$ is an externally controlled voltage. (We discuss circuits in greater detail in Section 3.1.) Substitute $I = dQ/dt$ in the second equation to obtain a single second-order o.d.e. for Q.

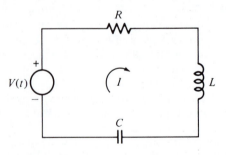

FIGURE 2.5

6. *Supply and Demand:* Assume, as in Exercise 25, Section 1.1, that the price of a product increases at a rate proportional to the excess demand (demand minus supply) and that demand in turn depends on price as a first-degree polynomial. Assume further that the supply of this product is controlled so that it increases at a rate proportional to the price.
   a. Differentiate the equation of Exercise 25, Section 1.1, to obtain an expression for d^2p/dt^2 in terms of dp/dt and ds/dt.

    b. Use the fact that ds/dt is proportional to p to rewrite this expression as a second-order o.d.e. for $p(t)$.

7. *A Floating Box:* A cube of wood 1 ft on each side, weighing 16 lb (so that its mass is 16/32 slugs), floats in the Charles River with the bottom of the box $x < 1$ ft below the surface of the water. According to the *principle of Archimedes,* the water buoys up the box by a force equal to the weight of the water it displaces. Write an o.d.e. for x,

    a. assuming that the only forces on the cube are the gravitational force ($=$ weight) downward and the buoyant force upward, and

    b. assuming that the cube is also subject to viscous damping with damping constant b lb/(ft/sec).

    Hint: The weight of the water displaced by the box is the product of the volume of the submerged part of the box with the density of water (62.5 lb/ft^3).

8. *An Anchored Floating Box:* Suppose the cube of Exercise 7 is anchored to the bottom of the river at a point where the water is 3 ft deep by a line that behaves like a spring of natural length $L = 2$ ft and spring constant $k = 22$ lb/ft. Set up an o.d.e. modeling the motion of the box,

    a. assuming no damping, and

    b. assuming viscous damping with constant b lb/(ft/sec).

9. *The Box Adrift:* The box and anchor in Exercise 8 are towed out to deeper water so that the anchor hangs suspended from the box but off the river bottom. The anchor weighs 16 lb and has a volume of 1/5 ft^3. Obtain an o.d.e. modeling the motion of the box (assuming no damping forces) as follows:

    a. By analyzing the forces on the box, obtain an expression in terms of x and d^2x/dt^2, for the amount $y = y(t)$ by which the spring is stretched.

    b. By analyzing the forces on the anchor, obtain a second-order o.d.e. for y.

    c. Substitute the expression from (a) into the equation of (b) to obtain a single fourth-order o.d.e. for x.

2.2 LINEAR DIFFERENTIAL EQUATIONS: A STRATEGY

In Section 1.3 we found that the general solution of a first-order linear o.d.e. has the form

$$x = ch(t) + p(t).$$

Note that $x = p(t)$ is itself a particular solution of the o.d.e. and that $x = ch(t)$ is the general solution of the related homogeneous equation. In this section we discuss a class of higher-order o.d.e.'s whose solutions follow a similar pattern.

Definition. *An nth-order o.d.e. is **linear** if it can be written in the form*

$$a_n(t)\frac{d^nx}{dt^n} + \cdots + a_1\frac{dx}{dt} + a_0(t)x = E(t).$$

The functions $a_n(t)$, . . . , $a_0(t)$ are called the **coefficients** *of the equation. The equation is* **homogeneous** *if $E(t) = 0$. The equation is* **normal** *on the interval I given by $\alpha < t < \beta$ provided the functions $a_n(t)$, . . . , $a_0(t)$ and $E(t)$ are continuous on I and $a_n(t)$ is never zero on I.*

The o.d.e.'s in Examples 2.1.1 through 2.1.4 are all linear. Let's look at some further examples.

▮ Example 2.2.1

Which of the following o.d.e.'s are linear? Which of the linear o.d.e.'s are homogeneous? Determine the largest intervals on which the linear o.d.e.'s are normal.

(1)
$$3 \frac{d^2x}{dt^2} - t \frac{dx}{dt} = t^2$$

(2)
$$t \frac{d^3x}{dt^3} + \frac{d^2x}{dt^2} = t^2$$

(3)
$$\frac{d^2x}{dt^2} + \frac{1}{t-1} x = 0$$

(4)
$$\frac{d^3x}{dt^3} - x \frac{dx}{dt} = t.$$

Equations (1), (2), and (3) are linear. The presence of the term $x(dx/dt)$ prevents equation (4) from being linear. Only equation (3) is homogeneous.

Equation (1) is normal on the entire real line $-\infty < t < \infty$. Equation (2) is not normal on any interval in which the leading coefficient t is zero. Thus, the largest intervals on which equation (2) is normal are the intervals $-\infty < t < 0$ and $0 < t < \infty$. In order to guarantee the continuity of the coefficients of equation (3), we must avoid $t = 1$. Thus, the largest intervals on which equation (3) is normal are the intervals $-\infty < t < 1$ and $1 < t < \infty$.

Among the simplest linear o.d.e.'s are those of the form

$$\frac{d^nx}{dt^n} = E(t).$$

These can be solved by integrating n times. Each integration will introduce a new constant of integration, so the general solution will involve n constants.

Specification of the initial values of the solution and its first $n - 1$ derivatives will determine these constants.

Example 2.2.2

Find the general solution of

(N)
$$\frac{d^3x}{dt^3} = \frac{12}{(t-1)^4}, \quad 1 < t < \infty.$$

Then find the specific solution of (N) that satisfies the initial condition

$$x(2) = 1, \quad x'(2) = 0, \quad x''(2) = 0.$$

We obtain the general solution of (N) by integrating three times:

$$\frac{d^2x}{dt^2} = \frac{-4}{(t-1)^3} + c_1$$

$$\frac{dx}{dt} = \frac{2}{(t-1)^2} + c_1 t + c_2$$

$$x = \frac{-1}{t-1} + c_1 \frac{t^2}{2} + c_2 t + c_3.$$

Substitution of the initial values for $x(2)$, $x'(2)$, and $x''(2)$ in the preceding three formulas yields three equations:

$$0 = -4 + c_1$$
$$0 = 2 + 2c_1 + c_2$$
$$1 = -1 + 2c_1 + 2c_2 + c_3.$$

The solution of these equations is

$$c_1 = 4, \quad c_2 = -10, \quad c_3 = 14.$$

We get the solution of (N) matching the given initial condition by substituting these values in the general solution:

$$x = \frac{-1}{t-1} + 2t^2 - 10t + 14.$$

Before we leave this example, we note that the general solution can be written in the form

$$x = H(t) + p(t),$$

where

$$H(t) = c_1 \frac{t^2}{2} + c_2 t + c_3 \quad \text{and} \quad p(t) = \frac{-1}{t-1}.$$

Here $x = p(t)$ is a particular solution of (N), and $x = H(t)$ is the general solution of the homogeneous o.d.e.

(H)
$$\frac{d^3 x}{dt^3} = 0$$

(check this).

Just as in Example 2.2.2, specification of the initial values of $x(t)$ and its first $n - 1$ derivatives determines a unique solution of an nth-order linear o.d.e.

Theorem: Existence and Uniqueness of Solutions of Linear O.D.E.'s. *Suppose $Lx = E(t)$ is an nth-order linear o.d.e. that is normal on an interval I. Let t_0 be a fixed value of t in I. Then, given any n real numbers $\alpha_0, \ldots, \alpha_{n-1}$, there exists a solution $x = \phi(t)$ of the o.d.e. that is defined for all t in I and that satisfies the initial condition*

$$x(t_0) = \alpha_0, \quad x'(t_0) = \alpha_1, \ldots, x^{(n-1)}(t_0) = \alpha_{n-1}.$$

Furthermore, if $x = \theta(t)$ is a solution of the o.d.e. that satisfies the same initial condition as $x = \phi(t)$, then $\theta(t) = \phi(t)$ for all t in I.

Our discussion of the solutions of linear o.d.e.'s will be aided by the introduction of some notation. In place of our usual notation for mth derivatives, we will use

$$D^m x = \frac{d^m x}{dt^m}.$$

We will extend this formalism to expressions that appear on the left sides of linear o.d.e.'s as follows:

$$[a_n(t) D^n + \cdots + a_1(t) D + a_0(t)]x = a_n(t) D^n x + \cdots + a_1(t) Dx + a_0(t)x$$

$$= a_n(t)\frac{d^n x}{dt^n} + \cdots + a_1(t)\frac{dx}{dt} + a_0(t)x.$$

An expression of the form $a_n(t) D^n + \cdots + a_1(t) D + a_0(t)$ is called a **linear differential operator.** Using operator notation, a typical linear o.d.e. has the form

$$[a_n(t) D^n + \cdots + a_1(t) D + a_0(t)]x = E(t).$$

Example 2.2.3

In operator notation, equations (1), (2), and (3) of Example 2.2.1 are

(1) $$(3D^2 - t D)x = t^2$$

(2) $$(t D^3 + D^2)x = t^2$$

(3) $$\left(D^2 - \frac{1}{t - 1}\right) x = 0.$$

Keep in mind that when we apply a linear differential operator $L = a_n(t)D^n + \cdots + a_0(t)$ to a function $x = \phi(t)$, we get a function $Lx = L[\phi(t)]$. Let's calculate some examples.

Example 2.2.4

Find $L(\sin 2t)$ and $L(e^{-4t})$, where $L = D^3 - t D^2 + t^3$.

$$L(\sin 2t) = D^3 \sin 2t - t D^2 \sin 2t + t^3 \sin 2t$$
$$= -8 \cos 2t + 4t \sin 2t + t^3 \sin 2t.$$
$$L(e^{-4t}) = D^3 e^{-4t} - t D^2 e^{-4t} + t^3 e^{-4t}$$
$$= (-64 - 16t + t^3)e^{-4t}.$$

Among the first properties we learned in calculus is that the derivative of a sum is the sum of the derivatives. This rule extends to an arbitrary linear differential operator

$$L = a_n(t)D^n + \cdots + a_1(t)D + a_0(t).$$

If we apply L to a sum, we get

$$L(x_1 + x_2) = a_n(t) D^n(x_1 + x_2) + \cdots + a_1(t) D(x_1 + x_2) + a_0(t)(x_1 + x_2)$$
$$= a_n(t)(D^n x_1 + D^n x_2) + \cdots + a_1(t)(Dx_1 + Dx_2) + a_0(t)(x_1 + x_2)$$
$$= a_n(t) D^n x_1 + \cdots + a_1(t) Dx_1 + a_0(t)x_1$$
$$\quad + a_n(t) D^n x_2 + \cdots + a_1(t) Dx_2 + a_0(t)x_2$$
$$= Lx_1 + Lx_2.$$

Fact: Principle of Superposition.   *If L is a linear differential operator, then*

$$L(x_1 + x_2) = Lx_1 + Lx_2.$$

Similarly, the fact that $D(cx) = c(Dx)$ for any constant c extends to linear differential operators.

Fact: Principle of Proportionality.   *If L is a linear differential operator and c is a constant, then*

$$L(cx) = c(Lx).$$

Let's now turn our attention to the linear o.d.e.

(N) $$\qquad\qquad a_n(t) \frac{d^n x}{dt^n} + \cdots + a_1(t) \frac{dx}{dt} + a_0(t)x = E(t).$$

If we set

$$L = a_n(t) D^n + \cdots + a_1(t)D + a_0(t),$$

then we can rewrite (N) as

(N) $$\qquad\qquad\qquad Lx = E(t).$$

The **related homogeneous equation** is

(H) $$\qquad\qquad\qquad Lx = 0.$$

For example, the homogeneous equations related to equations (1) and (2) of Example 2.2.3 are, respectively,

(H$_1$) $$\qquad\qquad\qquad (3D^2 - tD)x = 0$$

and

(H$_2$) $$(tD^3 + D^2)x = 0.$$

Let $x = p(t)$ be a solution of (N), and let $x = h(t)$ be a solution of (H). Then

$$Lp(t) = E(t) \quad \text{and} \quad Lh(t) = 0.$$

The principle of superposition applies to give

$$L[h(t) + p(t)] = Lh(t) + Lp(t) = 0 + E(t) = E(t).$$

Thus, $x = h(t) + p(t)$ is another solution of (N). If we let $h(t)$ range over all the solutions of (H), we get a collection of solutions of (N).

Suppose now that $x = f(t)$ is also a solution of (N). Then $Lf(t) = E(t)$. Let $h_1(t) = f(t) - p(t)$. Then

$$f(t) = h_1(t) + p(t)$$

and

$$Lh_1(t) = L[f(t) - p(t)] = Lf(t) - Lp(t) = E(t) - E(t) = 0.$$

Thus, $x = h_1(t)$ is a solution of (H), and $f(t)$ belongs to the collection we described in the previous paragraph.

We have obtained the following powerful information about the form of the general solution of a linear o.d.e.

Fact.   *If we find a particular solution $x = p(t)$ of the nonhomogeneous linear o.d.e.*

(N) $$Lx = E(t)$$

and if $x = H(t)$ is a formula that describes all solutions of the related homogeneous equation

(H) $$Lx = 0$$

then

$$x = H(t) + p(t)$$

describes all solutions of (N).

Example 2.2.5

Substitution will show that if

$$p(t) = te^t,$$

then $x = p(t)$ is a particular solution of the o.d.e.

(N) $$D^2x = (t + 2)e^t.$$

We can solve the related homogeneous equation

(H) $$D^2x = 0$$

by integrating twice. Its solution is $x = H(t)$, where

$$H(t) = c_1t + c_2.$$

The preceding fact tells us that the general solution of (N) is

$$x = H(t) + p(t)$$
$$= c_1t + c_2 + te^t.$$

We can now describe our strategy for solving a linear o.d.e. First find the general solution $x = H(t)$ of the related homogeneous equation. Then find a particular solution $x = p(t)$ of the nonhomogeneous equation. The general solution of the nonhomogeneous equation is $x = H(t) + p(t)$.

Let's summarize the major points of this section.

LINEAR DIFFERENTIAL EQUATIONS: A STRATEGY

An expression of the form

$$L = a_n(t)D^n + \cdots + a_1(t)D + a_0(t)$$

is called a **linear differential operator.** When L is applied to a function x, it gives a function Lx defined by the rule

$$Lx = a_n(t) \frac{d^n x}{dt^n} + \cdots + a_1(t) \frac{dx}{dt} + a_0(t)x.$$

Linear differential operators satisfy the following principles.

1. **Principle of superposition:** $L(x_1 + x_2) = Lx_1 + Lx_2$.
2. **Principle of proportionality:** $L(cx) = c(Lx)$.

A differential equation is **linear** if it can be written in the form $Lx = E(t)$, where L is a linear differential operator. The functions $a_n(t), \ldots, a_0(t)$ in the description of the operator are called the **coefficients** of the equation. The equation is **homogeneous** if $E(t) = 0$. The equation is **normal** on an interval I if the coefficients and $E(t)$ are continuous on I and the highest coefficient $a_n(t)$ is never zero on I.

If $x = p(t)$ is a particular solution of the linear o.d.e.

(N) $$Lx = E(t)$$

and if $x = H(t)$ is the general solution of the **related homogeneous equation**

then

(H) $$Lx = 0,$$

$$x = H(t) + p(t)$$

is the general solution of (N).

Existence and Uniqueness of Solutions:

Let $Lx = E(t)$ be an nth-order linear o.d.e. that is normal on an interval I and let t_0 be a fixed value of t in I. Given any n real numbers $\alpha_0, \ldots, \alpha_{n-1}$, there is exactly one solution of $Lx = E(t)$ that satisfies the initial condition

$$x(t_0) = \alpha_0, \quad x'(t_0) = \alpha_1, \ldots, x^{(n-1)}(t_0) = \alpha_{n-1}.$$

Note

A technicality

Our results in this chapter will depend on the existence and uniqueness theorem, which assumes we are dealing with an interval I on which the o.d.e. is normal and which provides information about solutions defined on I. *All statements in this chapter about general solutions are valid only on such intervals.* If an o.d.e. has a solution on an interval where the equation is *not* normal, then it can be obtained by piecing together solutions on the subintervals where the equation is normal (see Example 1.6.1, and Exercises 17–22 in Section 1.6).

EXERCISES

Which of the o.d.e.'s in Exercises 1 through 8 are linear? Rewrite each linear o.d.e. in operator notation. Of the linear equations, which are homogeneous?

1. $\dfrac{d^2x}{dt^2} - 5tx = t\dfrac{dx}{dt} - 25$

2. $\dfrac{d^2x}{dt^2} + 2t^2x^2 = 0$

3. $\left(\dfrac{dx}{dt}\right)^2 - 2t^2 = 0$

4. $\dfrac{d^3x}{dt^3} = -x \sin t$

5. $\dfrac{d^2x}{dt^2} = -x\dfrac{dx}{dt}$

6. $\dfrac{d^3x}{dt^3} - 3x = t\dfrac{d^2x}{dt^2}$

7. $\dfrac{d^4x}{dt^4} + 5t^3\dfrac{dx}{dt} = \sqrt{t^2 - 1}$

8. $x\dfrac{d^2x}{dt^2} + t\dfrac{dx}{dt} = t$

9. Which of the following linear o.d.e.'s are normal on the interval $0 < t < 2$?

   a. $(t - 1)\dfrac{d^2x}{dt^2} + t\dfrac{dx}{dt} - 5x = 3$

   b. $t\dfrac{d^2x}{dt^2} - e^t\dfrac{dx}{dt} + tx = \sin t$

   c. $5\dfrac{d^2x}{dt^2} - 3\dfrac{dx}{dt} + 2x = e^t - t$

   d. $3\dfrac{d^2x}{dt^2} + 5x = \csc \pi t$

10. Determine the largest intervals on which the given o.d.e. is normal.

    a. $(t^2 + 1)\dfrac{d^2x}{dt^2} + \dfrac{1}{t}\dfrac{dx}{dt} - e^tx = \sin t$

    b. $t^2\dfrac{d^2x}{dt^2} - 3\dfrac{dx}{dt} + x = t$

    c. $(t^2 - 1)\dfrac{d^2x}{dt^2} + 3x = 5$

    d. $5\dfrac{d^2x}{dt^2} - 3\dfrac{dx}{dt} + x = \sqrt{t - 1}$

In Exercises 11 through 15, find $Lf_i(t)$ for each given function $f_i(t)$.

11. $L = D^2 - 2D + 1;$ $f_1(t) = e^t,\ \ f_2(t) = 3e^{2t}$

12. $L = D^2 - 2D + 1;$ $f_1(t) = te^t,\ \ f_2(t) = \dfrac{5}{2}te^t$

13. $L = D^2 + 3D - 3;$ $f_1(t) = e^{2t},\ \ f_2(t) = t^3,\ \ f_3(t) = 5e^{2t} - 2t^3$

14. $L = D^3 - tD + 6;$ $f_1(t) = \sin 2t,\ \ f_2(t) = 1,\ \ f_3(t) = 1 - \sin 2t$

15. $L = D;$ $f_1(t) = t,\ \ f_2(t) = e^t,\ \ f_3(t) = te^t$

     Note that in this problem $L[f_1(t)f_2(t)] \neq L[f_1(t)]L[f_2(t)]$.

In Exercises 16 through 22:

(a) Find the general solution of the given o.d.e. by integrating several times.

(b) Find the solution that satisfies the given initial condition.

16. $\dfrac{d^2x}{dt^2} = \sin 2t,\ \ x(\pi) = 1,\ x'(\pi) = 0$

17. $\dfrac{d^2x}{dt^2} = \sin 2t,\ \ x(\pi) = 0,\ x'(\pi) = 1$

18. $\dfrac{d^2x}{dt^2} = \sin 2t,\ \ x(\pi) = 1,\ x'(\pi) = 2$

19. $\dfrac{d^3x}{dt^3} = te^t$, $x(0) = x'(0) = 0, x''(0) = 1$

20. $\dfrac{d^3x}{dt^3} = 16e^{2t}$, $x(0) = 1, x'(0) = 0, x''(0) = 24$

21. $\dfrac{d^4x}{dt^4} = 3t - 1$, $x(0) = x'(0) = x''(0) = x'''(0) = 0$

22. $\dfrac{d^4x}{dt^4} = 3t - 1$, $x(0) = x'(0) = 1, x''(0) = x'''(0) = -1$

In Exercises 23 through 27, you are given a nonhomogeneous equation (N) $Lx = E(t)$, the general solution $x = H(t)$ of the related homogeneous equation, and an expression involving one or more constants for a particular solution $x = p(t)$ of (N). In each problem:

(a) Find values of the constants for which $x = p(t)$ is a solution of (N).
(b) Find the general solution of (N).

23. $(D^2 - D - 2)x = -t + 4$; $H(t) = c_1e^{2t} + c_2e^{-t}$; $p(t) = At + B$
24. $(D^2 - D - 2)x = 3 \sin t$; $H(t) = c_1e^{2t} + c_2e^{-t}$; $p(t) = A \sin t + B \cos t$
25. $(D^3 - D^2 - 2D)x = 4$; $H(t) = c_1 + c_2e^{2t} + c_3e^{-t}$; $p(t) = At$
26. $(D^2 + 4)x = e^t + 1$; $H(t) = c_1 \sin 2t + c_2 \cos 2t$; $p(t) = Ae^t + B$
27. $(D^2 - 1)x = t^2$; $H(t) = c_1e^t + c_2e^{-t}$; $p(t) = A + Bt + Ct^2$

2.3 HOMOGENEOUS LINEAR EQUATIONS: THE WRONSKIAN

In Section 1.3 we saw that the general solution of a first-order homogeneous linear o.d.e. has the form $x = ch(t)$. The general solution of the homogeneous second-order o.d.e. in Example 2.2.5 was of the form $x = c_1h_1(t) + c_2h_2(t)$. In Example 2.2.2 we saw an example of a third-order homogeneous o.d.e. whose general solution had the form $x = c_1h_1(t) + c_2h_2(t) + c_3h_3(t)$. In this section we shall verify that this pattern is typical of normal homogeneous linear o.d.e.'s.

We begin by noting an important consequence of the principles of superposition and proportionality. Suppose $x = h_1(t), \ldots, x = h_k(t)$ are solutions of a homogeneous linear o.d.e. $Lx = 0$ and that $c_1, \ldots, c_k$ are constants. Then

$$L[c_1h_1(t) + c_2h_2(t) + \cdots + c_kh_k(t)]$$
$$= L[c_1h_1(t)] + L[c_2h_2(t)] + \cdots + L[c_kh_k(t)]$$
$$= c_1Lh_1(t) \quad + c_2Lh_2(t) \quad + \cdots + c_kLh_k(t)$$
$$= 0 \qquad\quad + 0 \qquad\quad + \cdots + 0 \qquad\quad = 0.$$

Thus, $x = c_1h_1(t) + \cdots + c_kh_k(t)$ is also a solution of $Lx = 0$. We refer to a function of the form $c_1h_1(t) + \cdots + c_kh_k(t)$ as a **linear combination** of $h_1(t), \ldots, h_k(t)$. Thus, we have the following.

Fact: *If $h_1(t), \ldots, h_k(t)$ are solutions of a homogeneous linear o.d.e. $Lx = 0$, then any linear combination of these functions is also a solution.*

Let $Lx = 0$ be an nth-order homogeneous linear o.d.e. that is normal on an interval I. If we have n solutions $h_1(t), \ldots, h_n(t)$ of $Lx = 0$, then we can view the formula

(S) $$x = c_1h_1(t) + c_2h_2(t) + \cdots + c_nh_n(t)$$

as describing a collection of solutions, with each choice of the constants $c_1, \ldots, c_n$ determining a different solution in the collection. If this is a *complete* collection of solutions, then the formula (S) is the general solution of (H). In this case, we say that the functions $h_1(t), \ldots, h_n(t)$ **generate** the general solution.

Our tool for deciding whether the collection of solutions described by (S) is complete will be the existence and uniqueness theorem for solutions of linear o.d.e.'s. We will choose a value of t_0 and check to see if we can match every initial condition at t_0

(I) $$x(t_0) = \alpha_0, \quad x'(t_0) = \alpha_1, \ldots, x^{(n-1)}(t_0) = \alpha_{n-1}$$

with a function in our collection. That is, we will check to see if we can always find constants $c_1, \ldots, c_n$ so that the function $x = c_1h_1(t) + \cdots + c_nh_n(t)$ satisfies (I). If there is an initial condition at t_0 that we can't match, then the "existence" part of the theorem says there is a solution that is not in our collection. Suppose, on the other hand, that we can match every initial condition with a function in our collection. If $x = \phi(t)$ is any solution at all, then it satisfies some initial condition. By "uniqueness," $\phi(t)$ must be equal to the function in the collection that satisfies the same initial condition. Thus, in this case, the collection is complete.

Example 2.3.1

The functions $h_1(t) = 1$ and $h_2(t) = e^{-t}$ are solutions of the homogeneous o.d.e.

(H) $$(D^2 + D)x = 0.$$

(Check this by substitution.) These functions generate a collection of solutions described by the formula

(S)
$$x = c_1h_1(t) + c_2h_2(t)$$
$$= c_1 + c_2e^{-t}.$$

Determine whether this is the general solution of (H).

Let's take $t_0 = 0$ and try to match every initial condition with a function in our collection. Since $n = 2$, the initial conditions we must match have the form

$$x(0) = \alpha_0, \quad x'(0) = \alpha_1.$$

For a function in the collection specified by (S),

$$x'(t) = -c_2e^{-t}.$$

Thus,

$$x(0) = c_1 + c_2$$
$$x'(0) = \quad\quad - c_2.$$

If we try to match the initial condition, we get a system of two algebraic equations

$$c_1 + c_2 = \alpha_0$$
$$- c_2 = \alpha_1$$

in the two unknowns c_1 and c_2. No matter what specific values we choose for α_0 and α_1, we can always solve these equations:

$$c_1 = \alpha_0 + \alpha_1, \quad c_2 = -\alpha_1.$$

Thus, by choosing c_1 and c_2 appropriately, we can match any initial condition at t_0. Formula (S) is the general solution of (H).

Example 2.3.2

The functions $h_1(t) = 1$ and $h_2(t) = t$ and $h_3(t) = 2t - 3$ are solutions of the homogeneous o.d.e.

(H)
$$(D^3 - 2D^2)x = 0.$$

(Check this by substitution.) These functions generate a collection of solutions described by the formula

(S)
$$x = c_1 h_1(t) + c_2 h_2(t) + c_3 h_3(t)$$
$$= c_1 + c_2 t + c_3(2t - 3).$$

Determine whether this is the general solution of (H).

Let's take $t_0 = 0$ and try to match every initial condition with a function in our collection. Since $n = 3$, the initial conditions we must match have the form

$$x(0) = \alpha_0, \quad x'(0) = \alpha_1, \quad x''(0) = \alpha_2.$$

For a function in the collection specified by (S),

$$x' = c_2 + 2c_3$$
$$x'' = 0.$$

If we try to match the initial condition, we get a system of three algebraic equations

$$c_1 \quad - 3c_3 = \alpha_0$$
$$c_2 + 2c_3 = \alpha_1$$
$$0 \ = \alpha_2.$$

We can't match any initial condition with $\alpha_2 \neq 0$. Thus, (S) is *not* the general solution of (H).

———————————————

Let $x = h_1(t), \ldots, x = h_n(t)$ be solutions of the nth-order homogeneous linear o.d.e.

(H)
$$Lx = 0,$$

which is normal on an interval I. These functions generate a collection of solutions described by the formula

(S)
$$x = c_1 h_1(t) + \cdots + c_n h_n(t).$$

Let t_0 be a value of t in I. To determine whether (S) is the general solution of (H), we try to match every initial condition at t_0

$$x(t_0) = \alpha_0, \quad x'(t_0) = \alpha_1, \ldots, x^{(n-1)}(t_0) = \alpha_{n-1}.$$

Substitution of our expression for x into the equations describing the initial condition leads to a system of n algebraic equations

(A)
$$
\begin{aligned}
c_1 h_1(t_0) + \cdots + c_n h_n(t_0) &= \alpha_0 \\
c_1 h_1'(t_0) + \cdots + c_n h_n'(t_0) &= \alpha_1 \\
\vdots \\
c_1 h_1^{(n-1)}(t_0) + \cdots + c_n h_n^{(n-1)}(t_0) &= \alpha_{n-1}
\end{aligned}
$$

in the n unknowns $c_1, \ldots, c_n$. Fortunately, we need only know if it is possible to solve this system of equations for all choices of specific values for the α's on the right-hand side. *We do not need to actually solve the equations.* The information we want can be obtained using Cramer's determinant test (see Appendix A). If the determinant of coefficients of (A) is not zero, then the system of equations can always be solved. In this case, (S) is the general solution. If the determinant of coefficients is zero, then there are values of $\alpha_0, \ldots, \alpha_{n-1}$ for which the system of equations does not have a solution. In this case, (S) is not the general solution.

The determinant formed from the n solutions and their first $n-1$ derivatives was introduced by the Polish mathematician H. Wronski (1811) and bears his name.

Definition: *The **Wronskian** of a list of n functions $h_1(t)$, $h_2(t)$, $\ldots$, $h_n(t)$ is the n × n determinant*

$$
W[h_1, h_2, \ldots, h_n](t) = \det
\begin{bmatrix}
h_1(t) & h_2(t) & \ldots & h_n(t) \\
h_1'(t) & h_2'(t) & \ldots & h_n'(t) \\
\vdots & \vdots & & \vdots \\
h_1^{(n-1)}(t) & h_2^{(n-1)}(t) & \ldots & h_n^{(n-1)}(t)
\end{bmatrix}.
$$

With this terminology, we can state our conclusion as follows.

Fact: Wronskian Test for Solutions of Homogeneous Linear O.D.E.'s.   *Suppose $Lx = 0$ is an nth-order homogeneous linear o.d.e. that is normal on an interval I. Let t_0 be a value of t in I. The functions $h_1(t), \ldots, h_n(t)$ generate the general solution of $Lx = 0$*

$$
x = c_1 h_1(t) + \cdots + c_n h_n(t)
$$

if and only if they are solutions of Lx = 0 and their Wronskian satisfies

$$W[h_1, \ldots, h_n](t_0) \neq 0.$$

Let's use the Wronskian test in some examples. Note that Examples 2.3.3 and 2.3.4 involve the same o.d.e.'s and functions as Examples 2.3.1 and 2.3.2. Note also that when applying the Wronskian test, we are free to choose t_0 to be any point in the interval I. In the examples, we will always choose t_0 to be a point at which functions that appear in the determinant are easy to evaluate.

Example 2.3.3

The two functions $h_1(t) = 1$ and $h_2(t) = e^{-t}$ are solutions of the second-order homogeneous o.d.e.

(H) $$(D^2 + D)x = 0.$$

Use the Wronskian test to determine whether these functions generate the general solution of (H).

The Wronskian of $h_1(t)$ and $h_2(t)$ is

$$W[1, e^{-t}](t) = \det \begin{bmatrix} 1 & e^{-t} \\ 0 & -e^{-t} \end{bmatrix}.$$

At $t_0 = 0$,

$$W[1, e^{-t}](0) = \det \begin{bmatrix} 1 & 1 \\ 0 & -1 \end{bmatrix} = -1 \neq 0.$$

Therefore, $h_1(t)$ and $h_2(t)$ generate the general solution

$$x = c_1 h_1(t) + c_2 h_2(t)$$
$$= c_1 + c_2 e^{-t}.$$

Example 2.3.4

The three functions $h_1(t) = 1$ and $h_2(t) = t$ and $h_3(t) = 2t - 3$ are solutions of the third-order homogeneous o.d.e.

(H) $$(D^3 - 2D^2)x = 0.$$

Use the Wronskian test to determine whether these functions generate the general solution of (H).

The Wronskian of $h_1(t)$, $h_2(t)$, and $h_3(t)$ is

$$W[1, t, 2t - 3](t) = \det \begin{bmatrix} 1 & t & 2t - 3 \\ 0 & 1 & 2 \\ 0 & 0 & 0 \end{bmatrix}.$$

At $t_0 = 0$,

$$W[1, t, 2t - 3](0) = \det \begin{bmatrix} 1 & 0 & -3 \\ 0 & 1 & 2 \\ 0 & 0 & 0 \end{bmatrix} = 0.$$

Thus, $h_1(t)$, $h_2(t)$, and $h_3(t)$ do not generate the general solution of (H).

We note that e^{2t} is also a solution of (H). The Wronskian of 1, t, and e^{2t} is

$$W[1, t, e^{2t}](t) = \det \begin{bmatrix} 1 & t & e^{2t} \\ 0 & 1 & 2e^{2t} \\ 0 & 0 & 4e^{2t} \end{bmatrix}.$$

At $t_0 = 0$,

$$W[1, t, e^{2t}](0) = \det \begin{bmatrix} 1 & 0 & 1 \\ 0 & 1 & 2 \\ 0 & 0 & 4 \end{bmatrix} = 4 \neq 0.$$

Thus, 1, t, and e^{2t} generate the general solution of (H)

$$x = c_1 + c_2 t + c_3 e^{2t}.$$

Example 2.3.5

Substitution will show that the two functions $h_1(t) = \cos 2t$ and $h_2(t) = 4 \sin^2 t - 2$ are solutions of the second-order homogeneous o.d.e.

(H) $(D^2 + 4)x = 0.$

Use the Wronskian test to determine whether these functions generate the general solution of (H).

The Wronskian of $h_1(t)$ and $h_2(t)$ is

$$W[\cos 2t, 4 \sin^2 t - 2](t) = \det \begin{bmatrix} \cos 2t & 4 \sin^2 t - 2 \\ -2 \sin 2t & 8 \sin t \cos t \end{bmatrix}.$$

At $t_0 = 0$,

$$W[\cos 2t, 4 \sin^2 t - 2](0) = \det \begin{bmatrix} 1 & -2 \\ 0 & 0 \end{bmatrix} = 0.$$

Therefore, $h_1(t)$ and $h_2(t)$ do not generate the general solution of (H).

We note that $\sin 2t$ is another solution of (H), and we leave it to you to check that the functions $\cos 2t$ and $\sin 2t$ generate the general solution of (H)

$$x = c_1 \cos 2t + c_2 \sin 2t.$$

Example 2.3.6

Find all solutions of

(H) $$(t^2 D^3 + 2t D^2 - 2D)x = 0, \quad 0 < t < \infty$$

that are of the form $x = t^\alpha$. Determine whether these functions generate the general solution of (H). Note that the o.d.e. is normal on the given interval, but not on any interval containing 0.

Substitution of $x = t^\alpha$ in (H) yields

$$t^2 \alpha(\alpha - 1)(\alpha - 2)t^{\alpha-3} + 2t\alpha(\alpha - 1)t^{\alpha-2} - 2\alpha t^{\alpha-1} = 0.$$

We can rewrite this equation as

$$\alpha(\alpha^2 - \alpha - 2)t^{\alpha-1} = 0.$$

Since we want this to hold for all values of t on the interval $0 < t < \infty$, we must have

$$\alpha(\alpha^2 - \alpha - 2) = 0.$$

Thus,

$$\alpha = 2, \quad \alpha = -1 \quad \text{or} \quad \alpha = 0.$$

Corresponding to these values of α, we have three solutions of our third-order homogeneous o.d.e. (H):

$$h_1(t) = t^2, \quad h_2(t) = t^{-1}, \quad h_3(t) = t^0 = 1.$$

The Wronskian of $h_1(t)$, $h_2(t)$, and $h_3(t)$ is

$$W[t^2, t^{-1}, 1](t) = \det \begin{bmatrix} t^2 & t^{-1} & 1 \\ 2t & -t^{-2} & 0 \\ 2 & 2t^{-3} & 0 \end{bmatrix}.$$

At $t_0 = 1$ (we can't use $t_0 = 0$ since it is outside the interval $0 < t < \infty$),

$$W[t^2, t^{-1}, 1](1) = \det \begin{bmatrix} 1 & 1 & 1 \\ 2 & -1 & 0 \\ 2 & 2 & 0 \end{bmatrix} = 6 \neq 0.$$

Thus, $h_1(t)$, $h_2(t)$, and $h_3(t)$ generate the general solution of (H)

$$x = c_1 t^2 + c_2 t^{-1} + c_3.$$

Suppose now that we are given an nth-order homogeneous linear o.d.e.

(H) $$Lx = 0$$

that is normal on an interval I. Let t_0 be a fixed value of t in I. Then the existence and uniqueness theorem guarantees that there is a solution $x = h_1(t)$ that satisfies the initial condition

$$x(t_0) = 1, \quad x'(t_0) = 0, \ldots, x^{(n-1)}(t_0) = 0.$$

There is also a solution $x = h_2(t)$ that satisfies the initial condition

$$x(t_0) = 0, \quad x'(t_0) = 1, \ldots, x^{(n-1)}(t_0) = 0.$$

Indeed, for each $i = 1, 2, \ldots, n$ there is a solution $x = h_i(t)$ that satisfies the initial condition

$$x(t_0) = \cdots = x^{(i-2)}(t_0) = 0, \quad x^{(i-1)}(t_0) = 1, \quad x^{(i)}(t_0) = \cdots = x^{(n-1)}(t_0) = 0.$$

At t_0, the Wronskian of these n functions is

$$W[h_1, h_2, \ldots, h_n](t_0) = \det \begin{bmatrix} 1 & 0 & \cdots & 0 \\ 0 & 1 & \cdots & 0 \\ \cdot & \cdot & & \cdot \\ \cdot & \cdot & & \cdot \\ \cdot & \cdot & & \cdot \\ 0 & 0 & \cdots & 1 \end{bmatrix} = 1 \neq 0.$$

Thus, $h_1(t)$, $h_2(t)$, $\ldots$, $h_n(t)$ generate the general solution of (H).

We have verified the pattern for solutions of normal homogeneous linear equations.

Fact: *The general solution of the nth-order normal homogeneous linear o.d.e. Lx = 0 is of the form*

$$x = c_1 h_1(t) + c_2 h_2(t) + \cdots + c_n h_n(t)$$

for a suitable choice of $h_1(t), \ldots, h_n(t)$.

Is it possible to find fewer than n functions that generate the general solution of an nth-order o.d.e. $Lx = 0$? Suppose that we have k solutions $h_1(t), \ldots, h_k(t)$, where $k < n$. Since the Wronskian test applies only if we have n solutions, let's throw in the extra solutions

$$h_{k+1}(t) = 0, \ldots, h_n(t) = 0.$$

The Wronskian of $h_1(t), \ldots, h_n(t)$ has at least one column (the last) consisting entirely of zeros. Thus, the Wronskian is zero (see Exercise 34 of Appendix A). By the Wronskian test, the collection of solutions described by the formula $x = c_1 h_1(t) + \cdots + c_n h_n(t)$ is not complete. But then neither is the collection of functions of the form $x = c_1 h_1(t) + \cdots + c_k h_k(t)$. Thus, we have the following.

Fact: *The general solution of an nth-order homogeneous normal linear o.d.e. cannot be generated by fewer than n solutions.*

We close with a summary.

HOMOGENEOUS LINEAR EQUATIONS: THE WRONSKIAN

Let $Lx = 0$ be an nth-order homogeneous linear o.d.e. that is normal on an interval I. The general solution of $Lx = 0$ is of the form

$$x = c_1 h_1(t) + c_2 h_2(t) + \cdots + c_n h_n(t)$$

for a suitable choice of $h_1(t), \ldots, h_n(t)$. The general solution cannot be generated by fewer than n functions.

To decide whether a given set of n solutions $h_1(t), \ldots, h_n(t)$ of $Lx = 0$ generates the general solution, choose a value of t_0 in the interval I and calculate the **Wronskian** of the solutions at this point:

$$W[h_1, \ldots, h_n](t_0) = \det \begin{bmatrix} h_1(t_0) & h_2(t_0) & \cdots & h_n(t_0) \\ h_1'(t_0) & h_2'(t_0) & \cdots & h_n'(t_0) \\ \cdot & \cdot & & \cdot \\ \cdot & \cdot & & \cdot \\ \cdot & \cdot & & \cdot \\ h_1^{(n-1)}(t_0) & h_2^{(n-1)}(t_0) & \cdots & h_n^{(n-1)}(t_0) \end{bmatrix}.$$

The general solution of $Lx = 0$ is $x = c_1 h_1(t) + \cdots + c_n h_n(t)$ if and only if $W[h_1, \ldots, h_n](t_0) \neq 0$.

Note

On the Wronskian of solutions of a homogeneous o.d.e.

Suppose the nth-order homogeneous linear o.d.e. $Lx = 0$ is normal on an interval I. Let $h_1(t), \ldots, h_n(t)$ be solutions of $Lx = 0$, and let $W(t)$ be the Wronskian of these functions. Let t_0 and t_0' be two values of t in I. If $W(t_0) \neq 0$, then the Wronskian test tells us that general solution of $Lx = 0$ is $x = c_1 h_1(t) + \cdots + c_n h_n(t)$. Since this is the general solution, the Wronskian test also tells us that $W(t_0') \neq 0$. Thus, *the Wronskian of a list of n solutions of an nth-order homogeneous linear o.d.e. is either always zero or is always nonzero on I.*

EXERCISES

In Exercises 1 through 14 you are given an nth-order linear differential operator L and n solutions $h_1(t), \ldots, h_n(t)$ of $Lx = 0$. Use the Wronskian test to determine whether these solutions generate the general solution of $Lx = 0$.

1. $L = D^2 - a^2$, $a \neq 0$; $h_1(t) = e^{at}$, $h_2(t) = e^{-at}$
2. $L = D^2 - 2aD + a^2$; $h_1(t) = e^{at}$, $h_2(t) = ae^{at}$
3. $L = D^2 - 2aD + a^2$; $h_1(t) = e^{at}$, $h_2(t) = te^{at}$
4. $L = D^2 + 4$; $h_1(t) = \sin 2t$, $h_2(t) = \sin t \cos t$
5. $L = D^2 + a^2$, $a \neq 0$; $h_1(t) = \sin at$, $h_2(t) = \cos at$
6. $L = tD^2 - D$, $t > 0$; $h_1(t) = 2$, $h_2(t) = t^2$
7. $L = t^2 D^2 + 4tD + 2$, $t > 0$; $h_1(t) = \dfrac{1}{t}$, $h_2(t) = \dfrac{1}{t^2}$
8. $L = D^3 + D^2$; $h_1(t) = e^{-t}$, $h_2(t) = t + 3e^{-t}$, $h_3(t) = t$
9. $L = D^3 - 4D$; $h_1(t) = e^{2t}$, $h_2(t) = e^{-2t}$, $h_3(t) = 1$

10. $L = D^4 + 5D^2 + 4$; $h_1(t) = \sin t$, $h_2(t) = \cos t$, $h_3(t) = \sin 2t$,

   $h_4(t) = \sin \left(\dfrac{\pi}{4} + t \right)$

11. $L = D^4 + 4D^3 + 6D^2 + 4D + 1$; $h_1(t) = e^{-t}$, $h_2(t) = te^{-t}$,
   $h_3(t) = t^2 e^{-t}$, $h_4(t) = t^3 e^{-t}$

12. $L = D^4 - 1$; $h_1(t) = \sin t$, $h_2(t) = \cos t$, $h_3(t) = e^t$, $h_4(t) = e^{-t}$

13. $L = D^4$; $h_1(t) = t^3 + t^2$, $h_2(t) = t^2 + 1$, $h_3(t) = t^3 - 1$, $h_4(t) = t$

14. $L = D^n$; $h_1(t) = 1$, $h_2(t) = t, \ldots, h_n(t) = t^{n-1}$

In Exercises 15 through 21:

(a) Find all solutions of the form $e^{\lambda t}$ or t^α.

(b) Determine whether the solutions found in (a) generate a complete collection of solutions.

15. $(D^2 - 1)x = 0$ 16. $(t^2 D^2 - tD)x = 0$, $t > 0$

17. $(tD^2 + 2D)x = 0$, $t > 0$ 18. $(D^3 + D^2 - D + 2)x = 0$

19. $(D^2 - 6D^2 + 11D - 6)x = 0$ 20. $(D^4 - 1)x = 0$

21. $(D^4 - 3D^2 - 4)x = 0$

In Exercises 22 through 29 you are given an nth-order o.d.e. and a formula describing a collection of solutions to the o.d.e. Show that this formula is *not* the general solution by finding an initial condition

$$x(t_0) = \alpha_0, \quad x'(t_0) = \alpha_1, \ldots, x^{(n-1)}(t_0) = \alpha_{n-1}$$

that cannot be matched by a function in the collection.

22. $(D^2 - 2D + 1)x = 0$; $x = c_1 e^t + c_2 e^{t+1}$

23. $D^3 x = 0$; $x = c_1(t + 1) + c_2(1 + t^2) + c_3(t^2 - t)$

24. $(D^3 + D)x = 0$; $x = c_1 \sin t + c_2 \cos t + c_3 \sin \left(t + \dfrac{\pi}{6} \right)$

25. $(D^3 + D^2)x = 0$; $x = c_1 e^{-t} + c_2$

26. $(D^3 + D^2)x = 0$; $x = c_1 e^{-t} + c_2(t + 2e^{-t}) + c_3 t$

27. $(D^3 - 2D^2 + D)x = -4e^{-t}$; $x = c_1 e^t + c_2 te^t + c_3(t - 1)e^t - e^{-t}$

28. $(D^4 - 1)x = 0$; $x = c_1 \sin t + c_2 \cos t + c_3 e^t$

29. $(D^4 - 16)x = 16$; $x = c_1 \sin 2t + c_2 \cos 2t + c_3 \sin t \cos t + c_4 e^{2t} - 1$

Two cautionary exercises:

30. Suppose $x = \phi(t)$ is a solution of the linear o.d.e. $Lx = E(t)$, where $E(t) \neq 0$. Show that $x = 2\phi(t) = \phi(t) + \phi(t)$ is *not* a solution of $Lx = E(t)$. Thus, linear combinations of solutions of *nonhomogeneous* equations need not be solutions.

31. Show that $x = \dfrac{1}{t^2 + 1}$ is a solution of the equation $\dfrac{dx}{dt} + 2tx^2 = 0$, whereas

   $x = \dfrac{2}{t^2 + 1} = \dfrac{1}{t^2 + 1} + \dfrac{1}{t^2 + 1}$ is not. Note that this o.d.e. is *not* linear.

Some more abstract problems:

32. a. Show that $W[t, t^2](t_0)$ is 0 when $t_0 = 0$ and is not 0 when $t_0 = 1$.
    b. Why doesn't this contradict the observation in Note 1?

33. a. Show that

$$p_n(r) = \det \begin{bmatrix} 1 & 1 & & 1 & 1 \\ a_1 & a_2 & & a_n & r \\ a_1^2 & a_2^2 & & a_n^2 & r^2 \\ \cdot & \cdot & \cdots & \cdot & \cdot \\ \cdot & \cdot & & \cdot & \cdot \\ \cdot & \cdot & & \cdot & \cdot \\ a_1^n & a_2^n & & a_n^n & r^n \end{bmatrix}$$

is a polynomial in r of degree at most n by expanding the determinant by minors along the last column.

    b. Show that $a_1, \ldots, a_n$ are roots of $p_n(r)$. [*Hint:* See Appendix A, Note 2(ii).]
    c. Conclude that $p_n(r) = A_n(r - a_1) \cdots (r - a_n)$ for some constant A_n.
    d. Use mathematical induction to show that if $a_1, a_2, \ldots, a_n$ are distinct, then $A_n \neq 0$.
    e. By taking $r = a_{n+1}$, show that if $a_1, \ldots, a_{n+1}$ are distinct, then

$$\det \begin{bmatrix} 1 & 1 & & 1 & 1 \\ a_1 & a_2 & & a_n & a_{n+1} \\ a_1^2 & a_2^2 & & a_n^2 & a_{n+1}^2 \\ \cdot & \cdot & \cdots & \cdot & \cdot \\ \cdot & \cdot & & \cdot & \cdot \\ \cdot & \cdot & & \cdot & \cdot \\ a_1^n & a_2^n & & a_n^n & a_{n+1}^n \end{bmatrix} \neq 0.$$

This determinant is called the **Vandermonde determinant.**

    f. Prove that if $a_1, \ldots, a_{n+1}$ are distinct, then $W[e^{a_1 t}, \ldots, e^{a_{n+1} t}](0) \neq 0$.

2.4   LINEAR INDEPENDENCE OF FUNCTIONS

In the last section we found that the Wronskian provides a test for determining whether solutions $h_1(t), \ldots, h_n(t)$ of a normal nth-order equation $Lx = 0$ generate the general solution. In this section we use this test to obtain another characterization of solutions that generate the general solution. This new characterization provides an alternative approach in cases where computation of the Wronskian is cumbersome. Our discussion will involve the notion of linear independence, the importance of which goes far beyond its use in this context.

Let's begin by taking another look at Example 2.3.4. In that example the three functions $h_1(t) = 1$, $h_2(t) = t$, and $h_3(t) = 2t - 3$ were solutions of the third-order equation $(D^3 - 2D^2)x = 0$, but did *not* generate the general solution. Note that $h_3(t) = -3h_1(t) + 2h_2(t)$, so

$$-3h_1(t) + 2h_2(t) - h_3(t) = 0.$$

We have found constants $c_1 = -3$, $c_2 = 2$, and $c_3 = -1$ so that

$$c_1 h_1(t) + c_2 h_2(t) + c_3 h_3(t) = 0$$

for all t. Of course, this last relationship would also hold if we took all the constants to be zero. The important thing is that we have found constants that are not all zero. We say these functions are linearly dependent.

Definition: *The functions $h_1(t), \ldots, h_n(t)$ are **linearly dependent** on the interval I if there exist constants $c_1, \ldots, c_n$, with at least one $c_i \neq 0$, so that*

$$c_1 h_1(t) + \cdots + c_n h_n(t) = 0$$

*for all t in I. The functions are **linearly independent** on I if the only constants for which this relationship holds for all t in I are $c_1 = c_2 = \cdots = c_n = 0$.*

Example 2.4.1

Show that the functions 1, t, and e^{2t} are linearly independent on any interval I.

Suppose

$$c_1 + c_2 t + c_3 e^{2t} = 0$$

for all t in I. If we differentiate this relationship twice, we see that

$$c_2 + 2c_3 e^{2t} = 0$$

and

$$4c_3 e^{2t} = 0$$

for all t in I. Since $e^{2t} \neq 0$, the last equation implies $c_3 = 0$. Substituting this value into the second equation yields $c_2 = 0$. The first equation now reads $c_1 = 0$. Thus, the functions are linearly independent on I.

Recall from Example 2.3.4 that these functions *do* generate the general solution of $(D^3 - 2D^2)x = 0$.

Example 2.4.2

Show that the functions 1, t, $\cos 2t$, and $\cos^2 t$ are linearly dependent on any interval I.

Since $\cos^2 t = (1 + \cos 2t)/2$,

$$\frac{1}{2}(1) + 0(t) + \frac{1}{2}\cos 2t + (-1)\cos^2 t = 0$$

for all t in I. Thus, the functions are linearly dependent on I.

Example 2.4.3

Show that the functions t^3 and $|t^3|$ are linearly independent on $-\infty < t < +\infty$.
Suppose

$$c_1 t^3 + c_2 |t^3| = 0$$

for all t. Then the relationship certainly holds for $t = 1$ and $t = -1$:

$$c_1(1) + c_2(1) = 0$$
$$c_1(-1) + c_2(1) = 0.$$

Adding these equations yields $c_2 = 0$; subtracting them yields $c_1 = 0$. Thus, the functions are linearly independent on $-\infty < t < +\infty$.

In Example 2.4.1, repeated differentiation of the original relationship led to a system of equations, which we used to show that the functions were independent. Let's try this in the general case. Suppose

$$c_1 h_1(t) + \cdots + c_n h_n(t) = 0$$

for all t in an interval I. Then

$$c_1 h_1'(t) + \cdots + c_n h_n'(t) \quad\quad = 0$$
$$\vdots$$
$$c_1 h_1^{(n-1)}(t) + \cdots + c_n h_n^{(n-1)}(t) = 0.$$

for all t in I. Substitution of a particular value of t, say $t = t_0$, yields a system of algebraic equations

$$c_1 h_1(t_0) + \cdots + c_n h_n(t_0) \quad = 0$$
$$c_1 h_1'(t_0) + \cdots + c_n h_n'(t_0) \quad = 0.$$

(A)

$$\vdots$$

$$c_1 h_1^{(n-1)}(t_0) + \cdots + c_n h_n^{(n-1)}(t_0) = 0.$$

Note that these equations always have at least one solution, namely

$$c_1 = c_2 = \cdots = c_n = 0.$$

If the determinant of coefficients, $W[h_1, \ldots, h_n](t_0)$, is not zero, then Cramer's determinant test tells us that the system of equations (A) has a *unique* solution. In this case, $c_1 = c_2 = \cdots = c_n = 0$ is the *only* solution, and the functions are independent.

Fact: Wronskian Test for Independence.   *If $W[h_1, \ldots, h_n](t_0) \neq 0$, then the functions $h_1(t), \ldots, h_n(t)$ are linearly independent on any interval I that contains t_0.*

Example 2.4.4

Show that the functions t and t^5 are linearly independent on $-\infty < t < +\infty$.
The Wronskian of these functions is

$$W[t, t^5](t) = \det \begin{bmatrix} t & t^5 \\ 1 & 5t^4 \end{bmatrix} = 4t^5.$$

Since $W[t, t^5](1) = 4 \neq 0$, the functions are independent.

Unfortunately, $h_1(t), \ldots, h_n(t)$ may be independent even if the determinant of coefficients of (A) is zero. The functions in Example 2.4.4 are independent even though their Wronskian is zero at $t_0 = 0$. The functions in Example 2.4.3 are independent on $-\infty < t < +\infty$ even though their Wronskian is *always* zero.
    Suppose, however, that $h_1(t), \ldots, h_n(t)$ are solutions of an nth-order homogeneous linear o.d.e. $Lx = 0$ that is normal on I. If $W[h_1, \ldots, h_n](t_0) = 0$, the second part of Cramer's determinant test tells us there are infinitely many solutions to the equations (A). Thus there are constants $c_1, \ldots, c_n$, with at least

one $c_i \neq 0$, such that these equations (A) hold. Let $x = c_1 h_1(t) + \cdots + c_n h_n(t)$. Then x is a solution of $Lx = 0$. The equations (A) tell us that x satisfies the same initial condition as 0. The "uniqueness" part of the existence and uniqueness theorem implies that $x = 0$—that is, $c_1 h_1(t) + \cdots + c_n h_n(t) = 0$ for all t in I. The functions are linearly dependent on I in this case.

   If we combine the result of the preceding paragraph with the Wronskian test for independence, we see that the solutions $h_1(t), \ldots, h_n(t)$ of an nth-order homogeneous linear o.d.e. $Lx = 0$ that is normal on I are linearly independent if and only if their Wronskian is not zero at t_0. The Wronskian test for solutions tells us that this last condition holds if and only if these functions generate the general solution of $Lx = 0$. Thus, we have the following fact.

Fact: *Let $h_1(t), \ldots, h_n(t)$ be solutions of the nth-order homogeneous linear o.d.e. $Lx = 0$ that is normal on an interval I. The general solution of $Lx = 0$ is $x = c_1 h_1(t) + \cdots + c_n h_n(t)$ if and only if $h_1(t), \ldots, h_n(t)$ are linearly independent on I.*

Example 2.4.5

Show that the general solution of

(H) $$(D^4 + 4D^3 + 6D^2 + 4D + 1)x = 0$$

is

(S) $$x = c_1 e^{-t} + c_2 t e^{-t} + c_3 t^2 e^{-t} + c_4 t^3 e^{-t}.$$

Substitution will show that the four functions e^{-t}, $t e^{-t}$, $t^2 e^{-t}$, and $t^3 e^{-t}$ are solutions of our fourth-order homogeneous o.d.e. (H). Suppose

(1) $$c_1 e^{-t} + c_2 t e^{-t} + c_3 t^2 e^{-t} + c_4 t^3 e^{-t} = 0$$

for all t. Since $e^{-t} \neq 0$, we can divide to get

(2) $$c_1 + c_2 t + c_3 t^2 + c_4 t^3 = 0.$$

We differentiate (2) three times to obtain three new relations:

(3) $$c_2 + 2c_3 t + 3c_4 t^2 = 0$$

(4) $$2c_3 + 6c_4 t = 0$$

(5) $$6c_4 = 0.$$

Equation (5) gives $c_4 = 0$, which substituted into (4) gives $c_3 = 0$. Substitution of these two values into (3) gives $c_2 = 0$, and finally (2) gives us $c_1 = 0$. Thus, the only way that (2), and hence (1), can hold is if

$$c_1 = c_2 = c_3 = c_4 = 0.$$

Our solutions of (H) are therefore independent, and (S) is the general solution.

Let's summarize.

LINEAR INDEPENDENCE OF FUNCTIONS

The functions $h_1(t), \ldots, h_n(t)$ are **linearly dependent** on the interval I if there exist constants $c_1, \ldots, c_n$, with at least one $c_i \neq 0$, so that

$$c_1 h_1(t) + \cdots + c_n h_n(t) = 0$$

for all t in I. The functions are **linearly independent** on I if the only constants for which this relationship holds for all t in I are $c_1 = c_2 = \cdots = c_n = 0$.

Let $Lx = 0$ be an nth-order homogeneous linear o.d.e. that is normal on I, and let t_0 be a value of t in I. Suppose $h_1(t), \ldots, h_n(t)$ are solutions of $Lx = 0$. Then the following are equivalent:

1. $h_1(t), \ldots, h_n(t)$ generate the general solution on I of $Lx = 0$.
2. $W[h_1, \ldots, h_n](t_0) \neq 0$.
3. $h_1(t), \ldots, h_n(t)$ are linearly independent on I.

EXERCISES

In Exercises 1 through 12, check the given functions for linear independence on $-\infty < t < \infty$.

1. $h_1(t) = 2t, \quad h_2(t) = 3t$
2. $h_1(t) = e^{2t}, \quad h_2(t) = e^{3t}$
3. $h_1(t) = \sin 5t, \quad h_2(t) = \cos 5t$
4. $h_1(t) = 1, \quad h_2(t) = t, \quad h_3(t) = t^2$
5. $h_1(t) = t + 1, \quad h_2(t) = t^2 + 1, \quad h_3(t) = t^2 - t$
6. $h_1(t) = t^2 + t, \quad h_2(t) = t^2 + 1, \quad h_3(t) = t^2 - 1$
7. $h_1(t) = te^t, \quad h_2(t) = (t + 1)e^t, \quad h_3(t) = te^{t+1}$
8. $h_1(t) = (t + 1)e^t, \quad h_2(t) = (t - 1)e^t, \quad h_3(t) = e^t$
9. $h_1(t) = \sin^2 t, \quad h_2(t) = \cos^2 t, \quad h_3(t) = \sin 2t$

10. $h_1(t) = \sin^2 t$, $h_2(t) = \cos^2 t$, $h_3(t) = \cos 2t$
11. $h_1(t) = e^t$, $h_2(t) = e^t \sin t$, $h_3(t) = e^t \cos t$
12. $h_1(t) = e^t$, $h_2(t) = te^t$, $h_3(t) = e^{2t}$, $h_4(t) = te^{2t}$

Some more abstract problems:

13. a. Show that if a, b, and c are distinct constants, then the functions e^{at}, e^{bt}, and e^{ct} are linearly independent.
    b. Show that the functions e^{at}, te^{at}, . . . , $t^{k-1}e^{at}$ are linearly independent.

14. a. Suppose we know that

$$h_1(0) = 1 \quad h_1(1) = 0 \quad h_1(2) = 0$$
$$h_2(0) = 0 \quad h_2(1) = 1 \quad h_2(2) = 0$$
$$h_3(0) = 0 \quad h_3(1) = 0 \quad h_3(2) = 1$$

Show that $h_1(t)$, $h_2(t)$, and $h_3(t)$ are linearly independent on the interval $-1 < t < 4$.

    b. Suppose we know that

$$g_1(0) = 0 \quad g_1(1) \neq 0 \quad g_1(2) \neq 0$$
$$g_2(0) \neq 0 \quad g_2(1) = 0 \quad g_2(2) \neq 0$$
$$g_3(0) \neq 0 \quad g_3(1) \neq 0 \quad g_3(2) = 0$$

Are $g_1(t)$, $g_2(t)$, and $g_3(t)$ necessarily independent on the interval $-1 < t < 4$?

15. a. Suppose the functions $h_1(t)$, . . . , $h_n(t)$ are linearly independent on an interval I. Show that $g_1(t) = e^t h_1(t)$, . . . , $g_n(t) = e^t h_n(t)$ are also linearly independent on I.
    b. Show that e^t in (a) can be replaced by any function that does not vanish on I.

16. a. Show that if $h_1(t)$, . . . , $h_n(t)$ are polynomials of degrees 1, 2, . . . , n, respectively, then they are linearly independent.
    b. Show that any collection of nonconstant polynomials of distinct degrees is independent.

*17. Show that a collection of more than $n + 1$ polynomials of degree at most n must be dependent.

18. Show that if $h_1(t)$, . . . , $h_n(t)$ are independent, then the two functions $g_1(t) = a_1 h_1(t) + \cdots + a_n h_n(t)$ and $g_2(t) = b_1 h_1(t) + \cdots + b_n h_n(t)$ are different, unless $a_1 = b_1$, . . . , $a_n = b_n$.

19. a. Show that if $h_1(t)$, . . . , $h_n(t)$ are linearly dependent, then $h_1'(t)$, . . . , $h_n'(t)$ are also linearly dependent.
    b. Find an example of two linearly independent functions $f(t)$, $g(t)$ whose derivatives are linearly dependent.

2.5 HOMOGENEOUS LINEAR EQUATIONS WITH CONSTANT COEFFICIENTS: REAL ROOTS

We saw in the last section that the problem of finding the general solution of an nth-order homogeneous linear o.d.e. boils down to finding n linearly independent solutions. In this section and the next we develop an algebraic way of finding such solutions for equations

(H) $$(a_n D^n + \cdots + a_1 D + a_0)x = 0$$

with *constant coefficients*.

The operator in (H) is a polynomial expression in D. We call the corresponding polynomial

$$P(r) = a_n r^n + \cdots + a_1 r + a_0$$

the **characteristic polynomial** of (H), and we denote the operator by $P(D)$:

$$P(D) = a_n D^n + \cdots + a_1 D + a_0.$$

We can take our algebraic notation further by making sense of multiplication of operators. We define the product $Q(D)F(D)$ to be the operator whose value at a function x is obtained by first applying $F(D)$ to x, and then applying $Q(D)$ to $F(D)x$:

$$[Q(D)F(D)]x = Q(D)[F(D)x].$$

Example 2.5.1

$$[(D^2 + D)(D - 1)]x = (D^2 + D)[(D - 1)x] = (D^2 + D)[Dx - x]$$
$$= D^2[Dx - x] + D[Dx - x] = D^3x - D^2x + D^2x - Dx$$
$$= D^3x - Dx = (D^3 - D)x.$$

This example illustrates a remarkable fact about polynomial expressions in D. The effect of applying first $F(D)$ to x and then $Q(D)$ to the result is exactly the same as applying their product *as polynomials* to x.

Fact: *If $P(r) = Q(r)F(r)$, then, for any function x, $P(D)x = Q(D)[F(D)x]$.*

Notice that if $P(r) = Q(r)F(r)$ and if $F(D)\phi(t) = 0$, then

$$P(D)\phi(t) = Q(D)[F(D)\phi(t)] = Q(D)[0] = 0.$$

Fact: *If $F(r)$ is a factor of $P(r)$, then any solution of $F(D)x = 0$ is also a solution of $P(D)x = 0$.*

We begin our search for solutions of $P(D)x = 0$ by looking for factors of $P(r)$. If the real number λ is a root of $P(r)$—that is, if $P(\lambda) = 0$—then $r - \lambda$ is a factor of $P(r)$. In this case, any solution of $(D - \lambda)x = 0$ will also be a solution of $P(D)x = 0$. In particular, $e^{\lambda t}$ is a solution of $P(D)x = 0$.

Fact: *If the real number λ is a root of $P(r)$, then $e^{\lambda t}$ is a solution of $P(D)x = 0$.*

Example 2.5.2

Solve $(D^3 - D)x = 0$.
The characteristic polynomial factors easily:

$$r^3 - r = r(r^2 - 1) = r(r - 1)(r + 1).$$

Corresponding to its roots, 0, 1, and -1, we get solutions, $e^{0t} = 1$, e^t, and e^{-t}. Since these three solutions are independent (check this), the general solution of our third-order o.d.e. is

$$x = c_1 + c_2 e^t + c_3 e^{-t}.$$

Now suppose that λ is a root of multiplicity k; that is, suppose $(r - \lambda)^k$ is the highest power of $r - \lambda$ that is a factor of $P(r)$. We know that $e^{\lambda t}$ is a solution of $P(D)x = 0$. Can we find any other solutions corresponding to λ? Let's look for some, using a trick we first used when dealing with first-order equations. Let's look for solutions of the form $e^{\lambda t}y$ where y is a function of t.
    The derivative of $e^{\lambda t}y$ is

$$D[e^{\lambda t}y] = e^{\lambda t} Dy + \lambda e^{\lambda t}y = e^{\lambda t}(D + \lambda)y.$$

Thus, we can pull out the factor $e^{\lambda t}$, provided we replace D by $D + \lambda$. Then

$$D^2[e^{\lambda t}y] = D[e^{\lambda t}(D + \lambda)y] = e^{\lambda t}[(D + \lambda)(D + \lambda)y] = e^{\lambda t}[(D + \lambda)^2 y].$$

If we continue this way, we get the general formula

$$D^m[e^{\lambda t}y] = e^{\lambda t}[(D + \lambda)^m y].$$

This formula extends easily to arbitrary polynomials in D.

Fact: Exponential Shift. $P(D)[e^{\lambda t}y] = e^{\lambda t}P(D + \lambda)y.$

Example 2.5.3

$$(D^2 - 4D + 5)[e^{2t} \cos t] = e^{2t}[(D + 2)^2 - 4(D + 2) + 5]\cos t$$
$$= e^{2t}(D^2 + 1)\cos t = e^{2t}(-\cos t + \cos t) = 0.$$

Example 2.5.4

$$(D - 1)^3[c_1 e^{2t} + c_2 t e^{2t}] = e^{2t}(D + 2 - 1)^3[c_1 + c_2 t]$$
$$= e^{2t}(D + 1)^3[c_1 + c_2 t]$$
$$= e^{2t}(D^3 + 3D^2 + 3D + 1)[c_1 + c_2 t]$$
$$= e^{2t}[(3c_2 + c_1) + c_2 t]$$
$$= (3c_2 + c_1)e^{2t} + c_2 t e^{2t}.$$

Let's apply the exponential shift in case $P(r) = Q(r)(r - \lambda)^k$. Here

$$P(D)[e^{\lambda t}y] = Q(D)(D - \lambda)^k[e^{\lambda t}y] = e^{\lambda t}[Q(D + \lambda)(D + \lambda - \lambda)^k y]$$
$$= e^{\lambda t}[Q(D + \lambda)D^k y].$$

If we take y to be a function whose kth derivative is 0, then $e^{\lambda t}y$ will be a solution of $P(D)x = 0$. In particular, taking $y = 1, t, \ldots, t^{k-1}$, we see that the functions $e^{\lambda t}, te^{\lambda t}, \ldots, t^{k-1}e^{\lambda t}$, are solutions of $P(D)x = 0$. What's more, an argument like the one in Example 2.4.5 will show they are independent (see Exercise 13, Section 2.4, and Exercise 27, Section 2.5).

Fact: *If the real number λ is a root of $P(r)$ of multiplicity k, then the k functions $e^{\lambda t}, te^{\lambda t}, \ldots, t^{k-1}e^{\lambda t}$ are linearly independent solutions of $P(D)x = 0$.*

Example 2.5.5

Solve $(D^3 - 3D^2 + 3D - 1)x = 0$.

The characteristic polynomial is $r^3 - 3r^2 + 3r - 1 = (r - 1)^3$. Corresponding to the triple root 1, we get three independent solutions, e^t, te^t, and $t^2 e^t$, of our third-order o.d.e. The general solution is

$$x = c_1 e^t + c_2 t e^t + c_3 t^2 e^t.$$

■ Example 2.5.6

Solve $(D - 2)^2(D - 1)^3 x = 0$.

The characteristic polynomial is $(r - 2)^2(r - 1)^3$. Corresponding to the double root 2, we get two independent solutions e^{2t} and te^{2t}. Corresponding to the triple root 1, we get three independent solutions, e^t, te^t, and $t^2 e^t$. We have found five solutions of our fifth-order o.d.e.

Are the solutions independent? Suppose

(1) $$c_1 e^{2t} + c_2 te^{2t} + c_3 e^t + c_4 te^t + c_5 t^2 e^t = 0$$

for all t. Note that e^t, te^t, and $t^2 e^t$ are solutions of $(D - 1)^3 x = 0$. Application of $(D - 1)^3$ to (1) will eliminate the terms involving these functions and will affect the first two terms as in Example 2.5.4.:

$$(3c_2 + c_1)e^{2t} + c_2 te^{2t} = 0.$$

Since e^{2t} and te^{2t} are independent,

$$3c_2 + c_1 = 0 \quad \text{and} \quad c_2 = 0.$$

Then

$$c_1 = c_2 = 0.$$

Now (1) reads

$$c_3 e^t + c_4 te^t + c_5 t^2 e^t = 0.$$

Since e^t, te^t, and $t^2 e^t$ are independent,

$$c_3 = c_4 = c_5 = 0.$$

Our solutions are independent. The general solution of the o.d.e. is

$$x = c_1 e^{2t} + c_2 te^{2t} + c_3 e^t + c_4 te^t + c_5 t^2 e^t.$$

The argument we used to show independence in the preceding example can be generalized to show that the solutions corresponding to different real roots are always independent.

Fact: *Associate functions to the polynomial $P(r)$ as follows: for each real root λ of $P(r)$, include the k functions $e^{\lambda t}, te^{\lambda t}, \ldots, t^{k-1}e^{\lambda t}$, where k is the multiplicity* *of* λ *as a root. These functions are linearly independent solutions of* $P(D)x = 0$.

Example 2.5.7

Solve $(D^3 - D^2 - 8D + 12)x = 0$.

The characteristic polynomial is $P(r) = r^3 - r^2 - 8r + 12$. We look for a root among the divisors of the constant term: ± 1 are not roots, but 2 is. This tells us that $r - 2$ is a factor of $P(r)$. Dividing, we see that $P(r) = (r - 2)(r^2 + r - 6)$. We can find the roots of the quadratic term by factoring or by the quadratic formula; these roots are 2 and -3. Thus, $P(r)$ has a double root 2 and a single root -3. Corresponding to these roots, we get three independent solutions, e^{2t}, te^{2t}, and e^{-3t}, of our third-order o.d.e. The general solution is

$$x = c_1 e^{2t} + c_2 t e^{2t} + c_3 e^{-3t}.$$

Example 2.5.8

Solve $(3D^5 - D^4 - 15D^3 + 5D^2 + 18D - 6)x = 0$.

The characteristic polynomial is

$$3r^5 - r^4 - 15r^3 + 5r^2 + 18r - 6 = 3\left(r - \frac{1}{3}\right)(r^2 - 3)(r^2 - 2)$$

(see Note 2 for a description of how we factored this polynomial). Corresponding to its roots, $1/3$, $\pm\sqrt{3}$, and $\pm\sqrt{2}$, we get five linearly independent solutions, $e^{t/3}$, $e^{\sqrt{3}t}$, $e^{-\sqrt{3}t}$, $e^{\sqrt{2}t}$, and $e^{-\sqrt{2}t}$, of our fifth-order equation. The general solution is

$$x = c_1 e^{t/3} + c_2 e^{\sqrt{3}t} + c_3 e^{-\sqrt{3}t} + c_4 e^{\sqrt{2}t} + c_5 e^{-\sqrt{2}t}.$$

Example 2.5.9

Solve the initial-value problem

$$(D^3 - 2D^2 + D)x = 0; \qquad x(0) = x'(0) = 0, \, x''(0) = 1.$$

The characteristic polynomial is $r^3 - 2r^2 + r = r(r - 1)^2$, so the general solution of the o.d.e. is

$$x = c_1 + c_2 e^t + c_3 t e^t.$$

Any function of this form satisfies

$$x' = (c_2 + c_3)e^t + c_3 t e^t$$

$$x'' = (c_2 + 2c_3)e^t + c_3 t e^t.$$

Substitution of our initial condition yields the equations

$$0 = c_1 + c_2$$

$$0 = \quad\quad c_2 + c_3$$

$$1 = \quad\quad c_2 + 2c_3$$

which we solve to find $c_3 = 1$, $c_2 = -1$, and $c_1 = 1$. The solution of our initial-value problem is

$$x = 1 - e^t + t e^t.$$

Let's summarize our facts about operators of the form $P(D)$.

HOMOGENEOUS EQUATIONS WITH CONSTANT COEFFICIENTS: REAL ROOTS

Associated to each polynomial $P(r) = a_n r^n + \cdots + a_1 r + a_0$ is an operator $P(D) = a_n D^n + \cdots + a_1 D + a_0$. Operators of this form satisfy the following:

1. If $P(r) = Q(r)F(r)$, then $P(D)x = Q(D)[F(D)x]$.
2. If $F(r)$ is a factor of $P(r)$, then any solution of $F(D)x = 0$ is also a solution of $P(D)x = 0$.
3. **Exponential Shift:** $P(D)[e^{\lambda t}y] = e^{\lambda t}[P(D + \lambda)y]$.
4. Associate functions to $P(r)$ as follows: for each real root λ of $P(r)$, include the k functions $e^{\lambda t}, t e^{\lambda t}, \ldots, t^{k-1} e^{\lambda t}$, where k is the multiplicity of λ as a root. These functions are linearly independent solutions of $P(D)x = 0$.

Notes

1.   A warning about variable-coefficient operators

The basis for our results was the observation that formal multiplication of polynomial expressions in D corresponds to successive application of these operators. In particular, the commutative law for multiplication holds for these operators: $Q(D)F(D)$ and $F(D)Q(D)$ multiply out to give the same operator.

We can still use the notion of successive application to define the product of two variable-coefficient operators, but the commutative law breaks down. For example,

$$[(D - t)D]x = (D - t)[Dx] = D^2x - tDx = (D^2 - tD)x,$$

but

$$[D(D - t)]x = D[(D - t)x] = D[Dx - tx] = D^2x - tDx - x = (D^2 - tD - 1)x.$$

Note that D is a factor of both $D(D - t)$ and $(D - t)D$ and that $x = 1$ is a solution of $Dx = 0$. Although $x = 1$ is also a solution of $[(D - t)D]x = 0$, it is *not* a solution of $[D(D - t)]x = 0$. This example illustrates why we can't expect a nice algebraic technique for solving general variable-coefficient equations.

2.   Finding roots of polynomials

The problem of finding roots of polynomials is a deep one. For polynomials of degree two, there is a formula; the roots of $ar^2 + br + c$ are $(-b \pm \sqrt{b^2 - 4ac})/2a$. Similar, but messier, formulas exist for finding the roots of polynomials of degrees three and four. No such formula exists for polynomials of degree $n > 4$.

Although trial and error plays a part in finding roots, there are some tricks worth knowing. Among these is one for finding *rational* roots of polynomials $P(r)$ with *integer* coefficients. If $\lambda = a/b$ is a root of $P(r)$, where a and b are integers from which we've canceled all common factors, then a is a divisor of the constant coefficient a_0 of $P(r)$ and b is a divisor of the highest coefficient a_n. Let's use this fact to help find the roots of

$$P(r) = 3r^5 - r^4 - 15r^3 + 5r^2 + 18r - 6$$

(as needed in Example 2.5.8).

The divisors of the constant coefficient of $P(r)$ are ± 1, ± 2, ± 3, and ± 6. The divisors of the highest coefficient are ± 1 and ± 3. The only possibilities for rational roots are ± 1, $\pm 1/3$, ± 2, $\pm 2/3$, ± 3, $\pm 3/3$, ± 6, and $\pm 6/3$. Substitution will show that ± 1 are not roots, but $1/3$ is. Then $r - 1/3$ is a factor. We divide to get

$$P(r) = \left(r - \frac{1}{3}\right)(3r^4 - 15r^2 + 18) = 3\left(r - \frac{1}{3}\right)(r^4 - 5r^2 + 6).$$

The fact that the fourth-degree factor is a quadratic in $y = r^2$ helps us to spot a factorization:

$$P(r) = 3\left(r - \frac{1}{3}\right)(r^2 - 3)(r^2 - 2).$$

The roots of $P(r)$ are $1/3$, $\pm\sqrt{3}$, and $\pm\sqrt{2}$.

EXERCISES

Find the general solution in Exercises 1 through 13.

1. $x'' - 5x = 0$ 2. $x'' - 2x' - 15x = 0$
3. $2x'' - 5x' = 0$ 4. $x'' + 4x' + 4x = 0$
5. $9x'' - 12x' + 4x = 0$ 6. $x''' - 2x'' - x' + 2x = 0$
7. $x^{(5)} - 2x^{(4)} + x^{(3)} = 0$ 8. $(D + 2)^3(D - 1)x = 0$
9. $D^2(D + 1)^3(D - 2)(3D + 5)(2D - 3)x = 0$
10. $(D^3 - 11D^2 + 31D - 21)x = 0$ 11. $(D^2 - 2)(D^2 + D - 1)x = 0$
12. $(4D^3 - 24D^2 + 35D - 12)x = 0$ 13. $(D^4 - 2D^2 + 1)x = 0$

Solve the initial-value problems in Exercises 14 through 17.

14. $15x'' - 2x' - x = 0;$ $x(0) = x'(0) = 1$
15. $x'' - 5x' = 0;$ $x(0) = x'(0) = 10$
16. $x''' + 3x'' + 3x' + x = 0;$ $x(0) = x'(0) = x''(0) = 0$
17. $(D - 1)^2(D + 2)x = 0;$ $x(0) = x'(0) = 0, \quad x''(0) = 9$

Exercises 18 through 21 refer to our models.

18. *A Damped Spring:*
    a. Find the general solution of the o.d.e. in Example 2.1.2, assuming $m = 5$ grams, $L = 3$ cm, $k = 60$ dynes/cm, and $b = 40$ dynes/(cm/sec).
    b. The mass in (a) starts at rest with the spring stretched to 5 cm. Find a formula for its position after t seconds.

19. *An LRC Circuit:* Using the o.d.e. derived in Exercise 5, Section 2.1, find a formula for the charge Q at time t seconds on the capacitor of an *LRC* circuit with no external voltage ($V(t) = 0$), resistance $R = 2$ ohms, capacitance $C = 1$ farad, and inductance $L = 1$ henry, given that charge $Q = 0$ and current $I = 1$ ampere at time $t = 0$.

20. *Punk Rock:* Consider the supply-and-demand model in Exercise 6, Section 2.1, under the following assumptions concerning the new album, *Annihilation*, by Dee and the Operators:
    a. If it were free, 5000 people per week would want a copy; every $1 in price reduces this number by 600.
    b. The contract that Dee and the Operators signed with Golden Groove Records contains an escalator clause: their commission (and hence the price of the record) is increased each week by 10¢ for every 100 fans who ask for, but can't get, the album.
    c. Every week the record production plant hires enough new people to increase their weekly output of copies by 50 for every dollar in price of the record.
    Using the model of Exercise 6, Section 2.1, predict what will happen to the price of the record after one year (52 weeks) if its initial price is $p = \$4.00$ and $dp/dt = 0$ at $t = 0$.

*21.   *Coupled Springs:*
    a.   Find a formula for the roots of the polynomial

$$P(r) = (mr^2 + br + k)^2 + k(mr^2 + br)$$

by means of the substitution $z = mr^2 + br$ and two uses of the quadratic formula.
    b.   Find the general solution of the o.d.e. modeling the damped coupled spring system in Exercise 4, Section 2.1, when $m = 1$, $k = 1$, and $b = 1 + \sqrt{5}$. [*Hint:* $(1 + \sqrt{5})^2 = 2(3 + \sqrt{5})$.]

Use the exponential shift to calculate $Lf_i(t)$ in Exercises 22 through 25.

22. $L = D^2 + 3D - 2;$ $f_1(t) = e^{2t},$ $f_2(t) = t^3 e^{2t}$
23. $L = (D + 1)^2;$ $f_1(t) = e^t,$ $f_2(t) = e^t \sin t,$ $f_3(t) = e^{-t} \sin t$
24. $L = D^2 + 4D + 5;$ $f_1(t) = te^{-2t} \cos t,$ $f_2(t) = te^{-2t} \sin t$
25. $L = [D^2 - 2\alpha D + (\alpha^2 + \beta^2)];$ $f_1(t) = e^{\alpha t} \cos \beta t,$ $f_2(t) = e^{\alpha t} \sin \beta t$

In Exercises 26 and 27, you are asked to use an argument similar to the one used in Example 2.5.7 to show the independence of given sets of functions.

26.   Show that if $m_1, \ldots, m_k$ are distinct, then $e^{m_1 t}, \ldots, e^{m_k t}$ are independent as follows: Suppose

$$c_1 e^{m_1 t} + \cdots + c_k e^{m_k t} = 0$$

for all t. Show that the operator

$$P_k(D) = (D - m_1)(D - m_2) \cdots (D - m_{k-1})$$

changes this equation into

$$c_k(m_k - m_1)(m_k - m_2) \cdots (m_k - m_{k-1})e^{m_k t} = 0$$

from which we can conclude that $c_k = 0$. What operator should we now use to show that $c_{k-1} = 0$?
27.   Show that the functions $e^{\lambda t}, te^{\lambda t}, \ldots, t^{k-1}e^{\lambda t}$ are independent as follows: Suppose $c_1 e^{\lambda t} + c_2 te^{\lambda t} + \cdots + c_k t^{k-1}e^{\lambda t} = 0$. Show that the operator $(D - \lambda)^{k-1}$ changes this equation into $e^{\lambda t}(k - 1)!c_k = 0$, from which we can conclude that $c_k = 0$. What operator should we now use to show that $c_{k-1} = 0$?

2.6   HOMOGENEOUS LINEAR EQUATIONS WITH CONSTANT COEFFICIENTS: COMPLEX ROOTS

In the last section we saw how to solve equations $P(D)x = 0$ when all the roots of the characteristic polynomial $P(r)$ are real. In this section we complete our treatment of homogeneous linear equations with constant coefficients by determining how to deal with complex roots.

Suppose $\alpha + \beta i$ is a root of $P(r)$, where α and β are real numbers and $\beta \neq 0$. Then $\alpha - \beta i$ is also a root, so

$$F(r) = r^2 - 2\alpha r + (\alpha^2 + \beta^2) = [r - (\alpha + \beta i)][r - (\alpha - \beta i)]$$

is a factor of $P(r)$. Any solution of $F(D)x = 0$ is also a solution of $P(D)x = 0$.

For the moment, let's ignore the fact that the roots are complex. Then the general solution of $F(D)x = 0$ is

$$x = k_1 e^{(\alpha + \beta i)t} + k_2 e^{(\alpha - \beta i)t}.$$

We can use **Euler's formula,**

$$e^{u + vi} = e^u(\cos v + i \sin v),$$

to rewrite our formula for x:

$$\begin{aligned} x &= k_1 e^{\alpha t}(\cos \beta t + i \sin \beta t) + k_2 e^{\alpha t}(\cos \beta t - i \sin \beta t) \\ &= (k_1 + k_2)e^{\alpha t} \cos \beta t + (k_1 i - k_2 i)e^{\alpha t} \sin \beta t \\ &= c_1 e^{\alpha t} \cos \beta t + c_2 e^{\alpha t} \sin \beta t. \end{aligned}$$

This formula involves two real-valued candidates for solutions.

Substitution (aided by the exponential shift formula) will show that the functions $e^{\alpha t} \cos \beta t$ and $e^{\alpha t} \sin \beta t$ are solutions of $F(D)x = 0$. Are they independent? Suppose

$$c_1 e^{\alpha t} \cos \beta t + c_2 e^{\alpha t} \sin \beta t = 0$$

for all t. Then, taking $t = 0$ gives

$$c_1 = 0.$$

Since $e^{\alpha t} \sin \beta t$ is not always 0, it now follows that

$$c_2 = 0.$$

We have found two independent solutions of $F(D)x = 0$.

Fact: *Suppose $\alpha \pm \beta i$ are roots of $P(r)$, where α and β are real and $\beta \neq 0$. Then $e^{\alpha t} \cos \beta t$ and $e^{\alpha t} \sin \beta t$ are linearly independent real-valued solutions of $P(D)x = 0$.*

Example 2.6.1

Solve $(D^2 + 4D + 5)x = 0$, which is the equation for the damped-spring system of Example 2.1.2, with $m = 1$, $b = 4$, $k = 5$.

The roots of the characteristic polynomial $r^2 + 4r + 5$ are $-2 \pm i$. Corresponding to these roots, we get two linearly independent solutions, $e^{-2t} \cos t$ and $e^{-2t} \sin t$, of our second-order o.d.e.. The general solution is

$$x = c_1 e^{-2t} \cos t + c_2 e^{-2t} \sin t.$$

Now suppose that $\alpha + \beta i$ is a root of $P(r)$ of multiplicity k. Then $\alpha - \beta i$ is also a root of multiplicity k. The polynomial

$$G(r) = [r^2 - 2\alpha + (\alpha^2 + \beta^2)]^k = [r - (\alpha + \beta i)]^k [r - (\alpha - \beta i)]^k$$

is a factor of $P(r)$. Any solution of $G(D)x = 0$ is also a solution of $P(D)x = 0$. Let's look at any example to see what solutions of $G(D)x = 0$ look like.

Example 2.6.2

Solve $(D^2 + 4D + 5)^2 x = 0$.

The roots of the characteristic polynomial are $-2 \pm i$, each of which has multiplicity 2. We know that $e^{-2t} \cos t$ and $e^{-2t} \sin t$ are linearly independent solutions. We need two more. By analogy with what worked when we had real roots of multiplicity 2, we guess that $te^{-2t} \cos t$ and $te^{-2t} \sin t$ are also solutions. We can verify this by substitution (aided, once more, by the exponential shift formula).

Are our four solutions independent? Suppose

$$c_1 e^{-2t} \cos t + c_2 e^{-2t} \sin t + c_3 te^{-2t} \cos t + c_4 te^{-2t} \sin t = 0$$

for all t. Applying $(D^2 + 4D + 5)$ to this equation gives

$$-2c_3 e^{-2t} \sin t + 2c_4 e^{-2t} \cos t = 0.$$

Since $e^{-2t} \sin t$ and $e^{-2t} \cos t$ are independent,

$$c_3 = c_4 = 0.$$

Now

$$c_1 e^{-2t} \cos t + c_2 e^{-2t} \sin t = 0,$$

so

$$c_1 = c_2 = 0.$$

Our solutions are independent. The general solution is

$$x = c_1 e^{-2t} \cos t + c_2 e^{-2t} \sin t + c_3 t e^{-2t} \cos t + c_4 t e^{-2t} \sin t.$$

The general case follows a similar pattern.

Fact: *Suppose that $\alpha \pm \beta i$ are roots of $P(r)$ of multiplicity k, where α and β are real numbers and $\beta \neq 0$. Then the $2k$ functions $e^{\alpha t} \cos \beta t$, $e^{\alpha t} \sin \beta t$, $te^{\alpha t} \cos \beta t$, $te^{\alpha t} \sin \beta t$, . . . , $t^{k-1} e^{\alpha t} \cos \beta t$, $t^{k-1} e^{\alpha t} \sin \beta t$ are linearly independent real-valued solutions of $P(D)x = 0$.*

Once again, the argument in Example 2.5.6 can be generalized to show that the solutions corresponding to distinct roots of $P(r)$ are independent. We can now solve any equation $P(D)x = 0$, *provided we can find the roots of $P(r)$.*

Fact: *Associate functions to the polynomial $P(r)$ as follows:*

1. *For each real root λ of $P(r)$, include the k functions $e^{\lambda t}, te^{\lambda t}, \ldots, t^{k-1} e^{\lambda t}$, where k is the multiplicity of λ as a root.*
2. *For each pair of complex roots $\alpha \pm \beta i$, include the $2k$ functions $e^{\alpha t} \cos \beta t$, $e^{\alpha t} \sin \beta t$, $te^{\alpha t} \cos \beta t$, $te^{\alpha t} \sin \beta t$, . . . , $t^{k-1} e^{\alpha t} \cos \beta t$, $t^{k-1} e^{\alpha t} \sin \beta t$, where k is the multiplicity of each of $\alpha \pm \beta i$ as a root.*

These functions are linearly independent solutions of $P(D)x = 0$.

Example 2.6.3

Solve $(D + 2)^3 D^2 (D^2 - D + 1)x = 0$.

The roots of the characteristic polynomial $(r + 2)^3 r^2 (r^2 - r + 1)$ are -2 (multiplicity 3), 0 (multiplicity 2), and $\dfrac{1}{2} \pm \dfrac{i \sqrt{3}}{2}$ (multiplicity 1). Corresponding to the roots, we get seven linearly independent solutions, e^{-2t}, te^{-2t}, $t^2 e^{-2t}$, 1, t, $e^{t/2} \cos(\sqrt{3}/2)t$, and $e^{t/2} \sin(\sqrt{3}/2)t$, of our seventh-order o.d.e. The general solution is

$$x = c_1 e^{-2t} + c_2 t e^{-2t} + c_3 t^2 e^{-2t} + c_4 + c_5 t$$
$$+ c_6 e^{t/2} \cos \frac{\sqrt{3}}{2} t + c_7 e^{t/2} \sin \frac{\sqrt{3}}{2} t.$$

Example 2.6.4

Solve $(D^8 - 8D^4 + 16)x = 0$.
The characteristic polynomial is

$$r^8 - 8r^4 + 16 = (r^4 - 4)^2 = [(r^2 - 2)(r^2 + 2)]^2.$$

Its roots are $\pm\sqrt{2}$, and $\pm\sqrt{2}i$, each of multiplicity 2. The general solution of our o.d.e. is

$$x = c_1 e^{\sqrt{2}t} + c_2 t e^{\sqrt{2}t} + c_3 e^{-\sqrt{2}t} + c_4 t e^{-\sqrt{2}t}$$
$$+ c_5 \cos\sqrt{2}t + c_6 \sin\sqrt{2}t + c_7 t \cos\sqrt{2}t + c_8 t \sin\sqrt{2}t.$$

In our next two examples we turn the tables. We are given a function $E(t)$ and are asked to find an operator $A(D)$ so that $A(D)E(t) = 0$. An operator that satisfies this condition is said to **annihilate** $E(t)$. Thus, saying $A(D)$ annihilates the function $E(t)$ is the same as saying $E(t)$ is a solution of $A(D)x = 0$.

Example 2.6.5

Find an operator $A(D)$ with constant coefficients that annihilates the function $E(t) = -3e^{3t} + 10te^{3t}$.

Any linear operator that annihilates both e^{3t} and te^{3t} will annihilate $E(t)$. We know that e^{3t} and te^{3t} are solutions of any homogeneous equation whose characteristic polynomial has 3 as a double root. The simplest such polynomial is $A(r) = (r - 3)^2$. Thus, $E(t)$ is annihilated by

$$A(D) = (D - 3)^2.$$

Example 2.6.6

Find an operator $A(D)$ with constant coefficients that annihilates the function $E(t) = 1 + 65e^t \cos 2t$.

We know that $1 = e^{0t}$ is a solution of any homogeneous equation whose characteristic polynomial has 0 as a root. Also, $e^t \cos 2t$ is a solution of any homogeneous equation whose characteristic polynomial has $1 \pm 2i$ as roots.

The simplest polynomial that has all three numbers as roots is

$$A(r) = r[r - (1 + 2i)][r - (1 - 2i)] = r(r^2 - 2r + 5).$$

Thus, $E(t)$ is annihilated by

$$A(D) = D(D^2 - 2D + 5).$$

We close with a description of our method for solving homogeneous linear equations with constant coefficients. This method was published by Leonhard Euler (1724).

HOMOGENEOUS LINEAR EQUATIONS WITH CONSTANT COEFFICIENTS

To solve a homogeneous linear o.d.e. with constant coefficients

(H) $$(a_n D^n + \cdots + a_1 D + a_0)x = 0:$$

1. Find all the roots of the characteristic polynomial

$$P(r) = a_n r^n + \cdots + a_1 r + a_0.$$

2. Obtain functions $h_1(t), \ldots, h_n(t)$ as follows:
   a. For each real root λ of $P(r)$, include on the list the k functions

   $$e^{\lambda t}, te^{\lambda t}, \ldots, t^{k-1}e^{\lambda t}$$

   where k is the multiplicity of λ as a root.
   b. For each pair of complex roots $\alpha \pm \beta i$, include on the list the $2k$ functions

   $$e^{\alpha t} \cos \beta t, e^{\alpha t} \sin \beta t, te^{\alpha t} \cos \beta t, te^{\alpha t} \sin \beta t,$$
   $$\ldots, t^{k-1}e^{\alpha t} \cos \beta t, t^{k-1}e^{\alpha t} \sin \beta t$$

   where k is the multiplicity of each of $\alpha \pm \beta i$ as a root.
3. The general solution of (H) is generated by these functions:

$$x = c_1 h_1(t) + \cdots + c_n h_n(t).$$

EXERCISES

Find the general solution in Exercises 1 through 12.

1. $9x'' + x = 0$
2. $x'' + 2x' + 5x = 0$
3. $x^{(4)} - 16x = 0$
4. $(4D^2 + 1)(D^2 + 2D + 2)x = 0$
5. $(D^2 + 1)(D^2 + D + 1)x = 0$
6. $(D^4 + 2D^2 + 1)x = 0$
7. $(D^3 - D^2 + 9D - 9)x = 0$
8. $(D^4 + 9D^2 + 20)x = 0$
9. $(D^4 - 1)x = 0$
10. $(D^4 - 2D^3 + 2D^2 - 2D + 1)x = 0$
11. $(D^2 + 1)^4 x = 0$
*12. $(D^4 + 2D^3 + 3D^2 + 2D + 1)x = 0$

Solve the initial-value problems in Exercises 13 through 18.

13. $16x'' + x = 0;$ $x(0) = 2, x'(0) = 9$
14. $x^{(4)} - 81x = 0;$ $x(0) = -2, x'(0) = 9, x''(0) = 18, x'''(0) = -27$
15. $(D^2 + 1)x = 0;$ $x(\pi) = 0, x'(\pi) = 0$
16. $(D^2 + 1)x = 0;$ $x(\pi) = 1, x'(\pi) = 1$
17. $(5D^2 + 2D + 1)x = 0;$ $x(0) = 0, x'(0) = 1$
18. $(D^3 - 2D^2 + 2D - 4)x = 0;$ $x(0) = 0, x'(0) = 4, x''(0) = 12$

Find an annihilator of smallest possible order for the given function in Exercises 19 through 24.

19. $e^{3t} - 5e^t$
20. $3 + te^t$
21. $t^2 e^{2t} - e^t + te^t$
22. $e^t + \sin 2t - 3$
23. $t \sin 2t$
24. $t^2 + e^t \sin 3t$

Exercises 25 through 29 refer to our models.

25. *A Damped Spring:*
    a. Find the general solution for the o.d.e. of Example 2.1.2, assuming $m = 2$ grams, $L = 7$ cm, $k = 5$ dynes/cm, and $b = 6$ dynes/(cm/sec).
    b. Find the specific solution that corresponds to a starting position at which the spring is 5 cm long and becoming longer at 1 cm/sec.
26. *A Vertical Spring:* Find the general solution of the o.d.e. modeling the vertical spring in Exercise 1, Section 2.1, if $w = 32$ and
    a. $b = 0, \ k = 1$
    b. $b = 3, \ k = 2$
    c. $b = 2, \ k = 1$
    d. $b = 2, \ k = 2$
27. *One Mass, Two Springs:* Find the general solution of the o.d.e. in Exercise 2, Section 2.1, assuming $m = 5$ grams, $k_1 = k_2 = 1/2$ dyne/cm, $L_1 + L_2 = B$, and $b = 2$ dynes/(cm/sec).
28. *Coupled Springs:* Find the general solution of the o.d.e. in Example 2.1.4, assuming $m_1 = m_2 = 1$ gram, $L_1 = 10$ cm, $L_2 = 3$ cm, $k_1 = 3$ dynes/cm, and $k_2 = 2$ dynes/cm.
29. *An LRC Circuit:* Find a formula, in terms of the inductance L and capacitance C, for the charge Q in the circuit of Exercise 5, Section 2.1, assuming $R = 0$, $V(t) = 0$, $Q(0) = Q_0$, and $I(0) = 0$.

30. *An Anchored Floating Box:*
   a. Find the general solution of the homogeneous equation related to the model of the anchored floating box in Exercise 8, Section 2.1, assuming damping constant $b = 5$.
   b. Find a constant solution of the nonhomogeneous equation (this is the equilibrium position of the box).
   c. Write a formula for the motion of the box in terms of its initial position x_0 and velocity v_0.

2.7 NONHOMOGENEOUS LINEAR EQUATIONS: UNDETERMINED COEFFICIENTS

Now that we have a method for solving homogeneous linear equations with constant coefficients, it is time to turn to the behavior of "forced" models, that is, to nonhomogeneous equations. Recall that if $x = p(t)$ is a particular solution of

(N) $$Lx = E(t)$$

and $x = H(t)$ is the general solution of the related homogeneous equation

(H) $$Lx = 0$$

then the general solution of (N) is

$$x = H(t) + p(t).$$

Once we can solve the homogeneous equation, it suffices to find just one function solving the nonhomogeneous equation.

In this section we describe a method for finding a particular solution of (N) when $L = P(D)$ has constant coefficients and $E(t)$ is itself the solution of a homogeneous equation with constant coefficients. If

$$A(D)E(t) = 0$$

we say (as in the previous section) that $A(D)$ **annihilates** $E(t)$. In this case any function that is a solution of

(N) $$P(D)x = E(t)$$

also satisfies

(H*) $$A(D)P(D)x = A(D)E(t) = 0.$$

Thus, we should look for our particular solution $x = p(t)$ among the solutions of the *homogeneous* equation (H*).

Since (H*) is homogeneous, we know how to get its general solution, $x = k_1 h_1(t) + \cdots + k_s h_s(t)$. Our particular solution is one of the functions described by this formula. That is,

$$p(t) = k_1 h_1(t) + \cdots + k_s h_s(t)$$

for some choice of the coefficients k_i. We can *treat this as a guess for* $x = p(t)$, *substitute into (N), and determine the coefficients.* However, we want to minimize our work. Note that if $h_i(t)$ is a solution of (H), then the term $k_i h_i(t)$ will vanish upon substitution into the left side of (N); this term will contribute nothing toward attaining the right side of (N). We therefore *obtain a simpler guess for p(t) by dropping terms that are themselves solutions of (H)*.

Example 2.7.1

Solve

(N) $$(D^2 - 2D - 3)x = 6 - 8e^t.$$

The characteristic polynomial of the related homogeneous equation

(H) $$(D^2 - 2D - 3)x = 0$$

is $(r - 3)(r + 1)$. Thus, the general solution of (H) is $x = H(t)$, where

$$H(t) = c_1 e^{3t} + c_2 e^{-t}.$$

The right side of (N) is a solution of any homogeneous equation whose characteristic polynomial has roots 0 and 1. In particular, it is annihilated by the operator $A(D) = D(D - 1)$. Thus, any function $x = p(t)$ that is a solution of (N) will also satisfy

(H*) $$D(D - 1)(D^2 - 2D - 3)x = 0.$$

The characteristic polynomial of (H*) is $r(r - 1)(r - 3)(r + 1)$, so $p(t)$ has to be of the form

$$p(t) = k_1 + k_2 e^t + k_3 e^{3t} + k_4 e^{-t}.$$

We get a simpler guess for $p(t)$ by crossing out the terms that satisfy (H):

$$p(t) = k_1 + k_2 e^t.$$

If we substitute our simpler guess for $x = p(t)$ into the left side of (N), we obtain

$$(D^2 - 2D - 3)(k_1 + k_2e^t) = -3k_1 - 4k_2e^t.$$

We want this to equal the right side of (N), $6 - 8e^t$. Thus,

$$k_1 = -2 \quad \text{and} \quad k_2 = 2.$$

Our particular solution is

$$p(t) = -2 + 2e^t.$$

The general solution of (N) is

$$x = H(t) + p(t) = c_1e^{3t} + c_2e^{-t} - 2 + 2e^t.$$

Example 2.7.2

Solve

(N) $$(D^2 - 4)x = -3e^{3t} + 10te^{3t}.$$

The general solution of the related homogeneous equation

(H) $$(D^2 - 4)x = 0$$

is $x = H(t)$, where

$$H(t) = c_1e^{2t} + c_2e^{-2t}.$$

The right side of (N) is annihilated by $(D - 3)^2$ (see Example 2.6.5), so any function $x = p(t)$ that is a solution of (N) also satisfies

(H*) $$(D - 3)^2(D^2 - 4)x = 0.$$

The characteristic polynomial of (H*) is $(r - 3)^2(r - 2)(r + 2)$, so $p(t)$ has to be of the form

$$p(t) = k_1e^{3t} + k_2te^{3t} + k_3e^{2t} + k_4e^{-2t}.$$

We get a simpler guess for $p(t)$ by dropping the terms that satisfy (H):

$$p(t) = k_1e^{3t} + k_2te^{3t}.$$

If we substitute our simpler guess for $x = p(t)$ into the left side of (N) and use the exponential shift, we obtain

$$(D^2 - 4)(k_1e^{3t} + k_2te^{3t}) = e^{3t}[(D + 3)^2 - 4](k_1 + k_2t)$$
$$= e^{3t}[D^2 + 6D + 5](k_1 + k_2t)$$
$$= (5k_1 + 6k_2)e^{3t} + 5k_2te^{3t}.$$

We want this to equal $-3e^{3t} + 10te^{3t}$, so

$$5k_1 + 6k_2 = -3, \quad 5k_2 = 10.$$

Then

$$k_1 = -3, \quad k_2 = 2.$$

Our particular solution is

$$p(t) = -3e^{3t} + 2te^{3t}.$$

The general solution of (N) is

$$x = H(t) + p(t) = c_1e^{2t} + c_2e^{-2t} - 3e^{3t} + 2te^{3t}.$$

In each of the preceding examples, the simplified guess for $x = p(t)$ was of the same form as the right side of (N), with specific constants replaced by undetermined coefficients. The following examples show that this is not always the case. The form of the simplified guess is sometimes more complicated than the right side of (N).

Example 2.7.3

Solve

(N) $$(D^2 - 4)x = e^t + 2e^{2t}.$$

The general solution of the related homogeneous equation

(H) $$(D^2 - 4)x = 0$$

is $x = H(t)$, where

$$H(t) = c_1e^{2t} + c_2e^{-2t}.$$

The right side of (N) is annihilated by $(D - 1)(D - 2)$, so a particular solution $x = p(t)$ of (N) will also satisfy

(H*) $$(D - 1)(D - 2)(D^2 - 4)x = 0.$$

The characteristic polynomial of (H*) is $(r - 1)(r - 2)^2(r + 2)$, so our particular solution has to be of the form

$$p(t) = k_1 e^t + k_2 e^{2t} + k_3 t e^{2t} + k_4 e^{-2t}.$$

We obtain a simpler guess for $p(t)$ by crossing out the terms that satisfy (H):

$$p(t) = k_1 e^t + k_3 t e^{2t}.$$

We substitute the simpler guess for $p(t)$ into the left side of (N):

$$\begin{aligned}
(D^2 - 4)p(t) &= (D^2 - 4)k_1 e^t + (D^2 - 4)k_3 t e^{2t} \\
&= -3k_1 e^t + e^{2t}[(D + 2)^2 - 4]k_3 t \\
&= -3k_1 e^t + e^{2t}(D^2 + 4D)k_3 t \\
&= -3k_1 e^t + 4k_3 e^{2t}.
\end{aligned}$$

We want this to equal $e^t + 2e^{2t}$, so

$$k_1 = -\frac{1}{3}, \quad k_3 = \frac{1}{2}.$$

Our particular solution is $x = p(t)$, where

$$p(t) = -\frac{1}{3} e^t + \frac{1}{2} t e^{2t}.$$

The general solution of (N) is

$$x = H(t) + p(t) = c_1 e^{2t} + c_2 e^{-2t} - \frac{1}{3} e^t + \frac{1}{2} t e^{2t}.$$

Example 2.7.4

Find the general solution of

(N) $$(D^2 + 4)x = \sin 2t$$

which describes a forced undamped-spring system as in Example 2.1.3, with

$m = 1$, $b = 0$, $k = 4$, and $E(t) = \sin 2t$. Find the specific solution satisfying the initial condition

$$x(0) = 0, \quad x'(0) = 0.$$

The related homogeneous equation is

(H) $$(D^2 + 4)x = 0.$$

Its characteristic polynomial $r^2 + 4$ has roots $\pm 2i$. The general solution of (H) is $x = H(t)$, where

$$H(t) = c_1 \cos 2t + c_2 \sin 2t.$$

The right side of (N) is annihilated by $(D^2 + 4)$, so a particular solution $x = p(t)$ of (N) will also satisfy

(H*) $$(D^2 + 4)^2 x = 0.$$

The roots $\pm 2i$ of the characteristic polynomial of (H*) both have multiplicity 2. Our particular solution has to be of the form

$$p(t) = k_1 \cos 2t + k_2 \sin 2t + k_3 t \cos 2t + k_4 t \sin 2t.$$

We obtain a simpler guess by crossing off the terms that satisfy (H):

$$p(t) = k_3 t \cos 2t + k_4 t \sin 2t.$$

We substitute the simpler guess into the left side of (N):

$$(D^2 + 4)(k_3 t \cos 2t + k_4 t \sin 2t) = -4k_3 \sin 2t + 4k_4 \cos 2t.$$

We want this to equal $\sin 2t$, so

$$k_3 = -\frac{1}{4}, \quad k_4 = 0.$$

Our particular solution is $x = p(t)$, where

$$p(t) = -\frac{t}{4} \cos 2t.$$

The general solution of (N) is

$$x = H(t) + p(t) = c_1 \cos 2t + c_2 \sin 2t - \frac{t}{4} \cos 2t.$$

To find the specific solution of (N) satisfying the given initial condition, we first differentiate the general solution to obtain

$$x' = -2c_1 \sin 2t + 2c_2 \cos 2t - \frac{1}{4} \cos 2t + \frac{t}{2} \sin 2t$$

so

$$x(0) = c_1 \quad \text{and} \quad x'(0) = 2c_2 - \frac{1}{4}.$$

Matching this to the initial condition, we find $c_1 = 0$ and $c_2 = 1/8$. The desired specific solution is

$$x = \frac{1}{8} \sin 2t - \frac{t}{4} \cos 2t.$$

Example 2.7.5

Solve

(N) $(D^2 - 4)x = 1 + 65e^t \cos 2t.$

The general solution of the related homogeneous equation

(H) $(D^2 - 4)x = 0$

is $x = H(t)$, where

$$H(t) = c_1 e^{2t} + c_2 e^{-2t}.$$

Since the right side of (N) is annihilated by $D(D^2 - 2D + 5)$ (see Example 2.6.6), a particular solution $x = p(t)$ of (N) will also satisfy

(H*) $D(D^2 - 2D + 5)(D^2 - 4)x = 0.$

The characteristic polynomial of (H*) is $r(r^2 - 2r + 5)(r - 2)(r + 2)$, so $p(t)$ has to be of the form

$$p(t) = k_1 + k_2 e^t \cos 2t + k_3 e^t \sin 2t + k_4 e^{2t} + k_5 e^{-2t}.$$

We get a simpler guess by crossing out the terms that satisfy (H):

$$p(t) = k_1 + k_2 e^t \cos 2t + k_3 e^t \sin 2t.$$

We substitute the simpler guess into the left side of (N):

$$(D^2 - 4)p(t) = (D^2 - 4)k_1 + (D^2 - 4)(k_2 e^t \cos 2t + k_3 e^t \sin 2t)$$
$$= -4k_1 + e^t(D^2 + 2D - 3)(k_2 \cos 2t + k_3 \sin 2t)$$
$$= -4k_1 + (-7k_2 + 4k_3)e^t \cos 2t + (-4k_2 - 7k_3)e^t \sin 2t.$$

We want this to equal $1 + 65e^t \cos 2t$, so

$$-4k_1 \qquad\qquad\qquad = 1$$
$$-7k_2 + 4k_3 = 65$$
$$-4k_2 - 7k_3 = 0.$$

Then

$$k_1 = -\frac{1}{4}, \quad k_2 = -7, \quad k_3 = 4.$$

Our particular solution is $x = p(t)$, where

$$p(t) = -\frac{1}{4} - 7e^t \cos 2t + 4e^t \sin 2t.$$

The general solution of (N) is

$$x = H(t) + p(t) = c_1 e^{2t} + c_2 e^{-2t} - \frac{1}{4} - 7e^t \cos 2t + 4e^t \sin 2t.$$

In order for the method we used in the examples to work, the right side of the o.d.e. must be a solution of a constant-coefficient linear homogeneous equation. But we know what all such solutions are. They are linear combinations of functions of the following types:

1.   constants
2.   exponentials: $e^{\lambda t}$
3.   trig functions: $\sin \beta t$, $\cos \beta t$
4.   exponentials times trig functions: $e^{\alpha t} \cos \beta t$, $e^{\alpha t} \sin \beta t$
5.   positive integer powers of t times any of these functions.

In brief, our method requires the right side of the o.d.e. to be a linear combination of nonnegative integer powers of t times exponentials times sines or cosines. We cannot use our method otherwise, because we will not be able to find a constant-coefficient annihilator for $E(t)$ if, say, $E(t)$ is $\sec t$ or $\ln t$ or even $1/(1 + e^t)$. We can summarize the method as follows.

UNDETERMINED COEFFICIENTS

To solve a constant-coefficient equation

(N) $$P(D)x = E(t)$$

where $E(t)$ is a linear combination of nonnegative integer powers of t times exponentials times sines or cosines:

1. Solve the related homogeneous equation

   (H) $$P(D)x = 0.$$

2. Find an **annihilator** $A(D)$ for the right side of (N):

   $$A(D)E(t) = 0.$$

3. Since a particular solution $x = p(t)$ of (N) also satisfies

   (H*) $$A(D)P(D)x = 0,$$

   we obtain a description of the form of $p(t)$ from the general solution of (H*).

4. Obtain a simpler guess for $p(t)$ by dropping those terms that satisfy (H).

5. Substitute the simpler guess into the left side of (N). (The exponential shift is often helpful in this step.)

6. Determine the coefficients that yield the required right side, $E(t)$.

7. The general solution of (N) is the function obtained in step 6 plus the general homogeneous solution from step 1.

EXERCISES

Find the general solution in Exercises 1 through 10.

1. $(D^2 + 2D + 1)x = 5 + t$

2. $(D - 1)^4 x = 2t + e^{-t}$

3. $(D^4 + 4D^3 + 4D^2)x = 2$

4. $(3D^2 + 2D - 1)x = 2\sin t$

5. $(D^2 + 1)x = \cos t$

6. $(4D^2 + 3D - 1)x = 25 - t^2$

7. $(D^2 + 6D + 10)x = 80e^t \sin t$

8. $(D^2 - D + 2)x = t^2 - 8e^{2t}$

9. $(9D^2 - 1)x = t \sin t$

10. $(D^5 - 4D^4 + 4D^3)x = 240t^2 + 4e^{2t}$

Solve the initial-value problems in Exercises 11 through 13.

11. $(D^2 + 1)x = \sin 3t$; $x(0) = x'(0) = 0$
12. $(D^3 + 5D^2 - 6D)x = 3e^t$; $x(0) = 1, x'(0) = 3/7, x''(0) = 6/7$
13. $(5D^2 + 2D + 1)x = 5t$; $x(0) = -1, x'(0) = 0$

Make a *simplified* guess for a particular solution in Exercises 14 through 16. Do *not* solve for the coefficients.

14. $(D - 1)^2(D^2 + 1)^3(D + 2)x = t^2e^{3t} + e^t + e^{-t}\sin 3t + t^4$
15. $(D - 3)(D^2 + D + 1)(D + 1)^3x = 3 - t^2 + te^{-t/2}\sin(t\sqrt{5}) + e^{3t}$
16. $(D + 2)^7(D^2 + 1)^6x = te^{-2t} + \sin t$

Exercises 17 through 21 refer to our models.

17. *A Floating Box:* Solve the o.d.e. modeling the floating box in Exercise 7, Section 2.1, assuming (a) no damping, (b) viscous damping with $b = 10$, and (c) viscous damping with $b = 15$.

18. *An LRC Circuit:* Solve the o.d.e. modeling the *LRC* circuit in Exercise 5, Section 2.1, assuming $C = 1$, $V(t) = 10 \sin 2t$, and (a) $L = 1$, $R = 1$; (b) $L = 4$, $R = 4$; (c) $L = 4$, $R = 0$.

19. *One Mass, Two Springs:* Show that in Exercise 2, Section 2.1, the distance from the mass to the left wall is the same as if the mass were attached to a single spring with constant $k = k_1 + k_2$ and natural length $L = [k_1L_1 + k_2(B - L_2)]/(k_1 + k_2)$, subject to no external forces.

20. *A Moving Spring:* Find the general solution of the o.d.e. modeling the moving spring in Exercise 3, Section 2.1, assuming (a) $b = 5$ and $k = 8$; (b) $b = 2$ and $k = 2$; (c) $b = 0$ and $k = 2$.

21. *The Box Adrift:* Find the general solution of the o.d.e. in Exercise 9c, Section 2.1.

Another useful substitution:

22. An o.d.e. of the form

(N) $(a_nt^nD^n + a_{n-1}t^{n-1}D^{n-1} + \cdots + a_1tD + a_0)x = E(t)$, $t > 0$

where $a_0, \ldots, a_n$ are constants, is known as an **equidimensional** or **Cauchy-Euler** equation. The substitution $s = \ln t$ (a change in time scale) transforms an equidimensional equation into a linear o.d.e. with constant coefficients.

a. Show that if $s = \ln t$, then

$$\frac{dx}{dt} = \frac{1}{t}\frac{dx}{ds} \quad \text{and} \quad \frac{d^2x}{dt^2} = \frac{1}{t^2}\left[\frac{d^2x}{ds^2} - \frac{dx}{ds}\right].$$

b. Use (a) to show that, when $n = 2$, the substitution $s = \ln t$ transforms (N) into the o.d.e.

$$a_2\frac{d^2x}{ds^2} + (a_1 - a_2)\frac{dx}{ds} + a_0x = E(e^s).$$

Use the substitution described in Exercise 22 to solve the following o.d.e.'s. Express your final answers in terms of t.

23. $t^2x'' - 2tx' + 2x = 8 + t$, $t > 0$ 24. $t^2x'' + 5tx' + 3x = \ln t$, $t > 0$
25. $t^2x'' + 5tx' + 4x = t^2 - 3t$, $t > 0$ 26. $t^2x'' + x = 3 - \ln t$, $t > 0$

2.8 NONHOMOGENEOUS LINEAR EQUATIONS: VARIATION OF PARAMETERS

In the previous section we discussed a method of finding a particular solution $x = p(t)$ of $Lx = E(t)$ that works whenever the equation has constant coefficients and $E(t)$ is of a special form. In this section we consider a method that does not require $E(t)$ to be of a special form and does not assume the equation has constant coefficients. It requires only that we know the general solution of the related homogeneous equation $Lx = 0$. This method is therefore much more general than the preceding one. Unfortunately, it requires us to solve n equations in n unknown functions and can lead to difficult integrals. The method is an extension of the method of varying parameters used in Section 1.3.

Let's begin by considering the second-order equation

$$[a_2(t)D^2 + a_1(t)D + a_0(t)]x = E(t).$$

We assume that the equation is normal on an interval I, and we look for solutions valid on I. We first divide by $a_2(t)$ to obtain an equation in **standard form**:

(N) $$[D^2 + b_1(t)D + b_0(t)]x = q(t).$$

Suppose we know that the general solution of the related homogeneous equation

(H) $$[D^2 + b_1(t)D + b_0(t)]x = 0$$

is

$$x = c_1 h_1(t) + c_2 h_2(t).$$

We look for a particular solution of (N) of the form $x = p(t)$, with

$$p(t) = c_1(t)h_1(t) + c_2(t)h_2(t).$$

In order to substitute our expression for x into (N), we must calculate $Dp(t)$ and $D^2 p(t)$. We begin with $Dp(t)$:

$$Dp(t) = c_1(t)h_1'(t) + c_2(t)h_2'(t) + [c_1'(t)h_1(t) + c_2'(t)h_2(t)].$$

This formula is quite messy; $D^2 p(t)$ will be even worse. Keep in mind, however, that we're not really interested in $c_1(t)$ and $c_2(t)$ for their own sake, but only as a means of finding $x = p(t)$. In general, many different choices of $c_1(t)$ and $c_2(t)$ will yield the same function $p(t)$. Thus, we are allowed some freedom in picking $c_1(t)$ and $c_2(t)$. Let's make a simplifying assumption. Let's assume that the term in brackets in our expression for Dx is zero:

(1) $$c_1'(t)h_1(t) + c_2'(t)h_2(t) = 0.$$

Our formula for $Dp(t)$ now reads

$$Dp(t) = c_1(t)h_1'(t) + c_2(t)h_2'(t).$$

Hence,

$$D^2p(t) = c_1(t)h_1''(t) + c_2(t)h_2''(t) + c_1'(t)h_1'(t) + c_2'(t)h_2'(t).$$

Substituting our expressions for $p(t)$, $Dp(t)$, and $D^2p(t)$ into (N) yields

$$c_1(t)[h_1''(t)+b_1(t)h_1'(t)+b_0(t)h_1(t)]+c_2(t)[h_2''(t)+b_1(t)h_2'(t)+b_0(t)h_2(t)]$$
$$+c_1'(t)h_1'(t)+c_2'(t)h_2'(t) = q(t).$$

Since $h_1(t)$ and $h_2(t)$ satisfy (H), the terms in brackets are zero. Thus, the last equation reads

(2) $$c_1'(t)h_1'(t) + c_2'(t)h_2'(t) = q(t).$$

Formulas (1) and (2) provide us with a system of two equations for two unknown functions, $c_1'(t)$ and $c_2'(t)$:

(V)
$$c_1'(t)h_1(t) + c_2'(t)h_2(t) = 0$$
$$c_2'(t)h_1'(t) + c_2'(t)h_2'(t) = q(t).$$

The determinant of coefficients of this system of equations is

$$\det \begin{bmatrix} h_1(t) & h_2(t) \\ h_1'(t) & h_2'(t) \end{bmatrix}.$$

This is just the Wronskian of $h_1(t)$ and $h_2(t)$. Since $h_1(t)$ and $h_2(t)$ generate the general solution of (H), the Wronskian test for solutions tells us that their Wronskian is never zero on I. Cramer's determinant test (Appendix A) tells us that the system of equations (V) always has a solution for $c_1'(t)$ and $c_2'(t)$. We can solve and integrate to find $c_1(t)$ and $c_2(t)$. These in turn determine $p(t) = c_1(t)h_1(t) + c_2(t)h_2(t)$.

Example 2.8.1

Solve $(4D^2 - 4D + 1)x = t^{1/2}e^{t/2}$, $0 < t < +\infty$.
We first put the equation in standard form:

(N) $$\left(D^2 - D + \frac{1}{4}\right) x = \frac{e^{1/2}e^{t/2}}{4}.$$

The characteristic polynomial of the related homogeneous equation

(H)
$$\left(D^2 - D + \frac{1}{4}\right) x = 0$$

is $\left(r - \frac{1}{2}\right)^2$. The general solution of (H) is $x = H(t)$, with

$$H(t) = c_1 e^{t/2} + c_2 t e^{t/2}.$$

We look for a particular solution of (N) in the form $x = p(t)$ with

$$p(t) = c_1(t) e^{t/2} + c_2(t) t e^{t/2}.$$

Here $h_1(t) = e^{t/2}$ and $h_2(t) = t e^{t/2}$, so $h_1'(t) = e^{t/2}/2$ and $h_2'(t) = (2 + t) e^{t/2}/2$. The system of equations (V) reads

(V)
$$c_1'(t) e^{t/2} + c_2'(t) t e^{t/2} = 0$$
$$c_1'(t) \frac{e^{t/2}}{2} + c_2'(t) \frac{(2 + t) e^{t/2}}{2} = \frac{t^{1/2} e^{t/2}}{4}.$$

We can eliminate $c_1'(t)$ from (V) by subtracting the first equation from twice the second. We get $2 c_2'(t) e^{t/2} = (t^{1/2} e^{t/2})/2$, or

$$c_2'(t) = \frac{1}{4} t^{1/2}.$$

Substituting this value into the first equation yields

$$c_1'(t) = -\frac{1}{4} t^{3/2}.$$

We integrate these formulas to obtain specific values of $c_1(t)$ and $c_2(t)$:

$$c_1(t) = -\frac{1}{10} t^{5/2}, \quad c_2(t) = \frac{1}{6} t^{3/2}.$$

Our particular solution is $x = p(t)$ with

$$p(t) = -\frac{1}{10} t^{5/2} e^{t/2} + \frac{1}{6} t^{3/2} t e^{t/2} = \frac{1}{15} t^{5/2} e^{t/2}.$$

The general solution of (N) is

$$x = H(t) + p(t) = c_1 e^{t/2} + c_2 t e^{t/2} + \frac{1}{15} t^{5/2} e^{t/2}.$$

Example 2.8.2

Solve

(N)
$$(D^2 + 1)x = \sec t, \quad -\frac{\pi}{2} < t < +\frac{\pi}{2}.$$

The equation is already in standard form. The general solution of the related homogeneous equation is $x = H(t)$, with

$$H(t) = c_1 \cos t + c_2 \sin t.$$

We look for a particular solution in the form $x = p(t)$ with

$$p(t) = c_1(t) \cos t + c_2(t) \sin t.$$

In this case the system of equations (V) reads

(V)
$$c_1'(t) \cos t + c_2'(t) \sin t = 0$$
$$-c_1'(t) \sin t + c_2'(t) \cos t = \sec t.$$

Cramer's rule provides formulas for $c_1'(t)$ and $c_2'(t)$:

$$c_1'(t) = \frac{\det \begin{bmatrix} 0 & \sin t \\ \sec t & \cos t \end{bmatrix}}{\det \begin{bmatrix} \cos t & \sin t \\ -\sin t & \cos t \end{bmatrix}} = \frac{-\sin t \sec t}{\cos^2 t + \sin^2 t} = -\frac{\sin t}{\cos t}$$

$$c_2'(t) = \frac{\det \begin{bmatrix} \cos t & 0 \\ -\sin t & \sec t \end{bmatrix}}{\det \begin{bmatrix} \cos t & \sin t \\ -\sin t & \cos t \end{bmatrix}} = \frac{\cos t \sec t}{\cos^2 t + \sin^2 t} = 1.$$

We integrate these formulas to get

$$c_1(t) = \ln|\cos t|, \quad c_2(t) = t.$$

Our particular solution is $x = p(t)$ with

$$p(t) = (\ln|\cos t|) \cos t + t \sin t.$$

The general solution is

$$x = H(t) + p(t) = c_1 \cos t + c_2 \sin t + (\ln|\cos t|) \cos t + t \sin t.$$

This method, which we have worked out for second-order equations, can be extended to apply to nth-order equations. Instead of making a single simplifying assumption, we make $n - 1$ simplifying assumptions. Substitution into the o.d.e. provides another equation. The details are quite messy, so we will content ourselves with a statement of the result.

Fact: *Suppose we are given an nth-order o.d.e. in **standard form**,*

(N) $$[D^n + b_{n-1}(t)D^{n-1} + \cdots + b_1(t)D + b_0(t)]x = q(t),$$

which is normal on an interval I. Suppose we know that the general solution of the related homogeneous equation is $x = H(t)$, with

$$H(t) = c_1 h_1(t) + \cdots + c_n h_n(t).$$

If $c_1(t), \ldots, c_n(t)$ are functions that satisfy the system of equations

(V)

$$
\begin{aligned}
c_1'(t)h_1(t) \quad &+ \cdots + c_n'(t)h_n(t) && = 0 \\
c_1'(t)h_1'(t) \quad &+ \cdots + c_n'(t)h_n'(t) && = 0 \\
& \qquad\quad \vdots \\
c_1'(t)h_1^{(n-2)}(t) &+ \cdots + c_n'(t)h_n^{(n-2)}(t) && = 0 \\
c_1'(t)h_1^{(n-1)}(t) &+ \cdots + c_n'(t)h_n^{(n-1)}(t) && = q(t)
\end{aligned}
$$

then

$$p(t) = c_1(t)h_1(t) + \cdots + c_n(t)h_n(t)$$

is a particular solution of (N).

The system of equations (V) is surprisingly easy to remember. It consists of equations for the unknowns $c_1'(t), \ldots, c_n'(t)$. The coefficients are just the entries of the Wronskian of $h_1(t), \ldots, h_n(t)$. The right sides are all zero, except the last, which is the right side of (N). Note, however, that *this assumes the o.d.e. is in standard form.*

Example 2.8.3

Solve $(t^2 D^3 + 2t D^2 - 2D)x = t^3, \quad 0 < t < +\infty.$
We divide by t^2 to obtain the standard form:

(N) $$\left(D^3 + \frac{2}{t}D^2 - \frac{2}{t^2}D\right)x = t.$$

We found the general solution of the related homogeneous equation in Example 2.3.6; it is $x = H(t)$, with

$$H(t) = c_1 t^2 + c_2 \frac{1}{t} + c_3.$$

We look for a particular solution in the form $x = p(t)$, with

$$p(t) = c_1(t) t^2 + c_2(t) \frac{1}{t} + c_3(t).$$

In this case the system of equations (V) reads

$$t^2 c_1'(t) + \frac{1}{t} c_2'(t) + c_3'(t) = 0$$

(V) $$2t c_1'(t) - \frac{1}{t^2} c_2'(t) \qquad\qquad = 0$$

$$2 c_1'(t) + \frac{2}{t^3} c_2'(t) \qquad\qquad = t.$$

If we multiply the last equation by t and subtract the second equation, we get $(3/t^2) c_2'(t) = t^2$, or

$$c_2'(t) = \frac{t^4}{3}.$$

Substituting this value into the second equation of (V) yields

$$c_1'(t) = \frac{t}{6}.$$

The first equation of (V) now gives

$$c_3'(t) = -\frac{t^3}{2}.$$

We integrate to get specific values for the $c_i(t)$'s:

$$c_1(t) = \frac{t^2}{12}, \quad c_2(t) = \frac{t^5}{15}, \quad c_3(t) = -\frac{t^4}{8}.$$

The function

$$p(t) = \frac{t^4}{12} + \frac{t^4}{15} - \frac{t^4}{8} = \frac{t^4}{40}$$

is a particular solution of (N). The general solution is

$$x = H(t) + p(t) = c_1 t^2 + c_2 \frac{1}{t} + c_3 + \frac{t^4}{40}.$$

A limited version of the method of variation of parameters was treated by L. Euler (1743). The systematic application for linear equations of all orders was worked out by Joseph Louis Lagrange (1774).

VARIATION OF PARAMETERS

To solve an nth-order nonhomogeneous linear equation

$$[a_n(t)D^n + \cdots + a_1(t)D + a_0(t)]x = E(t):$$

0. Divided by $a_n(t)$ to obtain an equation in **standard form:**

(N) $[D^n + \cdots + b_1(t)D + b_0(t)]x = q(t).$

1. Find the general solution of the related homogeneous equation

(H) $[D^n + \cdots + b_1(t)D + b_0(t)]x = 0.$

This has the form $x = H(t)$, with

$$H(t) = c_1 h_1(t) + \cdots + c_n h_n(t).$$

2. Solve the system of equations

$$c_1'(t)h_1(t) \quad + \cdots + c_n'(t)h_n(t) \quad = 0$$

$$c_1'(t)h_1'(t) \quad + \cdots + c_n'(t)h_n'(t) \quad = 0$$

(V)

$$\vdots$$

$$c_1'(t)h_1^{(n-2)}(t) + \cdots + c_n'(t)h_n^{(n-2)}(t) = 0$$

$$c_1'(t)h_1^{(n-1)}(t) + \cdots + c_n'(t)h_n^{(n-1)}(t) = q(t)$$

for the unknowns $c_1'(t), \ldots, c_n'(t)$.

3. Integrate to find specific values for $c_1(t), \ldots, c_n(t)$.

4. The function $p(t) = c_1(t)h_1(t) + \cdots + c_n(t)h_n(t)$ is a particular solution of (N).

5. The general solution of (N) is obtained by adding the particular solution found in step 4 to the general homogeneous solution found in step 1.

Notes

1.   A shortcut

In our examples we obtained specific values for each $c_i(t)$ by integrating the formulas for $c_i'(t)$, taking the constant of integration to be zero. If we include arbitrary constants c_i when we integrate, then substitution of the resulting functions into our formula for x yields the general solution of (N).

2.   Reduction of order

Sometimes a linear o.d.e. with variable coefficients $Lx = E(t)$ can be solved by a modified variation of parameters technique known as **reduction of order**. The method requires that we know one nonzero solution $x = h(t)$ of the related homogeneous o.d.e. $Lx = 0$. By analogy with variation of parameters, we guess that the general solution of $Lx = E(t)$ should have the form

$$x = k(t)h(t).$$

Substituting this guess for x into $Lx = E(t)$ leads to a linear equation of order $n - 1$ for $y = k'(t)$. With luck we can solve for y, integrate to find $k(t)$, and multiply by $h(t)$ to find x.

As an example, let's use this method to solve

$$(t^2 D^2 + tD - 1)x = t^2, \quad t > 0.$$

We observe that one solution of the related homogeneous o.d.e. is $h(t) = t$, and we look for solutions of the nonhomogeneous equation of the form $x = k(t)t$. Substitution into the o.d.e. yields

$$t^2 k''(t) + 3t^2 k'(t) = t^2.$$

If we set $y = k'(t)$, then this equation becomes

$$t^2 Dy + 3t^2 y = t^2.$$

The general solution of this first-order linear o.d.e. is

$$y = c_1 e^{-3t} + \frac{1}{3}.$$

Since $y = k'(t)$, integration gives

$$k(t) = -\frac{1}{3} c_1 e^{-3t} + \frac{1}{3} t + c_2.$$

We get the general solution by multiplying by t:

$$x = -\frac{1}{3} c_1 t e^{-3t} + c_2 t + \frac{1}{3} t^2.$$

EXERCISES

Find the general solution in Exercises 1 through 8 using variation of parameters.

1. $2x'' - 6x' + 4x = 6e^{2t}$

2. $4x'' - 4x' + x = \dfrac{8}{t^2} e^{t/2}, \quad t > 0$

3. $x'' + x = \tan t, \quad -\dfrac{\pi}{2} < t < \dfrac{\pi}{2}$

4. $x'' - 2x' + x = e^t \ln t, \quad t > 0$

5. $9x'' - 6x' + x = (te^t)^{1/3}$

6. $(2D - 1)(2D + 1)(D - 2)x = 15e^{2t}$

7. $(D - 1)^3 x = \dfrac{e^t}{t^3}, \quad t > 0$

8. $(D^3 + D)x = \sec^2 t, \quad -\dfrac{\pi}{2} < t < \dfrac{\pi}{2}$

Solve the initial-value problems in Exercises 9 through 11.

9. $x'' + x = \sec t, \quad -\dfrac{\pi}{2} < t < \dfrac{\pi}{2}; \quad x(0) = 1, x'(0) = 2$

10. $5x'' - 10x' + 5x = t^{1/5}e^t; \quad x(0) = x'(0) = 0$

11. $5x'' - 10x' + 5x = 60t^{-5}e^t; \quad x(\ln 2) = x'(\ln 2) = 0$

In Exercises 12 through 15, you are given the general solution $x = H(t)$ of the related homogeneous equation. Find the general solution of the given (nonhomogeneous) equation.

12. $[(t - 1)D^2 - tD + 1]x = (t - 1)^2 e^t, \quad t > 1; \qquad H(t) = c_1 t + c_2 e^t$

13. $[(t^2 + t)D^2 + (2 - t^2)D - (2 + t)]x = (t + 1)^2, \quad t > 0; \qquad H(t) = c_1 t^{-1} + c_2 e^t$

14. $(tD^3 + 3D^2)x = t^{-1/2}, \quad t > 0; \qquad H(t) = c_1 + c_2 t + c_3 t^{-1}$

15. $(t^3 D^3 + t^2 D^2 - 2tD + 2)x = t^{-1}, \quad t > 0; \qquad H(t) = c_1 t + c_2 t^2 + c_3 t^{-1}$

In Exercises 16 through 19, the general solution of the related homogeneous equation is generated by functions of the form t^α. Find the general solution of the given (nonhomogeneous) equation.

16. $(tD^2 - D)x = t^2, \quad t > 0$

17. $(t^2 D^2 + 4tD + 2)x = t^5, \quad t > 0$

18. $(t^3 D^3 - 3t^2 D^2 + 6tD - 6)x = t^{-1}, \quad t > 0$

19. $(tD^2 + 2D)x = \sqrt{t}, \quad t > 0$

In Exercises 20 through 24, the related homogeneous equation has a solution either of the form $h(t) = t^\alpha$ or of the form $h(t) = e^{\lambda t}$. Use reduction of order (Note 2) to find the general solution.

20. $[t^2 D^2 + 5tD + 4]x = 0, \quad t > 0$

21. $[t^3 D^2 + tD - 1]x = 0, \quad t > 0$

22. $[tD^2 - (t + 1)D + 1]x = 0, \quad t > 0$

23. $[(tD^2 - (t + 1)D + 1]x = 4, \quad t > 0$

24. $[tD^2 + (2t - 1)D + (t - 1)]x = e^{-t}, \quad t > 0$

Exercises 25 through 27 refer to our models.

25. *Springs:* Find the general solution of the o.d.e. modeling the spring of Example 2.1.3 if
    a. $m = 1$ gram, $b = 4$ dynes/(cm/sec), $k = 3$ dynes/cm, and $E(t) = e^{-t}$ dynes.
    b. $m = 2$ grams, $b = 0$, $k = 8$ dynes/cm, and $E(t) = 8 \sin^2 2t$ dynes.

26. *A Floating Box:* The floating box of Exercises 7 and 8, Section 2.1, has the anchor unhooked from the line (with $L = 2$ and $k = 22$), which remains attached to the box. The other end is pulled down so that at time t its depth is $2 + \sec 13t$ ft, $0 < t < \pi/2$. Set up an equation to model the motion of the box, assuming no damping, and find the general solution via variation of parameters.

27. *A Moving Spring:* Modify Exercise 3, Section 2.1, by assuming simply that at time t the elevator is $h(t)$ feet off the ground.
    a. Set up a second-order o.d.e. modeling the height z of the weight, assuming $b = 3$ and $k = 4$.
    b. Using variation of parameters, derive a formula (involving some integrals including $h(t)$, L, and two arbitrary constants) to predict the height z at any time $t > 0$.

2.9   BEHAVIOR OF SPRING MODELS

In this section we consider again the first three models of Section 2.1. We find the general solutions of their differential equations and interpret them in physical terms.

Example 2.9.1   Undamped Springs Revisited

The equation (from Example 2.1.1) for a mass m on a spring with constant k and no damping can be written in operator form as

$$(mD^2 + k)x = 0.$$

The characteristic polynomial $mr^2 + k$ has roots $r = \pm\sqrt{k/m}$ (remember that, for physical reasons, $k > 0$ and $m > 0$). If we set

$$\omega = \sqrt{\frac{k}{m}}$$

then the general solution is

$$x(t) = c_1 \cos \omega t + c_2 \sin \omega t.$$

To determine a specific solution (that is, one particular motion of the mass), we must specify the initial position and velocity:

$$x(0) = x_0 \quad \text{and} \quad x'(0) = v_0.$$

The solution of this initial-value problem, obtained by substituting the general solution into this initial condition and solving for c_1 and c_2, is

$$x(t) = x_0 \cos \omega t + \frac{v_0}{\omega} \sin \omega t.$$

If the mass starts at rest ($v_0 = 0$) from the equilibrium position ($x_0 = 0$), then the solution is $x(t) = 0$, expressing the fact that the mass remains at rest. If x_0 and v_0 are not both zero, we can rewrite the solution in **phase-amplitude form** as

$$x(t) = A \cos(\omega t - \alpha)$$

where

$$A = \sqrt{x_0^2 + \left(\frac{v_0}{\omega}\right)^2}$$

and α satisfies

$$\cos \alpha = \frac{x_0}{A} \quad \text{and} \quad \sin \alpha = \frac{v_0}{\omega A}$$

(see Exercise 1). The phase-amplitude form exhibits the solution as a regular oscillation about the equilibrium position (see Figure 2.6): A is the amplitude of the oscillation, and α is called the phase angle.

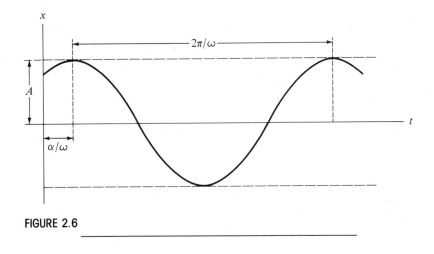

FIGURE 2.6

Example 2.9.2   Damped Springs Revisited

The equation of Example 2.1.2 modeling the motion of a mass m under the influence of a spring with constant k and damping with coefficient b has the operator form

$$(mD^2 + bD + k)x = 0.$$

The characteristic polynomial $mr^2 + br + k$ has roots

$$r = \frac{-b \pm \sqrt{b^2 - 4mk}}{2m}.$$

The nature of the solutions is determined by the sign of the discriminant $b^2 - 4mk$.

If $b^2 - 4mk < 0$, we say the motion is **underdamped.** In this case the roots are complex. If we set

$$\sigma = \frac{b}{2m} \quad \text{and} \quad \omega = \frac{\sqrt{4mk - b^2}}{2m},$$

then the roots are $r = -\sigma \pm \omega i$, so the general solution is

$$x(t) = c_1 e^{-\sigma t} \cos \omega t + c_2 e^{-\sigma t} \sin \omega t.$$

Substituting our formula for $x(t)$ into the initial condition

$$x(0) = x_0, \quad x'(0) = v_0,$$

we find that

$$x(t) = x_0 e^{-\sigma t} \cos \omega t + \frac{v_0 + \sigma x_0}{\omega} e^{-\sigma t} \sin \omega t.$$

We can rewrite this in phase-amplitude form as

$$x(t) = A e^{-\sigma t} \cos(\omega t - \alpha)$$

where

$$A = \sqrt{x_0^2 + \frac{(v_0 + \sigma x_0)^2}{\omega^2}}$$

and α satisfies

$$\cos \alpha = \frac{x_0}{A} \quad \text{and} \quad \sin \alpha = \frac{v_0 + \sigma x_0}{\omega A}.$$

The solution represents an oscillation about the equilibrium position whose amplitude dies down as $t \to \infty$ (see Figure 2.7). The number σ determines the rate at which the amplitude dies down.

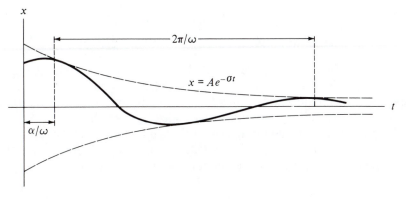

FIGURE 2.7

If $b^2 - 4mk > 0$, we say the motion is **overdamped.** In this case the characteristic roots are real and distinct. If we set

$$\sigma_1 = \frac{b - \sqrt{b^2 - 4mk}}{2m} \quad \text{and} \quad \sigma_2 = \frac{b + \sqrt{b^2 - 4mk}}{2m},$$

then the roots are $-\sigma_1$ and $-\sigma_2$, so the general solution is

$$x(t) = c_1 e^{-\sigma_1 t} + c_2 e^{-\sigma_2 t}.$$

Note that $0 < \sigma_1 < \sigma_2$, so both terms die down to 0 as $t \to \infty$. Substitution into the initial condition

$$x(0) = x_0, \quad x'(0) = v_0$$

gives

$$c_1 = \frac{\sigma_2 x_0 + v_0}{\sigma_2 - \sigma_1} \quad \text{and} \quad c_2 = -\frac{\sigma_1 x_0 + v_0}{\sigma_2 - \sigma_1}.$$

The solution crosses the equilibrium position at most once, at time

$$t_1 = \frac{1}{\sigma_2 - \sigma_1} \ln \left(-\frac{c_2}{c_1} \right) = \frac{1}{\sigma_2 - \sigma_1} \ln \left(\frac{\sigma_1 x_0 + v_0}{\sigma_2 x_0 + v_0} \right)$$

and has at most one relative extremum, at time

$$t_2 = \frac{1}{\sigma_2 - \sigma_1} \ln \left(-\frac{c_2}{c_1} \frac{\sigma_2}{\sigma_1} \right) = \frac{1}{\sigma_2 - \sigma_1} \ln \left[\left(\frac{\sigma_1 x_0 + v_0}{\sigma_2 x_0 + v_0} \right) \frac{\sigma_2}{\sigma_1} \right].$$

Note that in general, $t_2 > t_1$. We sketch three typical forms for the graph of the solution. Figure 2.8(a) depicts a solution with $x_0 > 0$ and $v_0 > 0$ (so that $t_1 < 0 < t_2$). Figure 2.8(b) depicts a solution with $x_0 > 0$ and $v_0 < -\sigma_2 x_0$ (so that $0 < t_1 < t_2$). Figure 2.8(c) depicts a solution with $x_0 > 0$ and $-\sigma_2 x_0 \leq v_0 < 0$ (so that t_1 and t_2 are both negative or both undefined). Physically, the first solution moves away from equilibrium, turns back at time t_2, and then approaches equilibrium asymptotically from the right; the second initially moves toward equilibrium, overshoots it (at time t_1), then turns around and approaches equilibrium asymptotically from the left; the third solution simply approaches equilibrium asymptotically from the right.

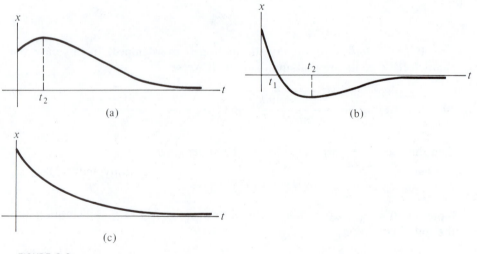

FIGURE 2.8

If $b^2 - 4mk = 0$, we say the motion is **critically damped.** In this case there is a real double root $r = -b/2m$. If we set

$$\sigma = \frac{b}{2m},$$

then the general solution is

$$x = c_1 e^{-\sigma t} + c_2 t e^{-\sigma t}.$$

The solution that satisfies the initial conditions

$$x(0) = x_0 \quad \text{and} \quad x'(0) = v_0$$

is

$$x = x_0 e^{-\sigma t} + (v_0 + \sigma x_0) t e^{-\sigma t}.$$

This solution again has at most one zero and at most one local extremum; they occur at times

$$t_1 = \frac{-x_0}{v_0 + \sigma x_0} \quad \text{and} \quad t_2 = \frac{v_0}{\sigma(v_0 + \sigma x_0)},$$

respectively. If $x_0 > 0$ and $v_0 > 0$, then $t_1 < 0 < t_2$ and the solution behaves roughly like Figure 2.8(a). If $x_0 > 0$ and $v_0 < -\sigma x_0$, then $t_2 > t_1 > 0$ and the graph resembles Figure 2.8(b). If $x_0 > 0$ and $-\sigma x_0 \leq v_0 < 0$, then t_1 and t_2 are both undefined or both negative and the solution behaves as in Figure 2.8(c).

The preceding discussion has indicated the kinds of unforced motion possible for a damped spring. We consider briefly three kinds of forced motion.

Example 2.9.3   Constant Forcing

When a mass m, hooked to a spring with constant k and damping with coefficient b, is subjected to a constant external force E (in the direction of increasing x)—for example, when the system hangs vertically and is acted on by gravity—the operator form of the equation is

$$(mD^2 + bD + k)x = E.$$

Applying the method of undetermined coefficients, we see that solutions have the form

$$x = \frac{E}{k} + h(t),$$

where $h(t)$ is a solution of the homogeneous (unforced) equation treated in Example 2.9.2. We saw that in the unforced case the motion either oscillated about, or tended asymptotically toward equilibrium. In the present case the graphs of all solutions are translated up by the constant amount E/k. This is the same effect as if the spring had natural length $L + E/k$. Thus, we can describe the effect of constant forcing by saying that *in the presence of a constant external force E, a damped spring acts the same as an unforced system with the same mass m, damping coefficient b, and spring constant k, but with the equilibrium length L replace by L + E/k.*

Example 2.9.4   Periodic Forcing of an Undamped Spring

When a mass m on a spring with constant k and no damping is subjected to an oscillating force $E(t) = E \sin \beta t$, the equation, in operator form, is

$$(mD^2 + k)x = E \sin \beta t.$$

We know from Example 2.9.1 that the general solution of the homogeneous (unforced) equation has the form $x = H(t)$, where

$$H(t) = c_1 \cos \omega t + c_2 \sin \omega t$$

with $\omega = \sqrt{k/m}$. We can use the method of undetermined coefficients to find a particular solution, and hence the general solution, of the forced equation. Note, however, that we have to handle the cases $\beta \neq \omega$ and $\beta = \omega$ separately. (Why?)

If $\beta \neq \omega$, the general (forced) solution is

$$x = \frac{-E}{m(\beta^2 - \omega^2)} \sin \beta t + c_1 \cos \omega t + c_2 \sin \omega t.$$

This is a superposition of regular oscillations with different amplitudes and frequencies, and in general it can be quite complicated. Note, however, that the solution is bounded by $|c_1| + |c_2| + |E/m(\beta^2 - \omega^2)|$, and that if β and ω are both integer multiples of some number γ, say $\beta = B\gamma$ and $\omega = W\gamma$, then x will be periodic, repeating itself after $2\pi/\gamma$.

When $\beta = \omega$, a particular solution is

$$p(t) = \frac{-E}{2\omega m} t \cos \omega t.$$

Note that this particular solution oscillates with increasing amplitude and becomes unbounded as $t \to \infty$ (Figure 2.9), whereas any solution of the homogeneous equation is bounded. Thus, *when $\beta = \omega$, every motion oscillates with*

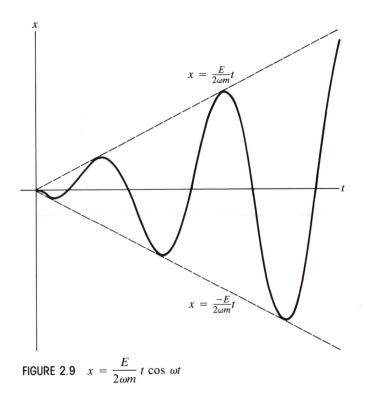

FIGURE 2.9 $x = \dfrac{E}{2\omega m} t \cos \omega t$

amplitude increasing without bound as $t \to \infty$. This phenomenon is known as **resonance.** If this phenomenon were achieved physically, the spring would snap once the oscillations became too big, and our model would not apply. In practice, there will be some friction present, and we will never have β *exactly* matching ω; it can be shown in these cases that resonant forcing leads to behavior with bounded but relatively large amplitude.

Example 2.9.5 Periodic Forcing of a Damped Spring

If $b \neq 0$, then the equation

$$(mD^2 + bD + k)x = E \sin \beta t$$

models a mass of m grams on a damped spring, subject to an oscillating force $E(t) = E \sin \beta t$. Each solution to this o.d.e. has the form

$$x = h(t) + p(t),$$

where $h(t)$ is a solution of the related homogeneous o.d.e., and $p(t)$ is of the form

$$p(t) = k_1 \sin \beta t + k_2 \cos \beta t.$$

(Why?) Although $h(t)$ will take different forms in the underdamped, over-damped, and critically damped cases, we always have

$$\lim_{t \to \infty} h(t) = 0.$$

We refer to $h(t)$ as the **transient term.** As time passes ($t \to \infty$), the behavior of the solution is determined by $p(t)$, which is called the **steady-state term.** Note that the steady-state term is periodic of the same period as the forcing term.

EXERCISES

1. Derive the values of A and α in the phase-amplitude form of the solution of Example 2.9.1, using the identity $\cos(\theta_1 - \theta_2) = \cos \theta_1 \cos \theta_2 + \sin \theta_1 \sin \theta_2$.

2. a. Verify the values of A and α in the phase-amplitude form of the solution in Example 2.9.2.
   b. Check the values of the times t_1 and t_2 in the overdamped case of Example 2.9.2.

3. *Springs:* Consider an unforced mass-spring system with $m = 1$ gram and $k = 100$ dynes/cm. For each of the following values of the damping constant, find the solution starting from equilibrium with velocity 1 cm/sec (increasing x) and graph it carefully for $0 \le t \le 1$ (you will need a calculator to do the numerical work).
   a. $b = 0$ b. $b = 12$ c. $b = 16$
   d. $b = 20$ e. $b = 52$

4. *More Springs:* Find the general solution of the o.d.e. modeling a 1-gram mass attached to a spring of natural length 10 cm and spring constant 1 dyne/cm, with no external forcing, for each of the following damping constants.
   a. $b = 1$ b. $b = 10$ c. $b = 2$

5. *Quantitative Predictions:* For each of the models in Exercise 4, predict the length of the spring after 10 seconds if the spring is initially
   i.   10 cm long and stretching at 5 cm/sec.
   ii. 9 cm long and compressing at 1 cm/sec.
   iii. 11 cm long and at rest.
   (You will need a calculator for the numerical work.)

6. *Qualitative Predictions:* For each of the models in Exercise 4, decide which of the following statements apply.
   i.   Unless the mass starts at the equilibrium position with zero velocity, it will move back and forth across equilibrium indefinitely.

    ii.   Given sufficient initial velocity at the equilibrium position, the spring will stretch indefinitely (until it breaks).

    iii.   The mass will, after sufficient time, stay forever within 1 angstrom (10^{-8} cm) of the equilibrium position.

7. *A Hanging Spring:* In Example 2.9.3 the equilibrium position of a system with constant forcing is independent of the mass. Yet if we hang heavier objects on a single spring, the equilibrium position changes. Why?

8. *An Undamped Spring:* What behavior should we expect in Example 2.9.4, as β approaches ω?

9. *A Damped Spring:* In Example 2.9.5, take $m = 1$, $b = 2$, $k = 2$, $\beta = 1$, and $E = 5$. Suppose that at the outset, $x(0) = 0$ and $x'(0) = 1$.

    a.   Find the transient term of the solution.

    b.   Find the steady-state term of the solution and write it in phase-amplitude form.

    c.   Sketch the graphs of the forcing function and of the steady-state term. Note that although these two graphs have the same period, they cross the t-axis at different times.

10. *Amplitude Modulation—Beats:*

    a.   Let α and β be distinct positive numbers. Find the solution of

$$(D^2 + \alpha^2)x = \cos \beta t$$

    that satisfies $x(0) = x'(0) = 0$.

    b.   Use the formulas for $\cos(\theta_1 \pm \theta_2)$, with $\theta_1 = \frac{1}{2}(\alpha + \beta)t$ and $\theta_2 = \frac{1}{2}(\alpha - \beta)t$, to rewrite the solution in the form

$$x = A(t) \sin \tfrac{1}{2}(\alpha + \beta)t$$

    where $A(t) = C \sin \frac{1}{2}(\alpha - \beta)t$.

Note that if $\alpha - \beta$ is small, then $\alpha + \beta$ will be relatively large. In this case $\sin \frac{1}{2}(\alpha + \beta)t$ oscillates with a high frequency while the amplitude $A(t)$ varies relatively slowly. The graph of x will resemble the curve in Figure 2.10. This phenomenon is referred to as *amplitude modulation* in electronics, and as *beats* in acoustics.

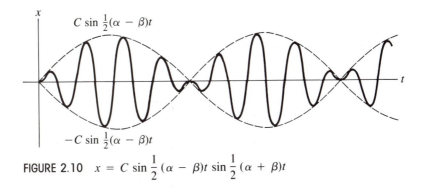

FIGURE 2.10 $x = C \sin \dfrac{1}{2}(\alpha - \beta)t \sin \dfrac{1}{2}(\alpha + \beta)t$

11. *An LRC Circuit:* Reinterpret the results of Example 2.9.2 in terms of the charge Q in the *LRC* circuit of Exercise 5, Section 2.1. What can you say about the

behavior of the *current* in the overdamped and underdamped cases? (*Hint:* Use phase-amplitude form for Q.)

12. *Another Circuit:* Suppose that the voltage in the circuit of Exercise 5, Section 2.1, is $V(t) = \sin \beta t$.
    a. Show that if $R \neq 0$, then the charge Q and the current I can each be written as the sum of a transient term that dies down to zero with time and a steady-state term that is periodic with the same period as the forcing term.
    b. What long-term behavior do the charge and current exhibit if $R = 0$ and if $\beta^2 = 1/LC$?

13. *A Floating Box:* A cubic box with 1-ft sides weighing w pounds (and hence of mass $w/32$ slugs) floats in the water as in Exercise 7, Section 2.1.
    a. Assuming no damping, find a general formula predicting the motion of the box.
    b. Find a formula to determine w from the frequency with which the box bobs up and down. [*Hint:* $x = A \cos(\lambda t - \alpha)$ has frequency $\lambda/2\pi$.]

2.10 ROTATIONAL MODELS

In this section we consider some mechanical models in which a body turns or bends instead of moving back and forth along a straight line. In these contexts it is convenient to reformulate the physical quantities of position, mass, and force into their rotational equivalents: angular displacement, moment of inertia, and moment.

Suppose a body of mass m is constrained to move in a circle of radius r (Figure 2.11). Its position is naturally described by an angular displacement θ from some reference position. An analysis of the motion based on forces has to take account of the forces that keep the body moving in a circle. These forces are taken care of automatically if we calculate the equations of motion in terms of moments and moment of inertia, which are defined as follows:

1. **Moment of inertia:** A point mass m traveling in a circle of radius r has moment of inertia

$$J = mr^2.$$

The moment of inertia for a continuously distributed mass is the integral of density times the square of the radius.

2. **Moment:** A force acting at radius r exerts a moment

$$M = F_{\text{tan}}r$$

where F_{tan} is the component of force tangent to the circle in the direction of increasing θ.

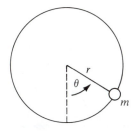

FIGURE 2.11

Newton's second law is expressed in terms of these quantities as

$$M = J\frac{d^2\theta}{dt^2}.$$

Example 2.10.1   A Twisted Shaft

A mass is attached to one end of a flexible shaft, the other end of which is fixed, and the contraption is mounted so that the mass can rotate (Figure 2.12). The mass is subject to two moments. The resistance of the shaft to twisting is reflected in a moment

$$M_{\text{shaft}} = -\kappa\theta$$

where θ is the angular displacement from an equilibrium position, and friction with the mounting mechanism or surrounding medium contributes

$$M_{\text{damp}} = -\beta\frac{d\theta}{dt}.$$

Combining these into Newton's law (in moment form), we obtain

$$J\frac{d^2\theta}{dt^2} = M_{\text{shaft}} + M_{\text{damp}} = -\kappa\theta - \beta\frac{d\theta}{dt}$$

or

$$(JD^2 + \beta D + \kappa)\theta = 0.$$

This is identical in form to the equation of an unforced damped spring, and the analysis of Examples 2.9.1 and 2.9.2 applies.

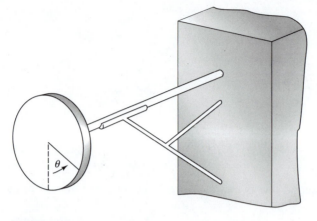

FIGURE 2.12

Example 2.10.2   The Simple Pendulum

A bob of mass m grams swings at the end of a rigid shaft of negligible weight that is L centimeters long; the other end of the shaft is attached to a support that allows it to rotate without friction (Figure 2.13). The force here is that of gravity, $G = gm$, where g is a constant ($g \approx 32$ ft/sec^2). If we measure angular displacement θ so that $\theta = 0$ when the shaft is hanging vertically, then the tangential component of the gravitational force has magnitude $G \sin \theta$. Thus, remembering that $r = L$ and gravity points down, we have

$$M_{\text{grav}} = -Lgm \sin \theta.$$

On the other hand, the moment of inertia is

$$J = mL^2.$$

Thus, the equation of motion is

$$mL^2 \frac{d^2\theta}{dt^2} = -Lgm \sin \theta$$

or

$$D^2\theta + \frac{g}{L} \sin \theta = 0.$$

This equation is nonlinear, and in fact the solution cannot be expressed in terms of elementary functions. However, for *small oscillations*—that is,

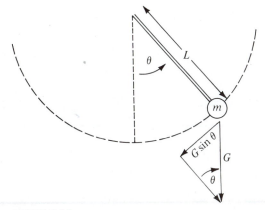

FIGURE 2.13

when the bob is not very far from the equilibrium $\theta = 0$—we can approximate with a linear equation. Recall from calculus that

$$\lim_{\theta \to 0} \frac{\sin \theta}{\theta} = 1$$

which means, when $|\theta|$ is small, that $\sin \theta$ is approximately equal to the numerical value of θ (in radians). Hence, we can replace $\sin \theta$ with θ in the o.d.e. to obtain the approximate, but linear, o.d.e.

$$\left(D^2 + \frac{g}{L} \right) \theta = 0.$$

This is identical to the equation in Example 2.9.1 modeling an undamped spring. Setting $\omega = \sqrt{g/L}$, we see that solutions oscillate periodically about equilibrium with a frequency of $\omega/2\pi$ cycles/sec. In particular, note that our approximation hypothesis—that $|\theta|$ is small—remains true in the approximate model for all time, provided the initial angular velocity is also small. (Compare Exercise 5 of this section.)

Example 2.10.3   Bending Beams

When a long, thin piece of metal (a thin wire several centimeters long or a steel beam many yards long) is supported at one or both ends and has various weights attached to it along its length, it bends (Figure 2.14). Suppose the beam is supported so that its undeflected shape is a horizontal line; then its deflected shape can be described by the graph of a function $y = -s(x)$, where x is the

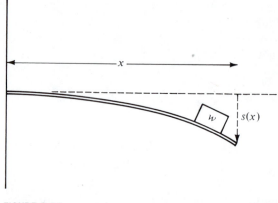

FIGURE 2.14

distance from one end of the beam and y is the vertical deflection of that point from its equilibrium position. Note that here, by contrast with our other examples, the independent variable is x, a distance, instead of time t.

The equations of elasticity theory are formulated in terms of the **turning moment** $\mu(x)$ of a given force relative to the point x. The turning moment is defined as the sum of the moments, with x as a center of rotation, of all forces acting to the right of x. If we have a weight distribution described by a function $w(x)$, then it can be shown that $w(x)$ and $\mu(x)$ are related by

$$\frac{d^2\mu(x)}{dx^2} = w(x).$$

On the other hand, elasticity theory says that the turning moment is proportional to the curvature of the beam. Recall from analytic geometry that the curvature of $s(x)$ is given by $s''(x)[1 + s'(x)^2]^{-3/2}$. Thus, the equation of elasticity theory is

$$\mu(x) = \frac{EIs''(x)}{[1 + s'(x)^2]^{3/2}},$$

where E and I are constants (E depends on the material of which the beam is composed, and I depends on the size and shape of a cross section).

We have two second-order o.d.e.'s: one for $\mu(x)$, given $w(x)$, and the other for $s(x)$, given $\mu(x)$. The second equation, however, is nonlinear; trying to solve for $s(x)$ in this form would lead to a horrible mess. But for *small deflections,* the values of $s(x)$ and $s'(x)$ will be small, and we can approximate the quantity $[1 + s'(x)^2]^{3/2}$ by the constant 1. This simplifies the second equation to

$$\mu(x) = EI\frac{d^2s}{dx^2}.$$

If we differentiate this twice and substitute into the relation between $\mu(x)$ and $w(x)$, we can eliminate the intermediate variable $\mu(x)$, which doesn't really interest us. This yields a fourth-order o.d.e. describing the deflection $s(x)$ in terms of the weight distribution $w(x)$:

$$EI \frac{d^4s}{dx^4} = w(x).$$

The general solution of this equation has the form

$$s(x) = H(x) + p(x)$$

where $p(x)$ is a particular solution depending on $w(x)$, and $H(x)$ is the general solution of the homogeneous equation

$$EI \frac{d^4H}{dx^4} = 0.$$

Thus,

$$s(x) = c_1 + c_2x + c_3x^2 + c_4x^3 + p(x).$$

If we have an evenly distributed weight

$$w(x) = w,$$

then by undetermined coefficients

$$p(x) = \frac{w}{24EI} x^4,$$

so

$$s(x) = c_1 + c_2x + c_3x^2 + c_4x^3 + \frac{w}{24EI} x^4.$$

In contrast to the earlier examples, for which the data naturally led to an initial-value problem, the natural data for the beam problem are values for $s(x)$ and/or its derivatives at each end of the beam. This is a **boundary-value problem.** For example, if the left end of the beam ($x = 0$) is embedded in the wall, $s(x)$ and $s'(x)$ are fixed at zero. If the right end ($x = L$) rests on a fulcrum support, we have $s(x) = 0$, $s'(x)$ unrestrained (because the beam is free to tilt), but $s''(x) = 0$ (because nothing is acting at the fulcrum to bend the beam). In all, we thus have four conditions: two at $x = 0$ ($s(0) = 0$, $s'(0) = 0$) and two at $x = L$ ($s(L) = 0$, $s''(L) = 0$). For the evenly distributed weight these boundary conditions give us

$$0 = s(0)\ \ = c_1$$

$$0 = s'(0)\ \ = c_2$$

$$0 = s(L)\ \ = c_1 + c_2L + c_3L^2 + c_4L^3 + \frac{w}{24EI}L^4$$

$$0 = s''(L) = 2c_3 + 6c_4L + \frac{w}{2EI}L^2$$

which we solve to find

$$s(x) = \frac{w}{48EI}(3L^2x^2 - 5Lx^3 + 2x^4).$$

In particular, if we set $L = 1$ and choose $w = 48EI$, we obtain

$$s(x) = 3x^2 - 5x^3 + 2x^4$$

for which the beam has the approximate shape in Figure 2.15.

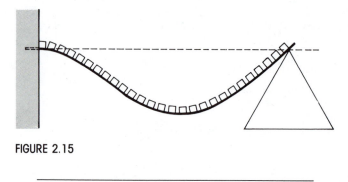

FIGURE 2.15

EXERCISES

1. *The Twisted Shaft:* What do the graphs in Figure 2.8 tell us about physical behavior when the equations model the shaft of Example 2.10.1?
2. *A Beam:* A beam 6 ft long, with constants $EI = 1$, has both ends embedded in walls. Find the shape of the beam if the weight distribution is $w(x) = 4$. In particular, how far below the horizontal will the beam be at its lowest point?
3. *Another Beam:* Repeat Exercise 2 under the assumption that each end of the beam rests on a fulcrum support.

4. *A Simple Pendulum:* Solve the linear o.d.e. in Example 2.10.2 to predict the motion of a 2-ft pendulum under each of the following initial conditions:
   a. $\theta(0) = 0$, $\theta'(0) = 2$ radians/sec
   b. $\theta(0) = 1/20$ radian, $\theta'(0) = 0$ radians/sec
   c. $\theta(0) = 1/20$ radian, $\theta'(0) = -2$ radians/sec
   Express your answers in phase amplitude form (see Example 2.9.1) and, in each case, determine the maximum value of θ and the frequency of the oscillation.

5. *A Balanced Stick:* Consider the variation of Example 2.10.2 obtained by turning the pendulum upside down (Figure 2.16).
   a. Find the appropriate linear approximation to the o.d.e. near the "balanced" equilibrium position.
   b. Is this linearization as useful, on physical grounds, as the linear approximation to the equation modeling the original pendulum? (*Hint:* Does the approximation hypothesis continue to hold?)

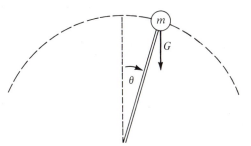

FIGURE 2.16

6. *A Damped Pendulum:* Suppose the pendulum of Example 2.10.2 encounters frictional resistance with moment $M_{damp} = -b\, d\theta/dt$.
   a. Obtain an approximate, but linear, o.d.e. for θ.
   b. What does the discussion in Example 2.9.2 tell us about the physical behavior of the pendulum in case $b^2 < 4gL^3m$?

7. *The Shaking Finger:* Consider the following simplified model of a muscle controlling a limb (see Figure 2.17). An arm of length $\sqrt{2}$ ft (and negligible weight) pivots at one end from the bottom of a vertical rod. The free end of the arm supports a weight of w lb and is attached to the vertical rod by a horizontal spring with constant $k = 5$ slugs/ft and natural length 1 ft. To simplify the mathematics, assume the spring is attached to a ring that can travel up and down the rod, so the spring remains horizontal.
   a. Add the tangential components of the gravitational and spring forces to find F_{tan}.
   b. Check that if no weight were supported ($w = 0$), the arm would rest at an angle of $\pi/4$ radians with the rod.
   c. Use the result of (a) to obtain an o.d.e. for θ.
   d. When θ is near $\pi/4$ radians, $\cos\theta \approx (1/\sqrt{2})[1 - (\theta - \pi/4)]$, $\sin\theta \approx (1/\sqrt{2})[1 + (\theta - \pi/4)]$, and $\cos\theta\sin\theta \approx 1/2$. Use these approximations to obtain an approximate linear equation modeling the arm.
   e. Predict the motion of the arm if the weight is $w = 1$ and the arm starts from rest with $\theta(0) = \pi/4$.

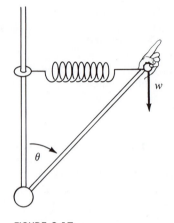

FIGURE 2.17

In the following exercise, which assumes some knowledge of geometric vectors, we derive the rotational form of Newton's second law of motion.

8. If a point moves counterclockwise on a circle of radius r centered at the origin, then its position at time t is described by the vector from the origin to the point:

$$\mathbf{R}(t) = r \cos\theta(t)\mathbf{i} + r \sin\theta(t)\mathbf{j}.$$

The vector

$$\mathbf{T}(t) = -\sin\theta(t)\mathbf{i} + \cos\theta(t)\mathbf{j}$$

is a unit vector tangent to the circle and points counterclockwise.

a. Differentiate $\mathbf{R}(t)$ twice (remembering that r is constant) to obtain a formula for the acceleration $\mathbf{a}(t) = \mathbf{R}''(t)$.

b. Show that the force $\mathbf{F}(t) = m\mathbf{a}(t)$ on the point can be written as the sum of a force $\mathbf{F}_1(t) = -m[\theta'(t)]^2\mathbf{R}(t)$ pointing toward the origin, and a tangential force $\mathbf{F}_2(t) = mr\theta''(t)\mathbf{T}(t)$.

c. Conclude that $F_{\text{tan}} = mr\theta''(t)$.

d. Use the expression for F_{tan} to check that $M = F_{\text{tan}}\, r$ and $J\theta''(t) = mr^2\theta''(t)$ are equal.

9. *A Derivation for the Bending Beam:*

a. Verify that if finitely many weights w_i, $i = 1, \ldots, k$, are placed at positions p_i, $i = 1, \ldots, k$, along a beam, then the turning moment relative to x (see Example 2.9.3) is

$$\mu(x) = \sum_{p_i > x} (p_i - x)w_i.$$

b. If the weight is continuously distributed along the beam according to the density function $w(p)$ (that is, the weight acting between points p and $q > p$ is given by $\int_p^q w(t)\, dt$), then use a partition of the beam into portions of length

Δp to obtain the approximation

$$\mu(x) \approx \sum_{p_i > x} (p_i - x) w(p_i) \, \Delta p$$

Now take the limit as $\Delta p \to 0$ to obtain

$$\mu(x) = \int_x^L (p - x) w(p) \, dp.$$

c. Differentiate this formula twice with respect to x, using the fundamental theorem of calculus, to obtain the formulas

$$\frac{d\mu}{dx} = -\int_x^L w(p) \, dp \quad \text{and} \quad \frac{d^2\mu}{dx^2} = w(x).$$

2.11   SYSTEMS OF DIFFERENTIAL EQUATIONS: ELIMINATION

The two-spring model in Example 2.1.4 led naturally to two o.d.e.'s in two unknowns, x_1 and x_2. By solving one equation for x_2 in terms of x_1 and its derivatives and substituting into the other, we obtained a single o.d.e. for x_1 to which we could apply the methods of this chapter. The same procedure enabled us to deal with several other models in the examples and exercises of Sections 2.1 and 2.10. In this section we discuss a general elimination method for handling two constant-coefficient o.d.e.'s in two unknowns. The method is based on the observation in Section 2.5 that constant-coefficient differential operators can be manipulated formally as if they were simple algebraic expressions. We illustrate with an example.

Example 2.11.1

Solve the system of o.d.e.'s

(S)
$$(2D + 8)x_1 + (D - 1)x_2 = 0$$
$$(D + 9)x_1 + \qquad Dx_2 = 9.$$

Let us treat this formally, as if it were an algebra problem. To eliminate x_2, we multiply the first equation by D and the second by $(D - 1)$:

$$D(2D + 8)x_1 + D(D - 1)x_2 = D(0) \qquad = 0$$
$$(D - 1)(D + 9)x_1 + (D - 1)Dx_2 = (D - 1)(9) = -9.$$

Now we subtract:

$$[D(2D + 8) - (D - 1)(D + 9)]x_1 = 9.$$

Simplifying the expression in brackets gives

$$(N_1) \qquad\qquad\qquad [D^2 + 9]x_1 = 9.$$

The general solution of (N_1) can be found by undetermined coefficients:

$$x_1(t) = c_1 \sin 3t + c_2 \cos 3t + 1.$$

In a similar way, we can eliminate x_1 from (S) by subtracting $(D + 9)$ times the first equation from $(2D + 8)$ times the second. We obtain

$$(N_2) \qquad\qquad\qquad [D^2 + 9]x_2 = 72.$$

The general solution of (N_2) (again using undetermined coefficients) is

$$x_2(t) = k_1 \sin 3t + k_2 \cos 3t + 8.$$

We now have expressions for x_1 and x_2 that together involve four constants. However, two of these can be eliminated by substituting both expressions into (S). To this end, we calculate each term of the second o.d.e.:

$$(D + 9)x_1 = (3c_1 + 9c_2) \cos 3t + (9c_1 - 3c_2) \sin 3t + 9$$
$$Dx_2 = 3k_1 \cos 3t - 3k_2 \sin 3t.$$

Thus, the second o.d.e. of (S) reads

$$(3c_1 + 9c_2 + 3k_1) \cos 3t + (9c_1 - 3c_2 - 3k_2) \sin 3t + 9 = 9.$$

Since $\cos 3t$ and $\sin 3t$ are linearly independent,

$$3c_1 + 9c_2 + 3k_1 = 0, \quad 9c_1 - 3c_2 - 3k_2 = 0.$$

We solve for k_1 and k_2 in terms of c_1 and c_2:

$$k_1 = -c_1 - 3c_2, \quad k_2 = 3c_1 - c_2.$$

You should check that substitution into the first equation of (S) yields the same expressions for k_1 and k_2.

The general solution of (S) is

$$x_1(t) = \qquad\qquad c_1 \sin 3t + \qquad\qquad c_2 \cos 3t + 1$$
$$x_2(t) = (-c_1 - 3c_2) \sin 3t + (3c_1 - c_2) \quad \cos 3t + 8$$

The method of this example can be applied to any system of the form

(S)
$$L_1x_1 + L_2x_2 = E_1(t)$$
$$L_3x_1 + L_4x_2 = E_2(t)$$

where each of L_1, L_2, L_3, and L_4 is a differential operator with constant coefficients (or a number). We can interpret our formal manipulations in terms of applying operators to various functions. If we apply L_4 to both sides of the first equation and L_2 to both sides of the second, we obtain

$$L_4L_1x_1 + L_4L_2x_2 = L_4E_1(t)$$
$$L_2L_3x_1 + L_2L_4x_2 = L_2E_2(t).$$

Note that $L_4L_2x_2 = L_2L_4x_2$. Thus, when we subtract, the x_2 terms cancel, leaving

(N₁)
$$(L_4L_1 - L_2L_3)x_1 = L_4E_1(t) - L_2E_2(t).$$

If $L_4L_1 - L_2L_3$ is a nonzero differential operator, we have an o.d.e. for x_1 whose general solution has the form

$$x_1(t) = c_1h_1(t) + c_2h_2(t) + \cdots + c_nh_n(t) + p_1(t).$$

Similarly, by applying L_3 to the first equation of (S) and L_1 to the second, we obtain

$$L_3L_1x_1 + L_3L_2x_2 = L_3E_1(t)$$
$$L_1L_3x_1 + L_1L_4x_2 = L_1E_2(t).$$

Subtracting the first equation from the second yields an o.d.e. for x_2:

(N₂)
$$(L_1L_4 - L_3L_2)x_2 = L_1E_2(t) - L_3E_1(t).$$

Note that the differential operators on the left in (N₁) and (N₂) are the same, so that the general solution of (N₂) is

$$x_2(t) = k_1h_1(t) + k_2h_2(t) + \cdots + k_nh_n(t) + p_2(t),$$

where $p_2(t)$ is a particular solution of (N₂), and $h_1(t), \ldots, h_n(t)$ are the same as in the solution of (N₁).

As in the method of undetermined coefficients in Section 2.7, the application of a differential operator to both sides of an equation gives a new o.d.e., which is satisfied by all solutions of the original equation. But again the new o.d.e. may also have other, extraneous solutions. These extra solutions are reflected in the presence of too many arbitrary constants in our expressions for x_1 and x_2. We can eliminate the extra constants by substituting both expressions into (S). *The general solution of (S) can always be written involving only n arbitrary constants, where n is the order of (N_1).*

Let us consider a few more examples.

Example 2.11.2

Solve the system of o.d.e.'s

$$2Dx_1 + Dx_2 = 7x_1 + 8x_2 + 50e^t$$

$$Dx_1 + 3Dx_2 = -4x_1 - 6x_2.$$

We first rearrange the terms, grouping together all terms involving x_1 and all terms involving x_2:

(S)
$$(2D - 7)x_1 + (D - 8)x_2 = 50e^t$$

$$(D + 4)x_1 + (3D + 6)x_2 = 0.$$

To eliminate x_2, we apply $3D + 6$ to the first o.d.e. and $D - 8$ to the second:

$$(3D + 6)(2D - 7)x_1 + (3D + 6)(D - 8)x_2 = (3D + 6)(50e^t) = 450e^t$$

$$(D - 8)(D + 4)x_1 + (D - 8)(3D + 6)x_2 = (D - 8)(0) = 0.$$

We subtract to get

$$[(3D + 6)(2D - 7) - (D - 8)(D + 4)]x_1 = 450e^t$$

or

(N_1)
$$(5D^2 - 5D - 10)x_1 = 450e^t.$$

The characteristic polynomial of (N_1) is $5r^2 - 5r - 10 = 5(r - 2)(r + 1)$. Two independent solutions of the related homogeneous o.d.e. are $h_1(t) = e^{2t}$ and $h_2(t) = e^{-t}$, and a particular solution of (N_1) is $p_1(t) = -45e^t$. Thus, the general solution of (N_1) is

$$x_1(t) = c_1 e^{2t} + c_2 e^{-t} - 45e^t.$$

Similarly, we eliminate x_1 from (S) by applying $D + 4$ to the first o.d.e. and $2D - 7$ to the second:

$$(D + 4)(2D - 7)x_1 + (D + 4)(D - 8)x_2 = (D + 4)(50e^t) = 250e^t$$

$$(2D - 7)(D + 4)x_1 + (2D - 7)(3D + 6)x_2 = (2D - 7)(0) = 0.$$

Subtracting the first of these equations from the second, we have

(N_2)
$$(5D^2 - 5D - 10)x_2 = -250e^t.$$

The general solution of (N_2) is

$$x_2(t) = k_1 e^{2t} + k_2 e^{-t} + 25e^t.$$

Since (N_1) is of order two, the general solution of (S) should involve only two arbitrary constants. Thus, two of the constants in our expressions for x_1 and x_2 must be eliminated by substitution into (S). We calculate

$$(2D - 7)x_1 = -3c_1 e^{2t} - 9c_2 e^{-t} + 225e^t$$

$$(D - 8)x_2 = -6k_1 e^{2t} - 9k_2 e^{-t} - 175e^t$$

and substitute into the first o.d.e. of (S) to get

$$(-3c_1 - 6k_1)e^{2t} + (-9c_2 - 9k_2)e^{-t} + 50e^t = 50e^t.$$

Hence, we must have $-3c_1 - 6k_1 = 0$ and $-9c_2 - 9k_2 = 0$, or

$$k_1 = -\frac{1}{2}c_1, \quad k_2 = -c_2.$$

The general solution of (S) is

$$x_1(t) = c_1 e^{2t} + c_2 e^{-t} - 45e^t$$

$$x_2(t) = -\frac{1}{2}c_1 e^{2t} - c_2 e^{-t} + 25e^t.$$

Example 2.11.3

Solve the system of o.d.e.'s

(S)
$$(D^2 - 1)x_1 + x_2 = 0$$
$$4(D - 1)x_1 + (D + 1)x_2 = 0.$$

To eliminate x_2, we apply $D + 1$ to the first equation,

$$(D + 1)(D^2 - 1)x_1 + (D + 1)x_2 = 0,$$

and then subtract the second o.d.e. We obtain the equation

(N$_1$) $[(D + 1)(D^2 - 1) - 4(D - 1)]x_1 = 0.$

Note that (N$_1$) is homogeneous. Its characteristic polynomial is

$$(r + 1)(r^2 - 1) - 4(r - 1) = (r - 1)[(r + 1)^2 - 4] = (r - 1)^2(r + 3)$$

so the general solution is

$$x_1(t) = c_1e^t + c_2te^t + c_3e^{-3t}.$$

If we eliminate x_1 from (S), by applying $4(D - 1)$ to the first o.d.e. and $D^2 - 1$ to the second and then subtracting, we find that x_2 also satisfies (N$_1$), so

$$x_2(t) = k_1e^t + k_2te^t + k_3e^{-3t}.$$

We expect only three arbitrary constants, but we have six. To substitute into the first o.d.e. of (S), we calculate

$$(D^2 - 1)x_1 = 2c_2e^t + 8c_3e^{-3t}.$$

Then

$$2c_2e^t + 8c_3e^{-3t} + k_1e^t + k_2te^t + k_3e^{-3t} = 0$$

so that

$$k_1 = -2c_2, \quad k_2 = 0, \quad k_3 = -8c_3.$$

The general solution of (S) is

$$x_1(t) = \quad c_1e^t + c_2te^t + \quad c_3e^{-3t}$$
$$x_2(t) = -2c_1e^t \quad\quad - 8c_3e^{-3t}.$$

If the operator $L_4L_1 - L_2L_3$ is zero, then the method fails. Just as in the case of an algebraic system, there may or may not be functions that satisfy both equations (see Exercises 20 and 21).

Although variants of this method can be applied to systems involving more than two unknowns, they become more and more cumbersome as the number of unknowns increases. For this reason a different approach, treated in Chapter

3, is more useful in handling larger systems. For two o.d.e.'s in two unknowns, the elimination method is the more direct way of obtaining the general solution.

SYSTEMS OF DIFFERENTIAL EQUATIONS VIA ELIMINATION

To solve the system of o.d.e.'s

(S)
$$L_1 x_1 + L_2 x_2 = E_1(t)$$
$$L_3 x_1 + L_4 x_2 = E_2(t)$$

where the L_i are linear differential operators with constant coefficients and $L_1 L_4 - L_2 L_3$ is a differential operator of order n:

1. Eliminate x_2 by applying to each equation in (S) the operator that appears before x_2 in the other, and subtracting. The general solution of the resulting o.d.e. has the form

$$x_1(t) = c_1 h_1(t) + \cdots + c_n h_n(t) + p_1(t).$$

2. Similarly, eliminate x_1 from (S) to obtain an o.d.e. for x_2. Its general solution has the form

$$x_2(t) = k_1 h_1(t) + \cdots + k_n h_n(t) + p_2(t)$$

where the functions $h_i(t)$ are the same as in step 1.

3. Substitute these expressions for $x_1(t)$ and $x_2(t)$ into (S) to determine the relations among the constants $k_1, \ldots, k_n$ and $c_1, \ldots, c_n$.

4. Write the general solution, which involves only n arbitrary constants.

Note

On elimination and Cramer's rule

If the L_i's, and $E_i(t)$'s in (S) were all numbers, with $L_1 L_4 - L_2 L_3 \neq 0$, then Cramer's rule (Appendix A) would apply to give

$$x_1 = \frac{\det \begin{bmatrix} E_1 & L_2 \\ E_2 & L_4 \end{bmatrix}}{\det \begin{bmatrix} L_1 & L_2 \\ L_3 & L_4 \end{bmatrix}} = \frac{L_4 E_1 - L_2 E_2}{L_4 L_1 - L_2 L_3}$$

$$x_2 = \frac{\det \begin{bmatrix} L_1 & E_1 \\ L_3 & E_2 \end{bmatrix}}{\det \begin{bmatrix} L_1 & L_2 \\ L_3 & L_4 \end{bmatrix}} = \frac{L_1 E_2 - L_3 E_1}{L_1 L_4 - L_2 L_3}.$$

Multiplication of both expressions by the denominator, $L_1L_4 - L_2L_3$, gives

$$(L_1L_4 - L_2L_3)x_1 = L_4E_1 - L_2E_2$$

$$(L_1L_4 - L_2L_3)x_2 = L_1E_2 - L_3E_1.$$

These equations are the same as (N_1) and (N_2), which we obtained earlier. Thus, *Cramer's rule provides a shortcut for arriving at the o.d.e.'s (N_1) and (N_2).* However, when taking determinants whose entries include operators and functions, we must take care to write the products in the correct order. For example, the numerator in the expression for x_1 is $L_4E_1(t) - L_2E_2(t)$, not $E_1(t)L_4 - L_2E_2(t)$.

For $n > 2$, Cramer's rule continues to describe the o.d.e.'s we need to solve in order to handle n o.d.e.'s in n unknowns by elimination. Since Cramer's rule requires the calculation of $n + 1$ determinants of size $n \times n$, its use involves a great deal of work. Furthermore, even after we have solved the o.d.e.'s for $x_1, \ldots, x_n$, we must substitute our expressions back into (S) in order to determine the relations among the various constants. This can be an enormous task.

EXERCISES

In Exercises 1 through 14, find the general solution.

1. $(D - 1)x_1 + (D + 3)x_2 = 2$
   $Dx_1 + (D - 1)x_2 = 0$

2. $Dx_1 = -5x_1 + 8x_2$
   $Dx_2 = -3x_1 + 5x_2$

3. $(3D + 3)x_1 + (2D - 1)x_2 = 0$
   $(D + 1)x_1 + (D - 4)x_2 = 0$

4. $Dx_1 = 4x_1 - 2x_2 + 4$
   $Dx_2 = 3x_1 - 3x_2$

5. $Dx_1 = x_2 + 10$
   $Dx_2 = -5x_1 - 4x_2$

6. $(D + 9)x_1 - 9x_2 = 3t + 1$
   $4x_1 + (D - 3)x_2 = t$

7. $(D + 2)x_1 + (D + 1)x_2 = 3e^{2t}$
   $(D - 7)x_1 - (D + 3)x_2 = 4$

8. $Dx_1 + Dx_2 + x_1 - 3x_2 - e^t = 0$
   $2Dx_1 + 3Dx_2 + 3x_1 - 8x_2 - 3e^t = 0$

9. $Dx_1 + Dx_2 = x_1 - x_2 + 1$
   $2Dx_1 + 4Dx_2 = -4x_1 + 3x_2 + 4$

10. $Dx_1 + Dx_2 = -x_1 + x_2 + \tan t$
    $3Dx_1 + 4Dx_2 = -4x_1 + 3x_2 + 4\tan t$

11. $(D^2 - 1)x_1 + (D + 1)x_2 = 0$
    $(D - 1)x_1 - Dx_2 = 3$

12. $(D - 2)x_1 + D^2x_2 = t$
    $(D^2 - 1)x_1 - D^3x_2 = 0$

13. $(D - 4)x_1 - Dx_2 = 32$
    $(D^3 - D)x_1 + (D + 4)x_2 = 0$

14. $(D - 2)x_1 + Dx_2 = 0$
    $(4D^2 - 8D)x_1 + x_2 = 2$

Solve the initial-value problems in Exercises 15 through 17.

15. $Dx_1 = -3x_1 - 2x_2 + 2$
    $Dx_2 = x_1 + e^t$ $x_1(0) = x_2(0) = 0$

16. $(D - 4)x_1 + (D - 1)x_2 = 3e^t$
    $-(D - 4)x_1 + (D + 2)x_2 = e^t$ $x_1(0) = x_2(0) = 1$

17. $(D + 2)x_1 + (2D + 3)x_2 = 6 + 16e^t$
    $(2D + 1)x_1 + Dx_2 = 3 + 8e^t$ $x_1(0) = 11, x_2(0) = 0$

Exercises 18 and 19 refer to our earlier models. (For more models, see Section 3.1.)

18. *Interacting Populations:*
    a. Solve the system of equations modeling the populations in Example 1.1.7.
    b. Assume that the populations start at $x_1(0) = 12$ million and $x_2(0) = 16$ million. Will they ever be equal?

19. *Supply and Demand:*
    a. Reformulate the price model of Exercise 6, Section 2.1, as a system of two o.d.e.'s for supply and price.
    b. What form will the general solution of this system take if the supplier is relatively insensitive to price (supply increases by very few units per dollar of price)?
    c. What form will the general solution take if the supplier is overzealous?
    d. Show that in either of these cases, the price will tend toward zero and the supply toward an "equilibrium" supply as t increases.

Exercises 20 and 21 deal with systems (S) for which $L_4L_1 - L_2L_3$ is zero.

20. By subtracting a (constant or operator) multiple of one equation from the other, show that each of the following systems has no solutions.
    a. $(D - 2)x_1 + Dx_2 = e^t$
       $(3D - 6)x_1 + 3Dx_2 = e^t$
    b. $(D - 1)x_1 + Dx_2 = t$
       $(D^2 - D)x_1 + D^2x_2 = 0$

21. Show that any differentiable choice for x_2 leads to a solution of
    $$2Dx_1 + Dx_2 = e^t$$
    $$6Dx_1 + 3Dx_2 = 3e^t.$$

REVIEW PROBLEMS

1. Let $L = tD^3 + 3D^2$.
   a. Find $L(1)$, $L(t)$, $L(1/t)$, and $L(t^3)$.
   b. Show that 1, t, $1/t$ are linearly independent on $0 < t$.
   c. Find the general solution of $Lx = 24t$ on $0 < t$.

Exercises 2 through 11 can be solved by the general methods discussed in the text. In Exercises 2 through 8, find the general solution.

2. $\dfrac{d^2x}{dt^2} + 2\dfrac{dx}{dt} - 3x = 3t + \sin 3t$

3. $9\dfrac{d^2x}{dt^2} - 6\dfrac{dx}{dt} + x = \dfrac{18}{t^3}e^{t/3}$, $t > 0$

4. $4\dfrac{d^3x}{dt^3} - 8\dfrac{d^2x}{dt^2} - \dfrac{dx}{dt} + 2x = 0$

5. $\dfrac{d^3x}{dt^3} + 3\dfrac{d^2x}{dt^2} + 3\dfrac{dx}{dt} + x = \dfrac{2}{t}e^{-t}$, $t > 0$

6. $(D - 7)^2(D + 1)x = e^{-t}$

7. $D^3(D^2 + 1)x = 3 + 2e^{-t}$

8. $(D - 1)(D^2 + 2D + 2)^3x = 3 + 2e^{-t}$

Solve the initial-value problems in Exercises 9 through 11.

9. $(2D^2 + 2D + 5)x = 0$; $x(0) = 2, x'(0) = -2$

10. $(D - 1)(D + 1)(D - 2)x = 6$; $x(0) = 1, x(0) = -1, x''(0) = 1$

11. $x'' - 2x' + x = 8e^{-t}$; $x(0) = x'(0) = 1$

In Exercises 12 through 23, find the general solution. Some of these problems require the use of special techniques discussed in the exercises of this chapter.

12. $2x'' - 4x' + 10x = 4e^t \sec 2t$, $-\dfrac{\pi}{4} < t < \dfrac{\pi}{4}$

13. $t^2 x'' + tx' + x = \dfrac{3}{t}$, $t > 0$ 14. $tx'' - x' - (t + 1)x = 0$, $t > 0$

15. $x'' + x = \sin t$ 16. $t\dfrac{d^3 x}{dt^3} - \dfrac{d^2 x}{dt^2} = 6t^2$, $t > 0$

17. $t^2 x'' - tx' + x = \dfrac{1}{t}$, $t > 0$ 18. $tx'' - (2t + 1)x' + 2x = t^2 e^{2t}$, $t > 0$

19. $4x'' - x = -6te^t + 8e^{t/2}$ 20. $t\dfrac{d^3 x}{dt^3} + 2\dfrac{d^2 x}{dt^2} = 0$

21. $Dx_1 = -4x_1 + 6x_2 + 2$ 22. $(D - 4)x_1 -$ $2x_2 = 4$
    $Dx_2 = -3x_1 + 5x_2 + e^t$ $3D^2 x_1 + (D^3 + 3D^2)x_2 = 0$

23. $Dx_1 +$ $x_2 = e^t$
    $-x_1 + (D + 1)x_2 = 0$

DETERMINANTS

The solution of systems of algebraic equations is at the core of many problems. In this appendix, we discuss determinants, a powerful tool for working with systems in which the number of equations equals the number of unknowns. In particular, we will see how to determine when such a system has solutions without actually finding the solutions themselves. Determinants play an important role in many areas of mathematics beyond the particular problems we address.

Let's begin by considering a system of two algebraic equations in two unknowns:

(A)
$$b_{11}u_1 + b_{12}u_2 = r_1$$
$$b_{21}u_1 + b_{22}u_2 = r_2.$$

Here the b_{ij}'s, r_1, and r_2 are given real numbers, and we wish to solve for the unknowns u_1 and u_2.

We multiply the first equation by b_{22} and the second by b_{12} to obtain

$$b_{22}b_{11}u_1 + b_{22}b_{12}u_2 = b_{22}r_1$$
$$b_{12}b_{21}u_1 + b_{12}b_{22}u_2 = b_{12}r_2.$$

Subtraction yields

$$(b_{11}b_{22} - b_{12}b_{21})u_1 = r_1b_{22} - b_{12}r_2.$$

Similarly, multiplying the first equation by b_{21}, the second by b_{11} and subtracting gives

$$(b_{11}b_{22} - b_{12}b_{21})u_2 = b_{11}r_2 - r_1b_{21}.$$

If $b_{11}b_{22} - b_{12}b_{21} \neq 0$, we can solve for u_1 and u_2:

$$u_1 = \frac{r_1 b_{22} - r_2 b_{12}}{b_{11}b_{22} - b_{12}b_{21}}, \quad u_2 = \frac{b_{11}r_2 - r_1 b_{21}}{b_{11}b_{22} - b_{12}b_{21}}.$$

We call $b_{11}b_{22} - b_{12}b_{21}$ the **determinant of coefficients** of (A) and write

$$\det \begin{bmatrix} b_{11} & b_{12} \\ b_{21} & b_{22} \end{bmatrix} = b_{11}b_{22} - b_{12}b_{21}.$$

The numerators of our expressions for u_1 and u_2 can also be expressed as determinants. We have

$$u_1 = \frac{\det \begin{bmatrix} r_1 & b_{12} \\ r_2 & b_{22} \end{bmatrix}}{\det \begin{bmatrix} b_{11} & b_{12} \\ b_{21} & b_{22} \end{bmatrix}}, \quad u_2 = \frac{\det \begin{bmatrix} b_{11} & r_1 \\ b_{21} & r_2 \end{bmatrix}}{\det \begin{bmatrix} b_{11} & b_{12} \\ b_{21} & b_{22} \end{bmatrix}}.$$

Example A.1

Solve the system of equations

$$2u_1 - 3u_2 = 7$$
$$5u_1 + 6u_2 = -1.$$

The determinant of coefficients is

$$\det \begin{bmatrix} 2 & -3 \\ 5 & 6 \end{bmatrix} = (2)(6) - (-3)(5) = 27.$$

Then

$$u_1 = \frac{\det \begin{bmatrix} 7 & -3 \\ -1 & 6 \end{bmatrix}}{27} = \frac{39}{27} = \frac{13}{9}$$

$$u_2 = \frac{\det \begin{bmatrix} 2 & 7 \\ 5 & -1 \end{bmatrix}}{27} = \frac{-37}{27}.$$

What if the determinant of coefficients is zero? In this case the answer depends on the values of r_1 and r_2. For certain values of r_1 and r_2 there will be no solutions. For all other values there will be infinitely many solutions.

■ Example A.2

Consider the system of equations

(A)
$$u_1 + 4u_2 = r_1$$
$$2u_1 + 8u_2 = r_2.$$

The determinant of coefficients is

$$\det \begin{bmatrix} 1 & 4 \\ 2 & 8 \end{bmatrix} = (1)(8) - (4)(2) = 0.$$

If we subtract twice the first equation from the second, we get

$$0 = r_2 - 2r_1.$$

This is impossible if $r_2 \neq 2r_1$; there are no solutions in this case. If $r_2 = 2r_1$, then $u_1 = r_1 - 4k$ and $u_2 = k$ will be a solution for each value of k; there are infinitely many solutions in this case.

With the right interpretation of determinants, the preceding observations can be extended to systems of n algebraic equations in n unknowns.

Fact: Cramer's Determinant Test.   *Given the system of n algebraic equations*

(A)
$$b_{11}u_1 + b_{12}u_2 + \cdots + b_{1n}u_n = r_1$$
$$b_{21}u_1 + b_{22}u_2 + \cdots + b_{2n}u_n = r_2$$
$$\vdots$$
$$b_{n1}u_1 + b_{n2}u_2 + \cdots + b_{nn}u_n = r_n$$

there is a number depending on the coefficients

$$\Delta = \det \begin{bmatrix} b_{11} & b_{12} & \cdots & b_{1n} \\ b_{11} & b_{12} & \cdots & b_{2n} \\ \cdot & \cdot & & \cdot \\ \cdot & \cdot & & \cdot \\ \cdot & \cdot & & \cdot \\ b_{n1} & b_{n2} & \cdots & b_{nn} \end{bmatrix}$$

*called the **determinant of coefficients**, with the following properties:*

1. *If $\Delta \neq 0$, then the system of equations (A) has a unique solution.*
2. *If $\Delta = 0$, then the system of equations (A) has no solutions for some values of $r_1, \ldots, r_n$ and has infinitely many solutions for all other values of $r_1, \ldots, r_n$.*

If $\Delta \neq 0$, we can use determinants to explicitly calculate the unique solution of (A) (see Note 3). However, in practice, Cramer's determinant test for large n is primarily a theoretical tool.

To make use of Cramer's test, we need to know how to compute the $n \times n$ determinant Δ. The method we'll use is called **expansion by minors:**

1. Associated to each entry b_{ij} of the $n \times n$ determinant is a smaller determinant, called the (i, j)th **minor.** It is the $(n - 1) \times (n - 1)$ determinant obtained by crossing out the row and column containing the given entry b_{ij}.

2. Also associated to each entry b_{ij} is a plus sign or a minus sign, obtained by forming a "checkerboard" pattern starting with $+$ in the upper left corner:

$$\begin{bmatrix} + & - & + & - & \cdot & \cdot & \cdot \\ - & + & - & + & \cdot & \cdot & \cdot \\ + & - & + & - & \cdot & \cdot & \cdot \\ \cdot & \cdot & \cdot & & & & \\ \cdot & \cdot & \cdot & & & & \\ \cdot & \cdot & \cdot & & & & \end{bmatrix}.$$

This sign can also be described by a formula. The sign associated to b_{ij} is $(-1)^{i+j}$.

3. The determinant is obtained by expanding along a given row or column. One fixes a row (or column) and adds the entries of the row (or column) times the associated sign times the associated minor:

$$\det = \sum (-1)^{i+j} (b_{ij}) \{(i, j)\text{th minor}\}.$$

There is a great deal of work in actually checking that calculations of a determinant using different rows or columns all give the same answer and that this answer is what we need to make Cramer's determinant test work. We'll take this on faith.

Example A.3

Calculate

$$\Delta = \det \begin{bmatrix} 1 & 2 & 3 & 4 \\ 2 & 0 & 3 & 1 \\ 3 & 2 & 1 & 0 \\ 5 & -2 & 2 & 0 \end{bmatrix}$$

We expand Δ by minors along the last column:

a. The sign associated with the 4 in the top right corner is $-$. We obtain the minor associated with this 4 by crossing out the top row and the last column of Δ:

$$\det \begin{bmatrix} 2 & 0 & 3 \\ 3 & 2 & 1 \\ 5 & -2 & 2 \end{bmatrix}$$

The corresponding summand in the expansion of Δ is

$$-4 \det \begin{bmatrix} 2 & 0 & 3 \\ 3 & 2 & 1 \\ 5 & -2 & 2 \end{bmatrix}$$

b. The sign associated with the 1 in the last column and second row of Δ is $+$. We obtain the associated minor by crossing out the last column and the second row of Δ. The corresponding summand is

$$+1 \det \begin{bmatrix} 1 & 2 & 3 \\ 3 & 2 & 1 \\ 5 & -2 & 2 \end{bmatrix}$$

c. The summands corresponding to the two 0's in the last column of Δ are both of the form

$$(\text{sign})(0)(\text{minor}) = 0.$$

Thus,

$$\Delta = -4 \det \begin{bmatrix} 2 & 0 & 3 \\ 3 & 2 & 1 \\ 5 & -2 & 2 \end{bmatrix} + \det \begin{bmatrix} 1 & 2 & 3 \\ 3 & 2 & 1 \\ 5 & -2 & 2 \end{bmatrix} + 0 + 0.$$

We calculate each of the 3 × 3 determinants in our expansion by expanding along the first row:

$$\det \begin{bmatrix} 2 & 0 & 3 \\ 3 & 2 & 1 \\ 5 & -2 & 2 \end{bmatrix} = +2 \det \begin{bmatrix} 2 & 1 \\ -2 & 2 \end{bmatrix} + 0 + 3 \det \begin{bmatrix} 3 & 2 \\ 5 & -2 \end{bmatrix}$$

$$= 2[(2)(2) - (1)(-2)] + 3[(3)(-2) - (2)(5)]$$

$$= -36$$

$$\det \begin{bmatrix} 1 & 2 & 3 \\ 3 & 2 & 1 \\ 5 & -2 & 2 \end{bmatrix} = +1 \det \begin{bmatrix} 2 & 1 \\ -2 & 2 \end{bmatrix} - 2 \det \begin{bmatrix} 3 & 1 \\ 5 & 2 \end{bmatrix} + 3 \det \begin{bmatrix} 3 & 2 \\ 5 & -2 \end{bmatrix}$$

$$= 6 - 2 - 48 = -44.$$

Substituting the values for the 3 × 3 determinants into our expression for Δ yields

$$\Delta = -4(-36) + (-44) = 100.$$

Example A.4

Determine whether the system of equations

(A)
$$\begin{aligned}
u_1 + 2u_2 + 3u_3 + 4u_4 &= r_1 \\
2u_1 \qquad + 3u_3 + \quad u_4 &= r_2 \\
3u_1 + 2u_2 + \quad u_3 \qquad &= r_3 \\
5u_1 - 2u_2 + 2u_3 \qquad &= r_4
\end{aligned}$$

always has a solution.

The determinant of coefficients is

$$\Delta = \det \begin{bmatrix} 1 & 2 & 3 & 4 \\ 2 & 0 & 3 & 1 \\ 3 & 2 & 1 & 0 \\ 5 & -2 & 2 & 0 \end{bmatrix}.$$

We found in Example A.3 that

$$\Delta = 100.$$

Since $\Delta \neq 0$, the system of equations (A) always has a solution.

Example A.5

Determine whether the system of equations

(A)
$$\begin{aligned}
3u_1 - \quad u_2 + \quad u_3 + \quad u_4 &= r_1 \\
2u_1 \qquad + \quad u_3 \qquad &= r_2 \\
8u_1 - 2u_2 + 3u_3 + 2u_4 &= r_3 \\
u_1 \qquad - \quad u_3 \qquad &= r_4
\end{aligned}$$

always has a solution.

The determinant of coefficients is

$$\Delta = \det \begin{bmatrix} 3 & -1 & 1 & 1 \\ 2 & 0 & 1 & 0 \\ 8 & -2 & 3 & 2 \\ 1 & 0 & -1 & 0 \end{bmatrix}$$

We expand Δ along the second column:

$$\Delta = -(-1) \det \begin{bmatrix} 2 & 1 & 0 \\ 8 & 3 & 2 \\ 1 & -1 & 0 \end{bmatrix} + 0 - (-2) \det \begin{bmatrix} 3 & 1 & 1 \\ 2 & 1 & 0 \\ 1 & -1 & 0 \end{bmatrix} + 0.$$

We expand each of the 3×3 determinants along the last column:

$$\Delta = -(-1) \left\{ -2 \det \begin{bmatrix} 2 & 1 \\ 1 & -1 \end{bmatrix} \right\} - (-2) \left\{ 1 \det \begin{bmatrix} 2 & 1 \\ 1 & -1 \end{bmatrix} \right\}$$

$$= (-2 + 2) \det \begin{bmatrix} 2 & 1 \\ 1 & -1 \end{bmatrix} = 0.$$

The system of equations (A) has no solutions for some values of $r_1, r_2, r_3,$ and r_4; it has infinitely many solutions for all other values.

In general, the expansion of an $n \times n$ determinant involves n minors, each of which is an $(n - 1) \times (n - 1)$ determinant. Each of these minors can be expanded in terms of $(n - 2) \times (n - 2)$ determinants, and so on. Eventually, the whole problem reduces to 2×2 determinants. As we saw in the examples, expansion along a row or column with zeros in it saves a lot of work, since we needn't calculate the corresponding minors. We list some other useful facts about determinants in the notes that follow our summary.

DETERMINANTS
Calculation of 2 × 2 Determinants

$$\det \begin{bmatrix} b_{11} & b_{12} \\ b_{21} & b_{22} \end{bmatrix} = b_{11}b_{22} - b_{12}b_{21}.$$

Calculation of n × n Determinants

$$\det \begin{bmatrix} b_{11} & \cdots & b_{1n} \\ \cdot & & \cdot \\ \cdot & & \cdot \\ \cdot & & \cdot \\ b_{n1} & \cdots & b_{nn} \end{bmatrix}.$$

Choose a row or column. Associated to each entry b_{ij} of the row or column is a sign, $(-1)^{i+j}$, and a **minor,** the $(n-1) \times (n-1)$ determinant obtained by crossing out the row and column containing the entry. The determinant is the sum along the chosen row or column of the terms consisting of entry times sign times minor.

Cramer's Determinant Test

Given the system of n algebraic equations for the n unknowns $u_1, \ldots, u_n$,

$$b_{11}u_1 + \cdots + b_{1n}u_n = r_1$$

(A)

$$b_{n1}u_1 + \cdots + b_{nn}u_n = r_n,$$

let Δ be the **determinant of coefficients,**

$$\Delta = \det \begin{bmatrix} b_{11} & \cdots & b_{1n} \\ \cdot & & \cdot \\ \cdot & & \cdot \\ \cdot & & \cdot \\ b_{n1} & \cdots & b_{nn} \end{bmatrix}.$$

1. If $\Delta \neq 0$, then the system of equations has a unique solution.
2. If $\Delta = 0$, then the system of equations has no solutions for some values of $r_1, \ldots, r_n$ and infinitely many solutions for all other values.

Notes

1. A warning

Some readers may have seen a "diagonal" method of calculating 3×3 determinants. This method does *not* extend to $n \times n$ determinants with $n > 3$.

2. On calculating determinants

The following properties of determinants are useful in simplifying calculations (for proofs, see Exercises 34 through 36):

Fact: Properties of Determinants

  i. *If all entries of a single row (or column) have a common factor, this factor can be pulled out in front of the determinant.*

ii.   *If a determinant has two identical rows (or columns), then the determinant is 0.*

iii. *If $i \neq j$, then replacement of row i by [(row i) + k (row j)] doesn't change the determinant:*

$$\det \begin{bmatrix} b_{11} & b_{12} & \cdots & b_{1n} \\ \cdot & \cdot & & \cdot \\ \cdot & \cdot & & \cdot \\ \cdot & \cdot & & \cdot \\ b_{i1} & b_{i2} & \cdots & b_{in} \\ \cdot & \cdot & & \cdot \\ \cdot & \cdot & & \cdot \\ \cdot & \cdot & & \cdot \\ b_{n1} & b_{n2} & \cdots & b_{nn} \end{bmatrix} = \det \begin{bmatrix} b_{11} & b_{12} & \cdots & b_{1n} \\ \cdot & \cdot & & \cdot \\ \cdot & \cdot & & \cdot \\ \cdot & \cdot & & \cdot \\ b_{i1} + kb_{j1} & b_{i2} + kb_{j2} & \cdots & b_{in} + kb_{jn} \\ \cdot & \cdot & & \cdot \\ \cdot & \cdot & & \cdot \\ \cdot & \cdot & & \cdot \\ b_{n1} & b_{n2} & \cdots & b_{nn} \end{bmatrix}.$$

The calculation of the 4×4 determinant in Example 2.4.5 can be simplified by the use of properties i and ii. We first factor out -1 from the second column; the resulting determinant is 0 since it has two identical columns:

$$\det \begin{bmatrix} 3 & -1 & 1 & 1 \\ 2 & 0 & 1 & 0 \\ 8 & -2 & 3 & 2 \\ 1 & 0 & -1 & 0 \end{bmatrix} = (-1) \det \begin{bmatrix} 3 & 1 & 1 & 1 \\ 2 & 0 & 2 & 0 \\ 8 & 2 & 3 & 2 \\ 1 & 0 & -1 & 0 \end{bmatrix} = (-1)(0) = 0.$$

We can simplify the calculation of the 4×4 determinant Δ in Example 2.4.3 by using property iii. We replace row 1 by [(row 1) $-$ 4(row 2)] and expand by minors along the last column:

$$\Delta = \det \begin{bmatrix} 1 & 2 & 3 & 4 \\ 2 & 0 & 3 & 1 \\ 3 & 2 & 1 & 0 \\ 5 & -2 & 2 & 0 \end{bmatrix} = \det \begin{bmatrix} -7 & 2 & -9 & 0 \\ 2 & 0 & 3 & 1 \\ 3 & 2 & 1 & 0 \\ 5 & -2 & 2 & 0 \end{bmatrix} = \det \begin{bmatrix} -7 & 2 & -9 \\ 3 & 2 & 1 \\ 5 & -2 & 2 \end{bmatrix}.$$

To calculate the 3×3 determinant, replace row 2 by [(row 2) $-$ (row 1)], and then replace row 3 by [(row 3) + (row 1)]:

$$\Delta = \det \begin{bmatrix} -7 & 2 & -9 \\ 10 & 0 & 10 \\ 5 & -2 & 2 \end{bmatrix} = \det \begin{bmatrix} -7 & 2 & -9 \\ 10 & 0 & 10 \\ -2 & 0 & -7 \end{bmatrix}.$$

We expand this last determinant by minors along the second column:

$$\Delta = -2 \det \begin{bmatrix} 0 & 10 \\ -2 & -7 \end{bmatrix} = (-2)(-50) = 100.$$

3.   **Cramer's rule**

When the determinant of coefficients Δ is not zero, we can calculate the unique solution of the system of equations (A) explicitly from the coefficients and right side, using determinants.

Fact: Cramer's Rule.   *Suppose the determinant of coefficients* Δ *of the system (A) is not zero. Let* Δ_i *be the determinant we obtain from* Δ *by replacing the entries of the ith column with the right side* $r_1, \ldots, r_n$:

$$\Delta_i = \det \begin{bmatrix} b_{11} & \cdots & b_{1i-1} & r_1 & b_{1i+1} & \cdots & b_{1n} \\ b_{21} & \cdots & b_{2i-1} & r_2 & b_{2i+1} & \cdots & b_{2n} \\ \cdot & & \cdot & \cdot & \cdot & & \cdot \\ \cdot & & \cdot & \cdot & \cdot & & \cdot \\ \cdot & & \cdot & \cdot & \cdot & & \cdot \\ b_{n1} & \cdots & b_{ni-1} & r_n & b_{ni+1} & \cdots & b_{nn} \end{bmatrix}.$$

Then the unique solution of the system (A) is given by the formulas

$$u_1 = \frac{\Delta_1}{\Delta}, \quad u_2 = \frac{\Delta_2}{\Delta}, \ldots, u_n = \frac{\Delta_n}{\Delta}.$$

Note that when $n = 2$, these are just the formulas we obtained at the beginning of this section.

EXERCISES

Calculate the determinants in Exercises 1 through 16.

1. $\det \begin{bmatrix} 1 & 2 \\ 3 & 4 \end{bmatrix}$

2. $\det \begin{bmatrix} 1 & -1 \\ 2 & 5 \end{bmatrix}$

3. $\det \begin{bmatrix} e^t & 1 \\ 1 & e^{-t} \end{bmatrix}$

4. $\det \begin{bmatrix} \sin t & \cos t \\ \cos t & -\sin t \end{bmatrix}$

5. $\det \begin{bmatrix} 0 & 1 & 0 \\ 5 & 2 & 7 \\ 5 & 9 & 3 \end{bmatrix}$

6. $\det \begin{bmatrix} 1 & 0 & -1 \\ 2 & 3 & 2 \\ 5 & 5 & 7 \end{bmatrix}$

7. $\det \begin{bmatrix} 1 & -1 & 2 \\ 2 & 3 & 2 \\ 6 & 8 & 8 \end{bmatrix}$

8. $\det \begin{bmatrix} e^t & \sin t & \cos t \\ e^t & \cos t & -\sin t \\ e^t & -\sin t & -\cos t \end{bmatrix}$

9. $\det \begin{bmatrix} e^t & e^{2t} & e^{-t} \\ e^t & 2e^{2t} & -e^{-t} \\ e^t & 4e^{2t} & e^{-t} \end{bmatrix}$

10. $\det \begin{bmatrix} 1 & 0 & 1 & 0 \\ 0 & -1 & -1 & 0 \\ 0 & 1 & 1 & -1 \\ 1 & 1 & 0 & -1 \end{bmatrix}$

11. $\det \begin{bmatrix} 1 & 2 & 1 & 3 \\ 2 & 0 & 5 & 0 \\ 0 & 1 & 3 & 7 \\ 0 & 5 & 0 & 5 \end{bmatrix}$

12. $\det \begin{bmatrix} 0 & -1 & 3 & 2 \\ 5 & 0 & 2 & 1 \\ 6 & 0 & 0 & 1 \\ 2 & 0 & 0 & 3 \end{bmatrix}$

13. $\det \begin{bmatrix} 1 & t & \cos 2t & \cos^2 t \\ 0 & 1 & -2\sin 2t & -2\cos t \sin t \\ 0 & 0 & -4\cos 2t & -2(\cos^2 t - \sin^2 t) \\ 0 & 0 & 8\sin 2t & 8\cos t \sin t \end{bmatrix}$

14. $\det \begin{bmatrix} 1 & t & t^2 & t^3 \\ 0 & 1 & 2t & 3t^2 \\ 0 & 0 & 2 & 6t \\ 0 & 0 & 0 & 6 \end{bmatrix}$

15. $\det \begin{bmatrix} 1 & 7 & -8 & 2 & 3 \\ 0 & 2 & 3 & 6 & -4 \\ 0 & 0 & 3 & 5 & 8 \\ 0 & 0 & 0 & 4 & -9 \\ 0 & 0 & 0 & 0 & 5 \end{bmatrix}$

16. $\det \begin{bmatrix} 0 & 0 & 0 & 1 & 0 \\ 1 & 3 & 2 & -3 & 3 \\ 0 & 2 & 0 & 2 & 0 \\ 1 & 5 & 3 & 5 & 4 \\ 1 & -2 & 2 & 6 & 5 \end{bmatrix}$

In each of Exercises 17 through 23, use Cramer's determinant test to determine whether the system has solutions for all values of the right side.

17. $x - y = a$
    $3x - 3y = b$

18. $x - y = a$
    $x + y = b$

19. $x + y = a$
    $3x - 3y = b$

20. $x + 2y = a$
    $2x + 4y = b$

21. $x - y + 3z = a$
    $x + y - 3z = b$
    $2x \quad - z = c$

22. $x - y + 3z = a$
    $x + y - 3z = b$
    $3x - y + 3z = c$

23. $x - y + u + v = a$
    $3x + 2y \quad = b$
    $x \quad - v = c$
    $y + 2u \quad = d$

In Exercises 24 through 27, decide whether the given system of equations has no solutions, a unique solution, or infinitely many solutions.

24. $x - y = 1$
    $3x - 3y = 3$

25. $x - y = 3$
    $3x - 3y = 1$

26. $x - y = 2$
    $x + y = 4$

27. $x + 2y = 1$
    $2x - y = 1$

In Exercises 28 through 32, use Cramer's rule (Note 3) to solve for x *only*.

28. $2x - 2y = 1$
    $x + y = 2$

29. $3x + 5y = 7$
    $x + 4y = 4$

30. $x + 2y - 3z = 1$
    $2x + 2y + 3z = 2$
    $x - y + z = 0$

31. $u + v + x = 2$
    $u - 3v + 2x = -1$
    $2u + 5v - 3x = 2$

32. $w + x + y - z = 5$
    $3w \quad + y \quad = 1$
    $w + x \quad + z = 2$
    $w \quad + y + z = 0$

Some more abstract problems:

33. Show that

$$
\det \begin{bmatrix} a_{11} & a_{12} & \cdots & a_{1n} \\ & a_{22} & \cdots & a_{2n} \\ & & \cdot & \cdot \\ & & \cdot & \cdot \\ 0\text{'s} & & \cdot & \cdot \\ & & & a_{nn} \end{bmatrix} = \det \begin{bmatrix} a_{11} & & & \\ a_{21} & a_{22} & & 0\text{'s} \\ \cdot & & \cdot & \\ \cdot & & & \cdot \\ \cdot & & & \cdot \\ a_{n1} & a_{n2} & \cdots & a_{nn} \end{bmatrix} = a_{11} a_{22} \cdots a_{nn}.
$$

34. a. Show that a factor common to all entries of a single row or column can be pulled out in front of the determinant. (*Hint:* Expand along this row or column.)

   b. Show that a determinant with a row or column of zeros must equal zero.

35. a. Show that interchanging adjacent rows (say row i and row $i + 1$) in a determinant reverses the sign. [*Hint:* Expand the original determinant along the ith row and the new one along the $(i + 1)$st row.]

   b. Show that interchanging any two rows in a determinant reverses the sign. [*Hint:* Use the result of (a) several times.]

   c. Show that a determinant with two equal rows is 0.

36. a. Suppose that each entry of the ith row of a determinant is a sum, say $a_{ij} = b_{ij} + c_{ij}$ for all j. Show that the determinant is the sum of the two determinants obtained by replacing the a_{ij}'s in that row by the b_{ij}'s and the c_{ij}'s, respectively.

   b. Use (a) together with the results of earlier exercises to show that the addition of a multiple of one row to a different row does not change the determinant.

LINEAR SYSTEMS OF DIFFERENTIAL EQUATIONS

3.1 SOME ELECTRICAL CIRCUIT MODELS

In this section we formulate some models of electrical circuits. As in Example 2.1.4, the description of these models involves several interrelated quantities. In the following sections we will learn to handle the resulting systems of o.d.e.'s directly, instead of reducing everything to a single o.d.e. This direct approach is particularly useful when dealing with more than two unknowns.

The quantities used to describe the state of an electrical circuit are currents (measured in amperes), charges (measured in coulombs), and voltage changes (measured in volts). Our variables will be currents and charges. We will derive the differential equations of our models from a voltage analysis, using the fundamental laws first formulated in the mid-nineteenth century by the German physicist G. R. Kirchhoff. **Kirchhoff's current law** states that *the total current entering any point of a circuit equals the total current leaving it.* **Kirchhoff's voltage law** states that *the sum of the voltage changes around any loop in a circuit is zero.*

The circuits we consider involve four kinds of elements, denoted symbolically as shown in Figure 3.1. For each element in a given circuit, we choose a positive direction for measuring the current through the element (we indicate our choices by arrows; for a voltage source we always choose the positive direction so that the arrow points from the − in the symbol to the +). As current flows through an element there is a voltage drop or rise (negative drop):

1. In a **resistor** the voltage drop is proportional to the current:

$$V_{\text{res}} = RI_{\text{res}}.$$

The positive constant R is called the **resistance** and is measured in ohms.

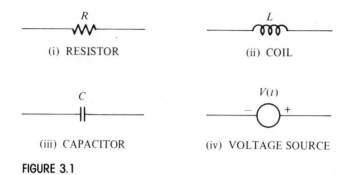

FIGURE 3.1

2. In a **coil** the voltage drop is proportional to the rate of change of the current:

$$V_{\text{coil}} = L\,\frac{dI_{\text{coil}}}{dt}.$$

The positive constant L is the **inductance** of the coil and is measured in henrys.

3. In a **capacitor** the voltage drop is proportional to the charge difference between two plates:

$$V_{\text{cap}} = \frac{1}{C}\,Q_{\text{cap}},$$

and the charge difference is related to the current by

$$I_{\text{cap}} = \frac{dQ_{\text{cap}}}{dt}.$$

The positive constant C is **capacitance** and is measured in farads.

4. A **voltage source** imposes an externally controlled (possibly varying) voltage drop

$$V_{\text{ext}} = -V(t).$$

In order to apply Kirchhoff's voltage law correctly, we have to measure the voltage changes in the loop consistently. We add the voltage changes clockwise around the loop. *If the positive direction for the current through an element agrees with the clockwise direction, we add the voltage drop as given in (1) to (4); if the directions disagree, we adjust the formula for the voltage drop by a sign change.*

Example 3.1.1   A Simple *LRC* Circuit

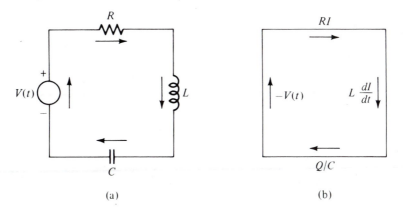

FIGURE 3.2

In Figure 3.2(a) we sketch the simplest circuit involving each of the four elements. The arrows indicate our choices for the positive directions of currents through the various elements. Kirchhoff's current law tells us that the current I is the same throughout the circuit:

$$I_{\text{res}} = I_{\text{coil}} = I_{\text{cap}} = I.$$

In Figure 3.2(b) we have sketched what relations (1) to (4) tell us about the voltage drops across the elements. If we apply Kirchhoff's voltage law, we obtain

$$0 = V_{\text{res}} + V_{\text{coil}} + V_{\text{cap}} + V_{\text{ext}}$$
$$= RI + L\frac{dI}{dt} + \frac{1}{C}Q - V(t).$$

We solve this equation for $DI = dI/dt$ and combine the result with the relation of Q to I from (3), to get the system of two o.d.e.'s

$$DQ = \qquad\qquad I$$
$$DI = \frac{-1}{LC}Q - \frac{R}{L}I + \frac{V(t)}{L}.$$

(We could, of course, substitute $I = DQ$ into the second equation to reduce this system to a single second-order o.d.e. We shall see that we can solve the system without doing this.)

Example 3.1.2   A Two-Loop Circuit

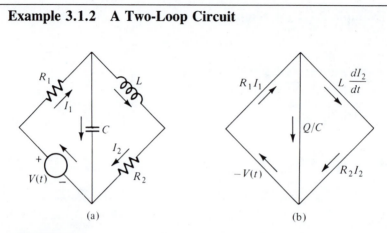

(a)   (b)

FIGURE 3.3

The wiring diagram in Figure 3.3(a) shows a circuit with two loops. The current I_1 through the left-hand resistor may differ from the current I_2 through the right-hand resistor. Kirchhoff's current law tells us that the current through the coil is $I_{coil} = I_2$ and that the current through the capacitor satisfies $I_1 = I_2 + I_{cap}$, so $I_{cap} = I_1 - I_2$.

The voltage analysis, sketched in Figure 3.3(b), leads to two equations (one for each loop):

$$R_1 I_1 + \frac{1}{C} Q - V(t) = 0$$

$$L D I_2 + R_2 I_2 - \frac{1}{C} Q = 0.$$

As before, Q is related to I_{cap} by differentiation:

$$D Q = I_{cap} = I_1 - I_2.$$

The second and third equations describe DQ and DI_2 in terms of I_1, I_2, and Q. We can solve the first equation for I_1:

$$I_1 = \frac{-1}{R_1 C} Q + \frac{V(t)}{R_1}.$$

Substituting this into the equation for DQ, we obtain the system of two o.d.e.'s

$$DQ = -\frac{1}{R_1 C} Q - I_2 + \frac{V(t)}{R_1}$$

$$DI_2 = \frac{1}{LC} Q - \frac{R_2}{L} I_2.$$

When there are several loops in a circuit, the analysis of currents can be aided by imagining a **mesh current** flowing clockwise through each loop. In the last example, we could imagine mesh currents I_1 and I_2 flowing through the left-hand and right-hand loops, respectively. *The actual current flowing through an element is the sum of the signed mesh currents for the loops containing the element; the sign is plus if the mesh current flows in the positive direction and minus otherwise.* (The reader should check that this works in the last example.)

Example 3.1.3   A Three-Loop Circuit

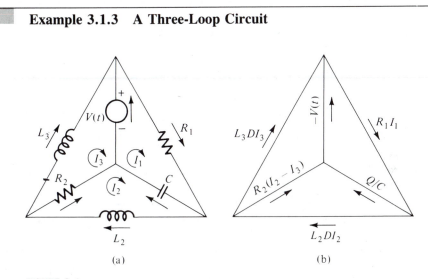

FIGURE 3.4

To handle the circuit sketched in Figure 3.4(a), we imagine a mesh current in each loop. The currents through the coils L_2 and L_3 and the resistor R_1 are equal to the mesh currents of the loops in which they lie. To obtain the currents through the elements common to two loops, we must add signed mesh currents; the current through the resistor R_2 is $I_2 - I_3$, and the current through the capacitor is $I_1 - I_2$.

The voltage analysis sketched in Figure 3.4(b) leads to three equations:

$$-V(t) + R_1 I_1 + \frac{1}{C} Q = 0$$

$$\frac{-1}{C} Q + L_2 \, DI_2 + R_2(I_2 - I_3) = 0$$

$$V(t) - R_2(I_2 - I_3) + L_3 \, DI_3 = 0.$$

We also have the equation for DQ:

$$DQ = I_1 - I_2.$$

The first voltage relation can be solved for I_1 to yield

$$I_1 = \frac{-1}{R_1 C} Q + \frac{V(t)}{R_1}.$$

Substituting this into the equation for DQ and combining with the second and third voltage relations, we come up with the system of three o.d.e.'s in Q, I_2, and I_3:

$$DQ = \frac{-1}{R_1 C} Q - I_2 \qquad\qquad + \frac{V(t)}{R_1}$$

$$DI_2 = \frac{1}{L_2 C} Q - \frac{R_2}{L_2} I_2 + \frac{R_2}{L_2} I_3$$

$$DI_3 = \qquad\qquad \frac{R_2}{L_3} I_2 - \frac{R_2}{L_3} I_3 - \frac{V(t)}{L_3}.$$

Example 3.1.4   Four Loops

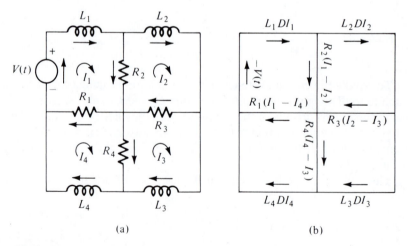

(a)   (b)

FIGURE 3.5

The complicated circuit of Figure 3.5(a) can be decomposed into four loops, with clockwise mesh currents I_j, $j = 1, 2, 3, 4$. Here there are no capacitors, hence no charges. The voltage analysis, sketched in Figure 3.5(b), leads directly to the system of four o.d.e.'s

$$L_1 DI_1 = -(R_1 + R_2)I_1 + \qquad R_2 I_2 + \qquad\qquad\qquad R_1 I_4 + V(t)$$

$$L_2 DI_2 = \qquad R_2 I_1 - (R_2 + R_3)I_2 + \qquad R_3 I_3$$

$$L_3 DI_3 = \qquad\qquad\qquad R_3 I_2 - (R_3 + R_4)I_3 + \qquad R_4 I_4$$

$$L_4 DI_4 = \qquad R_1 I_1 + \qquad\qquad\qquad R_4 I_3 - (R_1 + R_4)I_4$$

EXERCISES

Circuits: In Exercises 1 to 5, find a system of two first-order o.d.e.'s modeling the circuit of the indicated figure.

1. Figure 3.6; $R = 2$ ohms, $L_1 = L_2 = 1$ henry, and $V(t) = 4e^{-4t}$ volts.
2. Figure 3.7; $R_1 = R_2 = 4$ ohms, $L = 2$ henrys, $C = 2$ farads, and $V(t) = 4e^{-t/2}$ volts.
3. Figure 3.7; $R_1 = R_2 = 4$ ohms, $L = 2$ henrys, $C = 1$ farad, and $V(t) = 4e^{-t/2}$ volts.

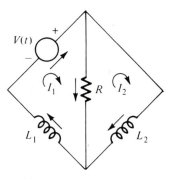

FIGURE 3.6

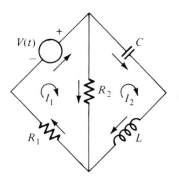

FIGURE 3.7

4. Figure 3.8; $R_1 = R_2 = 1$ ohm, $L = 2$ henrys, $C = 1$ farad, and $V(t) = 4e^{-t/2}$ volts.

5. Figure 3.9; $R_1 = R_2 = 1$ ohm, $C_1 = 3$ and $C_2 = 2$ farads, and $V(t) = 5$ volts.

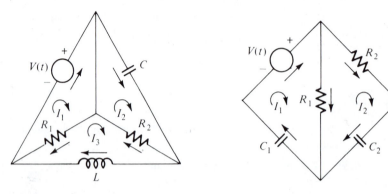

FIGURE 3.8   FIGURE 3.9

Other Models.:

6. *Supply and Demand:* The model in Exercise 6, Section 2.1, is most naturally modeled as a system of two first-order o.d.e.'s for price and supply. Find this system.

7. *Radioactive Decay:* A radioactive substance, A, decays into substance B. Substance B, although more stable than A, nevertheless decays into the stable substance C.
   a. Write a system of three first-order o.d.e.'s modeling the amounts of substances A, B, and C under the assumption that each of substances A and B decays at a rate proportional to the amount of that substance present.
   b. Do the same under the additional assumption that α grams per year of substance A are steadily added to the mixture while γ grams per year of substance C are steadily extracted.
   (*Note:* Assume that 1 gram of substance A decomposes into 1 gram of B, and 1 gram of B gives 1 gram of C.)

8. *Diffusion I:* When two solutions of a substance are separated by a permeable membrane, the *amount* of the substance that crosses the membrane in a given time is proportional to the difference between the *concentrations* of the neighboring solutions. The constant of proportionality is called the *permeability* of the membrane. Write a system of two first-order equations modeling the change in concentrations x_1 and x_2 of two saline solutions separated by a membrane with permeability P, assuming
   a. there are equal volumes $V_1 = V_2 = V$ of liquid on the two sides of the membrane.
   b. the two unequal volumes are V_1 and V_2, respectively.

9. *Diffusion II:* An organ has a double wall. The outer wall has a permeability P_1 to glucose, and the permeability of the inner wall is $0 < P_2 < P_1$. Let v and V denote, respectively, the (constant) volume between the two walls and the volume inside the organ. Denote the concentration of glucose in volume v by x and in volume V by y. The blood surrounding the organ has a steady concentration G of glucose. Write two first-order equations in x and y to model the distribution of glucose in the organ.

10. *Political Coexistence:* Two neighboring countries, Camponesa and Mandachuva, have strikingly different economic and social structures. Camponesa is ruled by a socialist military front that is nationalizing all large corporations and developing various social welfare programs. Mandachuva, ruled by a right-wing junta, is experiencing rapid economic growth, although unemployment and inflation are high. There is a net annual per capita birth rate in Camponesa of 15%, but the middle class is becoming uncomfortable with political developments and 4% of the population moves to Mandachuva each year. On the other hand, Mandachuva has an annual net per capita birth rate of 10%, but 3% of the population (unemployed laborers and leftist intellectuals) annually move to Camponesa. Write a system of two different equations modeling the demographic changes in Mandachuva and Camponesa. (Compare with Example 1.1.7.)

3.2 LINEAR SYSTEMS, MATRICES, AND VECTORS

In this section we introduce notation that helps us recognize parallels between the solutions of a single linear o.d.e. and the solutions of a linear system of o.d.e.'s. By the end of the chapter we hope you will agree that in this case a good notation, like a picture, is worth a thousand words.

A system of o.d.e.'s is **linear** if it can be written in the form

$$
\begin{aligned}
x_1' &= a_{11}x_1 + a_{12}x_2 + \cdots + a_{1n}x_n + E_1(t) \\
x_2' &= a_{21}x_1 + a_{22}x_2 + \cdots + a_{2n}x_n + E_2(t) \\
&\qquad\qquad\qquad \vdots \\
x_n' &= a_{n1}x_1 + a_{n2}x_2 + \cdots + a_{nn}x_n + E_n(t).
\end{aligned}
$$

(S)

The **coefficients** a_{ij} may be constants or functions of t. (As with a single o.d.e., we pay most attention to the constant-coefficient case.) The number n of unknowns is the **order** of the system. We will call the system **homogeneous** if $E_1(t) = E_2(t) = \cdots = E_n(t) = 0$.

The systems in Examples 3.1.1 to 3.1.4 were all linear. We will solve these systems (for specific choices of the resistances, inductances, capacitances, and voltages) in the examples and exercises of later sections. Linear systems that we can solve immediately can be obtained from single linear o.d.e.'s.

Example 3.2.1

Consider the second-order equation

(N) $$(D^2 - 2D - 3)x = 6 - 8e^t.$$

Suppose $x = x(t)$ is a solution of (N), and set

$$x_1 = x, \quad x_2 = x'.$$

Then we have

$$x_1' = x' = x_2,$$

and, since x satisfies (N),

$$x_2' = x'' = 2x' + 3x + 6 - 8e^t = 2x_2 + 3x_1 + 6 - 8e^t.$$

In other words, x_1 and x_2 satisfy the second-order system

(S$_N$)
$$x_1' = \quad\quad x_2$$
$$x_2' = 3x_1 + 2x_2 + 6 - 8e^t.$$

Conversely, if x_1 and x_2 satisfy (S$_N$), then

$$(D^2 - 2D - 3)x_1 = x_1'' - 2x_1' - 3x_1 = x_2' - 2x_2 - 3x_1 = 6 - 8e^t.$$

That is, $x = x_1$ satisfies (N). Thus, solutions of (N) yield solutions of (S$_N$), and vice versa.

We found the general solution of (N) in Example 2.7.1:

$$x = c_1 e^{3t} + c_2 e^{-t} - 2 + 2e^t.$$

The corresponding general solution of (S$_N$) is

$$x_1 = x = c_1 e^{3t} + c_2 e^{-t} - 2 + 2e^t$$
$$x_2 = x' = 3c_1 e^{3t} - c_2 e^{-t} \quad\quad + 2e^t.$$

Example 3.2.2

The process we used in the preceding example can be used to replace the nth-order linear equation

(N) $$(D^n + a_{n-1}D^{n-1} + \cdots + a_1D + a_0)x = E(t)$$

by an equivalent system. In this case, we introduce n unknowns

$$x_1 = x, \quad x_2 = x', \quad x_3 = x'', \ldots, x_n = x^{(n-1)}.$$

The single equation (N) is equivalent to the system

$$
\begin{aligned}
x_1' &= x_2 \\
x_2' &= x_3 \\
&\quad\cdot \\
(S_N) &\quad\cdot \\
&\quad\cdot \\
x_{n-1}' &= x_n \\
x_n' &= -a_0x_1 - a_1x_2 - \cdots - a_{n-1}x_n + E(t).
\end{aligned}
$$

Note that the order of (S_N) is the same as the order of (N) and that (S_N) is homogeneous precisely when (N) is.

The right side of a linear system (S) is determined by its coefficients and the n functions $E_1(t), \ldots, E_n(t)$. A solution of (S) consists of n functions $x_1(t), x_2(t), \ldots, x_n(t)$. Thus, we will be dealing with "arrays" of constants and functions. We need suitable notation and terminology for these arrays.

An $n \times m$ **matrix** is an array of the form

$$
B = \begin{bmatrix}
b_{11} & b_{12} & \cdots & b_{1m} \\
b_{21} & b_{22} & \cdots & b_{2m} \\
& & \cdot & \\
& & \cdot & \\
& & \cdot & \\
b_{n1} & b_{n2} & \cdots & b_{nm}
\end{bmatrix}.
$$

Note that an $n \times m$ matrix has n rows and m columns. The location of each **entry** b_{ij} is specified by two subscripts; the first indicates the row in which b_{ij} appears, and the second indicates the column. The entries may be constants or functions.

We refer to $n \times 1$ matrices with constant entries as **n-vectors**. Vectors are indicated by boldface letters. We use **0** to denote the n-vector whose entries are all 0.

An $n \times 1$ matrix whose entries are functions is an **n-vector valued function.** A vector valued function assigns a vector to each specific value of t. For example, the vector valued function

$$\mathbf{x}(t) = \begin{bmatrix} \sin t \\ 3t - 5 \\ t^2 + 2 \end{bmatrix}$$

assigns the 3-vector

$$\mathbf{x}(0) = \begin{bmatrix} 0 \\ -5 \\ 2 \end{bmatrix}$$

to the value $t = 0$.

We say that two $n \times m$ matrices B and C are **equal** provided each entry b_{ij} of B is equal to the corresponding entry c_{ij} of C. For example, the matrix equality

$$\begin{bmatrix} x_1 \\ x_2 \\ x_3 \end{bmatrix} = \begin{bmatrix} \sin t \\ 3t - 5 \\ t^2 + 2 \end{bmatrix}$$

means

$$x_1 = \sin t, \quad x_2 = 3t - 5, \quad \text{and} \quad x_3 = t^2 + 2.$$

The matrix equality

$$\begin{bmatrix} y_1 \\ y_2 \\ y_3 \end{bmatrix} = \mathbf{0}$$

means

$$y_1 = y_2 = y_3 = 0.$$

We obtain the **sum of two $n \times m$ matrices** by adding corresponding entries:

$$\begin{bmatrix} a_{11} & a_{12} & \cdots & a_{1m} \\ a_{21} & a_{22} & \cdots & a_{2m} \\ \vdots & \vdots & & \vdots \\ a_{n1} & a_{n2} & \cdots & a_{nm} \end{bmatrix} + \begin{bmatrix} b_{11} & b_{12} & \cdots & b_{1m} \\ b_{21} & b_{22} & \cdots & b_{2m} \\ \vdots & \vdots & & \vdots \\ b_{n1} & b_{n2} & \cdots & b_{nm} \end{bmatrix}$$

$$= \begin{bmatrix} (a_{11} + b_{11}) & (a_{12} + b_{12}) & \cdots & (a_{1m} + b_{1m}) \\ (a_{21} + b_{21}) & (a_{22} + b_{22}) & \cdots & (a_{2m} + b_{2m}) \\ \vdots & \vdots & & \vdots \\ (a_{n1} + b_{n1}) & (a_{n2} + b_{n2}) & \cdots & (a_{nm} + b_{nm}) \end{bmatrix}.$$

We obtain the **product of a number and an $n \times m$ matrix** by multiplying each entry of the matrix by the number:

$$
c \begin{bmatrix}
a_{11} & a_{12} & \cdots & a_{1m} \\
a_{21} & a_{22} & \cdots & a_{2m} \\
\vdots & \vdots & & \vdots \\
a_{n1} & a_{n2} & \cdots & a_{nm}
\end{bmatrix}
=
\begin{bmatrix}
ca_{11} & ca_{12} & \cdots & ca_{1m} \\
ca_{21} & ca_{22} & \cdots & ca_{2m} \\
\vdots & \vdots & & \vdots \\
ca_{n1} & ca_{n2} & \cdots & ca_{nm}
\end{bmatrix}.
$$

For example,

$$
-2 \begin{bmatrix} 1 & 2 \\ 0 & 3 \end{bmatrix}
+ 3 \begin{bmatrix} 4 & -1 \\ 1 & 0 \end{bmatrix}
= \begin{bmatrix} -2 & -4 \\ 0 & -6 \end{bmatrix}
+ \begin{bmatrix} 12 & -3 \\ 3 & 0 \end{bmatrix}
= \begin{bmatrix} 10 & -7 \\ 3 & -6 \end{bmatrix}
$$

and

$$
3 \begin{bmatrix} 0 \\ 1 \\ 2 \end{bmatrix}
+ \begin{bmatrix} 1 \\ 3 \\ 1 \end{bmatrix}
- 2 \begin{bmatrix} -2 \\ 0 \\ 1 \end{bmatrix}
= \begin{bmatrix} 0 \\ 3 \\ 6 \end{bmatrix}
+ \begin{bmatrix} 1 \\ 3 \\ 1 \end{bmatrix}
+ \begin{bmatrix} 4 \\ 0 \\ -2 \end{bmatrix}
= \begin{bmatrix} 5 \\ 6 \\ 5 \end{bmatrix}.
$$

We define the **product of an $n \times m$ matrix and an m-vector** by the rule

$$
\begin{bmatrix}
a_{11} & a_{12} & \cdots & a_{1m} \\
a_{21} & a_{22} & \cdots & a_{2m} \\
\vdots & \vdots & & \vdots \\
a_{n1} & a_{n2} & \cdots & a_{nm}
\end{bmatrix}
\begin{bmatrix} x_1 \\ x_2 \\ \vdots \\ x_m \end{bmatrix}
=
\begin{bmatrix}
a_{11}x_1 + a_{12}x_2 + \cdots + a_{1m}x_m \\
a_{21}x_1 + a_{22}x_2 + \cdots + a_{2m}x_m \\
\vdots \\
a_{n1}x_1 + a_{n2}x_2 + \cdots + a_{nm}x_m
\end{bmatrix}.
$$

Note that the product is an n-vector. The first (top) entry of the product is obtained by multiplying the entries of the first row of the matrix by the corresponding entries of the vector and then summing the products. Similarly, the ith entry of the product is obtained by "multiplying" the ith row of the matrix by the vector. For example,

$$
\begin{bmatrix}
3 & 2 & -1 \\
4 & 3 & 5 \\
0 & 1 & 0
\end{bmatrix}
\begin{bmatrix} 1 \\ 2 \\ -2 \end{bmatrix}
=
\begin{bmatrix}
3(1) + 2(2) + (-1)(-2) \\
4(1) + 3(2) + 5(-2) \\
0(1) + 1(2) + 0(-2)
\end{bmatrix}
=
\begin{bmatrix} 9 \\ 0 \\ 2 \end{bmatrix}.
$$

We define the **derivative of a vector valued function** by the rule

$$
D \begin{bmatrix} x_1(t) \\ x_2(t) \\ \cdot \\ \cdot \\ \cdot \\ x_n(t) \end{bmatrix} = \begin{bmatrix} x_1'(t) \\ x_2'(t) \\ \cdot \\ \cdot \\ \cdot \\ x_n'(t) \end{bmatrix}.
$$

For example,

$$
D \begin{bmatrix} \sin t \\ 3t - 5 \\ t^2 + 2 \end{bmatrix} = \begin{bmatrix} \cos t \\ 3 \\ 2t \end{bmatrix}.
$$

The equations that made up our system (S) can be replaced by an equivalent matrix equality, which we can simplify using the notions of vector addition and matrix multiplication:

$$
\begin{bmatrix} x_1' \\ x_2' \\ \cdot \\ \cdot \\ \cdot \\ x_n' \end{bmatrix} = \begin{bmatrix} a_{11}x_1 + a_{12}x_2 + \cdots + a_{1n}x_n + E_1(t) \\ a_{21}x_1 + a_{22}x_2 + \cdots + a_{2n}x_n + E_2(t) \\ \cdot \\ \cdot \\ \cdot \\ a_{n1}x_1 + a_{n2}x_2 + \cdots + a_{nn}x_n + E_n(t) \end{bmatrix}
$$

$$
= \begin{bmatrix} a_{11}x_1 + \cdots + a_{1n}x_n \\ a_{21}x_1 + \cdots + a_{2n}x_n \\ \cdot \\ \cdot \\ \cdot \\ a_{n1}x_1 + \cdots + a_{nn}x_n \end{bmatrix} + \begin{bmatrix} E_1(t) \\ E_2(t) \\ \cdot \\ \cdot \\ \cdot \\ E_n(t) \end{bmatrix}
$$

$$
= \begin{bmatrix} a_{11} & a_{12} & \cdots & a_{1n} \\ a_{21} & a_{22} & \cdots & a_{2n} \\ \cdot & \cdot & & \cdot \\ \cdot & \cdot & & \cdot \\ \cdot & \cdot & & \cdot \\ a_{n1} & a_{n2} & \cdots & a_{nn} \end{bmatrix} \begin{bmatrix} x_1 \\ x_2 \\ \cdot \\ \cdot \\ \cdot \\ x_n \end{bmatrix} + \begin{bmatrix} E_1(t) \\ E_2(t) \\ \cdot \\ \cdot \\ \cdot \\ E_n(t) \end{bmatrix}.
$$

If we set

$$\mathbf{x} = \begin{bmatrix} x_1 \\ x_2 \\ \cdot \\ \cdot \\ \cdot \\ x_n \end{bmatrix}, \quad A = \begin{bmatrix} a_{11} & a_{12} & \cdots & a_{1n} \\ a_{21} & a_{22} & \cdots & a_{2n} \\ \cdot & \cdot & & \cdot \\ \cdot & \cdot & & \cdot \\ \cdot & \cdot & & \cdot \\ a_{n1} & a_{n2} & \cdots & a_{nn} \end{bmatrix}, \quad \text{and} \quad \mathbf{E}(t) = \begin{bmatrix} E_1(t) \\ E_2(t) \\ \cdot \\ \cdot \\ \cdot \\ E_n(t) \end{bmatrix}$$

then our system (S) can be written in matrix form as

$$D\mathbf{x} = A\mathbf{x} + \mathbf{E}(t).$$

Example 3.2.3

The matrix form of the system

(S)
$$\begin{aligned} x_1' &= x_2 \\ x_2' &= 3x_1 + 2x_2 + 6 - 8e^t \end{aligned}$$

from Example 3.2.1 is $D\mathbf{x} = A\mathbf{x} + \mathbf{E}(t)$, where

$$\mathbf{x} = \begin{bmatrix} x_1 \\ x_2 \end{bmatrix}, \quad A = \begin{bmatrix} 0 & 1 \\ 3 & 2 \end{bmatrix}, \quad \text{and} \quad \mathbf{E}(t) = \begin{bmatrix} 0 \\ 6 - 8e^t \end{bmatrix}.$$

The general solution of (S) can be written in matrix form as

$$\mathbf{x} = \begin{bmatrix} c_1 e^{3t} + c_2 e^{-t} - 2 + 2e^t \\ 3c_1 e^{3t} - c_2 e^{-t} + 2e^t \end{bmatrix} = \begin{bmatrix} c_1 e^{3t} \\ 3c_1 e^{3t} \end{bmatrix} + \begin{bmatrix} c_2 e^{-t} \\ -c_2 e^{-t} \end{bmatrix} + \begin{bmatrix} -2 + 2e^t \\ 2e^t \end{bmatrix}$$

$$= c_1 \begin{bmatrix} e^{3t} \\ 3e^{3t} \end{bmatrix} + c_2 \begin{bmatrix} e^{-t} \\ -e^{-t} \end{bmatrix} + \begin{bmatrix} -2 + 2e^t \\ 2e^t \end{bmatrix}.$$

If we set

$$\mathbf{h}_1(t) = \begin{bmatrix} e^{3t} \\ 3e^{3t} \end{bmatrix}, \quad \mathbf{h}_2(t) = \begin{bmatrix} e^{-t} \\ -e^{-t} \end{bmatrix}, \quad \text{and} \quad \mathbf{p}(t) = \begin{bmatrix} -2 + 2e^t \\ 2e^t \end{bmatrix}$$

then the general solution of (S_N) is

$$\mathbf{x} = c_1 \mathbf{h}_1(t) + c_2 \mathbf{h}_2(t) + \mathbf{p}(t).$$

Note that $\mathbf{x} = \mathbf{p}(t)$ is a particular solution of (S_N). We leave it to you to check further (Exercise 18) that the general solution of the homogeneous system $D\mathbf{x} = A\mathbf{x}$ is $\mathbf{x} = c_1 \mathbf{h}_1(t) + c_2 \mathbf{h}_2(t)$. Note all the parallels between the solution of the system and the solution of the o.d.e.!

Example 3.2.4

The matrix form of the system (S_N) in Example 3.2.2 is $D\mathbf{x} = A\mathbf{x} + \mathbf{E}(t)$, where

$$\mathbf{x} = \begin{bmatrix} x_1 \\ x_2 \\ \cdot \\ \cdot \\ \cdot \\ x_{n-1} \\ x_n \end{bmatrix}, \quad A = \begin{bmatrix} 0 & 1 & 0 & \cdots & 0 \\ 0 & 0 & 1 & \cdots & 0 \\ \cdot & \cdot & \cdot & & \cdot \\ \cdot & \cdot & \cdot & & \cdot \\ \cdot & \cdot & \cdot & & \cdot \\ 0 & 0 & 0 & \cdots & 1 \\ -a_0 & -a_1 & -a_2 & \cdots & -a_{n-1} \end{bmatrix},$$

and

$$\mathbf{E}(t) = \begin{bmatrix} 0 \\ 0 \\ \cdot \\ \cdot \\ \cdot \\ 0 \\ E(t) \end{bmatrix}.$$

Example 3.2.5

The matrix form of the system in Example 3.1.3 is $D\mathbf{x} = A\mathbf{x} + \mathbf{E}(t)$, where

$$\mathbf{x} = \begin{bmatrix} Q \\ I_2 \\ I_3 \end{bmatrix}, \quad A = \begin{bmatrix} \dfrac{-1}{R_1 C} & -1 & 0 \\ \dfrac{1}{L_2 C} & -\dfrac{R_2}{L_2} & \dfrac{R_2}{L_2} \\ 0 & \dfrac{R_2}{L_3} & -\dfrac{R_2}{L_3} \end{bmatrix}, \quad \text{and} \quad \mathbf{E}(t) = \begin{bmatrix} \dfrac{V(t)}{R_1} \\ 0 \\ -\dfrac{V(t)}{L_3} \end{bmatrix}.$$

Example 3.2.6

Determine whether the vector valued functions

$$\mathbf{x}_1(t) = \begin{bmatrix} e^{2t} \\ 0 \end{bmatrix}, \quad \text{and} \quad \mathbf{x}_2(t) = \begin{bmatrix} e^t \\ 0 \end{bmatrix}$$

are solutions of $D\mathbf{x} = A\mathbf{x} + \mathbf{E}(t)$ where

$$A = \begin{bmatrix} 0 & 2 \\ -1 & 3 \end{bmatrix}, \quad \text{and} \quad \mathbf{E}(t) = \begin{bmatrix} e^t \\ e^t \end{bmatrix}.$$

We calculate

$$Dx_1(t) = \begin{bmatrix} 2e^{2t} \\ 0 \end{bmatrix}$$

and

$$Ax_1(t) + E(t) = \begin{bmatrix} 0 & 2 \\ -1 & 3 \end{bmatrix}\begin{bmatrix} e^{2t} \\ 0 \end{bmatrix} + \begin{bmatrix} e^t \\ e^t \end{bmatrix} = \begin{bmatrix} 0 \\ -e^{2t} \end{bmatrix} + \begin{bmatrix} e^t \\ e^t \end{bmatrix} = \begin{bmatrix} e^t \\ e^t - e^{2t} \end{bmatrix}.$$

Since these are unequal, $x_1(t)$ is not a solution of $Dx = Ax + E(t)$.
On the other hand,

$$Dx_2(t) = \begin{bmatrix} e^t \\ 0 \end{bmatrix}$$

and

$$Ax_1(t) + E(t) = \begin{bmatrix} 0 & 2 \\ -1 & 3 \end{bmatrix}\begin{bmatrix} e^t \\ 0 \end{bmatrix} + \begin{bmatrix} e^t \\ e^t \end{bmatrix} = \begin{bmatrix} 0 \\ -e^t \end{bmatrix} + \begin{bmatrix} e^t \\ e^t \end{bmatrix} = \begin{bmatrix} e^t \\ 0 \end{bmatrix}.$$

Since these are equal, x_2 is a solution of $Dx = Ax + E(t)$.

The following summary and notes include lists of important rules of matrix algebra and of differentiation of vector valued functions. We invite you to verify these rules (see Note 3 and Exercises 29 to 31).

LINEAR SYSTEMS, MATRICES, AND VECTORS

An $n \times m$ **matrix** is an array of the form

$$B = \begin{bmatrix} b_{11} & \cdots & b_{1m} \\ \cdot & & \cdot \\ \cdot & & \cdot \\ \cdot & & \cdot \\ b_{n1} & \cdots & b_{nm} \end{bmatrix}$$

The b_{ij}'s are called the **entries** of B. An $n \times 1$ matrix with constant entries is called an **n-vector**. We denote the n-vector all of whose entries are 0 by **0**. An $n \times 1$ matrix whose entries are functions is an **n-vector valued function**.

The **sum of two $n \times m$ matrices** is the $n \times m$ matrix whose entries are obtained by adding the corresponding entries of the two matrices. The **product of a number and an $n \times m$ matrix** is the new $n \times m$ matrix obtained by multiplying each entry of the old matrix by the number.

The **product of an $n \times m$ matrix and an m-vector** is the new n-vector given by the rule

$$\begin{bmatrix} a_{11} & \cdots & a_{1m} \\ \cdot & & \cdot \\ \cdot & & \cdot \\ \cdot & & \cdot \\ a_{n1} & \cdots & a_{nm} \end{bmatrix} \begin{bmatrix} x_1 \\ \cdot \\ \cdot \\ \cdot \\ x_m \end{bmatrix} = \begin{bmatrix} a_{11}x_1 + \cdots + a_{1m}x_m \\ \cdot \\ \cdot \\ \cdot \\ a_{n1}x_1 + \cdots + a_{nm}x_m \end{bmatrix}.$$

If $\mathbf{v}$ and $\mathbf{w}$ are m-vectors. A is an $n \times m$ matrix, and c is a number, then

(M1) $$A(\mathbf{v} + \mathbf{w}) = A\mathbf{v} + A\mathbf{w}$$

(M2) $$A(c\mathbf{v}) = c(A\mathbf{v}) = (cA)\mathbf{v}.$$

The **derivative of an n-vector valued function** is the new n-vector valued function obtained by differentiating each of the entries of the old one. If $\mathbf{x}$ and $\mathbf{y}$ are n-vector valued functions and c is a number, then

(D1) $$D(\mathbf{x} + \mathbf{y}) = D\mathbf{x} + D\mathbf{y}$$

(D2) $$D(c\mathbf{x}) = c(D\mathbf{x}).$$

A system of o.d.e.'s is **linear** if it can be written in the form

$$x_1' = a_{11}x_1 + \cdots + a_{1n}x_n + E_1(t)$$
$$\cdot$$
$$\cdot$$
$$\cdot$$
$$x_n' = a_{n1}x_1 + \cdots + a_{nn}x_n + E_n(t)$$

where the **coefficients** a_{ij} may be constants or functions of t. The matrix form of this system is

$$D\mathbf{x} = A\mathbf{x} + \mathbf{E}(t)$$

where

$$\mathbf{x} = \begin{bmatrix} x_1 \\ \cdot \\ \cdot \\ \cdot \\ x_n \end{bmatrix}, \quad A = \begin{bmatrix} a_{11} & \cdots & a_{1n} \\ \cdot & & \cdot \\ \cdot & & \cdot \\ \cdot & & \cdot \\ a_{n1} & \cdots & a_{nn} \end{bmatrix}, \quad \text{and} \quad \mathbf{E}(t) = \begin{bmatrix} E_1(t) \\ \cdot \\ \cdot \\ \cdot \\ E_n(t) \end{bmatrix}.$$

The **order** of the system is n. The system is **homogeneous** if $\mathbf{E}(t) = \mathbf{0}$.

Notes

1. Geometry and vectors

You may have already encountered vectors in the plane and in space as representations of forces or velocities in a physics course, or as an analytic device in multivariate calculus. Usually, we think of these as arrows that are free to move in position so long as they maintain the same direction and length. In the plane, such arrows can be described by their x and y components and are often written in the form $x\mathbf{i} + y\mathbf{j}$. If we identify $\mathbf{i}$ and $\mathbf{j}$, respectively, with the 2×1 matrices

$$\mathbf{i} = \begin{bmatrix} 1 \\ 0 \end{bmatrix} \quad \text{and} \quad \mathbf{j} = \begin{bmatrix} 0 \\ 1 \end{bmatrix},$$

then any 2-vector in our sense, written

$$\mathbf{v} = \begin{bmatrix} x \\ y \end{bmatrix},$$

corresponds uniquely to a free arrow in the plane, written

$$x\mathbf{i} + y\mathbf{j},$$

and vice versa. The pair of numbers in the matrix $\mathbf{v}$ can also be regarded as the coordinates of a point $P(x, y)$. The three interpretations are related as follows: The matrix $\mathbf{v}$ corresponds to the free arrow $x\mathbf{i} + y\mathbf{j}$ that, when moved so its tail is at the origin, has its head at the point $P(x, y)$ (see Figure 3.10). This visualization can be applied to vector operations: $c\mathbf{v}$ is an arrow with the same or opposite direction as $\mathbf{v}$ (depending on the sign of the number c) and with length $|c|$ times that of $\mathbf{v}$ [Figure 3.11(a)]; and $\mathbf{v} + \mathbf{w}$ is the diagonal of the parallelogram determined by $\mathbf{v}$ and $\mathbf{w}$ [Figure 3.11(b)].

A similar correspondence can be set up between 3-vectors

$$\mathbf{v} = \begin{bmatrix} x \\ y \\ z \end{bmatrix},$$

arrows, and points (x, y, z) in 3-space. The visualization of vector operations in 3-space is just like that for 2-vectors.

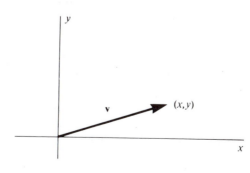

FIGURE 3.10

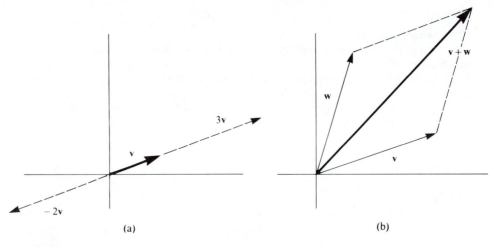

(a)   (b)

FIGURE 3.11

We can use our visualization of 2- and 3-vectors as arrows to picture vector valued functions $\mathbf{x} = \mathbf{x}(t)$. We regard $\mathbf{x}$ as a function that assigns to each time t an arrow $\mathbf{x}(t)$ from the origin. As time varies, the tip of the arrow will move along a path (Figure 3.12). We think of $\mathbf{x} = \mathbf{x}(t)$ as a parametrized curve describing the path.

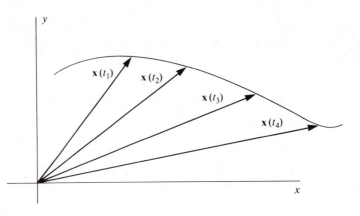

FIGURE 3.12

As an example, let's consider the 2-vector valued function

$$\mathbf{x}(t) = \begin{bmatrix} e^{2t} \\ -e^{4t} \end{bmatrix}.$$

Here $y = -e^{4t} = -(e^{2t})^2 = -x^2$. Thus, the path traced by $\mathbf{x}(t)$ lies on the parabola $y = -x^2$. Since $x = e^{2t} > 0$, we are only interested in the part of the parabola that lies to the right of the y axis. We have sketched the path in Figure 3.13. The arrow on the curve indicates the direction in which the point $\mathbf{x}(t)$ moves as t increases.

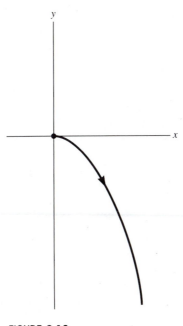

FIGURE 3.13

We can add $D\mathbf{x}$ to our picture as an arrow that points from the position $\mathbf{x}(t)$ on the path. Since the components of $D\mathbf{x}$ are the rates at which the components of $\mathbf{x}$ are changing, $D\mathbf{x}$ points in the direction of the motion (Figure 3.14). What's more, the length of $D\mathbf{x}$ tells us the speed at which $\mathbf{x}$ moves along the path. In this way, we can regard a system $D\mathbf{x} = A\mathbf{x} + \mathbf{E}(t)$ as relating the velocity $D\mathbf{x}$ of a moving object to its position $\mathbf{x}(t)$. A solution of the system is a parametrized curve describing the path of the object.

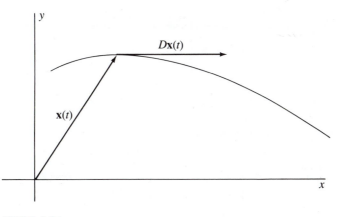

FIGURE 3.14

2. The laws of matrix algebra

Properties (M1) and (M2) in the summary show some analogies between the algebra of matrices and the algebra of numbers. In this note we list some further properties of matrices.

We begin with a list of laws satisfied by matrix addition and by multiplication of a number times a matrix. In this list, O represents the $n \times m$ matrix, all of whose entries are 0:

$$O = \begin{bmatrix} 0 & \cdots & 0 \\ \cdot & & \cdot \\ \cdot & & \cdot \\ \cdot & & \cdot \\ 0 & \cdots & 0 \end{bmatrix}.$$

This matrix is called the $n \times m$ **zero matrix**.

Fact: *If A, B, and C are $n \times m$ matrices, and if c and d are numbers, then*

(A1) $$(A + B) + C = A + (B + C)$$

(A2) $$A + B = B + A$$

(A3) $$A + O = A$$

(A4) $$A + (-1)A = O$$

(A5) $$c(A + B) = cA + cB$$

(A6) $$(c + d)A = cA + dA$$

(A7) $$c(dA) = (cd)A$$

(A8) $$1A = A.$$

Multiplication of a matrix and a vector satisfies the following list of laws. In this list I denotes the $m \times m$ matrix whose entries a_{ij} are 0 if $i \neq j$ and 1 if $i = j$:

$$I = \begin{bmatrix} 1 & 0 & \cdots & 0 \\ 0 & 1 & \cdots & 0 \\ \cdot & \cdot & & \cdot \\ \cdot & \cdot & & \cdot \\ \cdot & \cdot & & \cdot \\ 0 & 0 & \cdots & 1 \end{bmatrix}.$$

This matrix is called the $m \times m$ **identity matrix**.

Fact: *Suppose A and B are $n \times m$ matrices, **v** and **w** are m-vectors, and c is a number. Then*

(M1) $A(\mathbf{v} + \mathbf{w}) = A\mathbf{v} + A\mathbf{w}$ (M2) $A(c\mathbf{v}) = c(A\mathbf{v}) = (cA)\mathbf{v}$

(M3) $(A + B)\mathbf{v} = A\mathbf{v} + B\mathbf{v}$ (M4) $I\mathbf{v} = \mathbf{v}$

(M5) $A\mathbf{0} = \mathbf{0}$ (M6) $O\mathbf{v} = \mathbf{0}.$

3. Verifying the properties of matrix operations

The verification of properties (M1) to (M6), (A1) to (A8), and (D1) and (D2) follows a pattern that we illustrate by verifying (M1). Let

$$A = \begin{bmatrix} a_{11} & \cdots & a_{1m} \\ \cdot & & \cdot \\ \cdot & & \cdot \\ \cdot & & \cdot \\ a_{n1} & \cdots & a_{nm} \end{bmatrix}, \quad \mathbf{v} = \begin{bmatrix} v_1 \\ \cdot \\ \cdot \\ \cdot \\ v_m \end{bmatrix}, \quad \text{and} \quad \mathbf{w} = \begin{bmatrix} w_1 \\ \cdot \\ \cdot \\ \cdot \\ w_m \end{bmatrix}.$$

Then

$$A(\mathbf{v} + \mathbf{w}) = \begin{bmatrix} a_{11} & \cdots & a_{1m} \\ \cdot & & \cdot \\ \cdot & & \cdot \\ \cdot & & \cdot \\ a_{n1} & \cdots & a_{nm} \end{bmatrix} \begin{bmatrix} v_1 + w_1 \\ \cdot \\ \cdot \\ \cdot \\ v_m + w_m \end{bmatrix}$$

$$= \begin{bmatrix} a_{11}(v_1 + w_1) + \cdots + a_{1m}(v_m + w_m) \\ \cdot \\ \cdot \\ \cdot \\ a_{n1}(v_1 + w_1) + \cdots + a_{nm}(v_m + w_m) \end{bmatrix}$$

$$= \begin{bmatrix} (a_{11}v_1 + \cdots + a_{1m}v_m) + (a_{11}w_1 + \cdots + a_{1m}w_m) \\ \cdot \\ \cdot \\ \cdot \\ (a_{n1}v_1 + \cdots + a_{nm}v_m) + (a_{n1}w_1 + \cdots + a_{nm}w_m) \end{bmatrix}$$

$$= \begin{bmatrix} a_{11}v_1 + \cdots + a_{1m}v_m \\ \cdot \\ \cdot \\ \cdot \\ a_{n1}v_1 + \cdots + a_{nm}v_m \end{bmatrix} + \begin{bmatrix} a_{11}w_1 + \cdots + a_{1m}w_m \\ \cdot \\ \cdot \\ \cdot \\ a_{n1}w_1 + \cdots + a_{nm}w_m \end{bmatrix}$$

$$= A\mathbf{v} + A\mathbf{w}.$$

EXERCISES

1. Given $\mathbf{v} = \begin{bmatrix} 1 \\ 2 \end{bmatrix}$, $\mathbf{w} = \begin{bmatrix} -1 \\ 0 \end{bmatrix}$, $\mathbf{u} = \begin{bmatrix} 0 \\ 3 \end{bmatrix}$, $A = \begin{bmatrix} 1 & -1 \\ 2 & 3 \end{bmatrix}$, and $B = \begin{bmatrix} 1 & 2 \\ 0 & 1 \end{bmatrix}$, find

a. $\mathbf{v} + \mathbf{w}$ b. $3\mathbf{u}$ c. $2\mathbf{v} - 5\mathbf{w} + \mathbf{u}$

d. $A\mathbf{v}$ e. $A\mathbf{w}$ f. $A(3\mathbf{v} - \mathbf{w})$

g. $-3A$ h. $A - 2B$ i. $(A + B)\mathbf{v}$

2. Given $\mathbf{v} = \begin{bmatrix} 1 \\ 2 \\ 0 \\ 0 \end{bmatrix}$, $\mathbf{w} = \begin{bmatrix} -3 \\ 3 \\ 1 \\ -1 \end{bmatrix}$, and $A = \begin{bmatrix} 1 & 2 & 0 & 3 \\ 0 & 1 & -1 & 2 \\ 0 & 0 & 1 & 1 \\ 1 & -1 & 1 & 1 \end{bmatrix}$, find

a. $3\mathbf{v}$ b. $\mathbf{v} - 2\mathbf{w}$

c. $A\mathbf{v}$ d. $A\mathbf{v} - 2A\mathbf{w}$

e. $A - 3I$, where I is the 4×4 identity matrix (see Note 2)

Determine which of the systems in Exercises 3 to 9 are linear. For each linear system,

a.   determine whether the system is homogeneous,
b.   find its order, and
c.   write it in matrix form.

3. $\begin{aligned} x' &= -x + t + y \\ y' &= t - 2x \end{aligned}$

4. $\begin{aligned} x' &= 2x - 3xy \\ y' &= 3xy - 4y \end{aligned}$

5. $\begin{aligned} x' &= 5x - 6y \\ y' &= 2x + y \end{aligned}$

6. $\begin{aligned} x' &= -y - z \\ y' &= -x - z \\ z' &= -x - y \end{aligned}$

7. $\begin{aligned} x' &= -ty - z + t \\ y' &= -\frac{x}{t} - \frac{z}{t} + 1 \\ z' &= x - ty \end{aligned}$

8. $\begin{aligned} x' &= y^2 + z \\ y' &= x + z \\ z' &= x - z \end{aligned}$

9. $\begin{aligned} x' &= x + 3y + t^2 \\ y' &= 2x + y + t \\ z' &= x - t^2 + 3ty \end{aligned}$

In Exercises 10 to 18, you are given an o.d.e. (N). For each:

a.   Find the equivalent system (S_N).
b.   Find the general solution of (N) and use it to obtain the general solution of (S_N).
c.   Write (S_N) in matrix form.
d.   Write the general solution of (S_N) in the form $\mathbf{x} = c_1\mathbf{h}_1(t) + \cdots + c_n\mathbf{h}_n(t) + \mathbf{p}(t)$.

10. $(D^2 - 1) = 0$ 11. $(D^2 - 1)x = t$

12. $(D^2 + 1)x = 0$ 13. $(D^2 + 1)x = 1$

14. $(D - 1)^2(D + 1)x = 0$ 15. $(D - 1)^2(D + 1)x = 4$

16. $(D^3 - D)x = 0$ 17. $(D^3 - D)x = 1$

18. $(D^2 - 2D - 3)x = 0$

In Exercises 19 to 23, you are given A, $\mathbf{E}(t)$, and $\mathbf{x}_i(t)$.

a.   Find $D\mathbf{x}_i(t)$.
b.   Find $A\mathbf{x}_i(t) + \mathbf{E}(t)$.
c.   Determine whether $\mathbf{x}_i(t)$ is a solution of $D\mathbf{x} = A\mathbf{x} + \mathbf{E}(t)$.

19. $A = \begin{bmatrix} 1 & 2 \\ 0 & 3 \end{bmatrix}$, $\mathbf{E}(t) = \mathbf{0}$, $\mathbf{x}_1(t) = \begin{bmatrix} e^t \\ e^{3t} \end{bmatrix}$, $\mathbf{x}_2(t) = \begin{bmatrix} e^{3t} \\ e^{3t} \end{bmatrix}$

20. $A = \begin{bmatrix} 0 & 1 \\ -1 & 0 \end{bmatrix}$, $\mathbf{E}(t) = \begin{bmatrix} t \\ -1 \end{bmatrix}$, $\mathbf{x}_1(t) = \begin{bmatrix} \sin t \\ \cos t \end{bmatrix}$, $\mathbf{x}_2(t) = \begin{bmatrix} 0 \\ -t \end{bmatrix}$

21. $A = \begin{bmatrix} 0 & 1 \\ 0 & 0 \end{bmatrix}$, $\mathbf{E}(t) = 0$, $\mathbf{x}_1(t) = \begin{bmatrix} t \\ 1 \end{bmatrix}$, $\mathbf{x}_2(t) = \begin{bmatrix} 1 \\ 0 \end{bmatrix}$

22. $A = \begin{bmatrix} 1 & 2 & 0 \\ 0 & -1 & 0 \\ 0 & 0 & 2 \end{bmatrix}$, $\mathbf{E}(t) = 0$, $\mathbf{x}_1(t) = \begin{bmatrix} e^t \\ 0 \\ e^{2t} \end{bmatrix}$, $\mathbf{x}_2(t) = 2\mathbf{x}_1(t)$

23. $A = \begin{bmatrix} 1 & 2 & 0 \\ 0 & -1 & 0 \\ 0 & 0 & 2 \end{bmatrix}$, $\mathbf{E}(t) = \begin{bmatrix} -1 \\ 2 \\ -4 \end{bmatrix}$, $\mathbf{x}_1(t) = \begin{bmatrix} e^t - 3 \\ 2 \\ 2 \end{bmatrix}$, $\mathbf{x}_2(t) = 2\mathbf{x}_1(t)$

24. Each of the following vector valued functions determines a parametrized curve (see Note 1) that is easy to identify by solving for y in terms of x or for x in terms of y. Sketch the curve and indicate with an arrow the direction in which the point $\mathbf{x}(t)$ moves along the curve as t increases.

    a. $\mathbf{x} = \begin{bmatrix} e^t \\ e^{2t} \end{bmatrix}$ b. $\mathbf{x} = \begin{bmatrix} -e^{2t} \\ e^{2t} \end{bmatrix}$ c. $\mathbf{x} = \begin{bmatrix} -e^{2t} \\ -e^{2t} \end{bmatrix}$

    d. $\mathbf{x} = \begin{bmatrix} e^{3t} \\ 2e^t \end{bmatrix}$ e. $\mathbf{x} = \begin{bmatrix} e^{2t} \\ e^{3t} \end{bmatrix}$

25. The parametrized curve determined by each of the following vector valued functions (see Note 1) is a circle or an ellipse. Sketch the curve and indicate by an arrow whether the point $\mathbf{x}(t)$ moves around this curve in a clockwise or counterclockwise direction as t increases.

    a. $\mathbf{x} = \begin{bmatrix} \sin t \\ \cos t \end{bmatrix}$ b. $\mathbf{x} = \begin{bmatrix} \cos t \\ \sin t \end{bmatrix}$

    c. $\mathbf{x} = \begin{bmatrix} 2\sin 3t \\ 2\cos 3t \end{bmatrix}$ d. $\mathbf{x} = \begin{bmatrix} \sin t \\ 2\cos t \end{bmatrix}$

Several of the models in Chapter 2 led naturally to systems of o.d.e.'s involving second derivatives. We dealt with each of them by replacing the system by a single higher-order o.d.e. An alternative approach is to proceed as in Example 3.2.1. If an equation involves the second derivative of a variable, we introduce the first derivative as a new variable and replace the equation by two equations involving only first derivatives. Use this idea in Exercises 26 and 27 to obtain a linear system of o.d.e.'s for the given variables:

26. *Coupled Springs:* The model of Example 2.1.4; x_1, $v_1 = x_1'$, x_2, and $v_2 = x_2'$.

27. *A Moving Spring:* The model of Exercise 3, Section 2.1; the distance z of the weight from the ground, $v = z'$, and the distance y from the bottom of the elevator to the ground.

Some more abstract problems:

28. Suppose A is an $n \times n$ matrix, I is the $n \times n$ identity matrix (see Note 2), $\mathbf{v}$ is an n-vector, and λ is a number.

    a. Show that $A - \lambda I$ is the matrix obtained from A by subtracting λ from each of the diagonal entries a_{ii} while leaving the other entries alone.

    b. Show that $A\mathbf{v} - \lambda\mathbf{v} = (A - \lambda I)\mathbf{v}$.

29. Verify properties (A1) to (A8) of Note 2.

30. Verify properties (M2) to (M5) of Note 2.

31. Verify properties (D1) and (D2) of the summary.

32. a. Use some of the properties (M1), (M2), (D1), and (D2) to show that if $x = h_1(t)$ and $x = h_2(t)$ are solutions of the homogeneous system $Dx = Ax$ and c_1 and c_2 are constant scalars, then $x = c_1h_1(t) + c_2h_2(t)$ is also a solution.
     b. Does (a) remain true if c_1 and c_2 are replaced by scalar valued functions?
     c. Does (a) remain true if the system there is replaced by the nonhomogeneous system $Dx = Ax + E(t)$?

33. Use some of the properties in the summary to show that if $x = h(t)$ is a solution of the homogeneous system $Dx = Ax$, and $x = p(t)$ is a solution of the related nonhomogeneous system $Dx = Ax + E(t)$, then $x = p(t) + h(t)$ is also a solution of the nonhomogeneous system.

34. Use some of the properties in the summary to show that if $x = p(t)$ and $x = q(t)$ are both solutions of the same nonhomogeneous system $Dx = Ax + E(t)$, then $x = p(t) - q(t)$ is a solution of the related homogeneous system $Dx = Ax$.

35. Suppose A and B are $n \times m$ matrices with the property that $Av = Bv$ for every m-vector v. Show that the matrices A and B are equal. (*Hint:* What are Av and Bv in the special case that all entries of v except the jth are 0, and the jth entry of v is 1?)

3.3   LINEAR SYSTEMS OF O.D.E.'S: GENERAL PROPERTIES

We saw in Examples 3.2.1 and 3.2.2 that single linear o.d.e.'s can be replaced by equivalent systems. In Example 3.2.1 we solved such a system by first solving the single equation. If we learn to solve systems directly, we will be able to solve single equations by first solving the equivalent system. Thus our theory of systems includes the results of Chapter 2 as a special case. It should not surprise us that many of the features we see in solutions of systems will look familiar.

Suppose we are given an nth-order linear system of o.d.e.'s whose matrix form is

(S) $$Dx = Ax + E(t),$$

where

$$x = \begin{bmatrix} x_1 \\ \cdot \\ \cdot \\ \cdot \\ x_n \end{bmatrix}, \quad A = \begin{bmatrix} a_{11} & \cdots & a_{1n} \\ \cdot & & \cdot \\ \cdot & & \cdot \\ \cdot & & \cdot \\ a_{n1} & \cdots & a_{nn} \end{bmatrix}, \quad \text{and} \quad E(t) = \begin{bmatrix} E_1(t) \\ \cdot \\ \cdot \\ \cdot \\ E_n(t) \end{bmatrix}.$$

Let $\mathbf{x} = \mathbf{p}(t)$ be a solution of (S), and let $\mathbf{x} = \mathbf{h}(t)$ be a solution of the **related homogeneous system**

(H) $$D\mathbf{x} = A\mathbf{x}.$$

Then

$$D[\mathbf{h}(t) + \mathbf{p}(t] = D\mathbf{h}(t) + D\mathbf{p}(t)$$
$$= A\mathbf{h}(t) + [A\mathbf{p}(t) + \mathbf{E}(t)] = A[\mathbf{h}(t) + \mathbf{p}(t)] + \mathbf{E}(t).$$

Thus $\mathbf{x} = \mathbf{h}(t) + \mathbf{p}(t)$ is a solution of (S). If we let $\mathbf{h}(t)$ range over all solutions of (H), we get a collection of solutions of (S).

Suppose $\mathbf{x} = \mathbf{f}(t)$ is another solution of (S). Let $\mathbf{h}_1(t) = \mathbf{f}(t) - \mathbf{p}(t)$. Then

$$\mathbf{f}(t) = \mathbf{h}_1(t) + \mathbf{p}(t)$$

and

$$D\mathbf{h}_1(t) = D[\mathbf{f}(t) - \mathbf{p}(t)] = D\mathbf{f}(t) - D\mathbf{p}(t)$$
$$= [A\mathbf{f}(t) + \mathbf{E}(t)] - [A\mathbf{p}(t) + \mathbf{E}(t)] = A[\mathbf{f}(t) - \mathbf{p}(t)]$$
$$= A\mathbf{h}_1(t).$$

Thus $\mathbf{x} = \mathbf{h}_1(t)$ is a solution of (H), and $\mathbf{f}(t)$ belongs to the collection we described in the previous paragraph.

We have shown the following:

Fact: *If we find a particular solution $\mathbf{x} = \mathbf{p}(t)$ of the nonhomogeneous linear system (S), and if $\mathbf{x} = \mathbf{H}(t)$ is a formula that describes all solutions of the homogeneous system (H), then*

$$\mathbf{x} = \mathbf{H}(t) + \mathbf{p}(t)$$

describes all solutions of (S).

Our strategy for solving linear systems will be the same as our strategy for solving single linear equations (see Section 2.2). First we will find the general solution $\mathbf{x} = \mathbf{H}(t)$ of the related homogeneous system (H). Then we will look for a particular solution $\mathbf{x} = \mathbf{p}(t)$ of the nonhomogeneous system (S). The general solution of (S) will be $\mathbf{x} = \mathbf{H}(t) + \mathbf{p}(t)$.

Let's now take a look at what the general solution of (H) looks like. (You should look back at Section 2.3, noting all of the parallels between the properties of one homogeneous o.d.e. and the present situation.) Suppose we know that $\mathbf{x} = \mathbf{h}_1(t), \ldots, \mathbf{x} = \mathbf{h}_n(t)$ are solutions of (H) and that $c_1, \ldots, c_n$ are constants. Then

$$D[c_1\mathbf{h}_1(t) + \cdots + c_n\mathbf{h}_n(t)] = c_1 D\mathbf{h}_1(t) + \cdots + c_n D\mathbf{h}_n(t)$$
$$= c_1 A\mathbf{h}_1(t) + \cdots + c_n A\mathbf{h}_n(t)$$
$$= A[c_1\mathbf{h}_1(t) + \cdots + c_n\mathbf{h}_n(t)].$$

Thus $\mathbf{x} = c_1\mathbf{h}_1(t) + \cdots + c_n\mathbf{h}_n(t)$ is a solution of (H). We refer to a vector valued function of this form as a **linear combination** of $\mathbf{h}_1(t), \ldots, \mathbf{h}_n(t)$. Thus we have the following:

Fact: *If $\mathbf{h}_1(t), \ldots, \mathbf{h}_n(t)$ are solutions of (H), then any linear combination of these vector valued functions is also a solution of (H).*

If we find some solutions of (H), then we can use this fact to generate a collection of solutions. The following theorem will provide us with a criterion for deciding whether the collection we get is complete.

Theorem: Existence and Uniqueness of Solutions of Linear Systems. *Suppose $D\mathbf{x} = A\mathbf{x} + \mathbf{E}(t)$ is an nth-order linear system of o.d.e.'s. Assume the entries $E_i(t)$ of $\mathbf{E}(t)$ and the coefficients a_{ij} are continuous on an interval I. Let t_0 be a fixed value of t in I. Then, given any n-vector $\mathbf{v}$, there exists a solution $\mathbf{x} = \boldsymbol{\phi}(t)$ of the system, which is defined for all t in I and which satisfies the initial condition*

$$\mathbf{x}(t_0) = \mathbf{v}.$$

Furthermore, if $\mathbf{x} = \boldsymbol{\psi}(t)$ is a solution of the system that satisfies the same initial condition as $\mathbf{x} = \boldsymbol{\phi}(t)$, then $\boldsymbol{\phi}(t) = \boldsymbol{\psi}(t)$ for all t in I.

To decide whether a given collection of solutions is complete, we try to match every initial condition at t_0 with a function in the collection. *If there is some initial condition that we cannot match*, then the existence part of the theorem says that *at least one solution has been omitted* from the collection. On the other hand, *if every initial condition can be matched, then the collection is complete*, since any solution satisfies some initial condition at t_0 and by uniqueness must therefore agree with a function in our collection.

To determine whether a formula $\mathbf{x} = c_1\mathbf{h}_1(t) + \cdots + c_n\mathbf{h}_n(t)$ giving solutions of the nth-order homogeneous system (H) is the general solution, we must determine whether we can match each initial condition. That is, we must determine whether the equation

(M) $$c_1\mathbf{h}_1(t_0) + c_2\mathbf{h}_2(t_0) + \cdots + c_n\mathbf{h}_n(t_0) = \mathbf{v}$$

can be solved for $c_1, \ldots, c_n$ for all choices of $\mathbf{v}$. Now each $\mathbf{h}_i(t)$ is an n-vector valued function:

$$
\mathbf{h}_1(t) = \begin{bmatrix} h_{11}(t) \\ h_{21}(t) \\ \cdot \\ \cdot \\ \cdot \\ h_{n1}(t) \end{bmatrix}, \quad
\mathbf{h}_2(t) = \begin{bmatrix} h_{12}(t) \\ h_{22}(t) \\ \cdot \\ \cdot \\ \cdot \\ h_{n2}(t) \end{bmatrix}, \ldots,
\mathbf{h}_n(t) = \begin{bmatrix} h_{1n}(t) \\ h_{2n}(t) \\ \cdot \\ \cdot \\ \cdot \\ h_{nn}(t) \end{bmatrix}.
$$

The vector $\mathbf{v}$ is an n-vector:

$$
\mathbf{v} = \begin{bmatrix} v_1 \\ v_2 \\ \cdot \\ \cdot \\ \cdot \\ v_n \end{bmatrix}.
$$

Equation (M) is merely a matrix formulation of the system of algebraic equations

(M')
$$
\begin{aligned}
c_1 h_{11}(t_0) + c_2 h_{12}(t_0) + \cdots + c_n h_{1n}(t_0) &= v_1 \\
c_1 h_{21}(t_0) + c_2 h_{22}(t_0) + \cdots + c_n h_{2n}(t_0) &= v_2 \\
&\;\cdot \\
&\;\cdot \\
&\;\cdot \\
c_1 h_{n1}(t_0) + c_2 h_{n2}(t_0) + \cdots + c_n h_{nn}(t_0) &= v_n.
\end{aligned}
$$

Cramer's determinant test tells us that we can solve this system for all choices of $v_1, \ldots, v_n$ if and only if the determinant of coefficients is not zero.

We define the **Wronskian** of $\mathbf{h}_1(t), \mathbf{h}_2(t), \ldots, \mathbf{h}_n(t)$ to be the determinant of the $n \times n$ matrix whose columns are $\mathbf{h}_1(t), \mathbf{h}_2(t), \ldots, \mathbf{h}_n(t)$:

$$
W[\mathbf{h}_1, \mathbf{h}_2, \ldots, \mathbf{h}_n](t) = \det \begin{bmatrix} h_{11}(t)\ h_{12}(t), & \cdots & h_{1n}(t) \\ h_{21}(t)\ h_{22}(t), & \cdots & h_{2n}(t) \\ \cdot & \cdot & \cdot \\ \cdot & \cdot & \cdot \\ \cdot & \cdot & \cdot \\ h_{n1}(t)\ h_{n2}(t), & \cdots & h_{nn}(t) \end{bmatrix}
$$

The determinant of coefficients of (M') is the Wronskian evaluated at t_0. Thus we can state the conclusion of the preceding paragraph as follows:

Fact: *Suppose the coefficients a_{ij} of the nth-order linear homogeneous system (H) are continuous on an interval I. Let $h_1(t), \ldots, h_n(t)$ be solutions of (H), and let t_0 be a fixed value of t in I. The general solution of (H) is*

$$x = c_1 h_1(t) + \cdots + c_n h_n(t)$$

if and only if

$$W[h_1, \ldots, h_n](t_0) \neq 0.$$

Example 3.3.1

The equations of the simple *LRC* circuit of Example 3.1.1 with inductance $L = 1$ henry, resistance $R = 3$ ohms, capacitance $C = 1/2$ farad, and no voltage source ($V(t) = 0$) can be written in the form $Dx = Ax$, where

$$x = \begin{bmatrix} Q \\ I \end{bmatrix} \quad \text{and} \quad A = \begin{bmatrix} 0 & 1 \\ -2 & -3 \end{bmatrix}.$$

Let

$$h_1(t) = \begin{bmatrix} e^{-t} \\ -e^{-t} \end{bmatrix} \quad \text{and} \quad h_2(t) = \begin{bmatrix} e^{-2t} \\ -2e^{-2t} \end{bmatrix}.$$

Then

$$Ah_1(t) = \begin{bmatrix} 0 & 1 \\ -2 & -3 \end{bmatrix} \begin{bmatrix} e^{-t} \\ -e^{-t} \end{bmatrix} = \begin{bmatrix} -e^{-t} \\ e^{-t} \end{bmatrix} = Dh_1(t)$$

and

$$Ah_2(t) = \begin{bmatrix} 0 & 1 \\ -2 & -3 \end{bmatrix} \begin{bmatrix} e^{-2t} \\ -2e^{-2t} \end{bmatrix} = \begin{bmatrix} -2e^{-2t} \\ 4e^{-2t} \end{bmatrix} = Dh_2(t).$$

Thus, $h_1(t)$ and $h_2(t)$ are solutions of the second-order system $Dx = Ax$. The Wronskian of $h_1(t)$ and $h_2(t)$ is

$$W[h_1, h_2](t) = \det \begin{bmatrix} e^{-t} & e^{-2t} \\ -e^{-t} & -2e^{-2t} \end{bmatrix}.$$

Thus

$$W[h_1, h_2](0) = \det \begin{bmatrix} 1 & 1 \\ -1 & -2 \end{bmatrix} = -1 \neq 0.$$

Since the Wronskian is not zero at $t_0 = 0$, $h_1(t)$ and $h_2(t)$ generate the general solution of $D\mathbf{x} = A\mathbf{x}$:

$$\mathbf{x} = c_1 \begin{bmatrix} e^{-t} \\ -e^{-t} \end{bmatrix} + c_2 \begin{bmatrix} e^{-2t} \\ -2e^{-2t} \end{bmatrix}.$$

In terms of the charge and current in the circuit model, this reads

$$Q = c_1 e^{-t} + c_2 e^{-2t}$$
$$I = -c_1 e^{-t} - 2c_2 e^{-2t}.$$

Example 3.3.2

The equations of the three-loop circuit of Example 3.1.3 with $V(t) = 0$, $R_1 = R_2 = 1$, $L_2 = L_3 = 1$, and $C = 1$ can be written $D\mathbf{x} = A\mathbf{x}$, where

$$\mathbf{x} = \begin{bmatrix} Q \\ I_2 \\ I_3 \end{bmatrix} \quad \text{and} \quad A = \begin{bmatrix} -1 & -1 & 0 \\ 1 & -1 & 1 \\ 0 & 1 & 1 \end{bmatrix}.$$

Direct substitution will show that

$$\mathbf{h}_1(t) = \begin{bmatrix} (1+t)e^{-t} \\ -e^{-t} \\ -(1+t)e^{-t} \end{bmatrix}, \quad \mathbf{h}_2(t) = \begin{bmatrix} (1-t)e^{-t} \\ e^{-t} \\ -(1-t)e^{-t} \end{bmatrix}, \quad \mathbf{h}_3(t) = \begin{bmatrix} (3+t)e^{-t} \\ -e^{-t} \\ -(3+t)e^{-t} \end{bmatrix}$$

are solutions of this third-order system. If we evaluate the Wronskian of these functions at $t = 0$, we get

$$W[\mathbf{h}_1, \mathbf{h}_2, \mathbf{h}_3](0) = \det \begin{bmatrix} 1 & 1 & 3 \\ -1 & 1 & -1 \\ -1 & -1 & -3 \end{bmatrix} = 0,$$

so that $\mathbf{h}_1$, $\mathbf{h}_2$, $\mathbf{h}_3$ do not generate the general solution.

On the other hand, we leave it to the reader to check that

$$\mathbf{k}(t) = \begin{bmatrix} (2 - t^2)e^{-t} \\ 2te^{-t} \\ t^2 e^{-t} \end{bmatrix}$$

is a solution of $D\mathbf{x} = A\mathbf{x}$ and that $\mathbf{h}_1(t)$, $\mathbf{h}_2(t)$, and $\mathbf{k}(t)$ generate the general solution,

$$\mathbf{x} = c_1 \begin{bmatrix} (1+t)e^{-t} \\ -e^{-t} \\ -(1+t)e^{-t} \end{bmatrix} + c_2 \begin{bmatrix} (1-t)e^{-t} \\ e^{-t} \\ -(1-t)e^{-t} \end{bmatrix} + c_3 \begin{bmatrix} (2-t^2)e^{-t} \\ 2te^{-t} \\ t^2 e^{-t} \end{bmatrix}.$$

In terms of charge and current, this reads

$$Q = (c_1 + c_2 + 2c_3)e^{-t} + (c_1 - c_2)te^{-t} - c_3t^2e^{-t}$$
$$I_2 = (-c_1 + c_2)e^{-t} + 2c_3te^{-t}$$
$$I_3 = (-c_1 - c_2)e^{-t} + (-c_1 + c_2)te^{-t} + c_3t^2e^{-t}.$$

Example 3.3.3

If we add to the circuit of Example 3.3.1 a voltage source with

$$V(t) = 10 \cos t$$

then the equations read $D\mathbf{x} = A\mathbf{x} + \mathbf{E}(t)$, where

$$\mathbf{x} = \begin{bmatrix} Q \\ I \end{bmatrix}, \quad A = \begin{bmatrix} 0 & 1 \\ -2 & -3 \end{bmatrix}, \quad \text{and} \quad \mathbf{E}(t) = \begin{bmatrix} 0 \\ 10 \cos t \end{bmatrix}.$$

We saw in Example 3.3.1 that the general solution of the homogeneous system $D\mathbf{x} = A\mathbf{x}$ (no voltage source) is $\mathbf{x} = \mathbf{H}(t)$, where

$$\mathbf{H}(t) = c_1 \begin{bmatrix} e^{-t} \\ -e^{-t} \end{bmatrix} + c_2 \begin{bmatrix} e^{-2t} \\ -2e^{-2t} \end{bmatrix}.$$

Let

$$\mathbf{p}(t) = \begin{bmatrix} \cos t + 3 \sin t \\ 3 \cos t - \sin t \end{bmatrix}.$$

Then

$$A\mathbf{p}(t) + \mathbf{E}(t) = \begin{bmatrix} 0 & 1 \\ -2 & -3 \end{bmatrix} \begin{bmatrix} \cos t + 3 \sin t \\ 3 \cos t - \sin t \end{bmatrix} + \begin{bmatrix} 0 \\ 10 \cos t \end{bmatrix}$$

$$= \begin{bmatrix} 3 \cos t - \sin t \\ -11 \cos t - 3 \sin t \end{bmatrix} + \begin{bmatrix} 0 \\ 10 \cos t \end{bmatrix} = \begin{bmatrix} 3 \cos t - \sin t \\ -\cos t - 3 \sin t \end{bmatrix}$$

$$= D\mathbf{p}(t).$$

Thus, $\mathbf{p}(t)$ is a solution of $D\mathbf{x} = A\mathbf{x} + \mathbf{E}(t)$, and the general solution is

$$\mathbf{x} = \mathbf{H}(t) + \mathbf{p}(t) = c_1 \begin{bmatrix} e^{-t} \\ -e^{-t} \end{bmatrix} + c_2 \begin{bmatrix} e^{-2t} \\ -2e^{-2t} \end{bmatrix} + \begin{bmatrix} \cos t + 3 \sin t \\ 3 \cos t - \sin t \end{bmatrix}.$$

In terms of charge and current

$$Q = \quad c_1 e^{-t} + \quad c_2 e^{-2t} + \quad \cos t + 3 \sin t$$
$$I = -c_1 e^{-t} - 2c_2 e^{-2t} + 3 \cos t - \quad \sin t.$$

In each of the preceding examples, the general solution of an nth-order homogeneous system had the form $\mathbf{x} = c_1\mathbf{h}_1(t) + \cdots + c_n\mathbf{h}_n(t)$. Let's now check that this is always the case.

The existence and uniqueness theorem tells us that an nth-order homogeneous system has solutions that match each initial condition at t_0. In particular, there are solutions $\mathbf{h}_1(t), \ldots, \mathbf{h}_n(t)$ that match the initial conditions

$$\mathbf{h}_1(t_0) = \begin{bmatrix} 1 \\ 0 \\ 0 \\ \cdot \\ \cdot \\ \cdot \\ 0 \end{bmatrix}, \quad \mathbf{h}_2(t_0) = \begin{bmatrix} 0 \\ 1 \\ 0 \\ \cdot \\ \cdot \\ \cdot \\ 0 \end{bmatrix}, \ldots, \mathbf{h}_n(t_0) = \begin{bmatrix} 0 \\ 0 \\ 0 \\ \cdot \\ \cdot \\ \cdot \\ 1 \end{bmatrix}.$$

Then

$$W[\mathbf{h}_1, \ldots, \mathbf{h}_n](t_0) = \det \begin{bmatrix} 1 & 0 & \cdots & 0 \\ 0 & 1 & \cdots & 0 \\ 0 & 0 & \cdots & 0 \\ \cdot & \cdot & & \cdot \\ \cdot & \cdot & & \cdot \\ \cdot & \cdot & & \cdot \\ 0 & 0 & \cdots & 1 \end{bmatrix} = 1 \neq 0.$$

These solutions generate a complete collection of solutions. Furthermore, as in Section 2.3, the general solution cannot be generated by fewer than n solutions.

Fact: *Suppose the coefficients of the nth-order linear homogeneous system (H) are continuous on an interval I. Then the general solution of (H) has the form*

$$\mathbf{x} = c_1\mathbf{h}_1(t) + \cdots + c_n\mathbf{h}_n(t)$$

for a suitable choice of $\mathbf{h}_1(t), \ldots, \mathbf{h}_n(t)$. The general solution cannot be generated by fewer than n solutions.

Let's summarize.

LINEAR SYSTEMS OF O.D.E.'s: GENERAL PROPERTIES

Given an nth-order linear system of o.d.e.'s

$$\text{(S)} \qquad\qquad D\mathbf{x} = A\mathbf{x} + \mathbf{E}(t).$$

The **related homogeneous system** is

$$\text{(H)} \qquad\qquad D\mathbf{x} = A\mathbf{x}.$$

Suppose the entries a_{ij} of A and $E_i(t)$ of $\mathbf{E}(t)$ are continuous on an interval I. Then the following are true:

1. The general solution of (S) is of the form

$$\mathbf{x} = \mathbf{H}(t) + \mathbf{p}(t)$$

where $\mathbf{x} = \mathbf{H}(t)$ is the general solution of (H) and $\mathbf{x} = \mathbf{p}(t)$ is a particular solution of (S).

2. The general solution of (H) is of the form $\mathbf{x} = \mathbf{H}(t)$, where

$$\mathbf{H}(t) = c_1\mathbf{h}_1(t) + \cdots + c_n\mathbf{h_n}(t)$$

for a suitable choice of $\mathbf{h}_1(t), \ldots, \mathbf{h}_n(t)$.

3. Let t_0 be a fixed value of t in I. A collection of solutions of (H) or (S) is complete if and only if every initial condition $\mathbf{x}(t_0) = \mathbf{v}$ is matched by some solution in the collection.

4. Let t_0 be a fixed value of t in I and let $\mathbf{h}_1(t), \ldots, \mathbf{h}_n(t)$ be solutions of (H). The general solution of (H) is $\mathbf{x} = c_1\mathbf{h}_1(t) + \cdots + c_n\mathbf{h}_n(t)$ if and only if $W[\mathbf{h}_1, \ldots, \mathbf{h}_n](t_0) \neq 0$. Here $W[\mathbf{h}_1, \ldots, \mathbf{h}_n](t)$, the **Wronskian** of $\mathbf{h}_1, \ldots, \mathbf{h}_n$, is the determinant of the $n \times n$ matrix whose columns are $\mathbf{h}_1, \ldots, \mathbf{h}_n$.

Notes

1.   A technicality

Our results depend heavily on the existence and uniqueness theorem, which assumes that the coefficients a_{ij} and $E_i(t)$ are continuous on an interval I and which provides information about solutions defined on I. *All statements in this and later sections about general solutions are valid only on such intervals. The values t_0 that* *we use to test for completeness of a collection of solutions will always be in the interval I.*

2. On the new and old Wronskians

If $x = h(t)$ is a solution of the single nth-order linear homogeneous equation $(a_n D^n + a_{n-1} D^{n-1} + \cdots + a_0)x = 0$, then

$$\mathbf{h}(t) = \begin{bmatrix} h(t) \\ h'(t) \\ \cdot \\ \cdot \\ \cdot \\ h^{(n-1)}(t) \end{bmatrix}$$

is a solution of the equivalent system (see Example 3.2.2). If we start with n solutions $h_1(t), \ldots, h_n(t)$ of the o.d.e., then we get n solutions $\mathbf{h}_1(t), \ldots, \mathbf{h}_n(t)$ of the system. The Wronskian of $\mathbf{h}_1(t), \ldots, \mathbf{h}_n(t)$ is

$$W[\mathbf{h}_1, \ldots, \mathbf{h}_n](t) = \det \begin{bmatrix} h_1(t) & \cdots & h_n(t) \\ \cdot & & \cdot \\ \cdot & & \cdot \\ \cdot & & \cdot \\ h_1^{(n-1)}(t) & \cdots & h_n^{(n-1)}(t) \end{bmatrix}.$$

This is the same as $W[h_1, \ldots, h_n](t)$ as we defined it in Chapter 2.

3. Fundamental matrices

Suppose the vector valued functions

$$\mathbf{h}_1(t) = \begin{bmatrix} h_{11}(t) \\ \cdot \\ \cdot \\ \cdot \\ h_{n1}(t) \end{bmatrix}, \quad \mathbf{h}_2(t) = \begin{bmatrix} h_{12}(t) \\ \cdot \\ \cdot \\ \cdot \\ h_{n2}(t) \end{bmatrix}, \ldots, \mathbf{h}_n(t) = \begin{bmatrix} h_{1n}(t) \\ \cdot \\ \cdot \\ \cdot \\ h_{nn}(t) \end{bmatrix}$$

generate all solutions of the homogeneous system $D\mathbf{x} = A\mathbf{x}$. Then every solution can be written in the form

$$\mathbf{x} = c_1 \mathbf{h}_1(t) + c_2 \mathbf{h}_2(t) + \cdots + c_n \mathbf{h}_n(t)$$

$$= \begin{bmatrix} c_1 h_{11}(t) + c_2 h_{12}(t) + \cdots + c_n h_{1n}(t) \\ \cdot \\ \cdot \\ \cdot \\ c_1 h_{n1}(t) + c_2 h_{n2}(t) + \cdots + c_n h_{nn}(t) \end{bmatrix}.$$

Note that this last expression can be written in the matrix form

$$\mathbf{x} = \Phi(t)\mathbf{C}$$

where

$$\Phi(t) = \begin{bmatrix} h_{11}(t) & h_{12}(t) & \cdots & h_{1n}(t) \\ & & & \\ \cdot & \cdot & & \cdot \\ \cdot & \cdot & & \cdot \\ \cdot & \cdot & & \cdot \\ h_{n1}(t) & h_{n2}(t) & \cdots & h_{nn}(t) \end{bmatrix}$$

is the $n \times n$ matrix whose columns are $\mathbf{h}_1(t), \ldots, \mathbf{h}_n(t)$, and

$$\mathbf{C} = \begin{bmatrix} c_1 \\ \cdot \\ \cdot \\ \cdot \\ c_n \end{bmatrix}$$

is a constant n-vector whose entries are the coefficients of the linear combination. The matrix $\Phi(t)$ is called a **fundamental matrix** of $D\mathbf{x} = A\mathbf{x}$. Note that $\Phi(t)$ is the matrix whose determinant we compute to find the Wronskian of $\mathbf{h}_1(t), \ldots, \mathbf{h}_n(t)$:

$$W[\mathbf{h}_1, \ldots, \mathbf{h}_n](t) = \det \Phi(t).$$

4. A warning

Statement 3 in the summary applies to homogeneous and nonhomogeneous systems alike: A collection of solutions is complete if and only if every possible initial condition is matched by some member of the collection. However, the Wronskian test (statement 4) applies only to solutions of homogeneous equations. Of course, in the expression $\mathbf{x} = \mathbf{H}(t) + \mathbf{p}(t)$, one can use the Wronskian to check that $\mathbf{H}(t)$ represents a general solution of (H) and then use statement 1 to conclude that $\mathbf{H}(t) + \mathbf{p}(t)$ is the general solution of (S).

EXERCISES

In Exercises 1 to 10, you are given a matrix A and a list of vector valued functions $\mathbf{h}_1(t), \ldots, \mathbf{h}_k(t)$. For each problem decide (a) whether the functions $\mathbf{x} = \mathbf{h}_i(t)$ are solutions of the homogeneous system $D\mathbf{x} = A\mathbf{x}$ and (b) whether the ones that are solutions generate the general solution of $D\mathbf{x} = A\mathbf{x}$.

1. $A = \begin{bmatrix} -3 & -2 \\ 1 & 0 \end{bmatrix};$ $\mathbf{h}_1(t) = \begin{bmatrix} 2e^{-2t} \\ -e^{-2t} \end{bmatrix},$ $\mathbf{h}_2(t) = \begin{bmatrix} e^{-t} \\ -e^{-t} \end{bmatrix}$

2. $A = \begin{bmatrix} 0 & -1 \\ 4 & 0 \end{bmatrix};$ $\mathbf{h}_1(t) = \begin{bmatrix} \cos 2t \\ 2 \sin 2t \end{bmatrix},$ $\mathbf{h}_2(t) = \begin{bmatrix} \sin 2t \\ -2 \cos 2t \end{bmatrix}$

3. $A = \begin{bmatrix} 0 & 1 \\ -4 & 0 \end{bmatrix}$; $\mathbf{h}_1(t) = \begin{bmatrix} \cos 2t \\ -2 \sin 2t \end{bmatrix}$, $\mathbf{h}_2(t) = \begin{bmatrix} \sin 2t \\ -2 \cos 2t \end{bmatrix}$

4. $A = \begin{bmatrix} 5 & -3 \\ 3 & -5 \end{bmatrix}$; $\mathbf{h}_1(t) = \begin{bmatrix} 3e^{4t} + e^{-4t} \\ e^{4t} + 3e^{-4t} \end{bmatrix}$, $\mathbf{h}_2(t) = \begin{bmatrix} 3e^{4t} - e^{-4t} \\ e^{4t} - 3e^{-4t} \end{bmatrix}$

5. $A = \begin{bmatrix} 5 & -3 \\ 3 & -5 \end{bmatrix}$; $\mathbf{h}_1(t) = \begin{bmatrix} 3e^{4t} \\ e^{4t} \end{bmatrix}$, $\mathbf{h}_2(t) = \begin{bmatrix} 3e^{-4t} \\ e^{-4t} \end{bmatrix}$

6. $A = \begin{bmatrix} 1 & 1 \\ 1 & 1 \end{bmatrix}$; $\mathbf{h}_1(t) = \begin{bmatrix} 1 \\ 1 \end{bmatrix}$, $\mathbf{h}_2(t) = \begin{bmatrix} 1 \\ -1 \end{bmatrix}$, $\mathbf{h}_3(t) = \begin{bmatrix} 1 + e^{2t} \\ -1 + e^{2t} \end{bmatrix}$

7. $A = \begin{bmatrix} -1 & -1 & 0 \\ 0 & -1 & 0 \\ 0 & 1 & -1 \end{bmatrix}$; $\mathbf{h}_1(t) = \begin{bmatrix} e^{-t} \\ 0 \\ -e^{-t} \end{bmatrix}$, $\mathbf{h}_2(t) = \begin{bmatrix} te^{-t} \\ -e^{-t} \\ -te^{-t} \end{bmatrix}$,

$$\mathbf{h}_3(t) = \begin{bmatrix} (3t - 2)e^{-t} \\ -3e^{-t} \\ -(3t - 2)e^{-t} \end{bmatrix}$$

8. $A = \begin{bmatrix} 2 & 0 & 4 \\ 0 & 2 & 0 \\ -1 & 0 & 2 \end{bmatrix}$; $\mathbf{h}_1(t) = \begin{bmatrix} 2e^{2t} \cos 2t \\ e^{2t} \\ -e^{2t} \sin 2t \end{bmatrix}$, $\mathbf{h}_2(t) = \begin{bmatrix} 2e^{2t} \sin 2t \\ e^{2t} \\ e^{2t} \cos 2t \end{bmatrix}$,

$$\mathbf{h}_3(t) = \begin{bmatrix} 2e^{2t} (\sin 2t + \cos 2t) \\ 2e^{2t} \\ 2e^{2t} (\cos 2t - \sin 2t) \end{bmatrix}$$

9. $A = \begin{bmatrix} 2 & 0 & 4 \\ 0 & 2 & 0 \\ -1 & 0 & 2 \end{bmatrix}$; $\mathbf{h}_1(t) = \begin{bmatrix} 2e^{2t} \cos 2t \\ e^{2t} \\ -e^{2t} \sin 2t \end{bmatrix}$, $\mathbf{h}_2(t) = \begin{bmatrix} 2e^{2t} \sin 2t \\ e^{2t} \\ e^{2t} \cos 2t \end{bmatrix}$,

$$\mathbf{h}_3(t) = \begin{bmatrix} 0 \\ e^{2t} \\ 0 \end{bmatrix}$$

10. $A = \begin{bmatrix} 0 & 8 & 0 & 0 \\ -2 & 0 & 0 & 0 \\ 0 & 0 & 0 & 2 \\ 0 & 0 & -8 & 0 \end{bmatrix}$; $\mathbf{h}_1(t) = \begin{bmatrix} 2 \sin 4t \\ \cos 4t \\ 0 \\ 0 \end{bmatrix}$, $\mathbf{h}_2(t) = \begin{bmatrix} 2 \cos 4t \\ -\sin 4t \\ 0 \\ 0 \end{bmatrix}$,

$$\mathbf{h}_3(t) = \begin{bmatrix} 0 \\ 0 \\ \sin 4t \\ 2 \cos 4t \end{bmatrix}, \quad \mathbf{h}_4(t) = \begin{bmatrix} 0 \\ 0 \\ \cos 4t \\ -2 \sin 4t \end{bmatrix}$$

In Exercises 11 to 15, you are given a matrix A, a vector valued function $\mathbf{E}(t)$, and formulas describing a collection of solutions of the nonhomogeneous system $D\mathbf{x} = A\mathbf{x} + \mathbf{E}(t)$. In each case decide whether the collection is complete.

11. $A = \begin{bmatrix} -3 & -2 \\ 1 & 0 \end{bmatrix}$, $\mathbf{E}(t) = \begin{bmatrix} 2e^{-t} \\ -e^{-t} \end{bmatrix}$; $\begin{cases} x_1 = 2c_1e^{-2t} + c_2e^{-t} \\ x_2 = -c_1e^{-2t} - c_2e^{-t} + e^{-t} \end{cases}$

12. $A = \begin{bmatrix} 0 & -1 \\ 4 & 0 \end{bmatrix}$, $\mathbf{E}(t) = \begin{bmatrix} 0 \\ -5e^t \end{bmatrix}$; $\begin{cases} x_1 = c_1 \cos 2t + c_2 \sin 2t + e^t \\ x_2 = -2c_2 \cos 2t + 2c_1 \sin 2t - e^t \end{cases}$

13. $A = \begin{bmatrix} 5 & -3 & 0 \\ 3 & -5 & 0 \\ 0 & 1 & 2 \end{bmatrix}$, $E(t) = \begin{bmatrix} 0 \\ 0 \\ 4 \end{bmatrix}$; $\begin{cases} x_1 = 6c_1e^{4t} - 2c_2e^{-4t} \\ x_2 = 2c_1e^{4t} - 6c_2e^{-4t} \\ x_3 = c_1e^{4t} + c_2e^{-4t} - 2 \end{cases}$

14. $A = \begin{bmatrix} 5 & -3 & 0 \\ 3 & -5 & 0 \\ 0 & 1 & 2 \end{bmatrix}$, $E(t) = \begin{bmatrix} 0 \\ 0 \\ 4 \end{bmatrix}$; $\begin{cases} x_1 = (6c_1 + 6c_3)e^{4t} + (-2c_2 + 2c_3)e^{-4t} \\ x_2 = (2c_1 + 2c_3)e^{4t} + (-6c_2 + 6c_3)e^{-4t} \\ x_3 = (c_1 + c_3)e^{4t} + (c_2 - c_3)e^{-4t} - 2 \end{cases}$

15. $A = \begin{bmatrix} 5 & -3 & 0 \\ 3 & -5 & 0 \\ 0 & 1 & 2 \end{bmatrix}$, $E(t) = \begin{bmatrix} 0 \\ 0 \\ 4 \end{bmatrix}$; $\begin{cases} x_1 = 6c_1e^{4t} - 2c_2e^{-4t} \\ x_2 = 2c_1e^{4t} - 6c_2e^{-4t} \\ x_3 = c_1e^{4t} + c_2e^{-4t} + c_3e^{2t} - 2 \end{cases}$

3.4   LINEAR INDEPENDENCE OF VECTORS

In the previous section we obtained a Wronskian test for determining whether solutions $h_1(t), \ldots, h_n(t)$ of an nth-order homogeneous linear system (H) generate the general solution. In this section we obtain another test. The new test involves a concept that will be important when we try to build generating sets of solutions.

Recall that the solutions $h_1(t), \ldots, h_n(t)$ of (H) generate the general solution provided we can always solve the equation

$$c_1h_1(t_0) + \cdots + c_nh_n(t_0) = v$$

for the unknowns $c_1, \ldots, c_n$. This is really a statement about the *initial vectors* $h_1(t_0), \ldots, h_n(t_0)$. Let's take a look at the vectors that arise in this way in Example 3.3.2. In that example, $h_1(t)$, $h_2(t)$, and $h_3(t)$ do not generate the general solution. Since t_0 is 0, the vectors in question are

$$h_1(0) = \begin{bmatrix} 1 \\ -1 \\ -1 \end{bmatrix}, \quad h_2(0) = \begin{bmatrix} 1 \\ 1 \\ -1 \end{bmatrix}, \quad \text{and} \quad h_3(0) = \begin{bmatrix} 3 \\ -1 \\ -3 \end{bmatrix}.$$

Note that $h_3(0) = 2h_1(0) + h_2(0)$, so

$$2h_1(0) + h_2(0) - h_3(0) = 0.$$

We have found constants $c_1 = 2$, $c_2 = 1$, and $c_3 = -1$ so that

$$c_1h_1(0) + c_2h_2(0) + c_3h_3(0) = 0.$$

Of course, this last relationship would also be true if we took all the constants to be zero. The important thing is that we have found constants that are not all zero. We say these vectors are linearly dependent.

Definition: *The n-vectors* $v_1, \ldots, v_k$ *are* **linearly dependent** *if there exist constants* $c_1, \ldots, c_k$ *with at least one* $c_i \neq 0$ *so that*

$$c_1 v_1 + \cdots + c_k v_k = 0.$$

The vectors are **linearly independent** *if the only constants for which this relationship holds are* $c_1 = c_2 = \cdots = c_k = 0$.

Note that the vectors $v_1, \ldots, v_k$ all have to be n-vectors (otherwise we couldn't form the linear combination $c_1 v_1 + \cdots + c_k v_k$). We do *not* require that $k = n$.

Example 3.4.1

Check for independence:

$$v_1 = \begin{bmatrix} 1 \\ -1 \\ -1 \end{bmatrix}, \quad v_2 = \begin{bmatrix} 1 \\ 1 \\ -1 \end{bmatrix}, \quad v_3 = \begin{bmatrix} 2 \\ 0 \\ 0 \end{bmatrix}.$$

A typical linear combination of v_1, v_2, and v_3 looks like

$$c_1 v_1 + c_2 v_2 + c_3 v_3 = \begin{bmatrix} c_1 + c_2 + 2c_3 \\ -c_1 + c_2 \\ -c_1 - c_2 \end{bmatrix}.$$

We wish to know which values of c_1, c_2, and c_3 give 0. Equating corresponding entries of the linear combination and 0 gives us the equations

$$c_1 + c_2 + 2c_3 = 0$$
$$-c_1 + c_2 \qquad = 0$$
$$-c_1 - c_2 \qquad = 0.$$

Adding the first and last equations, we obtain $2c_3 = 0$, so $c_3 = 0$. The last two equations force $c_1 = c_2 = 0$. Thus, the only values of c_1, c_2, and c_3 for which these equations hold are

$$c_1 = c_2 = c_3 = 0.$$

Hence, the vectors v_1, v_2, and v_3 are independent.

Note that v_1, v_2, and v_3 are the initial values at $t = 0$ of the vector valued functions $h_1(t)$, $h_2(t)$, and $k(t)$ in Example 3.3.2. These functions *do* generate the general solution of the homogeneous system (H) in that example.

Example 3.4.2

Check for independence:

$$\mathbf{v}_1 = \begin{bmatrix} 1 \\ 0 \\ 0 \\ 0 \end{bmatrix}, \quad \mathbf{v}_2 = \begin{bmatrix} 0 \\ -1 \\ 0 \\ 0 \end{bmatrix}, \quad \mathbf{v}_3 = \begin{bmatrix} 2 \\ 3 \\ 0 \\ 0 \end{bmatrix}.$$

A typical linear combination of these vectors looks like

$$c_1 \mathbf{v}_1 + c_2 \mathbf{v}_2 + c_3 \mathbf{v}_3 = \begin{bmatrix} c_1 & + 2c_3 \\ -c_2 & + 3c_3 \\ 0 & \\ 0 & \end{bmatrix}.$$

We wish to know which values of c_1, c_2, and c_3 give $\mathbf{0}$. Equating corresponding entries of the linear combination and $\mathbf{0}$ gives us the equations

$$c_1 \qquad + 2c_3 = 0$$
$$-c_2 + 3c_3 = 0$$
$$0 = 0$$
$$0 = 0.$$

The constants c_1, c_2, and c_3 will satisfy these equations if and only if

$$c_1 = -2c_3 \quad \text{and} \quad c_2 = 3c_3.$$

We *can* find constants that are not all 0 and that satisfy these equations. For example, we could take

$$c_3 = 1, \quad c_1 = -2, \quad c_2 = 3.$$

The vectors are linearly dependent.

Example 3.4.3

Check for independence

$$\mathbf{v}_1 = \begin{bmatrix} 1 \\ 0 \\ 0 \end{bmatrix}, \quad \mathbf{v}_2 = \begin{bmatrix} 1 \\ 1 \\ 0 \end{bmatrix}, \quad \mathbf{v}_3 = \begin{bmatrix} 1 \\ 2 \\ 0 \end{bmatrix}, \quad \mathbf{v}_4 = \begin{bmatrix} 2 \\ -2 \\ 2 \end{bmatrix}.$$

A typical linear combination looks like

$$c_1\mathbf{v}_1 + c_2\mathbf{v}_2 + c_3\mathbf{v}_3 + c_4\mathbf{v}_4 = \begin{bmatrix} c_1 + c_2 + c_3 + 2c_4 \\ c_2 + 2c_3 - 2c_4 \\ 2c_4 \end{bmatrix}.$$

Equating corresponding entries of the linear combination and **0** gives us the equations

$$c_1 + c_2 + c_3 + 2c_4 = 0$$
$$c_2 + 2c_3 - 2c_4 = 0$$
$$2c_4 = 0.$$

The constants c_1, c_2, c_3, and c_4 solve these equations if and only if

$$c_4 = 0$$
$$c_2 = -2c_3 + 2c_4 \qquad = -2c_3$$
$$c_1 = -c_2 - c_3 - 2c_4 = 2c_3 - c_3 = c_3.$$

We *can* find constants that are not all zero and that satisfy these equations. For example, we could take

$$c_4 = 0, \quad c_3 = 1, \quad c_2 = -2, \quad c_1 = 1.$$

The vectors are linearly dependent.

Example 3.4.4

Check for independence:

$$\mathbf{v}_1 = \begin{bmatrix} 2 \\ 0 \\ 0 \\ 2 \end{bmatrix}, \quad \mathbf{v}_2 = \begin{bmatrix} 2 \\ 0 \\ 3 \\ 0 \end{bmatrix}, \quad \mathbf{v}_3 = \begin{bmatrix} 1 \\ 1 \\ 0 \\ 0 \end{bmatrix}.$$

A typical linear combination looks like

$$c_1\mathbf{v}_1 + c_2\mathbf{v}_2 + c_3\mathbf{v}_3 = \begin{bmatrix} 2c_1 + 2c_2 + c_3 \\ c_3 \\ 3c_2 \\ 2c_1 \end{bmatrix}.$$

Equating corresponding second, third, and fourth entries of the linear combination and $\mathbf{0}$ yields $c_3 = 0$, $3c_2 = 0$, and $2c_1 = 0$. The only solution is

$$c_1 = c_2 = c_3 = 0.$$

The vectors are linearly independent.

In general, to determine whether a given collection $\mathbf{v}_1, \ldots, \mathbf{v}_k$ of n-vectors is linearly independent, we look for values of $c_1, \ldots, c_k$ for which the linear combination $c_1\mathbf{v}_1 + \cdots + c_k\mathbf{v}_k$ is $\mathbf{0}$. If we set corresponding entries of the linear combination and $\mathbf{0}$ equal, we get n algebraic equations in the k unknowns $c_1, \ldots, c_k$. The coefficients of these equations are just the entries of the vectors $\mathbf{v}_1, \ldots, \mathbf{v}_k$. These equations always have at least one solution, $c_1 = \cdots = c_k = 0$. The vectors are independent if and only if this is the *only* solution.

In Section 3.6 we discuss a systematic method for solving systems of algebraic equations. For now, let's recall that *if $k = n$* (that is, if the number of unknowns is equal to the number of equations), then Cramer's determinant test applies. If the determinant of coefficients is not zero, then the unique solution is $c_1 = c_2 = \cdots = c_n = 0$; the vectors $\mathbf{v}_1, \ldots, \mathbf{v}_n$ are linearly independent in this case. If the determinant of coefficients is zero, then there will be infinitely many other solutions; the vectors are linearly dependent in this case. Thus, we have the following:

Fact: *The n-vectors $\mathbf{v}_1, \ldots, \mathbf{v}_n$ are linearly independent if and only if $\det V \neq 0$, where V is the $n \times n$ matrix whose columns are $\mathbf{v}_1, \ldots, \mathbf{v}_n$.*

If the vectors $\mathbf{v}_1, \ldots, \mathbf{v}_n$ are initial vectors,

$$\mathbf{v}_1 = \mathbf{h}_1(t_0), \ldots, \mathbf{v}_n = \mathbf{h}_n(t_0)$$

then $\det V$ is just $W[\mathbf{h}_1, \ldots, \mathbf{h}_n](t_0)$. Thus $\mathbf{v}_1, \ldots, \mathbf{v}_n$ are linearly independent if and only if this Wronskian is not zero. If we combine this with the Wronskian test from the previous section, we get the following:

Fact: *Suppose $\mathbf{h}_1(t), \ldots, \mathbf{h}_n(t)$ are solutions of the nth-order homogeneous linear system of o.d.e.'s $D\mathbf{x} = A\mathbf{x}$. Suppose the coefficients a_{ij} are continuous on an interval I and t_0 is a fixed value of t in I. The general solution of $D\mathbf{x} = A\mathbf{x}$ is $\mathbf{x} = c_1\mathbf{h}_1(t) + \cdots + c_n\mathbf{h}_n(t)$ if and only if the initial vectors $\mathbf{h}_1(t_0), \ldots, \mathbf{h}_n(t_0)$ are linearly independent.*

This new test determines the strategy we will follow when we look for the general solution of a homogeneous system with constant coefficients. We will start by looking for solutions with linearly independent initial vectors. If we find n of these immediately, then they will generate the general solution. If we find fewer than n, we will look for additional ones to obtain to a larger set of solutions with linearly independent initial vectors. We will continue until we get n linearly independent solutions.

Let's summarize.

LINEAR INDEPENDENCE OF VECTORS

The n-vectors $\mathbf{v}_1, \ldots, \mathbf{v}_k$ are **linearly dependent** if there exist constants $c_1, \ldots, c_k$ with at least one $c_i \neq 0$ so that

$$c_1\mathbf{v}_1 + \cdots + c_k\mathbf{v}_k = \mathbf{0}.$$

The vectors are **linearly independent** if the only constants for which this relationship holds are $c_1 = \cdots = c_k = 0$.

To check whether a given set of vectors

$$\mathbf{v}_1 = \begin{bmatrix} v_{11} \\ \cdot \\ \cdot \\ \cdot \\ v_{n1} \end{bmatrix}, \quad \mathbf{v}_2 = \begin{bmatrix} v_{12} \\ \cdot \\ \cdot \\ \cdot \\ v_{n2} \end{bmatrix}, \ldots, \mathbf{v}_k = \begin{bmatrix} v_{1k} \\ \cdot \\ \cdot \\ \cdot \\ v_{nk} \end{bmatrix}$$

is independent, we set corresponding coefficients of $c_1\mathbf{v}_1 + \cdots + c_k\mathbf{v}_k$ and $\mathbf{0}$ equal to obtain a system of algebraic equations:

$$c_1 v_{11} + c_2 v_{12} + \cdots + c_k v_{1k} = 0$$
$$\cdot$$
$$\cdot$$
$$\cdot$$
$$c_1 v_{n1} + c_2 v_{n2} + \cdots + c_k v_{nk} = 0.$$

If $c_1 = \cdots = c_k = 0$ is the only solution of these equations, then the vectors are linearly independent. Otherwise, they are linearly dependent.

Suppose $\mathbf{h}_1(t), \ldots, \mathbf{h}_n(t)$ are solutions of the nth-order linear system of o.d.e.'s $D\mathbf{x} = A\mathbf{x}$. Suppose the coefficients a_{ij} are continuous on an interval I and t_0 is a fixed value in I. Then the general solution of $D\mathbf{x} = A\mathbf{x}$ is $\mathbf{x} = c_1\mathbf{h}_1(t) + \cdots + c_n\mathbf{h}_n(t)$ if and only if the initial vectors $\mathbf{h}_1(t_0), \ldots, \mathbf{h}_n(t_0)$ are linearly independent.

Notes

1. On the number of solutions needed to generate a general solution

We've seen that the general solution of an nth-order homogeneous system $Dx = Ax$ cannot be generated by fewer than n solutions. Suppose, on the other hand, that we are given $k > n$ solutions $\mathbf{h}_1(t), \ldots, \mathbf{h}_k(t)$ that generate the general solution. Then every initial condition $\mathbf{x}(t_0) = \mathbf{v}$ can be matched by a function of the form $\mathbf{x} = c_1\mathbf{h}_1(t) + \cdots + c_k\mathbf{h}_k(t)$. This says that every n-vector $\mathbf{v}$ can be expressed as a linear combination of the vectors $\mathbf{v}_1 = \mathbf{h}_1(t_0), \ldots, \mathbf{v}_k = \mathbf{h}_k(t_0)$. Since $k > n$, there is a j so that every n-vector $\mathbf{v}$ can still be expressed as a linear combination of $\mathbf{v}_1, \ldots, \mathbf{v}_{j-1}, \mathbf{v}_{j+1}, \ldots, \mathbf{v}_k$ (see Exercise 17). The corresponding solutions $\mathbf{h}_1(t), \ldots, \mathbf{h}_{j-1}(t), \mathbf{h}_{j+1}(t), \ldots, \mathbf{h}_k(t)$ still generate the general solution of $Dx = Ax$, since every initial condition can be matched by some linear combination of these functions. This means that *a generating set of $k > n$ solutions is redundant*, since we can generate the general solution without using all of them.

2. Visualizing independent 3-vectors

In this note we use our visualization of 3-vectors (Note 1, Section 3.2) to obtain a picture of independence. Since any set of vectors that contains $\mathbf{0}$ or that consists of four or more 3-vectors is dependent (Exercise 11 of this section and Exercise 24, Section 3.6), we may restrict our attention to consideration of two or three nonzero 3-vectors.

Two nonzero 3-vectors $\mathbf{v}_1$ and $\mathbf{v}_2$ are independent provided $\mathbf{v}_2$ is not a multiple of $\mathbf{v}_1$ (Exercise 13). Any multiple of $\mathbf{v}_1$ lies on the line through the origin determined by $\mathbf{v}_1$ [see Figure 3.15(a)]. Thus $\mathbf{v}_1$ and $\mathbf{v}_2$ are independent provided $\mathbf{v}_2$ does not lie on this line.

Three nonzero 3-vectors $\mathbf{v}_1$, $\mathbf{v}_2$, and $\mathbf{v}_3$ are independent provided $\mathbf{v}_2$ is not a multiple of $\mathbf{v}_1$ and $\mathbf{v}_3$ is not a linear combination of $\mathbf{v}_1$ and $\mathbf{v}_2$ (Exercise 13). As we've just seen, the first of these conditions tells us that $\mathbf{v}_1$ and $\mathbf{v}_2$ determine different lines through the origin. The parallelogram rule shows that any linear combination of $\mathbf{v}_1$ and $\mathbf{v}_2$ lies in the plane through the origin determined by these vectors [see Figure 3.15(b)]. Thus the second condition implies that $\mathbf{v}_3$ points out of this plane (see Figure 3.16).

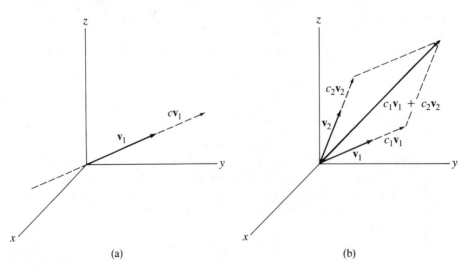

(a)   (b)

FIGURE 3.15

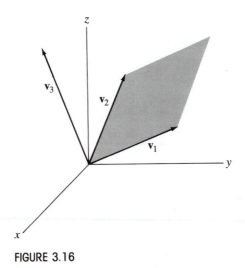

FIGURE 3.16

EXERCISES

In Exercises 1 to 10 check the given set of vectors for linear independence.

1. $\mathbf{v}_1 = \begin{bmatrix} 1 \\ 2 \end{bmatrix}$, $\mathbf{v}_2 = \begin{bmatrix} 2 \\ 1 \end{bmatrix}$

2. $\mathbf{v}_1 = \begin{bmatrix} 3 \\ -1 \end{bmatrix}$, $\mathbf{v}_2 = \begin{bmatrix} 12 \\ -4 \end{bmatrix}$

3. $\mathbf{v}_1 = \begin{bmatrix} 3 \\ -2 \end{bmatrix}$, $\mathbf{v}_2 = \begin{bmatrix} 1 \\ 1 \end{bmatrix}$, $\mathbf{v}_3 = \begin{bmatrix} 2 \\ 2 \end{bmatrix}$

4. $\mathbf{v}_1 = \begin{bmatrix} 1 \\ 0 \\ 1 \end{bmatrix}$, $\mathbf{v}_2 = \begin{bmatrix} 1 \\ 1 \\ -1 \end{bmatrix}$, $\mathbf{v}_3 = \begin{bmatrix} 3 \\ 1 \\ 1 \end{bmatrix}$

5. $\mathbf{v}_1 = \begin{bmatrix} 1 \\ 0 \\ 1 \end{bmatrix}$, $\mathbf{v}_2 = \begin{bmatrix} 1 \\ 1 \\ -1 \end{bmatrix}$, $\mathbf{v}_3 = \begin{bmatrix} 1 \\ 0 \\ 0 \end{bmatrix}$

6. $\mathbf{v}_1 = \begin{bmatrix} 1 \\ 0 \\ 1 \\ 1 \end{bmatrix}$, $\mathbf{v}_2 = \begin{bmatrix} 1 \\ 0 \\ 0 \\ 3 \end{bmatrix}$, $\mathbf{v}_3 = \begin{bmatrix} 1 \\ 1 \\ 0 \\ 0 \end{bmatrix}$

7. $\mathbf{v}_1 = \begin{bmatrix} 1 \\ 0 \\ 1 \\ 1 \end{bmatrix}$, $\mathbf{v}_2 = \begin{bmatrix} 1 \\ 0 \\ 0 \\ 3 \end{bmatrix}$, $\mathbf{v}_3 = \begin{bmatrix} 1 \\ 0 \\ 2 \\ -2 \end{bmatrix}$

8. $\mathbf{v}_1 = \begin{bmatrix} 1 \\ 2 \\ 3 \\ 4 \\ 1 \end{bmatrix}$, $\mathbf{v}_2 = \begin{bmatrix} 2 \\ 1 \\ 4 \\ 3 \\ 2 \end{bmatrix}$, $\mathbf{v}_3 = \begin{bmatrix} -1 \\ 1 \\ -1 \\ 1 \\ -1 \end{bmatrix}$

9. $\mathbf{v}_1 = \begin{bmatrix} 1 \\ 1 \\ 1 \\ 1 \\ 1 \end{bmatrix}$, $\mathbf{v}_2 = \begin{bmatrix} 0 \\ 1 \\ 1 \\ 1 \\ 1 \end{bmatrix}$, $\mathbf{v}_3 = \begin{bmatrix} 0 \\ 0 \\ 1 \\ 1 \\ 1 \end{bmatrix}$, $\mathbf{v}_4 = \begin{bmatrix} 0 \\ 0 \\ 0 \\ 1 \\ 1 \end{bmatrix}$, $\mathbf{v}_5 = \begin{bmatrix} 0 \\ 0 \\ 0 \\ 0 \\ 1 \end{bmatrix}$

10. $\mathbf{v}_1 = \begin{bmatrix} 1 \\ 1 \\ 1 \\ 1 \\ 0 \end{bmatrix}$, $\mathbf{v}_2 = \begin{bmatrix} 2 \\ 0 \\ 3 \\ 2 \\ 1 \end{bmatrix}$, $\mathbf{v}_3 = \begin{bmatrix} 2 \\ 0 \\ 4 \\ 2 \\ 0 \end{bmatrix}$, $\mathbf{v}_4 = \begin{bmatrix} 1 \\ 0 \\ 2 \\ 1 \\ 0 \end{bmatrix}$

Some more abstract questions:

11. Show that any set of n-vectors that includes $\mathbf{0}$ is dependent.

12. Show that if two vectors in a set are equal, then the vectors are dependent.

13. a. Show that if $\mathbf{v}_j = a_1\mathbf{v}_1 + \cdots + a_{j-1}\mathbf{v}_{j-1}$, then $\mathbf{v}_1, \ldots, \mathbf{v}_j, \ldots, \mathbf{v}_k$ are dependent.
    b. Show that if $\mathbf{v}_1, \ldots, \mathbf{v}_k$ are dependent, then some $\mathbf{v}_j$ can be written as a linear combination of the preceding ones, $\mathbf{v}_j = a_1\mathbf{v}_1 + \cdots + a_{j-1}\mathbf{v}_{j-1}$.

14. Show that if $\mathbf{v}_1, \ldots, \mathbf{v}_k$ are independent, then the set obtained by deleting $\mathbf{v}_i$ is also independent.

15. Suppose $\mathbf{v}_1, \ldots, \mathbf{v}_k$ are n-vectors and, for each i, the vector $\mathbf{v}_i$ has a nonzero entry in a position where all the other vectors $\mathbf{v}_1, \ldots, \mathbf{v}_{i-1}, \mathbf{v}_{i+1}, \ldots, \mathbf{v}_k$ have zero entries. Show that $\mathbf{v}_1, \ldots, \mathbf{v}_k$ are linearly independent.

16. a. Suppose $\mathbf{v}_1, \ldots, \mathbf{v}_n$ are n linearly independent n-vectors. Show that every n-vector $\mathbf{v}$ can be written as a linear combination of $\mathbf{v}_1, \ldots, \mathbf{v}_n$:

    $$\mathbf{v} = a_1\mathbf{v}_1 + \cdots + a_n\mathbf{v}_n.$$

    (*Hint:* View this as a system of equations for the unknowns $a_1, \ldots, a_n$. What can you say about the determinant of coefficients?)
    b. Prove that any set $\mathbf{v}_1, \ldots, \mathbf{v}_n, \mathbf{v}$ consisting of $n + 1$ different n-vectors must be linearly dependent.
    c. Does a set of $k \leq n$ different n-vectors have to be independent?

17. Suppose that every n-vector $\mathbf{v}$ can be expressed as a linear combination of the vectors $\mathbf{v}_1, \ldots, \mathbf{v}_k$, where $k > n$. Use the results of Exercises 16(b) and 13(b) to show that for some j, every n-vector $\mathbf{v}$ can be expressed as a linear combination of the vectors $\mathbf{v}_1, \ldots, \mathbf{v}_{j-1}, \mathbf{v}_{j+1}, \ldots, \mathbf{v}_k$, that is, without using $\mathbf{v}_j$.

3.5 HOMOGENEOUS SYSTEMS, EIGENVALUES, AND EIGENVECTORS

We now have methods for determining whether given solutions of a homogeneous linear system

(H) $D\mathbf{x} = A\mathbf{x}$

generate the general solution. In this section we begin our discussion of a method for finding solutions in case the system has *constant coefficients*.

In Chapter 2 we saw that if λ is a root of the polynomial $P(r)$, then $x = e^{\lambda t}$ is a solution of the o.d.e. $P(D)x = 0$. The equivalent system (see Example 3.2.2) has a solution

$$
\mathbf{x} =
\begin{bmatrix}
x \\
x' \\
\cdot \\
\cdot \\
\cdot \\
x^{(n-1)}
\end{bmatrix}
=
\begin{bmatrix}
e^{\lambda t} \\
\lambda e^{\lambda t} \\
\cdot \\
\cdot \\
\cdot \\
\lambda^{n-1} e^{\lambda t}
\end{bmatrix}
= e^{\lambda t}
\begin{bmatrix}
1 \\
\lambda \\
\cdot \\
\cdot \\
\cdot \\
\lambda^{n-1}
\end{bmatrix}.
$$

Let's look for solutions to (H) of the form

$$\mathbf{x} = e^{\lambda t}\mathbf{v}.$$

Note that the initial vector at $t = 0$ of such a solution will be $\mathbf{x}(0) = \mathbf{v}$. Since we want to eventually have solutions with linearly independent initial vectors, we certainly want $\mathbf{v} \neq \mathbf{0}$. If we substitute $\mathbf{x} = e^{\lambda t}\mathbf{v}$ into (H), we get

$$\lambda e^{\lambda t}\mathbf{v} = e^{\lambda t}A\mathbf{v}.$$

Cancellation of $e^{\lambda t}$ yields

$$\lambda\mathbf{v} = A\mathbf{v}.$$

If we find a constant λ and a vector $\mathbf{v}$ with this property, then $\mathbf{x} = e^{\lambda t}\mathbf{v}$ will be a solution of (H).

Definition: *Let A be an n $\times$ n matrix with constant entries. We say the number λ is an **eigenvalue** of A provided there exists a nonzero vector $\mathbf{v}$ such that*

$$A\mathbf{v} = \lambda\mathbf{v}.$$

*Any nonzero vector with this property is then called an **eigenvector** of A corresponding to λ.*

Note that an eigen*value* λ can be zero, but $\mathbf{0}$ is *not* allowed as an eigen*vector*.

With this terminology we can state our observation about solutions of (H) as follows.

Fact: *If λ is an eigenvalue of A and $\mathbf{v}$ is an eigenvector of A corresponding to λ, then $\mathbf{x} = e^{\lambda t}\mathbf{v}$ is a solution of $D\mathbf{x} = A\mathbf{x}$.*

Example 3.5.1

Let

$$A = \begin{bmatrix} 2 & 1 \\ 0 & 2 \end{bmatrix} \quad \text{and} \quad \mathbf{v} = \begin{bmatrix} 1 \\ 0 \end{bmatrix}.$$

Then

$$A\mathbf{v} = \begin{bmatrix} 2 \\ 0 \end{bmatrix} = 2\mathbf{v}.$$

Thus 2 is an eigenvalue of A, and $\mathbf{v}$ is an eigenvector of A corresponding to the eigenvalue 2. The vector valued function

$$\mathbf{x} = e^{2t}\mathbf{v} = \begin{bmatrix} e^{2t} \\ 0 \end{bmatrix}$$

is a solution of $D\mathbf{x} = A\mathbf{x}$.

In order to make use of the preceding observation, we need a procedure for calculating the eigenvalues and eigenvectors of a given matrix

$$A = \begin{bmatrix} a_{11} & a_{12} & \cdots & a_{1n} \\ a_{21} & a_{22} & \cdots & a_{2n} \\ \cdot & \cdot & & \cdot \\ \cdot & \cdot & & \cdot \\ \cdot & \cdot & & \cdot \\ a_{n1} & a_{n2} & \cdots & a_{nn} \end{bmatrix}$$

An eigenvalue of A is a number λ for which the equation

$$A\mathbf{v} = \lambda\mathbf{v}$$

has a nonzero solution

$$\mathbf{v} = \begin{bmatrix} v_1 \\ v_2 \\ \cdot \\ \cdot \\ \cdot \\ v_n \end{bmatrix}.$$

Now

$$Av = \begin{bmatrix} a_{11}v_1 + a_{12}v_2 + \cdots + a_{1n}v_n \\ a_{21}v_1 + a_{22}v_2 + \cdots + a_{2n}v_n \\ \vdots \\ a_{n1}v_1 + a_{n2}v_2 + \cdots + a_{nn}v_n \end{bmatrix} \quad \text{and} \quad \lambda v = \begin{bmatrix} \lambda v_1 \\ \lambda v_2 \\ \vdots \\ \lambda v_n \end{bmatrix}.$$

Equating corresponding entries and regrouping terms, we get the system of algebraic equations

(C)

$$
\begin{array}{rcl}
(a_{11} - \lambda)v_1 + a_{12}v_2 + \cdots + a_{1n}v_n &=& 0 \\
a_{21}v_1 + (a_{22} - \lambda)v_2 + \cdots + a_{2n}v_n &=& 0 \\
\vdots & & \\
a_{n1}v_1 + a_{n2}v_2 + \cdots + (a_{nn} - \lambda)v_n &=& 0.
\end{array}
$$

An eigenvalue of A is a number λ for which the equations (C) have a solution other than the trivial solution $v_1 = \cdots = v_n = 0$.

The system (C) can be rewritten in matrix form as

(C)
$$(A - \lambda I)v = 0$$

where $A - \lambda I$ is the matrix obtained from A by subtracting λ from each of the diagonal entries a_{ii} of A:

$$A - \lambda I = \begin{bmatrix} a_{11} - \lambda & a_{12} & \cdots & a_{1n} \\ a_{21} & a_{22} - \lambda & \cdots & a_{2n} \\ \vdots & \vdots & & \vdots \\ a_{n1} & a_{n2} & \cdots & a_{nn} - \lambda \end{bmatrix}.$$

Note that this matrix is the same as the matrix obtained by subtracting from A the product of the number λ with the $n \times n$ identity matrix I, which we introduced in Note 2 of Section 3.2 (see Exercise 28 in Section 3.2). Cramer's determinant test tells us that (C) has nontrivial solutions if and only if the determinant of coefficients, $\det(A - \lambda I)$, is zero. Thus we have the following:

Fact: *The number λ is an eigenvalue of A if and only if $\det(A - \lambda I) = 0$.*

Example 3.5.2

Find the eigenvalues of

$$A = \begin{bmatrix} -1 & -1 & 0 \\ 1 & -\frac{3}{2} & \frac{3}{2} \\ 0 & 1 & -1 \end{bmatrix}.$$

We expand $\det(A - \lambda I)$ along the first column:

$$\det(A - \lambda I) = \det \begin{bmatrix} -1 - \lambda & -1 & 0 \\ 1 & -\frac{3}{2} - \lambda & \frac{3}{2} \\ 0 & 1 & -1 - \lambda \end{bmatrix}$$

$$= -(1 + \lambda) \det \begin{bmatrix} -\frac{3}{2} - \lambda & \frac{3}{2} \\ 1 & -1 - \lambda \end{bmatrix} - \det \begin{bmatrix} -1 & 0 \\ 1 & -1 - \lambda \end{bmatrix}$$

$$= -(1 + \lambda)\left(\lambda^2 + \frac{5}{2}\lambda + 1\right)$$

$$= -(1 + \lambda)(2 + \lambda)\left(\frac{1}{2} + \lambda\right).$$

The eigenvalues of A are the values that make this zero,

$$\lambda = -1, \quad \lambda = -2, \quad \text{and} \quad \lambda = -\frac{1}{2}.$$

Example 3.5.3

Find the eigenvalues of

$$A = \begin{bmatrix} 2 & 1 & 0 \\ 0 & 2 & 0 \\ 0 & 0 & 2 \end{bmatrix}.$$

Here

$$\det(A - \lambda I) = \det \begin{bmatrix} 2 - \lambda & 1 & 0 \\ 0 & 2 - \lambda & 0 \\ 0 & 0 & 2 - \lambda \end{bmatrix} = (2 - \lambda)^3.$$

The only eigenvalue of A is

$$\lambda = 2.$$

Example 3.5.4

Find the eigenvalues of

$$A = \begin{bmatrix} 1 & 0 & 4 \\ 0 & 3 & 0 \\ 1 & 0 & 1 \end{bmatrix}.$$

Here

$$\det(A - \lambda I) = \det \begin{bmatrix} 1 - \lambda & 0 & 4 \\ 0 & 3 - \lambda & 0 \\ 1 & 0 & 1 - \lambda \end{bmatrix} = (3 - \lambda) \det \begin{bmatrix} 1 - \lambda & 4 \\ 1 & 1 - \lambda \end{bmatrix}$$

$$= (3 - \lambda)(\lambda^2 - 2\lambda - 3) = (3 - \lambda)(\lambda - 3)(\lambda + 1).$$

The eigenvalues of A are

$$\lambda = 3 \quad \text{and} \quad \lambda = -1.$$

In each of our examples, $\det(A - \lambda I)$ was a polynomial in λ. This is true in general (see Exercises 23 and 24).

Fact: *If A is an $n \times n$ matrix, then $\det(A - \lambda I)$ is a polynomial in λ of degree n.*

We refer to $\det(A - \lambda I)$ as the **characteristic polynomial** of A. The eigenvalues of A are the roots of the characteristic polynomial.

Once we have found a specific eigenvalue λ, we find eigenvectors corresponding to λ by finding nonzero solutions of $(A - \lambda I)\mathbf{v} = \mathbf{0}$. Note that this system will have infinitely many solutions (why?), so there will be infinitely many eigenvectors corresponding to λ.

Example 3.5.5

The eigenvalues of

$$A = \begin{bmatrix} 1 & 0 & 4 \\ 0 & 3 & 0 \\ 1 & 0 & 1 \end{bmatrix}$$

are 3 and -1 (see Example 3.5.4). Find an eigenvector corresponding to each eigenvalue.

To find an eigenvector corresponding to -1, we solve the equation $(A - (-1)I)\mathbf{v} = \mathbf{0}$, which is just

$$\begin{bmatrix} 2 & 0 & 4 \\ 0 & 4 & 0 \\ 1 & 0 & 2 \end{bmatrix} \mathbf{v} = \mathbf{0}.$$

Written out in terms of the entries v_1, v_2, and v_3 of $\mathbf{v}$, this reads

$$2v_1 \quad + 4v_3 = 0$$
$$4v_2 \quad\quad = 0$$
$$v_1 \quad + 2v_3 = 0.$$

If we solve these equations for v_1 and v_2, we get

$$v_1 = -2v_3, \quad v_2 = 0.$$

Any choice of v_3, say $v_3 = a$, will lead to a solution:

$$v_1 = -2a, \quad v_2 = 0, \quad v_3 = a.$$

In vector terms, the solutions are the vectors of the form

$$\mathbf{v} = \begin{bmatrix} -2a \\ 0 \\ a \end{bmatrix} = a \begin{bmatrix} -2 \\ 0 \\ 1 \end{bmatrix}.$$

Any nonzero vector of this form will be an eigenvector. In particular, taking $a = 1$,

$$\mathbf{v} = \begin{bmatrix} -2 \\ 0 \\ 1 \end{bmatrix}$$

is an eigenvector corresponding to -1.

To find an eigenvector corresponding to 3, we must solve $(A - 3I)\mathbf{w} = \mathbf{0}$. Written out in terms of the entries of $\mathbf{w}$, this reads

$$-2w_1 + 4w_3 = 0$$
$$0 = 0$$
$$w_1 - 2w_3 = 0.$$

If we solve for w_1, we get

$$w_1 = 2w_3.$$

Any choices of w_2 and w_3, say $w_2 = a$ and $w_3 = b$, will lead to a solution:

$$w_1 = 2b, \quad w_2 = a, \quad w_3 = b.$$

In vector terms, the solutions are the vectors of the form

$$\mathbf{w} = \begin{bmatrix} 2b \\ a \\ b \end{bmatrix} = a \begin{bmatrix} 0 \\ 1 \\ 0 \end{bmatrix} + b \begin{bmatrix} 2 \\ 0 \\ 1 \end{bmatrix}.$$

Any nonzero vector of this form is an eigenvector. In particular, the vectors

$$\mathbf{w}_1 = \begin{bmatrix} 0 \\ 1 \\ 0 \end{bmatrix} \qquad \text{(take } a = 1 \text{ and } b = 0\text{)}$$

and

$$\mathbf{w}_2 = \begin{bmatrix} 2 \\ 0 \\ 1 \end{bmatrix} \qquad \text{(take } a = 0 \text{ and } b = 1\text{)}$$

are eigenvectors corresponding to 3. Note that these two vectors are linearly independent.

In the next section we describe a systematic approach to solving systems of algebraic equations. This approach will be extremely useful in finding eigenvectors. In Sections 3.7 to 3.11 we elaborate on our observation about eigenvalues, eigenvectors, and the corresponding solutions of homogeneous systems of o.d.e.'s to find special solutions that generate the general solution.

Example 3.5.6

Solve $D\mathbf{x} = A\mathbf{x}$, where

$$A = \begin{bmatrix} -3 & -1 & 0 \\ 2 & 0 & 0 \\ 0 & 0 & 0 \end{bmatrix}.$$

This system models the three-loop circuit of Example 3.1.3 with $V(t) = 0$, $R_2 = 0$, $R_1 = 1/3$, $L_2 = L_3 = 1/2$, and $C = 1$.

The characteristic polynomial, $\det(A - \lambda I) = -\lambda(\lambda^2 + 3\lambda + 2)$, has three roots

$$\lambda = -2, -1, 0.$$

To find an eigenvector of A for $\lambda = -2$, we must solve $[A - (-2)I]\mathbf{v} = \mathbf{0}$, or

$$
\begin{aligned}
-v_1 - v_2 &= 0 \\
2v_1 + 2v_2 &= 0 \\
2v_3 &= 0.
\end{aligned}
$$

The last equation forces $v_3 = 0$, and the other two give $v_1 = -v_2$. Any nonzero choice of v_2 leads to an eigenvector; for example, $v_2 = 1$ gives the eigenvector

$$\mathbf{v} = \begin{bmatrix} -1 \\ 1 \\ 0 \end{bmatrix}$$

and a corresponding solution of $D\mathbf{x} = A\mathbf{x}$

$$\mathbf{h}_1(t) = e^{-2t}\mathbf{v} = \begin{bmatrix} -e^{-2t} \\ e^{-2t} \\ 0 \end{bmatrix}.$$

To find an eigenvector corresponding to $\lambda = -1$, we solve the equation $[A - (-1)I]\mathbf{w} = \mathbf{0}$, or

$$
\begin{aligned}
-2w_1 - w_2 &= 0 \\
2w_1 + w_2 &= 0 \\
w_3 &= 0.
\end{aligned}
$$

Here the last equation forces $w_3 = 0$, and the other two give $w_1 = -w_2/2$. Any nonzero choice of w_2 leads to an eigenvector; for example, $w_2 = 2$ gives

$$\mathbf{w} = \begin{bmatrix} -1 \\ 2 \\ 0 \end{bmatrix}$$

and a corresponding solution to $D\mathbf{x} = A\mathbf{x}$

$$\mathbf{h}_2(t) = e^{-t}\mathbf{w} = \begin{bmatrix} -e^{-t} \\ 2e^{-t} \\ 0 \end{bmatrix}.$$

Finally, we look for an eigenvector corresponding to $\lambda = 0$, that is, a nonzero solution of $A\mathbf{u} = \mathbf{0}$. This amounts to

$$-3u_1 - u_2 = 0$$
$$2u_1 = 0$$
$$0 = 0.$$

The second equation forces $u_1 = 0$, which then forces $u_2 = 0$ in the first; there is no restriction on u_3. The choice $u_3 = 1$ gives the eigenvector

$$\mathbf{u} = \begin{bmatrix} 0 \\ 0 \\ 1 \end{bmatrix}$$

and a corresponding solution of $D\mathbf{x} = A\mathbf{x}$

$$\mathbf{h}_3(t) + e^{0t}\mathbf{u} = \mathbf{u} = \begin{bmatrix} 0 \\ 0 \\ 1 \end{bmatrix}.$$

The reader can check that the vectors $\mathbf{h}_1(0) = \mathbf{v}$, $\mathbf{h}_2(0) = \mathbf{w}$, and $\mathbf{h}_3(0) = \mathbf{u}$ are linearly independent. It follows that the general solution of our third-order system $D\mathbf{x} = A\mathbf{x}$ is

$$\mathbf{x} = c_1\mathbf{h}_1(t) + c_2\mathbf{h}_2(t) + c_3\mathbf{h}_3(t) = c_1 \begin{bmatrix} -e^{-2t} \\ e^{-2t} \\ 0 \end{bmatrix} + c_2 \begin{bmatrix} -e^{-t} \\ 2e^{-t} \\ 0 \end{bmatrix} + c_3 \begin{bmatrix} 0 \\ 0 \\ 1 \end{bmatrix}.$$

In terms of currents and charges, this is

$$Q = -c_1e^{2t} - c_2e^{-t}$$
$$I_2 = c_1e^{-2t} + 2c_2e^{-t}$$
$$I_3 = c_3.$$

The absence of the resistance R_2 in this example means that the equation for DI_3 does not involve Q or I_2, and the equations for DQ and DI_2 do not involve I_3. In effect, the one-loop subcircuit carrying the current I_3 does not interact with the two-loop subcircuit carrying Q and I_2 (and hence I_1). The solutions $c_3\mathbf{h}_3(t)$, corresponding to the eigenvalue $\lambda = 0$, are precisely the solutions for which the two-loop subcircuit is dormant ($Q = I_2 = 0$). The solutions $c_1\mathbf{h}_1(t) + c_2\mathbf{h}_2(t)$ are the solutions for which the one-loop subcircuit is dormant ($I_3 = 0$).

The currents in the two-loop subcircuit do interact, via the capacitor. However, there is a different way of measuring the state of the circuit using variables (related to our eigenvectors) whose action is independent: in terms of the variables $J_1 = 2Q + I_2$, $J_2 = Q + I_2$, and I_3, the differential equations describing our circuit read

$$DJ_1 = -2J_1, \quad DJ_2 = -J_2, \quad DI_3 = 0.$$

The solutions $c_1\mathbf{h}_1(t)$, corresponding to the eigenvalue $\lambda = -2$, are the solutions for which $J_2 = I_3 = 0$. The solutions $c_2\mathbf{h}_2(t)$, corresponding to $\lambda = -1$, are the solutions for which $J_1 = I_3 = 0$.

HOMOGENEOUS SYSTEMS, EIGENVALUES, AND EIGENVECTORS

Let A be an $n \times n$ matrix with constant entries. The number λ is an **eigenvalue** of A provided there exists a nonzero vector $\mathbf{v}$ such that $A\mathbf{v} = \lambda\mathbf{v}$. Any nonzero vector with this property is then called an **eigenvector** of A corresponding to λ.

Let $A - \lambda I$ be the matrix obtained from A by subtracting λ from each of the diagonal entries a_{ii} of A. The eigenvalues of A are the roots of the **characteristic polynomial,** $\det(A - \lambda I)$. The eigenvectors of A corresponding to λ are the nonzero solutions of

(C) $$(A - \lambda I)\mathbf{v} = \mathbf{0}.$$

If λ is an eigenvalue of A and $\mathbf{v}$ is an eigenvector of A corresponding to λ, then $\mathbf{x} = e^{\lambda t}\mathbf{v}$ is a solution of $D\mathbf{x} = A\mathbf{x}$.

Notes

1. On the old and new characteristic polynomial

In Section 2.6 we referred to $P(r) = r^n + a_{n-1}r^{n-1} + \cdots + a_0$ as the characteristic polynomial of the equation $(D^n + a_{n-1}D^{n-1} + \cdots + a_0)x = 0$. The matrix of coefficients of the equivalent system is

$$A = \begin{bmatrix} 0 & 1 & 0 & \cdots & 0 \\ 0 & 0 & 1 & \cdots & 0 \\ \cdot & \cdot & \cdot & & \cdot \\ \cdot & \cdot & \cdot & & \cdot \\ \cdot & \cdot & \cdot & & \cdot \\ 0 & 0 & 0 & \cdots & 1 \\ -a_0 & -a_1 & -a_2 & \cdots & -a_{n-1} \end{bmatrix}.$$

The characteristic polynomial of this matrix is $(-1)^n P(\lambda)$.

2. Visualizing the solutions corresponding to an eigenvector

The search for eigenvectors in the case of a constant-coefficient homogeneous system

(H) $$D\mathbf{x} = A\mathbf{x}$$

can be viewed as a search for straight-line solutions. If $\mathbf{v}$ is a nonzero constant vector, then the parametrized curve $x(t) = t\mathbf{v}$ describes a straight line through the origin parallel to $\mathbf{v}$. For any differentiable function $f(t)$ the parametrized curve

$$\mathbf{x} = f(t)\mathbf{v}$$

describes the same line (or a portion of it), but traversed with speed $f'(t)$. To find solutions that travel along straight lines through the origin, we substitute $\mathbf{x}(t) = f(t)\mathbf{v}$ into (H); this leads to

$$f'(t)\mathbf{v} = A[f(t)\mathbf{v}] = f(t)A\mathbf{v}.$$

Since the two sides are the same, $A\mathbf{v}$ must be parallel (or opposite) to $\mathbf{v}$; that is

$$A\mathbf{v} = \lambda\mathbf{v}$$

for some real number λ. In other words, $\mathbf{v}$ has to be an eigenvector of A. If we assume the existence of λ (the eigenvalue), then equating coefficients of $\mathbf{v}$ in

$$f'(t)\mathbf{v} = f(t)A\mathbf{v} = f(t)\lambda\mathbf{v}$$

gives the single first-order o.d.e.

$$f'(t) = f(t)\lambda.$$

The general solution of this o.d.e. is

$$f(t) = ce^{\lambda t}.$$

Thus, we can reformulate the relation between eigenvectors of A and solutions of (H) by saying that *a solution* $\mathbf{x} = \mathbf{x}(t)$ *of (H) that moves along a straight line through the origin must have the form*

$$\mathbf{x}(t) = ce^{\lambda t}\mathbf{v}$$

where **v** *is an eigenvector of A corresponding to the real eigenvalue* λ.

EXERCISES

In Exercises 1 to 6 find (a) the characteristic polynomial and (b) the eigenvalues of the given matrix A.

1. $A = \begin{bmatrix} 0 & 2 \\ -1 & 3 \end{bmatrix}$

2. $A = \begin{bmatrix} 1 & 1 \\ 1 & 1 \end{bmatrix}$

3. $A = \begin{bmatrix} 1 & 1 \\ 3 & 1 \end{bmatrix}$

4. $A = \begin{bmatrix} 1 & -1 \\ 1 & 3 \end{bmatrix}$

5. $A = \begin{bmatrix} 2 & 1 & -2 \\ -3 & 0 & 4 \\ -2 & -1 & 4 \end{bmatrix}$

6. $A = \begin{bmatrix} 0 & -2 & 2 \\ 1 & 3 & -2 \\ 2 & 4 & -3 \end{bmatrix}$

In Exercises 7 to 11 find the eigenvalues of A, and for each eigenvalue find a corresponding eigenvector.

7. $A = \begin{bmatrix} 1 & 1 \\ 3 & -1 \end{bmatrix}$

8. $A = \begin{bmatrix} -3 & 1 \\ -1 & -1 \end{bmatrix}$

9. $A = \begin{bmatrix} 1 & 1 & 1 \\ 0 & 2 & -1 \\ 0 & 0 & -3 \end{bmatrix}$

10. $A = \begin{bmatrix} 1 & -1 & -1 \\ 0 & 0 & -1 \\ 0 & 0 & 1 \end{bmatrix}$

11. $A = \begin{bmatrix} 1 & -1 & -1 \\ 0 & -1 & -1 \\ 0 & 0 & 1 \end{bmatrix}$

In Exercises 12 to 14 you are given a matrix A and an eigenvector of A. Find

a.   the eigenvalue λ to which **v** corresponds,

b.   the associated solution of $D\mathbf{x} = A\mathbf{x}$.

12. $A = \begin{bmatrix} 1 & -3 \\ -3 & 1 \end{bmatrix}$, $\mathbf{v} = \begin{bmatrix} 1 \\ -1 \end{bmatrix}$

13. $A = \begin{bmatrix} 1 & -3 \\ -3 & 1 \end{bmatrix}$, $\mathbf{v} = \begin{bmatrix} 1 \\ 1 \end{bmatrix}$

14. $A = \begin{bmatrix} 2 & -1 \\ -2 & 1 \end{bmatrix}$, $\mathbf{v} = \begin{bmatrix} 1 \\ 2 \end{bmatrix}$

In Exercises 15 to 18 find the general solution of $D\mathbf{x} = A\mathbf{x}$ (which is generated by solutions corresponding to eigenvectors of A).

15. A as in Exercises 12 and 13

16. A as in Exercise 7

17. A as in Exercise 9

18. A as in Exercise 10

In Exercises 19 and 20, (a) find all solutions of $D\mathbf{x} = A\mathbf{x}$ corresponding to eigenvectors of A and (b) show that these do *not* generate the general solution, by exhibiting an initial condition not satisfied by the solutions these generate.

19. A as in Exercise 8   20. A as in Exercise 11

Some more abstract questions:

21.   Suppose $\mathbf{x} = f(t)\mathbf{v}$ is a solution of the homogeneous system $D\mathbf{x} = A\mathbf{x}$ with $\mathbf{v} \neq \mathbf{0}$ and $f(t)$ is a real-valued function that is not identically zero. Show that $f(t)$ must have the form $f(t) = ce^{\lambda t}$, where $c \neq 0$, λ is an eigenvalue of A, and $\mathbf{v}$ is a corresponding eigenvector. [*Hint:* If $\mathbf{v} \neq \mathbf{0}$ and $g(t)\mathbf{v} = h(t)\mathbf{u}$, then $\mathbf{u} = \lambda\mathbf{v}$ for some λ and $g(t) = \lambda h(t)$.]

22.   Suppose A is an $n \times n$ matrix in one of the two forms that follow:

$$\text{upper triangular:} \quad A = \begin{bmatrix} a_{11} & a_{12} & \cdots & a_{1n} \\ & a_{22} & \cdots & a_{2n} \\ & & \ddots & \vdots \\ 0\text{'s} & & & \vdots \\ & & & a_{nn} \end{bmatrix}$$

$$\text{lower triangular:} \quad A = \begin{bmatrix} a_{11} & & & \\ a_{21} & a_{22} & & 0\text{'s} \\ \vdots & & \ddots & \\ \vdots & & & \\ a_{n1} & \cdots & & a_{nn} \end{bmatrix}.$$

   a.   Show that the determinant of A equals the product of its diagonal entries: $\det A = a_{11}a_{22} \cdots a_{nn}$.

   b.   Show that the eigenvalues of A are the same as the diagonal entries of A: $\lambda = a_{11}, \lambda = a_{22}, \dots, \lambda = a_{nn}$.

23.   Show that the characteristic polynomial of a 3×3 matrix is a polynomial of degree 3.

*24. a.   Suppose B is a $k \times k$ matrix, each of whose entries is either a number or a term of the form $a_{ij} - \lambda$ (λ a variable). By induction on k show that if the number of entries of B involving λ is $p \le k$, then $\det B$ is a polynomial in λ of degree at most k. (*Hint:* Expand along a row containing an entry that involves λ, and count.)

   b.   Use (a) to show that the characteristic polynomial of an $n \times n$ matrix is a polynomial of degree n.

3.6   SYSTEMS OF ALGEBRAIC EQUATIONS: ROW REDUCTION

Many of our methods require us to solve algebraic systems of equations. If a system consists of n equations in n unknowns and if the determinant of coefficients of the system is not zero, then Cramer's rule (Note 3, Appendix A) provides us with a formula for the unique solution. The systems that arose in the last two sections often failed to satisfy these conditions. Indeed the systems $(A - \lambda I)\mathbf{v} = \mathbf{0}$ that we have to solve to find eigenvectors are precisely the ones whose determinant of coefficients is zero. In this section we describe a systematic approach to solving algebraic systems of equations. Even when Cramer's rule can be used, our new approach is usually more efficient.

We will be dealing with systems of the form

(E)

$$b_{11}u_1 + b_{12}u_2 + \cdots + b_{1m}u_m = k_1$$
$$b_{21}u_1 + b_{22}u_2 + \cdots + b_{2m}u_m = k_2$$
$$\vdots$$
$$b_{n1}u_1 + b_{n2}u_2 + \cdots + b_{nm}u_m = k_n$$

The coefficients b_{ij} and the k_i's will be given, and we will want to solve for the unknowns $u_1, \ldots, u_m$.

To begin with, we note that (E) can be written in matrix form as

$$B\mathbf{u} = \mathbf{k}$$

where

$$B = \begin{bmatrix} b_{11} & b_{12} & \cdots & b_{1m} \\ b_{21} & b_{22} & \cdots & b_{2m} \\ & & \vdots & \\ b_{n1} & b_{n2} & \cdots & b_{nm} \end{bmatrix}, \quad \mathbf{u} = \begin{bmatrix} u_1 \\ u_2 \\ \vdots \\ u_m \end{bmatrix}, \quad \text{and} \quad \mathbf{k} = \begin{bmatrix} k_1 \\ k_2 \\ \vdots \\ k_n \end{bmatrix}.$$

The information essential for describing the system is contained in the **augmented matrix**

$$[B \mid \mathbf{k}] = \begin{bmatrix} b_{11} & b_{12} & \cdots & b_{1m} & k_1 \\ b_{21} & b_{22} & \cdots & b_{2m} & k_2 \\ & & \vdots & & \\ b_{n1} & b_{n2} & \cdots & b_{nm} & k_n \end{bmatrix}.$$

The vertical line before the last column of this matrix is there just as a reminder that this is the augmented matrix and *not* the coefficient matrix B.

Our general approach to solving linear algebraic systems will be to replace the given system by a simpler system that has the same solutions. We will do this by systematically eliminating variables from the equations of the system.

Since the augmented matrix carries all the information needed to describe the system, we will work with it rather than with the equations themselves.

If we start with a system (E) and perform any of the following operations on the equations, then we get a system that has the same solutions as (E):

1. Adding multiples of one fixed equation to one or more of the other equations.
2. Multiplying an equation by a nonzero number.
3. Rearranging the order of the equations.

Each of these operations affects the augmented matrix of (E). The corresponding matrix operations are called **row operations:**

1. Adding multiples of a fixed row to one or more of the other rows.
2. Multiplying a row by a nonzero constant.
3. Rearranging the order of the rows.

We say that two matrices are **row equivalent** if we can get from one to the other by a sequence of row operations. Since the operations do not affect the solutions of the corresponding systems, we have the following:

Fact: *If* $[B \mid \mathbf{k}]$ *and* $[B' \mid \mathbf{k}']$ *are row equivalent, then the systems* $B\mathbf{u} = \mathbf{k}$ *and* $B'\mathbf{u} = \mathbf{k}'$ *have the same solutions.*

Let's look at some examples. In each example we will start with the augmented matrix of the system and perform a sequence of row operations to obtain a row-equivalent matrix whose system is easier to solve. We aim for a matrix that has a lot of zeros. In particular, we want as many zeros as possible in the lower left. We use the notation $[B \mid \mathbf{k}] \xrightarrow[R_j \to R_j + cR_i]{} [B' \mid \mathbf{k}']$ to indicate that we get $[B' \mid \mathbf{k}']$ from $[B \mid \mathbf{k}]$ by replacing (row j) by {(row j) + c(row i)}, and $[B \mid \mathbf{k}] \xrightarrow[R_i \to cR_i]{} [B' \mid \mathbf{k}']$ to indicate that we get $[B' \mid \mathbf{k}']$ by multiplying (row i) by c. $[B \mid \mathbf{k}] \xrightarrow[\text{rearr}]{} [B' \mid \mathbf{k}']$ indicates that we get $[B' \mid \mathbf{k}]$ by rearranging the order of the rows of $[B \mid \mathbf{k}]$.

Example 3.6.1

Solve

$$
\begin{aligned}
u_1 + u_2 \quad + u_4 &= 3 \\
u_1 + 2u_2 \quad\quad &= 1 \\
u_3 + u_4 &= 2 \\
2u_1 + 2u_2 \quad + 3u_4 &= 3.
\end{aligned}
$$

(E)

Starting with the augmented matrix, we perform a sequence of row operations to introduce zeros:

$$\left[\begin{array}{cccc|c} 1 & 1 & 0 & 1 & 3 \\ 1 & 2 & 0 & 0 & 1 \\ 0 & 0 & 1 & 1 & 2 \\ 2 & 2 & 0 & 3 & 3 \end{array}\right] \xrightarrow[R_4 \to R_4 - 2R_1]{R_2 \to R_2 - R_1} \left[\begin{array}{cccc|c} 1 & 1 & 0 & 1 & 3 \\ 0 & 1 & 0 & -1 & -2 \\ 0 & 0 & 1 & 1 & 2 \\ 0 & 0 & 0 & 1 & -3 \end{array}\right]$$

$$\xrightarrow[\substack{R_1 \to R_1 - R_4 \\ R_2 \to R_2 + R_4 \\ R_3 \to R_3 - R_4}]{} \left[\begin{array}{cccc|c} 1 & 1 & 0 & 0 & 6 \\ 0 & 1 & 0 & 0 & -5 \\ 0 & 0 & 1 & 0 & 5 \\ 0 & 0 & 0 & 1 & -3 \end{array}\right] \xrightarrow[R_1 \to R_1 - R_2]{} \left[\begin{array}{cccc|c} 1 & 0 & 0 & 0 & 11 \\ 0 & 1 & 0 & 0 & -5 \\ 0 & 0 & 1 & 0 & 5 \\ 0 & 0 & 0 & 1 & -3 \end{array}\right].$$

The last matrix is the augmented matrix of the system

$$\begin{aligned} u_1 & & = & 11 \\ & u_2 & = & -5 \\ & & u_3 = & 5 \\ & & u_4 = & -3. \end{aligned}$$

These equations are simply the statement of the (unique) solution of (E), which can be written in vector form as

$$\mathbf{u} = \left[\begin{array}{c} 11 \\ -5 \\ 5 \\ -3 \end{array}\right].$$

Example 3.6.2

Solve

(E)
$$\begin{aligned} 2u_1 + u_2 + 4u_3 + 3u_4 & = 3 \\ 2u_2 + 2u_4 + 5u_5 & = 8 \\ -u_1 + 4u_2 - 2u_3 + 3u_4 + 6u_5 & = 6 \\ u_1 + 2u_3 + u_4 - 2u_5 & = -2. \end{aligned}$$

We start with the augmented matrix and perform row operations to introduce zeros:

$$\begin{bmatrix} 2 & 1 & 4 & 3 & 0 & | & 3 \\ 0 & 2 & 0 & 2 & 5 & | & 8 \\ -1 & 4 & -2 & 3 & 6 & | & 6 \\ 1 & 0 & 2 & 1 & -2 & | & -2 \end{bmatrix} \xrightarrow[\substack{R_1 \to R_1 - 2R_4 \\ R_3 \to R_3 + R_4}]{} \begin{bmatrix} 0 & 1 & 0 & 1 & 4 & | & 7 \\ 0 & 2 & 0 & 2 & 5 & | & 8 \\ 0 & 4 & 0 & 4 & 4 & | & 4 \\ 1 & 0 & 2 & 1 & -2 & | & -2 \end{bmatrix}$$

$$\xrightarrow[\substack{R_2 \to R_2 - 2R_1 \\ R_3 \to R_3 - 4R_1}]{} \begin{bmatrix} 0 & 1 & 0 & 1 & 4 & | & 7 \\ 0 & 0 & 0 & 0 & -3 & | & -6 \\ 0 & 0 & 0 & 0 & -12 & | & -24 \\ 1 & 0 & 2 & 1 & -2 & | & -2 \end{bmatrix} \xrightarrow[\substack{R_2 \to -\frac{1}{3}R_2 \\ R_3 \to -\frac{1}{12}R_3}]{} \begin{bmatrix} 0 & 1 & 0 & 1 & 4 & | & 7 \\ 0 & 0 & 0 & 0 & 1 & | & 2 \\ 0 & 0 & 0 & 0 & 1 & | & 2 \\ 1 & 0 & 2 & 1 & -2 & | & -2 \end{bmatrix}$$

$$\xrightarrow[\substack{R_1 \to R_1 - 4R_2 \\ R_3 \to R_3 - R_2 \\ R_4 \to R_4 + 2R_2}]{} \begin{bmatrix} 0 & 1 & 0 & 1 & 0 & | & -1 \\ 0 & 0 & 0 & 0 & 1 & | & 2 \\ 0 & 0 & 0 & 0 & 0 & | & 0 \\ 1 & 0 & 2 & 1 & 0 & | & 2 \end{bmatrix} \xrightarrow[\text{rearr}]{} \begin{bmatrix} 1 & 0 & 2 & 1 & 0 & | & 2 \\ 0 & 1 & 0 & 1 & 0 & | & -1 \\ 0 & 0 & 0 & 0 & 1 & | & 2 \\ 0 & 0 & 0 & 0 & 0 & | & 0 \end{bmatrix}.$$

The last matrix represents the system

$$\begin{aligned} u_1 \quad + 2u_3 + u_4 \quad &= \quad 2 \\ u_2 \quad\quad + u_4 \quad &= -1 \\ u_5 &= \quad 2 \\ 0 &= \quad 0. \end{aligned}$$

We solve these for u_1, u_2, and u_5 in terms of u_3 and u_4:

$$u_1 = 2 - 2u_3 - u_4, \quad u_2 = -1 - u_4, \quad u_5 = 2.$$

Any choice of u_3 and u_4, say $u_3 = a$ and $u_4 = b$, will lead to a solution of (E):

$$u_1 = 2 - 2a - b, \quad u_2 = -1 - b, \quad u_3 = a, \quad u_4 = b, \quad u_5 = 2.$$

In vector form this reads

$$\mathbf{u} = \begin{bmatrix} 2 - 2a - b \\ -1 - b \\ a \\ b \\ 2 \end{bmatrix} = \begin{bmatrix} 2 \\ -1 \\ 0 \\ 0 \\ 2 \end{bmatrix} + a \begin{bmatrix} -2 \\ 0 \\ 1 \\ 0 \\ 0 \end{bmatrix} + b \begin{bmatrix} -1 \\ -1 \\ 0 \\ 1 \\ 0 \end{bmatrix}.$$

Example 3.6.3

Solve

(E)
$$\begin{aligned} u_1 + 2u_2 + 2u_3 &= 0 \\ 2u_1 + 5u_2 + 5u_3 &= 0 \\ 2u_1 + 2u_2 + 2u_3 &= 1. \end{aligned}$$

We perform a sequence of row operations:

$$\begin{bmatrix} 1 & 2 & 2 & | & 0 \\ 2 & 5 & 5 & | & 0 \\ 2 & 2 & 2 & | & 1 \end{bmatrix} \xrightarrow[\substack{R_2 \to R_2 - 2R_1 \\ R_3 \to R_3 - 2R_1}]{} \begin{bmatrix} 1 & 2 & 2 & | & 0 \\ 0 & 1 & 1 & | & 0 \\ 0 & -2 & -2 & | & 1 \end{bmatrix} \xrightarrow[\substack{R_1 \to R_1 - 2R_2 \\ R_3 \to R_3 + 2R_2}]{} \begin{bmatrix} 1 & 0 & 0 & | & 0 \\ 0 & 1 & 1 & | & 0 \\ 0 & 0 & 0 & | & 1 \end{bmatrix}.$$

The last matrix is the augmented matrix of the system

$$\begin{aligned} u_1 &= 0 \\ u_2 + u_3 &= 0 \\ 0 &= 1. \end{aligned}$$

Since the last equation of this system is impossible, the system, and hence also (E), has no solutions.

In each of these examples the matrix at the end of the sequence of row operations has the following properties:

1. Any row consisting entirely of zeros is at the bottom.
2. The first nonzero entry of each nonzero row is 1. We refer to these as **corner** (or pivot) entries.
3. The corner entry of each nonzero row is further to the right than the corner entries of the preceding rows.
4. The corner entries are the only nonzero entries in their columns.

Any matrix with these four properties is said to be **reduced.**

Schematically, the nonzero entries of a reduced matrix lie in a kind of stair-step pattern. The entry in the corner of each step is 1. Below and to the left of the corners, all the entries are zero. The entries above and to the right of the corners can be anything, except that the corner entries are the only nonzero entries in their columns.

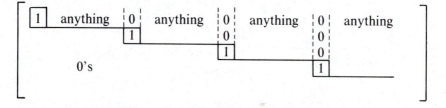

Starting with the augmented matrix $[B \mid \mathbf{k}]$ of (E), we can always obtain a reduced matrix $[B^* \mid \mathbf{k}^*]$ by a sequence of row operations. In fact, $[B^* \mid \mathbf{k}^*]$ is uniquely determined by $[B \mid \mathbf{k}]$.

Fact: *Each matrix [B | k] is row equivalent to exactly one reduced matrix* [B* | k*].

The process of finding [B* | k*] is called **reduction.** The fact that [B* | k*] is unique means we needn't perform the row operations in any particular order. At each stage, any row operation that is convenient can be performed without affecting the final answer. The system B*u = k* has the same solutions as (E). What's more, it is easy to solve.

Example 3.6.4

The eigenvalues of

$$A = \begin{bmatrix} -1 & -1 & 0 \\ 1 & -\frac{3}{2} & \frac{3}{2} \\ 0 & 1 & -1 \end{bmatrix}$$

are $\lambda = -1, -2, -1/2$ (see Example 3.5.2). Find the eigenvectors of A corresponding to each of these eigenvalues.

To find eigenvectors of A corresponding to $\lambda = -1$, we must solve $[A - (-1)I]v = 0$. We reduce the augmented matrix $[A + I \mid 0]$:

$$\begin{bmatrix} 0 & -1 & 0 & \mid & 0 \\ 1 & -\frac{1}{2} & \frac{3}{2} & \mid & 0 \\ 0 & 1 & 0 & \mid & 0 \end{bmatrix} \xrightarrow[\substack{R_2 \to R_2 - \frac{1}{2}R_1 \\ R_3 \to R_3 + R_1}]{} \begin{bmatrix} 0 & -1 & 0 & \mid & 0 \\ 1 & 0 & \frac{3}{2} & \mid & 0 \\ 0 & 0 & 0 & \mid & 0 \end{bmatrix} \xrightarrow[\substack{R_1 \to -R_1 \\ \text{and rearr}}]{} \begin{bmatrix} 1 & 0 & \frac{3}{2} & \mid & 0 \\ 0 & 1 & 0 & \mid & 0 \\ 0 & 0 & 0 & \mid & 0 \end{bmatrix}.$$

The equations of the corresponding system

$$v_1 \quad\quad + \tfrac{3}{2}v_3 = 0$$
$$v_2 \quad\quad\quad = 0$$
$$0 \quad = 0$$

can be solved for the corner variables, v_1 and v_2, in terms of the noncorner variable v_3:

$$v_1 = -\tfrac{3}{2}v_3, \quad v_2 = 0.$$

Any choice of v_3, say $v_3 = 2a$, leads to a solution. In vector terms, the solutions are the vectors of the form

$$\mathbf{v} = \begin{bmatrix} -3a \\ 0 \\ 2a \end{bmatrix} = a \begin{bmatrix} -3 \\ 0 \\ 2 \end{bmatrix}.$$

The nonzero vectors of this form are the eigenvectors for $\lambda = -1$.

To find the eigenvectors for $\lambda = -2$, we solve $[A - (-2)I]\mathbf{w} = \mathbf{0}$. The augmented matrix $[A + 2I \mid \mathbf{0}]$ can be reduced as follows:

$$
\begin{bmatrix}
1 & -1 & 0 & | & 0 \\
1 & \frac{1}{2} & \frac{3}{2} & | & 0 \\
0 & 1 & 1 & | & 0
\end{bmatrix}
\xrightarrow[\substack{R_1 \to R_1 + R_3 \\ R_2 \to R_2 - \frac{1}{2}R_3}]{}
\begin{bmatrix}
1 & 0 & 1 & | & 0 \\
1 & 0 & 1 & | & 0 \\
0 & 1 & 1 & | & 0
\end{bmatrix}
\xrightarrow[\substack{R_2 \to R_2 - R_1 \\ \text{and rearr}}]{}
\begin{bmatrix}
1 & 0 & 1 & | & 0 \\
0 & 1 & 1 & | & 0 \\
0 & 0 & 0 & | & 0
\end{bmatrix}.
$$

The equations corresponding to the last matrix are

$$
\begin{aligned}
w_1 \quad\;\; + w_3 &= 0 \\
w_2 + w_3 &= 0 \\
0 &= 0.
\end{aligned}
$$

Any choice of the noncorner variable, w_3, say $w_3 = a$, leads to a solution. These solutions can be written in the form

$$
\mathbf{w} =
\begin{bmatrix}
-a \\
-a \\
a
\end{bmatrix}
= a
\begin{bmatrix}
-1 \\
-1 \\
1
\end{bmatrix}.
$$

The eigenvectors for $\lambda = -2$ are the nonzero vectors of this form.

To find eigenvectors for $\lambda = -1/2$, we solve $[A - (-\frac{1}{2})I]\mathbf{u} = \mathbf{0}$ by reducing the augmented matrix $[A + \frac{1}{2}I \mid \mathbf{0}]$:

$$
\begin{bmatrix}
-\frac{1}{2} & -1 & 0 & | & 0 \\
1 & -1 & \frac{3}{2} & | & 0 \\
0 & 1 & -\frac{1}{2} & | & 0
\end{bmatrix}
\xrightarrow[\substack{R_1 \to R_1 + R_3 \\ R_2 \to R_2 + R_3}]{}
\begin{bmatrix}
-\frac{1}{2} & 0 & -\frac{1}{2} & | & 0 \\
1 & 0 & 1 & | & 0 \\
0 & 1 & -\frac{1}{2} & | & 0
\end{bmatrix}
$$

$$
\xrightarrow[\substack{R_1 \to R_1 + \frac{1}{2}R_2 \\ \text{and rearr}}]{}
\begin{bmatrix}
1 & 0 & 1 & | & 0 \\
0 & 1 & -\frac{1}{2} & | & 0 \\
0 & 0 & 0 & | & 0
\end{bmatrix}.
$$

The corresponding equations, solved for the corner variables u_1 and u_2 in terms of the noncorner variable u_3, give

$$
u_1 = -u_3, \quad u_2 = \tfrac{1}{2}u_3.
$$

The eigenvectors of A with eigenvalue $\lambda = -1/2$ are the nonzero vectors of the form

$$
\mathbf{u} =
\begin{bmatrix}
-2a \\
a \\
2a
\end{bmatrix}
= a
\begin{bmatrix}
-2 \\
1 \\
2
\end{bmatrix}.
$$

Example 3.6.5

Given that 2 is an eigenvalue of

$$A = \begin{bmatrix} 2 & 0 & 0 \\ -6 & 8 & -2 \\ -9 & 9 & -1 \end{bmatrix}$$

describe the eigenvectors of A corresponding to $\lambda = 2$.

We reduce the augmented matrix of $(A - 2I)\mathbf{v} = \mathbf{0}$:

$$\begin{bmatrix} 0 & 0 & 0 & | & 0 \\ -6 & 6 & -2 & | & 0 \\ -9 & 9 & -3 & | & 0 \end{bmatrix} \xrightarrow[\substack{R_2 \to -\frac{1}{6}R_2 \\ \text{and rearr}}]{} \begin{bmatrix} 1 & -1 & \frac{1}{3} & | & 0 \\ -9 & 9 & -3 & | & 0 \\ 0 & 0 & 0 & | & 0 \end{bmatrix}$$

$$\xrightarrow[R_2 \to R_2 + 9R_1]{} \begin{bmatrix} 1 & -1 & \frac{1}{3} & | & 0 \\ 0 & 0 & 0 & | & 0 \\ 0 & 0 & 0 & | & 0 \end{bmatrix}.$$

If we solve the nontrivial equation of the corresponding system for the corner variable v_1 in terms of the noncorner variables, v_2 and v_3, we get

$$v_1 = v_2 - \tfrac{1}{3}v_3.$$

Any choice of the noncorner variables, say $v_2 = a$ and $v_3 = 3b$, leads to a solution. The eigenvectors are the nonzero vectors of the form

$$\mathbf{v} = \begin{bmatrix} a - b \\ a \\ 3b \end{bmatrix} = a \begin{bmatrix} 1 \\ 1 \\ 0 \end{bmatrix} + b \begin{bmatrix} -1 \\ 0 \\ 3 \end{bmatrix}.$$

Let's summarize.

SYSTEMS OF ALGEBRAIC EQUATIONS: ROW REDUCTION

Two matrices are **row equivalent** if we can get from one to the other by a sequence of **row operations:**

1. Adding multiples of one fixed row to one or more of the other rows.
2. Multiplying a row by a nonzero constant.
3. Rearranging the order of the rows.

> ### A matrix is **reduced** provided:
>
> 1. Any row consisting entirely of zeros is at the bottom.
> 2. The first nonzero entry, or **corner** entry, of each nonzero row is 1.
> 3. The corner entry of each nonzero row is further to the right than the corner entries of the preceding rows.
> 4. The corner entries are the only nonzero entries in their columns.
>
> To solve $B\mathbf{u} = \mathbf{k}$, we first **reduce** the **augmented matrix** $[B \mid \mathbf{k}]$—that is, we perform a sequence of row operations to obtain a reduced matrix $[B^* \mid \mathbf{k}^*]$. We then solve $B^*\mathbf{u} = \mathbf{k}^*$. The solutions of $B\mathbf{u} = \mathbf{k}$ are the same as the solutions of $B^*\mathbf{u} = \mathbf{k}^*$.

Notes

1. A warning

The row operation of type 1 allows us to alter several rows at once, using a single row R_i, by adding or subtracting multiples of R_i to each of the other rows. We urge you to be careful when doing this. It is tempting to try to do several such operations at once, using a different row to alter each of the original rows of the matrix simultaneously. However, if you are not careful, this can lead to subtracting a row from itself or some other operation equivalent to multiplying a row by zero, which is *not* an allowable operation.

2. On the solutions of $B^*\mathbf{u} = \mathbf{k}^*$

If $[B^* \mid \mathbf{k}^*]$ is reduced, then the location of the corners determines whether $B^*\mathbf{u} = \mathbf{k}^*$ has no solutions, a unique solution, or infinitely many solutions. If there is a corner in the last column, as in Example 3.6.3, then there are no solutions. If each column except the last has a corner, as in Example 3.6.1, then $\mathbf{u} = \mathbf{k}^*$ is the unique solution. If the last column does not have a corner and there are i other columns without corners, as in Example 3.6.2, then it is possible to find vectors $\mathbf{w}, \mathbf{v}_1, \ldots, \mathbf{v}_i$ so that the solutions are the vectors of the form $\mathbf{u} = \mathbf{w} + a_1\mathbf{v}_1 + \cdots + a_i\mathbf{v}_i$. There are infinitely many solutions in this case.

It is often possible to determine how many solutions a system $B\mathbf{u} = \mathbf{k}$ has without completing the reduction. If, for example, one of the rows of an intermediate matrix is $[0 \quad 0 \quad \cdots \quad 0 \mid 1]$, then this will also be true of the reduced matrix. There are no solutions in this case.

3. Row operations and determinants

Row operations can be useful when calculating the determinant of an $n \times n$ matrix M (see Note 2 in Appendix A). However, it is *not* true that row-equivalent matrices necessarily have the same determinant. The following table shows the effect on the determinant of each of our row operations:

$$M \xrightarrow[R_j \to R_j + cR_i]{} M' \qquad \det M' = \det M$$

$$M \xrightarrow[R_j \to cR_j]{} M' \qquad \det M' = c(\det M)$$

$$M \xrightarrow[\text{rearr}]{} M' \qquad \det M' = (-1)^j(\det M), \text{ where } j \text{ is the number of interchanges of rows it would take to arrive at the rearrangement.}$$

EXERCISES

In Exercises 1 to 4 reduce the given matrix.

1. $\begin{bmatrix} 1 & 3 & -1 & 1 \\ 0 & 1 & 0 & 0 \\ 1 & 1 & 1 & 1 \\ 2 & 2 & 2 & 2 \end{bmatrix}$

2. $\begin{bmatrix} 1 & 1 & 1 & 1 \\ 1 & 1 & 0 & 1 \\ 0 & 1 & 1 & 1 \\ 1 & 1 & 1 & 1 \end{bmatrix}$

3. $\begin{bmatrix} 1 & 0 & 2 & 3 \\ 2 & 1 & 4 & 5 \\ 1 & 1 & 2 & 2 \\ 0 & 1 & 0 & 1 \end{bmatrix}$

4. $\begin{bmatrix} 0 & 2 & 2 & 1 \\ 0 & 1 & 1 & 3 \\ 0 & 4 & 4 & -1 \\ 0 & 0 & 0 & 6 \end{bmatrix}$

In Exercise 5 to 9 find all solutions of the given system of equations (if any exist). Express your answer (a) as separate parametric equations for the variables and (b) as a linear combination of vectors.

5.
$$\begin{aligned}
x + 2y - z + 4w - u &= 0 \\
2x - y + 3z - w + u &= 0 \\
-x + y - z + 2w + u &= 0 \\
x + y - 4z + 2w &= 0 \\
-x + 3z + 2w - u &= 0
\end{aligned}$$

6.
$$\begin{aligned}
x_1 + 2x_2 + 3x_3 - 2x_4 &= 0 \\
3x_1 - 7x_2 - 4x_3 + 7x_4 &= 0 \\
4x_1 - 3x_2 + x_3 + 3x_4 &= 0 \\
x_1 + 3x_2 + 4x_3 - 3x_4 &= 0
\end{aligned}$$

7.
$$\begin{aligned}
x - y + z &= 1 \\
2x + y + 2z &= 5 \\
x + 2y + z &= 4
\end{aligned}$$

8.
$$\begin{aligned}
x_1 + 2x_2 + x_3 - x_4 - x_5 &= 2 \\
2x_1 + 2x_2 + 2x_3 - 3x_4 - 2x_5 &= 1 \\
-x_1 - x_3 + 2x_4 + x_5 &= 0
\end{aligned}$$

9.
$$\begin{aligned}
x_1 + x_2 - x_3 + x_4 &= 1 \\
x_1 - x_2 + x_3 + x_4 &= 0 \\
x_1 + x_2 - x_3 - x_4 &= 1 \\
x_1 + x_2 + x_3 + x_4 &= 0
\end{aligned}$$

In Exercises 10 to 13 solve $B\mathbf{u} = \mathbf{k}$ for $\mathbf{u}$ (if possible):

10. $B = \begin{bmatrix} 1 & -1 & 1 \\ 0 & 1 & 1 \\ 1 & 1 & 1 \end{bmatrix}$, $\mathbf{k} = \begin{bmatrix} 1 \\ 2 \\ 3 \end{bmatrix}$

11. $B = \begin{bmatrix} 1 & 2 & 3 & 0 \\ 1 & -1 & -3 & 1 \\ 2 & 1 & 0 & 1 \end{bmatrix}$, $\mathbf{k} = \begin{bmatrix} 1 \\ 1 \\ 1 \end{bmatrix}$

12. $B = \begin{bmatrix} 1 & 2 & 3 & 0 \\ 1 & -1 & -3 & 1 \\ 2 & 1 & 0 & 1 \end{bmatrix}$, $\mathbf{k} = \begin{bmatrix} 1 \\ 1 \\ 2 \end{bmatrix}$

13. $B = \begin{bmatrix} 1 & 2 & 3 \\ 1 & -1 & 2 \\ 2 & 1 & 5 \\ 0 & 3 & 1 \end{bmatrix}$, $\mathbf{k} = \begin{bmatrix} 1 \\ 2 \\ 3 \\ -1 \end{bmatrix}$

In Exercises 14 to 19 find the eigenvectors corresponding to each of the eigenvalues of A, where A is the matrix in the indicated exercise from Section 3.5:

14. Exercise 1

15. Exercise 2

16. Exercise 3

17. Exercise 4

18. Exercise 5

19. Exercise 6

We have encountered algebraic systems of equations before. In Exercises 20 and 21 use row reduction to help solve the algebraic systems that arise in solving the problems.

20. Check for independence:

a. $\mathbf{v}_1 = \begin{bmatrix} 1 \\ 2 \\ -1 \end{bmatrix}$, $\mathbf{v}_2 = \begin{bmatrix} 0 \\ -1 \\ 3 \end{bmatrix}$, $\mathbf{v}_3 = \begin{bmatrix} 1 \\ 4 \\ 5 \end{bmatrix}$

b. $\mathbf{v}_1 = \begin{bmatrix} 1 \\ 3 \\ -1 \\ 1 \end{bmatrix}$, $\mathbf{v}_2 = \begin{bmatrix} -1 \\ 2 \\ 2 \\ 1 \end{bmatrix}$, $\mathbf{v}_3 = \begin{bmatrix} 1 \\ 4 \\ 1 \\ 3 \end{bmatrix}$, $\mathbf{v}_4 = \begin{bmatrix} 1 \\ -1 \\ 0 \\ 1 \end{bmatrix}$

c. $\mathbf{v}_1 = \begin{bmatrix} 1 \\ 0 \\ 1 \\ 2 \end{bmatrix}$, $\mathbf{v}_2 = \begin{bmatrix} 1 \\ -1 \\ 0 \\ 1 \end{bmatrix}$, $\mathbf{v}_3 = \begin{bmatrix} 1 \\ 2 \\ 3 \\ 4 \end{bmatrix}$

21. Find the general solution and the specific solution satisfying the given initial conditions:

a. $(D - 1)(D + 1)(D^2 + 1)x = 0$; $x(0) = x'(0) = 1, x''(0) = x'''(0) = 2$
b. $(D - 1)^2(D + 1)^2 x = 0$; $x(0) = x'(0) = 1, x''(0) = x'''(0) = 0$

Some more abstract questions:

22. Show that a reduced matrix with fewer rows than columns ($n \times m, n < m$) must have some columns without corners.

23. Use Exercise 22 to show that a system of n equations in $m > n$ unknowns, and with zero right-hand side, $B\mathbf{u} = \mathbf{0}$, always has nonzero solutions.

24. Use the fact proved in Exercise 23 to show that any set of $m > n$ different n-vectors must be linearly dependent.

25. Give an example to show that the assumption of a zero right-hand side in Exercise 23 is necessary; that is, find a system of n equations in $m > n$ unknowns, $B\mathbf{u} = \mathbf{v}$, with no solutions.

3.7 HOMOGENEOUS SYSTEMS WITH CONSTANT COEFFICIENTS: REAL ROOTS

In Section 3.4 we saw that the problem of solving an nth-order homogeneous system $D\mathbf{x} = A\mathbf{x}$ boils down to finding n solutions with linearly independent initial vectors. In Section 3.5 we saw that if $\mathbf{v}$ is an eigenvector of A corresponding to the eigenvalue λ, then $\mathbf{h}(t) = e^{\lambda t}\mathbf{v}$ is a solution with initial vector $\mathbf{h}(0) = \mathbf{v}$. If we find n *linearly independent* eigenvectors $\mathbf{v}_1, \ldots, \mathbf{v}_n$ corresponding to the eigenvalues $\lambda_1, \ldots, \lambda_n$, respectively, then the associated solutions

$$\mathbf{h}_1(t) = e^{\lambda_1 t}\mathbf{v}_1, \ldots, \mathbf{h}_n(t) = e^{\lambda_n t}\mathbf{v}_n$$

will generate the general solution of $D\mathbf{x} = A\mathbf{x}$:

$$\mathbf{x} = c_1\mathbf{h}_1(t) + \cdots + c_n\mathbf{h}_n(t).$$

Example 3.7.1

Solve $D\mathbf{x} = A\mathbf{x}$, where

$$A = \begin{bmatrix} 1 & 0 & 4 \\ 0 & 3 & 0 \\ 1 & 0 & 1 \end{bmatrix}.$$

We found the eigenvalues and corresponding eigenvectors of A in Examples 3.5.4 and 3.5.5. The vector

$$\mathbf{v} = \begin{bmatrix} -2 \\ 0 \\ 1 \end{bmatrix}$$

is an eigenvector of A corresponding to -1. Associated to this vector is a solution of $D\mathbf{x} = A\mathbf{x}$:

$$\mathbf{h}_1(t) = e^{-t}\mathbf{v} = \begin{bmatrix} -2e^{-t} \\ 0 \\ e^{-t} \end{bmatrix}.$$

The vectors

$$\mathbf{w}_1 = \begin{bmatrix} 0 \\ 1 \\ 0 \end{bmatrix} \quad \text{and} \quad \mathbf{w}_2 = \begin{bmatrix} 2 \\ 0 \\ 1 \end{bmatrix}$$

are linearly independent eigenvectors corresponding to 3. Associated to these vectors are two more solutions:

$$\mathbf{h}_2(t) = e^{3t}\mathbf{w}_1 = \begin{bmatrix} 0 \\ e^{3t} \\ 0 \end{bmatrix} \quad \text{and} \quad \mathbf{h}_3(t) = e^{3t}\mathbf{w}_2 = \begin{bmatrix} 2e^{3t} \\ 0 \\ e^{3t} \end{bmatrix}.$$

If the initial vectors $\mathbf{v}$, $\mathbf{w}_1$, and $\mathbf{w}_2$ are linearly independent, then $\mathbf{h}_1(t)$, $\mathbf{h}_2(t)$, and $\mathbf{h}_3(t)$ generate the general solution of our third-order system.

To determine whether $\mathbf{v}$, $\mathbf{w}_1$, and $\mathbf{w}_2$ are independent, we look for the values of c_1, c_2, and c_3 for which

(I)
$$c_1\mathbf{v} + c_2\mathbf{w}_1 + c_3\mathbf{w}_2 = \mathbf{0}.$$

Since $\mathbf{v}$, $\mathbf{w}_1$, and $\mathbf{w}_2$ are eigenvectors of A corresponding to -1, 3, and 3, respectively, then

$$A\mathbf{v} = -\mathbf{v}, \quad A\mathbf{w}_1 = 3\mathbf{w}_1, \quad \text{and} \quad A\mathbf{w}_2 = 3\mathbf{w}_2.$$

Multiplication of (I) by A yields

(II) $$-c_1\mathbf{v} + 3c_2\mathbf{w}_1 + 3c_3\mathbf{w}_2 = \mathbf{0}.$$

If we subtract 3 times (I) from (II), we get $-4c_1\mathbf{v} = \mathbf{0}$. Since $\mathbf{v} \neq \mathbf{0}$,

$$c_1 = 0.$$

Substitution into (I) yields $c_2\mathbf{w}_1 + c_3\mathbf{w}_2 = \mathbf{0}$. Since $\mathbf{w}_1$ and $\mathbf{w}_2$ are linearly independent,

$$c_2 = c_3 = 0.$$

Thus, $\mathbf{v}$, $\mathbf{w}_1$, and $\mathbf{w}_2$ are linearly independent.

The general solution of $D\mathbf{x} = A\mathbf{x}$ is

$$\mathbf{x} = c_1\mathbf{h}_1(t) + c_2\mathbf{h}_2(t) + c_3\mathbf{h}_3(t)$$

$$= c_1 \begin{bmatrix} -2e^{-t} \\ 0 \\ e^{-t} \end{bmatrix} + c_2 \begin{bmatrix} 0 \\ e^{3t} \\ 0 \end{bmatrix} + c_3 \begin{bmatrix} 2e^{3t} \\ 0 \\ e^{3t} \end{bmatrix}.$$

The argument we used in this example to show independence can be extended to show that a list of eigenvectors of A corresponding to several eigenvalues will be independent provided the eigenvectors corresponding to each single eigenvalue are independent. Combining this with our earlier observations, we have the following:

Fact: *Let A be an $n \times n$ matrix with constant entries. Associate a list of vector valued functions to A as follows: For each eigenvalue λ of A, find as many linearly independent eigenvectors corresponding to λ as possible, and associate to each such eigenvector $\mathbf{v}$ the function $e^{\lambda t}\mathbf{v}$. These vector valued functions are solutions of $D\mathbf{x} = A\mathbf{x}$ with linearly independent initial vectors. In particular, if we find n such functions $\mathbf{h}_1(t), \ldots, \mathbf{h}_n(t)$, then the general solution of $D\mathbf{x} = A\mathbf{x}$ is $\mathbf{x} = c_1\mathbf{h}_1(t) + \cdots + c_n\mathbf{h}_n(t)$.*

One case in which we can be sure that this process will yield n functions occurs when A has n *distinct* eigenvalues. In this case, each eigenvalue will contribute one solution toward the n we need.

Example 3.7.2

Solve $D\mathbf{x} = A\mathbf{x}$, where

$$A = \begin{bmatrix} -1 & -1 & 0 \\ 1 & -\frac{3}{2} & \frac{3}{2} \\ 0 & 1 & -1 \end{bmatrix}.$$

This represents the three-loop circuit of Example 3.1.3 with $V(t) = 0$, $R_1 = 2$, $R_2 = 3$, $L_2 = 2$, $L_3 = 3$, and $C = 1/2$.

We found the eigenvalues and eigenvectors of A in Examples 3.5.2 and 3.6.4. The vectors

$$\mathbf{v} = \begin{bmatrix} -3 \\ 0 \\ 2 \end{bmatrix}, \quad \mathbf{w} = \begin{bmatrix} -1 \\ -1 \\ 1 \end{bmatrix}, \quad \text{and} \quad \mathbf{u} = \begin{bmatrix} -2 \\ 1 \\ 2 \end{bmatrix}$$

are eigenvectors corresponding to the eigenvalues -1, -2, and $-1/2$, respectively. Associated to these eigenvectors are the three solutions

$$\mathbf{h}_1(t) = e^{-t}\mathbf{v}, \quad \mathbf{h}_2(t) = e^{-2t}\mathbf{w}, \quad \mathbf{h}_3(t) = e^{-t/2}\mathbf{u}$$

with independent initial vectors. The general solution of our third-order system is therefore

$$\mathbf{x} = c_1\mathbf{h}_1(t) + c_2\mathbf{h}_2(t) + c_3\mathbf{h}_3(t)$$

$$= c_1 \begin{bmatrix} -3e^{-t} \\ 0 \\ 2e^{-t} \end{bmatrix} + c_2 \begin{bmatrix} -e^{-2t} \\ -e^{-2t} \\ e^{-2t} \end{bmatrix} + c_3 \begin{bmatrix} -2e^{-t/2} \\ e^{-t/2} \\ 2e^{-t/2} \end{bmatrix}.$$

In terms of the currents and charges in the circuit of Example 3.1.3, this is

$$\begin{aligned} Q(t) &= -3c_1e^{-t} - c_2e^{-2t} - 2c_3e^{-t/2} \\ I_2(t) &= \qquad\quad - c_2e^{-2t} + c_3e^{-t/2} \\ I_3(t) &= \quad 2c_1e^{-t} + c_2e^{-2t} + 2c_3e^{-t/2}. \end{aligned}$$

Example 3.7.3

Find the solution of

$$\text{(H)} \qquad \begin{aligned} x_1' &= x_1 - x_2 \\ x_2' &= \quad - x_2 + 3x_3 \\ x_3' &= -x_1 + x_2 \end{aligned}$$

that satisfies the initial condition

$$x_1(0) = x_2(0) = 0, \quad x_3(0) = 1.$$

This system can be written in matrix terms as $D\mathbf{x} = A\mathbf{x}$, where

$$\mathbf{x} = \begin{bmatrix} x_1 \\ x_2 \\ x_3 \end{bmatrix} \quad \text{and} \quad A = \begin{bmatrix} 1 & -1 & 0 \\ 0 & -1 & 3 \\ -1 & 1 & 0 \end{bmatrix}.$$

The characteristic polynomial of A is

$$\det(A - \lambda I) = \det \begin{bmatrix} 1 - \lambda & -1 & 0 \\ 0 & -1 - \lambda & 3 \\ -1 & 1 & -\lambda \end{bmatrix}$$

$$= -\lambda^3 + 4\lambda = -\lambda(\lambda - 2)(\lambda + 2).$$

The eigenvalues of A are

$$\lambda = 0, 2, -2.$$

To find eigenvectors of A, we reduce $[A - 0I \mid \mathbf{0}]$, $[A - 2I \mid \mathbf{0}]$, and $[A - (-2)I \mid \mathbf{0}]$. The row-equivalent reduced matrices are

$$\begin{bmatrix} 1 & 0 & -3 & | & 0 \\ 0 & 1 & -3 & | & 0 \\ 0 & 0 & 0 & | & 0 \end{bmatrix}, \quad \begin{bmatrix} 1 & 0 & 1 & | & 0 \\ 0 & 1 & -1 & | & 0 \\ 0 & 0 & 0 & | & 0 \end{bmatrix}, \quad \text{and} \quad \begin{bmatrix} 1 & 0 & 1 & | & 0 \\ 0 & 1 & 3 & | & 0 \\ 0 & 0 & 0 & | & 0 \end{bmatrix},$$

respectively. Each of the algebraic systems corresponding to the reduced matrices has a single noncorner variable. If we take the noncorner variable to be 1, we obtain eigenvectors.

$$\mathbf{v} = \begin{bmatrix} 3 \\ 3 \\ 1 \end{bmatrix}, \quad \mathbf{w} = \begin{bmatrix} -1 \\ 1 \\ 1 \end{bmatrix}, \quad \text{and} \quad \mathbf{u} = \begin{bmatrix} -1 \\ -3 \\ 1 \end{bmatrix}$$

corresponding to 0, 2, and -2, respectively. The three solutions associated to these vectors generate the general solution of our third-order system (H). In matrix terms, the general solution is

$$\mathbf{x} = c_1 \mathbf{v} + c_2 e^{2t} \mathbf{w} + c_3 e^{-2t} \mathbf{u}$$

$$= c_1 \begin{bmatrix} 3 \\ 3 \\ 1 \end{bmatrix} + c_2 \begin{bmatrix} -e^{2t} \\ e^{2t} \\ e^{2t} \end{bmatrix} + c_3 \begin{bmatrix} -e^{2t} \\ -3e^{-2t} \\ e^{-2t} \end{bmatrix}.$$

In matrix terms, the initial condition is

$$\mathbf{x}(0) = \begin{bmatrix} 0 \\ 0 \\ 1 \end{bmatrix}.$$

If we equate corresponding entries of $\mathbf{x}(0)$ and the given initial vector, we obtain the algebraic system of equations

$$3c_1 - c_2 - c_3 = 0$$
$$3c_1 + c_2 - 3c_3 = 0$$
$$c_1 + c_2 + c_3 = 1.$$

We reduce the augmented matrix of this algebraic system

$$\begin{bmatrix} 3 & -1 & -1 & | & 0 \\ 3 & 1 & -3 & | & 0 \\ 1 & 1 & 1 & | & 1 \end{bmatrix} \rightarrow \begin{bmatrix} 1 & 0 & 0 & | & \frac{1}{4} \\ 0 & 1 & 0 & | & \frac{3}{8} \\ 0 & 0 & 1 & | & \frac{3}{8} \end{bmatrix}$$

to determine the solution

$$c_1 = \tfrac{1}{4}, \quad c_2 = \tfrac{3}{8}, \quad c_3 = \tfrac{3}{8}.$$

The solution of (H) satisfying the given initial condition is

$$\mathbf{x} = \tfrac{1}{4} \begin{bmatrix} 3 \\ 3 \\ 1 \end{bmatrix} + \tfrac{3}{8} \begin{bmatrix} -e^{2t} \\ e^{2t} \\ e^{2t} \end{bmatrix} + \tfrac{3}{8} \begin{bmatrix} -e^{-2t} \\ -3e^{-2t} \\ e^{-2t} \end{bmatrix}.$$

Written parametrically, the solution is

$$x_1 = \tfrac{3}{4} - \tfrac{3}{8}e^{2t} - \tfrac{3}{8}e^{-2t}$$
$$x_2 = \tfrac{3}{4} + \tfrac{3}{8}e^{2t} - \tfrac{9}{8}e^{-2t}$$
$$x_3 = \tfrac{1}{4} + \tfrac{3}{8}e^{2t} + \tfrac{3}{8}e^{-2t}.$$

In Example 3.7.1 we had fewer than n distinct eigenvalues, but we still found n solutions with independent initial vectors. This was because we were able to find two linearly independent eigenvectors corresponding to a single eigenvalue. Let's take another look at our method for finding eigenvectors to see how many independent eigenvectors we can expect to find corresponding to a single eigenvalue.

The eigenvectors of A for the eigenvalue λ are the nonzero solutions of the algebraic system

(C) $(A - \lambda I)\mathbf{v} = \mathbf{0}.$

We solve (C) by reducing the augmented matrix and expressing the corner variables in terms of the noncorner variables. Suppose there are k noncorner variables. Then each choice of values $a_1, \ldots, a_k$ for these variables leads to a solution of (C) in the form

$$\mathbf{v} = a_1\mathbf{v}_1 + \cdots + a_k\mathbf{v}_k,$$

where $\mathbf{v}_i$ is the solution obtained by setting the ith noncorner variable equal to 1 and the other noncorner variables equal to zero. The eigenvectors $\mathbf{v}_1, \ldots, \mathbf{v}_k$ are linearly independent (Exercise 15, Section 3.4) and cannot be increased to a larger independent set of eigenvectors for λ.

   Although the process just described is by no means the only way of finding k linearly independent eigenvectors corresponding to λ, it results in as large a set of independent eigenvectors as one can find for λ. Furthermore, the number k is at most equal to the multiplicity of λ as a root of the characteristic polynomial.

Fact: *Let λ be a root of $\det(A - \lambda I)$ with multiplicity m. Suppose that when we reduce the augmented matrix $[A - \lambda I \mid \mathbf{0}]$ of the algebraic system*

(C) $(A - \lambda I)\mathbf{v} = \mathbf{0}$

we obtain a matrix whose system has k noncorner variables. For $j = 1, \ldots, k$, let $\mathbf{v}_j$ be the solution of (C) obtained by taking the jth noncorner variable to be 1 while taking all the other noncorner variables to be 0. Then

1. *$\mathbf{v}_1, \ldots, \mathbf{v}_k$ are linearly independent eigenvectors for λ*
2. *every eigenvector $\mathbf{v}$ for λ is a linear combination of $\mathbf{v}_1, \ldots, \mathbf{v}_k$*
3. *k is the largest possible number of independent eigenvectors for λ*
4. *$1 \leq k \leq m$.*

Example 3.7.4

Solve $D\mathbf{x} = A\mathbf{x}$, where

$$A = \begin{bmatrix} -2 & 1 & 0 & 1 \\ 1 & -2 & 1 & 0 \\ 0 & 1 & -2 & 1 \\ 1 & 0 & 1 & -2 \end{bmatrix}.$$

This represents the four-loop circuit of Example 3.1.4 with $V(t) = 0$, every $L_i = 1$, and every $R_i = 1$.

The characteristic polynomial of A is $\det(A - \lambda I) = (\lambda + 2)^2(\lambda + 4)\lambda$. There are three eigenvalues:

$$\lambda = 0, \quad \lambda = -4, \quad \text{and} \quad \lambda = -2.$$

Note that $\lambda = -2$ is a double root.

To find eigenvectors for $\lambda = 0$, we reduce $[A \mid \mathbf{0}]$ to obtain

$$\left[\begin{array}{cccc|c} 1 & 0 & 0 & -1 & 0 \\ 0 & 1 & 0 & -1 & 0 \\ 0 & 0 & 1 & -1 & 0 \\ 0 & 0 & 0 & 0 & 0 \end{array}\right].$$

Thus, an eigenvector for $\lambda = 0$ is

$$\mathbf{v} = \begin{bmatrix} 1 \\ 1 \\ 1 \\ 1 \end{bmatrix}.$$

The corresponding solution (remember $e^0 = 1$) is

$$\mathbf{h}_1(t) = \begin{bmatrix} 1 \\ 1 \\ 1 \\ 1 \end{bmatrix}.$$

To find an eigenvector for $\lambda = -4$, we reduce $[A + 4I \mid \mathbf{0}]$ to get

$$\left[\begin{array}{cccc|c} 1 & 0 & 0 & 1 & 0 \\ 0 & 1 & 0 & -1 & 0 \\ 0 & 0 & 1 & 1 & 0 \\ 0 & 0 & 0 & 0 & 0 \end{array}\right].$$

An eigenvector for $\lambda = -4$ is

$$\mathbf{w} = \begin{bmatrix} -1 \\ 1 \\ -1 \\ 1 \end{bmatrix}.$$

The associated solution of $D\mathbf{x} = A\mathbf{x}$ is

$$\mathbf{h}_2(t) = \begin{bmatrix} -e^{-4t} \\ e^{-4t} \\ -e^{-4t} \\ e^{-4t} \end{bmatrix}.$$

Finally, we look for eigenvectors for $\lambda = -2$. We reduce $[A + 2I \mid \mathbf{0}]$ to get

$$\begin{bmatrix} 1 & 0 & 1 & 0 & 0 \\ 0 & 1 & 0 & 1 & 0 \\ 0 & 0 & 0 & 0 & 0 \\ 0 & 0 & 0 & 0 & 0 \end{bmatrix}.$$

The corresponding equations have two noncorner variables (the last two). If in turn we set each noncorner variable equal to 1 while setting the other equal to zero, we get two independent eigenvectors,

$$\mathbf{u}_1 = \begin{bmatrix} -1 \\ 0 \\ 1 \\ 0 \end{bmatrix} \quad \text{and} \quad \mathbf{u}_2 = \begin{bmatrix} 0 \\ -1 \\ 0 \\ 1 \end{bmatrix}.$$

The associated solutions of $D\mathbf{x} = A\mathbf{x}$,

$$\mathbf{h}_3(t) = \begin{bmatrix} -e^{-2t} \\ 0 \\ e^{-2t} \\ 0 \end{bmatrix} \quad \text{and} \quad \mathbf{h}_4(t) = \begin{bmatrix} 0 \\ -e^{-2t} \\ 0 \\ e^{-2t} \end{bmatrix},$$

have independent initial vectors.

The general solution of our fourth-order system is

$$\mathbf{x} = c_1\mathbf{h}_1(t) + c_2\mathbf{h}_2(t) + c_3\mathbf{h}_3(t) + c_4\mathbf{h}_4(t)$$

$$= c_1 \begin{bmatrix} 1 \\ 1 \\ 1 \\ 1 \end{bmatrix} + c_2 \begin{bmatrix} -e^{-4t} \\ e^{-4t} \\ -e^{-4t} \\ e^{-4t} \end{bmatrix} + c_3 \begin{bmatrix} -e^{-2t} \\ 0 \\ e^{-2t} \\ 0 \end{bmatrix} + c_4 \begin{bmatrix} 0 \\ -e^{-2t} \\ 0 \\ e^{-2t} \end{bmatrix}.$$

In terms of currents in our model, we have

$$I_1 = c_1 - c_2e^{-4t} - c_3e^{-2t}$$
$$I_2 = c_1 + c_2e^{-4t} - c_4e^{-2t}$$
$$I_3 = c_1 + c_2e^{-4t} - c_3e^{-2t}$$
$$I_4 = c_1 + c_2e^{-4t} - c_4e^{-2t}.$$

The method described in the summary will yield the general solution of the system $Dx = Ax$ provided the following are true:

1. Each root of the characteristic polynomial of A is a real number.
2. For each eigenvalue λ of A, we can find as many independent eigenvectors corresponding to λ as the multiplicity of λ as a root of the characteristic polynomial.

In the next three sections, we discuss methods for solving systems for which one or both of these conditions fail.

HOMOGENEOUS SYSTEMS WITH CONSTANT COEFFICIENTS: REAL ROOTS

To solve $Dx = Ax$ where A is an $n \times n$ matrix:

1. Find the roots of the characteristic polynomial $\det(A - \lambda I)$.
2. For each real root λ find as many linearly independent eigenvectors corresponding to λ as possible. (One way to do this is to reduce $[A - \lambda I \mid 0]$ and then list the solutions of the corresponding algebraic system obtained by in turn taking each of the noncorner variables to be 1 while taking all other noncorner variables to be 0.) Associate to each of these eigenvectors v the solution $e^{\lambda t} v$ of $Dx = Ax$.
3. If by combining the solutions of $Dx = Ax$ associated to the various roots we obtain n vector valued functions $h_1(t), \ldots, h_n(t)$, then the general solution of $Dx = Ax$ is $x = c_1 h_1(t) + \cdots + c_n h_n(t)$.

EXERCISES

In Exercises 1 to 5 find the general solution of the system $Dx = Ax$, where A is the matrix considered earlier in the exercises indicated.

1. Exercise 1, Section 3.5 and Exercise 14, Section 3.6
2. Exercise 2, Section 3.5 and Exercise 15, Section 3.6
3. Exercise 3, Section 3.5 and Exercise 16, Section 3.6
4. Exercise 5, Section 3.5 and Exercise 18, Section 3.6
5. Exercise 6, Section 3.5 and Exercise 19, Section 3.6

In Exercises 6 to 13 find (a) the general solution of the given system of o.d.e.'s, and (b) the specific solution satisfying the given initial conditions.

6. $Dx = Ax$, where $A = \begin{bmatrix} -3 & 1 \\ -2 & 0 \end{bmatrix}$; $x(0) = \begin{bmatrix} 0 \\ 1 \end{bmatrix}$

7. $\begin{aligned} x_1' &= x_1 + 2x_2 \\ x_2' &= 2x_1 + x_2 \end{aligned}$ $x_1(0) = 1, x_2(0) = 3$

8. $x_1' = \qquad -x_2$
   $x_2' = 2x_1 + \quad 3x_2 \qquad x_1(0) = 1, x_2(0) = 3$

9. $D\mathbf{x} = A\mathbf{x}$, where $A = \begin{bmatrix} 1 & 1 & 0 \\ 1 & 1 & 0 \\ 0 & 0 & -1 \end{bmatrix}$; $\qquad \mathbf{x}(0) = \begin{bmatrix} 2 \\ 4 \\ 2 \end{bmatrix}$

10. $D\mathbf{x} = A\mathbf{x}$, where $A = \begin{bmatrix} 1 & 1 & 0 \\ 1 & 1 & 0 \\ 1 & 1 & 0 \end{bmatrix}$; $\qquad \mathbf{x}(0) = \begin{bmatrix} 2 \\ 4 \\ 2 \end{bmatrix}$

11. $x_1' = \quad -x_1 + \quad x_2$
    $x_2' = -6x_1 + 4x_2$
    $x_3' = \qquad\qquad x_2 - x_3 \qquad x_1(0) = 1, x_2(0) = 2, x_3(0) = 3$

12. $D\mathbf{x} = A\mathbf{x}$, where $A = \begin{bmatrix} 1 & 0 & 1 & 0 \\ 0 & 2 & 0 & 2 \\ 0 & 0 & -1 & 1 \\ 0 & 0 & 0 & 4 \end{bmatrix}$; $\qquad \mathbf{x}(0) = \begin{bmatrix} 2 \\ 16 \\ 5 \\ 15 \end{bmatrix}$

13. $x_1' = -4x_1 \qquad\qquad + 6x_4$
    $x_2' = \qquad 2x_2$
    $x_3' = \qquad\qquad - 2x_3 + 3x_4 \qquad x_1(0) = x_2(0) = x_3(0) = 1, x_4(0) = 5$
    $x_4' = -5x_1 \qquad\qquad + 7x_4$

Exercises 14 to 18 refer to our earlier models.

14. *LRC Circuit:* Consider the *LRC* circuit of Example 3.1.1 with no external voltage, resistance 5 ohms, inductance $\frac{1}{2}$ henry, and capacitance $\frac{1}{8}$ farad.
    a. Write the system of o.d.e.'s modeling the circuit.
    b. Find the general solution.
    c. Predict the charge Q after 1 second, given that initially there is no current and $Q = 1$ (use a table of exponentials or a calculator for this last part).

15. *Two-Loop Circuit:* Consider the two-loop circuit of Example 3.1.2 with no external voltage, $R_1 = 2$ ohms, $R_2 = 3$ ohms, $L = 6$ henrys, and $C = \frac{1}{6}$ farad. Find
    a. The general solution.
    b. The solution with $I_2 = 1$ and $I_1 = -3$ at time $t = 0$.

16. *Radioactive Decay:*
    a. Find the general solution of the system modeling the radioactive decay chain of Exercise 7, Section 3.1, when no extra material is added or extracted.
    b. Recall from Exercise 23, Section 1.2, the relation between half-life and rate of decay. Suppose that zigonium has a half-life of 3 years and decomposes into martinium-123. This in turn has a half-life of 15 years and decomposes into carbon. Starting from an initial sample of 1 kg ($= 10^3$ gram) of zigonium, find the amount of each substance present after 15 years.

17. *Diffusion I:* Find the general solution of the system modeling the diffusion problem of Exercise 8, Section 3.1, in case
    a. $V_1 = V_2 = 5, P = 0.1$
    b. $V_1 = 5, V_2 = 10, P = 0.1$

18. *Political Coexistence:*
    a. Find the general solution of the system modeling the problem of Exercise 10, Section 3.1.
    b. Assume that the population of Camponesa is now 12 million and the population

of Mandachuva is 16 million. Will there ever be a time when the two populations are equal?

Some more abstract problems:

19. Suppose that $v_1, \ldots, v_k$ are independent eigenvectors for A corresponding to the eigenvalues $\lambda_1, \ldots, \lambda_k$, respectively. Suppose that $v = c_1 v_1 + \cdots + c_k v_k$ is also an eigenvector for A, corresponding to λ. By applying A to the expression for v and regrouping terms, show that

$$c_1(\lambda - \lambda_1)v_1 + c_2(\lambda - \lambda_2)v_2 + \cdots + c_k(\lambda - \lambda_k)v_k = 0.$$

Use the independence of the v_i's to conclude that for each i, either $c_i = 0$ or $\lambda_i = \lambda$.

20. Use the result of Exercise 19 to show that if $v_1, \ldots, v_n$ is a collection of eigenvectors for A such that the eigenvectors corresponding to each particular eigenvalue are independent, then the whole collection is independent.

3.8 HOMOGENEOUS SYSTEMS WITH CONSTANT COEFFICIENTS: COMPLEX ROOTS

If the characteristic polynomial of A has complex roots, then the methods of the last section do not provide enough solutions to generate the general solution of $Dx = Ax$. We faced a similar situation in Section 2.6. In that case we proceeded formally with the method we had developed for real roots, ignoring the fact that the roots were complex. We then rewrote our solution using Euler's formula,

$$e^{u + iv} = e^u(\cos v + i \sin v)$$

and found real-valued solutions that were part of our generating set for the general solution. In this section we describe a similar approach for systems.

By analogy with real roots, we refer to complex roots of the characteristic polynomial of A as **complex eigenvalues** of A. Since our matrices A have *real* entries, their characteristic polynomials have *real* coefficients. Thus, complex eigenvalues come in conjugate pairs, $\alpha \pm \beta i$.

We refer to a nonzero vector with complex entries that satisfies $[A - (\alpha + \beta i)I]v = 0$ as an **eigenvector** of A corresponding to $\alpha + \beta i$. Although the examples we saw in Section 3.6 all involved real coefficients, the methods of that section are perfectly valid for systems of algebraic equations with complex coefficients. We can use those methods to find eigenvectors corresponding to complex eigenvalues.

Let's see how proceeding formally, ignoring the fact that our roots are complex, works for a system of order 2.

Example 3.8.1

Solve $D\mathbf{x} = A\mathbf{x}$, where

$$A = \begin{bmatrix} -1 & -1 \\ 4 & -1 \end{bmatrix}.$$

This represents the two-loop circuit of Example 3.1.2 with $V(t) = 0$, $C = 1/4$, $L = 1$, $R_1 = 4$, and $R_2 = 1$.

The characteristic polynomial of A is $\det(A - \lambda I) = \lambda^2 + 2\lambda + 5$, so A has the pair of complex eigenvalues

$$\lambda = -1 + 2i \quad \text{and} \quad \lambda = -1 - 2i.$$

To find an eigenvector corresponding to $\lambda = -1 + 2i$, we reduce the augmented matrix $[A - (-1 + 2i)I \mid \mathbf{0}]$:

$$\begin{bmatrix} -2i & -1 & 0 \\ 4 & -2i & 0 \end{bmatrix} \xrightarrow[\substack{R_2 \to \frac{1}{4}R_2 \\ \text{and rearr}}]{} \begin{bmatrix} 1 & -i/2 & 0 \\ -2i & -1 & 0 \end{bmatrix} \xrightarrow[R_2 \to R_2 + 2iR_1]{} \begin{bmatrix} 1 & -i/2 & 0 \\ 0 & 0 & 0 \end{bmatrix}.$$

If we set the noncorner variable of the corresponding algebraic system to be 2, we obtain the eigenvector

$$\mathbf{v} = \begin{bmatrix} i \\ 2 \end{bmatrix}.$$

Associated to this vector is a complex "solution" of $D\mathbf{x} = A\mathbf{x}$,

$$\mathbf{g}_1(t) = e^{(-1+2i)t}\mathbf{v}.$$

We can use Euler's formula to rewrite $\mathbf{g}_1(t)$ in terms of its real and imaginary parts:

$$\mathbf{g}_1(t) = e^{-t}(\cos 2t + i \sin 2t)\begin{bmatrix} i \\ 2 \end{bmatrix} = \begin{bmatrix} e^{-t}(-\sin 2t + i \cos 2t) \\ e^{-t}(2 \cos 2t + 2i \sin 2t) \end{bmatrix}$$

$$= \begin{bmatrix} -e^{-t} \sin 2t \\ 2e^{-t} \cos 2t \end{bmatrix} + i \begin{bmatrix} e^{-t} \cos 2t \\ 2e^{-t} \sin 2t \end{bmatrix}.$$

To find an eigenvector for the second complex eigenvalue, we reduce $[A - (-1 - 2i)I \mid \mathbf{0}]$ to get

$$\begin{bmatrix} 1 & i/2 & 0 \\ 0 & 0 & 0 \end{bmatrix}.$$

We see that

$$\mathbf{w} = \begin{bmatrix} -i \\ 2 \end{bmatrix}$$

is an eigenvector of A corresponding to $\lambda = -1 - 2i$. Associated to $\mathbf{w}$ is another complex "solution,"

$$\mathbf{g}_2(t) = e^{(-1-2i)t}\mathbf{w} = \begin{bmatrix} -e^{-t}\sin 2t \\ 2e^{-t}\cos 2t \end{bmatrix} - i\begin{bmatrix} e^{-t}\cos 2t \\ 2e^{-t}\sin 2t \end{bmatrix}.$$

Proceeding formally, we obtain the "general solution" of $D\mathbf{x} = A\mathbf{x}$:

$$\begin{aligned}
\mathbf{x} &= k_1\mathbf{g}_1(t) + k_2\mathbf{g}_2(t) \\
&= (k_1 + k_2)\begin{bmatrix} -e^{-t}\sin 2t \\ 2e^{-t}\cos 2t \end{bmatrix} + i(k_1 - k_2)\begin{bmatrix} e^{-t}\cos 2t \\ 2e^{-t}\sin 2t \end{bmatrix} \\
&= c_1\begin{bmatrix} -e^{-t}\sin 2t \\ 2e^{-t}\cos 2t \end{bmatrix} + c_2\begin{bmatrix} e^{-t}\cos 2t \\ 2e^{-t}\sin 2t \end{bmatrix}.
\end{aligned}$$

This expression involves two real vector valued candidates for solutions:

$$\mathbf{h}_1(t) = \begin{bmatrix} -e^{-t}\sin 2t \\ 2e^{-t}\cos 2t \end{bmatrix} \quad \text{and} \quad \mathbf{h}_2(t) = \begin{bmatrix} e^{-t}\cos 2t \\ 2e^{-t}\sin 2t \end{bmatrix}.$$

Direct substitution will show that $\mathbf{h}_1(t)$ and $\mathbf{h}_2(t)$ are indeed solutions of the system $D\mathbf{x} = A\mathbf{x}$. Their initial vectors

$$\mathbf{h}_1(0) = \begin{bmatrix} 0 \\ 2 \end{bmatrix} \quad \text{and} \quad \mathbf{h}_2(0) = \begin{bmatrix} 1 \\ 0 \end{bmatrix}$$

are linearly independent, so the real general solution of our second-order system is

$$\mathbf{x} = c_1\mathbf{h}_1(t) + c_2\mathbf{h}_2(t) = c_1\begin{bmatrix} -e^{-t}\sin 2t \\ 2e^{-t}\cos 2t \end{bmatrix} + c_2\begin{bmatrix} e^{-t}\cos 2t \\ 2e^{-t}\sin 2t \end{bmatrix}.$$

In terms of current and charge in our model, the solution is

$$\begin{aligned}
Q &= -c_1e^{-t}\sin 2t + c_2e^{-t}\cos 2t \\
I &= 2c_1e^{-t}\cos 2t + 2c_2e^{-t}\sin 2t.
\end{aligned}$$

Note that both $\mathbf{h}_1(t)$ and $\mathbf{h}_2(t)$ appear in our expression for the first complex "solution" $\mathbf{g}_1(t)$, obtained from $\lambda = -1 + 2i$; $\mathbf{h}_1(t)$ is the "real part" and $\mathbf{h}_2(t)$ the "imaginary part" of $\mathbf{g}_1(t)$.

The major features of this example are quite typical.

Fact: *Let $\alpha \pm \beta i$ be complex eigenvalues of the real $n \times n$ matrix A. Let $\mathbf{v}$ be an eigenvector of A corresponding to $\alpha + \beta i$. Then $\mathbf{h}_1(t) = \mathrm{Re}(e^{(\alpha + \beta i)t}\mathbf{v})$ and $\mathbf{h}_2(t) = \mathrm{Im}(e^{(\alpha + \beta i)t}\mathbf{v})$ are solutions of $D\mathbf{x} = A\mathbf{x}$ with linearly independent initial vectors.*

In stating this fact, we have used the convention that if r and s are real, then $\mathrm{Re}(r + is) = r$ and $\mathrm{Im}(r + is) = s$. Note that *we need only work with one of the two eigenvalues $\alpha \pm \beta i$ in order to find two solutions.*

In general, the characteristic polynomial of A could have mixed real and complex roots, any of which could be a multiple root. A modification of our previous methods, along the line suggested by the preceding fact, helps us build toward the general solution of $D\mathbf{x} = A\mathbf{x}$.

Fact: *Let A be an $n \times n$ matrix with real, constant entries. Associate real vector valued functions to A as follows:*

1.   *For each real eigenvalue λ, find as many linearly independent eigenvectors corresponding to λ as possible. To each such eigenvector $\mathbf{v}$, associate $e^{\lambda t}\mathbf{v}$.*
2.   *For each pair of complex eigenvalues $\lambda = \alpha \pm \beta i$, find as many linearly independent eigenvectors corresponding to one of the eigenvalues, $\alpha + \beta i$, as possible. To each such eigenvector $\mathbf{v}$, associate the two functions $\mathrm{Re}(e^{(\alpha + \beta i)t}\mathbf{v})$ and $\mathrm{Im}(e^{(\alpha + \beta i)t}\mathbf{v})$.*

These functions are solutions of $D\mathbf{x} = A\mathbf{x}$ with linearly independent initial vectors. In particular, if we find n such functions $\mathbf{h}_1(t), \ldots, \mathbf{h}_n(t)$, then the general solution of $D\mathbf{x} = A\mathbf{x}$ is $\mathbf{x} = c_1\mathbf{h}_1(t) + \cdots + c_n\mathbf{h}_n(t)$.

Example 3.8.2

Solve $D\mathbf{x} = A\mathbf{x}$, where

$$A = \begin{bmatrix} -1 & -1 & 0 \\ 2 & -1 & 1 \\ 0 & 1 & -1 \end{bmatrix}.$$

This represents the three-loop circuit of Example 3.1.3 with $V(t) = 0$, $C = 1/2$, $L_2 = L_3 = 1$, $R_1 = 2$, and $R_2 = 1$.

The characteristic polynomial is $\det(A - \lambda I) = -(1 + \lambda)(\lambda^2 + 2\lambda + 2)$, so the eigenvalues of A are

$$\lambda = -1, \quad \lambda = -1 \pm i.$$

We find eigenvectors for $\lambda = -1$ by reducing $[A - (-1)I \mid \mathbf{0}]$:

$$\begin{bmatrix} 0 & -1 & 0 & 0 \\ 2 & 0 & 1 & 0 \\ 0 & 1 & 0 & 0 \end{bmatrix} \rightarrow \begin{bmatrix} 1 & 0 & \frac{1}{2} & 0 \\ 0 & 1 & 0 & 0 \\ 0 & 0 & 0 & 0 \end{bmatrix}.$$

The choice $v_3 = -2$ leads to an eigenvector for $\lambda = -1$,

$$\mathbf{v} = \begin{bmatrix} 1 \\ 0 \\ -2 \end{bmatrix}$$

and the associated solution of $D\mathbf{x} = A\mathbf{x}$,

$$\mathbf{h}_1(t) = e^{-t}\mathbf{v} = \begin{bmatrix} e^{-t} \\ 0 \\ -2e^{-t} \end{bmatrix}.$$

We choose *one* of the pair of complex eigenvalues, say $\lambda = -1 + i$, and look for the corresponding eigenvectors by reducing $[A - (-1 + i)I \mid \mathbf{0}]$:

$$\begin{bmatrix} -i & -1 & 0 & 0 \\ 2 & -i & 1 & 0 \\ 0 & 1 & -i & 0 \end{bmatrix} \rightarrow \begin{bmatrix} 1 & 0 & 1 & 0 \\ 0 & 1 & -i & 0 \\ 0 & 0 & 0 & 0 \end{bmatrix}.$$

The complex vector

$$\begin{bmatrix} -1 \\ i \\ 1 \end{bmatrix}$$

is an eigenvector for $\lambda = -1 + i$. The corresponding complex solution of the system $D\mathbf{x} = A\mathbf{x}$ is

$$e^{(-1+i)t} \begin{bmatrix} -1 \\ i \\ 1 \end{bmatrix} = e^{-t}(\cos t + i \sin t) \begin{bmatrix} -1 \\ i \\ 1 \end{bmatrix}$$

$$= \begin{bmatrix} -e^{-t} \cos t \\ -e^{-t} \sin t \\ e^{-t} \cos t \end{bmatrix} + i \begin{bmatrix} -e^{-t} \sin t \\ e^{-t} \cos t \\ e^{-t} \sin t \end{bmatrix}.$$

The real and imaginary parts of this complex solution,

$$\mathbf{h}_2(t) = \begin{bmatrix} -e^{-t} \cos t \\ -e^{-t} \sin t \\ e^{-t} \cos t \end{bmatrix} \quad \text{and} \quad \mathbf{h}_3(t) = \begin{bmatrix} -e^{-t} \sin t \\ e^{-t} \cos t \\ e^{-t} \sin t \end{bmatrix}$$

are solutions with independent initial vectors.

The general solution of our third-order system is

$$\mathbf{x} = c_1\mathbf{h}_1(t) + c_2\mathbf{h}_2(t) + c_3\mathbf{h}_3(t)$$

$$= c_1\begin{bmatrix} e^{-t} \\ 0 \\ -2e^{-t} \end{bmatrix} + c_2\begin{bmatrix} -e^{-t}\cos t \\ -e^{-t}\sin t \\ e^{-t}\cos t \end{bmatrix} + c_3\begin{bmatrix} -e^{-t}\sin t \\ e^{-t}\cos t \\ e^{-t}\sin t \end{bmatrix}.$$

In terms of charges and currents, this reads

$$Q = \quad c_1e^{-t} - c_2e^{-t}\cos t - c_3e^{-t}\sin t$$

$$I_2 = \qquad\quad - c_2e^{-t}\sin t + c_3e^{-t}\cos t$$

$$I_3 = -2c_1e^{-t} + c_2e^{-t}\cos t + c_3e^{-t}\sin t.$$

Example 3.8.3

Solve $D\mathbf{x} = A\mathbf{x}$, where

$$A = \begin{bmatrix} 0 & -2 & 0 & 0 \\ 2 & 0 & 0 & 0 \\ 0 & 0 & 0 & -4 \\ 0 & 0 & 1 & 0 \end{bmatrix}.$$

The characteristic polynomial of A is $\det(A - \lambda I) = (\lambda^2 + 4)^2$. The roots are $\pm 2i$, each of which has multiplicity 2.

We choose *one* of the pair of complex eigenvalues, say $2i$, and look for corresponding eigenvectors by reducing $[A - 2iI \mid \mathbf{0}]$:

$$\begin{bmatrix} -2i & -2 & 0 & 0 & | & 0 \\ 2 & -2i & 0 & 0 & | & 0 \\ 0 & 0 & -2i & -4 & | & 0 \\ 0 & 0 & 1 & -2i & | & 0 \end{bmatrix} \rightarrow \begin{bmatrix} 1 & -i & 0 & 0 & | & 0 \\ 0 & 0 & 1 & -2i & | & 0 \\ 0 & 0 & 0 & 0 & | & 0 \\ 0 & 0 & 0 & 0 & | & 0 \end{bmatrix}.$$

The algebraic system corresponding to this matrix has two noncorner variables (the second and fourth). If in turn we set each of these to be 1 while setting the other to be zero, we obtain two independent eigenvectors corresponding to $2i$:

$$\mathbf{v}_1 = \begin{bmatrix} i \\ 1 \\ 0 \\ 0 \end{bmatrix} \quad \text{and} \quad \mathbf{v}_2 = \begin{bmatrix} 0 \\ 0 \\ 2i \\ 1 \end{bmatrix}.$$

Associated to these vectors are two complex solutions:

$$e^{2it}\mathbf{v}_1 = (\cos 2t + i \sin 2t) \begin{bmatrix} i \\ 1 \\ 0 \\ 0 \end{bmatrix} = \begin{bmatrix} -\sin 2t \\ \cos 2t \\ 0 \\ 0 \end{bmatrix} + i \begin{bmatrix} \cos 2t \\ \sin 2t \\ 0 \\ 0 \end{bmatrix}$$

and

$$e^{2it}\mathbf{v}_2 = (\cos 2t + i \sin 2t) \begin{bmatrix} 0 \\ 0 \\ 2i \\ 1 \end{bmatrix} = \begin{bmatrix} 0 \\ 0 \\ -2 \sin 2t \\ \cos 2t \end{bmatrix} + i \begin{bmatrix} 0 \\ 0 \\ 2 \cos 2t \\ \sin 2t \end{bmatrix}.$$

The real and imaginary parts of these complex solutions provide us with four solutions with independent initial vectors.

The general solution of our fourth-order system is

$$\mathbf{x} = c_1 \begin{bmatrix} -\sin 2t \\ \cos 2t \\ 0 \\ 0 \end{bmatrix} + c_2 \begin{bmatrix} \cos 2t \\ \sin 2t \\ 0 \\ 0 \end{bmatrix} + c_3 \begin{bmatrix} 0 \\ 0 \\ -2 \sin 2t \\ \cos 2t \end{bmatrix} + c_4 \begin{bmatrix} 0 \\ 0 \\ 2 \cos 2t \\ \sin 2t \end{bmatrix}.$$

The method described in the summary will yield the general solution of the system $D\mathbf{x} = A\mathbf{x}$ provided that for each eigenvalue (real or complex) we can find as many independent eigenvectors as the multiplicity of that eigenvalue as a root of the characteristic polynomial. In the next two sections we discuss systems for which this condition fails to hold.

HOMOGENEOUS SYSTEMS WITH CONSTANT COEFFICIENTS: COMPLEX ROOTS

To solve $D\mathbf{x} = A\mathbf{x}$ where A is an $n \times n$ matrix:

1. Find the roots of the characteristic polynomial $\det(A - \lambda I)$.

2. For each real root λ find as many linearly independent eigenvectors corresponding to λ as possible. Assign to each of these eigenvectors $\mathbf{v}$ the solution $e^{\lambda t}\mathbf{v}$.

3. For each pair of complex roots $\alpha \pm \beta i$, work with one of the roots as if it were real to obtain complex solutions. Associate to the pair of eigenvalues the real and imaginary parts of these complex solutions.

If by combining the solutions associated to the various roots we obtain n vector valued functions $h_1(t), \ldots, h_n(t)$, then the general solution of the system $D\mathbf{x} = A\mathbf{x}$ is $\mathbf{x} = c_1 h_1(t) + \cdots + c_n h_n(t)$.

EXERCISES

In Exercises 1 to 8 find the general solution of $D\mathbf{x} = A\mathbf{x}$.

1. $A = \begin{bmatrix} 0 & 1 \\ -1 & 0 \end{bmatrix}$

2. $A = \begin{bmatrix} 3 & -2 \\ 2 & 3 \end{bmatrix}$

3. $A = \begin{bmatrix} 1 & 0 & -1 \\ 0 & 2 & 0 \\ 1 & 0 & 1 \end{bmatrix}$

4. $A = \begin{bmatrix} 1 & -2 & 2 \\ 0 & 1 & 1 \\ -1 & -2 & 4 \end{bmatrix}$

5. $A = \begin{bmatrix} 0 & 1 & -2 \\ -1 & -1 & 2 \\ 0 & 0 & -1 \end{bmatrix}$

6. $A = \begin{bmatrix} 1 & 1 & 0 & 0 \\ 3 & -1 & 0 & 0 \\ 0 & 0 & 0 & 1 \\ 0 & 0 & -1 & 0 \end{bmatrix}$

7. $A = \begin{bmatrix} 0 & 1 & 1 & -1 \\ 0 & -1 & 0 & 1 \\ -5 & -5 & -2 & 0 \\ 0 & -4 & 0 & -1 \end{bmatrix}$

8. $A = \begin{bmatrix} 0 & 1 & 0 & 0 & 0 & 0 \\ -1 & 0 & 0 & 0 & 0 & 0 \\ 0 & 0 & 0 & -1 & 0 & 0 \\ 0 & 0 & 1 & 0 & 0 & 0 \\ 0 & 0 & 0 & 0 & 0 & -1 \\ 0 & 0 & 0 & 0 & -1 & 0 \end{bmatrix}$

In Exercises 9 to 12 find (a) the general solution and (b) the specific solution satisfying the given initial conditions.

9. $D\mathbf{x} = A\mathbf{x}$, where $A = \begin{bmatrix} 1 & -1 \\ 1 & 1 \end{bmatrix}$; $\mathbf{x}(0) = \begin{bmatrix} 2 \\ 3 \end{bmatrix}$.

10. $\begin{aligned} x_1' &= 5x_1 - 4x_2 \\ x_2' &= 10x_1 - 7x_2 \end{aligned}$ $x_1(0) = 70, \, x_2(0) = 10$

11. $\begin{aligned} x_1' &= \quad\;\; - x_2 \\ x_2' &= 3x_1 + 2x_2 \end{aligned}$ $x_1(0) = 2, \, x_2(0) = 1$

12. $\begin{aligned} x_1' &= x_1 + 2x_2 \qquad\;\; + x_4 \\ x_2' &= -x_1 - x_2 + x_3 \\ x_3' &= \qquad\qquad x_3 + x_4 \\ x_4' &= \qquad\quad - 2x_3 - x_4 \end{aligned}$ $x_1(0) = x_2(0) = x_3(0) = 0, \, x_4(0) = 4$

Exercises 13 to 15 refer to our earlier models.

13. *LRC Circuit:* Find the general solution of the system modeling the *LRC* circuit of Example 3.1.1 with $V(t) = 0$, $R = 1$ ohm, $C = 1$ farad, and $L = 1$ henry.

14. *Two-Loop Circuit:*
    a. Find the general solution of the system modeling the two-loop circuit of Example 3.1.2 with $V(t) = 0$, $R_1 = R_2 = 1$ ohm, $L = 1$ henry, and $C = 1$ farad.

b.   Find a formula for the current I_{cap} through the capacitor at time t, given that $I_2 = 1$ and $I_1 = 2$ amperes when $t = 0$.

15.   *Coupled Springs:* Find the general solution of the system of Exercise 26, Section 3.2, assuming $m_1 = m_2 = 1$ gram, $L_1 = 10$ cm, $L_2 = 3$ cm, $k_1 = 3$ dynes/cm, and $k_2 = 2$ dynes/cm. (Compare this with Exercise 28, Section 2.6.)

3.9   DOUBLE ROOTS AND MATRIX PRODUCTS

We have seen how to calculate the general solution of $Dx = Ax$ provided that for each eigenvalue λ the number of independent eigenvectors matches the multiplicity m of λ as a root of the characteristic polynomial of A. In this section and the next, we discuss a general method for associating m solutions with independent initial vectors to the eigenvalue λ. This method works even when there are fewer than m independent eigenvectors. The discussion in this section focuses on the simplest case, when λ is a real root of multiplicity two. Once the machinery for this case is set up, the extension to complex roots and higher multiplicities (in the next section) is quite straightforward.

Let's start by investigating a system of order two that is equivalent to a single o.d.e.

Example 3.9.1

Recall from Section 3.2 that the o.d.e.

(H) $(D^2 - 2D + 1)x = 0$

can be replaced by an equivalent system. Introducing two unknowns

$$x_1 = x \quad \text{and} \quad x_2 = x',$$

we obtain the system

(S_H) $Dx = Ax$

where

$$\mathbf{x} = \begin{bmatrix} x_1 \\ x_2 \end{bmatrix} \quad \text{and} \quad A = \begin{bmatrix} 0 & 1 \\ -1 & 2 \end{bmatrix}.$$

We leave it to you to check that the characteristic polynomial of A is $(\lambda - 1)^2$ and that all eigenvectors are multiples of a single eigenvector.

Although the method of the preceding two sections will not yield the general solution of (S$_H$), we can find it by first solving (H). Since the general solution of (H) is $x = c_1e^t + c_2te^t$, the general solution of (S$_H$) can be written as

$$\mathbf{x} = \begin{bmatrix} x \\ x' \end{bmatrix} = \begin{bmatrix} c_1e^t + c_2te^t \\ c_1e^t + c_2(t + 1)e^t \end{bmatrix}$$
$$= c_1\mathbf{h}_1(t) + c_2\mathbf{h}_2(t),$$

where

$$\mathbf{h}_1(t) = e^t \left(\begin{bmatrix} 1 \\ 1 \end{bmatrix} + t \begin{bmatrix} 0 \\ 0 \end{bmatrix} \right) \quad \text{and} \quad \mathbf{h}_2(t) = e^t \left(\begin{bmatrix} 0 \\ 1 \end{bmatrix} + t \begin{bmatrix} 1 \\ 1 \end{bmatrix} \right).$$

Based on this example, we guess that whenever λ is a double root of $\det(A - \lambda I)$, we can find solutions to $D\mathbf{x} = A\mathbf{x}$ of the form

(1) $$\mathbf{h}(t) = e^{\lambda t}(\mathbf{v}_0 + t\mathbf{v}_1).$$

Substitution of $\mathbf{x} = \mathbf{h}(t)$ into $D\mathbf{x} = A\mathbf{x}$ yields the equation

$$\lambda e^{\lambda t}(\mathbf{v}_0 + t\mathbf{v}_1) + e^{\lambda t}\mathbf{v}_1 = e^{\lambda t}(A\mathbf{v}_0 + tA\mathbf{v}_1).$$

If we divide by $e^{\lambda t}$ and subtract $\lambda(\mathbf{v}_0 + t\mathbf{v}_1)$ from both sides, we get

$$\mathbf{v}_1 = (A\mathbf{v}_0 - \lambda\mathbf{v}_0) + t(A\mathbf{v}_1 - \lambda\mathbf{v}_1).$$

This can be rewritten [see Exercise 28(b), Section 3.2] as

(2) $$\mathbf{v}_1 = (A - \lambda I)\mathbf{v}_0 + t(A - \lambda I)\mathbf{v}_1.$$

When $t = 0$, this reads

(3) $$\mathbf{v}_1 = (A - \lambda I)\mathbf{v}_0,$$

so that $\mathbf{v}_1$ will be determined as soon as we find $\mathbf{v}_0$. If we subtract (3) from (2) and take $t = 1$, we get

$$(A - \lambda I)\mathbf{v}_1 = \mathbf{0}.$$

By substituting (3) into this equation, we eliminate $\mathbf{v}_1$ and obtain

(4) $$(A - \lambda I)[(A - \lambda I)\mathbf{v}_0] = \mathbf{0}.$$

This tells us where to look for $\mathbf{v}_0$. We want a vector with the property that when we multiply it by $A - \lambda I$ and then multiply the resulting vector by $A - \lambda I$, we get the zero vector.

If we were able to rewrite (4) as a problem involving a single known matrix M times the unknown vector $\mathbf{v}_0$

$$M\mathbf{v}_0 = \mathbf{0},$$

then we could solve it by reduction. Luckily it is possible to do this in a very general setting.

It turns out that the effect of multiplying a k-vector $\mathbf{v}$ by an $m \times k$ matrix C, and then multiplying the resulting m-vector $C\mathbf{v}$ by an $n \times m$ matrix B—that is, the calculation of the n-vector $B(C\mathbf{v})$—can be rewritten as multiplication of $\mathbf{v}$ by a single $n \times k$ matrix BC called the **product** of B and C. To calculate BC, we think of C as made up from its columns

$$\mathbf{c}_1 = \begin{bmatrix} c_{11} \\ \cdot \\ \cdot \\ \cdot \\ c_{m1} \end{bmatrix}, \quad \mathbf{c}_2 = \begin{bmatrix} c_{12} \\ \cdot \\ \cdot \\ \cdot \\ c_{m2} \end{bmatrix}, \ldots, \mathbf{c}_k = \begin{bmatrix} c_{1k} \\ \cdot \\ \cdot \\ \cdot \\ c_{mk} \end{bmatrix}.$$

We take the product of the $n \times m$ matrix B with each of these m-vectors to obtain n-vectors

$$B\mathbf{c}_1, \quad B\mathbf{c}_2, \ldots, B\mathbf{c}_k.$$

These n-vectors form the columns of the product BC:

$$BC = B[\mathbf{c}_1 \cdots \mathbf{c}_k] = [B\mathbf{c}_1 \cdots B\mathbf{c}_k].$$

With this definition we have the following (see Exercise 20).

Fact: *Suppose B is an $n \times m$ matrix, C is an $m \times k$ matrix, and $\mathbf{v}$ is a k-vector. Then $(BC)\mathbf{v} = B(C\mathbf{v})$.*

Example 3.9.2

Let

$$B = \begin{bmatrix} 1 & 2 & 3 & 0 \\ 1 & 0 & -1 & 2 \\ 0 & 1 & 0 & 1 \end{bmatrix} \quad \text{and} \quad C = \begin{bmatrix} 1 & 2 \\ 0 & 1 \\ 1 & 0 \\ -1 & 0 \end{bmatrix}.$$

Then C has two columns

$$\mathbf{c}_1 = \begin{bmatrix} 1 \\ 0 \\ 1 \\ -1 \end{bmatrix} \quad \text{and} \quad \mathbf{c}_2 = \begin{bmatrix} 2 \\ 1 \\ 0 \\ 0 \end{bmatrix}.$$

To obtain the columns of BC, we multiply the columns of C by B:

$$B\mathbf{c}_1 = \begin{bmatrix} 1 & 2 & 3 & 0 \\ 1 & 0 & -1 & 2 \\ 0 & 1 & 0 & 1 \end{bmatrix} \begin{bmatrix} 1 \\ 0 \\ 1 \\ -1 \end{bmatrix} = \begin{bmatrix} 4 \\ -2 \\ -1 \end{bmatrix},$$

$$B\mathbf{c}_2 = \begin{bmatrix} 1 & 2 & 3 & 0 \\ 1 & 0 & -1 & 2 \\ 0 & 1 & 0 & 1 \end{bmatrix} \begin{bmatrix} 2 \\ 1 \\ 0 \\ 0 \end{bmatrix} = \begin{bmatrix} 4 \\ 2 \\ 1 \end{bmatrix}.$$

Thus

$$BC = \begin{bmatrix} 4 & 4 \\ -2 & 2 \\ -1 & 1 \end{bmatrix}.$$

Note that *the product of two matrices is only defined if the number of columns of the first is the same as the number of rows of the second.* Since C has two columns and B has three rows, the product in the opposite order CB is not defined for the matrices in this example.

Example 3.9.3

Let

$$B = \begin{bmatrix} 2 & 1 \\ 1 & 0 \end{bmatrix} \quad \text{and} \quad C = \begin{bmatrix} 1 & 2 \\ 0 & -1 \end{bmatrix}.$$

To obtain the columns of BC, we multiply the columns $\mathbf{c}_1$ and $\mathbf{c}_2$ of C by B:

$$B\mathbf{c}_1 = \begin{bmatrix} 2 & 1 \\ 1 & 0 \end{bmatrix} \begin{bmatrix} 1 \\ 0 \end{bmatrix} = \begin{bmatrix} 2 \\ 1 \end{bmatrix}, \quad B\mathbf{c}_2 = \begin{bmatrix} 2 & 1 \\ 1 & 0 \end{bmatrix} \begin{bmatrix} 2 \\ -1 \end{bmatrix} = \begin{bmatrix} 3 \\ 2 \end{bmatrix}.$$

Thus

$$BC = \begin{bmatrix} 2 & 3 \\ 1 & 2 \end{bmatrix}.$$

These matrices can also be multiplied in the opposite order. We obtain the columns of CB by multiplying the columns $\mathbf{b}_1$ and $\mathbf{b}_2$ of B by C:

$$Cb_1 = \begin{bmatrix} 1 & 2 \\ 0 & -1 \end{bmatrix} \begin{bmatrix} 2 \\ 1 \end{bmatrix} = \begin{bmatrix} 4 \\ -1 \end{bmatrix}, \quad Cb_2 = \begin{bmatrix} 1 & 2 \\ 0 & -1 \end{bmatrix} \begin{bmatrix} 1 \\ 0 \end{bmatrix} = \begin{bmatrix} 1 \\ 0 \end{bmatrix}.$$

Thus

$$CB = \begin{bmatrix} 4 & 1 \\ -1 & 0 \end{bmatrix}.$$

Note that $CB \neq BC$. Thus, *even if the products CB and BC are both defined, they need not be equal.*

In order to rewrite equation (4), we need to form the product of the $n \times n$ matrix $B = (A - \lambda I)$ with itself. Just as with numbers, we write the product of an $n \times n$ matrix with itself as a **square**:

$$B^2 = BB.$$

Higher powers are obtained by successively multiplying by B:

$$B^3 = BB^2, \quad B^4 = BB^3, \quad \text{etc.}$$

Example 3.9.4

Let

$$B = \begin{bmatrix} -1 & 1 \\ -1 & 1 \end{bmatrix}.$$

We obtain the columns of $B^2 = BB$ by multiplying the columns of B by B:

$$\begin{bmatrix} -1 & 1 \\ -1 & 1 \end{bmatrix} \begin{bmatrix} -1 \\ -1 \end{bmatrix} = \begin{bmatrix} 0 \\ 0 \end{bmatrix}, \quad \begin{bmatrix} -1 & 1 \\ -1 & 1 \end{bmatrix} \begin{bmatrix} 1 \\ 1 \end{bmatrix} = \begin{bmatrix} 0 \\ 0 \end{bmatrix}.$$

Thus

$$B^2 = \begin{bmatrix} 0 & 0 \\ 0 & 0 \end{bmatrix}.$$

Note that all the entries of the product BB are zero, even though none of the entries of the factors is zero.

Example 3.9.5

Let

$$B = \begin{bmatrix} 0 & 0 & 0 & 0 \\ 1 & 0 & 0 & 0 \\ 0 & 0 & 0 & 1 \\ 0 & 0 & 0 & 2 \end{bmatrix}.$$

Then

$$B^2 = BB = \begin{bmatrix} 0 & 0 & 0 & 0 \\ 1 & 0 & 0 & 0 \\ 0 & 0 & 0 & 1 \\ 0 & 0 & 0 & 2 \end{bmatrix} \begin{bmatrix} 0 & 0 & 0 & 0 \\ 1 & 0 & 0 & 0 \\ 0 & 0 & 0 & 1 \\ 0 & 0 & 0 & 2 \end{bmatrix} = \begin{bmatrix} 0 & 0 & 0 & 0 \\ 0 & 0 & 0 & 0 \\ 0 & 0 & 0 & 2 \\ 0 & 0 & 0 & 4 \end{bmatrix}$$

and

$$B^3 = BB^2 = \begin{bmatrix} 0 & 0 & 0 & 0 \\ 1 & 0 & 0 & 0 \\ 0 & 0 & 0 & 1 \\ 0 & 0 & 0 & 2 \end{bmatrix} \begin{bmatrix} 0 & 0 & 0 & 0 \\ 0 & 0 & 0 & 0 \\ 0 & 0 & 0 & 2 \\ 0 & 0 & 0 & 4 \end{bmatrix} = \begin{bmatrix} 0 & 0 & 0 & 0 \\ 0 & 0 & 0 & 0 \\ 0 & 0 & 0 & 4 \\ 0 & 0 & 0 & 8 \end{bmatrix}.$$

Using the notation of matrix powers, we can rewrite equation (4) as

$$(A - \lambda I)^2 \mathbf{v}_0 = \mathbf{0}.$$

If $\mathbf{v}_0$ is a solution of this equation, and if $\mathbf{v}_1 = (A - \lambda I)\mathbf{v}_0$ [as in equation (3)], then substitution of these vectors into equation (1) will yield a solution of $D\mathbf{x} = A\mathbf{x}$.

Fact: *If* $(A - \lambda I)^2 \mathbf{v} = \mathbf{0}$, *then*

$$\mathbf{h}(t) = e^{\lambda t}(\mathbf{v} + t[A - \lambda I]\mathbf{v})$$

is a solution of $D\mathbf{x} = A\mathbf{x}$ *with initial vector* $\mathbf{h}(0) = \mathbf{v}$.

We refer to nonzero solutions of $(A - \lambda I)^2 \mathbf{v} = \mathbf{0}$ as **generalized eigenvectors** of A corresponding to the eigenvalue λ. If we find independent generalized eigenvectors, then we can use the preceding fact to obtain solutions of $D\mathbf{x} = A\mathbf{x}$ with independent initial vectors.

Let's see how this works for the system of Example 3.9.1.

Example 3.9.6

The characteristic polynomial of

$$A = \begin{bmatrix} 0 & 1 \\ -1 & 2 \end{bmatrix}$$

has a double root $\lambda = 1$. With this choice of λ,

$$A - \lambda I = \begin{bmatrix} -1 & 1 \\ -1 & 1 \end{bmatrix}.$$

We calculated the square of this matrix in Example 3.9.3:

$$(A - \lambda I)^2 = \begin{bmatrix} 0 & 0 \\ 0 & 0 \end{bmatrix}.$$

In this case, *every* vector is a solution of $(A - \lambda I)^2 \mathbf{v} = \mathbf{0}$. In particular, the vectors

$$\mathbf{v} = \begin{bmatrix} 1 \\ 1 \end{bmatrix} \quad \text{and} \quad \mathbf{w} = \begin{bmatrix} 0 \\ 1 \end{bmatrix}$$

are independent generalized eigenvectors. The solutions associated to these vectors,

$$\mathbf{h}_1(t) = e^{\lambda t}(\mathbf{v} + t[A - \lambda I]\mathbf{v})$$

$$= e^{\lambda t}\left(\begin{bmatrix} 1 \\ 1 \end{bmatrix} + t\begin{bmatrix} -1 & 1 \\ -1 & 1 \end{bmatrix}\begin{bmatrix} 1 \\ 1 \end{bmatrix}\right)$$

$$= e^{\lambda t}\left(\begin{bmatrix} 1 \\ 1 \end{bmatrix} + t\begin{bmatrix} 0 \\ 0 \end{bmatrix}\right)$$

$$\mathbf{h}_2(t) = e^{\lambda t}(\mathbf{w} + t[A - \lambda I]\mathbf{w})$$

$$= e^{\lambda t}\left(\begin{bmatrix} 0 \\ 1 \end{bmatrix} + t\begin{bmatrix} -1 & 1 \\ -1 & 1 \end{bmatrix}\begin{bmatrix} 0 \\ 1 \end{bmatrix}\right)$$

$$= e^{\lambda t}\left(\begin{bmatrix} 0 \\ 1 \end{bmatrix} + t\begin{bmatrix} 1 \\ 1 \end{bmatrix}\right)$$

are the solutions that generated the general solution in Example 3.9.1.

If we have a system with more than one eigenvalue, then, as in Sections 3.7 and 3.8, we work with each individually and combine the resulting solutions.

Example 3.9.7

Find the general solution to $D\mathbf{x} = A\mathbf{x}$, where

$$A = \begin{bmatrix} -1 & 1 & 4 \\ -2 & 2 & 4 \\ -1 & 0 & 4 \end{bmatrix}.$$

The characteristic polynomial of A is $(2 - \lambda)^2(1 - \lambda)$, so $\lambda = 2$ is a double root and $\lambda = 1$ is a simple root.

To find the generalized eigenvectors corresponding to the double root $\lambda = 2$, we must solve

$$(A - 2I)^2\mathbf{v} = \mathbf{0}.$$

We first square $A - 2I$:

$$(A - 2I)^2 = \begin{bmatrix} -3 & 1 & 4 \\ -2 & 0 & 4 \\ -1 & 0 & 2 \end{bmatrix} \begin{bmatrix} -3 & 1 & 4 \\ -2 & 0 & 4 \\ -1 & 0 & 2 \end{bmatrix} = \begin{bmatrix} 3 & -3 & 0 \\ 2 & -2 & 0 \\ 1 & -1 & 0 \end{bmatrix}.$$

We then reduce the augmented matrix $[(A - 2I)^2 \mid \mathbf{0}]$:

$$\begin{bmatrix} 3 & -3 & 0 & | & 0 \\ 2 & -2 & 0 & | & 0 \\ 1 & -1 & 0 & | & 0 \end{bmatrix} \rightarrow \begin{bmatrix} 1 & -1 & 0 & | & 0 \\ 0 & 0 & 0 & | & 0 \\ 0 & 0 & 0 & | & 0 \end{bmatrix}.$$

The vectors

$$\mathbf{v} = \begin{bmatrix} 0 \\ 0 \\ 1 \end{bmatrix} \quad \text{and} \quad \mathbf{w} = \begin{bmatrix} 1 \\ 1 \\ 0 \end{bmatrix}$$

are independent solutions of the corresponding algebraic system of equations. Associated to these generalized eigenvectors are two solutions of $D\mathbf{x} = A\mathbf{x}$:

$$\mathbf{h}_1(t) = e^{2t}(\mathbf{v} + t[A - 2I]\mathbf{v})$$

$$= e^{2t}\left(\begin{bmatrix} 0 \\ 0 \\ 1 \end{bmatrix} + t \begin{bmatrix} -3 & 1 & 4 \\ -2 & 0 & 4 \\ -1 & 0 & 2 \end{bmatrix} \begin{bmatrix} 0 \\ 0 \\ 1 \end{bmatrix} \right)$$

$$= e^{2t}\left(\begin{bmatrix} 0 \\ 0 \\ 1 \end{bmatrix} + t \begin{bmatrix} 4 \\ 4 \\ 2 \end{bmatrix} \right) = \begin{bmatrix} 4te^{2t} \\ 4te^{2t} \\ (1 + 2t)e^{2t} \end{bmatrix}$$

$$h_2(t) = e^{2t}(\mathbf{w} + t[A - 2I]\mathbf{w})$$

$$= e^{2t}\left(\begin{bmatrix} 1 \\ 1 \\ 0 \end{bmatrix} + t \begin{bmatrix} -3 & 1 & 4 \\ -2 & 0 & 4 \\ -1 & 0 & 2 \end{bmatrix} \begin{bmatrix} 1 \\ 1 \\ 0 \end{bmatrix}\right)$$

$$= e^{2t}\left(\begin{bmatrix} 1 \\ 1 \\ 0 \end{bmatrix} + t \begin{bmatrix} -2 \\ -2 \\ -1 \end{bmatrix}\right) = \begin{bmatrix} (1 - 2t)e^{2t} \\ (1 - 2t)e^{2t} \\ -te^{2t} \end{bmatrix}.$$

We can find the eigenvectors for $\lambda = 1$ by reducing $[A - I \mid \mathbf{0}]$:

$$\begin{bmatrix} -2 & 1 & 4 & \mid & 0 \\ -2 & 1 & 4 & \mid & 0 \\ -1 & 0 & 3 & \mid & 0 \end{bmatrix} \rightarrow \begin{bmatrix} 1 & 0 & -3 & \mid & 0 \\ 0 & 1 & -2 & \mid & 0 \\ 0 & 0 & 0 & \mid & 0 \end{bmatrix}.$$

Every eigenvector for $\lambda = 1$ is a multiple of

$$\mathbf{u} = \begin{bmatrix} 3 \\ 2 \\ 1 \end{bmatrix},$$

and a solution associated to $\lambda = 1$ is

$$\mathbf{h}_3(t) = e^t\mathbf{u} = \begin{bmatrix} 3e^t \\ 2e^t \\ e^t \end{bmatrix}.$$

Since the initial vectors

$$\mathbf{h}_1(0) = \begin{bmatrix} 3 \\ 2 \\ 1 \end{bmatrix}, \quad \mathbf{h}_2(0) = \begin{bmatrix} 0 \\ 0 \\ 1 \end{bmatrix}, \quad \mathbf{h}_3(0) = \begin{bmatrix} 1 \\ 1 \\ 0 \end{bmatrix}$$

are independent (check this), the general solution of our third-order system $D\mathbf{x} = A\mathbf{x}$ is

$$\mathbf{x} = c_1\mathbf{h}_1(t) + c_2\mathbf{h}_2(t) + c_3\mathbf{h}_3(t)$$

$$= c_1 \begin{bmatrix} 3e^t \\ 2e^t \\ e^t \end{bmatrix} + c_2 \begin{bmatrix} 4te^{2t} \\ 4te^{2t} \\ (1 + 2t)e^{2t} \end{bmatrix} + c_3 \begin{bmatrix} (1 - 2t)e^{2t} \\ (1 - 2t)e^{2t} \\ -te^{2t} \end{bmatrix}.$$

The method we have outlined relies on our being able to find two linearly independent generalized eigenvectors for each eigenvalue of multiplicity 2. This is guaranteed by a general theorem of linear algebra, which we state more explicitly in the next section.

MATRIX PRODUCTS AND DOUBLE ROOTS

The **product** of an $n \times m$ matrix B and an $m \times k$ matrix C is the $n \times k$ matrix whose columns are the products of B with the columns of C:

$$BC = B[\mathbf{c}_1 \cdots \mathbf{c}_k] = [B\mathbf{c}_1 \cdots B\mathbf{c}_k].$$

The effect of multiplying a k-vector by BC is the same as multiplying the vector by C and then multiplying the resulting vector by B:

$$(BC)\mathbf{v} = B(C\mathbf{v}).$$

The **square** of an $n \times n$ matrix is its product with itself, and we obtain **higher powers** by successively multiplying by the matrix:

$$B^2 = BB, \quad B^3 = BB^2, \quad B^4 = BB^3, \quad \text{etc.}$$

If λ is a double root of the characteristic polynomial $\det(A - \lambda I)$, then a **generalized eigenvector** corresponding to the eigenvalue λ is a vector $\mathbf{v} \neq \mathbf{0}$ satisfying

$$(A - \lambda I)^2 \mathbf{v} = \mathbf{0}.$$

There are always two linearly independent generalized eigenvectors corresponding to a double eigenvalue of A. Associated to each of these generalized eigenvectors is a solution of $D\mathbf{x} = A\mathbf{x}$:

$$\mathbf{h}(t) = e^{\lambda t}(\mathbf{v} + t[A - \lambda I]\mathbf{v}).$$

Notes

1. On the entries of the product of two matrices

We have described the product BC of an $n \times m$ matrix B and an $m \times k$ matrix C by describing its columns. Sometimes it helps to have a description of the entry in the ith row and jth column of BC.

If $\mathbf{c}_j$ is the jth column of C, then $B\mathbf{c}_j$ is the jth column of BC. The entry in the ith row and jth column of BC is the same as the ith entry down in the vector $B\mathbf{c}_j$. To get this, we multiply the entries in the ith row of B by the corresponding entries in $\mathbf{c}_j$ and sum the products. Thus BC is the $n \times k$ matrix whose entry in the ith row and jth column is

$$b_{i1}c_{1j} + b_{i2}c_{2j} + \cdots + b_{in}c_{nj}.$$

2. **On the properties of matrix multiplication**

Examples 3.9.2 and 3.9.3 illustrate ways in which matrix multiplication differs from multiplication of numbers. The following list of basic properties of matrix multiplication demonstrates some of the similarities between these operations. In this list we use I_n to denote the $n \times n$ identity matrix and $O_{m \times k}$ to denote the $m \times k$ zero matrix (see Note 2 in Section 3.2).

Fact: *Suppose that A and A' are $n \times m$ matrices, B and B' are $m \times k$ matrices, C is a $k \times r$ matrix, and c is a number. Then*

1. $(AB)C = A(BC)$
2. $I_n A = A$ and $AI_m = A$
3. $(cA)B = c(AB) = A(cB)$
4. $AO_{m \times k} = O_{n \times k}$ and $O_{r \times n} A = O_{r \times m}$
5. $A(B + B') = AB + AB'$
6. $(A + A')B = AB + A'B$.

3. **Eigenvectors are generalized eigenvectors**

Suppose that λ is a double eigenvalue of A. If $\mathbf{v}$ is an eigenvector of A corresponding to λ, then

$$(A - \lambda I)^2 \mathbf{v} = (A - \lambda I)[(A - \lambda I)\mathbf{v}] = (A - \lambda I)\mathbf{0} = \mathbf{0}.$$

Thus $\mathbf{v}$ is also a generalized eigenvector of A. The solution of $D\mathbf{x} = A\mathbf{x}$ associated to $\mathbf{v}$ by virtue of the fact that $\mathbf{v}$ is a generalized eigenvector is

$$\mathbf{h}(t) = e^{\lambda t}(\mathbf{v} + t[A - \lambda I]\mathbf{v}) = e^{\lambda t}(\mathbf{v} + t\mathbf{0}) = e^{\lambda t}\mathbf{v},$$

which is the same as the solution associated to $\mathbf{v}$ by virtue of the fact that $\mathbf{v}$ is an eigenvector.

EXERCISES

1. Let $B = \begin{bmatrix} 3 & -1 & 0 \\ 1 & 0 & 1 \\ -1 & 0 & 0 \end{bmatrix}$, $C = \begin{bmatrix} 1 & 1 & 0 \\ 0 & 1 & -1 \\ 1 & 0 & 1 \end{bmatrix}$, and $\mathbf{v} = \begin{bmatrix} 1 \\ 2 \\ -1 \end{bmatrix}$.

    Find the indicated matrices.

    a. BC b. $(BC)\mathbf{v}$ c. $B(C\mathbf{v})$
    d. CB e. B^2 f. B^3
    g. $(BC)^2$ h. $B^2 C^2$ i. $(B + C)^2$

2. Let $A = \begin{bmatrix} 0 & 0 & 3 & 0 \\ 1 & 3 & 0 & -1 \\ 0 & 0 & 3 & 0 \\ 1 & 0 & -1 & 3 \end{bmatrix}$, $B = \begin{bmatrix} 0 & 1 \\ 1 & 0 \\ 0 & 1 \\ 0 & 1 \end{bmatrix}$, $C = \begin{bmatrix} 1 & 3 \\ 0 & -1 \\ 1 & 0 \\ 0 & 0 \end{bmatrix}$,

and $F = \begin{bmatrix} 1 & 0 & -1 & 0 \\ 0 & 1 & 0 & 0 \end{bmatrix}$.

Find the indicated matrices.

a. AB b. AC c. $A(B - C)$

d. FA e. BF f. FB

g. $(FA)B$ h. $F(AB)$ i. $(AB)F$

In Exercises 3 and 4 you are given a matrix A, an eigenvalue λ, and a generalized eigenvector v. Find the associated solution of $Dx = Ax$.

3. $A = \begin{bmatrix} 0 & 1 \\ -1 & 2 \end{bmatrix}$, $\lambda = 1$, $v = \begin{bmatrix} 1 \\ 0 \end{bmatrix}$

4. $A = \begin{bmatrix} -1 & 1 & 4 \\ -2 & 2 & 4 \\ -1 & 0 & 4 \end{bmatrix}$, $\lambda = 2$, $v = \begin{bmatrix} 1 \\ 1 \\ 1 \end{bmatrix}$

In Exercises 5 to 10 find the general solution of $Dx = Ax$.

5. A as in Exercise 4, Section 3.5 (and Exercise 17, Section 3.6)

6. A as in Exercise 8, Section 3.5

7. A as in Exercise 11, Section 3.5

8. $A = \begin{bmatrix} 2 & -1 & -4 \\ 0 & 2 & -4 \\ 0 & 1 & -2 \end{bmatrix}$

9. $A = \begin{bmatrix} 2 & 0 & 1 & 0 \\ 0 & 1 & 0 & 1 \\ 0 & 0 & 2 & 1 \\ 0 & -1 & 0 & 1 \end{bmatrix}$

10. $\begin{bmatrix} -2 & 0 & 0 & 0 \\ 0 & -2 & 2 & -1 \\ 0 & 0 & -4 & 9 \\ 0 & 0 & -4 & 8 \end{bmatrix}$

In Exercises 11 to 14 find (a) the general solution and (b) the specific solution satisfying the given condition.

11. $Dx = Ax$, $A = \begin{bmatrix} 0 & 1 \\ -4 & 4 \end{bmatrix}$; $x(0) = \begin{bmatrix} 3 \\ 4 \end{bmatrix}$

12. $\begin{aligned} x_1' &= \quad\quad x_2 \\ x_2' &= -x_1 - 2x_2 \end{aligned}$ $x_1(0) = 3,\ x_2(0) = 4$

13. $\begin{aligned} x_1' &= 2x_1 - \ x_2 \\ x_2' &= \ x_1 + 4x_2 \end{aligned}$ $x_1(0) = 3,\ x_2(0) = 4$

14. $\begin{aligned} x_1' &= 3x_1 - \ x_2 \quad\ - 4x_4 \\ x_2' &= \quad\quad 3x_2 \quad\ - 4x_4 \\ x_3' &= \quad\quad\quad 2x_3 \\ x_4' &= \quad\quad\ x_2 \quad\ - \ x_4 \end{aligned}$ $x_1(0) = x_2(0) = x_3(0) = 1,\ x_4(0) = 0$

Exercises 15 and 16 refer to our earlier models.

15. *LRC Circuit:* Find the general solution of the system modeling the *LRC* circuit of Example 3.1.1 with $V(t) = 0$, $R = 2$ ohms, $C = 1$ farad, and $L = 1$ henry.

16. *Two-Loop Circuit:*

a. Find the general solution of the system modeling the two-loop circuit of Example 3.1.2 with $V(t) = 0$, $R_1 = 1$, $R_2 = 3$ ohms, $C = 1$ farad, and $L = 1$ henry.

    b. Find the specific solution satisfying $Q(0) = 1$ coulomb and $I_2(0) = 2$ amperes.

Some more abstract problems:

17. Suppose that A is an $n \times m$ matrix.
    a. Show that if X is a matrix for which AX and XA are both defined, then X must be $m \times n$.
    b. Show that if $A^2 = AA$ is defined, then $m = n$.

18. Let A and B be $n \times n$ matrices.
    a. Use properties (5) and (6) from Note 2 to show that
$$(A + B)^2 = A^2 + AB + BA + B^2.$$
    b. Find an example of 2×2 matrices A and B for which
$$(A + B)^2 \neq A^2 + 2AB + B^2.$$

19. a. Show that any 2×2 matrix of the form $X = \begin{bmatrix} 0 & a \\ 0 & 0 \end{bmatrix}$ or $X = \begin{bmatrix} 0 & 0 \\ a & 0 \end{bmatrix}$ satisfies
$$X^2 = \begin{bmatrix} 0 & 0 \\ 0 & 0 \end{bmatrix}.$$
    b. Are there any other 2×2 matrices that satisfy this equation?

20. Suppose B is an $n \times m$ matrix, C is an $m \times k$ matrix, and $\mathbf{v}$ is a k-vector. Show that $(BC)\mathbf{v} = B(C\mathbf{v})$. (*Hint:* Use the description of BC given in Note 1.)

21. Verify properties (1) through (6) from Note 2. (*Hint:* You may use properties (M1) to (M6) from Note 2, Section 3.2, as well as Exercise 20 with B and C replaced by A and B.)

22. Suppose A is an $n \times n$ matrix. Show that if $(A - \lambda I)^2 \mathbf{v} = \mathbf{0}$ but $(A - \lambda I)\mathbf{v} \neq \mathbf{0}$, then $\mathbf{v}_1 = (A - \lambda I)\mathbf{v}$ is an eigenvector of A.

23. An important fact about matrix multiplication is that the determinant of a product of $n \times n$ matrices is the same as the product of the determinants:
$$\det(AB) = (\det A)(\det B).$$
Verify this formula when

    a. $A = \begin{bmatrix} 1 & 2 \\ 0 & 3 \end{bmatrix}$, $B = \begin{bmatrix} -1 & 0 \\ 1 & 3 \end{bmatrix}$

    b. $A = \begin{bmatrix} 1 & 0 & 3 \\ -1 & 2 & 0 \\ 1 & 1 & 0 \end{bmatrix}$, $B = \begin{bmatrix} 1 & 1 & 1 \\ 3 & 0 & 0 \\ -1 & 2 & 5 \end{bmatrix}$

    c. $A = \begin{bmatrix} a_{11} & a_{12} \\ a_{21} & a_{22} \end{bmatrix}$, $B = \begin{bmatrix} b_{11} & b_{12} \\ b_{21} & b_{22} \end{bmatrix}$

24. The **Cayley-Hamilton theorem** says that if the $n \times n$ matrix A has characteristic polynomial $p(\lambda) = \det(A - \lambda I) = b_n\lambda^n + \cdots + b_1\lambda + b_0$, then the matrix $p(A) = b_nA^n + \cdots + b_1A + b_0I_n$ is equal to the $n \times n$ zero matrix. Verify this fact for

    a. $A = \begin{bmatrix} 1 & 2 \\ -1 & 3 \end{bmatrix}$ b. $A = \begin{bmatrix} 1 & 0 & 3 \\ -1 & 2 & 0 \\ 1 & 1 & 0 \end{bmatrix}$ c. $A = \begin{bmatrix} a_{11} & a_{12} \\ a_{21} & a_{22} \end{bmatrix}$

3.10 HOMOGENEOUS SYSTEMS WITH CONSTANT COEFFICIENTS: MULTIPLE ROOTS

In this section we complete our treatment of homogeneous systems with constant coefficients by discussing (real or complex) eigenvalues of arbitrary multiplicity. Our method for dealing with multiple roots is a straightforward generalization of the methods we used for real double roots.

In Section 3.9 we associated to a double root λ solutions of $D\mathbf{x} = A\mathbf{x}$ of the form $\mathbf{h}(t) = e^{\lambda t}(\mathbf{v}_0 + t\mathbf{v}_1)$. In general, if λ is a root of multiplicity m, we look for solutions of the form

$$(1) \qquad \mathbf{h}(t) = e^{\lambda t}(\mathbf{v}_0 + t\mathbf{v}_1 + t^2\mathbf{v}_2 + \cdots + t^{m-1}\mathbf{v}_{m-1}).$$

Substitution of $\mathbf{x} = \mathbf{h}(t)$ into $D\mathbf{x} = A\mathbf{x}$ yields the equation

$$\lambda e^{\lambda t}(\mathbf{v}_0 + t\mathbf{v}_1 + \cdots + t^{m-1}\mathbf{v}_{m-1}) + e^{\lambda t}(\mathbf{v}_1 + 2t\mathbf{v}_2 + \cdots + (m-1)t^{m-2}\mathbf{v}_{m-1})$$
$$= e^{\lambda t}(A\mathbf{v}_0 + tA\mathbf{v}_1 + \cdots + t^{m-1}A\mathbf{v}_{m-1}).$$

If we divide by $e^{\lambda t}$ and subtract $\lambda(\mathbf{v}_0 + \cdots + t^{m-1}\mathbf{v}_{m-1})$ from both sides, we get

$$\mathbf{v}_1 + 2t\mathbf{v}_2 + \cdots + (m-1)t^{m-2}\mathbf{v}_{m-1}$$
$$= (A\mathbf{v}_0 - \lambda\mathbf{v}_0) + t(A\mathbf{v}_1 - \lambda\mathbf{v}_1) + \cdots + t^{m-1}(A\mathbf{v}_{m-1} - \lambda\mathbf{v}_{m-1})$$
$$= (A - \lambda I)\mathbf{v}_0 + t(A - \lambda I)\mathbf{v}_1 + \cdots + t^{m-1}(A - \lambda I)\mathbf{v}_{m-1}.$$

Comparing corresponding powers of t on both sides, we obtain the equations

$$(A - \lambda I)\mathbf{v}_0 = \mathbf{v}_1$$
$$(A - \lambda I)\mathbf{v}_1 = 2\mathbf{v}_2$$

$$(2) \qquad \qquad \cdot$$
$$\cdot$$
$$\cdot$$

$$(A - \lambda I)\mathbf{v}_{m-2} = (m-1)\mathbf{v}_{m-1}$$
$$(A - \lambda I)\mathbf{v}_{m-1} = \mathbf{0}.$$

Starting from the first equation of (2) and successively multiplying by $A - \lambda I$ yields

$$(A - \lambda I)\mathbf{v}_0 = \mathbf{v}_1$$
$$(A - \lambda I)^2\mathbf{v}_0 = (A - \lambda I)\mathbf{v}_1 = 2\mathbf{v}_2$$
$$(A - \lambda I)^3\mathbf{v}_0 = 2(A - \lambda I)\mathbf{v}_2 = 3\cdot2\mathbf{v}_3$$

$$(3) \qquad \qquad \cdot$$
$$\cdot$$
$$\cdot$$

$$(A - \lambda I)^{m-1}\mathbf{v}_0 = (m-1)!\mathbf{v}_{m-1}$$
$$(A - \lambda I)^m\mathbf{v}_0 = (m-1)!(A - \lambda I)\mathbf{v}_{m-1} = \mathbf{0}.$$

The last equation of (3) tells us where to look for $\mathbf{v}_0$. Once we find $\mathbf{v}_0$, we can use the earlier equations to find the other $\mathbf{v}_j$'s; you should check that

$$\mathbf{v}_j = \frac{1}{j!}(A - \lambda I)^j \mathbf{v}_0.$$

Substitution of these vectors into (1) will yield a solution of $D\mathbf{x} = A\mathbf{x}$.

Fact: *If* $(A - \lambda I)^m \mathbf{v} = \mathbf{0}$, *then*

$$\mathbf{h}(t) = e^{\lambda t}\left(\mathbf{v} + t[A - \lambda I]\mathbf{v} + \frac{1}{2}t^2[A - \lambda I]^2\mathbf{v}\right.$$

$$\left. + \cdots + \frac{1}{(m-1)!}t^{m-1}[A - \lambda I]^{m-1}\mathbf{v}\right)$$

is a solution of $D\mathbf{x} = A\mathbf{x}$ *with initial vector* $\mathbf{h}(0) = \mathbf{v}$.

Just as when the multiplicity is 2, we refer to nonzero solutions of $(A - \lambda I)^m\mathbf{v} = \mathbf{0}$ as **generalized eigenvectors** of A corresponding to λ. Independent generalized eigenvectors yield solutions with independent initial vectors.

Example 3.10.1

Solve $D\mathbf{x} = A\mathbf{x}$, where

$$A = \begin{bmatrix} -1 & -1 & 0 \\ 1 & -1 & 1 \\ 0 & 1 & -1 \end{bmatrix}.$$

This represents the three-loop circuit of Example 3.1.3 with $V(t) = 0$, $C = 1$, $R_1 = R_2 = 1$, and $L_2 = L_3 = 1$.

The characteristic polynomial of A is $-(\lambda + 1)^3$, so -1 is a root of multiplicity 3. To find solutions associated to $\lambda = -1$, we first calculate

$$A - (-1)I = \begin{bmatrix} 0 & -1 & 0 \\ 1 & 0 & 1 \\ 0 & 1 & 0 \end{bmatrix}, \quad [A - (-1)I]^2 = \begin{bmatrix} -1 & 0 & -1 \\ 0 & 0 & 0 \\ 0 & 0 & 1 \end{bmatrix}$$

and

$$[A - (-1)I]^3 = \begin{bmatrix} 0 & 0 & 0 \\ 0 & 0 & 0 \\ 0 & 0 & 0 \end{bmatrix}.$$

Every 3-vector satisfies $[A - (-1)I]^3\mathbf{v} = \mathbf{0}$. In particular, the vectors

$$\mathbf{v} = \begin{bmatrix} 1 \\ 0 \\ 0 \end{bmatrix}, \quad \mathbf{w} = \begin{bmatrix} 0 \\ 1 \\ 0 \end{bmatrix}, \quad \text{and} \quad \mathbf{u} = \begin{bmatrix} 0 \\ 0 \\ 1 \end{bmatrix}$$

are independent generalized eigenvectors. Associated to these vectors are three solutions of $D\mathbf{x} = A\mathbf{x}$ with independent initial vectors:

$$\mathbf{h}_1(t) = e^{-t}\left(\mathbf{v} + t[A - (-1)I]\mathbf{v} + \frac{1}{2}t^2[A - (-1)I]^2\mathbf{v} \right)$$

$$= e^{-t}\left(\begin{bmatrix} 1 \\ 0 \\ 0 \end{bmatrix} + t\begin{bmatrix} 0 \\ 1 \\ 0 \end{bmatrix} + \frac{1}{2}t^2\begin{bmatrix} -1 \\ 0 \\ 1 \end{bmatrix} \right) = \begin{bmatrix} (1 - \frac{1}{2}t^2)e^{-t} \\ te^{-t} \\ \frac{1}{2}t^2e^{-t} \end{bmatrix}$$

$$\mathbf{h}_2(t) = e^{-t}\left(\mathbf{w} + t[A - (-1)I]\mathbf{w} + \frac{1}{2}t^2[A - (-1)I]^2\mathbf{w} \right)$$

$$= e^{-t}\left(\begin{bmatrix} 0 \\ 1 \\ 0 \end{bmatrix} + t\begin{bmatrix} -1 \\ 0 \\ 1 \end{bmatrix} + \frac{1}{2}t^2\begin{bmatrix} 0 \\ 0 \\ 0 \end{bmatrix} \right) = \begin{bmatrix} -te^{-t} \\ e^{-t} \\ te^{-t} \end{bmatrix}$$

$$\mathbf{h}_3(t) = e^{-t}\left(\mathbf{u} + t[A - (-1)I]\mathbf{u} + \frac{1}{2}t^2[A - (-1)I]^2\mathbf{u} \right)$$

$$= e^{-t}\left(\begin{bmatrix} 0 \\ 0 \\ 1 \end{bmatrix} + t\begin{bmatrix} 0 \\ 1 \\ 0 \end{bmatrix} + \frac{1}{2}t^2\begin{bmatrix} -1 \\ 0 \\ 1 \end{bmatrix} \right) = \begin{bmatrix} \frac{1}{2}t^2e^{-t} \\ te^{-t} \\ (1 + \frac{1}{2}t^2)e^{-t} \end{bmatrix}.$$

The general solution of $D\mathbf{x} = A\mathbf{x}$ is

$$\mathbf{x} = c_1\mathbf{h}_1(t) + c_2\mathbf{h}_2(t) + c_3\mathbf{h}_3(t),$$

which, in terms of charge and current, is

$$Q = c_1(1 - \tfrac{1}{2}t^2)e^{-t} - c_2te^{-t} - \tfrac{1}{2}c_3t^2e^{-t}$$

$$I_2 = c_1te^{-t} \qquad\qquad + c_2e^{-t} + c_3te^{-t}$$

$$I_3 = \tfrac{1}{2}c_1t^2e^{-t} \qquad\qquad + c_2te^{-t} + c_3(1 + \tfrac{1}{2}t^2)e^{-t}.$$

Note that this system of o.d.e.'s is the same as that of Example 3.3.2. However, the generating solutions we have found here are different from those given there. As a result, the two expressions for the general solution look different. They are equivalent, however, in the sense that each expression yields every solution exactly once as the parameters c_1, c_2, and c_3 range over all possible real values.

If $\alpha \pm \beta_i$ are complex roots of the characteristic polynomial, then we can work with one of the roots as if it were real to obtain complex solutions of $D\mathbf{x} = A\mathbf{x}$. The real and imaginary parts of these complex solutions are real solutions.

Fact: *Suppose that λ is a complex root of $\det(A - \lambda I)$ and $\mathbf{w}_1, \ldots, \mathbf{w}_j$ are independent complex solutions of $(A - \lambda I)^m \mathbf{w} = \mathbf{0}$. For $i = 1, 2, \ldots, j$ let*

$$\mathbf{h}_i(t) = e^{\lambda t} \left(\mathbf{w}_i + t[A - \lambda I]\mathbf{w}_i + \frac{1}{2} t^2 [A - \lambda I]^2 \mathbf{w}_i \right.$$

$$\left. + \cdots + \frac{1}{(m-1)!} t^{m-1}[A - \lambda I]^{m-1} \mathbf{w}_i \right).$$

Then $\text{Re}(\mathbf{h}_1(t)), \text{Im}(\mathbf{h}_1(t)), \ldots, \text{Re}(\mathbf{h}_j(t)), \text{Im}(\mathbf{h}_j(t))$ *are solutions of $D\mathbf{x} = A\mathbf{x}$ with independent initial vectors.*

Example 3.10.2

Solve $D\mathbf{x} = A\mathbf{x}$, where

$$A = \begin{bmatrix} 0 & -2 & 0 & 0 \\ 2 & 0 & 4 & 0 \\ 0 & 0 & 0 & -4 \\ 0 & 0 & 1 & 0 \end{bmatrix}.$$

The characteristic polynomial of A is $(\lambda^2 + 4)^2$. The roots are $\pm 2i$, each of which has multiplicity 2.

To find solutions associated to $\lambda = 2i$, we first calculate

$$A - 2iI = \begin{bmatrix} -2i & -2 & 0 & 0 \\ 2 & -2i & 4 & 0 \\ 0 & 0 & -2i & -4 \\ 0 & 0 & 1 & -2i \end{bmatrix}$$

and

$$(A - 2iI)^2 = \begin{bmatrix} -8 & 8i & -8 & 0 \\ -8i & -8 & -16i & -16 \\ 0 & 0 & -8 & 16i \\ 0 & 0 & -4i & -8 \end{bmatrix}.$$

We solve $(A - 2iI)^2 \mathbf{w} = \mathbf{0}$ by reducing the augmented matrix:

$$[(A - 2iI)^2 \mid \mathbf{0}] \rightarrow \begin{bmatrix} 1 & -i & 0 & 2i & 0 \\ 0 & 0 & 1 & -2i & 0 \\ 0 & 0 & 0 & 0 & 0 \\ 0 & 0 & 0 & 0 & 0 \end{bmatrix}.$$

The vectors

$$\mathbf{w}_1 = \begin{bmatrix} i \\ 1 \\ 0 \\ 0 \end{bmatrix} \quad \text{and} \quad \mathbf{w}_2 = \begin{bmatrix} -2i \\ 0 \\ 2i \\ 1 \end{bmatrix}$$

are independent complex generalized eigenvectors. Associated to these vectors are two complex solutions:

$$\mathbf{h}_1(t) = e^{2it}(\mathbf{w}_1 + t[A - 2iI]\mathbf{w}_1)$$

$$= (\cos 2t + i \sin 2t)\left(\begin{bmatrix} i \\ 1 \\ 0 \\ 0 \end{bmatrix} + t \begin{bmatrix} 0 \\ 0 \\ 0 \\ 0 \end{bmatrix}\right)$$

$$= \begin{bmatrix} -\sin 2t \\ \cos 2t \\ 0 \\ 0 \end{bmatrix} + i \begin{bmatrix} \cos 2t \\ \sin 2t \\ 0 \\ 0 \end{bmatrix}$$

$$\mathbf{h}_2(t) = e^{2it}(\mathbf{w}_2 + t[A - 2iI]\mathbf{w}_2)$$

$$= (\cos 2t + i \sin 2t)\left(\begin{bmatrix} -2i \\ 0 \\ 2i \\ 1 \end{bmatrix} + t \begin{bmatrix} -4 \\ 4i \\ 0 \\ 0 \end{bmatrix}\right)$$

$$= \begin{bmatrix} 2 \sin 2t - 4t \cos 2t \\ -4t \sin 2t \\ -2 \sin 2t \\ \cos 2t \end{bmatrix} + i \begin{bmatrix} -2 \cos 2t - 4t \sin 2t \\ 4t \cos 2t \\ 2 \cos 2t \\ \sin 2t \end{bmatrix}.$$

The real and imaginary parts of these solutions provide us with four solutions with independent initial vectors. The general solution of our fourth-order system $D\mathbf{x} = A\mathbf{x}$ is

$$\mathbf{x} = c_1 \begin{bmatrix} -\sin 2t \\ \cos 2t \\ 0 \\ 0 \end{bmatrix} + c_2 \begin{bmatrix} \cos 2t \\ \sin 2t \\ 0 \\ 0 \end{bmatrix} + c_3 \begin{bmatrix} 2 \sin 2t - 4t \cos 2t \\ -4t \sin 2t \\ -2 \sin 2t \\ \cos 2t \end{bmatrix}$$

$$+ c_4 \begin{bmatrix} -2 \cos 2t - 4t \sin 2t \\ 4t \cos 2t \\ 2 \cos 2t \\ \sin 2t \end{bmatrix}.$$

As usual, if we have a system with more than one eigenvalue, then we work with each individually and combine the resulting solutions.

Fact: *The solutions of* $D\mathbf{x} = A\mathbf{x}$ *that we get by combining solutions associated to various eigenvalues will have independent initial vectors as long as the solutions associated to each single eigenvalue or pair of complex eigenvalues have independent initial vectors.*

Solve $D\mathbf{x} = A\mathbf{x}$, where

$$A = \begin{bmatrix} 2 & 0 & 0 & 0 \\ 1 & 2 & 0 & 0 \\ 0 & 0 & 2 & 1 \\ 0 & 0 & 0 & 4 \end{bmatrix}.$$

The characteristic polynomial is $(2 - \lambda)^3(4 - \lambda)$. The root $\lambda = 2$ has multiplicity 3, and $\lambda = 4$ has multiplicity 1.

We calculated the first three powers of $A - 2I$ in Example 3.9.4:

$$A - 2I = \begin{bmatrix} 0 & 0 & 0 & 0 \\ 1 & 0 & 0 & 0 \\ 0 & 0 & 0 & 1 \\ 0 & 0 & 0 & 2 \end{bmatrix}, \quad (A - 2I)^2 = \begin{bmatrix} 0 & 0 & 0 & 0 \\ 0 & 0 & 0 & 0 \\ 0 & 0 & 0 & 2 \\ 0 & 0 & 0 & 4 \end{bmatrix},$$

$$(A - 2I)^3 = \begin{bmatrix} 0 & 0 & 0 & 0 \\ 0 & 0 & 0 & 0 \\ 0 & 0 & 0 & 4 \\ 0 & 0 & 0 & 8 \end{bmatrix}.$$

We reduce $[(A - 2I)^3 \mid \mathbf{0}]$,

$$[(A - 2I)^3 \mid \mathbf{0}] \rightarrow \begin{bmatrix} 0 & 0 & 0 & 0 & 0 \\ 0 & 0 & 0 & 0 & 0 \\ 0 & 0 & 0 & 0 & 0 \\ 0 & 0 & 0 & 1 & 0 \end{bmatrix},$$

and read off three independent generalized eigenvectors

$$\begin{bmatrix} 1 \\ 0 \\ 0 \\ 0 \end{bmatrix}, \quad \begin{bmatrix} 0 \\ 1 \\ 0 \\ 0 \end{bmatrix}, \quad \text{and} \quad \begin{bmatrix} 0 \\ 0 \\ 1 \\ 0 \end{bmatrix}.$$

The associated solutions are

$$\mathbf{h}_1(t) = e^{2t}\left(\begin{bmatrix} 1 \\ 0 \\ 0 \\ 0 \end{bmatrix} + t\begin{bmatrix} 0 \\ 1 \\ 0 \\ 0 \end{bmatrix} + \frac{1}{2}t^2\begin{bmatrix} 0 \\ 0 \\ 0 \\ 0 \end{bmatrix}\right) = \begin{bmatrix} e^{2t} \\ te^{2t} \\ 0 \\ 0 \end{bmatrix}$$

$$\mathbf{h}_2(t) = e^{2t}\left(\begin{bmatrix} 0 \\ 1 \\ 0 \\ 0 \end{bmatrix} + t\begin{bmatrix} 0 \\ 0 \\ 0 \\ 0 \end{bmatrix} + \frac{1}{2}t^2\begin{bmatrix} 0 \\ 0 \\ 0 \\ 0 \end{bmatrix}\right) = \begin{bmatrix} 0 \\ e^{2t} \\ 0 \\ 0 \end{bmatrix}$$

$$\mathbf{h}_3(t) = e^{2t}\left(\begin{bmatrix} 0 \\ 0 \\ 1 \\ 0 \end{bmatrix} + t\begin{bmatrix} 0 \\ 0 \\ 0 \\ 0 \end{bmatrix} + \frac{1}{2}t^2\begin{bmatrix} 0 \\ 0 \\ 0 \\ 0 \end{bmatrix}\right) = \begin{bmatrix} 0 \\ 0 \\ e^{2t} \\ 0 \end{bmatrix}.$$

To find a solution associated to $\lambda = 4$, we reduce $[A - 4I \mid \mathbf{0}]$:

$$\begin{bmatrix} -2 & 0 & 0 & 0 & | & 0 \\ 1 & -2 & 0 & 0 & | & 0 \\ 0 & 0 & -2 & 1 & | & 0 \\ 0 & 0 & 0 & 0 & | & 0 \end{bmatrix} \rightarrow \begin{bmatrix} 1 & 0 & 0 & 0 & | & 0 \\ 0 & 1 & 0 & 0 & | & 0 \\ 0 & 0 & 1 & -\frac{1}{2} & | & 0 \\ 0 & 0 & 0 & 0 & | & 0 \end{bmatrix}.$$

The vector

$$\begin{bmatrix} 0 \\ 0 \\ 1 \\ 2 \end{bmatrix}$$

is an eigenvector with associated solution

$$\mathbf{h}_4(t) = e^{4t}\begin{bmatrix} 0 \\ 0 \\ 1 \\ 2 \end{bmatrix} = \begin{bmatrix} 0 \\ 0 \\ e^{4t} \\ 2e^{4t} \end{bmatrix}.$$

If we combine the solutions associated to $\lambda = 2$ and $\lambda = 4$, we obtain four solutions with independent initial vectors. Since $D\mathbf{x} = A\mathbf{x}$ has order 4, the general solution is

$$\mathbf{x} = c_1\mathbf{h}_1(t) + c_2\mathbf{h}_2(t) + c_3\mathbf{h}_3(t) + c_4\mathbf{h}_4(t)$$

$$= c_1\begin{bmatrix} e^{2t} \\ te^{2t} \\ 0 \\ 0 \end{bmatrix} + c_2\begin{bmatrix} 0 \\ e^{2t} \\ 0 \\ 0 \end{bmatrix} + c_2\begin{bmatrix} 0 \\ 0 \\ e^{2t} \\ 0 \end{bmatrix} + c_4\begin{bmatrix} 0 \\ 0 \\ e^{4t} \\ 2e^{4t} \end{bmatrix}.$$

Example 3.10.4

Solve $D\mathbf{x} = A\mathbf{x}$, where

$$A = \begin{bmatrix} 1 & 0 & 0 & 0 & 0 & 0 \\ 0 & 1 & 0 & 0 & 1 & 0 \\ 0 & 0 & 0 & 0 & 0 & 1 \\ 1 & 0 & 0 & 1 & 0 & 0 \\ 0 & 0 & 0 & 1 & 1 & 0 \\ 0 & 0 & 0 & 0 & 0 & 0 \end{bmatrix}.$$

The characteristic polynomial is $(1 - \lambda)^4 \lambda^2$. The root $\lambda = 1$ has multiplicity 4, and $\lambda = 0$ has multiplicity 2.

The first four powers of $A - I$ are

$$(A - I) = \begin{bmatrix} 0 & 0 & 0 & 0 & 0 & 0 \\ 0 & 0 & 0 & 0 & 1 & 0 \\ 0 & 0 & -1 & 0 & 0 & 1 \\ 1 & 0 & 0 & 0 & 0 & 0 \\ 0 & 0 & 0 & 1 & 0 & 0 \\ 0 & 0 & 0 & 0 & 0 & -1 \end{bmatrix}, (A - I)^2 = \begin{bmatrix} 0 & 0 & 0 & 0 & 0 & 0 \\ 0 & 0 & 0 & 1 & 0 & 0 \\ 0 & 0 & 1 & 0 & 0 & -2 \\ 0 & 0 & 0 & 0 & 0 & 0 \\ 1 & 0 & 0 & 0 & 0 & 0 \\ 0 & 0 & 0 & 0 & 0 & 1 \end{bmatrix},$$

$$(A - I)^3 = \begin{bmatrix} 0 & 0 & 0 & 0 & 0 & 0 \\ 1 & 0 & 0 & 0 & 0 & 0 \\ 0 & 0 & -1 & 0 & 0 & 3 \\ 0 & 0 & 0 & 0 & 0 & 0 \\ 0 & 0 & 0 & 0 & 0 & 0 \\ 0 & 0 & 0 & 0 & 0 & -1 \end{bmatrix}, (A - I)^4 = \begin{bmatrix} 0 & 0 & 0 & 0 & 0 & 0 \\ 0 & 0 & 0 & 0 & 0 & 0 \\ 0 & 0 & 1 & 0 & 0 & -4 \\ 0 & 0 & 0 & 0 & 0 & 0 \\ 0 & 0 & 0 & 0 & 0 & 0 \\ 0 & 0 & 0 & 0 & 0 & 1 \end{bmatrix}.$$

The vectors

$$\begin{bmatrix} 1 \\ 0 \\ 0 \\ 0 \\ 0 \\ 0 \end{bmatrix}, \begin{bmatrix} 0 \\ 1 \\ 0 \\ 0 \\ 0 \\ 0 \end{bmatrix}, \begin{bmatrix} 0 \\ 0 \\ 0 \\ 1 \\ 0 \\ 0 \end{bmatrix}, \text{ and } \begin{bmatrix} 0 \\ 0 \\ 0 \\ 0 \\ 1 \\ 0 \end{bmatrix}$$

are independent solutions of $(A - I)^4 \mathbf{v} = \mathbf{0}$. The solution associated to the first of these generalized eigenvectors is

$$\mathbf{h}_1(t) = e^t \left(\begin{bmatrix} 1 \\ 0 \\ 0 \\ 0 \\ 0 \\ 0 \end{bmatrix} + t \begin{bmatrix} 0 \\ 0 \\ 0 \\ 1 \\ 0 \\ 0 \end{bmatrix} + \frac{1}{2} t^2 \begin{bmatrix} 0 \\ 0 \\ 0 \\ 0 \\ 1 \\ 0 \end{bmatrix} + \frac{1}{3 \cdot 2} t^3 \begin{bmatrix} 0 \\ 1 \\ 0 \\ 0 \\ 0 \\ 0 \end{bmatrix} \right) = \begin{bmatrix} e^t \\ \frac{1}{6} t^3 e^t \\ 0 \\ t e^t \\ \frac{1}{2} t^2 e^t \\ 0 \end{bmatrix}.$$

We leave it to you to check that the solutions associated to the other generalized eigenvectors are

$$
\mathbf{h}_2(t) = \begin{bmatrix} 0 \\ e^t \\ 0 \\ 0 \\ 0 \\ 0 \end{bmatrix}, \quad
\mathbf{h}_3(t) = \begin{bmatrix} 0 \\ \frac{1}{2}t^2 e^t \\ 0 \\ e^t \\ te^t \\ 0 \end{bmatrix}, \quad
\mathbf{h}_4(t) = \begin{bmatrix} 0 \\ te^t \\ 0 \\ 0 \\ e^t \\ 0 \end{bmatrix}.
$$

To find solutions associated to $\lambda = 0$, we calculate

$$
(A - 0I)^2 = A^2 = \begin{bmatrix}
1 & 0 & 0 & 0 & 0 & 0 \\
0 & 1 & 0 & 1 & 2 & 0 \\
0 & 0 & 0 & 0 & 0 & 0 \\
2 & 0 & 0 & 1 & 0 & 0 \\
1 & 0 & 0 & 2 & 1 & 0 \\
0 & 0 & 0 & 0 & 0 & 0
\end{bmatrix}
$$

and reduce $[A^2 \mid \mathbf{0}]$:

$$
[A^2 \mid \mathbf{0}] \rightarrow \left[\begin{array}{cccccc|c}
1 & 0 & 0 & 0 & 0 & 0 & 0 \\
0 & 1 & 0 & 0 & 0 & 0 & 0 \\
0 & 0 & 0 & 1 & 0 & 0 & 0 \\
0 & 0 & 0 & 0 & 1 & 0 & 0 \\
0 & 0 & 0 & 0 & 0 & 0 & 0 \\
0 & 0 & 0 & 0 & 0 & 0 & 0
\end{array}\right].
$$

The vectors

$$
\begin{bmatrix} 0 \\ 0 \\ 1 \\ 0 \\ 0 \\ 0 \end{bmatrix} \quad \text{and} \quad \begin{bmatrix} 0 \\ 0 \\ 0 \\ 0 \\ 0 \\ 1 \end{bmatrix}
$$

are generalized eigenvectors with associated solutions (remember $e^{0t} = 1$)

$$
\mathbf{h}_5(t) = \begin{bmatrix} 0 \\ 0 \\ 1 \\ 0 \\ 0 \\ 0 \end{bmatrix} \quad \text{and} \quad \mathbf{h}_6(t) = \begin{bmatrix} 0 \\ 0 \\ t \\ 0 \\ 0 \\ 1 \end{bmatrix}.
$$

The general solution of $D\mathbf{x} = A\mathbf{x}$ is

$$\mathbf{x} = c_1\mathbf{h}_1(t) + c_2\mathbf{h}_2(t) + c_3\mathbf{h}_3(t) + c_4\mathbf{h}_4(t) + c_5\mathbf{h}_5(t) + c_6\mathbf{h}_6(t)$$

$$= c_1\begin{bmatrix} e^t \\ \frac{1}{6}t^3e^t \\ 0 \\ te^t \\ \frac{1}{2}t^2e^t \\ 0 \end{bmatrix} + c_2\begin{bmatrix} 0 \\ e^t \\ 0 \\ 0 \\ 0 \\ 0 \end{bmatrix} + c_3\begin{bmatrix} 0 \\ \frac{1}{2}t^2e^t \\ 0 \\ e^t \\ te^t \\ 0 \end{bmatrix} + c_4\begin{bmatrix} 0 \\ te^t \\ 0 \\ 0 \\ e^t \\ 0 \end{bmatrix} + c_5\begin{bmatrix} 0 \\ 0 \\ 1 \\ 0 \\ 0 \\ 0 \end{bmatrix} + c_6\begin{bmatrix} 0 \\ t \\ 0 \\ 0 \\ 0 \\ 1 \end{bmatrix}.$$

We can now find the general solution of $D\mathbf{x} = A\mathbf{x}$ provided that we can find m independent generalized eigenvectors corresponding to each root of $\det(A - \lambda I)$ with multiplicity m. The following result from linear algebra guarantees that this is always possible.

Theorem: If λ is a root of $\det(A - \lambda I)$ *with multiplicity m, then the equation* $(A - \lambda I)^m\mathbf{v} = \mathbf{0}$ *has m linearly independent solutions for* $\mathbf{v}$.

HOMOGENEOUS SYSTEMS WITH CONSTANT COEFFICIENTS

To solve $D\mathbf{x} = A\mathbf{x}$ where A is an $n \times n$ matrix:

1. Find the roots of the characteristic polynomial $\det(A - \lambda I)$ and their multiplicities.
2. For each real root λ of multiplicity m, find m linearly independent solutions of $(A - \lambda I)^m\mathbf{v} = \mathbf{0}$. Associate to each of these **generalized eigenvectors** the solution of $D\mathbf{x} = A\mathbf{x}$:

$$\mathbf{h}(t) = e^{\lambda t}\left(\mathbf{v} + t[A - \lambda I]\mathbf{v} + \frac{1}{2}t^2[A - \lambda I]^2\mathbf{v}\right.$$

$$\left. + \cdots + \frac{1}{(m-1)!}t^{m-1}[A - \lambda I]^{m-1}\mathbf{v}\right).$$

3. For each pair of complex roots $\alpha \pm \beta i$, work with one of the roots as if it were real to obtain complex solutions. Associate to the pair of eigenvalues the real and imaginary parts of these complex solutions.
4. Combine the solutions associated to the various roots to obtain n-vector valued functions $\mathbf{h}_1(t), \ldots, \mathbf{h}_n(t)$ that generate the general solution

$$\mathbf{x} = c_1\mathbf{h}_1(t) + \cdots + c_n\mathbf{h}_n(t).$$

Note

On finding generalized eigenvectors

   If λ is a root of $\det(A - \lambda I)$ with multiplicity m, then we can always find m independent generalized eigenvectors by calculating $(A - \lambda I)^m$ and reducing $[(A - \lambda I)^m \mid \mathbf{0}]$. If A is a large matrix and λ is a root of high multiplicity, then this may be a messy calculation. In this case we can try to find the generalized eigenvectors without calculating the mth power of $A - \lambda I$.

   We've already seen (in Section 3.7) that if there are m independent eigenvectors, the associated solutions can be obtained without calculating any powers of $A - \lambda I$ at all. If $A - \lambda I$ has j independent eigenvectors, where $1 < j < m$, then the following theorem tells us that we can stop short of calculating the mth power of $A - \lambda I$.

Theorem: Suppose λ *is a root of* $\det(A - \lambda I)$ *with multiplicity m and that there are j independent eigenvectors corresponding to* λ. *Then every generalized eigenvector corresponding to* λ *is a solution of the equation* $(A - \lambda I)^{m-j+1}\mathbf{v} = \mathbf{0}$.

   Note that if the number j of independent eigenvectors is relatively high, this theorem tells us that to find the generalized eigenvectors we need only calculate a relatively low power of $A - \lambda I$. If, on the other hand, $j = 1$, the theorem tells us that every generalized eigenvector is a solution of $(A - \lambda I)^m\mathbf{v} = \mathbf{0}$; in this case there is nothing gained by first finding an eigenvector.

   Exercise 16 describes another approach that is effective when the number of independent eigenvectors is relatively high. Exercises 17 and 18 are useful when dealing with systems for which this number is relatively low.

EXERCISES

In Exercises 1 and 2 you are given a matrix A, an eigenvalue λ, and a generalized eigenvector $\mathbf{v}$. Find the associated solution of $D\mathbf{x} = A\mathbf{x}$.

1. $A = \begin{bmatrix} -1 & -1 & 0 \\ 1 & -1 & 1 \\ 0 & 1 & -1 \end{bmatrix}$, $\lambda = -1$, $\mathbf{v} = \begin{bmatrix} 1 \\ 0 \\ 0 \end{bmatrix}$

2. $A = \begin{bmatrix} 1 & 0 & 0 & 0 & 0 & 0 \\ 0 & 1 & 0 & 0 & 1 & 0 \\ 0 & 0 & 0 & 0 & 0 & 1 \\ 1 & 0 & 0 & 1 & 0 & 0 \\ 0 & 0 & 0 & 1 & 1 & 0 \\ 0 & 0 & 0 & 0 & 0 & 0 \end{bmatrix}$, $\lambda = 1$, $\mathbf{v} = \begin{bmatrix} 6 \\ -1 \\ 0 \\ 2 \\ 0 \\ 0 \end{bmatrix}$

In Exercises 3 to 12 find the general solution of $D\mathbf{x} = A\mathbf{x}$.

3. $A = \begin{bmatrix} 0 & 1 & 0 \\ 0 & 0 & 1 \\ 1 & -3 & 3 \end{bmatrix}$

4. $A = \begin{bmatrix} 1 & 0 & 0 \\ 1 & 0 & 1 \\ 1 & -1 & 2 \end{bmatrix}$

5. $A = \begin{bmatrix} 2 & 0 & -1 & 1 \\ 0 & 0 & -1 & 0 \\ 0 & -1 & 0 & 0 \\ -1 & -1 & 0 & 0 \end{bmatrix}$
6. $A = \begin{bmatrix} 1 & 1 & 0 & 0 \\ 0 & 1 & 1 & 0 \\ 0 & 0 & 1 & -1 \\ 0 & 0 & 0 & 2 \end{bmatrix}$

7. $A = \begin{bmatrix} 0 & 2 & 0 & 0 \\ -2 & 0 & 1 & 0 \\ 0 & 0 & 0 & 2 \\ 0 & 0 & -2 & 0 \end{bmatrix}$
8. $A = \begin{bmatrix} 2 & 1 & 0 & 0 \\ 0 & 1 & 1 & 0 \\ -1 & -1 & 0 & 0 \\ 0 & 0 & 0 & 1 \end{bmatrix}$

9. $A = \begin{bmatrix} 2 & 0 & 0 & 0 \\ 0 & 2 & 0 & 0 \\ 1 & 0 & 1 & 1 \\ 1 & 0 & -1 & 3 \end{bmatrix}$
10. $A = \begin{bmatrix} 1 & 1 & 0 & 0 \\ -1 & 1 & 1 & 0 \\ 0 & 0 & 1 & -1 \\ 0 & 0 & 1 & 1 \end{bmatrix}$

11. $A = \begin{bmatrix} 0 & 0 & 0 & 0 & 1 \\ 1 & -1 & 0 & -1 & 1 \\ 1 & 0 & -1 & -1 & 0 \\ 0 & 0 & 0 & 0 & 1 \\ 1 & 0 & 0 & -1 & 0 \end{bmatrix}$
12. $A = \begin{bmatrix} 2 & 1 & 0 & 0 & 0 & 0 \\ 0 & 2 & 1 & 0 & 0 & 0 \\ 0 & 0 & 2 & 1 & 0 & 0 \\ 0 & 0 & 0 & 2 & 0 & 0 \\ 0 & 0 & 0 & 0 & 0 & 1 \\ 0 & 0 & 0 & 0 & -2 & -2 \end{bmatrix}$

In Exercises 13 and 14 find (a) the general solution and (b) the specific solution satisfying the given initial condition.

13. $\begin{aligned} x_1' &= -x_1 + x_2 \\ x_2' &= \quad\;\; - x_2 \\ x_3' &= \qquad\qquad - x_3 + x_4 \\ x_4' &= \qquad\qquad\qquad - x_4 \end{aligned}$ $x_1(0) = x_2(0) = 1,\ x_3(0) = -1,\ x_4(0) = 0$

14. $\begin{aligned} x_1' &= -x_1 + x_2 \\ x_2' &= \quad\;\; - x_2 + x_3 \\ x_3' &= \qquad\qquad - x_3 + x_4 \\ x_4' &= \qquad\qquad\qquad - x_4 \end{aligned}$ $x_1(0) = 1,\ x_2(0) = x_3(0) = 0,\ x_4(0) = -1$

Some more abstract problems:

15. Suppose λ is a root of $\det(A - \lambda I)$ with multiplicity m and that $\mathbf{v}$ is a solution of $(A - \lambda I)^k \mathbf{v} = \mathbf{0}$ for some $k \le m$. Show that $\mathbf{v}$ is a generalized eigenvector for A with associated solution

$$\mathbf{h}(t) = e^{\lambda t}\left(\mathbf{v} + t[A - \lambda I]\mathbf{v} + \frac{1}{2}t^2[A - \lambda I]^2\mathbf{v} + \cdots + \frac{1}{(k-1)!}t^{k-1}[A - \lambda I]^{k-1}\mathbf{v} \right)$$

Note, as a special case, that any eigenvector $\mathbf{v}$ is a generalized eigenvector with associated solution $\mathbf{h}(t) = e^{\lambda t}\mathbf{v}$.

16. If $\mathbf{v}$ is a solution of $(A - \lambda I)^k \mathbf{v} = \mathbf{0}$ for a low power k, then the associated solution of $D\mathbf{x} = A\mathbf{x}$ is easy to calculate (see Exercise 15). One approach to finding generalized eigenvectors is to start by finding as many independent eigenvectors as possible. If this is not enough, then we combine these eigenvectors with solutions of $(A - \lambda I)^2 \mathbf{v} = \mathbf{0}$, making sure that the expanded list is independent. We continue this way, solving $(A - \lambda I)^k \mathbf{v} = \mathbf{0}$ for $k = 3, 4$, etc., until we have found enough independent generalized eigenvectors. Use this approach to solve the system of the indicated exercise.

   a. Exercise 4   b. Exercise 5   c. Exercise 9

17. Suppose λ is a root of $\det(A - \lambda I)$ with multiplicity m and that there is an integer $k \leq m$ so that

$$(A - \lambda I)^k \mathbf{v} = \mathbf{0} \quad \text{but} \quad (A - \lambda I)^{k-1} \mathbf{v} \neq \mathbf{0}.$$

Then we refer to the vectors

$$\mathbf{w}_1 = \mathbf{v}, \quad \mathbf{w}_2 = (A - \lambda I)\mathbf{v}, \ldots, \mathbf{w}_k = (A - \lambda I)^{k-1}\mathbf{v}$$

as a **string of generalized eigenvectors.**

a. Show that $\mathbf{w}_1, \ldots, \mathbf{w}_k$ are indeed generalized eigenvectors and that the associated solutions are

$$\mathbf{h}_1(t) = e^{\lambda t}\left(\mathbf{w}_1 + t\mathbf{w}_2 + \cdots + \frac{1}{(k-2)!}t^{k-2}\mathbf{w}_{k-1} + \frac{1}{(k-1)!}t^{k-1}\mathbf{w}_k\right)$$

$$\mathbf{h}_2(t) = e^{\lambda t}\left(\mathbf{w}_2 + t\mathbf{w}_3 + \cdots + \frac{1}{(k-2)!}t^{k-2}\mathbf{w}_k\right)$$

.
.
.

$$\mathbf{h}_k(t) = e^{\lambda t}\mathbf{w}_k.$$

b. Show $\mathbf{w}_1, \ldots, \mathbf{w}_k$ are independent: Suppose $c_1\mathbf{w}_1 + \cdots + c_k\mathbf{w}_k = \mathbf{0}$. Multiply this equation by $(A - \lambda I)^{k-1}$ to obtain the equation $c_1\mathbf{w}_1 = \mathbf{0}$ from which you can conclude $c_1 = 0$. Now multiply the original equation by $(A - \lambda I)^{k-2}$ to show that $c_2 = 0$, etc.

18. A theorem from linear algebra states that *if λ is a root of $\det(A - \lambda I)$ of multiplicity m, and if every eigenvector corresponding to λ is a multiple of one fixed eigenvector, then it is possible to find a single string of m generalized eigenvectors* (see Exercise 17). The solutions associated to this string are all determined once we have calculated the vectors in the string. Use this fact to obtain the general solution of the system in the indicated exercise.

a. Exercise 6   b. Exercise 11   c. Exercise 12

3.11   NONHOMOGENEOUS SYSTEMS

We turn now to the problem of solving a nonhomogeneous system

$$(S) \qquad\qquad D\mathbf{x} = A\mathbf{x} + \mathbf{E}(t).$$

Recall that the general solution of (S) will be of the form

$$\mathbf{x} = \mathbf{H}(t) + \mathbf{p}(t)$$

where $\mathbf{x} = \mathbf{H}(t)$ is the general solution of the related homogeneous system

$$(H) \qquad\qquad D\mathbf{x} = A\mathbf{x}$$

and $\mathbf{x} = \mathbf{p}(t)$ is a particular solution of (S). Our technique of solution will rely on having $H(t)$. We will find a particular solution by extending to systems the method of variation of parameters that we used in Section 2.8.

If A is an $n \times n$ matrix, then the general solution of (H) is of the form $x = H(t)$ where

$$H(t) = c_1 h_1(t) + \cdots + c_n h_n(t)$$

for a suitable choice of $h_1(t), \ldots, h_n(t)$. By analogy with what we did in Section 2.8, we look for a solution $x = p(t)$ with

$$p(t) = c_1(t) h_1(t) + \cdots + c_n(t) h_n(t).$$

When we substitute this formula into (S), the left side $Dp(t)$, is a sum of terms of the form

$$D[c_i(t) h_i(t)] = c_i'(t) h_i(t) + c_i(t) h_i'(t).$$

The first part of the right side, $Ap(t)$, is a sum of terms of the form

$$A[c_i(t) h_i(t)] = c_i(t) A h_i(t).$$

Since $h_i(t)$ is a solution of (H), we know that $A h_i(t) = h_i'(t)$. Then

$$A[c_i(t) h_i(t)] = c_i(t) h_i'(t).$$

These terms appear on both sides. Cancellation leaves

(V) $$c_1'(t) h_1(t) + \cdots + c_n'(t) h_n(t) = E(t).$$

Thus we have the following:

Fact: *Let* $x = c_1 h_1(t) + \cdots + c_n h_n(t)$ *be the general solution of (H). If* $c_1'(t), \ldots, c_n'(t)$ *satisfy*

(V) $$c_1'(t) h_1(t) + \cdots + c_n'(t) h_n(t) = E(t)$$

then $p(t) = c_1(t) h_1(t) + \cdots + c_n(t) h_n(t)$ *is a solution of (S).*

If we equate corresponding entries of the left and right sides of (V), we obtain an algebraic system of n equations in the n unknowns $c_1'(t), \ldots, c_n'(t)$. The matrix of coefficients of this system has columns $h_1(t), \ldots, h_n(t)$, and its determinant is the Wronskian of these functions, $W(t) = W[h_1, \ldots, h_n](t)$. Since $h_1(t), \ldots, h_n(t)$ generate the general solution of (H), $W(t_0) \neq 0$ for all t_0. Cramer's determinant test tells us we can always solve the system (V). Once we find $c_1'(t), \ldots, c_n'(t)$, we can integrate to find $c_1(t), \ldots, c_n(t)$. These in turn determine $p(t)$.

Cramer's rule is one way of solving (V). We can also find the solution by reducing the augmented matrix of the system. This augmented matrix will be $[\Phi(t) \mid E(t)]$, where $\Phi(t)$ is the matrix whose columns are $h_1(t), \ldots , h_n(t)$.

Example 3.11.1

Solve

(S) $$Dx = Ax + E(t),$$

where

$$A = \begin{bmatrix} 1 & 0 & 4 \\ 0 & 3 & 0 \\ 1 & 0 & 1 \end{bmatrix} \quad \text{and} \quad E(t) = \begin{bmatrix} -2e^t \\ 9t \\ e^t \end{bmatrix}.$$

We found the general solution of the related homogeneous system in Example 3.7.1. It is $x = H(t)$, where

$$H(t) = c_1 \begin{bmatrix} -2e^{-t} \\ 0 \\ e^{-t} \end{bmatrix} + c_2 \begin{bmatrix} 0 \\ e^{3t} \\ 0 \end{bmatrix} + c_3 \begin{bmatrix} 2e^{3t} \\ 0 \\ e^{3t} \end{bmatrix}.$$

We look for a particular solution of (S) of the form $x = p(t)$, with

$$p(t) = c_1(t) \begin{bmatrix} -2e^{-t} \\ 0 \\ e^{-t} \end{bmatrix} + c_2(t) \begin{bmatrix} 0 \\ e^{3t} \\ 0 \end{bmatrix} + c_3(t) \begin{bmatrix} 2e^{3t} \\ 0 \\ e^{3t} \end{bmatrix}.$$

To find functions $c_1(t)$, $c_2(t)$, and $c_3(t)$ that work, we first solve

(V) $$c_1'(t) \begin{bmatrix} -2e^{-t} \\ 0 \\ e^{-t} \end{bmatrix} + c_2'(t) \begin{bmatrix} 0 \\ e^{3t} \\ 0 \end{bmatrix} + c_3'(t) \begin{bmatrix} 2e^{3t} \\ 0 \\ e^{3t} \end{bmatrix} = \begin{bmatrix} -2e^t \\ 9t \\ e^t \end{bmatrix}$$

for $c_1'(t)$, $c_2'(t)$, and $c_3'(t)$, by reducing the augmented matrix:

$$\begin{bmatrix} -2e^{-t} & 0 & 2e^{3t} & \mid & -2e^t \\ 0 & e^{3t} & 0 & \mid & 9t \\ e^{-t} & 0 & e^{3t} & \mid & e^t \end{bmatrix} \xrightarrow{R_1 \to R_1 + 2R_3} \begin{bmatrix} 0 & 0 & 4e^{3t} & \mid & 0 \\ 0 & e^{3t} & 0 & \mid & 9t \\ e^{-t} & 0 & e^{3t} & \mid & e^t \end{bmatrix}$$

$$\xrightarrow{R_3 \to R_3 - \frac{1}{4}R_1} \begin{bmatrix} 0 & 0 & 4e^{3t} & \mid & 0 \\ 0 & e^{3t} & 0 & \mid & 9t \\ e^{-t} & 0 & 0 & \mid & e^t \end{bmatrix} \xrightarrow[\substack{R_2 \to e^{-3t}R_2 \\ R_3 \to e^t R_3}]{R_1 \to (e^{-3t}/4)R_1} \begin{bmatrix} 0 & 0 & 1 & \mid & 0 \\ 0 & 1 & 0 & \mid & 9te^{-3t} \\ 1 & 0 & 0 & \mid & e^{2t} \end{bmatrix}$$

$$\xrightarrow{\text{rearr}} \begin{bmatrix} 1 & 0 & 0 & \mid & e^{2t} \\ 0 & 1 & 0 & \mid & 9te^{-3t} \\ 0 & 0 & 1 & \mid & 0 \end{bmatrix}.$$

The solution of the algebraic system corresponding to the reduced matrix is

$$c_1'(t) = e^{2t}, \quad c_2'(t) = 9te^{-3t}, \quad c_3'(t) = 0.$$

Any functions with these derivatives will yield a particular solution. We can take

$$c_1(t) = \frac{e^{2t}}{2}, \quad c_2(t) = -3te^{-3t} - e^{-3t}, \quad c_3(t) = 0$$

to obtain the particular solution

$$\mathbf{p}(t) = \frac{e^{2t}}{2} \begin{bmatrix} -2e^{-t} \\ 0 \\ e^{-t} \end{bmatrix} + (-3te^{-3t} - e^{-3t}) \begin{bmatrix} 0 \\ e^{3t} \\ 0 \end{bmatrix} = \begin{bmatrix} -e^{t} \\ -3t - 1 \\ e^{t}/2 \end{bmatrix}.$$

The general solution of (S) is

$$\mathbf{x} = \mathbf{H}(t) + \mathbf{p}(t) = c_1 \begin{bmatrix} -2e^{-t} \\ 0 \\ e^{-t} \end{bmatrix} + c_2 \begin{bmatrix} 0 \\ e^{3t} \\ 0 \end{bmatrix} + c_3 \begin{bmatrix} 2e^{3t} \\ 0 \\ e^{3t} \end{bmatrix} + \begin{bmatrix} -e^{t} \\ -3t - 1 \\ e^{t}/2 \end{bmatrix}.$$

Example 3.11.2

Solve

(S) $$\qquad\qquad D\mathbf{x} = A\mathbf{x} + \mathbf{E}(t),$$

where

$$A = \begin{bmatrix} -1 & -1 \\ 4 & -1 \end{bmatrix} \quad \text{and} \quad \mathbf{E}(t) = \begin{bmatrix} 2e^{-t} \\ 0 \end{bmatrix}.$$

This system describes the two-loop circuit of Example 3.1.2 with $C = 1/4$, $L = 1$, $R_1 = 4$, $R_2 = 1$, and a varying voltage source with $V(t) = 8e^{-t}$.

We solved the related homogeneous system, corresponding to no voltage source, in Example 3.8.1. The general solution was $\mathbf{x} = \mathbf{H}(t)$, where

$$\mathbf{H}(t) = c_1 \begin{bmatrix} -e^{-t} \sin 2t \\ 2e^{-t} \cos 2t \end{bmatrix} + c_2 \begin{bmatrix} e^{-t} \cos 2t \\ 2e^{-t} \sin 2t \end{bmatrix}.$$

We look for a particular solution of (S) in the form

$$\mathbf{p}(t) = c_1(t) \begin{bmatrix} -e^{-t} \sin 2t \\ 2e^{-t} \cos 2t \end{bmatrix} + c_2(t) \begin{bmatrix} e^{-t} \cos 2t \\ 2e^{-t} \sin 2t \end{bmatrix}.$$

To find functions $c_1(t)$ and $c_2(t)$ that work, we solve

(V) $\quad c_1'(t) \begin{bmatrix} -e^{-t}\sin 2t \\ 2e^{-t}\cos 2t \end{bmatrix} + c_2'(t) \begin{bmatrix} e^{-t}\cos 2t \\ 2e^{-t}\sin 2t \end{bmatrix} = \begin{bmatrix} 2e^{-t} \\ 0 \end{bmatrix}$

for $c_1'(t)$ and $c_2'(t)$, using Cramer's rule:

$$c_1'(t) = \frac{\det \begin{bmatrix} 2e^{-t} & e^{-t}\cos 2t \\ 0 & 2e^{-t}\sin 2t \end{bmatrix}}{\det \begin{bmatrix} -e^{-t}\sin 2t & e^{-t}\cos 2t \\ 2e^{-t}\cos 2t & 2e^{-t}\sin 2t \end{bmatrix}} = -2\sin 2t.$$

$$c_2'(t) = \frac{\det \begin{bmatrix} -e^{-t}\sin 2t & 2e^{-t} \\ 2e^{-t}\cos 2t & 0 \end{bmatrix}}{\det \begin{bmatrix} -e^{-t}\sin 2t & e^{-t}\cos 2t \\ 2e^{-t}\cos 2t & 2e^{-t}\sin 2t \end{bmatrix}} = 2\cos 2t.$$

We can take

$$c_1(t) = \cos 2t \quad \text{and} \quad c_2(t) = \sin 2t$$

to obtain the particular solution

$$\mathbf{p}(t) = \cos 2t \begin{bmatrix} -e^{-t}\sin 2t \\ 2e^{-t}\cos 2t \end{bmatrix} + \sin 2t \begin{bmatrix} e^{-t}\cos 2t \\ 2e^{-t}\sin 2t \end{bmatrix} = \begin{bmatrix} 0 \\ 2e^{-t} \end{bmatrix}.$$

The general solution of (S) is

$$\mathbf{x} = \mathbf{H}(t) + \mathbf{p}(t) = c_1 \begin{bmatrix} -e^{-t}\sin 2t \\ 2e^{-t}\cos 2t \end{bmatrix} + c_2 \begin{bmatrix} e^{-t}\cos 2t \\ 2e^{-t}\sin 2t \end{bmatrix} + \begin{bmatrix} 0 \\ 2e^{-t} \end{bmatrix}.$$

Written in terms of currents and charges in the circuit, this reads

$$Q = -c_1 e^{-t}\sin 2t + c_2 e^{-t}\cos 2t$$

$$I_2 = 2c_1 e^{-t}\cos 2t + 2c_2 e^{-t}\sin 2t + 2e^{-t}.$$

Note that in our particular solution $\mathbf{p}(t)$ (i.e., $c_1 = c_2 = 0$), the initial condition of an uncharged capacitor and current of 2 amperes in the second loop leads to a perpetually uncharged capacitor and to currents I_1 and I_2 moving in opposite directions with equal (diminishing) magnitudes that cancel out at the capacitor.

Example 3.11.3

Find the solution of the system

$$
\begin{aligned}
DQ &= -2Q - I_2 && + e^{-3t} \\
DI_2 &= 2Q - I_2 + I_3 \\
DI_3 &= 2I_2 - 2I_3 - 2e^{-3t}
\end{aligned}
$$

(S)

satisfying the initial conditions

$$Q(0) = 1, \quad I_2(0) = 0, \quad I_3(0) = 0.$$

This represents the circuit of Example 3.1.3 with $R_1 = R_2 = 2$, $L_2 = 2$, $L_3 = 1$, $C = 1/4$, and $V(t) = 2e^{-3t}$.

In matrix notation, (S) is $D\mathbf{x} = A\mathbf{x} + \mathbf{E}(t)$, where

$$
\mathbf{x} = \begin{bmatrix} Q \\ I_2 \\ I_3 \end{bmatrix}, \quad
A = \begin{bmatrix} -2 & -1 & 0 \\ 2 & -1 & 1 \\ 0 & 2 & -2 \end{bmatrix}, \quad \text{and} \quad
\mathbf{E}(t) = \begin{bmatrix} e^{-3t} \\ 0 \\ -2e^{-3t} \end{bmatrix}.
$$

To solve the related homogeneous system (H), we find the characteristic polynomial of A:

$$
\det(A - \lambda I) = \det \begin{bmatrix} -2 - \lambda & -1 & 0 \\ 2 & -1 - \lambda & 1 \\ 0 & 2 & -2 - \lambda \end{bmatrix} = -(\lambda + 2)^2(\lambda + 1).
$$

Thus, the eigenvalues of A are $\lambda = -1$ and $\lambda = -2$.

To find the eigenvectors for $\lambda = -1$, we reduce $[A + I \mid \mathbf{0}]$:

$$
\begin{bmatrix} -1 & -1 & 0 & | & 0 \\ 2 & 0 & 1 & | & 0 \\ 0 & 2 & -1 & | & 0 \end{bmatrix} \rightarrow
\begin{bmatrix} 1 & 0 & \frac{1}{2} & | & 0 \\ 0 & 1 & -\frac{1}{2} & | & 0 \\ 0 & 0 & 0 & | & 0 \end{bmatrix}.
$$

This gives the eigenvector

$$
\mathbf{v} = \begin{bmatrix} 1 \\ -1 \\ -2 \end{bmatrix}
$$

and corresponding solution of (H)

$$
\mathbf{h}_1(t) = e^{-t}\mathbf{v} = \begin{bmatrix} e^{-t} \\ -e^{-t} \\ -2e^{-t} \end{bmatrix}.
$$

To find the generalized eigenvectors for $\lambda = -2$, we reduce $[(A + 2I)^2 \mid \mathbf{0}]$:

$$[(A + 2I)^2 \mid \mathbf{0}] = \begin{bmatrix} -2 & -1 & 1 & 0 \\ 2 & 1 & 1 & 0 \\ 4 & 2 & 2 & 0 \end{bmatrix} \rightarrow \begin{bmatrix} 1 & \frac{1}{2} & \frac{1}{2} & 0 \\ 0 & 0 & 0 & 0 \\ 0 & 0 & 0 & 0 \end{bmatrix}.$$

We read off two independent solutions of $(A + 2I)^2\mathbf{w} = \mathbf{0}$

$$\mathbf{w}_1 = \begin{bmatrix} -1 \\ 0 \\ 2 \end{bmatrix} \quad \text{and} \quad \mathbf{w}_2 = \begin{bmatrix} -1 \\ 2 \\ 0 \end{bmatrix}$$

and obtain the associated solutions of (H):

$$\mathbf{h}_2(t) = e^{-2t}(\mathbf{w}_1 + t[A + 2I]\mathbf{w}_1) = e^{-2t} \begin{bmatrix} -1 \\ 0 \\ 2 \end{bmatrix} = \begin{bmatrix} -e^{-2t} \\ 0 \\ 2e^{-2t} \end{bmatrix}$$

$$\mathbf{h}_3(t) = e^{-2t}(\mathbf{w}_2 + t[A + 2I]\mathbf{w}_2) = e^{-2t} \left(\begin{bmatrix} -1 \\ 2 \\ 0 \end{bmatrix} + t \begin{bmatrix} -2 \\ 0 \\ 4 \end{bmatrix} \right)$$

$$= \begin{bmatrix} -(2t + 1)e^{-2t} \\ 2e^{-2t} \\ 4te^{-2t} \end{bmatrix}.$$

The general solution of the related homogeneous system is $\mathbf{x} = \mathbf{H}(t)$, where

$$\mathbf{H}(t) = c_1\mathbf{h}_1(t) + c_2\mathbf{h}_2(t) + c_3\mathbf{h}_3(t)$$

$$= c_1 \begin{bmatrix} e^{-t} \\ -e^{-t} \\ -2e^{-t} \end{bmatrix} + c_2 \begin{bmatrix} -e^{-2t} \\ 0 \\ 2e^{-2t} \end{bmatrix} + c_3 \begin{bmatrix} -(2t + 1)e^{-2t} \\ 2e^{-2t} \\ 4te^{-2t} \end{bmatrix}.$$

We look for a particular solution of (S) in the form

$$\mathbf{p}(t) = c_1(t)\mathbf{h}_1(t) + c_2(t)\mathbf{h}_2(t) + c_3(t)\mathbf{h}_3(t).$$

To find functions $c_1(t)$, $c_2(t)$, $c_3(t)$ that work, we solve

$$\text{(V)} \quad \begin{aligned} e^{-t}c_1'(t) - e^{-2t}c_2'(t) - (2t + 1)e^{-2t}c_3'(t) &= e^{-3t} \\ -e^{-t}c_1'(t) \qquad\qquad + 2e^{-2t}c_3'(t) &= 0 \\ -2e^{-t}c_1'(t) + 2e^{-2t}c_2'(t) + 4te^{-2t}c_3'(t) &= -2e^{-3t}. \end{aligned}$$

Using row reduction or Cramer's rule, we find

$$c_1'(t) = 0, \quad c_2'(t) = -e^{-t}, \quad c_3'(t) = 0.$$

Integration gives

$$c_1(t) = 0, \quad c_2(t) = e^{-t}, \quad c_3(t) = 0$$

so a particular solution of (S) is

$$\mathbf{p}(t) = 0\mathbf{h}_1(t) + e^{-t}\mathbf{h}_2(t) + 0\mathbf{h}_3(t) = \begin{bmatrix} -e^{-3t} \\ 0 \\ 2e^{-3t} \end{bmatrix}.$$

The general solution of (S) is

$$\mathbf{x} = \mathbf{H}(t) + \mathbf{p}(t)$$

$$= c_1 \begin{bmatrix} e^{-t} \\ -e^{-t} \\ -2e^{-t} \end{bmatrix} + c_2 \begin{bmatrix} -e^{-2t} \\ 0 \\ 2e^{-2t} \end{bmatrix} + c_3 \begin{bmatrix} -(2t+1)e^{-2t} \\ 2e^{-2t} \\ 4te^{-2t} \end{bmatrix} + \begin{bmatrix} -e^{-3t} \\ 0 \\ 2e^{-3t} \end{bmatrix}.$$

To find the specific solution satisfying the desired initial condition, we substitute $t = 0$ and set the resulting vector equal to the required initial vector:

$$\mathbf{v}(0) = c_1 \begin{bmatrix} 1 \\ -1 \\ -2 \end{bmatrix} + c_2 \begin{bmatrix} -1 \\ 0 \\ 2 \end{bmatrix} + c_3 \begin{bmatrix} -1 \\ 2 \\ 0 \end{bmatrix} + \begin{bmatrix} -1 \\ 0 \\ 2 \end{bmatrix} = \begin{bmatrix} 1 \\ 0 \\ 0 \end{bmatrix}.$$

This leads to the equations

$$
\begin{aligned}
c_1 - c_2 - c_3 &= 2 \\
-c_1 \qquad + 2c_3 &= 0 \\
-2c_1 + 2c_2 \qquad &= -2
\end{aligned}
$$

with solutions

$$c_1 = -2, \quad c_2 = -3, \quad c_3 = -1.$$

The solution of the initial-value problem is

$$\mathbf{x} = -2\mathbf{h}_1(t) - 3\mathbf{h}_2(t) - \mathbf{h}_3(t) + \mathbf{p}(t).$$

Written in terms of charges and currents, this is

$$Q = -2e^{-t} + (4 + 2t)e^{-2t} - e^{-3t}$$
$$I_2 = 2e^{-t} - 2e^{-2t}$$
$$I_3 = 4e^{-t} - (6 + 4t)e^{-2t} + 2e^{-3t}.$$

Let's summarize.

NONHOMOGENEOUS SYSTEMS

To solve the nth-order system

(S) $$D\mathbf{x} = A\mathbf{x} + \mathbf{E}(t)$$

given that the general solution of the related homogeneous system $D\mathbf{x} = A\mathbf{x}$ is $\mathbf{x} = \mathbf{H}(t)$, with

$$\mathbf{H}(t) = c_1\mathbf{h}_1(t) + \cdots + c_n\mathbf{h}_n(t),$$

we look for a particular solution of (S) of the form $\mathbf{x} = \mathbf{p}(t)$, with

(P) $$\mathbf{p}(t) = c_1(t)\mathbf{h}_1(t) + \cdots + c_n(t)\mathbf{h}_n(t).$$

To find functions $c_1(t), \ldots, c_n(t)$ that work, we solve

(V) $$c_1'(t)\mathbf{h}_1(t) + \cdots + c_n'(t)\mathbf{h}_n(t) = \mathbf{E}(t)$$

for $c_1'(t), \ldots, c_n'(t)$. We then integrate to find $c_1(t), \ldots, c_n(t)$. Substitution into (P) yields a particular solution of (S). The general solution of (S) is

$$\mathbf{x} = \mathbf{H}(t) + \mathbf{p}(t).$$

Notes

1. A shortcut

In our examples we obtained specific values for each $c_i(t)$ by integrating the formulas for $c_i'(t)$, taking the constants of integration to be zero. If we include arbitrary constants c_i when we integrate, then substitution of the resulting functions into formula (P) yields the general solution of (S).

2. On the old and new methods of variation of parameters

We have seen (Example 3.2.2) that an nth-order o.d.e. $Lx = E(t)$ can be replaced by an equivalent system $Dx = Ax + E(t)$. The forcing term of this system is just

$$\mathbf{E}(t) = \begin{bmatrix} 0 \\ \cdot \\ \cdot \\ \cdot \\ 0 \\ E(t) \end{bmatrix}.$$

If the general solution of the related homogeneous equation $Lx = 0$ is $x = H(t)$, where $H(t) = c_1 h_1(t) + \cdots + c_n h_n(t)$, then the general solution of the homogeneous system $Dx = Ax$ is $\mathbf{x} = \mathbf{H}(t)$, where

$$\mathbf{H}(t) = c_1 \begin{bmatrix} h_1(t) \\ h_1'(t) \\ \cdot \\ \cdot \\ \cdot \\ h_1^{(n-1)}(t) \end{bmatrix} + \cdots + c_n \begin{bmatrix} h_n(t) \\ h_n'(t) \\ \cdot \\ \cdot \\ \cdot \\ h_n^{(n-1)}(t) \end{bmatrix}.$$

If we use variation of parameters to look for a particular solution of the non-homogeneous system $Dx = Ax + E(t)$, then we must solve the algebraic system of equations

$$c_1'(t) \begin{bmatrix} h_1(t) \\ h_1'(t) \\ \cdot \\ \cdot \\ \cdot \\ h_1^{(n-1)}(t) \end{bmatrix} + \cdots + c_n'(t) \begin{bmatrix} h_n(t) \\ h_n'(t) \\ \cdot \\ \cdot \\ \cdot \\ h_n^{(n-1)}(t) \end{bmatrix} = \begin{bmatrix} 0 \\ 0 \\ \cdot \\ \cdot \\ \cdot \\ E(t) \end{bmatrix}.$$

This system of equations is *exactly the same* as the system we would have to solve if we used variation of parameters to look for a particular solution of $Lx = E(t)$.

EXERCISES

In Exercises 1 to 12 find the general solution of $Dx = Ax + E(t)$. (Note: You solved $Dx = Ax$ in the exercise indicated.)

1. $A = \begin{bmatrix} 0 & 2 \\ -1 & 3 \end{bmatrix}$, $\mathbf{E}(t) = \begin{bmatrix} e^t \\ e^t \end{bmatrix}$ (Exercise 1, Section 3.7)

2. $A = \begin{bmatrix} 1 & 1 \\ 1 & 1 \end{bmatrix}$, $\mathbf{E}(t) = \begin{bmatrix} 0 \\ t \end{bmatrix}$ (Exercise 2, Section 3.7)

3. $A = \begin{bmatrix} 1 & 1 \\ 3 & 1 \end{bmatrix}$, $\mathbf{E}(t) = \begin{bmatrix} 1 \\ 1 \end{bmatrix}$ (Exercise 3, Section 3.7)

4. $A = \begin{bmatrix} 1 & -1 \\ 1 & 3 \end{bmatrix}$, $\mathbf{E}(t) = \begin{bmatrix} e^{2t} \\ 0 \end{bmatrix}$ (Exercise 5, Section 3.9)

5. $A = \begin{bmatrix} 2 & 1 & -2 \\ -3 & 0 & 4 \\ -2 & -1 & 4 \end{bmatrix}$, $\mathbf{E}(t) = \begin{bmatrix} e^t \\ e^t \\ e^t \end{bmatrix}$ (Exercise 4, Section 3.7)

6. $\hat{A} = \begin{bmatrix} 0 & -2 & 2 \\ 1 & 3 & -2 \\ 2 & 4 & -3 \end{bmatrix}$, $\mathbf{E}(t) = \begin{bmatrix} 2e^t - 2t \\ t \\ e^t \end{bmatrix}$ (Exercise 5, Section 3.7)

7. $A = \begin{bmatrix} 1 & -2 & 2 \\ 0 & 1 & 1 \\ -1 & -2 & 4 \end{bmatrix}$, $\mathbf{E}(t) = \begin{bmatrix} 0 \\ e^{2t} \\ 0 \end{bmatrix}$ (Exercise 4, Section 3.8)

8. $A = \begin{bmatrix} 1 & 0 & -1 \\ 0 & 2 & 0 \\ 1 & 0 & 1 \end{bmatrix}$, $\mathbf{E}(t) = \begin{bmatrix} e^t \\ e^{2t} \sin t \\ 0 \end{bmatrix}$ (Exercise 3, Section 3.8)

9. $A = \begin{bmatrix} -3 & 1 \\ -1 & -1 \end{bmatrix}$, $\mathbf{E}(t) = \begin{bmatrix} 2te^{-2t} \\ te^{-2t} \end{bmatrix}$ (Exercise 6, Section 3.9)

10. $A = \begin{bmatrix} 3 & -2 \\ 2 & 3 \end{bmatrix}$, $\mathbf{E}(t) = \begin{bmatrix} 2e^{3t} \\ 2e^{3t} \end{bmatrix}$ (Exercise 2, Section 3.8)

11. $A = \begin{bmatrix} 1 & 0 & 0 \\ 1 & 0 & 1 \\ 1 & -1 & 2 \end{bmatrix}$, $\mathbf{E}(t) = \begin{bmatrix} 0 \\ 0 \\ 1 \end{bmatrix}$ (Exercise 4, Section 3.10)

12. $A = \begin{bmatrix} 0 & 0 & 0 & 0 & 1 \\ 1 & -1 & 0 & -1 & 1 \\ 1 & 0 & -1 & -1 & 0 \\ 0 & 0 & 0 & 0 & 1 \\ 1 & 0 & 0 & -1 & 0 \end{bmatrix}$, $\mathbf{E}(t) = \begin{bmatrix} t \\ 1 \\ 0 \\ t \\ 1 \end{bmatrix}$ (Exercise 11, Section 3.10)

In Exercises 13 to 16 find (a) the general solution and (b) the specific solution satisfying the given initial conditions.

13. $\begin{aligned} x_1' &= -2x_1 + 2x_2 + e^{3t} \\ x_2' &= -4x_1 + 4x_2 \end{aligned}$ $x_1(0) = x_2(0) = \dfrac{2}{3}$

14. $D\mathbf{x} = A\mathbf{x} + \mathbf{E}(t)$, $A = \begin{bmatrix} 2 & 1 & 0 \\ 0 & 2 & 0 \\ 1 & 0 & 2 \end{bmatrix}$, $\mathbf{E}(t) = \begin{bmatrix} e^{2t} \\ 0 \\ 0 \end{bmatrix}$, $\mathbf{x}(0) = \begin{bmatrix} 1 \\ 2 \\ 3 \end{bmatrix}$

15. $\begin{aligned} x_1' &= \quad\;\; - 3x_2 + 2x_3 + e^{2t} \\ x_2' &= 3x_1 \quad\quad\;\; - 3x_3 + 3 \\ x_3' &= \quad\quad\quad\quad\; 2x_3 + e^{2t} \end{aligned}$ $x_1(0) = 1, \; x_2(0) = 2, \; x_3(0) = 3$

16. $D\mathbf{x} = A\mathbf{x} + \mathbf{E}(t)$, $A = \begin{bmatrix} 2 & 0 & 0 \\ -1 & 5 & -2 \\ -1 & 3 & 0 \end{bmatrix}$, $\mathbf{E}(t) = \begin{bmatrix} e^t \\ 0 \\ e^t \end{bmatrix}$, $\mathbf{x}(0) = \begin{bmatrix} 1 \\ 2 \\ 1 \end{bmatrix}$

Exercises 17 to 21 refer to our models.

17. *Two-Loop Circuit:* Find the general solution of the system modeling the two-loop circuit of Example 3.1.2 with $V(t) = e^{-t}$ volts, $R_1 = R_2 = 1$ ohm, $L = 1$ henry, and $C = 1$ farad. (*Note:* See Exercise 14, Section 3.8.)

18. *Four-Loop Circuit:* Find the general solution of the system modeling the four-loop circuit of Example 3.1.4 with each $R_1 = 1$ ohm, each $L_i = 1$ henry, and $V(t) = 4$ volts. (*Note:* See Example 3.7.4.)

*19. *Radioactive Decay:* Professor Adam D. Kay is performing an experiment involving the radioactive chain of Exercise 7, Section 3.1. There are now 500 grams of each substance, and substances A and B decay at the rates of A/2 and B/2, respectively. The experiment requires the removal of substance C at the rate of 100 grams per year and allows the addition of substance A at a rate of α grams per year. It is crucial that α be chosen so that the amounts of substances A, B, and C never hit 0. Professor Kay guesses that $\alpha = 250$ is big enough.

   a. Find formulas for the amounts of substances A, B, and C at time t. [*Note:* See Exercise 16(a), Section 3.7.]
   b. By investigating the minima of the functions found in (a), decide whether Professor Kay's guess will work.

20. *Diffusion II:* Find the general solution of the system in Exercise 9, Section 3.1, with $P_1 = 0.2$, $P_2 = 0.1$, $v = 20$, $V = 10$, and $G = 0.15$. What happens to x and y as $t \to \infty$?

21. *Supply and Demand:*
   a. Find the general solution of the system in Exercise 6, Section 3.1, under the following assumptions:
      i. If the product were free, 5000 people would buy it; every \$1 in price reduces this number by 600.
      ii. The price increases 10¢ per week for every 100 people who want the product but can't get it.
      iii. The weekly production is increased by 50 for every dollar in price of the product.
   b. Suppose that initially the price is $p_0 = \$4$ and the supply and demand are equal ($s_0 = w_0$). Predict the price, supply, and demand after 52 weeks. (Compare this with Exercise 20, Section 2.7.)

REVIEW PROBLEMS

In Exercises 1 to 12 find
a. the general solution and
b. the specific solution satisfying the given initial condition.

1. $\begin{aligned} x_1' &= 5x_1 - 2x_2 + 4e^{3t} \\ x_2' &= 4x_1 + x_2 + 4e^{3t} \end{aligned}$ $x_1(0) = -1, x_2(0) = 6$

2. $\begin{aligned} x_1' &= 4x_1 - 5x_2 - 4e^t \\ x_2' &= x_1 - 2x_2 + 12e^t \end{aligned}$ $x_1(0) = 11, x_2(0) = 7$

3. $\begin{aligned} x_1' &= 3x_1 - 4x_2 + e^{-t} + 1 \\ x_2' &= 4x_1 - 5x_2 + 1 \end{aligned}$ $x_1(0) = 2, x_2(0) = 1$

4. $\begin{aligned} x_1' &= 2x_1 + x_2 + e^{3t} \\ x_2' &= x_2 + e^{2t} \end{aligned}$ $x_1(0) = 3, x_2(0) = 2$

5. $\begin{aligned} x_1' &= 4x_1 \\ x_2' &= 2x_1 + 2x_2 - 4x_3 + 5e^{3t} \\ x_3' &= x_1 - 2x_3 + 5e^{3t} \end{aligned}$ $x_1(0) = 6, x_2(0) = 0, x_3(0) = 1$

6. $\begin{aligned} x_1' &= 3x_1 - x_2 + x_3 \\ x_2' &= 2x_1 + x_2 + x_3 \\ x_3' &= x_1 \quad\quad + 2x_3 \end{aligned}$ $x_1(0) = 2, x_2(0) = 8, x_3(0) = 4$

7. $\begin{aligned} x_1' &= 2x_1 \quad\quad\quad\quad + 2e^{2t} \\ x_2' &= x_1 + 2x_2 + 2x_3 + 2e^{4t} \\ x_3' &= x_1 \quad\quad + 4x_3 - e^{2t} \end{aligned}$ $x_1(0) = -2, x_2(0) = 7, x_3(0) = -1$

8. $\begin{aligned} x_1' &= -2x_1 + 2x_2 - 2x_3 + 2 \\ x_2' &= -4x_1 + 2x_2 \quad\quad + e^{2t} \\ x_3' &= \quad\quad\quad\quad 2x_3 + e^{2t} \end{aligned}$ $x_1(0) = 3, x_2(0) = x_3(0) = 1$

9. $\begin{aligned} x_1' &= 2x_1 \quad\quad + x_3 + te^{-2t} \\ x_2' &= \quad 2x_2 + x_3 \\ x_3' &= \quad\quad\quad 2x_3 + e^{-2t} \end{aligned}$ $x_1(0) = 0, x_2(0) = x_3(0) = 1$

10. $\begin{aligned} x_1' &= x_1 \quad + x_3 \\ x_2' &= \quad x_2 \quad\quad + x_4 \\ x_3' &= \quad\quad\quad x_3 \\ x_4' &= x_1 \quad\quad\quad + x_4 \end{aligned}$ $x_1(0) = x_2(0) = 0, x_3(0) = x_4(0) = 1$

11. $\begin{aligned} x_1' &= x_1 + x_2 - x_3 + x_4 + 4e^{2t} \\ x_2' &= \quad - x_2 \quad\quad\quad + 2e^{2t} \\ x_3' &= \quad - 2x_2 + x_3 - 2x_4 + 4e^{2t} \\ x_4' &= \quad\quad\quad\quad - x_4 + 2e^{2t} \end{aligned}$ $x_1(0) = 4, x_2(0) = 5, x_3(0) = 6, x_4(0) = 1$

12. $\begin{aligned} x_1' &= \quad 3x_2 \\ x_2' &= -3x_1 \quad + 3x_3 \\ x_3' &= \quad\quad\quad 3x_4 \\ x_4' &= \quad\quad - 3x_3 \end{aligned}$ $x_1(0) = 0, x_2(0) = 2, x_3(0) = 2, x_4(0) = 4$

13. *Circuits:* Find the general solution of the system modeling the circuit of the indicated exercise from Section 3.1.

    a. Exercise 1   b. Exercise 2   c. Exercise 3
    d. Exercise 4   e. Exercise 5

14. *Moving Springs:* Find the general solution of the system of Exercise 27, Section 3.2, assuming

    a. $b = 5$ and $k = 8$ b. $b = 2$ and $k = 2$ c. $b = 0$ and $k = 2$
    (Compare this with Exercise 20, Section 2.9.)

4

QUALITATIVE THEORY OF SYSTEMS OF O.D.E.'s

4.1 INTERACTING POPULATIONS

In the preceding three chapters, we have developed an effective technique for finding formulas to express the solutions of a linear o.d.e. or system of o.d.e.'s with constant coefficients as functions of time. Unfortunately, no such general technique exists for handling most *nonlinear* equations or systems—nor even for linear systems whose coefficients vary with time. In this chapter we will develop further the point of view introduced for first-order o.d.e.'s in Section 1.8: we try to obtain *qualitative* information about the long-term behavior of solutions to a system directly from a study of the equations, without trying to explicitly "solve" them. In this section, we will study systems modeling the growth of two interacting populations

$$\frac{dx}{dt} = f(x, y)$$

$$\frac{dy}{dt} = g(x, y)$$

by analyzing the signs of $f(x, y)$ and $g(x, y)$ in various regions of the xy- plane. This is very closely analogous to our study of first-order equations in Section 1.8. The following sections will address the role of eigenvalues in studying the behavior of linear systems and the applications of this to the behavior of solutions of nonlinear systems.

Suppose that we have two **competing species** whose sizes are denoted $x(t)$ and $y(t)$. Assume that in the absence of competition, the growth of the species would be governed by the equations

$$\frac{dx}{dt} = a_1 x - b_1 x^2$$

$$\frac{dy}{dt} = a_2 y - b_2 y^2,$$

where each a_i and b_i is a positive constant (compare Examples 1.1.6 and 1.8.1). However, these two species compete for the same resources. Assume that the effects of this competition can be accounted for by subtracting a term that is proportional to the product of the sizes of the two populations. This leads to a nonlinear system of o.d.e.'s of the form

(S)
$$\frac{dx}{dt} = a_1 x - b_1 x^2 - c_1 xy$$

$$\frac{dy}{dt} = a_2 y - b_2 y^2 - c_2 xy,$$

where each c_i is positive. Since $x(t)$ and $y(t)$ represent populations, we are only interested in solutions with

$$x \geq 0 \quad \text{and} \quad y \geq 0.$$

We will view solutions $x = x(t)$, $y = y(t)$ of (S) as parametric descriptions of curves in the xy- plane. As t varies, the point $(x(t), y(t))$ moves along this curve. Rather than attempting to find an accurate picture of the curve, our goal in this section will be to determine the general direction of this motion.

Our expression for dx/dt factors as follows:

$$\frac{dx}{dt} = x(a_1 - b_1 x - c_1 y).$$

Thus dx/dt is zero at any point on the $y - $ axis

$$x = 0$$

and at any point on the line

(L$_1$)
$$a_1 - b_1 x - c_1 y = 0.$$

This line forms a triangle with the x- and y-axes (see Figure 4.1a). Inside this triangle dx/dt is positive since it is the product of two positive factors. Thus x is increasing inside the triangle. At any point in the first quadrant that lies outside this triangle and off the y-axis, dx/dt is negative since it is the product of one positive and one negative factor. Thus x is decreasing at these points.

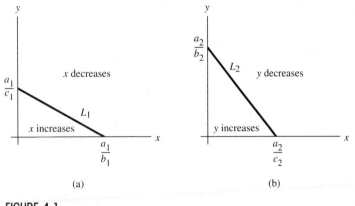

(a)   (b)

FIGURE 4.1

Our expression for dy/dt also factors:

$$\frac{dy}{dt} = y(a_2 - b_2y - c_2x).$$

Thus dy/dt is zero on the x-axis

$$y = 0$$

and on the line

(L₂) $$a_2 - b_2y - c_2x = 0.$$

This line also forms a triangle with the x- and y-axes (see Figure 4.1b). Inside this triangle, dy/dt is positive so y is increasing. At points outside this triangle and off the x-axis, dy/dt is negative so y is decreasing.

Let's look at some specific examples to see how we can combine our information about x and y.

Example 4.1.1

Use the signs of dx/dt and dy/dt to investigate the long-term behavior of the species modeled by the equations

(S)
$$\frac{dx}{dt} = 2x - 2x^2 - 5xy$$

$$\frac{dy}{dt} = y - y^2 - 2xy.$$

In this case,

$$\frac{dx}{dt} = x(2 - 2x - 5y)$$

$$\frac{dy}{dt} = y(1 - y - 2x).$$

Thus dx/dt is zero along the y-axis ($x = 0$) and along the line

(L₁) $2 - 2x - 5y = 0.$

Line L_1 separates the points where x is increasing from those where x is decreasing. Similarly, dy/dt is zero along the x-axis ($y = 0$) and along the line

(L₂) $1 - y - 2x = 0.$

Line L_2 separates the points where y is increasing from those where y is decreasing. Lines L_1 and L_2 divide the first quadrant into four regions (see Figure 4.2). In the region below both lines, x and y are both increasing. In the region above both lines, x and y are both decreasing. In the region below the first line and above the second, x is increasing and y is decreasing. In the fourth region, y is increasing and x is decreasing. The arrows in Figure 4.3 indicate the general direction in which solutions move from points within these regions, on the lines, and on the axes.

Note that there are four special points:

$$(0, 0), \quad (0, 1), \quad (1, 0), \quad \text{and} \quad \left(\frac{3}{8}, \frac{1}{4}\right).$$

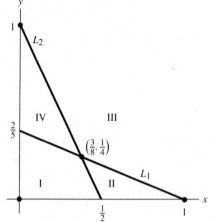

Region I: x increases, y increases
Region II: x increases, y decreases
Region III: x decreases, y decreases
Region IV: x decreases y increases

FIGURE 4.2

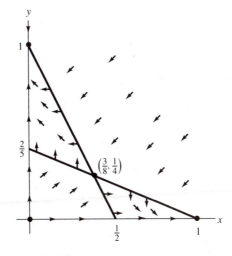

FIGURE 4.3

At each of these points, dx/dt and dy/dt are both zero. Thus a solution that starts at any of these points, remains stationary. If we follow the directions indicated by the arrows, we see that all other solutions appear to approach one or another of these four points as t goes to infinity. Most of the solutions appear to approach either $(0, 1)$ or $(1, 0)$. Thus the most likely long-term outcome appears to be that one of the populations becomes extinct while the other approaches 1.

Example 4.1.2

Use the signs of dx/dt and dy/dt to investigate the long-term behavior of the species modeled by the equations

(S)
$$\frac{dx}{dt} = 2x - 2x^2 - xy$$

$$\frac{dy}{dt} = y - y^2 - 2xy.$$

In this case,

$$\frac{dx}{dt} = x(2 - 2x - y)$$

$$\frac{dy}{dt} = y(1 - y - 2x).$$

Thus dx/dt is zero along the y-axis, and along the line

(L$_1$) $2 - 2x - y = 0.$

Similarly, dy/dt is zero along the x-axis and along the line

(L$_2$) $1 - y - 2x = 0.$

These lines divide the first quadrant into three regions (see Figure 4.4). Above the first line, x and y are both decreasing. Below the second line, x and y are both increasing. Between the two lines, x is increasing and y is decreasing. The arrows in Figure 4.5 show the general direction in which solutions move from points in these three regions, on the lines, and on the axes.

   Here there are three points at which both dx/dt and dy/dt are zero:

$$(0, 0), \quad (0, 1), \quad \text{and} \quad (1, 0).$$

A solution that starts at any of these points remains stationary. If we follow the directions indicated by the arrows, we see that all other solutions appear to approach one or another of these three points as t approaches infinity. If we start with $x = 0$ and $y \neq 0$, then y approaches 1. All other solutions eventually appear to approach the point $(1, 0)$. Thus the most likely long-term outcome for our populations appears to be that x approaches 1 and y approaches 0 as t approaches infinity. Note that unless species x is absent at the outset, it appears that species y will eventually become extinct.

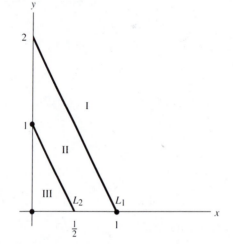

Region I: x decreases, y decreases
Region II: x increases, y decreases
Region III: x increases, y increases

FIGURE 4.4

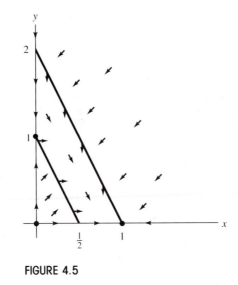

FIGURE 4.5

In each of the preceding examples, we were able to make reasonable predictions about the long-term behavior of the population. Such predictions are not always so easy when the populations consist of **a predator and its prey.** Assume that in the absence of any predators, the growth of one species is governed by the equation

$$\frac{dx}{dt} = a_1 x - b_1 x^2,$$

where a_1 and b_1 are positive constants. In the absence of its prey, the population of the other species decreases, as modeled by the equation

$$\frac{dy}{dt} = -a_2 y,$$

where a_2 is a positive constant. However, the second species preys on the first. To account for this we must subtract a term from dx/dt, reflecting the number of prey killed by the predator. At the same time, we assume the predator population increases in proportion to the number of prey eaten by the predators, so we add a term to dy/dt. We assume the number of prey killed and eaten by predators is proportional to the product of the two population sizes. This results in a system of the form

$$\frac{dx}{dt} = a_1 x - b_1 x^2 - c_1 xy$$

$$\frac{dt}{dt} = -a_2 y + c_2 xy,$$

where each of the constants c_i is positive. In the following example, we investigate the behavior of populations modeled by a system of this form.

Example 4.1.3

Use the signs of dx/dt and dy/dt to investigate the behavior of the populations modeled by the equations

$$\frac{dx}{dt} = x - x^2 - xy$$

$$\frac{dy}{dt} = -y + 2xy.$$

Here

$$\frac{dx}{dt} = x(1 - x - y)$$

$$\frac{dy}{dt} = y(-1 + 2x).$$

Thus dx/dt is zero along the y-axis and along the line

$$1 - x - y = 0.$$

Similarly, dy/dt is zero along the x-axis and along the line

$$-1 + 2x = 0.$$

As before, we can determine whether $x(t)$ and $y(t)$ are increasing or decreasing in each of the regions determined by these lines (see Figure 4.6). The arrows in Figure 4.7 show the general direction in which solutions move from points in these regions, on the lines and on the axes.

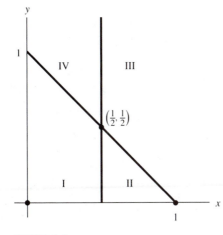

Region I: x increases, y decreases
Region II: x increases, y increases
Region III: x decreases, y increases
Region IV: x decreases y decreases

FIGURE 4.6

There are three points at which both dx/dt and dy/dt are zero:

$$(0, 0), \quad (1, 0), \quad \text{and} \quad \left(\frac{1}{2}, \frac{1}{2}\right).$$

A solution that starts at one of these points remains stationary. Solutions starting along the y-axis approach $(0, 0)$ as t goes to infinity. Solutions starting along the x-axis, but not at the origin, approach $(1, 0)$.

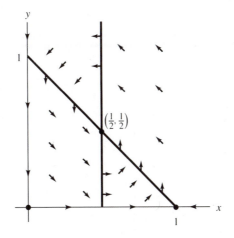

FIGURE 4.7

The arrows in Figure 4.7 are consistent with several very different long-term possibilities for solutions that start at points off the axes. It might happen that the point $P(t) = (x(t), y(t))$ spirals out away from $Q = (1/2, 1/2)$. On the other hand, $P(t)$ might spiral in towards Q. Intermediate between these two cases, is the possibility that $P(t)$ follows a closed curve that forever circles around Q. The tools we have used here are too crude to distinguish among these possibilities.

In the preceding examples, the points where dx/dt and dy/dt were both zero were of special interest. We refer to such a point as an **equilibrium** of the system (S). If we start at an equilibrium, then neither x nor y changes, so we remain there. In Example 4.1.1, the equilibria were $(0, 0)$, $(0, 1)$, $(1, 0)$, and $(3/8, 1/4)$. In Example 4.1.2 the equilibria were $(0, 0)$, $(0, 1)$, and $(1, 0)$. In Example 4.1.3 the equilibria were $(0, 0)$, $(1, 0)$, $(1/2, 1/2)$. In Section 4.3 we will see how approximation of nonlinear systems (like the ones in this section) by linear systems can lead to a more detailed picture of the solutions near the equilibria.

EXERCISES

In the following exercises, use the signs of dx/dt and dy/dt to investigate the behavior of those solutions of the given system that satisfy $x \geq 0$ and $y \geq 0$. If possible, make reasonable predictions about the long-term behavior of these solutions. Note that the systems in Exercises 1 through 5 model competing species, and the systems in Exercises 6 through 8 model predator-prey populations. In Exercises 9 and 10, the curves along which dx/dt and dy/dt are zero include curves that are not straight lines.

1. $\dfrac{dx}{dt} = 4x - 2x^2 - xy$

    $\dfrac{dy}{dt} = 3y - xy - y^2$

2. $\dfrac{dx}{dt} = 3x - x^2 - xy$

    $\dfrac{dy}{dt} = 4y - 2xy - y^2$

3. $\dfrac{dx}{dt} = 2x - 2x^2 - xy$

    $\dfrac{dy}{dt} = 3y - xy - y^2$

4. $\dfrac{dx}{dt} = 3x - x^2 - xy$

    $\dfrac{dy}{dt} = 2y - 2xy - y^2$

5. $\dfrac{dx}{dt} = 2x - 2x^2 - xy$

    $\dfrac{dy}{dt} = 4y - 2xy - y^2$

6. $\dfrac{dx}{dt} = 4x - 2x^2 - xy$

    $\dfrac{dy}{dt} = -3y + xy$

7. $\dfrac{dx}{dt} = 2x - 4x^2 - xy$

    $\dfrac{dy}{dt} = -3y + 7xy$

8. $\dfrac{dx}{dt} = 4x - 2x^2 - xy$

    $\dfrac{dy}{dt} = -y + xy$

9. $\dfrac{dx}{dt} = xy - x^3 + x$

   $\dfrac{dy}{dt} = 3y + xy - y^2$

10. $\dfrac{dx}{dt} = 5x + x^2 - xy$

    $\dfrac{dy}{dt} = y^2 - x^2y + y$

4.2 PHASE PORTRAITS OF LINEAR SYSTEMS

In this section we obtain a geometric representation of the solutions to a second-order, linear, homogeneous, constant-coefficient system

$$\frac{dx}{dt} = a_1 x + b_1 y$$

$$\frac{dy}{dt} = a_2 x + b_2 y.$$

In matrix form, this is

$$\mathbf{Dx} = A\mathbf{x},$$

where

$$\mathbf{x} = \begin{bmatrix} x \\ y \end{bmatrix} \quad \text{and} \quad A = \begin{bmatrix} a_1 & b_1 \\ a_2 & b_2 \end{bmatrix}.$$

Although our discussion focuses on second-order systems, the ideas and arguments are more general and can be used to get a picture of solutions of higher-order systems (see Note 2). The heart of our discussion is our observation from Note 1 of Section 3.2 that a vector valued function

$$\mathbf{x}(t) = \begin{bmatrix} x(t) \\ y(t) \end{bmatrix}$$

can be thought of as a parametrized curve. If $\mathbf{x}(t)$ is a solution of a system, then we refer to the corresponding curve as an **integral curve** of the system. Our first goal in picturing the solutions of a given system will be to sketch its integral curves. Of course an integral curve gives less information than an explicit formula for a solution; from a formula we can determine not only the points on the integral curve but also the time at which each point is reached. Our second goal will be to recapture some of this information by indicating with arrows the direction in which the integral curve is traversed. These directed integral curves form what is called the **phase portrait** of the system. We will see from the phase portraits that a great deal of qualitative information about the behavior of solutions can be obtained from knowledge of the eigen-values alone.

We begin our discussion by noting that if $\mathbf{x}(t)$ is a solution of $D\mathbf{x} = A\mathbf{x}$ and if t_0 is a fixed value of t, then $\mathbf{x}(t + t_0)$ is also a solution. However, these solutions determine the same integral curve: If $\mathbf{x}(t)$ arrives at a point when $t = t_1$, then $\mathbf{x}(t + t_0)$ arrives at the same point when $t = t_1 - t_0$. On the other hand, if $\mathbf{x}_1$ and $\mathbf{x}_2$ are solutions of $D\mathbf{x} = A\mathbf{x}$ that arrive at the same point at different times, then $\mathbf{x}_2(t) = \mathbf{x}_1(t + t_0)$ for a suitable choice of t_0 (Exercise 29), so $\mathbf{x}_1$ and $\mathbf{x}_2$ determine the same integral curve. Thus we have

Fact: *Distinct integral curves of the constant-coefficient system $D\mathbf{x} = A\mathbf{x}$ do not intersect.*

We also note that $\mathbf{x} = \mathbf{0}$ is a solution of $D\mathbf{x} = A\mathbf{x}$. As in Section 1.8, we refer to a constant solution of a system as an **equilibrium.** Thus $\mathbf{0}$ is an equi-librium of $D\mathbf{x} = A\mathbf{x}$. The integral curve determined by this solution consists of a single point at the origin and must always be included in the phase portrait of $D\mathbf{x} = A\mathbf{x}$.

Example 4.2.1

Sketch the phase portrait of $D\mathbf{x} = A\mathbf{x}$, where

$$A = \begin{bmatrix} 1 & 1 \\ 3 & -1 \end{bmatrix}.$$

The eigenvalues of A and associated eigenvectors are

$$\lambda_1 = -2, \quad \mathbf{v}_1 = \begin{bmatrix} 1 \\ -3 \end{bmatrix}, \quad \text{and} \quad \lambda_2 = 2, \quad \mathbf{v}_2 = \begin{bmatrix} 1 \\ 1 \end{bmatrix}$$

(see Exercise 7, Section 3.5). Corresponding to these eigenvalues and eigen-vectors are solutions

$$\mathbf{x}_1 = e^{\lambda_1 t}\mathbf{v}_1 \quad \text{and} \quad \mathbf{x}_2 = e^{\lambda_2 t}\mathbf{v}_2.$$

As t varies, these vector valued functions will cover all positive multiples of $\mathbf{v}_1$ and $\mathbf{v}_2$. Thus the integral curves corresponding to these solutions are the half-lines through the origin determined by the eigenvectors. The negatives of these functions are also solutions, so the half-lines opposite to the eigenvectors are also integral curves. Since the origin itself is an integral curve, our phase portrait includes the complete lines through the origin determined by the eigenvectors [Figure 4.8(a)]. Note, however, that each of these lines is made up of three integral curves: the origin and two half-lines. The direction of motion along these lines is determined by the signs of the eigenvalues. Since $\lambda_1 = -2$ is negative, $\mathbf{x}_1(t)$ and $-\mathbf{x}_1(t)$ both approach the origin as t gets larger. On the other hand, since $\lambda_2 = 2$ is positive, the points $\mathbf{x}_2(t)$ and $-\mathbf{x}_2(t)$ move out from the origin as t increases. In Figure 4.8(b) we have added arrows to show the directions of motion along these straight-line integral curves.

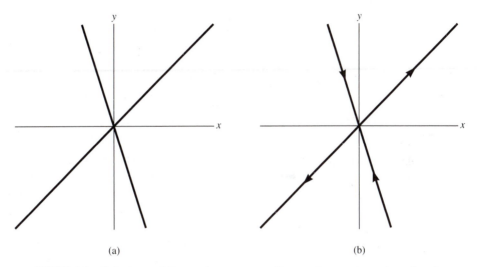

(a)   (b)

FIGURE 4.8 Solutions of $D\mathbf{x} = A\mathbf{x}$ corresponding to eigenvectors, $\lambda_1 < 0 < \lambda_2$

All other solutions of $D\mathbf{x} = A\mathbf{x}$ are of the form

$$\mathbf{x}(t) = ae^{\lambda_1 t}\mathbf{v}_1 + be^{\lambda_2 t}\mathbf{v}_2$$
$$= ae^{-2t}\mathbf{v}_1 + be^{2t}\mathbf{v}_2,$$

where a and b are nonzero constants. As t increases, the first term in this sum approaches zero and the second grows. This means that the integral curve determined by $\mathbf{x}(t)$ is asymptotic, as $t \to +\infty$, to the line determined by $\mathbf{v}_2$. On the other hand, to describe fully the integral curve, we should also trace it backward. As $t \to -\infty$, the first term in the sum grows and the second approaches zero, so the integral curve is asymptotic to the line determined by the eigenvector $\mathbf{v}_1$. These observations allow us to interpolate the rest of the phase portrait (Figure 4.9).

What does this phase portrait tell us about the behavior of solutions to $D\mathbf{x} = A\mathbf{x}$? We have the equilibrium at the origin, and two radial lines. Solutions along one of these lines tend toward the origin. Solutions along the other line, and all other solutions, grow without bound. By analogy with the terminology we used in Section 1.8, the presence of solutions near the origin that move

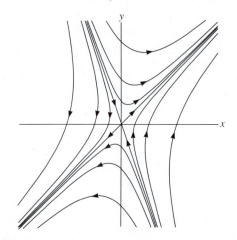

FIGURE 4.9   Phase portrait for $D\mathbf{x} = A\mathbf{x}$, $\lambda_1 < 0 < \lambda_2$

away from it would lead us to call the origin "unstable." However, since some of the solutions near the origin move toward the origin, we cannot call it a "repeller."

Figure 4.9 is a typical phase portrait for a second-order system $D\mathbf{x} = A\mathbf{x}$ when A has eigenvalues of opposite signs. The following are the key observations that led to this portrait.

Fact: *If $\mathbf{v}$ is an eigenvector of A corresponding to the eigenvalue $\lambda \neq 0$, then the phase portrait of $D\mathbf{x} = A\mathbf{x}$ includes the line through the origin determined by $\mathbf{v}$. This line is made up of three integral curves: the origin and two half-lines.*

Fact: *If A is a 2×2 matrix with real eigenvalues of opposite sign, $\lambda_1 < 0 < \lambda_2$, then the phase portrait of $D\mathbf{x} = A\mathbf{x}$ includes the two lines through the origin determined by eigenvectors $\mathbf{v}_1$ and $\mathbf{v}_2$ corresponding to λ_1 and λ_2, respectively. Every other integral curve is asymptotic as $t \to -\infty$ to the line determined by $\mathbf{v}_1$, and as $t \to +\infty$ to the line determined by $\mathbf{v}_2$.*

Example 4.2.2

Sketch the phase portrait of $D\mathbf{x} = A\mathbf{x}$, where

$$A = \begin{bmatrix} 4 & 1 \\ 3 & 2 \end{bmatrix}.$$

The eigenvalues of A and associated eigenvectors are

$$\lambda_1 = 1, \quad \mathbf{v}_1 = \begin{bmatrix} 1 \\ -3 \end{bmatrix}, \quad \text{and} \quad \lambda_2 = 5, \quad \mathbf{v}_2 = \begin{bmatrix} 1 \\ 1 \end{bmatrix}.$$

The eigenvectors are the same as in Example 4.2.1. However, in this example, the eigenvalues are both positive. Thus motion along both of the lines through the origin determined by the eigenvectors is radially outward (Figure 4.10).

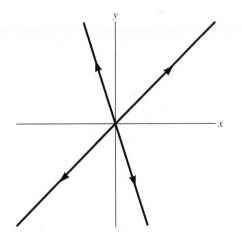

FIGURE 4.10   Solutions of $D\mathbf{x} = A\mathbf{x}$ corresponding to eigenvectors, $0 < \lambda_1 < \lambda_2$

As before, any solution that does not lie on these radial lines can be written in the form

$$\mathbf{x}(t) = ae^{\lambda_1 t}\mathbf{v}_1 + be^{\lambda_2 t}\mathbf{v}_2$$
$$= ae^{t}\mathbf{v}_1 + be^{5t}\mathbf{v}_2$$

where a and b are nonzero constants. This time, however, both terms in this sum increase without bound as t increases. To determine the *relative* rates at which the terms grow, we rewrite the expression for $\mathbf{x}(t)$ by factoring out the exponential of the higher eigenvalue:

$$\mathbf{x}(t) = e^{\lambda_2 t}(ae^{(\lambda_1 - \lambda_2)t}\mathbf{v}_1 + b\mathbf{v}_2)$$
$$= e^{5t}(ae^{-4t}\mathbf{v}_1 + b\mathbf{v}_2).$$

Inside the parentheses, the first term goes to zero as $t \to +\infty$ and the second stays constant. This means that $e^{-5t}\mathbf{x}(t)$ approaches $b\mathbf{v}_2$. Now $e^{-5t}\mathbf{x}(t)$ is always parallel to $\mathbf{x}(t)$, and $b\mathbf{v}_2$ is parallel to $\mathbf{v}_2$. Thus we can conclude that as $t \to \infty$, the slope of $\mathbf{x}(t)$ approaches the slope of the line determined by $\mathbf{v}_2$. A similar analysis shows (Exercise 31) that as $t \to -\infty$, the slope of $\mathbf{x}(t)$ approaches the slope of the line determined by $\mathbf{v}_1$. Of course, $\mathbf{x}(t)$ approaches $\mathbf{0}$ as $t \to -\infty$, so the integral curve enters the origin tangent to the line determined by $\mathbf{v}_1$.

The interpolated integral curves are sketched in Figure 4.11. Since every integral curve moves away from the origin (except the integral curve consisting of the origin itself), it would be natural to call the origin a "repeller."

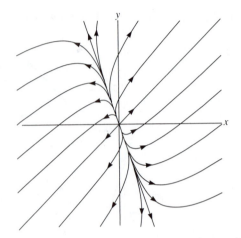

FIGURE 4.11   Phase portrait for $D\mathbf{x} = A\mathbf{x}$, $0 < \lambda_1 < \lambda_2$

Figure 4.11 is a typical phase portrait for $D\mathbf{x} = A\mathbf{x}$ when A has distinct positive real eigenvalues. If A has distinct *negative* eigenvalues, then a similar argument (Exercise 32) yields a phase portrait like the one in this figure but with the arrows reversed.

Fact: *If A is a 2 × 2 matrix with real nonzero eigenvalues $\lambda_1 < \lambda_2$ of the same sign, then the phase portrait of $D\mathbf{x} = A\mathbf{x}$ includes two lines through the origin determined by eigenvectors $\mathbf{v}_1$ and $\mathbf{v}_2$ corresponding to λ_1 and λ_2, respectively. The slope of every other integral curve approaches the slope of the line determined by $\mathbf{v}_1$ as $t \to -\infty$ and approaches the slope of the line determined by $\mathbf{v}_2$ as $t \to \infty$.*

If A is a 2 × 2 matrix with two equal nonzero eigenvalues, $\lambda_1 = \lambda_2 \neq 0$, and if there are two independent eigenvectors corresponding to this double eigenvalue, then *every* nonzero 2-vector is an eigenvector [Exercise 34(a)]. In this case the phase portrait includes all lines through the origin. The arrows point inward if the eigenvalue is negative (Figure 4.12) and outward if the eigenvalue is positive.

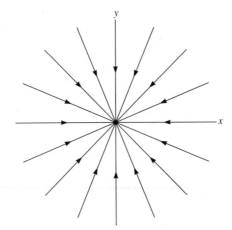

FIGURE 4.12 Phase portrait for $D\mathbf{x} = A\mathbf{x}$, $\lambda_1 = \lambda_2 < 0$ and A has two independent eigenvectors

The argument of Example 4.2.2 can be modified to deal with the case when A has two equal nonzero eigenvalues but only one independent eigenvector corresponding to this eigenvalue. In this case we find (Exercise 33) that the phase portrait resembles Figure 4.13(a) or Figure 4.13(b), with the arrows pointing inward (as shown) if the eigenvalue is negative and outward if it is positive.

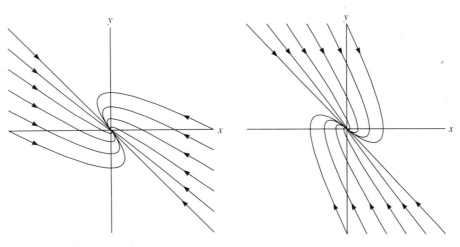

(a) Counterclockwise motion   (b) Clockwise motion

FIGURE 4.13 Phase portraits for $D\mathbf{x} = A\mathbf{x}$, $\lambda_1 = \lambda_2 < 0$, A has only one independent eigenvector

Fact: *Suppose A is a 2 × 2 matrix with two equal nonzero eigenvalues,* $\lambda_1 = \lambda_2 \neq 0$. *If A has two independent eigenvectors, then the phase portrait of* $D\mathbf{x} = A\mathbf{x}$ *consists of radial lines through the origin. If every eigenvector of A is a multiple of one eigenvector* $\mathbf{v}$, *then the integral curves of* $D\mathbf{x} = A\mathbf{x}$ *that do not lie on the line determined by* $\mathbf{v}$ *have slopes that approach the slope of this line as* $t \to \pm\infty$.

▪ Example 4.2.3

Sketch the phase portrait of $D\mathbf{x} = A\mathbf{x}$, where

$$A = \begin{bmatrix} 0 & 1 \\ -1 & -2 \end{bmatrix}.$$

The only eigenvalue of A is $\lambda = -1$, and each eigenvector is a multiple of

$$\mathbf{v} = \begin{bmatrix} -1 \\ 1 \end{bmatrix}.$$

Since λ is negative, motion along the line determined by $\mathbf{v}$ is inward. As we've just seen, the slopes of all other integral curves approach the slope of this line as $t \to \pm\infty$. This leaves two possibilities for the phase portrait [Figure 4.13(a and b)].

To determine which portrait is correct, we need to find the direction of motion at a point off the line through $\mathbf{v}$. Let's use the point one unit up the y-axis. When an integral curve $\mathbf{x}(t)$ reaches this point,

$$\mathbf{x}(t) = \begin{bmatrix} 0 \\ 1 \end{bmatrix},$$

so

$$D\mathbf{x} = A\mathbf{x} = \begin{bmatrix} 0 & 1 \\ -1 & -2 \end{bmatrix}\begin{bmatrix} 0 \\ 1 \end{bmatrix} = \begin{bmatrix} 1 \\ -2 \end{bmatrix}.$$

Since the top coordinate of $D\mathbf{x}$ is the rate of change of the x-coordinate, we see that at this point,

$$\frac{dx}{dt} = 1 > 0.$$

This means that the motion along the integral curve at this point is toward the right. Thus the correct phase portrait is the one in Figure 4.13(b).

Our eventual goal is to use information about the phase portraits of linear systems to obtain information about nonlinear systems (Sections 4.3 through 4.5). The cases involving real eigenvalues that arise in that context are the ones we have discussed so far. Note, however, that we have not yet considered any matrices that have zero as an eigenvalue.

Example 4.2.4

Sketch the phase portrait for $D\mathbf{x} = A\mathbf{x}$ where

$$A = \begin{bmatrix} 1 & 1 \\ 1 & 1 \end{bmatrix}.$$

The eigenvalues of A and associated eigenvectors are

$$\lambda_1 = 0, \quad \mathbf{v}_1 = \begin{bmatrix} 1 \\ -1 \end{bmatrix}, \quad \text{and} \quad \lambda_2 = 2, \quad \mathbf{v}_2 = \begin{bmatrix} 1 \\ 1 \end{bmatrix}$$

(see Exercise 2, Section 3.5, and Exercise 15, Section 3.6). Any multiple of $\mathbf{v}_1$ satisfies $A(c\mathbf{v}_1) = \mathbf{0}$, so each point on the line determined by $\mathbf{v}_1$ is an equilibrium point. Any other solution has the form

$$\mathbf{x}(t) = a\mathbf{v}_1 + be^{2t}\mathbf{v}_2,$$

where a and b are nonzero constants. The integral curve determined by this solution is a half-line parallel to $\mathbf{v}_2$. As $t \to -\infty$, $\mathbf{x}(t)$ approaches the line determined by $\mathbf{v}_1$, and as $t \to +\infty$, $\mathbf{x}(t)$ grows without bound. The integral curves for $D\mathbf{x} = A\mathbf{x}$ are sketched in Figure 4.14.

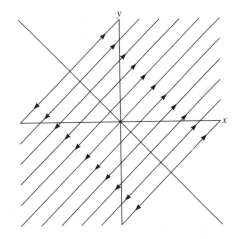

FIGURE 4.14 Phase portrait for $D\mathbf{x} = A\mathbf{x}$, $\lambda_1 = 0$, $\lambda_2 > 0$

The phase portrait in Figure 4.14 is typical when zero is an eigenvalue of A and the other eigenvalue is positive. If zero is an eigenvalue and the other eigenvalue is negative, then the phase portrait of $D\mathbf{x} = A\mathbf{x}$ will look like the portrait in this figure with the arrows reversed.

We leave it for you to check (Exercise 35) that if both of the eigenvalues of A are zero, and if every eigenvector is a multiple of one fixed eigenvector $\mathbf{v}$, then the phase portrait of $D\mathbf{x} = A\mathbf{x}$ is as in Figure 4.15; this is referred to as "shear flow." If there are two independent eigenvectors corresponding to zero, then $A = 0$ and every point is an equilibrium point [Exercise 34(b)].

Fact: *Suppose A is a nonzero 2×2 matrix and 0 is an eigenvalue of A. Then:*

a.   *Every point on the line L determined by an eigenvector for 0 is an equilibrium point.*

b.   *If 0 is the only eigenvalue, then every other integral curve is parallel to L; points on opposite sides of L move in opposite directions.*

c.   *If $\lambda \neq 0$ is another eigenvalue with eigenvector $\mathbf{v}$, then every integral curve off L is a half-line parallel to $\mathbf{v}$, and with one endpoint on L; motion along these integral curves is toward L if $\lambda < 0$ and away from L if $\lambda > 0$.*

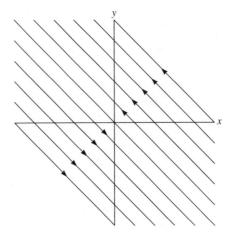

FIGURE 4.15   Phase portrait for $D\mathbf{x} = A\mathbf{x}$, $\lambda_1 = \lambda_2 = 0$, A has only one independent eigenvector

Let's turn now to the case when A is a 2×2 matrix with complex eigenvalues $\alpha \pm \beta i$—that is, to the case when A has no real eigenvalues. Recall from Note 1 in Section 3.2 that the vector $D\mathbf{x}(t)$ points from the point $\mathbf{x}(t)$ in the direction of motion along the integral curve. Since A has no real eigenvalues, this vector cannot be parallel to (or opposite to) the vector $\mathbf{x}(t)$. Thus the direction of motion is never radial. As a consequence, we have the following.

Fact: *If the eigenvalues of A are complex, then every integral curve winds perpetually around the origin.*

In the plane this can happen in three ways: spiraling in, spiraling out, or following closed curves (e.g., circles or ellipses).

When the eigenvalues of A are $\alpha \pm \beta i$, the solutions to $D\mathbf{x} = A\mathbf{x}$ involve terms of the form $e^{\alpha t} \cos \beta t$ and $e^{\alpha t} \sin \beta t$. In particular, $e^{\alpha t}$ is a factor in every term of the solution. Now, if $\alpha < 0$, then as $t \to +\infty$ this factor goes to zero (dominating any other terms that may occur), so every solution tends to the origin. On the other hand, if $\alpha > 0$, then this factor grows without bound as $t \to +\infty$. Thus we have the following.

Fact: *If the 2 × 2 matrix A has complex eigenvalues $\alpha \pm \beta i$ with $\alpha \neq 0$, then the integral curves of $D\mathbf{x} = A\mathbf{x}$ are spirals. If $\alpha < 0$, the curves spiral in to the origin; if $\alpha > 0$, they spiral out.*

Example 4.2.5

Sketch the phase portrait of $D\mathbf{x} = A\mathbf{x}$, where

$$A = \begin{bmatrix} -0.5 & -1 \\ 1 & -0.5 \end{bmatrix}.$$

The eigenvalues of A are $-0.5 \pm i$. Since the real part of these eigenvalues is negative, the integral curves spiral inward. There are two ways in which this

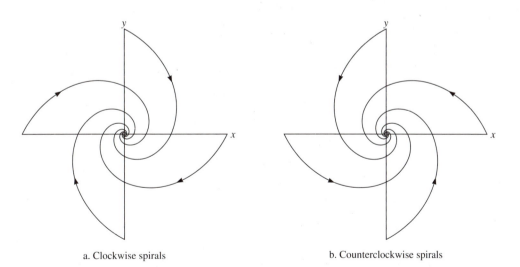

a. Clockwise spirals   b. Counterclockwise spirals

FIGURE 4.16   Phase portraits for $D\mathbf{x} = A\mathbf{x}$, $\lambda_j = \alpha \pm \beta_i$, $\alpha < 0$

can happen: clockwise spirals [Figure 4.16(a), page 343] or counterclockwise spirals [Figure 4.16(b)].

To determine the correct phase portrait, let's find the direction of motion at the point one unit up along the y-axis. At this point

$$Dx = Ax = \begin{bmatrix} -0.5 & -1 \\ 1 & -0.5 \end{bmatrix} \begin{bmatrix} 0 \\ 1 \end{bmatrix} = \begin{bmatrix} -1 \\ -0.5 \end{bmatrix}.$$

From the top entry we see that dx/dt is negative, so motion along the integral curve is toward the left at this point. Thus the spirals are counterclockwise, and the correct phase portrait is the one in Figure 4.16(b).

Finally, we turn to the case of pure imaginary eigenvalues $\pm \beta i$. We have already noted that the integral curves must wind around the origin. When A had complex roots with positive real part, the integral curves spiraled out; when the real part was negative, the curves spiraled in. Since the pure imaginary case is squarely between these, we expect to see the intermediate behavior— that is, periodic motion about the origin along closed curves. This intuition is correct. In fact, the closed curves are either ellipses or circles.

Fact: *If A has pure imaginary eigenvalues $\pm \beta i$, then the integral curves of $Dx = Ax$ are closed loops (ellipses or circles) around the origin.*

The phase portrait of a system $Dx = Ax$ with pure imaginary eigenvalues will look like one of the portraits in Figure 4.17(a and b). To decide, in any

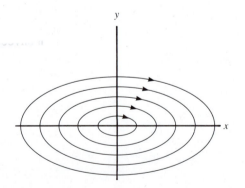

FIGURE 4.17   (a) Phase portraits for $Dx = Ax$, $\lambda_j = \alpha + \beta_i$, clockwise motion

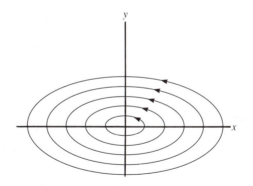

FIGURE 4.17 (b) Phase portraits for $D\mathbf{x} = A\mathbf{x}$, $\lambda_j = \alpha + \beta_i$, counterclockwise motion

given situation, whether the arrows should point clockwise or counterclockwise, we need only check the direction at one point away from the origin. The ellipses in Figure 4.17 have their major axes along the x-axis. In general, the major axes can lie on any line through the origin. We will not attempt to determine their orientation.

PHASE PORTRAITS OF LINEAR SYSTEMS

Any solution $\mathbf{x}(t)$ of the system $D\mathbf{x} = A\mathbf{x}$ can be thought of as a parametrized curve. We refer to these curves as **integral curves** of the system. By adding arrows indicating the direction in which the integral curves are traversed, we obtain the **phase portrait** of the system.

A constant solution of $D\mathbf{x} = A\mathbf{x}$ is referred to as an **equilibrium.** One such equilibrium is $\mathbf{x} = \mathbf{0}.$

If λ is a real eigenvalue of A with associated eigenvector $\mathbf{v}$, then the line through the origin determined by $\mathbf{v}$ is always part of the phase portrait of $D\mathbf{x} = A\mathbf{x}$. If $\lambda \neq 0$, then this line consists of three integral curves: the origin and two half-lines. If $\lambda = 0$, then each point on this line is an equilibrium.

Suppose A is a 2×2 matrix with distinct real eigenvalues λ_1 and λ_2, and let $\mathbf{v}_1$ and $\mathbf{v}_2$ be the corresponding eigenvectors. The properties of the integral curves that do not lie on the lines determined by the eigenvectors are determined by the signs of the eigenvalues. If the eigenvalues have opposite signs, then these curves are asymptotic to the line corresponding to the lower eigenvalue as $t \to -\infty$ and to the line corresponding to the higher eigenvalue as $t \to +\infty$. If the eigenvalues have the same sign, then these curves have slopes approaching the slopes of these lines. If one of the eigenvalues is zero, then these curves are lines parallel to the line determined by the other eigenvector.

Suppose A is a 2×2 matrix with only one eigenvalue λ, and let $\mathbf{v}$ be a corresponding eigenvector. If every eigenvector is a multiple of $\mathbf{v}$, then the integral curves that do not lie on the line determined by $\mathbf{v}$ have slopes approaching the slope of this line. If there are two independent eigenvectors, then either $\lambda \neq 0$ and the phase portrait consists of the radial lines through the origin, or $\lambda = 0$ and every point is an equilibrium.

Suppose A is a 2×2 matrix with complex eigenvalues $\alpha \pm \beta i$. If $\alpha > 0$, then the integral curves spiral out from the origin. If $\alpha < 0$, then the integral curves spiral in to the origin. If $\alpha = 0$, then the integral curves are circles or ellipses around the origin.

Notes

1. **Phase plane portraits of second-order o.d.e.'s**

   Recall from Section 3.2 that the second-order o.d.e.

$$D^2x + a_1\, Dx + a_0x = 0$$

can be replaced by an equivalent system. If we set $y = dx/dt$, then this system is $D\mathbf{x} = A\mathbf{x}$, where

$$\mathbf{x} = \begin{bmatrix} x \\ y \end{bmatrix} \quad \text{and} \quad A = \begin{bmatrix} 0 & 1 \\ -a_0 & -a_1 \end{bmatrix}.$$

We can apply the methods of this section to this system to obtain information about the behavior of the solutions to the equation (see Exercises 13 to 18).

Note that the direction of motion in phase portraits that arise this way is always easy to find. The y-coordinate is dx/dt. Thus at points above the x-axis, dx/dt is positive, and at points below the x-axis, dx/dt is negative. This means that motion along the integral curves is always clockwise.

2. **Higher-order systems**

   We will consider two examples that illustrate how the ideas we have developed can be used to help visualize solutions to higher-order systems. Our examples are third-order systems, so the integral curves are parametrized curves in 3-space.

   First let's consider the system $D\mathbf{x} = A\mathbf{x}$ with

$$A = \begin{bmatrix} -1 & 0 & 0 \\ 0 & -2 & 0 \\ 0 & 0 & 2 \end{bmatrix}.$$

The eigenvalues and associated eigenvectors are

$$\lambda_1 = -1, \quad \mathbf{v}_1 = \begin{bmatrix} 1 \\ 0 \\ 0 \end{bmatrix}; \quad \lambda_2 = -2, \quad \mathbf{v}_2 = \begin{bmatrix} 0 \\ 1 \\ 0 \end{bmatrix}; \quad \lambda_3 = +2, \quad \mathbf{v}_3 = \begin{bmatrix} 0 \\ 0 \\ 1 \end{bmatrix}.$$

The lines determined by these eigenvectors are the three coordinate axes. From the signs of the eigenvalues, we see that motion along the x- and y-axes is radially inward, and motion along the z-axis is radially outward. Any integral curve that lies in the xz-plane (or in the yz-plane) is a combination of solutions corresponding to eigenvalues of opposite signs. The integral curves that lie in the xy-plane are combinations of solutions corresponding to negative eigenvalues. What can we say about an integral curve $\mathbf{x}(t)$ that does not lie in the coordinate planes? Reasoning as in Example 4.2.1, we see that as $t \to \infty$, the direction of $\mathbf{x}(t)$ approaches the direction of the line corresponding to the highest eigenvalue (in this case, the z-axis). To determine the behavior of $\mathbf{x}(t)$ as $t \to -\infty$, we must reason as in Example 4.2.2; we find that the direction of $\mathbf{x}(t)$ approaches the direction of the line corresponding to the lowest eigenvalue (the y-axis). We have sketched some typical integral curves in Figure 4.18.

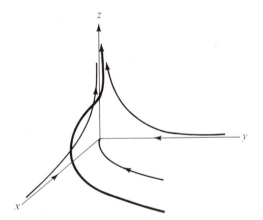

FIGURE 4.18

Let's next consider $D\mathbf{x} = A\mathbf{x}$, with

$$A = \begin{bmatrix} -2 & 1 & 0 \\ -1 & -2 & 0 \\ 0 & 0 & 1 \end{bmatrix}.$$

In this example there is one real eigenvalue $\lambda = 1$ with eigenvector along the z axis, and a pair of complex eigenvalues $\lambda = -2 \pm i$ whose corresponding solutions stay in the xy-plane (check this). Since the real eigenvalue is positive, motion along the z-axis is radially outward. In the xy-plane solutions spiral in toward the origin in a counterclockwise direction (why?) As $t \to +\infty$, an integral curve that does not lie in the xy-

plane will asymptotically approach the z axis. As $t \to -\infty$, this curve will approach the xy-plane. We have sketched some typical integral curves in Figure 4.19.

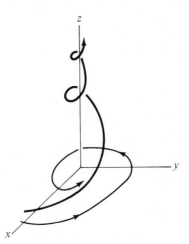

FIGURE 4.19

EXERCISES

In Exercises 1 to 12 find the eigenvalues and any real eigenvectors of A, and use this information to sketch the phase portrait of the system $D\mathbf{x} = A\mathbf{x}$.

1. $A = \begin{bmatrix} 1 & 1 \\ -3 & -1 \end{bmatrix}$

2. $A = \begin{bmatrix} 1 & 1 \\ -4 & -3 \end{bmatrix}$

3. $A = \begin{bmatrix} 2 & 1 \\ -3 & -1 \end{bmatrix}$

4. $A = \begin{bmatrix} 1 & 1 \\ -2 & 2 \end{bmatrix}$

5. $A = \begin{bmatrix} 5 & -6 \\ 5 & -2 \end{bmatrix}$

6. $A = \begin{bmatrix} -2 & 1 \\ 3 & 0 \end{bmatrix}$

7. $A = \begin{bmatrix} -2 & -1 \\ -1 & 2 \end{bmatrix}$

8. $A = \begin{bmatrix} -2 & 1 \\ -3 & 1 \end{bmatrix}$

9. $A = \begin{bmatrix} 5 & 4 \\ -2 & -1 \end{bmatrix}$

10. $A = \begin{bmatrix} -1 & 1 \\ -9 & 5 \end{bmatrix}$

11. $A = \begin{bmatrix} 1 & 5 \\ 1 & -3 \end{bmatrix}$

12. $A = \begin{bmatrix} 1 & 1 \\ -5 & -1 \end{bmatrix}$

13. $A = \begin{bmatrix} -4 & 2 \\ -3 & 1 \end{bmatrix}$

14. $A = \begin{bmatrix} -2 & 2 \\ -1 & 1 \end{bmatrix}$

15. $A = \begin{bmatrix} 3 & 0 \\ 0 & 3 \end{bmatrix}$

16. $A = \begin{bmatrix} 3 & -7 \\ 4 & -8 \end{bmatrix}$

In Exercises 17 to 22 sketch the phase portrait of the system equivalent to the given second-order o.d.e.

17. $(D^2 - 2D - 3)x = 0$ 18. $(4D^2 + 4D + 1)x = 0$

19. $(D^2 - 4D + 5)x = 0$ 20. $(D^2 + 3)x = 0$

21. $(D^2 - 3D + 1)x = 0$ 22. $(4D^2 + 4D - 3)x = 0$

In Exercises 23 and 24 the solutions corresponding to the two complex eigenvectors of A stay in the xy-plane. The solutions corresponding to the third eigenvector lie on the z-axis. Sketch several typical integral curves (see Note 2).

23. $A = \begin{bmatrix} -2 & 1 & 0 \\ -1 & -2 & 0 \\ 0 & 0 & -1 \end{bmatrix}$ 24. $A = \begin{bmatrix} 0 & 2 & 0 \\ -2 & 0 & 0 \\ 0 & 0 & 2 \end{bmatrix}$

Exercises 25 to 27 refer to our models.

25. *Springs:* Show that if m and b are positive, then the phase portrait of the system equivalent to $(mD^2 + bD + k)x = 0$ resembles
   a. Figure 4.17(a) in the undamped case ($b = 0$),
   b. Figure 4.16(a) in the underdamped case ($b \neq 0$, $b^2 - 4mk < 0$),
   c. Figure 4.11, with reversed arrows, in the overdamped case ($b^2 - 4mk > 0$), and
   d. Figure 4.13 in the critically damped case ($b^2 - 4mk = 0$).

26. *Circuits:* Show that if $R \neq 0$ and $V(t) = 0$, then the eigenvalues of the system in Example 3.1.1 are either negative or complex with negative real parts. What do the phase portraits associated with systems of these types tell you about the long-term behavior of the charge and current?

27. *More Circuits:* Repeat Exercise 26 for the system in Example 3.1.2, assuming that $R_2 \neq 0$ and $V(t) = 0$.

More abstract exercises.

28. Suppose that $\mathbf{x} = \mathbf{c}$ is a constant solution of the nonhomogeneous system $D\mathbf{x} = A\mathbf{x} + \mathbf{E}$, where $\mathbf{E}$ is a constant vector.
   a. Show that $A\mathbf{c} = -\mathbf{E}$.
   b. Show that $\mathbf{x}$ is a solution of $D\mathbf{x} = A\mathbf{x} + \mathbf{E}$ if and only if $\mathbf{y} = \mathbf{x} - \mathbf{c}$ is a solution of $D\mathbf{y} = A\mathbf{y}$.
   c. How are the phase portraits of these two systems related?

29. Suppose that $\mathbf{x}_1(t)$ and $\mathbf{x}_2(t)$ are solutions of $D\mathbf{x} = A\mathbf{x}$, and that $\mathbf{x}_1(t_1) = \mathbf{x}_2(t_2)$: Let $a = t_1 - t_2$.
   a. Show that $\mathbf{x}_2(t)$ and $\mathbf{x}_1(t + a)$ satisfy the same initial condition at time $t = t_2$.
   b. What does the existence and uniqueness theorem tell you about $\mathbf{x}_2(t)$ and $\mathbf{x}_1(t + a)$?

30. Use the result of the previous exercise to show that if an integral curve of $D\mathbf{x} = A\mathbf{x}$ reaches the same point at two different times, then the solution is periodic.

31. Suppose that $\mathbf{x} = ae^{\lambda_1 t}\mathbf{v}_1 + be^{\lambda_2 t}\mathbf{v}_2$, where $0 < \lambda_1 < \lambda_2$. Show that as $t \to -\infty$, the slope of $\mathbf{x}(t)$ approaches the slope of the line determined by $\mathbf{v}_1$. (*Hint:* Factor $e^{\lambda_1 t}$ out of the expression for $\mathbf{x}$.)

32. Suppose A is a 2×2 matrix with distinct negative eigenvalues $\lambda_1 < \lambda_2 < 0$. Show that the phase portrait of $D\mathbf{x} = A\mathbf{x}$ resembles the portrait in Figure 4.11 with the arrows reversed.

33. Suppose that A is a 2×2 matrix with only one eigenvalue λ, and that every eigenvector is a multiple of one fixed eigenvector $\mathbf{v}$.
     a. Show that the general solution of $D\mathbf{x} = A\mathbf{x}$ has the form

$$\mathbf{x} = e^{\lambda t}(a\mathbf{v} + bt\mathbf{v} + b\mathbf{w}) = te^{\lambda t}\left(\frac{1}{t}(a\mathbf{v} + \mathbf{w}) + b\mathbf{v}\right)$$

   for a suitable choice of $\mathbf{w}$.
     b. Check that the slope of $x(t)$ approaches the slope of the line through the orgin determined by $\mathbf{v}$ as $t \to \pm\infty$.
     c. Check that if the motion along the curve $\mathbf{x}(t)$ at the point one unit up the y-axis is toward the right (resp. left), then the motion at the point one unit down the y-axis is toward the left (resp. right).

34. Suppose that A is a 2×2 matrix with eigenvalue λ and that there are two independent eigenvectors corresponding to λ.
     a. Use the result of Exercise 16(a) in Section 3.4 to show that every 2-vector is an eigenvector of A.
     b. Use the result of Exercise 35 of Section 3.2 to show that if $\lambda = 0$, then $A = 0$ and every point is an equilibrium point of $D\mathbf{x} = A\mathbf{x}$.

35. Show that if both eigenvalues of the 2×2 matrix A are zero, and if every eigenvector of A is a multiple of one fixed eigenvector, then the phase portrait of $D\mathbf{x} = A\mathbf{x}$ resembles the portrait in Figure 4.15.

4.3   LINEARIZATION AND STABILITY OF EQUILIBRIA

In this section we continue our exploration of phase portraits for systems of o.d.e.'s. In particular, we extend to systems the notions of stability that we discussed in Section 1.8. We will see how our knowledge of phase portraits for linear systems can help us obtain information about the nonlinear case. This approach is the most effective means of studying the behavior of the many nonlinear systems that cannot be solved exactly.

As before, we would like to deal with systems for which distinct integral curves do not intersect. To guarantee this property, we require our system to be **autonomous,** which is to say that the derivative of each variable is a function of the values of the variables but not of time. The typical second-order autonomous system can be written in the form

(S) $$\frac{dx}{dt} = f(x, y), \quad \frac{dy}{dt} = g(x, y).$$

We assume the functions f and g are sufficiently well behaved that each initial-value problem has a unique solution. As usual, we use vector notation for the solutions of (S):

$$\mathbf{x} = \begin{bmatrix} x(t) \\ y(t) \end{bmatrix}.$$

The analysis of the behavior of solutions of a system (S) begins with the determination of the constant solutions. Such a solution $\mathbf{x}(t) = \mathbf{c}$ is called an **equilibrium** of the system. Equilibria of (S) occur at points where the coordinates have derivative zero simultaneously.

Fact: *The equilibria of (S) occur at the points in the plane where*

$$f(x, y) = g(x, y) = 0.$$

Suppose now that $\mathbf{x} = \mathbf{c}$ is an equilibrium of (S), where

$$\mathbf{c} = \begin{bmatrix} a \\ b \end{bmatrix}.$$

We say that $\mathbf{x} = \mathbf{c}$ is an **attractor** if there is a circle with center at the point (a, b) so that any solution $x(t)$ that starts from within the circle approaches $\mathbf{c}$ as $t \to \infty$. An attractor $\mathbf{x} = \mathbf{c}$ is **stable** in the sense that all solutions starting near $\mathbf{c}$ remain close to $\mathbf{x} = \mathbf{c}$. If, on the other hand, there is a circle centered at (a, b) with the property that some solutions starting near $\mathbf{c}$ eventually leave and stay out of the circle, then we say that $\mathbf{x} = \mathbf{c}$ is **unstable**. In particular, if *all* solutions starting inside the circle (except $\mathbf{x} = \mathbf{c}$) eventually leave and stay out of the circle, then we say that $\mathbf{x} = \mathbf{c}$ is a **repeller**. A look back at the discussion in Section 4.2 should convince you that when (S) is linear, the stability of the equilibrium $\mathbf{x} = \mathbf{0}$ is determined by the eigenvalues.

Fact: *The constant zero function* $\mathbf{x} = \mathbf{0}$ *is an equilibrium of the linear system* $D\mathbf{x} = A\mathbf{x}$.

1. *If every eigenvalue of A has negative real part, then* $\mathbf{x} = \mathbf{0}$ *is an attractor.*
2. *If every eigenvalue of A has positive real part, then* $\mathbf{x} = \mathbf{0}$ *is a repeller.*
3. *The presence of eigenvalues with real parts of opposite sign, or the presence of zero or pure imaginary eigenvalues, means that* $\mathbf{x} = \mathbf{0}$ *is neither an attractor nor a repeller.*

Our approach to analyzing an equilibrium of a nonlinear system (S) will be to **linearize** the system—that is, we will replace the given system by a linear system that approximates the given system near the equilibrium. In describing the approximating system, we will use f_x and f_y to denote the partial derivatives $\partial f / \partial x$ and $\partial f / \partial y$, respectively. Recall from multidimensional calculus that if f

and g have continuous partials, then we can write

$$f(x, y) = f(a, b) + f_x(a, b)(x - a) + f_y(a, b)(y - b) + \phi_1(x, y)$$

$$g(x, y) = g(a, b) + g_x(a, b)(x - a) + g_y(a, b)(y - b) + \phi_2(x, y),$$

where ϕ_1 and ϕ_2 are functions satisfying

$$\lim_{(x,y)\to(a,b)} \frac{\phi_i(x, y)}{[(x - a)^2 + (y - b)^2]^{1/2}} = 0, \quad i = 1, 2.$$

The conditions on the functions ϕ_i say that these quantities are equal to a small fraction of the distance of (x, y) from (a, b), with the fraction approaching zero as we get closer to (a, b). An immediate consequence of this is that if

$$\mathbf{c} = \begin{bmatrix} a \\ b \end{bmatrix}$$

is an equilibrium of the system

(S) $$\frac{dx}{dt} = f(x, y), \quad \frac{dy}{dt} = g(x, y)$$

(so that $f(a, b) = g(a, b) = 0$), then the linear system

$$\frac{dx}{dt} = f_x(a, b)(x - a) + f_y(a, b)(y - b)$$

$$\frac{dy}{dt} = g_x(a, b)(x - a) + g_y(a, b)(y - b)$$

is a close approximation to (S) near $\mathbf{c}$. Note that if we set $\mathbf{y} = \mathbf{x} - \mathbf{c}$, and

$$A = \begin{bmatrix} f_x(a, b) & f_y(a, b) \\ g_x(a, b) & g_y(a, b) \end{bmatrix}$$

then $D\mathbf{y} = D\mathbf{x}$, so that the linear system approximating (S) can be rewritten in the form

(L) $$D\mathbf{y} = A\mathbf{y}.$$

We will refer to A as the **linearization matrix** of (S) near $\mathbf{c}$.

The extent to which this approximation shows up in the phase portrait is summarized in the following important theorem (which holds in all dimensions).

Proofs of this result were published independently by the American mathematician Philip Hartman (1964) and the Soviet mathematician David Grobman (1965).

Hartman-Grobman Theorem: *If the linearization matrix A has no zero or pure imaginary eigenvalues, then the phase portrait for (S) near the equilibrium **c** can be obtained from the phase portrait of the linear system (L) via a continuous change of coordinates.*

We will not explain in detail the phrase "continuous change of coordinates." Instead, we will rely on an intuitive picture of the effect of such a change as a deformation of one of the portraits in Section 4.2, changing straight lines into curves, and circles or ellipses into other loops, but not "breaking" any curves. In particular, this means that *if A has no zero or pure imaginary eigenvalues, then the stability properties of the equilibrium **c** of (S) are the same as those of the equilibrium **0** of (L).*

Let's see what this theorem tells us in some examples. The systems considered in these examples are the same as those in Section 4.1, but here we allow the possibility that x and y are negative.

Example 4.3.1

Discuss the stability of the equilibria of the system

(S)

$$\frac{dx}{dt} = 2x - 2x^2 - 5xy$$

$$\frac{dy}{dt} = y - y^2 - 2xy.$$

Here

$$f(x, y) = 2x - 2x^2 - 5xy = x(2 - 2x - 5y)$$

$$g(x, y) = y - y^2 - 2xy = y(1 - y - 2x).$$

The equilibria of (S) occur at points where

$$x(2 - 2x - 5y) = 0$$

and simultaneously

$$y(1 - y - 2x) = 0.$$

At these points

(1) $x = 0$ or $2 - 2x - 5y = 0$

and simultaneously

(2) $y = 0$ or $1 - y - 2x = 0.$

To find the equilibria, we solve each of the equations in (1) simultaneously with each of the equations in (2). This yields four points

$$(0, 0), \quad (0, 1), \quad (1, 0), \quad \left(\frac{3}{8}, \frac{1}{4}\right).$$

To find the linearizations at the equilibria, we first calculate

$$f_x = 2 - 4x - 5y, \quad f_y = -5x$$

$$g_x = -2y, \qquad\qquad g_y = 1 - 2y - 2x.$$

We obtain the linearization matrices by substituting the coordinates of the equilibria into these formulas. The linearization matrices and their eigenvalues are as follows:

$$A_{(0,0)} = \begin{bmatrix} 2 & 0 \\ 0 & 1 \end{bmatrix}; \quad \lambda_1 = 1, \quad \lambda_2 = 2,$$

$$A_{(0,1)} = \begin{bmatrix} -3 & 0 \\ -2 & -1 \end{bmatrix}; \quad \lambda_1 = -3, \quad \lambda_2 = -1,$$

$$A_{(1,0)} = \begin{bmatrix} -2 & -5 \\ 0 & -1 \end{bmatrix}; \quad \lambda_1 = -2, \quad \lambda_2 = -1,$$

$$A_{(\frac{3}{8},\frac{1}{4})} = \begin{bmatrix} -\frac{3}{4} & -\frac{15}{8} \\ -\frac{1}{2} & -\frac{1}{4} \end{bmatrix}; \quad \lambda_1 = -\frac{3}{2}, \quad \lambda_2 = \frac{1}{2}.$$

Since none of the eigenvalues are zero or pure imaginary, the Hartman-Grobman theorem tells us that the phase portrait of (S) near each equilibrium (a, b) is similar to the phase portrait of the linearized system $D\mathbf{y} = A_{(a,b)}\mathbf{y}$. We have sketched the phase portraits of the linearized systems in Figure 4.20. Since the eigenvalues of $A_{(0,0)}$ are both positive, (0, 0) is a repeller. The eigenvalues of $A_{(0,1)}$ and $A_{(1,0)}$ are negative, so (0, 1) and (1, 0) are both attractors. The two eigenvalues of $A_{(\frac{3}{8},\frac{1}{4})}$ have opposite signs, so (3/8, 1/4) is unstable but is not a repeller.

In Figure 4.21, we have sketched the phase portrait of (S) directly, using graphical methods similar to those used in Section 1.7. Notice the resemblance

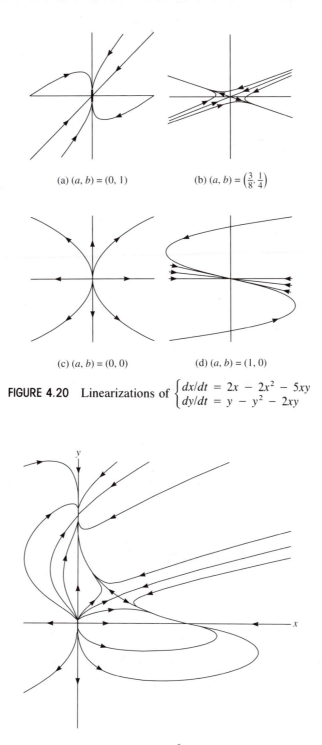

(a) $(a, b) = (0, 1)$ (b) $(a, b) = \left(\frac{3}{8}, \frac{1}{4}\right)$

(c) $(a, b) = (0, 0)$ (d) $(a, b) = (1, 0)$

FIGURE 4.20 Linearizations of $\begin{cases} dx/dt = 2x - 2x^2 - 5xy \\ dy/dt = y - y^2 - 2xy \end{cases}$

FIGURE 4.21 Phase portrait of $\begin{cases} dx/dt = 2x - 2x^2 - 5xy \\ dy/dt = y - y^2 - 2xy \end{cases}$

near the equilibria to the portraits in Figure 4.20. Note, however, that some of the straight lines in the portraits of the linearizations have become curved.

Figure 4.21 suggests that each integral curve that starts off the coordinate axes is contained in a single quadrant. We can verify this observation by noting that the axes are themselves made up of integral curves and that distinct integral curves do not intersect.

Figure 4.21 confirms the predictions we made in Example 4.1.1, particularly the fact that most solutions in the first quadrant tend toward one of the two attractors $(0, 1)$ and $(1, 0)$. In fact the only solutions in the first quadrant that do not tend toward the attractors are the solutions starting from the other two equilibria, and solutions starting along two exceptional integral curves, one coming in to $(3/8, 1/4)$ from the repeller at $(0, 0)$, and the other coming in to $(3/8, 1/4)$ from infinity. These exceptional integral curves divide the first quadrant into two regions. Above the curves, all solutions tend toward the attractor $(0, 1)$, and below them, all solutions tend toward the attractor $(1, 0)$. Each of the two integral curves that separate these regions is called a **separatrix**.

If we follow solutions in the first quadrant backward in time, we see that most solutions either approach the repeller at the origin, or are unbounded as $t \to -\infty$. The solutions approaching the repeller as $t \to -\infty$ are again separated from the ones that go to infinity by two exceptional integral curves, going from $(3/8, 1/4)$ to the two attractors. These integral curves are also referred to as separatrices of the system.

Comparison of Figure 4.21 with the linearization of the system at $(3/8, 1/4)$ in Figure 4.20 suggests that the eigenvectors are tangent to the separatrices at $(3/8, 1/4)$. This turns out to be true, but it is beyond the scope of this book to justify this fact.

The behavior of solutions in the second, third, and fourth quadrants is simpler. Solutions in the second quadrant that start above the x-axis approach the attractor $(0, 1)$ as $t \to \infty$ and approach the repeller $(0, 0)$ as $t \to -\infty$. Similarly, solutions in the fourth quadrant that start to the right of the y-axis approach $(1, 0)$ as $t \to \infty$ and approach $(0, 0)$ as $t \to -\infty$. Solutions in the third quadrant are unbounded as $t \to \infty$ and approach $(0, 0)$ as $t \to -\infty$.

Example 4.3.2

Discuss the stability of the equilibria of the system

(S)
$$\frac{dx}{dt} = 2x - 2x^2 - xy$$
$$\frac{dy}{dt} = y - y^2 - 2xy.$$

The equilibria of (S) occur at the three points

$$(0, 0), \quad (0, 1), \quad (1, 0).$$

To find the linearizations at these points, we calculate

$$f_x = 2 - 4x - y, \quad f_y = -x$$
$$g_x = -2y, \quad\quad\quad g_y = 1 - 2y - 2x.$$

The linearization matrices and their eigenvalues are as follows:

$$A_{(0,0)} = \begin{bmatrix} 2 & 0 \\ 0 & 1 \end{bmatrix}; \quad \lambda_1 = 1, \quad \lambda_2 = 2,$$

$$A_{(0,1)} = \begin{bmatrix} 1 & 0 \\ -2 & -1 \end{bmatrix}; \quad \lambda_1 = -1, \quad \lambda_2 = 1,$$

$$A_{(1,0)} = \begin{bmatrix} -2 & -1 \\ 0 & -1 \end{bmatrix}; \quad \lambda_1 = -2, \quad \lambda_2 = -1.$$

Since none of the eigenvalues are zero or pure imaginary, the Hartman-Grobman theorem tells us that the phase portrait of (S) near each equilibrium (a, b) is similar to the phase portrait of the linearized system $Dy = A_{(a,b)}y$. We have sketched the phase portraits of the linearized systems in Figure 4.22. Since the eigenvalues of $A_{(0,0)}$ are both positive, $(0, 0)$ is a repeller. The eigenvalues of $A_{(1,0)}$ are both negative, so $(1, 0)$ is an attractor. The two eigenvalues of $A_{(0,1)}$ have opposite signs, so $(0, 1)$ is unstable but is not a repeller.

In Figure 4.23, we have sketched the phase portrait of (S) directly. Notice the resemblance near the equilibria to the portraits in Figures 4.22. Note, once more, that some of the straight lines in the portraits of the linearizations have become curved. As before, each integral curve that starts off the coordinate axes is contained in a single quadrant.

We see from Figure 4.23 that most solutions in the first and second quadrants either approach the attractor $(1, 0)$ as $t \to \infty$ or are unbounded as $t \to \infty$. The curves exhibiting these two behaviors are separated by two separatrices along the positive y-axis, coming in to $(0, 1)$. Most solutions in the first and second quadrants approach the repeller $(0, 0)$ as $t \to -\infty$ or are unbounded as $t \to -\infty$. The separatrices that separate the curves exhibiting these behaviors are an integral curve that goes from $(0, 1)$ to the attractor $(1, 0)$ and an integral curve going out from $(0, 1)$ to infinity. As before, the separatrices are tangent to the eigenvectors of the linearization at $(0, 1)$.

Solutions in the third quadrant are unbounded as $t \to \infty$ and approach $(0, 0)$ as $t \to -\infty$. Solutions that start to the right of the y-axis in the fourth quadrant approach $(1, 0)$ as $t \to \infty$ and either approach $(0, 0)$ or become unbounded as $t \to -\infty$.

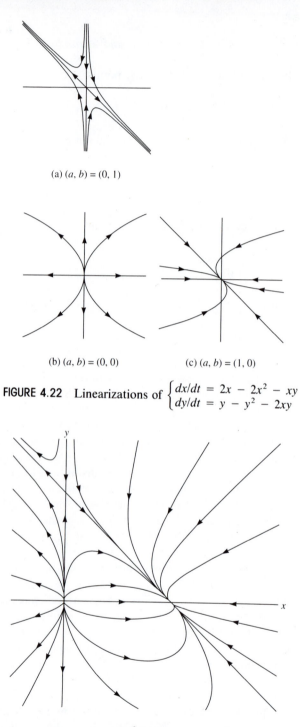

(a) $(a, b) = (0, 1)$

(b) $(a, b) = (0, 0)$ (c) $(a, b) = (1, 0)$

FIGURE 4.22   Linearizations of $\begin{cases} dx/dt = 2x - 2x^2 - xy \\ dy/dt = y - y^2 - 2xy \end{cases}$

FIGURE 4.23   Phase portrait of $\begin{cases} dx/dt = 2x - 2x^2 - xy \\ dy/dt = y - y^2 - 2xy \end{cases}$

Example 4.3.3

Discuss the stability of the equilibria of the system

(S)
$$\frac{dx}{dt} = x - x^2 - xy$$

$$\frac{dy}{dt} = -y + 2xy.$$

The equilibria of (S) occur at the three points

$$(0, 0), \quad (1, 0), \quad \left(\frac{1}{2}, \frac{1}{2}\right).$$

To find the linearizations at these points, we calculate

$$f_x = 1 - 2x - y, \quad f_y = -x$$
$$g_x = 2y, \qquad\qquad g_y = -1 + 2x.$$

The linearization matrices and their eigenvalues are as follows:

$$A_{(0,0)} = \begin{bmatrix} 1 & 0 \\ 0 & -1 \end{bmatrix}; \quad \lambda_1 = -1, \quad \lambda_2 = 1,$$

$$A_{(1,0)} = \begin{bmatrix} -1 & -1 \\ 0 & 1 \end{bmatrix}; \quad \lambda_1 = -1, \quad \lambda_2 = 1,$$

$$A_{(\frac{1}{2},\frac{1}{2})} = \begin{bmatrix} -\frac{1}{2} & -\frac{1}{2} \\ 1 & 0 \end{bmatrix}; \quad \lambda_1 = \frac{-1 + i\sqrt{7}}{4}, \quad \lambda_2 = \frac{-1 - i\sqrt{7}}{4}.$$

Here again, none of the eigenvalues are zero or pure imaginary, so the phase portrait near an equilibrium of (S) resembles the phase portrait of the corresponding linearization. We have sketched the phase portraits of the linearizations in Figure 4.24. Because each of $A_{(0,0)}$ and $A_{(1,0)}$ has eigenvalues of opposite signs, (0, 0) and (1, 0) are unstable but are not repellers. Since $A_{(\frac{1}{2},\frac{1}{2})}$ has complex roots with negative real parts, (1/2, 1/2) is an attractor.

In Figure 4.25, we have sketched the phase portrait of (S) directly. Notice the resemblance near the equilibria to the portraits in Figure 4.24.

We see from Figure 4.25 that most solutions in the first quadrant spiral in toward the attractor at (1/2, 1/2) as $t \to \infty$. The exceptions lie along the coordinate axes: solutions along the y-axis approach the origin as $t \to \infty$, and solutions along the positive x-axis approach (1, 0) as $t \to \infty$. As $t \to -\infty$, most solutions in the first quadrant are unbounded. The exceptions are the integral curve along the x-axis joining (0, 0) and (1, 0), and an integral curve leaving (1, 0) and spiraling in to (1/2, 1/2).

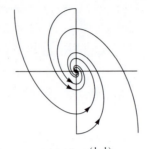

(a) $(a, b) = \left(\frac{1}{2}, \frac{1}{2}\right)$

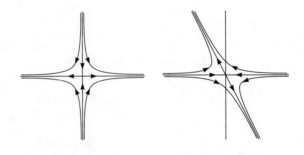

(b) $(a, b) = (0, 0)$ (c) $(a, b) = (1, 0)$

FIGURE 4.24   Linearizations of $\begin{cases} dx/dt = x - x^2 - xy \\ dy/dt = -y + 2xy \end{cases}$

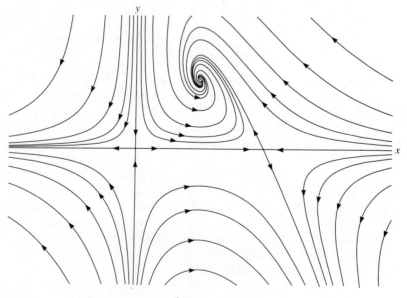

FIGURE 4.25   Phase portrait of $\begin{cases} dx/dt = x - x^2 - xy \\ dy/dt = -y + 2xy \end{cases}$

Most solutions in the second, third, and fourth quadrants are unbounded as $t \to \infty$ and as $t \to -\infty$. The exceptions are the integral curves along the axes and an integral curve going from $(1, 0)$ to infinity. Again, the exceptional integral curves are separatrices, tangent to the eigenvectors of the linearizations at $(0, 0)$ and $(1, 0)$.

Note that the Hartman-Grobman theorem has two serious limitations. First, it tells us nothing when the linearization has zero or pure imaginary eigenvalues (see Exercise 24). Second, when it does apply, it provides information only about the behavior of solutions near the equilibria. In order to make global predictions, we need to know more. The global predictions that we made in the preceding examples relied on graphs made using other methods. Nevertheless, the use of the theorem to investigate the stability of equilibria is a good start toward understanding the long-term behavior of solutions.

We've seen that when the linearization matrix A has no zero or pure imaginary eigenvalues, the stability of an equilibrium can be decided from the signs of the real parts of the eigenvalues. Negative conclusions can sometimes be drawn even when some of the eigenvalues are zero or pure imaginary. We will formulate this result for systems in n variables:

$$\frac{dx_1}{dt} = f_1(x_1, \ldots, x_n)$$

(S)
$$\cdot$$
$$\cdot$$
$$\cdot$$

$$\frac{dx_n}{dt} = f_n(x_1, \ldots, x_n)$$

As when $n = 2$, we write solutions of (S) as n-vector valued functions, and we take limits componentwise. Equilibria, or constant solutions, occur at points where all of the f_i's are simultaneously zero. The definitions of attractors and repellers remain the same, except that we need to replace the word "circle" in these definitions by the word "ball." By a **ball** with center

$$\mathbf{c} = \begin{bmatrix} c_1 \\ \cdot \\ \cdot \\ \cdot \\ c_n \end{bmatrix}$$

and radius ϵ we mean all those points

$$\mathbf{x} = \begin{bmatrix} x_1 \\ \cdot \\ \cdot \\ \cdot \\ x_n \end{bmatrix}$$

that satisfy the inequality

$$[(x_1 - c_1)^2 + \cdots + (x_n - c_n)^2]^{1/2} < \epsilon.$$

In this more general context the linearization matrix for (S) near **c** is

$$A = \begin{bmatrix} \partial f_1/\partial x_1 & \cdots & \partial f_1/\partial x_n \\ & & \\ \cdot & & \cdot \\ \cdot & & \cdot \\ \partial f_n/\partial x_1 & \cdots & \partial f_n/\partial x_n \end{bmatrix},$$

where the partials are all evaluated at **c**. To investigate the stability of the equilibrium **c**, we calculate the eigenvalues of A and apply the following test. Note, however, that the test does not apply if the real parts of all the eigenvalues are less than or equal to zero but are not all negative.

Fact: Linearization Stability Test: *Suppose that* **c** *is an equilibrium of (S). Let A be the linearization matrix for (S) near* **c**.

1. *If the real parts of* **all** *the eigenvalues of A are negative, then* **c** *is an attractor.*
2. *If the real parts of* **at least one** *eigenvalue of A is positive, then* **c** *is unstable.*
3. *If the real parts of* **all** *the eigenvalues of A are positive, then* **c** *is a repeller.*

Let's apply the linearization stability test to one more example.

Example 4.3.4

Use the linearization stability test to discuss the stability of the equilibria of

$$\frac{dx}{dt} = xy + 2y + z^2$$

$$\frac{dy}{dt} = x - y$$

$$\frac{dz}{dt} = (x - y)^2 - 3z.$$

The equilibria occur at points where

$$xy + 2y + z^2 = 0$$

$$x - y = 0$$

$$(x - y)^2 - 3z = 0.$$

From the second equation, we see that $x = y$. From the third equation, we now see that $z = 0$. Substitution of $x = y$ and $z = 0$ into the first equation leads to two values of x, $x = 0$, and $x = -2$. Thus there are two equilibria

$$(0, 0, 0) \quad \text{and} \quad (-2, -2, 0).$$

To find the linearizations, we first calculate the partials

$$\frac{\partial f_1}{\partial x} = y, \qquad \frac{\partial f_1}{\partial y} = x + 2, \qquad \frac{\partial f_1}{\partial z} = 2z$$

$$\frac{\partial f_2}{\partial x} = 1, \qquad \frac{\partial f_2}{\partial y} = -1, \qquad \frac{\partial f_2}{\partial z} = 0$$

$$\frac{\partial f_3}{\partial x} = 2(x - y), \quad \frac{\partial f_3}{\partial y} = -2(x - y), \quad \frac{\partial f_3}{\partial z} = -3.$$

The linearization matrices and their eigenvalues are as follows:

$$A_{(0,0,0)} = \begin{bmatrix} 0 & 2 & 0 \\ 1 & -1 & 0 \\ 0 & 0 & -3 \end{bmatrix}; \quad \lambda_1 = -3, \quad \lambda_2 = -2, \quad \lambda_3 = 1,$$

$$A_{(-2,-2,0)} = \begin{bmatrix} -2 & 0 & 0 \\ 1 & -1 & 0 \\ 0 & 0 & -3 \end{bmatrix}; \quad \lambda_1 = -3, \quad \lambda_2 = -2, \quad \lambda_3 = -1.$$

Since one of the eigenvalues of $A_{(0,0,0)}$ is positive, $(0, 0, 0)$ is unstable. Since the other eigenvalues are negative, $(0, 0, 0)$ is not a repeller. The eigenvalues of $A_{(-2,-2,0)}$ are all negative, so $(-2, -2, 0)$ is an attractor.

Let's summarize our observations.

LINEARIZATION AND STABILITY OF EQUILIBRIA

A constant solution $x = c$ of a system of o.d.e.'s is called an **equilibrium**. Equilibria occur at those points at which all of the functions on the right-hand side of the system are simultaneously zero.

If $x = c$ is an equilibrium of a system (S), then the **linearization matrix** for (S) near c is the matrix A of partial derivatives of the right-hand sides of (S), evaluated at c.

Hartman-Grobman Theorem

If the eigenvalues of A include no zeros or pure imaginary numbers, then the actual phase portrait of the original system (S) near the equilibrium **c** is obtained from the phase portrait of the **linearized system** $D\mathbf{y} = A\mathbf{y}$ by a continuous change of variables.

Linearization Stability Test

1. If all the eigenvalues of A have negative real part, then all solutions of (S) that start near **c** tend to **c**. In this case **c** is an **attractor**.

2. If some eigenvalue of A has positive real part, then some solutions eventually escape from the neighborhood of **c**. In this case $\mathbf{x} = \mathbf{c}$ is **unstable**.

3. If all eigenvalues of A have positive real part, then all solutions (other than $\mathbf{x} = \mathbf{c}$) eventually escape from the neighborhood of **c**, so **c** is a **repeller**.

EXERCISES

In Exercises 1 to 7 determine whether the equilibrium $\mathbf{x} = \mathbf{0}$ of the linear system $D\mathbf{x} = A\mathbf{x}$ is stable or unstable.

1. $A = \begin{bmatrix} 1 & 0 & 3 \\ 3 & 2 & -2 \\ 1 & 0 & -1 \end{bmatrix}$

2. $A = \begin{bmatrix} 1 & 5 & 0 \\ 0 & 2 & 0 \\ 1 & 7 & 1 \end{bmatrix}$

3. $A = \begin{bmatrix} 1 & -2 & 0 \\ 2 & -3 & 0 \\ 5 & 2 & -3 \end{bmatrix}$

4. $A = \begin{bmatrix} -1 & 0 & 0 \\ 1 & -1 & -1 \\ 2 & 1 & -1 \end{bmatrix}$

5. $A = \begin{bmatrix} 1 & 0 & -1 \\ 3 & 2 & 5 \\ 2 & 0 & -1 \end{bmatrix}$

6. $A = \begin{bmatrix} 1 & 0 & 0 \\ 3 & 0 & 2 \\ 1 & 0 & 1 \end{bmatrix}$

7. $A = \begin{bmatrix} 1 & 2 & 3 \\ 0 & -1 & -1 \\ 0 & 1 & -1 \end{bmatrix}$

In Exercises 8 through 23, find the equilibria and determine their stability. Decide whether each equilibrium is an attractor, a repeller, or neither. Note that the systems in Exercises 8 through 17 are the same as those in Exercises 1 through 10 of Section 4.1, but here we do not restrict attention to solutions for which x and y are nonnegative.

8. $\dfrac{dx}{dt} = 4x - 2x^2 - xy$

    $\dfrac{dy}{dt} = 3y - xy - y^2$

9. $\dfrac{dx}{dt} = 3x - x^2 - xy$

    $\dfrac{dy}{dt} = 4y - 2xy - y^2$

10. $\dfrac{dx}{dt} = 2x - 2x^2 - xy$

    $\dfrac{dy}{dt} = 3y - xy - y^2$

11. $\dfrac{dx}{dt} = 3x - x^2 - xy$

    $\dfrac{dy}{dt} = 2y - 2xy - y^2$

12. $\dfrac{dx}{dt} = 2x - 2x^2 - xy$

$\dfrac{dy}{dt} = 4y - 2xy - y^2$

13. $\dfrac{dx}{dt} = 4x - 2x^2 - xy$

$\dfrac{dy}{dt} = -3y + xy$

14. $\dfrac{dx}{dt} = 2x - 4x^2 - xy$

$\dfrac{dy}{dt} = -3y + 7xy$

15. $\dfrac{dx}{dt} = 4x - 2x^2 - xy$

$\dfrac{dy}{dt} = -y + xy$

16. $\dfrac{dx}{dt} = xy - x^3 + x$

$\dfrac{dy}{dt} = 3y + xy - y^2$

17. $\dfrac{dx}{dt} = 5x + x^2 - xy$

$\dfrac{dy}{dt} = y^2 - x^2y + y$

18. $\dfrac{dx}{dt} = y$

$\dfrac{dy}{dt} = (x^2 - 1)y - x$

19. $\dfrac{dx}{dt} = x - y$

$\dfrac{dy}{dt} = -2x + x^2$

20. $\dfrac{dx}{dt} = 2x - 2x^2 + xy$

$\dfrac{dy}{dt} = y - y^2 + 2xy$

21. $\dfrac{dx}{dt} = 2x - 2x^2 + 5xy$

$\dfrac{dy}{dt} = y - 2y^2 + 2xy$

22. $\dfrac{dx}{dt} = -2x + xy + z^2$

$\dfrac{dy}{dt} = x - y$

$\dfrac{dz}{dt} = (x - y)^2 - z$

23. $\dfrac{dx}{dt} = 4x + xy$

$\dfrac{dy}{dt} = x + z$

$\dfrac{dz}{dt} = -2y + z$

24. a. Check that $\mathbf{x} = \mathbf{0}$ is an equilibrium of

(S)
$$\dfrac{dx}{dt} = ax^3 + y$$
$$\dfrac{dy}{dt} = -x + ay^3$$

and that the linearization matrix of (S) near $\mathbf{0}$ has pure imaginary eigenvalues. (Note that this means that the integral curves of the linearized system are closed loops around the origin.)

  b. The distance r from a point to the origin satisfies $r^2 = x^2 + y^2$, so

$$\dfrac{dr}{dt} = \dfrac{1}{r}\left(x\dfrac{dx}{dt} + y\dfrac{dy}{dt}\right).$$

Substitute the values of dx/dt and dy/dt from (S) into this formula to obtain an expression for the rate of change of the distance from the origin to the points on an integral curve of (S).

  c. Show that the sign of dr/dt is the same as the sign of a. Note that this means that if $a > 0$, then points on an integral curve of (S) move away from the origin with time. If $a < 0$, then these points move toward the origin. In either event the curves are *not* closed loops.

4.4   CONSTANTS OF MOTION

The Hartman-Grobman theorem discussed in the preceding section tells us that in many cases the solutions of a nonlinear system near an equilibrium point mimic the solutions of the linearization of the system there. However, the result does not apply if the linearization matrix has any pure imaginary (or zero) eigenvalues. Furthermore, even if it does apply, it tells us only about *local* behavior: if the equilibrium is unstable, then solutions which get away from the equilibrium are no longer governed by the linearization. In this section, we shall see how knowledge of a function that is constant along solutions can yield a great deal of data about the phase portrait.

Let's begin with a familiar example, the unforced, undamped pendulum (see Example 2.10.2).

■   Example 4.4.1   The Undamped Pendulum

The motion of an unforced, undamped pendulum is modeled by the o.d.e.

$$\frac{d^2\theta}{dt^2} + \frac{g}{L} \sin\theta = 0.$$

The substitution

$$x = \theta \quad \text{and} \quad y = \frac{d\theta}{dt}$$

yields a system equivalent to the o.d.e.

$$\text{(S)} \qquad \frac{dx}{dt} = y$$

$$\frac{dy}{dt} = -\frac{g}{L} \sin x.$$

We first note that the equilibria of (S) occur when

$$0 = y$$

$$0 = -\frac{g}{L} \sin x,$$

which is to say, at the points

$$\mathbf{c}_n = \begin{bmatrix} \pi n \\ 0 \end{bmatrix},$$

where n is an integer. The linearization matrix at $\mathbf{c}_n$ is

$$A_n = \begin{bmatrix} 0 & 1 \\ -\dfrac{g}{L}\cos(\pi n) & 0 \end{bmatrix}.$$

When n is odd, $\cos(\pi n) = -1$ so that

$$A_n = \begin{bmatrix} 0 & 1 \\ \dfrac{g}{L} & 0 \end{bmatrix}, \quad n \text{ odd}.$$

This matrix has eigenvalues $\pm\sqrt{g/L}$. Thus, the Hartman-Grobman theorem applies to tell us that solutions near the equilibrium resemble Figure 4.9.

On the other hand, when n is even, $\cos(\pi n) = 1$ so that

$$A_n = \begin{bmatrix} 0 & 1 \\ -\dfrac{g}{L} & 0 \end{bmatrix}, \quad n \text{ even}.$$

This matrix has pure imaginary eigenvalues $\pm i\sqrt{g/L}$. All solutions of the linearized system are periodic, circling around the equilibrium, but the Hartman-Grobman theorem does not allow us to conclude anything about the solutions to (S).

To study the phase portrait of (S), we will analyze the energy of the pendulum. In problems such as this, where the force on the object depends only on its position (and not on its velocity or on time), the *potential energy* is defined, up to an additive constant, as the negative of an antiderivative of the force:

$$U(x) = -\int -\frac{g}{L}\sin x\, dx = -\frac{g}{L}\cos x.$$

The *kinetic energy* is half the square of the velocity

$$T(y) = \frac{1}{2}y^2,$$

so the *total energy* is

$$E(x, y) = U(x) + T(y) = -\frac{g}{L}\cos x + \frac{1}{2}y^2.$$

For each solution of (S),

$$x = x(t), \quad y = y(t),$$

the energy along the solution is measured by the composite function

$$E(t) = E(x(t), y(t)).$$

We calculate the derivative of $E(t)$ using the chain rule:

$$\frac{dE}{dt} = \frac{\partial E}{\partial x}\frac{dx}{dt} + \frac{\partial E}{\partial y}\frac{dy}{dt}$$

$$= \frac{g}{L}\sin x\,\frac{dx}{dt} + y\,\frac{dy}{dt}.$$

Along the solution, $dx/dt = y$ and $dy/dt = -(g/L)\sin x$. Thus

$$\frac{dE}{dt} = \left(\frac{g}{L}\sin x\right)y + y\left(-\frac{g}{L}\sin x\right) = 0.$$

This shows that the energy is constant along the solution.

Since the energy $E(x, y)$ is constant along solutions, the solutions must lie on the level curves $E(x, y) = c$. We have sketched the level curves

$$-\frac{g}{L}\cos x + \frac{1}{2}y^2 = c$$

for various values of c in Figure 4.26 (using a computer). There are three distinct kinds of level curves. For $c > g/L$, the level curve has two branches, symmetrically placed with respect to the x-axis; these are the undulating curves toward the outside of the figure. For $c = g/L$, the level curve consists of the graphs of two functions that go through the equilibria at odd multiples of π. Within the "blips" formed by these two graphs are concentric closed curves (not circles) representing energy levels $-g/L < c < g/L$ and nesting down to the equilibria at even multiples of π ($c = -g/L$). There are no level curves with $c < -g/L$ (why?).

This picture tells us a great deal about the behavior of the solutions to the system. At sufficiently high energy levels ($c > g/L$), $x = \theta$ is steadily increasing (if $y > 0$) or steadily decreasing (if $y < 0$). These solutions are

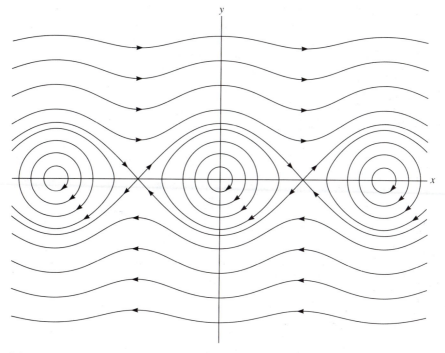

FIGURE 4.26

unbounded as $t \to \pm\infty$. Physically, the pendulum performs complete rotations (counterclockwise if $y > 0$, and clockwise if $y < 0$).

At lower energy levels ($-g/L < c < g/L$), the solutions are periodic. Here the angle $\theta = x$ oscillates about the equilibrium at the nearest even multiple of π and the velocity y oscillates about 0. Physically, the pendulum swings back and forth about the "straight down" position.

At the threshold energy $c = g/L$, there are two kinds of integral curves: equilibria at odd multiples of π, and arcs joining adjacent pairs of these equilibria. An equilibrium at an odd multiple of π represents the pendulum perfectly balanced in a "straight up" position. If the pendulum starts with angle θ_0 and with its initial velocity y_0 chosen so that the point (θ_0, y_0) lies on an arc in the phase plane connecting two upright equilibria, then the pendulum will continue to swing toward the upright position, approaching it in the limit as $t \to \infty$ without ever actually reaching it.

In Example 4.4.1, the hypothesis of the Hartman-Grobman theorem fails to hold, yet the phase portrait near each equilibrium still looks like the phase portrait of the corresponding linearized system. This will not always be the case. In some examples the linearized system may have closed periodic curves (corresponding to pure imaginary eigenvalues of the linearization matrix), whereas the system itself may have integral curves that spiral in or out from the equilibrium (see Exercise 24, Section 4.3). The point is that when the linearized system has zero or pure imaginary eigenvalues, the phase portrait must be analyzed directly, without relying on the linearized system.

The preceding example illustrates one approach to dealing directly with a nonlinear system. A function $E(x, y)$ of two variables is called a **constant of motion** of a system if it is constant along every solution. If $E(x, y)$ and its partial derivatives are continuous, then as in the example, the chain rule gives us a test for constants of motion:

Fact: *Suppose that the function $E(x, y)$ and its partials, $\partial E/\partial x$ and $\partial E/\partial y$, are continuous. Then $E(x, y)$ is a constant of motion for the system*

(S)
$$\frac{dx}{dt} = f(x, y)$$
$$\frac{dy}{dt} = g(x, y)$$

if and only if

$$f(x, y)\,\frac{\partial E}{\partial x} + g(x, y)\,\frac{\partial E}{\partial y} = 0$$

identically in x and y.

Geometrically, we have the following

Fact: *If $E(x, y)$ is a constant of motion for the system (S), then the integral curves of (S) lie on the level curves of E.*

In the example, the equilibria of the system are critical points of the constant of motion (that is, points where $\partial E/\partial x$ and $\partial E/\partial y$ are both zero). When this happens, the nature of the critical point can tell us a great deal about the solutions of the system.

Let (x_0, y_0) be an equilibrium of the system (S) and let $E(x, y)$ be a constant of motion for (S). Suppose that $E(x, y)$ has a strict local minimum at the point (x_0, y_0)—that is, suppose that for some $\epsilon > 0$

$$E(x_0, y_0) < E(x, y)$$

for all points (x, y) within distance ϵ of (x_0, y_0) but distinct from it. Let $c_0 = E(x_0, y_0)$. If c is a value slightly above c_0, then the level curve $E(x, y) = c$ is

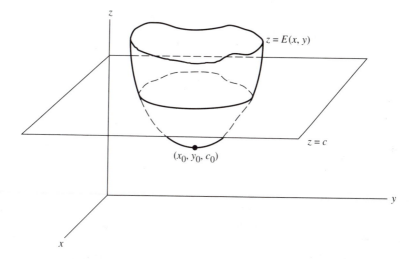

FIGURE 4.27   Graph of E near an extremum

a closed curve (see Figure 4.27). Since the integral curves of (S) must stay on the level curves of E, the phase portrait near (x_0, y_0) will consist of closed curves surrounding (x_0, y_0) (see Figure 4.28). Thus (x_0, y_0) is a stable equilibrium. In fact, if some disc about (x_0, y_0) contains no other equilibria of the system, then each integral curve near (x_0, y_0) is a closed curve around (x_0, y_0), representing a solution that oscillates about the equilibrium point periodically. This resembles the picture corresponding to a linear system with pure imaginary eigenvalues.

You should convince yourself that a similar phase portrait occurs if the constant of motion has a strict local *maximum* at (x_0, y_0). Of course, in this case the closed curves around (x_0, y_0) are level curves $E(x, y) = c$ with c slightly *below* $c_0 = E(x_0, y_0)$.

You may recall from Calculus that there is a second-partials test for analyzing the nature of a critical point of a function $E(x, y)$. The key to this test

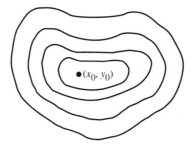

FIGURE 4.28   Level curves near an extremum

is the **discriminant** of $E(x, y)$ which is the function defined by

$$\Delta(x, y) = \frac{\partial^2 E}{\partial^2 x} \frac{\partial^2 E}{\partial^2 y} - \left(\frac{\partial^2 E}{\partial x \, \partial y}\right)^2.$$

If (x_0, y_0) is a critical point of $E(x, y)$ and if $\Delta(x_0, y_0) \neq 0$, then we can decide the behavior of $E(x, y)$ near (x_0, y_0) as follows.

Fact: Second Partials Test   *Let $E(x, y)$ be a function that has continuous first and second partial derivatives. Suppose that (x_0, y_0) is a point at which $\partial E/\partial x = 0 = \partial E/\partial y$.*

a.   *If $\Delta(x_0, y_0) > 0$, then $E(x, y)$ has a strict local extremum at (x_0, y_0): it is a maximum if $\partial^2 E/\partial^2 x < 0$ at (x_0, y_0), and a minimum if $\partial^2 E/\partial^2 x > 0$ at (x_0, y_0).*

b.   *If $\Delta(x_0, y_0) < 0$, then $E(x, y)$ has neither a local maximum nor a local minimum at (x_0, y_0). In this case, we say that (x_0, y_0) is a **saddle point** of $E(x, y)$.*

c.   *If $\Delta(x_0, y_0) = 0$, then no conclusion can be drawn. In this case, we say that (x_0, y_0) is a **degenerate critical point**.*

The second case of the second-partials test can be described in greater detail. Suppose that (x_0, y_0) is a critical point of E with $\Delta(x_0, y_0) < 0$. Let $c_0 = E(x_0, y_0)$. Then near (x_0, y_0), the level set $E(x, y) = c_0$ consists of two curves crossing at (x_0, y_0) and therefore separating the plane near (x_0, y_0) into four "quadrants." If c is slightly above or slightly below c_0, then the level set $E(x, y) = c$ has two branches, and resembles a hyperbola asymptotic to the two curves through (x_0, y_0) (see Figure 4.29). The graph of such a function is sketched in Figure 4.30, and explains why case (b) is called a "saddle point."

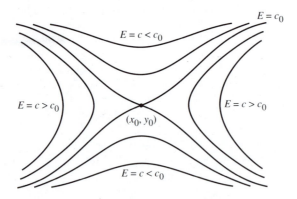

$E = c_0$

$E = c < c_0$

$E = c > c_0$

$E = c > c_0$

(x_0, y_0)

$E = c < c_0$

FIGURE 4.29   Level curves near a saddle point

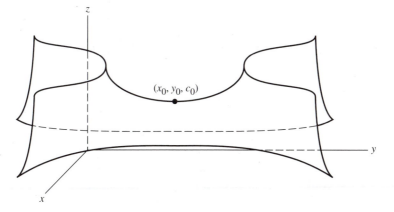

FIGURE 4.30   Graph of E near a saddle point

Now suppose that $E(x, y)$ is a constant of motion of a system (S) and that (x_0, y_0) is both an equilibrium of (S) and a saddle point of E. The integral curves of the system near (x_0, y_0) must lie on the level curves in Figure 4.29. If there are no other equilibria near (x_0, y_0), then we can obtain the phase portrait near (x_0, y_0) from Figure 4.29, by adding arrows to indicate the direction of motion. Note the resemblance of this picture to the phase portrait of a linear system with eigenvalues of opposite signs.

Our observations about equilibria of (S) and critical points of (E) can be summarized as follows.

Fact: *Let*

$$\frac{dx}{dt} = f(x, y)$$

(S)

$$\frac{dy}{dt} = g(x, y)$$

be a system of o.d.e.'s and let $E(x, y)$ be a constant of motion of (S) that has continuous first and second partial derivatives. Suppose that (x_0, y_0) is an equilibrium of (S) and that there is a neighborhood of (x_0, y_0) that does not contain any other equilibria of (S). Then

a.   *If (x_0, y_0) is a critical point of E with $\Delta(x_0, y_0) > 0$, then the equilibrium of (S) at (x_0, y_0) is surrounded by closed integral curves, representing solutions that oscillate periodically about equilibrium. In particular, the equilibrium is stable.*

b. *If (x_0, y_0) is a critical point of E with $\Delta(x_0, y_0) < 0$, then the equilibrium at (x_0, y_0) is a saddle point. If $c_0 = E(x_0, y_0)$, then the integral curves in the level set $E(x, y) = c_0$ are separatrices tending toward the equilibrium as $t \to \infty$ or as $t \to -\infty$; all other integral curves near (x_0, y_0) leave a disc around (x_0, y_0) as $t \to \infty$ and as $t \to -\infty$. In particular, the equilibrium is unstable.*

We illustrate this fact with an example coming from population models.

Example 4.4.2   Volterra-Lotka Equations

We discussed models for two populations consisting of a predator and its prey in Section 4.1. The simplest such model is given by the system

(S)
$$\frac{dx}{dt} = a_1x - c_1xy$$
$$\frac{dy}{dt} = -a_2y + c_2xy,$$

where a_1, a_2, c_1, and c_2 are positive constants. Since we are dealing with populations, we restrict attention to points where $x \geq 0$ and $y \geq 0$.

The equilibria of (S) occur at two points: $(0, 0)$ and $(a_2/c_2, a_1/c_1)$. The linearization matrix at $(0, 0)$ is

$$A_{(0,0)} = \begin{bmatrix} a_1 & 0 \\ 0 & -a_2 \end{bmatrix},$$

which has eigenvectors along the coordinate axes, with eigenvalues a_1 and $-a_2$. The Hartman-Grobman theorem applies to tell us that the phase portrait near $(0, 0)$ resembles the linearizations at $(0, 0)$ in Figures 4.20 and 4.22. However, the linearization matrix at $(a_2/c_2, a_1/c_1)$ is

$$A(a_2/c_2,\ a_1/c_1) = \begin{bmatrix} 0 & \dfrac{-c_1a_2}{c_2} \\ \dfrac{c_2a_1}{c_1} & 0 \end{bmatrix},$$

which has pure imaginary eigenvalues $\pm i\sqrt{a_1a_2}$. The Hartman-Grobman theorem does not apply at this equilibrium.

To search for a constant of motion, we note that the integral curves have slope

$$\frac{dy}{dx} = \frac{dy/dt}{dx/dt} = \frac{(-a_2 + c_2 x)y}{(a_1 - c_1 y)x}.$$

This equation can be solved by separation of variables to give us

$$(x^{a_2} e^{-c_2 x})(y^{a_1} e^{-c_1 y}) = C.$$

Thus, a constant of motion is given by the function

$$E(x, y) = (x^{a_2} e^{-c_2 x})(y^{a_1} e^{-c_1 y}).$$

You can check that

$$\frac{\partial E}{\partial x} = \left(\frac{a_2}{x} - c_2 \right) E(x, y) \quad \text{and} \quad \frac{\partial E}{\partial y} = \left(\frac{a_1}{y} - c_1 \right) E(x, y)$$

so that

$$\frac{\partial^2 E}{\partial x^2} = \left[\left(\frac{a_2}{x} - c_2 \right)^2 - \frac{a_2}{x^2} \right] E(x, y), \quad \frac{\partial^2 E}{\partial y^2} = \left[\left(\frac{a_1}{y} - c_1 \right)^2 - \frac{a_1}{y^2} \right] E(x, y),$$

and

$$\frac{\partial^2 E}{\partial y \, \partial x} = \left(\frac{a_2}{x} - c_2 \right) \left(\frac{a_1}{y} - c_1 \right) E(x, y).$$

The points along the axes are degenerate critical points of E. The only other critical point of E occurs at $(a_2/c_2, a_1/c_1)$. The discriminant there is

$$\Delta \, (a_2/c_2, a_1/c_1) = \frac{c_1^2 c_2^2}{a_1 a_2} \, [E \, (a_2/c_2, a_1/c_1)]^2 > 0.$$

Thus $(a_2/c_2, a_1/c_1)$ is a stable equilibrium, surrounded nearby by periodic orbits.
A more extensive investigation of the level sets of E would show that *every* integral curve in the first quadrant and off the axes is a closed curve surrounding the equilibrium at $(a_2/c_2, a_1/c_1)$. If both species are present at the outset, then this model predicts periodic fluctuation about an equilibrium value.

Let's summarize our observations about constants of motion.

CONSTANTS OF MOTION

A function $E(x, y)$ is said to be a **constant of motion** for a system of o.d.e.'s

(S)
$$\frac{dx}{dt} = f(x, y)$$
$$\frac{dy}{dt} = g(x, y),$$

provided that along each integral curve of (S)

$$x = x(t), \quad y = y(t)$$

we have

$$\frac{d}{dt} E(x(t), y(t)) = 0.$$

If $E(x, y)$ is a constant of motion for (S), then the integral curves of (S) lie on the level curves of E.

Assume that $E(x, y)$ is a constant of motion with continuous first and second partial derivatives. The **discriminant** of E is the function

$$\Delta(x, y) = \frac{\partial^2 E}{\partial^2 x} \frac{\partial^2 E}{\partial^2 y} - \left(\frac{\partial^2 E}{\partial x \, \partial y} \right)^2.$$

Assume that (x_0, y_0) is a critical point of $E(x, y)$, that the discriminant is not zero at this point, that (x_0, y_0) is an equilibrium of (S), and that there is a neighborhood of (x_0, y_0) containing no other equilibria of (S).

1. If (x_0, y_0) is a strict local extremum of E—that is, if $\Delta(x_0, y_0) > 0$— then (x_0, y_0) is a stable equilibrium of (S). In this case, the phase portrait of (S) near (x_0, y_0) consists of closed curves surrounding this point.

2. If (x_0, y_0) is a saddle point of $E(x, y)$—that is, if $\Delta(x_0, y_0) < 0$—then (x_0, y_0) is an unstable equilibrium, but not a repeller. In this case, a pair of separatrices approaches (x_0, y_0) as $t \to \infty$ and another approaches (x_0, y_0) as $t \to -\infty$. These four curves separate the region of the plane near (x_0, y_0) into four "quadrants." All solutions that start inside these quadrants leave the region as $t \to \infty$ and as $t \to -\infty$.

EXERCISES

In Exercises 1 through 8, you are given a system of o.d.e.'s (S), and a function $E(x, t)$. For each problem:

a. Verify that E is a constant of motion for (S).

b. Find the equilibria of (S).

c. Find the critical points of E, and classify each as a maximum, minimum, or saddle point.

d. Classify each equilibrium of (S) as stable or unstable.

1. $\dfrac{dx}{dt} = -6x - 10y$

   $E(x, y) = 5x^2 + 6xy + 5y^2$

   $\dfrac{dy}{dt} = 10x + 6y$

2. $\dfrac{dx}{dt} = x^2y + y^3$

   $E(x, y) = y^2 - x^2$

   $\dfrac{dy}{dt} = x^3 + xy^2$

3. $\dfrac{dx}{dt} = 2xy - x^2 - x$

   $E(x, y) = x^2y - xy^2 + xy$

   $\dfrac{dy}{dt} = 2xy - y^2 + y$

4. $\dfrac{dx}{dt} = 3y^2 - 1$

   $E(x, y) + x^2 + 2y^3 - 2y$

   $\dfrac{dy}{dt} = -x$

5. $\dfrac{dx}{dt} = 4 - 2y$

   $E(x, y) = x^2 - 2x + y^2 - 4y$

   $\dfrac{dy}{dt} = 2x - 2$

6. $\dfrac{dx}{dt} = 2xy$

   $E(x, y) = xe^{x - y^2}$

   $\dfrac{dy}{dt} = 1 + x$

7. $\dfrac{dx}{dt} = -\sin x \cos y$

   $E(x, y) = \sin x \sin y$

   $\dfrac{dy}{dt} = \cos x \sin y$

8. $\dfrac{dx}{dt} = (x - x^3 - 2)2y$

   $E(x, y) = (x^3 - x)(y^2 + 1) + 2y^2$

   $\dfrac{dy}{dt} = (1 - 3x^2)(y^2 + 1)$

9. a. Show that the function $E(x, y) = ax^2 + 2bxy + cy^2$ is a constant of motion for the system $D\mathbf{x} = A\mathbf{x}$ where

$$A = \begin{bmatrix} b & c \\ -a & -b \end{bmatrix}.$$

 b. If $ac \neq b^2$, show that the only critical point of E is at the origin.
 c. Show that the discriminant of E at the origin is $\Delta = ac - b^2$.
 d. What can you conclude about the stability of the origin as an equilibrium of $D\mathbf{x} = A\mathbf{x}$?
 e. Verify your conclusion by calculating the eigenvalues of A.

10. Suppose $\int f(x)\, dx = F(x) + c$, and consider the system

(S)

$$\frac{dx}{dt} = y$$

$$\frac{dy}{dt} = -f(x)$$

equivalent to the second-order o.d.e. $x'' + f(x) = 0$.
 a. Show that $E(x, y) = F(x) + y^2/2$ is a constant of motion for (S).
 b. Show that if $F(x)$ has a minimum at x_0, then E has a minimum at $(x_0, 0)$.
 c. Show that if $F(x)$ has a maximum at x_0, then $(x_0, 0)$ is a saddle point for E.
 d. Show that every critical point of E has the form $(x_0, 0)$, with $f(x_0) = 0$.

11. Use the results of Exercise 10 to locate and characterize the stability of the equilibria of the systems equivalent to the following second-order o.d.e.'s:
 a. $x'' = x$ b. $x'' = -x$ c. $x'' = x^2 - 1$ d. $x'' = -x^3$
 e. $x'' = x^3 - x$

12. Let $E(x, y)$ be a function with continuous first and second partial derivatives. Show that E is a constant of motion for the system

$$\frac{dx}{dt} = \frac{\partial E}{\partial y}$$

$$\frac{dy}{dt} = -\frac{\partial E}{\partial x}.$$

Systems of this form are called *Hamiltonian*. This and its higher-dimensional analogues are a standard form for systems modeling a conservative physical system.

13. For each of the following functions $E(x, y)$, write down its corresponding Hamiltonian system, identify its critical points, and classify them as stable or unstable.
 a. $E(x, y) = \cos x + y^2/2$ b. $E(x, y) = xy$
 c. $E(x, y) = x^2 + y^2$ d. $E(x, y) = x^4 - 2x^2 + y^2$

4.5   LYAPUNOV FUNCTIONS

Our analysis in Section 4.4 of the undamped pendulum relied on the fact that the energy was constant along solution curves. Most physical phenomena are dissipative: they tend to lose energy to their surroundings through friction, radiation, or other processes. Often, such dissipation has a stabilizing effect on the system. The stability of dissipative systems motivated the Russian mathematician and engineer A. M. Lyapunov (1857–1918) to analyze functions $E(x, y)$ that are nonincreasing along solutions of the system.

Definition: *A function $E(x, y)$ is a* **Lyapunov function** *for the system of o.d.e.'s*

(S)
$$\frac{dx}{dt} = f(x, y)$$

$$\frac{dy}{dt} = g(x, y)$$

if, along each solution of (S)

$$x = x(t), \quad y = y(t)$$

we have

$$\frac{d}{dt} E(x(t), y(t)) \leq 0.$$

Note that this definition allows the possibility that there are values of t for which

$$\frac{d}{dt} E(x(t), y(t)) = 0.$$

In order to avoid messy special cases, we will assume that this can only happen for an *interval* of time values if the solution is an equilibrium. This assumption guarantees that as we move along a non-equilibrium solution, $E(x(t), y(t))$ is strictly decreasing over any interval of time. As in our discussion of constants of motion, we will assume that (x_0, y_0) is an equilibrium of the system of o.d.e.'s (S), that there is a neighborhood of (x_0, y_0) that contains no other equilibria of (S), and that (x_0, y_0) is a nondegenerate critical point of E.

Suppose now that (x_0, y_0) is a strict local minimum of the Lyapunov function E. As before, the level sets $E(x, y) = c$ for values of c slightly above $c_0 = E(x_0, y_0)$ are closed curves surrounding (x_0, y_0). These curves bound disc-like regions that close down on (x_0, y_0) as c gets smaller. But note that now, when a point on a non-equilibrium solution reaches a level curve $E(x, y) = c$, instead of following the level curve, it must cross it and move *inside* the region bounded by this curve, to points where $E(x, y)$ has a lower value. In fact, the value of E must continue to decrease along the integral curve, approaching the minimum value $c_0 = E(x_0, y_0)$ in the limit (see Figure 4.31). This means that

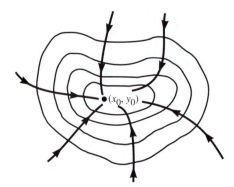

FIGURE 4.31   Level curves and integral curves near a local minimum of a Lyapunov function

every integral curve near (x_0, y_0) must approach this point as $t \to \infty$. Thus the equilibrium is an attractor, and in particular, it is stable.

For a Lyapunov function, unlike a constant of motion, the situation is radically different at a maximum of E. Again, the level sets $E(x, y) = c$ near a strict local maximum point (x_0, y_0) form closed curves surrounding (x_0, y_0), but this time they correspond to values of c slightly below the maximum value $c_0 = E(x_0, y_0)$. This means that as c decreases, the level sets $E(x, y) = c$ move *out* from (x_0, y_0). When a point on a non-equilibrium solution reaches one of these level curves, it must move out of the region bounded by the level curve. Thus, the equilibrium is a repeller and is, of course, unstable. We note that as $t \to -\infty$, a point moving along a non-equilibrium solution will approach (x_0, y_0).

The analysis of the integral curves near a saddle point of E is somewhat more delicate. It turns out that the phase portrait looks much like that near a

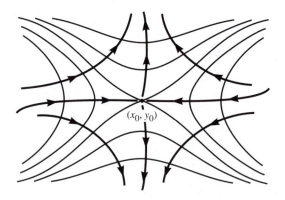

FIGURE 4.32   Level curves and integral curves near a saddle point of a Lyapunov function

saddle point of a constant of motion, but the separatrices and hyperbola-like integral curves cross the (hyperbola-like) level sets of E transversally instead of following them (Figure 4.32).

Thus we have the following

Fact: *Let*

(S)

$$\frac{dx}{dt} = f(x, y)$$

$$\frac{dy}{dt} = g(x, y)$$

be a system of o.d.e.'s and let $E(x, y)$ be a Lyapunov function for (S) that has continuous first and second partials. Assume that the only way in which $dE(x(t), y(t))/dt$ can remain zero along a solution for an interval of time, is for the solution to be an equilibrium. Suppose that (x_0, y_0) is an equilibrium of (S) and that there is a neighborhood of (x_0, y_0) that does not contain any other equilibria of (S). Then

a. *If (x_0, y_0) is a strict local minimum of E, then it is an attractor of (S).*
b. *If (x_0, y_0) is a strict local maximum of E, then it is a repeller of (S).*
c. *If (x_0, y_0) is a saddle point of E, then it is unstable, but not a repeller.*

We illustrate the use of Lyapunov functions in an important example. In this example, the local analysis near the equilibria could also be carried out using the Hartman-Grobman theorem. However, the Lyapunov function allows us to obtain global information.

Example 4.5.1   The Damped Pendulum

The motion of an unforced pendulum subject to viscous damping is modeled by the o.d.e.

$$\frac{d^2\theta}{dt^2} + b\frac{d\theta}{dt} + \frac{g}{L}\sin\theta = 0,$$

where $b > 0$. The substitution $x = \theta$ and $y = d\theta/dt$ leads to the system

(S)

$$\frac{dx}{dt} = y$$

$$\frac{dy}{dt} = -\frac{g}{L}\sin x - by.$$

If we check the total energy

$$E(x, y) = \frac{y^2}{2} - \frac{g}{L} \cos x$$

along a solution of (S), we see that the presence of the friction term turns E into a Lyapunov function:

$$\frac{dE}{dt} = \left(\frac{g}{L} \sin x \right) y + y \left(-\frac{g}{L} \sin x - by \right) = -by^2 \leq 0.$$

As before, the equilibria of (S) occur at the points

$$c_n = \begin{bmatrix} n\pi \\ 0 \end{bmatrix},$$

where n is an integer. The only points where $dE/dt = 0$ are points on the x-axis, $y = 0$. If a *non-equilibrium* solution curve starts at a point on the x-axis, then at that point $dy/dt \neq 0$; the solution immediately leaves the x-axis. Thus, the only way for dE/dt to remain zero along a solution for an interval of time is for the solution to be an equilibrium.

You can check that

$$\frac{\partial^2 E}{\partial x^2} = \Delta(x, y) = \frac{g}{L} \cos x.$$

If n is even, then Δ and $\partial^2 E/\partial x^2$ are both positive at c_n, so E has a strict local minimum at this equilibrium. Thus c_n is an attractor when n is even. If n is odd, then Δ is negative at c_n, so E has a saddle point at this equilibrium. Thus c_n is unstable, but not a repeller, when n is odd.

The level sets $E(x, y) = c$ are the same as those we had in Figure 4.26: the level sets with $c > g/L$ are the undulating curves toward the outside of the figure; the level set with $c = g/L$ consists of two functions that cross at c_n, n odd; and for $-g/L < c < g/L$, we have closed curves nesting down to the minima at c_n. However, this time, the energy decreases along the non-equilibrium integral curves, so the integral curves cross the level curves of E. Inside a "blip" formed by the level curves with $c = g/L$, all solutions eventually approach the attractor c_n (n even) at the center of the blip. This corresponds to the pendulum going through successively narrower swings, approaching rest in the straight-down position as $t \to \infty$. Non-equilibrium solutions that start on the level set $E(x, y) = g/L$ have decreasing energy; these solutions enter the blips and are attracted by one of the attractors at c_n, n even. A non-equilibrium solution that starts on a level set $E(x, y) = c$ with $c > g/L$ must have E decreasing until it either enters a blip and is attracted by an attractor, or else approaches one of the saddles at c_n, n odd.

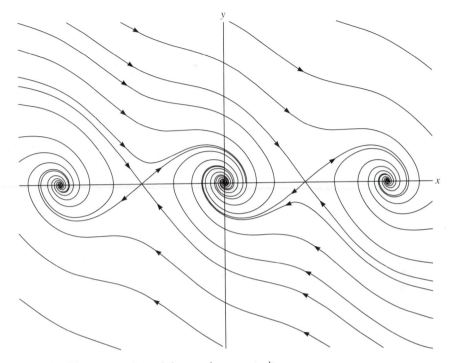

FIGURE 4.33   Damped pendulum—phase portrait

We know from our general discussion that there are two incoming separatrices at the saddle points, and two outgoing ones. In the case of the undamped pendulum, the separatrices lie along the level set $E(x, y) = g/L$. In the case of the damped pendulum, the energy decreases along the outgoing separatrices, so these curves immediately fall into a blip and approach the neighboring attractor at $c_{n \pm 1}$. If we follow the incoming separatrices *backward*, the energy increases; these separatrices cross the level sets $E(x, y) = c$ with $c > g/L$ and escape to ∞. The incoming separatrices represent the situation in which the pendulum approaches the upright equilibrium as $t \rightarrow \infty$; if we start with slightly more energy, the pendulum continues to move beyond the upright position and approaches the next downward equilibrium, whereas a pendulum that starts with slightly less energy fails to make it to the upright equilibrium and falls back to the previous downward one.

We have sketched the phase portrait in Figure 4.33 (using a computer). You should verify that our predictions based on the analysis of the Lyapunov function E are borne out by this picture.

The primary difficulty with Lyapunov functions is finding them in the first place. If one is simply given a system of o.d.e.'s out of the blue, then it is virtually impossible (barring luck) to decide whether the system has a Lyapunov function, and if so, to find one. Nevertheless, certain problems have natural Lyapunov functions, and these functions can be quite useful in analyzing the phase portrait.

We summarize the method outlined in this section.

LYAPUNOV FUNCTIONS

The function $E(x, y)$ is said to be a **Lyapunov function** for the system of o.d.e.'s

$$\frac{dx}{dt} = f(x, y)$$

(S)

$$\frac{dy}{dt} = g(x, y),$$

provided that along each integral curve of (S)

$$x = x(t), \quad y = y(t)$$

we have

$$\frac{d}{dt} E(x(t), y(t)) \leq 0.$$

Assume that $E(x, y)$ is a Lyapunov function for (S), and that the only way in which dE/dt can remain zero along a solution for an interval of time, is for the solution to be an equilibrium. Assume that (x_0, y_0) is a critical point of $E(x, y)$, that the discriminant Δ of E is not zero at this point, that (x_0, y_0) is an equilibrium of (S), and that there is a neighborhood of (x_0, y_0) containing no other equilibria of (S).

a.   If (x_0, y_0) is a relative minimum for E—that is, if Δ and $\partial^2 E/\partial^2 x$ are both positive at (x_0, y_0)—then (x_0, y_0) is an attractor for (S).

b.   If (x_0, y_0) is a relative maximum for E—that is, if $\Delta > 0$ and $\partial^2 E/\partial^2 x < 0$ at (x_0, y_0)—then (x_0, y_0) is a repeller for (S).

c.   If (x_0, y_0) is a saddle point for E—that is, if $\Delta(x_0, y_0) < 0$—then (x_0, y_0) is unstable, but is not a repeller. In this case, a pair of separatrices approaches (x_0, y_0) as $t \to \infty$ and another approaches (x_0, y_0) as $t \to -\infty$. These four curves separate the region of the plane near (x_0, y_0) into "quadrants." All solutions that start inside these quadrants leave the region as $t \to \infty$ and as $t \to -\infty$.

EXERCISES

In Exercises 1 through 8, you are given a system (S) of o.d.e.'s and a function $E(x, y)$. (Note that the functions are the same as those in Exercises 1 through 8 of Section 4.4.) For each problem:

a.   Verify that E is a Lyapunov function for (S).

b.   Find the equilibrium points of (S), and classify each as an attractor, repeller, or neither.

1. $\dfrac{dx}{dt} = -10x - 6y$

                $E(x, y) = 5x^2 + 6xy + 5y^2$

    $\dfrac{dy}{dt} = -6x - 10y$

2. $\dfrac{dx}{dt} = x^3 + xy^2$

                $E(x, y) = y^2 - x^2$

    $\dfrac{dy}{dt} = -x^2y - y^3$

3. $\dfrac{dx}{dt} = y^2 - 2xy - y$

                $E(x, y) = x^2y - xy^2 + xy$

    $\dfrac{dy}{dt} = 2xy - x^2 - x$

4. $\dfrac{dx}{dt} = 1 - 2x - 3y^2$

                $E(x, y) = x^2 + 2y^3 - 2y$

    $\dfrac{dy}{dt} = 2 + x - 6y^2$

5. $\dfrac{dx}{dt} = 2y - x - 3$

                $E(x, y) = x^2 - 2x + y^2 - 4y$

    $\dfrac{dy}{dt} = 4 - 2x - y$

6. $\dfrac{dx}{dt} = -x^3 - x^2 - 2xy$

                $E(x, y) = xe^{x-y^2}$

    $\dfrac{dy}{dt} = 2x^3y - x - 1$

7. $\dfrac{dx}{dt} = \sin x \cos y - \cos x \sin y$

                $E(x, y) = \sin x \sin y$

    $\dfrac{dy}{dt} = -\sin x \cos y - \cos x \sin y$

8. $\dfrac{dx}{dt} = 1 - 3x^2$

                $E(x, y) = (x^3 - x)(y^2 + 1) + 2y^2$

    $\dfrac{dy}{dt} = \dfrac{(x - 2 - x^3)y}{y^2 + 1}$

9. a. Check that $\mathbf{x} = \mathbf{0}$ is an equilibrium of

(S)
$$\frac{dx}{dt} = ax^3 + y$$

$$\frac{dy}{dt} = -x + ay^3$$

and that the linearization matrix of (S) near $\mathbf{0}$ has pure imaginary eigenvalues, so that the Hartman-Grobman theorem does not apply.

b. Show that if $a > 0$, then the function $E(x, y) = -(x^2 + y^2)$ is a Lyapunov function for (S).

c. Show that if $a < 0$, then the function $E(x, y) = x^2 + y^2$ is a Lyapunov function for (S).

d. In each of these cases, classify the equilibrium $\mathbf{x} = \mathbf{0}$ as an attractor, a repeller, or neither.

Compare Exercise 24, Section 4.3.

10. Suppose $\int f(x) \, dx = F(x) + c$, and consider the system

(S)
$$\frac{dx}{dt} = y$$

$$\frac{dy}{dt} = -f(x) - g(y)$$

equivalent to the second-order o.d.e. $x'' + g(x') + f(x) = 0$. Show that if $g(y)$ and y never have the same sign, then $E(x, y) = F(x) + y^2/2$ is a Lyapunov function for (S).

11. Recall that the *gradient* of a function $F(x, y)$ is the vector

$$\nabla F(x, y) = \begin{bmatrix} \dfrac{\partial F}{\partial x} \\[2mm] \dfrac{\partial F}{\partial y} \end{bmatrix}.$$

A *gradient system* is a system of o.d.e.'s which can be written in the form

(G) $D\mathbf{x} = \nabla F(x, y)$

for some function $F(x, y)$.

a. Show that $E(x, y) = -F(x, y)$ is a Lyapunov function for (G).

b. Show that the system

$$\frac{dx}{dt} = f(x, y)$$

$$\frac{dy}{dt} = g(x, y)$$

is a gradient system if and only if $\partial f/\partial y = \partial g/\partial x$.

12. Use the results of Exercise 11 to decide which of the following systems are gradient systems. For each gradient system, find a Lyapunov function; locate the equilibria; and classify them as attractors, repellers, or neither.

a. $\dfrac{dx}{dt} = 10x - 6y$

   $\dfrac{dy}{dt} = -6x + 10y$

b. $\dfrac{dx}{dt} = 3(x + y)^2 + 2x + 1 - 2y$

   $\dfrac{dy}{dt} = 3(x + y)^2 + 2y + 1 - 2x$

c. $\dfrac{dx}{dt} = 4(y^2 + 1)(x^3 - x)$

   $\dfrac{dy}{dt} = 2y(x^4 - 2x^2 + 2)$

*13. Given a function $E(x, y)$, and a system of o.d.e.'s,

$$\frac{dx}{dt} = f(x, y)$$

$$\frac{dy}{dt} = g(x, y),$$

show that E is a Lyapunov function for (S) if and only if there exist functions $\alpha(x, y)$ and $\beta(x, y)$, with $\alpha(x, y) \geq 0$, such that

$$f(x, y) = -\alpha(x, y) \frac{\partial E}{\partial x} - \beta(x, y) \frac{\partial E}{\partial y}$$

$$g(x, y) = \beta(x, y) \frac{\partial E}{\partial x} - \alpha(x, y) \frac{\partial E}{\partial y}.$$

4.6 LIMIT CYCLES AND CHAOS

The linearization of nonlinear systems near equilibria and the analysis of constants of motion or Lyapunov functions are major theoretical tools for studying the long-term behavior of solutions to systems of o.d.e.'s. In this section, we consider two classic examples that show that the study of equilibria alone is not enough to describe the long-term behavior of solutions to nonlinear systems.

Example 4.6.1 The Van der Pol Equation

The second-order o.d.e.

$$\frac{d^2x}{dx^2} + \epsilon(x^2 - 1)\frac{dx}{dt} + x = 0$$

was derived in 1920 by the Dutch electrical engineer B. Van der Pol as a model for the changing current in a triode vacuum tube. We will take $\epsilon = 1$ and consider the equivalent system (obtained by introducing the new variable $y = dx/dt$)

(S)
$$\frac{dx}{dt} = y$$
$$\frac{dy}{dt} = (1 - x^2)y - x.$$

We first note that the only equilibrium is at the origin. We obtain the linearization of (S) by taking partial derivatives:

$$f_x = 0, \qquad\qquad f_y = 1$$
$$g_x = -1 - 2xy, \quad g_y = 1 - x^2.$$

Evaluation at the origin yields the linearization matrix

$$A = \begin{bmatrix} 0 & 1 \\ -1 & 1 \end{bmatrix}.$$

The eigenvalues of A are $(1 \pm i\sqrt{3})/2$. Since the real part is positive, the origin is a repeller for $Dy = Ay$. Indeed, solutions of $Dy = Ay$ spiral out clockwise (why?) from the origin (see Figure 4.34). By the Hartman-Grobman theorem the phase portrait of (S) is similar to the one for $Dy = Ay$, and $x = 0$ is a repeller for (S).

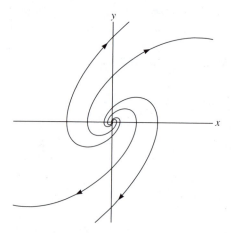

FIGURE 4.34   Van der Pol equation with $\epsilon = 1$, linearization near (0,0)

It is tempting to predict that solutions grow without bound as they spiral out from the origin. On the other hand, experimentally, the current in an unforced triode is observed to settle down to periodic behavior. To account for this in the model, we look beyond the local linearization, taking account of global considerations. Note that if we set $\epsilon = 0$ instead of $\epsilon = 1$, then the effect on the system (S) is to eliminate the term $(1 - x^2)y$ from the second equation, thereby obtaining a linear system describing simple (circular) rotation. We can regard the extra term in (S) as imposing upon the rotational motion a vertical component, proportional to the height y and to $1 - x^2$. Note that for $|x| < 1$, the $1 - x^2$ factor is positive, but as $|x|$ grows beyond 1, it becomes increasingly negative. Now let's follow the variation of this vertical component around a circle of very large radius (say, $x^2 + y^2 = 100$). We see that in a relatively narrow band about the y-axis, $-1 \le x \le 1$, the component is outward (up if $y > 0$, down if $y < 0$). On the other hand, outside this band, the component is inward; what's more, since the radius of the circle is large, the magnitude of this inward component is fairly large except where y is very small. Thus over most of the circle, there is a strong inward component superimposed on the rotation (Figure 4.35). Although this intuitive analysis has gaps, it certainly suggests that we should expect solutions with $\mathbf{x}(0)$ far from the origin to wind, on average, inward. (This can be made more precise and rigorous by various technical devices.) On the other hand, our local analysis tells us that solutions starting near the origin spiral outward. It follows that we should expect, at some intermediate distance from the origin, that some solution will adopt the intermediate behavior, winding about the origin along a closed loop. This expectation is fulfilled; the resulting *limit cycle* is shown in Figure 4.36, obtained via graphical means.

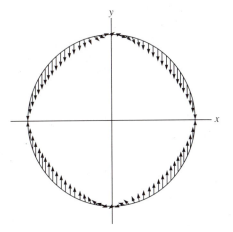

FIGURE 4.35

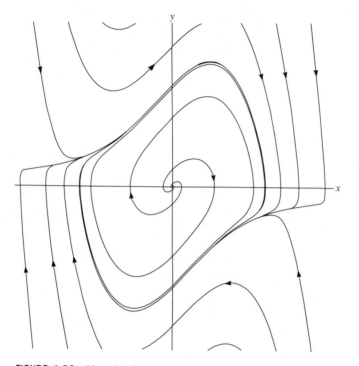

FIGURE 4.36   Van der Pol equation with $\epsilon = 1$, phase portrait

The Van der Pol equation illustrates the importance of determining whether or not a system of o.d.e.'s has any closed integral curves representing periodic solutions. In checking for such curves, it is helpful to know that they must surround equilibria.

Theorem: *If the system of o.d.e.'s*

(S)

$$\frac{dx}{dt} = f(x, y)$$

$$\frac{dy}{dt} = g(x, y)$$

has a closed integral curve, then the curve surrounds an equilibrium of (S). If there is exactly one equilibrium surrounded by the curve, then the linearization at this equilibrium cannot have eigenvalues of opposite sign.

In the following example, we use this theorem to show that the system in Example 4.3.1 cannot have any closed integral curves (without using the computer-drawn phase portrait).

Example 4.6.2

Show that the system

(S)
$$\frac{dx}{dt} = 2x - 2x^2 - 5xy$$

$$\frac{dy}{dt} = y - y^2 - 2xy$$

has no closed integral curves.

The origin, and the positive and negative halves of the axes are integral curves for (S). If (S) has a closed integral curve C, then it cannot intersect the axes. Thus C lies entirely in the interior of one of the four quadrants. The theorem tells us that C must surround an equilibrium. Since (3/8, 1/4) is the only equilibrium that does not lie on an axis, this point is the only equilibrium surrounded by C. We saw in Example 4.3.1 that the linearization at (3/8, 1/4) has eigenvalues with opposite signs. Thus, by the theorem, there cannot be any closed integral curves.

The following theorem will enable us to show that the system in Example 4.3.3 has no closed integral curves. We leave the proof of this theorem to you (Exercise 13). Recall that a region of the plane is simply connected if it contains the interior of each simple closed curve that lies in the region.

Theorem: *Let $f(x, y)$ and $g(x, y)$ be functions with continuous first partial derivatives. Suppose that $\partial f/\partial x + \partial g/\partial y$ is always positive or always negative in a simply connected domain of the plane. Then the system of o.d.e.'s*

(S)
$$\frac{dx}{dt} = f(x, y)$$

$$\frac{dy}{dt} = g(x, y)$$

has no closed integral curves in this domain.

▮   **Example 4.6.3**

Show that the system

$$\frac{dx}{dt} = x - x^2 - xy$$

(S)

$$\frac{dy}{dt} = -y + 2xy$$

has no closed integral curves.

As in Example 4.6.2, a closed integral curve C would have to be contained in the interior of one of the quadrants and would have to surround an equilibrium. The only equilibrium that does not lie on an axis is (1/2, 1/2), so C has to lie in the first quadrant. In the interior of the first quadrant,

$$\frac{\partial}{\partial x}(x - x^2 - xy) + \frac{\partial}{\partial y}(-y + 2xy) = -y < 0.$$

Thus (S) cannot have any closed integral curves.

_____

The preceding two theorems are useful in ruling out the existence of closed integral curves. The following theorem gives us a way of proving the existence of such a curve.

Poincaré-Bendixson Theorem: *Suppose that $f(x, y)$ and $g(x, y)$ have continuous first partial derivatives. Let R be a bounded region of the plane that contains its boundary and does not contain any equilibria of the system of o.d.e.'s*

$$\frac{dx}{dt} = f(x, y)$$

(S)

$$\frac{dy}{dt} = g(x, y).$$

If (S) has a solution that lies in R for all $t \geq t_0$, then R contains a closed integral curve for (S).

Note that if R is a region satisfying the conditions of this theorem, then R contains a closed integral curve, which in turn surrounds an equilibrium. Since R contains no equilibria, it must have a hole.

Example 4.6.4

Show that the system of o.d.e.'s

(S)
$$\frac{dx}{dt} = -y + x - x^3 - 3xy^2$$

$$\frac{dy}{dt} = x + y - x^2y - 3y^3$$

has a closed integral curve.

As a point moves along an integral curve of (S), its distance r from the origin satisfies

$$r^2 = x^2 + y^2$$

so

$$\frac{dr}{dt} = \frac{1}{r}\left(x\frac{dx}{dt} + y\frac{dy}{dt}\right)$$

$$= \frac{1}{r}(x^2 + y^2)(1 - x^2 - 3y^2)$$

$$= r(1 - x^2 - 3y^2) = r(1 - r^2 - 2y^2).$$

On the circle about (0, 0) of radius 1/2, $y^2 < 1/4$, so

$$\frac{dr}{dt} = \frac{1}{2}\left(1 - \frac{1}{4} - 2y^2\right) > 0.$$

A solution that reaches this circle must move outward, and cannot cross the circle again. On the other hand, on the circle of radius 3/2 about (0, 0),

$$\frac{dr}{dt} = \frac{3}{2}\left(1 - \frac{9}{4} - 2y^2\right) < 0.$$

A solution that reaches this circle must move inward, and cannot cross the circle again. It follows that *any* integral curve that starts from a point between the two circles (see Figure 4.37) will remain in this region. By the Poincaré-Bendixson theorem, (S) has a closed integral curve that lies in this region.

Note that solutions that start near the circle $r = 1/2$ will move outward toward the closed integral curve C. Solutions that start near $r = 3/2$ will move inward toward C.

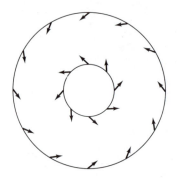

FIGURE 4.37   Finding a closed integral curve

The analysis we have carried out so far in this section is strictly two-dimensional. In particular, the Poincaré-Bendixson theorem does not hold for systems of order higher than 2; the limiting behavior of solutions to a nonlinear system of order 3 or more can be far more complex than the equilibria and limit cycles we have seen for second-order systems. This is illustrated by the following classic example.

Example 4.6.5   The Lorenz Equations

In 1963, E. Lorenz, a mathematician and meteorologist at M.I.T., derived a simple model for certain modes of oscillation in the atmosphere. For certain specific values of the parameters in his model, we obtain the system

(S)

$$\frac{dx}{dt} = -10x + 10y$$

$$\frac{dy}{dt} = 25x - y - xz$$

$$\frac{dz}{dt} = -\frac{8}{3}z + xy.$$

This system has three equilibria:

$$\mathbf{c}_0 = \begin{bmatrix} 0 \\ 0 \\ 0 \end{bmatrix}, \quad \mathbf{c}_1 = \begin{bmatrix} 8 \\ 8 \\ 24 \end{bmatrix}, \quad \text{and} \quad \mathbf{c}_2 = \begin{bmatrix} -8 \\ -8 \\ 24 \end{bmatrix}.$$

Let's investigate the stability of these equilibria.

The linearization matrix for (S) near $c_0 = 0$ is

$$A_0 = \begin{bmatrix} -10 & 10 & 0 \\ 25 & -1 & 0 \\ 0 & 0 & -\frac{8}{3} \end{bmatrix}.$$

This matrix has three real eigenvalues:

$$\lambda_1 = \frac{-11 - \sqrt{1081}}{2}, \quad \lambda_2 = -\frac{8}{3}, \quad \text{and} \quad \lambda_3 = \frac{-11 + \sqrt{1081}}{2}.$$

Since there are two negative eigenvalues, and one positive one, c_0 is an unstable equilibrium but is not a repeller. We can use the Hartman-Grobman theorem (for systems of any order) to tell us more about the behavior of solutions near the origin. The phase diagram of the linearized system $Dy = Ay$ is similar to the one in Figure 4.18 (Section 4.2): Linear combinations of the solutions corresponding to λ_1 and λ_2 lie in a plane and tend to the origin, but all other solutions go to infinity, approaching the line determined by an eigenvector corresponding to λ_3. Correspondingly, there will be a two-dimensional surface on which solutions to (S) will approach the origin, and all other solutions will move away from the origin.

The linearization matrix for (S) at c_1 is

$$A_1 = \begin{bmatrix} -10 & 10 & 0 \\ 1 & -1 & -8 \\ 8 & 8 & -\frac{8}{3} \end{bmatrix}.$$

The characteristic equation of this matrix is

$$3\lambda^3 + 41\lambda^2 + 280\lambda + 3840 = 0.$$

This equation has a real root

$$\lambda_1 \approx -13.6825$$

and complex roots $\alpha \pm \beta i$ where

$$\alpha \approx 7.98035 \times 10^{-3} \quad \text{and} \quad \beta \approx 8.78814.$$

Since λ_1 is negative, and the real part of the complex roots is positive, c_1 is unstable, but is not a repeller. Although we won't carry out the analysis in detail, we note that solutions of $Dy = Ay$ that lie on the line determined by λ_1

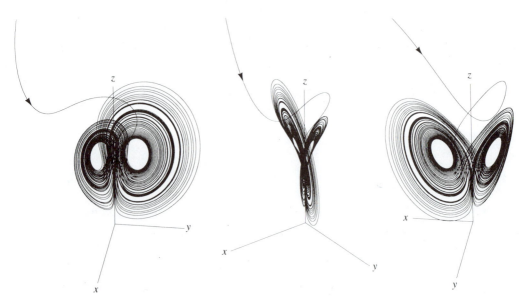

FIGURE 4.38   A single integral curve for the Lorenz equations from three perspectives

approach the origin, but all other solutions escape. The Hartman-Grobman theorem tells us to expect similar behavior for solutions of (S) near c_1 (although the curve along which solutions approach c_1 may no longer be a line). Thus this equilibrium is "more unstable" than the origin.

Finally, we note that if in (S) we simultaneously replace x and y by $-x$ and $-y$ respectively (leaving z alone), then the system is unchanged. This means that the phase portrait of (S) is symmetric about the z-axis. Since this symmetry interchanges c_2 and c_1, the phase portrait near c_2 is just like that near c_1.

Although all three equilibria are unstable, we have seen that none of them is a repeller. Where do most solutions go? Lorenz showed that (with a few exceptions beyond those we have found) most solutions tend toward a "strange attractor" that is neither an equilibrium nor a limit cycle, but rather a complicated set of solutions inside which the long-term behavior is extremely sensitive to initial conditions (see Figure 4.38). Lorenz pointed out that this chaotic behavior in atmospheric models suggests the theoretical impossibility of long-range weather prediction.

Let's summmarize our observations.

LIMIT CYCLES AND CHAOS

Given the system of o.d.e.'s

(S)

$$\frac{dx}{dt} = f(x, y)$$

$$\frac{dy}{dt} = g(x, y)$$

where $f(x, y)$ and $g(x, y)$ have continuous first partial derivatives:

1. Any closed integral curve C of (S) surrounds an equilibrium. If C surrounds exactly one equilibrium, then the linearization at this equilibrium cannot have eigenvalues of opposite sign.

2. If $\partial f/\partial x + \partial g/\partial y$ is always positive or always negative in a simply connected domain, then (S) has no closed integral curves in this domain.

3. **Poincaré-Bendixson Theorem:** Let R be a bounded region of the plane that contains its boundary and does not contain any equilibrium of (S). If (S) has a solution that lies in R for all $t \geq t_0$, then R contains a closed integral curve for (S).

Warning:

These results do *not* hold in general (even in a modified form) for nonlinear systems of degree 3 or more.

EXERCISES

In Exercises 1 through 8, show that the given system of o.d.e.'s does not have any closed integral curves.

1. $\dfrac{dx}{dt} = 3x - x^2 - xy$

    Compare Exercise 2, Section 4.1 and Exercise 9, Section 4.3.

    $\dfrac{dy}{dt} = 4y - 2xy - y^2$

2. $\dfrac{dx}{dt} = 2x - 2x^2 - xy$

    Compare Exercise 3, Section 4.1 and Exercise 10, Section 4.3.

    $\dfrac{dy}{dt} = 3y - xy - y^2$

3. $\dfrac{dx}{dt} = 2x - 2x^2 - xy$

   Compare Exercise 5, Section 4.1 and Exercise 12, Section 4.3.

   $\dfrac{dy}{dt} = 4y - 2xy - y^2$

4. $\dfrac{dx}{dt} = 4x - 2x^2 - xy$

   Compare Exercise 6, Section 4.1 and Exercise 13, Section 4.3.

   $\dfrac{dy}{dt} = -2y + xy$

5. $\dfrac{dx}{dt} = 2x - 4x^2 - xy$

   Compare Exercise 7, Section 4.1 and Exercise 14, Section 4.3.

   $\dfrac{dy}{dt} = -3y + 7xy$

6. $\dfrac{dx}{dt} = 5x + x^2 - xy$

   Compare Exercise 10, Section 4.1 and Exercise 17, Section 4.3.

   $\dfrac{dy}{dt} = y^2 - x^2y + y$

7. $\dfrac{dx}{dt} = y$

   $\dfrac{dy}{dt} = e^x y - x$

8. $\dfrac{dx}{dt} = y$

   $\dfrac{dy}{dt} = 2y - \sin(\pi x)y + x^2 - 1$

In Exercises 9 through 12, use the Poincaré-Bendixson theorem (applied to the region between two appropriately chosen circles centered at the origin) to show that the given system of o.d.e.'s has a closed integral curve.

9. $\dfrac{dx}{dt} = y$

   $\dfrac{dy}{dt} = -x + y(4 - 2x^2 - 3y^2)$

10. $\dfrac{dx}{dt} = y + x(4 - 2x^2 - y^2)$

    $\dfrac{dy}{dt} = -x + y(4 - 2x^2 - y^2)$

11. $\dfrac{dx}{dt} = y$

    $\dfrac{dy}{dt} = -x + (3 - e^{x^2 + y^2})$

12. $\dfrac{dx}{dt} = -y + x(3 - x^2 - y^2 - x\cos x)$

    $\dfrac{dy}{dt} = -x + y(3 - x^2 - y^2 - x\cos x)$

13. Suppose that $f(x, y)$ and $g(x, y)$ have continuous first partial derivatives in a simply connected domain D in the plane.

    a. Assume that the system of o.d.e.'s

    $$\text{(S)} \qquad \begin{aligned} \frac{dx}{dt} &= f(x, y) \\[2mm] \frac{dy}{dt} &= g(x, y) \end{aligned}$$

    has a closed integral curve C with equations $x = x(t), y = y(t)$. Show that

    $$\oint_C [f(x, y)\, dy - g(x, y)\, dx] = 0.$$

b. Green's theorem tells us that if C is a simple closed curve in D, and if R is the region enclosed by C, then

$$\oint_C [f(x, y) \, dy - g(x, y) \, dx] = \iint_R \left[\frac{\partial f}{\partial x} + \frac{\partial g}{\partial y} \right] dx \, dy.$$

Show that if $\partial f/\partial x + \partial g/\partial y$ is always positive or always negative in D, then (S) cannot have any closed integral curves.

REVIEW PROBLEMS

In Problems 1 through 8, sketch the phase portrait of $D\mathbf{x} = A\mathbf{x}$ and determine the stability of the equilibrium $\mathbf{x} = \mathbf{0}$. Is this equilibrium an attractor, a repeller, or neither?

1. $A = \begin{bmatrix} 7 & -2 \\ 10 & -2 \end{bmatrix}$.

2. $A = \begin{bmatrix} -4 & 3 \\ -2 & 1 \end{bmatrix}$.

3. $A = \begin{bmatrix} -3 & 2 \\ -2 & 1 \end{bmatrix}$.

4. $A = \begin{bmatrix} 1 & 1 \\ 6 & 0 \end{bmatrix}$.

5. $A = \begin{bmatrix} 1 & -5 \\ 1 & -1 \end{bmatrix}$.

6. $A = \begin{bmatrix} 1 & -5 \\ 1 & 1 \end{bmatrix}$.

7. $A = \begin{bmatrix} -1 & -5 \\ 1 & -1 \end{bmatrix}$.

8. $A = \begin{bmatrix} 5 & -2 \\ 6 & -2 \end{bmatrix}$.

In Problems 9 through 16, you are given a system of o.d.e.'s. Find the equilibria and determine their stability. State whether each of the equilibria is an attractor, a repeller, or neither. Note that the systems in Problems 9 through 11 model damped pendula with varying choices of g/L and b. The systems in Problems 12 and 13 were obtained from the Van der Pol equation by taking ϵ to be 2 and 3, respectively.

9. $\dfrac{dx}{dt} = y$

$\dfrac{dy}{dt} = -4 \sin x - 5y$

10. $\dfrac{dx}{dt} = y$

$\dfrac{dy}{dt} = -4 \sin x - 4y$

11. $\dfrac{dx}{dt} = y$

$\dfrac{dy}{dt} = -5 \sin x - 4y$

12. $\dfrac{dx}{dt} = y$

$\dfrac{dy}{dt} = 2(1 - x^2)y - x$

13. $\dfrac{dx}{dt} = y$

$\dfrac{dy}{dt} = 3(1 - x^2)y - x$

14. $\dfrac{dx}{dt} = -2x + xy$

$\dfrac{dy}{dt} = x - y$

15. $\dfrac{dx}{dt} = x^2 + y$

$\dfrac{dy}{dt} = 2x + y$

16. $\dfrac{dx}{dt} = 2x - x^2 - y$

$\dfrac{dy}{dt} = 2x + x^2$

17. The equation modeling an unforced spring with constant $k > 0$ and damping with constant $b > 0$

$$(m D^2 + b D + k)x = 0$$

is equivalent to the system

$$\frac{dx}{dt} = y$$

$$\frac{dy}{dt} = -\frac{k}{m}x - \frac{b}{m}y.$$

a. Show that $(0, 0)$ is the only equilibrium of this system.
b. Show that this equilibrium is an attractor.

18. The *LRC* circuit in Example 3.1.1 is modeled by the system

$$\frac{dQ}{dt} = I$$

$$\frac{dI}{dt} = \frac{-1}{LC}Q - \frac{R}{L}I + \frac{V(t)}{L},$$

where L, C, and R are positive constants. Assume that the external voltage is a constant

$$V(t) = V.$$

a. Show that this system has only one equilibrium.
b. Show that this equilibrium is an attractor.

19. The equilibria of the system modeling the damped pendulum

(S)
$$\frac{dx}{dt} = y$$

$$\frac{dy}{dt} = -\frac{g}{L}\sin x - by.$$

occur at the points where $y = 0$ and $x = n\pi$ with n an integer. Use the Hartman-Grobman theorem to show the following:
a. If n is even, then the equilibrium at $y = 0$, $x = n\pi$ is an attractor.
b. If n is odd, then the equilibrium at $y = 0$, $x = n\pi$ is unstable, but is not a repeller.

20. Let ϵ be a constant.
a. Show that the only equilibrium of the system equivalent to the Van der Pol equation

$$\frac{dx}{dt} = y$$

$$\frac{dy}{dt} = \epsilon(1 - x^2)y - x$$

is the point $x = 0$, $y = 0$.
b. Show that if $\epsilon > 0$, then this equilibrium is a repeller.
c. Show that if $\epsilon < 0$, then this equilibrium is an attractor.

In Problems 21 through 23, you are given a system of o.d.e.'s modeling two populations. Since we are dealing with populations, you need only consider solutions for which $x \geq 0$ and $y \geq 0$.
a. Use the signs of dx/dt and dy/dt to investigate the direction in which solutions move as t increases.
b. Determine the stability of all equilibria. State whether each of the equilibria is an attractor, a repeller, or neither.
c. Show that the system does not have any closed integral curves.
d. Discuss the probable long-term behavior of solutions.

21. $\dfrac{dx}{dt} = 3x - x^2 - 3xy$

    $\dfrac{dy}{dt} = -2y + 2xy + y^2$

23. $\dfrac{dx}{dt} = 4x - 4x^2 - xy$

    $\dfrac{dy}{dt} = 8y - 4xy - y^2$

22. $\dfrac{dx}{dt} = 2x - 2x^2 - xy$

    $\dfrac{dy}{dt} = 3y - xy - 3y^2$

In Problems 24 and 25, you are given a system of o.d.e.'s (S), and a function $E(x, t)$. For each problem:
a. Verify that E is a constant of motion for (S).
b. Find the equilibria of (S).
c. Find the critical points of E, and classify each as a maximum, minimum, or saddle point.
d. Classify each equilibrium of (S) as stable or unstable.

24. $\dfrac{dx}{dt} = x^2 - 3x$

    $\dfrac{dy}{dt} = 3y - 2xy$

    $E(x, y) = 3xy - x^2y$

25. $\dfrac{dx}{dt} = 2y - x$

    $\dfrac{dy}{dt} = 1 + y$

    $E(x, y) = x - y^2 + xy$

In Problems 26 and 27, you are given a system of o.d.e.'s (S), and a function $E(x, t)$. (Note that the functions are the same as in Problems 24 and 25.) For each problem:
a. Verify that E is a Lyapunov function for (S).
b. Find the equilibria of (S), and classify each as an attractor, repeller, or neither.

26. $\dfrac{dx}{dt} = (x^2 - 3x) + (2xy - 3y)$

    $\dfrac{dy}{dt} = (x^2 - 3x) - (2xy - 3y)$

    $E(x, y) = 3xy - x^2y$

27. $\dfrac{dx}{dt} = x - 3y - 1$

    $\dfrac{dy}{dt} = y - x - 1$

    $E(x, y) = x - y^2 + xy$

Show that the systems in Problems 28 and 29 do not have any closed integral curves.

28. $\dfrac{dx}{dt} = 4x^3 - 4y^2$

$\dfrac{dy}{dt} = 8y - 4x^5$

29. $\dfrac{dx}{dt} = 4x^2 - 4y^2$

$\dfrac{dy}{dt} = -8xy + x^2y + y^3 + y$

Use the Poincaré-Bendixson theorem to show that the systems in Problems 30 and 31 have closed integral curves.

30. $\dfrac{dx}{dt} = y$

$\dfrac{dy}{dt} = -x + y(4 - 2x^2 - 3y^2)$

31. $\dfrac{dx}{dt} = -y + x(4 - x^2 - y^2 - e^{-x})$

$\dfrac{dy}{dt} = x + y(4 - x^2 - y^2 - e^{-x})$

C H A P T E R

5

THE LAPLACE TRANSFORM

5.1 OLD MODELS FROM A NEW VIEWPOINT

Most of our methods so far have aimed at finding *all* solutions of an o.d.e. or system. To find the specific solution satisfying given conditions, we have had to find the general solution first and then solve for the coefficients to fit our initial data. Although the general solution gives us a better picture of the possible patterns of behavior for a model, concrete problems often require only a specific solution. The technique we consider in this chapter solves initial-value problems directly, without requiring the general solution. An added bonus from this method is its effectiveness in handling forcing terms defined ''in pieces.''

In this section we reexamine several models from earlier chapters to see how initial-value problems and forcing terms defined in pieces arise. Note that every problem we pose can be solved by methods from Chapters 1 to 3. However, by the end of the chapter it should be clear that the Laplace transform method often leads to specific solutions much more simply than our old methods.

Example 5.1.1 Controlled Immigration

In Section 1.1 we considered several specific models for populations with given growth and immigration rates. In general, a natural per capita growth rate $g(t)$ together with an immigration rate $E(t)$ is modeled by the first-order o.d.e.

(N) $$Dx = g(t)x + E(t).$$

Suppose the population of Mania has a natural per capita growth rate of 5% per year. From January 1965 to January 1975, government policy allowed immigration into Mania at the rate of $(1/1000)e^{t/20}$ million per year (t measured in years since 1965). Then an abrupt change in policy led to the sealing of all borders and an end to immigration. Since January 1965, when the population was 5 million, the Maniac government has kept the census figures secret. We would like to estimate the population of Mania in January 1985.

To use our old methods, we break the problem into two parts. First, we consider equation (N) with $g(t) = 1/20$ and $E(t) = (1/1000)e^{t/20}$:

(N₁) $$\left(D - \frac{1}{20}\right) x = \frac{1}{1000} e^{t/20}.$$

We find the general solution of (N₁) and then determine the specific solution satisfying the initial condition $x(0) = 5$. You can check that the solution of this initial-value problem is

$$x(t) = 5e^{t/20} + \frac{t}{1000} e^{t/20}.$$

This describes the population from January 1965 to January 1975. In particular, the population in January 1975 was

$$x(10) = 5e^{1/2} + \frac{1}{100} e^{1/2} \approx 8.26 \text{ million.}$$

Now, to study the population since January 1975, we consider equation (N) with $g(t) = 1/20$ and $E(t) = 0$:

(N₂) $$\left(D - \frac{1}{20}\right) x = 0.$$

For a prediction of the 1985 population, we must find the specific solution of (N₂) satisfying the initial condition $x(10) = 8.26$. The solution of this initial-value problem is

$$x(t) = 8.26e^{(t-10)/20}.$$

The population in January 1985 was

$$x(20) = 8.26e^{1/2} \approx 13.62.$$

The Laplace transform gives the same prediction (see Exercise 34, Section 5.5) by attacking directly the o.d.e. that models the behavior of the Maniac population:

(N₃) $$\left(D - \frac{1}{20}\right) x = \begin{cases} \dfrac{1}{1000} e^{t/20}, & 0 \le t < 10 \\[2mm] 0, & t > 10 \end{cases}$$

with initial condition

$$x(0) = 5.$$

The right side of (N_3) is a typical example of a function defined in pieces; its graph is sketched in Figure 5.1.

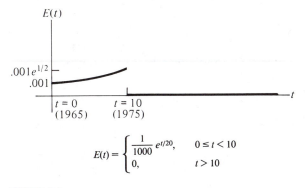

$$E(t) = \begin{cases} \dfrac{1}{1000}\, e^{t/20}, & 0 \le t < 10 \\ 0, & t > 10 \end{cases}$$

FIGURE 5.1

Example 5.1.2   Damped Forced Springs

Recall from Section 2.1 that a mass m attached to a horizontal spring with constant k, encountering damping b, and driven by the external force $E(t)$ is modeled by the equation

$$(mD^2 + bD + k)\, x = E(t).$$

Specific physical problems are naturally formulated as initial-value problems. For example, if the mass is pulled out p units from equilibrium and let go, the initial data are

$$x(0) = p, \quad x'(0) = 0.$$

If the mass starts from the equilibrium position but is nudged so that the spring is contracting at a rate q, the initial data are

$$x(0) = 0, \quad x'(0) = -q.$$

Certain physical solutions lead to forcing terms defined in pieces. If the mass is constrained to move along a shaft that is spun (see Figure 5.2), then the forcing term comes from centrifugal force. Assume the shaft is at rest until time t_1, spins steadily faster between times t_1 and t_2 (when the force reaches

FIGURE 5.2

F), and continues spinning at a constant angular velocity from time t_2 on. Then

$$E(t) = \begin{cases} 0, & t < t_1 \\ F(t - t_1)/(t_2 - t_1), & t_1 < t < t_2 \\ F & t > t_2 \end{cases}$$

(see Figure 5.3). Note that our old methods would require us to solve three successive initial-value problems to predict the long-term behavior of this model.

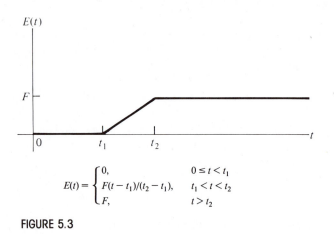

$$E(t) = \begin{cases} 0, & 0 \leq t < t_1 \\ F(t-t_1)/(t_2 - t_1), & t_1 < t < t_2 \\ F, & t > t_2 \end{cases}$$

FIGURE 5.3

Example 5.1.3   An *LRC* Circuit

The circuit sketched in Figure 5.4 has one loop involving an inductance L, a resistance R, a capacitance C, and a time-dependent voltage $V(t)$. The charge Q on the capacitor is modeled (see Section 3.1 and Exercise 5 in Section

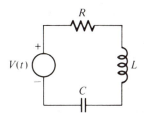

FIGURE 5.4

2.1) by the o.d.e.

$$\left(LD^2 + RD + \frac{1}{C} \right) Q = V(t).$$

An initial charge Q_0 and initial current I_0 are formulated as the initial conditions

$$Q(0) = Q_0, \quad Q'(0) = I_0.$$

If the circuit is initially unforced, but is plugged into an alternating voltage source $V(t) = E \sin \beta t$ at time t_1, then $V(t)$ is defined in pieces by

$$V(t) = \begin{cases} 0, & t < t_1 \\ E \sin \beta t, & t > t_1. \end{cases}$$

The graph of this function is sketched in Figure 5.5.

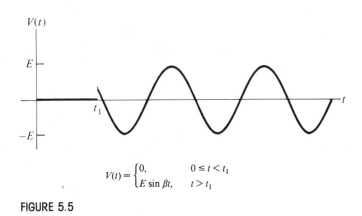

$$V(t) = \begin{cases} 0, & 0 \leq t < t_1 \\ E \sin \beta t, & t > t_1 \end{cases}$$

FIGURE 5.5

Although alternating current or voltage is often modeled by sinusoidal functions, a different model is provided by the **square wave**, sketched in Figure

5.6. The square wave is defined in many pieces by

$$V(t) = \begin{cases} +1 & \text{if } n < t < n + 1, \quad n \text{ even} \\ -1 & \text{if } n < t < n + 1, \quad n \text{ odd.} \end{cases}$$

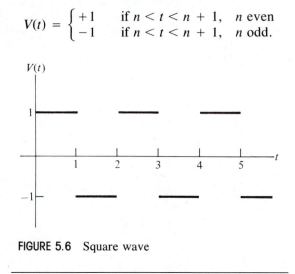

FIGURE 5.6   Square wave

Our final examples deal with systems of o.d.e.'s.

Example 5.1.4   A Multiloop Circuit with a Switch

The circuit in Figure 5.7 is that of Example 3.1.3, except that a switch is provided to allow us to change the voltage source from $V_1(t)$ (position ''a'') to $V_2(t)$ (position ''b''). To model a circuit that has the switch at ''a'' until time

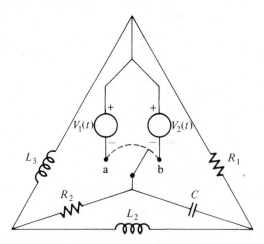

FIGURE 5.7

t_1 and at "b" from then on, we can use the equations of Example 3.1.3,

$$DQ = -\frac{1}{R_1 C} Q - I_2 \qquad\qquad + \frac{V(t)}{R_1}$$

$$DI_2 = \frac{1}{L_2 C} Q - \frac{R_2}{L_2} I_2 + \frac{R_2}{L_2} I_3$$

$$DI_3 = \qquad\qquad \frac{R_2}{L_3} I_2 - \frac{R_2}{L_3} I_3 - \frac{V(t)}{L_3},$$

with $V(t)$ defined in pieces by

$$V(t) = \begin{cases} V_1(t), & t < t_1 \\ V_2(t), & t > t_1. \end{cases}$$

To apply the methods of Chapter 3 to this circuit, we would have to solve two initial-value problems, each involving a third-order system. The Laplace transform method will let us handle only one initial-value problem involving a single system.

Example 5.1.5   Forced Coupled Springs

The physical setup in Figure 5.8 is that of Example 2.1.4, but with an external force $E(t)$ pulling the mass m_2. The force analysis in Example 2.1.4 is easily modified to take account of this extra force and leads to two second-order o.d.e.'s,

$$m_1 D^2 x_1 = -(k_1 + k_2)x_1 + k_2 x_2$$
$$m_2 D^2 x_2 = \qquad k_2 x_1 - k_2 x_2 + E(t).$$

Recall that in Section 2.1 we worked hard to change this system into a single fourth-order o.d.e. for x_1. An adaptation of the methods of Chapter 3

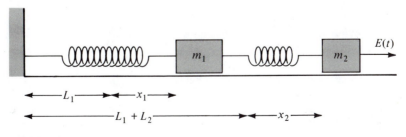

FIGURE 5.8

would require that we introduce two new variables ($v_1 = Dx_1$ and $v_2 = Dx_2$) and study an equivalent system of four first-order o.d.e.'s (for x_1, v_1, x_2, and v_2). The Laplace transform requires no preliminary changes; it can be applied directly to a system of higher-order o.d.e.'s.

EXERCISES

Your final answer to each problem should consist of an o.d.e. (or system of o.d.e.'s) together with a set of initial conditions at $t = 0$ for the variable(s) in the problem.

1. *A Savings Account:* Suppose the savings account of Exercise 19, Section 1.1, was opened with an initial deposit of $1000. After two years the income from the other investment increased to $500 per year. Set up an initial-value problem modeling the growth of the account.

2. *Ice Water:* Suppose that when $t = 0$ the water in a tank is at 70°F and the surroundings are at 40°F. Assume that the water loses heat according to Newton's law of cooling (Section 1.4) with constant of proportionality $\gamma = 1/10$. Suppose the temperature of the surroundings decreases at 2° per hour until it reaches 30°F and remains at 30°F thereafter. Write an initial-value problem to model the temperature of the water.

3. *A Tipped Spring:* Suppose a 16-lb weight is attached to an undamped horizontal spring system with constant $k = 2$ lb/ft. At $t = 0$ the spring is compressed $\frac{1}{2}$ ft and released. After 2 seconds, the system is tipped so it hangs vertically. Write an initial-value problem for the amount $x = x(t)$ by which the spring is stretched. Be sure to take into account that once the system is tipped, gravity acts against the spring.

4. *A Vertical Spring:* Suppose a 16-lb weight is hung from a vertical spring with natural length $L = 3$ ft and constant $k = 4$ lb/ft. The only forces acting on the weight are gravity, the restoring force of the spring, and a damping force with constant $b = 3$ lb/(ft/sec). Denote the amount by which the spring is stretched by $x = x(t)$.
   a. Find the equilibrium position of the spring. (*Hint:* In this position, the restoring force exactly balances the gravitational force.)
   b. Suppose that at $t = 0$ the mass is raised $\frac{1}{2}$ ft above the equilibrium position and released. Set up an initial-value problem modeling the behavior of the weight.
   c. Suppose that at $t = 0$ the mass is $\frac{1}{2}$ ft below the equilibrium position and moving upward at $\frac{1}{2}$ ft per second. Set up an initial-value problem modeling the behavior of the weight.

5. *Another LRC Circuit:* The voltage source in the circuit of Example 5.1.3 is turned on at $t = 0$, at which time there is no current or charge. The voltage increases at 1 volt per second until it reaches 10 volts; it stays at 10 volts thereafter. Find an o.d.e. for the charge Q on the capacitor and formulate an initial-value problem modeling this situation.

6. *An LRC Circuit with a Switch:* The switch in the circuit of Figure 5.9 allows us to change the voltage source from $V_1(t)$ to $V_2(t)$. At $t = 0$ there is no charge or current, and $V_1(t)$ is connected. After 2 seconds, the voltage source is changed to $V_2(t)$ and, 3 seconds later, back to $V_1(t)$. Find an o.d.e. for the charge Q on the capacitor and formulate an initial-value problem to model the behavior of the circuit.

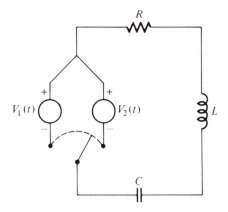

FIGURE 5.9

7. *A Moving Spring:* Suppose that at $t = 0$ the weight suspended from the elevator of Exercise 3, Section 2.1, is at rest, even with the ground floor. Suppose further that the spring has natural length $L = 10$ ft and spring constant $k = 2$ lb/ft and that the damping constant is $b = 2$ lb/(ft/sec). Set up an initial-value problem to model the height of the weight if at $t = 0$ the bottom of the elevator is (a) 16 ft above ground level or (b) 20 ft above ground level.

8. *Fishing:* A fisherman attaches a baited hook to the bottom of the floating box of Exercise 7, Section 2.1, pushes the box down so its bottom is $\frac{1}{4}$ ft from the surface of the water, and releases it. After 3 seconds, a fish bites and pulls straight down with a force of 2 lb. Find an initial-value problem modeling the motion of the box, assuming no damping.

Exercises 9 and 10 involve systems of o.d.e.'s.

9. *A Two-Loop Circuit:* Suppose that at $t = 0$ the currents through the resistors of the circuit in Example 3.1.2 are $I_1(0) = I_2(0) = 0$. Find $Q(0)$ if $R_1 = 1$ ohm, $R_2 = 3$ ohms, $L = 1$ henry, $C = 1$ farad, and $V(t) = 10$ volts, and formulate the initial-value problem modeling this circuit.

10. *Radioactive Decay:* Suppose the radioactive substances A and B decay into substances B and C at the rates of A/2 and B/10, as in Exercise 7, Section 3.1. At $t = 0$, there are 500 grams of each of the three substances. Each year, 250 grams of substance A are steadily added to the mix. During the first year, 100 grams of substance C are extracted. From then on, 125 grams per year of substance C are extracted. Write an initial-value problem modeling the amounts of substances A, B, and C.

5.2   DEFINITIONS AND BASIC CALCULATIONS

In this section we introduce the Laplace transform, which will enable us in later sections to solve initial-value problems directly without requiring the general solution. Philosophically, the Laplace transform technique is analogous to the method of undetermined coefficients considered in Section 2.7. There, we solved a nonhomogeneous equation by applying an appropriate operator (an annihilator of the forcing term) to both sides. This changed the original problem into another (a higher-order homogeneous equation), which we solved using algebra (factoring the new characteristic polynomial). We then used the solution of the new problem to find the solution of the original one. Similarly, the Laplace transform is an operator that, when applied to both sides of a (linear, constant-coefficient) differential equation or system with initial conditions, yields an algebra problem. We will use the solution of this algebra problem to find the desired solution of the initial-value o.d.e. problem.

Definition: *The **Laplace transform** assigns to the function f(t) a new function, F(s), defined by the formula*

$$F(s) = \int_0^\infty e^{-st} f(t) \, dt.$$

F(s) is called the Laplace transform of f(t), and we write

$$F(s) = \mathcal{L}[f(t)].$$

At first sight, the formula defining $F(s) = \mathcal{L}[f(t)]$ is rather imposing. It involves two variables, an infinite limit of integration, and a complicated integrand. But let's take a closer look at each feature separately.

1.   The formula defines a *function F(s)*—that is, it assigns a number $F(s)$ to each numerical value of s. This means that *s acts like a constant inside the integral sign.*

2.   Since the upper limit of integration is infinite, we have an *improper integral.* Recall from calculus that this is interpreted as a limit of proper integrals:

$$\int_0^\infty e^{-st} f(t) \, dt = \lim_{h \to \infty} \left[\int_0^h e^{-st} f(t) \, dt \right].$$

3.   Even though the integrand may seem complicated at first, the formulas for the transforms of the functions we most often deal with turn out to be quite reasonable. What's more, we only have to derive each of them once.

To give some feeling for the Laplace transform, let's calculate some specific examples.

Example 5.2.1

Find $\mathcal{L}[e^{\lambda t}]$.
The definition tells us that

$$\mathcal{L}[e^{\lambda t}] = \int_0^\infty e^{-st}e^{\lambda t}\, dt = \lim_{h \to \infty} \int_0^h e^{-(s-\lambda)t}\, dt.$$

We obtain different values for the integral in the last expression, depending on whether or not $s - \lambda = 0$:

$$\mathcal{L}[e^{\lambda t}] = \lim_{h \to \infty} \begin{cases} h & \text{if } s - \lambda = 0 \\[2ex] -\dfrac{e^{-(s-\lambda)h}}{s-\lambda} + \dfrac{1}{s-\lambda} & \text{if } s - \lambda \neq 0. \end{cases}$$

This limit makes sense only for $s - \lambda > 0$. Thus, the function $F(s)$ is defined only for $s > \lambda$, and

$$F(s) = \mathcal{L}[e^{\lambda t}] = \frac{1}{s - \lambda} \qquad \text{for } s > \lambda.$$

Note that when $\lambda = 0$ this reads

$$\mathcal{L}[1] = \frac{1}{s} \qquad \text{for } s > 0.$$

Example 5.2.2

Find $\mathcal{L}[\cos \beta t]$, $\beta \neq 0$.
The definition says

$$\mathcal{L}[\cos \beta t] = \int_0^\infty e^{-st} \cos \beta t\, dt = \lim_{h \to \infty} \int_0^h e^{-st} \cos \beta t\, dt.$$

To evaluate the integral, we use integration by parts twice (first with $u = e^{-st}$ and $dv = \cos \beta t\, dt$, then with $U = e^{-st}$ and $dV = \sin \beta t\, dt$):

$$\int_0^h e^{-st} \cos \beta t\, dt = \frac{e^{-st} \sin \beta t}{\beta} \bigg]_0^h + \frac{s}{\beta} \int_0^h e^{-st} \sin \beta t\, dt$$

$$= \frac{e^{-st} \sin \beta t}{\beta} \bigg]_0^h - \frac{s e^{-st} \cos \beta t}{\beta^2} \bigg]_0^h - \frac{s^2}{\beta^2} \int_0^h e^{-st} \cos \beta t\, dt.$$

We solve this equation for the integral:

$$\int_0^h e^{-st} \cos \beta t \, dt = \frac{\beta e^{-st} \sin \beta t}{s^2 + \beta^2} \bigg]_0^h - \frac{s e^{-st} \cos \beta t}{s^2 + \beta^2} \bigg]_0^h$$

$$= \frac{e^{-sh}[\beta \sin (\beta h) - s \cos (\beta h)]}{s^2 + \beta^2} + \frac{s}{s^2 + \beta^2}$$

Taking limits as $h \to \infty$, we get

$$\mathcal{L}[\cos \beta t] = \frac{s}{s^2 + \beta^2} \qquad \text{for } s > 0.$$

Example 5.2.3

Find $\mathcal{L}[\sin \beta t]$, $\beta \neq 0$.

We could do this problem by mimicking the previous example. But instead, note that the first integration by parts in that example gave the equation

$$\int_0^h e^{-st} \cos \beta t \, dt = \frac{e^{-st} \sin \beta t}{\beta} \bigg]_0^h + \frac{s}{\beta} \int_0^h e^{-st} \sin \beta t \, dt,$$

which we can solve for the integral on the right-hand side (after carrying out the indicated evaluation of the first term):

$$\int_0^h e^{-st} \sin \beta t \, dt = \frac{-\beta}{s} \frac{e^{-sh} \sin \beta h}{\beta} + \frac{\beta}{s} \int_0^h e^{-st} \cos \beta t \, dt.$$

As $h \to \infty$, the left side of this equation approaches $\mathcal{L}[\sin \beta t]$; if $s > 0$, then the first term on the right goes to 0 and the second tends to $(\beta/s)\mathcal{L}[\cos \beta t]$. Hence

$$\mathcal{L}[\sin \beta t] = \frac{\beta}{s} \mathcal{L}[\cos \beta t] = \frac{\beta}{s} \frac{s}{s^2 + \beta^2} = \frac{\beta}{s^2 + \beta^2}, \qquad s > 0.$$

Example 5.2.4

Find $\mathcal{L}[t^n]$ when n is a positive integer.

By definition,

$$\mathcal{L}[t^n] = \int_0^\infty e^{-st} t^n \, dt = \lim_{h \to \infty} \int_0^h e^{-st} t^n \, dt.$$

Integration by parts (with $u = t^n$ and $dv = e^{-st} \, dt$) gives

$$\int_0^h e^{-st}\, t^n\, dt = \frac{-h^n e^{-sh}}{s} + \frac{n}{s} \int_0^h e^{-st} t^{n-1}\, dt.$$

As $h \to \infty$, the left side approaches $\mathscr{L}[t^n]$. If $s > 0$, then by L'Hôpital's rule the first term on the right approaches 0 while the second approaches $(n/s)\mathscr{L}[t^{n-1}]$. Thus

$$\mathscr{L}[t^n] = \frac{n}{s}\, \mathscr{L}[t^{n-1}] \qquad \text{for } s > 0.$$

Repeated use of this reduction formula ultimately gives us a formula for $\mathscr{L}[t^n]$ in terms of $\mathscr{L}[t^0] = \mathscr{L}[1]$, which we already know from Example 5.2.1:

$$\mathscr{L}[t^n] = \frac{n}{s}\, \mathscr{L}[t^{n-1}] = \frac{n(n-1)}{s^2}\, \mathscr{L}[t^{n-2}] = \cdots = \frac{n!}{s^n}\, \mathscr{L}[1]$$

$$= \frac{n!}{s^{n+1}} \qquad \text{for } s > 0.$$

Example 5.2.5

Find $\mathscr{L}[5e^{2t} - t^3]$.
The definition says

$$\mathscr{L}[5e^{2t} - t^3] = \int_0^{\infty} e^{-st}(5e^{2t} - t^3)\, dt.$$

We could use integration by parts to evaluate this integral, but familiar properties of the integral allow us to break it up instead:

$$\mathscr{L}[5e^{2t} - t^3] = 5 \int_0^{\infty} e^{-st} e^{2t}\, dt - \int_0^{\infty} e^{-st} t^3\, dt.$$

The integrals on the right side are themselves Laplace transforms—$\mathscr{L}[e^{2t}]$ and $\mathscr{L}[t^3]$, respectively. We can use the results of Examples 5.2.1 and 5.2.4 to find them:

$$\mathscr{L}[5e^{2t} - t^3] = 5\mathscr{L}[e^{2t}] - \mathscr{L}[t^3] = \frac{5}{s-2} - \frac{6}{s^4}.$$

Note that $\mathscr{L}[e^{2t}]$ is defined for $s > 2$, and $\mathscr{L}[t^3]$ is defined for $s > 0$. Thus, our new formula will be valid for all values of s that are higher than both 0 and 2—that is, for $s > 2$.

The observation contained in the previous example applies to the transforms of linear combinations of any known functions. We state it formally:

Fact: *The Laplace transform is linear. That is, for any two functions $f_1(t)$ and $f_2(t)$ and constants c_1 and c_2,*

$$\mathcal{L}[c_1 f_1(t) + c_2 f_2(t)] = c_1 \mathcal{L}[f_1(t)] + c_2 \mathcal{L}[f_2(t)].$$

Example 5.2.5 illustrates how linearity, together with knowledge of the transforms of a few basic functions, enables us to calculate the transforms of their linear combinations. Let's look at another such example.

Example 5.2.6

Find $\mathcal{L}[3 - e^{-3t} + 5 \sin 2t]$.
By linearity,

$$\mathcal{L}[3 - e^{-3t} + 5 \sin 2t] = 3\mathcal{L}[1] - \mathcal{L}[e^{-3t}] + 5\mathcal{L}[\sin 2t].$$

Using the results of Examples 5.2.1 and 5.2.3, we get

$$\mathcal{L}[3 - e^{-3t} + 5 \sin 2t] = \frac{3}{s} - \frac{1}{s+3} + \frac{10}{s^2+4} \qquad \text{for } s > 0.$$

The Laplace transform will allow us to deal easily with o.d.e.'s whose forcing terms are defined in pieces. We next calculate a specific example; a general, efficient method for such examples will be worked out in Section 5.5

Example 5.2.7

Find $\mathcal{L}[u_a(t)]$ where $a > 0$, and

$$u_a(t) = \begin{cases} 0 & \text{if } 0 \le t < a \\ 1 & \text{if } a \le t. \end{cases}$$

The definition of $\mathcal{L}$ says

$$\mathcal{L}[u_a(t)] = \int_0^{\infty} e^{-st} u_a(t)\, dt.$$

We calculate this integral in pieces, one for each part of the formula defining $u_a(t)$:

$$\mathcal{L}[u_a(t)] = \int_0^a e^{-st}u_a(t)\ dt + \int_a^{\infty} e^{-st}u_a(t)\ dt$$

$$= \int_0^a e^{-st}0\ dt + \int_a^{\infty} e^{-st}1\ dt$$

$$= 0 + \lim_{h \to \infty} \int_a^h e^{-st}\ dt$$

$$= \lim_{h \to \infty} \begin{cases} h - a & \text{if } s = 0 \\ \dfrac{-e^{-sh}}{s} + \dfrac{e^{-sa}}{s} & \text{if } s \neq 0 \end{cases}$$

$$= \frac{e^{-sa}}{s} \quad \text{for } s > 0.$$

The last step of the Laplace transform method for solving initial-value problems will require us to retrieve a function from its Laplace transform. To conclude this section, let's use this new perspective to look back at the formulas we have derived so far.

Definition: *If $F(s) = \mathcal{L}[f(t)]$, then we say that $f(t)$ is an **inverse Laplace transform** of $F(s)$, and write*

$$f(t) = \mathcal{L}^{-1}[F(s)].$$

The relation between $\mathcal{L}$ and $\mathcal{L}^{-1}$ is like the relation between differentiation and integration. Just as a table of integrals starts from a "backward" reading of differentiation formulas, so an inverse transform table begins with transform formulas read backward. We list here some Laplace transform formulas (obtained from the results of Examples 5.2.1 to 5.2.4), together with the corresponding inverse Laplace transform formulas.

$$\mathcal{L}[e^{\lambda t}] = \frac{1}{s - \lambda} \qquad\qquad \mathcal{L}^{-1}\left[\frac{1}{s - \lambda}\right] = e^{\lambda t}$$

$$\mathcal{L}[1] = \frac{1}{s} \qquad\qquad\qquad \mathcal{L}^{-1}\left[\frac{1}{s}\right] = 1$$

$$\mathcal{L}[\cos \beta t] = \frac{s}{s^2 + b^2} \qquad\qquad \mathcal{L}^{-1}\left[\frac{s}{s^2 + \beta^2}\right] = \cos \beta t$$

$$\mathcal{L}\left[\frac{1}{\beta} \sin \beta t\right] = \frac{1}{s^2 + \beta^2} \qquad \mathcal{L}^{-1}\left[\frac{1}{s^2 + \beta^2}\right] = \frac{1}{\beta} \sin \beta t$$

$$\mathcal{L}\left[\frac{t^{n-1}}{(n - 1)!}\right] = \frac{1}{s^n} \qquad\qquad \mathcal{L}^{-1}\left[\frac{1}{s^n}\right] = \frac{t^{n-1}}{(n - 1)!}$$

Recall also the linearity of $\mathcal{L}$: If $\mathcal{L}[f_1(t)] = F_1(s)$ and $\mathcal{L}[f_2(t)] = F_2(s)$, then $\mathcal{L}[c_1 f_1(t) + c_2 f_2(t)] = c_1 F_1(s) + c_2 F_2(s)$. Read backward, this gives

$$\mathcal{L}^{-1}[c_1 F_1(s) + c_2 F_2(s)] = c_1 f_1(t) + c_2 f_2(t) = c_1 \mathcal{L}^{-1}[F_1(s)] + c_2 \mathcal{L}^{-1}[F_2(s)].$$

Thus, $\mathcal{L}^{-1}$, like $\mathcal{L}$, is *linear!* Just as above for $\mathcal{L}$, if we know the inverse transforms of a few basic functions, linearity lets us find the inverse transforms of their linear combinations.

Example 5.2.8

Find $\mathcal{L}^{-1}\left[\dfrac{5}{s+1} - \dfrac{6}{s^2+4} + \dfrac{1}{s^4}\right]$.

Using linearity and the list of inverse transforms, we have

$$\mathcal{L}^{-1}\left[\frac{5}{s+1} - \frac{6}{s^2+4} + \frac{1}{s^4}\right]$$

$$= 5\mathcal{L}^{-1}\left[\frac{1}{s+1}\right] - 6\mathcal{L}^{-1}\left[\frac{1}{s^2+4}\right] + \mathcal{L}^{-1}\left[\frac{1}{s^4}\right]$$

$$= 5e^{-t} - 3\sin 2t + \frac{1}{6}t^3.$$

The manipulation rules that we obtain in the following sections will greatly increase our repertoire of transforms. However, the effective use of the Laplace transform depends on knowing well the basic formulas listed in the following summary.

BASIC FORMULAS FOR THE LAPLACE TRANSFORM

Definitions

$$\mathcal{L}[f(t)] = \int_0^\infty e^{-st} f(t)\, dt$$

$$\mathcal{L}^{-1}[F(s)] = f(t) \quad \text{provided} \quad F(s) = \mathcal{L}[f(t)]$$

Linearity

$$\mathcal{L}[c_1 f_1(t) + c_2 f_2(t)] = c_1 \mathcal{L}[f_1(t)] + c_2 \mathcal{L}[f_2(t)]$$

$$\mathcal{L}^{-1}[c_1 F_1(s) + c_2 F_2(s)] = c_1 \mathcal{L}^{-1}[F_1(s)] + c_2 \mathcal{L}^{-1}[F_2(s)]$$

Basic Transforms and Inverse Transforms

$$\mathcal{L}[e^{\lambda t}] = \frac{1}{s - \lambda} \qquad \mathcal{L}^{-1}\left[\frac{1}{s - \lambda}\right] = e^{\lambda t}$$

$$\mathcal{L}[1] = \frac{1}{s} \qquad \mathcal{L}^{-1}\left[\frac{1}{s}\right] = 1$$

$$\mathcal{L}[t^n] = \frac{n!}{s^{n+1}} \qquad \mathcal{L}^{-1}\left[\frac{1}{s^n}\right] = \frac{t^{n-1}}{(n-1)!}$$

$$\mathcal{L}[\cos \beta t] = \frac{s}{s^2 + \beta^2} \qquad \mathcal{L}^{-1}\left[\frac{s}{s^2 + \beta^2}\right] = \cos \beta t$$

$$\mathcal{L}[\sin \beta t] = \frac{\beta}{s^2 + \beta^2} \qquad \mathcal{L}^{-1}\left[\frac{1}{s^2 + \beta^2}\right] = \frac{1}{\beta} \sin \beta t$$

The Laplace transform is named after the French mathematician Marquis Pierre-Simon de Laplace, in whose masterwork on probability ("Théorie Analytique des Probabilités," 1812) the integral expression defining $\mathcal{L}[f(t)]$ appears.

Notes

1. On the domain of $\mathcal{L}$

What is the domain of $\mathcal{L}$? That is, for what functions $f(t)$ does the improper integral in the definition of $\mathcal{L}[f(t)]$ converge for at least some values of s? An exhaustive answer to this question would involve technicalities beyond the scope of this book. However, we can formulate fairly general conditions on $f(t)$ that do ensure that $\mathcal{L}[f(t)]$ is defined.

There are two kinds of difficulties that can occur in trying to define the improper integral

$$\int_0^\infty e^{-st}f(t)\, dt = \lim_{h \to \infty} \int_0^h e^{-st}f(t)\, dt.$$

The first is that for some h, the proper definite integral

$$\int_0^\infty e^{-st}f(t)\, dt$$

may be undefined. Recall from calculus that $f(t)$ is **piecewise continuous** on $0 \le t \le h$ provided it is continuous there, except possibly at a finite number of points $t_1, \ldots, t_k$ at which the two one-sided limits

$$\lim_{t \to t_i^-} f(t) \quad \text{and} \quad \lim_{t \to t_i^+} f(t)$$

exist but are unequal. (The functions defined in pieces that came up in Section 5.1 and 5.2 are all examples of such functions.) If $f(t)$ is piecewise continuous on $0 \le t \le h$, then $\int_0^h f(t)\, dt$ exists. Thus, the first problem does not arise if $f(t)$ is piecewise continuous on every interval $0 \le t \le h$.

The second possible problem is that, even though the proper integrals exist, they may not tend to a finite limit as $h \to \infty$. One way to avoid this problem is to require that the integrand approach 0 rapidly as $t \to \infty$. To make this precise, we say a function $f(t)$ is of **exponential order** if there is a constant c so that

$$\lim_{t \to \infty} e^{-ct} f(t) = 0.$$

When this limit condition holds for a given value of c, then one can show that for $s > c$, the proper integrals defining the Laplace transform do converge. We can therefore formulate the following general criterion:

Fact: *If $f(t)$ is piecewise continuous on each interval $0 \le t \le h$ and of exponential order, then $\mathcal{L}[f(t)] = \int_0^\infty e^{-st} f(t)\, dt$ will be defined for all $s > c$.*

Although all the examples we consider in this chapter are piecewise continuous and of exponential order (see Exercises 27 and 28), there are functions that are not included in this class but that still have Laplace transforms (see Exercise 26).

2.   The linearity of $\mathcal{L}$—a technicality

Recall from calculus that two functions are equal provided (i) they have the same domain and (ii) they assign the same value to each element of the domain. Now, if we let $f_1(t) = e^t + 1$ and $f_2(t) = e^t$, then

$$\mathcal{L}[f_1(t)] = \frac{1}{s-1} + \frac{1}{s}, \qquad s > 1$$

and

$$\mathcal{L}[f_2(t)] = \frac{1}{s-1}, \qquad s > 1.$$

Hence

$$\mathcal{L}[f_1(t)] - \mathcal{L}[f_2(t)] = \frac{1}{s}, \qquad s > 1.$$

On the other hand, $f_1(t) - f_2(t) = 1$, so

$$\mathcal{L}[f_1(t) - f_2(t)] = \mathcal{L}[1] = \frac{1}{s}, \qquad s > 0.$$

Even though $\mathcal{L}[f_1(t)] - \mathcal{L}[f_2(t)]$ and $\mathcal{L}[f_1(t) - f_2(t)]$ assign the same value to every $s > 1$, they have different domains and so are not equal in the above sense.

When we say the Laplace transform is linear, we are using a different notion of equality for functions. We consider two functions $F(s)$ and $G(s)$ to be equal provided they agree eventually, that is, provided $F(s) = G(s)$ for all s larger than some fixed value s_0.

3. **The definition of $\mathscr{L}^{-1}$—another technicality**

Our definition tells us $\mathscr{L}^{-1}[F(s)] = f(t)$ provided $\mathscr{L}[f(t)] = F(s)$. Unfortunately, there do exist different functions $f(t)$ and $g(t)$ with the same Laplace transform. Thus it is possible to have $f(t) = \mathscr{L}^{-1}[F(s)]$ and $g(t) = \mathscr{L}^{-1}[F(s)]$ even though $f(t) \neq g(t)$. This is why we called $f(t)$ *an* inverse transform, rather than *the* inverse transform, of $F(s)$.

One way this difficulty can arise is if $f(t)$ and $g(t)$ agree except at a number of isolated points. Fortunately, for the functions we will be dealing with this is the *only* way the problem can arise.

Theorem (Lerch): *If $f(t)$ and $g(t)$ are piecewise continuous and of exponential order, and if $\mathscr{L}[f(t)] = \mathscr{L}[g(t)]$, then $f(t) = g(t)$ for all $t > 0$, except possibly at points where one or both of $f(t)$ and $g(t)$ are discontinuous.*

Note in particular that if $f(t)$ and $g(t)$ are both continuous and $\mathscr{L}[f(t)] = \mathscr{L}[g(t)]$, then $f(t) = g(t)$ for all $t > 0$. Thus a given function $F(s)$ can have at most one *continuous* inverse transform defined on $(0, \infty)$.

EXERCISES

In Exercises 1 through 7, calculate $F(s) = \mathscr{L}[f(t)]$ directly from the definition and indicate the values of s for which the integral defining $F(s)$ converges.

1. $f(t) = e^{4t}$

2. $f(t) = e^{-t}$

3. $f(t) = t^2$

4. $f(t) = te^{3t}$

5. $f(t) = \sin 2t$

6. $f(t) = e^t \sin 2t$

7. $f(t) = \begin{cases} 1 & \text{if } t < 3 \\ 0 & \text{if } t > 3 \end{cases}$

In Exercises 8 through 16, calculate $F(s) = \mathscr{L}[f(t)]$ using the linearity of $\mathscr{L}$ together with the basic formulas summarized at the end of this section.

8. $f(t) = e^{-t}$

9. $f(t) = t^4$

10. $f(t) = \sin 2t$

11. $f(t) = t^2 - 7 + \cos 2t$

12. $f(t) = -3t + e^{-3t} - 5 \sin 6t$

13. $f(t) = 3 + 12t + 42t^3 - 3e^{2t}$

14. $f(t) = e^{2t+3}$

15. $f(t) = (t + 1)(t + 2)$

16. $f(t) = \sin\left(t + \dfrac{\pi}{6}\right)$ (*Hint:* Use trig identities.)

In Exercises 17 through 24, calculate $f(t) = \mathscr{L}[F(s)]$ using the linearity of $\mathscr{L}^{-1}$ together with the basic formulas summarized at the end of this section.

17. $F(s) = \dfrac{1}{s-2}$

18. $F(s) = \dfrac{2}{s-1}$

19. $F(s) = \dfrac{1}{2s-1}$

20. $F(s) = \dfrac{1}{s+2}$

21. $F(s) = \dfrac{2}{s^5}$

22. $F(s) = \dfrac{s}{s^2+4}$

23. $F(s) = \dfrac{1}{s^2+3}$

24. $F(s) = \dfrac{3}{s^2+1} - \dfrac{20}{s^4} + \dfrac{3}{s}$

More advanced problems:

25. Is the Laplace transform of a product, $\mathcal{L}[f(t)g(t)]$, the same as the product of the transforms, $\mathcal{L}[f(t)]\mathcal{L}[g(t)]$? (*Hint:* Try some examples.)

26. Work through the following determination of $\mathcal{L}[t^{-1/2}]$:
    a. Show that for $s > 0$ the substitution $y = \sqrt{st}$ gives

$$\mathcal{L}[t^{-1/2}] = \int_0^\infty t^{-1/2} e^{-st}\, dt = \frac{2}{\sqrt{s}} \int_0^\infty e^{-y^2}\, dy.$$

    b. Rewrite the square of this integral in the form

$$(\mathcal{L}[t^{-1/2}])^2 = \frac{4}{s} \int_0^\infty e^{-x^2}\, dx \int_0^\infty e^{-y^2}\, dy = \frac{4}{s} \int_0^\infty \int_0^\infty e^{-x^2-y^2}\, dx\, dy.$$

    c. Convert this double integral to an integral in polar coordinates and evaluate to obtain

$$(\mathcal{L}[t^{-1/2}])^2 = \frac{4}{s} \int_0^{\pi/2} \int_0^\infty e^{-r^2} r\, dr\, d\theta = \frac{\pi}{s}.$$

    d. Conclude that $\mathcal{L}[t^{-1/2}] = \sqrt{\pi/s}$ for $s > 0$. Note that $t^{-1/2}$ is not piecewise continuous on any interval of the form $[0, h]$, since $\lim_{t\to 0^+} t^{-1/2}$ does not exist.

Exercises 27 and 28 together with the results of Chapter 2 guarantee that all solutions of any homogeneous constant-coefficient linear o.d.e. are of exponential order.

27. a. Work through the following demonstration that if n is a nonnegative integer, then $f(t) = t^n e^{\alpha t} \cos \beta t$ is of exponential order:
    i. Show that if $c = \alpha + 1$, then $|e^{-ct} f(t)| \le t^n e^{-t}$ for $t > 0$.
    ii. Show that $\lim_{t\to\infty}(t^n e^{-t}) = 0$. (*Hint:* Use L'Hôpital's rule to handle the case $n > 0$.)
    iii. Conclude that $\lim_{t\to\infty} e^{-ct} f(t) = 0$.
    b. Repeat part (a) for $f(t) = t^n e^{\alpha t} \sin \beta t$.

28. Show that if $f(t)$ and $g(t)$ are of exponential order, then so is any linear combination $h(t) = af(t) + bg(t)$. (*Hint:* Suppose $\lim_{t\to\infty} e^{-\alpha t} f(t) = 0 = \lim_{t\to\infty} e^{-\beta t} g(t)$. Show that $\lim_{t\to\infty} e^{-ct} h(t) = 0$, where c is the larger of α and β.)

29. Which of the following functions are
     a. of exponential order and
     b. piecewise continuous on every interval of the form $[0, h]$?
         i. $1/t$ ii. e^{t^2} iii. e^{-t^2}
         iv. $\arctan t$ v. $\ln t$ vi. $t^{-1/2}$

30. Show that the integral defining $\mathcal{L}[e^{t^2}]$ does not converge for any s.
     (*Hint:* $t^2 - st > 0$ for $t > s$.)

31. Show that $f(t)$ is of exponential order if and only if there exist constants a and b
     and a value t_0 so that

     $$|f(t)| \le ae^{bt} \quad \text{for all } t \ge t_0.$$

5.3 THE LAPLACE TRANSFORM AND INITIAL-VALUE PROBLEMS

In this section we will see how the Laplace transform is used to solve initial-value problems.

The first step in solving an o.d.e. with constant coefficients

$$a_n D^n x + a_{n-1} D^{n-1}x + \cdots + a_0 x = g(t)$$

using the Laplace transform is to apply $\mathcal{L}$ to both sides. Since $\mathcal{L}$ is linear, we get

$$a_n \mathcal{L}[D^n x] + a_{n-1}\mathcal{L}[D^{n-1}x] + \cdots + a_0\mathcal{L}[x] = \mathcal{L}[g(t)].$$

We already know how to find the right-hand term, $\mathcal{L}[g(t)]$, for many choices of $g(t)$. However, we need a way of dealing with the terms $\mathcal{L}[D^k x]$ appearing on the left side. We will try to express all terms of this form in terms of the transform $\mathcal{L}[x]$ of our unknown.

Let's start by considering $\mathcal{L}[Dx] = \mathcal{L}[x'(t)]$. By definition,

$$\mathcal{L}[x'(t)] = \int_0^\infty e^{-st}x'(t)\, dt = \lim_{h\to\infty} \int_0^h e^{-st}x'(t)\, dt.$$

Integration by parts (with $u = e^{-st}$ and $dv = x'(t)\, dt$) gives

$$\mathcal{L}[x'(t)] = \lim_{h\to\infty} \left(e^{-st}x(t)\Big]_0^h + s\int_0^h e^{-st}x(t)\, dt \right)$$

$$= \lim_{h\to\infty} e^{-sh}x(h) - x(0) + s\mathcal{L}[x(t)].$$

For the functions we are interested in (see Note 1, Section 5.2),

$$\lim_{h \to \infty} e^{-sh} x(h) = 0$$

as long as s is sufficiently large. Thus we have the following.

Fact: First Differentiation Formula ($k = 1$). $\mathscr{L}[Dx] = s\mathscr{L}[x] - x(0)$.

We consider how this formula is used in an extremely simple problem.

Example 5.3.1

Solve the initial-value problem

$$Dx = t, \quad x(0) = 2.$$

If we apply $\mathscr{L}$ to both sides of the o.d.e., we have

$$\mathscr{L}[Dx] = \mathscr{L}[t].$$

Using the formula obtained in Example 5.2.4 for $\mathscr{L}[t]$ and the first differentiation formula for $\mathscr{L}[Dx]$, we rewrite this equation as

$$s\mathscr{L}[x] - x(0) = \frac{1}{s^2}.$$

Since we require $x(0)$ to be 2,

$$s\mathscr{L}[x] - 2 = \frac{1}{s^2}.$$

We solve for $\mathscr{L}[x]$:

$$\mathscr{L}[x] = \frac{1}{s^3} + \frac{2}{s}.$$

Now we take inverse transforms to find x:

$$x = \mathscr{L}^{-1}\left[\frac{1}{s^3} + \frac{2}{s}\right] = \mathscr{L}^{-1}\left[\frac{1}{s^3}\right] + 2\mathscr{L}^{-1}\left[\frac{1}{s}\right]$$

$$= \frac{1}{2}t^2 + 2.$$

To deal with o.d.e.'s of order higher than 1, we must find formulas for terms of the form $\mathcal{L}[D^k x]$ with $k > 1$. For $k = 2$ we note that $D^2 x = Dx'(t)$, so

$$\mathcal{L}[D^2 x] = \mathcal{L}[Dx'(t)].$$

But now we can use the first differentiation formula for $k = 1$ to rewrite $\mathcal{L}[Dx'(t)]$ in terms of $\mathcal{L}[x'(t)]$:

$$\mathcal{L}[D^2 x] = s\mathcal{L}[x'(t)] - x'(0).$$

Of course, we already know how to express $\mathcal{L}[x'(t)]$ in terms of $\mathcal{L}[x]$; substituting into the preceding equation, we find

$$\begin{aligned}\mathcal{L}[D^2 x] &= s\{s\mathcal{L}[x] - x(0)\} - x'(0) \\ &= s^2 \mathcal{L}[x] - sx(0) - x'(0).\end{aligned}$$

Repeated application of this process yields the following general formula:

Fact: First Differentiation Formula

$$\mathcal{L}[D^k x] = s^k \mathcal{L}[x] - s^{k-1}x(0) - s^{k-2}x'(0) - \cdots - x^{(k-1)}(0).$$

Example 5.3.2

Solve the initial-value problem

$$D^3 x - D^2 x = 0, \quad x(0) = x'(0) = x''(0) = 3.$$

If we apply $\mathcal{L}$ to both sides of our o.d.e., we get

$$\mathcal{L}[D^3 x] - \mathcal{L}[D^2 x] = \mathcal{L}[0] = 0.$$

By the differentiation formula,

$$\begin{aligned}\mathcal{L}[D^3 x] &= s^3 \mathcal{L}[x] - s^2 x(0) - sx'(0) - x''(0) \\ &= s^3 \mathcal{L}[x] - 3s^2 - 3s - 3\end{aligned}$$

and

$$\begin{aligned}\mathcal{L}[D^2 x] &= s^2 \mathcal{L}[x] - sx(0) - x'(0) \\ &= s^2 \mathcal{L}[x] - 3s - 3.\end{aligned}$$

Substitution into the transformed o.d.e. gives

$$(s^3 - s^2)\mathcal{L}[x] - 3s^2 = 0$$

which we solve for $\mathcal{L}[x]$:

$$\mathcal{L}[x] = \frac{3s^2}{s^3 - s^2} = \frac{3}{s - 1}.$$

Then

$$x = \mathcal{L}^{-1}\left[\frac{3}{s - 1}\right] = 3\mathcal{L}^{-1}\left[\frac{1}{s - 1}\right]$$

$$= 3e^t.$$

These examples illustrate the three steps in solving an initial-value problem by Laplace transforms. First we *transform the o.d.e., incorporating the initial data by means of the first differentiation formula.* Then we *solve algebraically for $\mathcal{L}[x]$ in terms of s.* Finally we *obtain x as the inverse transform of $\mathcal{L}[x]$.*

In the examples so far, the third step was unusually easy. Most of the time we will need to rewrite $\mathcal{L}[x]$ in order to recognize its inverse transform. This is done by means of partial fractions decomposition of quotients of polynomials. A reminder of how such decompositions look is given in Note 2. We illustrate their use with three examples.

Example 5.3.3

Solve the initial-value problem

$$Dx - x = 2 \sin t, \quad x(0) = 0.$$

We transform both sides of the o.d.e.:

$$\mathcal{L}[Dx] - \mathcal{L}[x] = \mathcal{L}[2 \sin t]$$

$$(s\mathcal{L}[x] - x(0)) - \mathcal{L}[x] = \frac{2}{s^2 + 1}$$

Now, we solve for $\mathcal{L}[x]$:

$$\mathcal{L}[x] = \frac{2}{(s - 1)(s^2 + 1)}$$

so that

$$x = \mathcal{L}^{-1}\left[\frac{2}{(s - 1)(s^2 + 1)}\right].$$

To find this inverse transform, we look for the partial fractions decomposition of $\mathcal{L}[x]$, which is of the form

$$\frac{2}{(s - 1)(s^2 + 1)} = \frac{A}{s - 1} + \frac{Bs + C}{s^2 + 1}.$$

To find the values of the constants A, B, and C, we multiply both sides by $(s - 1)(s^2 + 1)$:

$$\begin{aligned}
2 &= A(s^2 + 1) + (Bs + C)(s - 1) \\
&= (A + B)s^2 + (-B + C)s + (A - C)
\end{aligned}$$

and equate the coefficients of s^2, s, and 1 on either side of the equation:

$$\begin{aligned}
A + B &= 0 \\
-B + C &= 0 \\
A \qquad - C &= 2.
\end{aligned}$$

The solution of this algebraic system is

$$A = 1, \quad B = C = -1.$$

Hence

$$\frac{2}{(s - 1)(s^2 + 1)} = \frac{1}{s - 1} + \frac{-s - 1}{s^2 + 1}.$$

This allows us to use the inverse transform formulas from the last section to find x:

$$\begin{aligned}
x &= \mathcal{L}^{-1}\left[\frac{2}{(s - 1)(s^2 + 1)}\right] = \mathcal{L}^{-1}\left[\frac{1}{s - 1} + \frac{-s - 1}{s^2 + 1}\right] \\
&= \mathcal{L}^{-1}\left[\frac{1}{s - 1}\right] - \mathcal{L}^{-1}\left[\frac{s}{s^2 + 1}\right] - \mathcal{L}^{-1}\left[\frac{1}{s^2 + 1}\right] \\
&= e^t - \cos t - \sin t.
\end{aligned}$$

Example 5.3.4

Solve the initial-value problem

$$D^2x - x = 0, \quad x(0) = 3, x'(0) = 1.$$

We transform both sides of the o.d.e.:

$$\mathcal{L}[D^2x] - \mathcal{L}[x] = 0$$
$$(s^2\mathcal{L}[x] - sx(0) - x'(0)) - \mathcal{L}[x] = 0$$
$$(s^2\mathcal{L}[x] - 3s - 1) - \mathcal{L}[x] = 0.$$

Solving for $\mathcal{L}[x]$, we have

$$\mathcal{L}[x] = \frac{3s + 1}{s^2 - 1} = \frac{3s + 1}{(s - 1)(s + 1)}.$$

Thus, we know that

$$x = \mathcal{L}^{-1}\left[\frac{3s + 1}{(s - 1)(s + 1)}\right].$$

To find this inverse transform, we look for the partial fractions decomposition of $\mathcal{L}[x]$, which has the form

$$\frac{3s + 1}{(s - 1)(s + 1)} = \frac{A}{s - 1} + \frac{B}{s + 1}.$$

We multiply by $(s - 1)(s + 1)$:

$$3s + 1 = A(s + 1) + B(s - 1) = (A + B)s + (A - B)$$

and equate coefficients:

$$A + B = 3$$
$$A - B = 1.$$

This gives

$$A = 2, \quad B = 1.$$

Thus

$$\frac{3s + 1}{(s - 1)(s + 1)} = \frac{2}{s - 1} + \frac{1}{s + 1}.$$

and we can find x:

$$x = \mathscr{L}^{-1}\left[\frac{2}{s - 1} + \frac{1}{s + 1}\right] = 2e^t + e^{-t}.$$

Example 5.3.5

Solve the initial-value problem

$$D^2x - 2Dx = 4, \quad x(0) = -1, x'(0) = 2.$$

We transform the o.d.e.:

$$\mathscr{L}[D^2x - 2Dx] = \frac{4}{s}$$

$$(s^2\mathscr{L}[x] + s - 2) - 2(s\mathscr{L}[x] + 1) = \frac{4}{s}$$

Solving for $\mathscr{L}[x]$, we have

$$(s^2 - 2s)\mathscr{L}[x] = \frac{4}{s} - s + 4 = \frac{-s^2 + 4s + 4}{s}$$

$$\mathscr{L}[x] = \frac{-s^2 + 4s + 4}{s^2(s - 2)}.$$

The partial fractions decomposition of the right side is

$$\frac{-s^2 + 4s + 4}{s^2(s - 2)} = \frac{-3}{s} + \frac{-2}{s^2} + \frac{2}{s - 2}$$

so that

$$x = \mathscr{L}^{-1}\left[\frac{-3}{s} - \frac{2}{s^2} + \frac{2}{s - 2}\right] = -3 - 2t + 2e^{2t}.$$

Let's summarize the method.

THE LAPLACE TRANSFORM AND INITIAL-VALUE PROBLEMS

To solve an o.d.e. with constant coefficients, subject to initial conditions at $t = 0$:

1. Transform both sides of the o.d.e., incorporating the initial data by means of the **first differentiation formula:**

$$\mathscr{L}[D^k x] = s^k \mathscr{L}[x] - s^{k-1} x(0) - s^{k-2} x'(0) - \cdots - x^{(k-1)}(0).$$

2. Solve algebraically for $\mathscr{L}[x]$ in terms of s.
3. Obtain x as the inverse Laplace transform of $\mathscr{L}[x]$. In this last step we often use the linearity of $\mathscr{L}^{-1}$ and partial fractions.

Notes

1. The first differentiation formula—a technicality

A careful look at our argument for the first differentiation formula when $k = 1$ shows that we assumed $x(t)$ and $x'(t)$ were reasonably well behaved (e.g., we assumed $\lim_{h \to \infty} e^{-sh} x(h) = 0$ for large values of s). Since the general formula is obtained by applying this case to successive derivatives of x, the formula is valid only if these derivatives are also well behaved. A more careful statement of the formula would read as follows:

Theorem: *Suppose $x(t)$, $x'(t)$, . . . , $x^{(k-1)}(t)$ are continuous and of exponential order, and suppose $x^{(k)}(t)$ is piecewise continuous and of exponential order. Then*

$$\mathscr{L}[D^k x] = s^k \mathscr{L}[x] - s^{k-1} x(0) - \cdots - x^{(k-1)}(0).$$

We will be using the method described in the summary to solve nth-order linear constant-coefficient o.d.e.'s whose forcing terms are piecewise continuous and of exponential order. The solutions we obtain for such o.d.e.'s will satisfy the hypotheses of the preceding theorem (with $k = n$).

2. Partial fractions

Each polynomial $q(s)$ with real coefficients can, at least in theory, be factored as a number (the leading coefficient) times a product of irreducible polynomials of two kinds: **linear factors** $s - a$, where a is a real root of the polynomial and **irreducible quadratic factors** $s^2 + bs + c$ (with $b^2 < 4c$), corresponding to pairs of complex roots. If $p(s)$ is a polynomial whose degree is strictly less than the degree of $q(s)$, then the rational expression $p(s)/q(s)$ can be written as a sum according to the following rules:

i. If $(s - a)^m$ is the highest power of $s - a$ that divides $q(s)$, then the sum should include terms of the form

$$\frac{A_1}{s - a} + \frac{A_2}{(s - a)^2} + \cdots + \frac{A_m}{(s - a)^m}.$$

ii. If $(s^2 + bs + c)^m$ is the highest power of the irreducible quadratic $s^2 + bs + c$ that divides $q(s)$, then the sum should include terms of the form

$$\frac{B_1 s + C_1}{(s^2 + bs + c)} + \frac{B_2 s + C_2}{(s^2 + bs + c)^2} + \cdots + \frac{B_m s + C_m}{(s^2 + bs + c)^m}.$$

In general, we obtain the **partial fractions decomposition** of a rational expression $p(s)/q(s)$, once we know how to factor $q(s)$, by first using long division to rewrite the original quotient as a polynomial plus a new quotient whose numerator has degree less than the denominator $q(s)$, and then writing this new quotient as a sum of terms of the form (i) and (ii), just described, corresponding to all the factors of $q(s)$.

3. The first differentiation formula and inverse transforms

Sometimes we can use a restatement of the first differentiation formula in place of partial fractions to calculate inverse transforms. To obtain this restatement, first note that the function

$$g(t) = \int_0^t f(t)\, dt$$

satisfies

$$Dg(t) = f(t) \quad \text{and} \quad g(0) = 0.$$

Taking Laplace transforms, we get

$$s\mathcal{L}[g(t)] - 0 = \mathcal{L}[f(t)]$$

$$\mathcal{L}[g(t)] = \frac{1}{s}\,\mathcal{L}[f(t)].$$

Setting $F(s) = \mathcal{L}[f(t)]$, we see that

$$\mathcal{L}\left[\int_0^t f(t)\, dt\right] = \frac{1}{s} F(s)$$

or, reformulating this in terms of $\mathcal{L}^{-1}$,

$$\mathcal{L}^{-1}\left[\frac{1}{s} F(s)\right] = \int_0^t \mathcal{L}^{-1}[F(s)]\, dt.$$

For example, we see that

$$\mathcal{L}^{-1}\left[\frac{1}{s(s^2 + \beta^2)}\right] = \int_0^t \mathcal{L}^{-1}\left[\frac{1}{s^2 + \beta^2}\right] dt = \frac{1}{\beta}\int_0^t \sin \beta t \, dt$$

$$= -\frac{1}{\beta^2}\cos \beta t + \frac{1}{\beta^2}.$$

Repeated application of the formula for $\mathcal{L}^{-1}[F(s)/s]$ gives us the following.

Fact: First Differentiation Formula—Inverse Transform Version

$$\mathcal{L}^{-1}\left[\frac{1}{s^k}F(s)\right] = \underbrace{\int_0^t \int_0^t \cdots \int_0^t}_{k \text{ integrals}} \mathcal{L}^{-1}[F(s)] \, dt \cdots dt.$$

For example,

$$\mathcal{L}^{-1}\left[\frac{1}{s^3(s-2)}\right] = \int_0^t \int_0^t \int_0^t \mathcal{L}^{-1}\left[\frac{1}{s-2}\right] dt \, dt \, dt$$

$$= \int_0^t \int_0^t \int_0^t e^{2t} \, dt \, dt \, dt = \int_0^t \int_0^t \left(\frac{e^{2t}}{2} - \frac{1}{2}\right) dt \, dt$$

$$= \int_0^t \left(\frac{e^{2t}}{4} - \frac{1}{2} - \frac{1}{4}\right) dt = \frac{e^{2t}}{8} - \frac{t^2}{4} - \frac{t}{4} - \frac{1}{8}.$$

Of course, this formula should not be used unless $\mathcal{L}^{-1}[F(s)]$ is both easy to find and easy to integrate.

4. Initial conditions at nonzero time

The Laplace transform of an o.d.e. depends on the initial conditions that hold at $t = 0$. To solve an o.d.e.

(N) $(a_n D^n + \cdots + a_0)x = E(t),$

with conditions imposed at time $t = u$, $u \neq 0$, we can restate the problem in terms of the time $\tau = t - u$ elapsed since imposition of the given conditions. The substitution $t = \tau + u$ changes the o.d.e. (N) into

(N') $(a_n D^n + \cdots + a_0)y = E(\tau + u).$

If $y = y(\tau)$ is a solution of (N') with certain conditions at $\tau = 0$, then $x = y(t - u)$ is a solution of (N) with the same conditions imposed at $t = u$. This observation should be used to do Exercises 26 through 29.

EXERCISES

In Exercises 1 through 6, use the first differentiation formula to find an expression for $\mathcal{L}[x]$, where x is the solution of the given initial-value problem.

1. $(D - 1)x = 0;$ $x(0) = -3$
2. $(D - 1)x = e^{3t};$ $x(0) = 3$

3. $(D^2 - 1)x = e^{2t}$; $x(0) = x'(0) = 0$
4. $(D^2 - 1)x = e^{2t}$; $x(0) = 0, x'(0) = 1$
5. $(D^2 + 1)x = \cos 3t$; $x(0) = x'(0) = 0$
6. $(D^2 + 1)x = \cos 3t$; $x(0) = 0, x'(0) = 3$

In Exercises 7 through 9, use the fact that $x = f(t)$ is the solution of the given initial-value problem to find $\mathcal{L}[f(t)]$.

7. $f(t) = \cos \beta t$; $(D^2 + \beta^2)x = 0$, $x(0) = 1, x'(0) = 0$
8. $f(t) = t \cos \beta t$; $(D^4 + 2\beta^2 D^2 + \beta^4)x = 0$, $x(0) = 0, x'(0) = 1, x''(0) = 0$,
    $x'''(0) = -3\beta^2$
9. $f(t) = (t + 2)e^{-t}$; $(D^2 + 2D + 1)x = 0$, $x(0) = 2, x'(0) = -1$

Find the inverse transforms of the functions in Exercises 10 through 15.

10. $\dfrac{1}{(s + 1)(s - 3)}$

11. $\dfrac{1}{s(s + 1)(s - 3)}$

12. $\dfrac{s + 4}{s^2 + 4s + 3}$

13. $\dfrac{1}{s^4 - 1}$

14. $\dfrac{1}{(s^2 + 1)(s^2 + 4)}$

15. $\dfrac{s + 1}{(s^2 + 4)s}$

Use the Laplace transform to solve the initial-value problems in Exercises 16 through 25.

16. $x'' - 2x' - 3x = 0$; $x(0) = 2, x'(0) = -2$
17. $x'' - 2x' - 3x = 1$; $x(0) = 2, x'(0) = -2$
18. $(D^2 + 4)x = \sin t$; $x(0) = x'(0) = 0$
19. $(D^2 + 4)x = e^t$; $x(0) = x'(0) = 0$
20. $(D^2 + 4)x = t$; $x(0) = -1, x'(0) = 0$
21. $(D^2 + 2D - 3)x = 5e^{2t}$; $x(0) = 2, x'(0) = 3$
22. $(D^3 - 4D)x = 0$; $x(0) = 4, x'(0) = x''(0) = 8$
23. $(D^3 - 4D)x = e^t$; $x(0) = 5, x'(0) = x''(0) = 9$
24. $(D^4 - 1)x = 0$; $x(0) = 1, x'(0) = x''(0) = x'''(0) = 0$
25. $(D^4 - 1)x = 1$; $x(0) = 1, x'(0) = x''(0) = x'''(0) = 0$

Use the substitution $\tau = t - u$ (discussed in Note 4) and the Laplace transform to solve the initial-value problems in Exercises 26 through 29.

26. $(D^2 - 1)x = 3t$; $x(1) = x'(1) = 0$
27. $(D^2 - 1)x = 3t$; $x(1) = x'(1) = 2$
28. $(D^2 - 1)x = e^{3t}$; $x(1) = x'(1) = 0$
29. $(D^2 - 1)x = \sin t$; $x(\pi) = x'(\pi) = 0$

In Exercises 30 through 34, which refer to our models, solve the initial-value problem of the indicated exercise.

30. *A Mixing Problem:* Exercise 24, Section 1.1 (compare Exercise 22, Section 1.2)
31. *Cooling Water:* Exercise 8(b), Section 1.4

32. *A Vertical Spring:* Exercise 4(b), Section 5.1

33. *Another Vertical Spring:* Exercise 4(c), Section 5.1

34. *A Circuit:* Exercise 5, Section 2.1, with $L = 1$, $C = 1/9$, $R = 0$, $V(t) = \sin 2t$, and subject to $Q(0) = 1$, $I(0) = -1$

More abstract problems:

35. Suppose $f(t)$ is a polynomial of degree n. Show that

$$\mathscr{L}[f(t)] = \frac{f(0)}{s} + \frac{f'(0)}{s^2} + \cdots + \frac{f^{(n)}(0)}{s^{n+1}}.$$

[*Hint:* Use the first differentiation formula, or else express the coefficients of $f(t)$ in terms of the initial values $f^{(i)}(0)$.]

36. a. Suppose $x = h(t)$ is a solution of the homogeneous o.d.e. $P(D)x = 0$, where $P(D) = a_2 D^2 + a_1 D + a_0$. Show that $\mathscr{L}[H(t)]$ has the form $g(s)/P(s)$, where $g(s) = b_1 s + b_0$ for some constants b_0, b_1.

  b. What is the analogous form for $\mathscr{L}[H(t)]$ if $x = h(t)$ is a solution of the o.d.e. $P(D)x = 0$ with $P(D) = a_n D^n + \cdots + a_1 D + a_0$?

37. a. Suppose $q(s) = (s - m_1)(s - m_2) \cdots (s - m_k)$ with $m_1, \ldots, m_k$ distinct. The partial fractions decomposition of $p(s)/q(s)$ is

$$\frac{p(s)}{q(s)} = \frac{A_1}{s - m_1} + \frac{A_2}{s - m_2} + \cdots + \frac{A_k}{s - m_k}.$$

Multiply both sides of this equation by $s - m_1$ and substitute $s = m_1$ to obtain an expression for A_1. Interpret this in terms of p and the factorization of q.

  b. Show that your expression in part (a) is the same as $A_1 = p(m_1)/q'(m_1)$.

38. a. Suppose $q(s) = (s - m_1)^P(s - m_2) \cdots (s - m_k)$ with $m_1, \ldots, m_k$ distinct and $p \geq 1$. The partial fractions decomposition of $p(s)/q(s)$ is

$$\frac{p(s)}{q(s)} = \frac{B_1}{s - m_1} + \frac{B_2}{(s - m_1)^2} + \cdots + \frac{B_p}{(s - m_1)^P} + \frac{A_2}{s - m_2} + \cdots + \frac{A_k}{s - m_k}.$$

Multiply both sides of this equation by $(s - m_1)^P$ to obtain an equation that we will call (*).

  b. Substitute $s = m_1$ in (*) to obtain an expression for B_p.

  c. Differentiate both sides of (*) r times, $r < p$, and substitute $s = m_1$ to obtain an expression for B_{p-r}.

  d. Do the expressions for B_{p-r}, $r = 0, \ldots, p - 1$ obtained in parts (b) and (c) remain valid when $m_i = m_j$ for some $i \neq j$ with $i, j > 1$?

5.4 FURTHER PROPERTIES OF THE LAPLACE TRANSFORM AND INVERSE TRANSFORM

We saw in the last section how the Laplace transform is used to solve initial-value problems. In this section we develop some rules of manipulation that will greatly extend our repertoire of Laplace transforms and make its use more efficient.

We begin by considering the effect of the Laplace transform on a function of the form $e^{\alpha t} f(t)$, assuming we know $F(s) = \mathcal{L}[f(t)]$. The definition says

$$\mathcal{L}[e^{\alpha t} f(t)] = \int_0^\infty e^{-st} e^{\alpha t} f(t) \, dt = \int_0^\infty e^{-(s-\alpha)t} f(t) \, dt.$$

This last integral is the value at $s - \alpha$ of the Laplace transform of $f(t)$—that is, it is $F(s - \alpha)$. In other words, we have the following.

Fact: First Shift Formula. *If* $\mathcal{L}[f(t)] = F(s)$, *then*

$$\mathcal{L}[e^{\alpha t} f(t)] = F(s - \alpha).$$

Example 5.4.1

Find $\mathcal{L}[t^3 e^{2t}]$.

The first shift formula tells us that $\mathcal{L}[t^3 e^{2t}]$ is obtained from $\mathcal{L}[t^3]$ by replacing s by $s - 2$. Since

$$\mathcal{L}[t^3] = F(s) = \frac{6}{s^4}$$

this means

$$\mathcal{L}[t^3 e^{2t}] = F(s - 2) = \frac{6}{(s - 2)^4}.$$

Example 5.4.2

Find $\mathcal{L}[e^{-t} \cos 3t]$.

$\mathcal{L}[e^{-t} \cos 3t]$ is obtained from

$$\mathcal{L}[\cos 3t] = F(s) = \frac{s}{s^2 + 9}$$

by replacing s with $s - (-1) = s + 1$. Thus

$$\mathcal{L}[e^{-t}\cos 3t] = F(s + 1) = \frac{s + 1}{(s + 1)^2 + 9} = \frac{s + 1}{s^2 + 2s + 10}.$$

Of course, every transform formula can be rewritten as an inverse transform formula. We can rewrite the first shift formula as

$$\mathcal{L}^{-1}[F(s - \alpha)] = e^{\alpha t}\mathcal{L}^{-1}[F(s)].$$

If we replace s with $s + \alpha$, we obtain a formula that is easier to work with.

Fact: First Shift Formula—Inverse Version

$$\mathcal{L}^{-1}[F(s)] = e^{\alpha t}\mathcal{L}^{-1}[F(s + \alpha)].$$

This formula says that we can substitute $s + \alpha$ for s *inside* the transform, provided we put the exponential factor $e^{\alpha t}$ *outside*.

Example 5.4.3

Find $\mathcal{L}^{-1}\left[\dfrac{3}{(s - 2)^5}\right]$.

We know how to inverse-transform powers of s; in the present case, we note that the substitution of $s + 2$ for s would turn our problem into one of this type. The shift formula tells us how to carry out this substitution:

$$\mathcal{L}^{-1}\left[\frac{3}{(s - 2)^5}\right] = e^{2t}\mathcal{L}^{-1}\left[\frac{3}{(s + 2 - 2)^5}\right] = e^{2t}\mathcal{L}^{-1}\left[\frac{3}{s^5}\right]$$

$$= \frac{1}{8}e^{2t}t^4.$$

Example 5.4.4

Find $\mathcal{L}^{-1}\left[\dfrac{2}{(s + 4)^3}\right]$.

The inverse version of the first shift formula says

$$\mathcal{L}^{-1}\left[\frac{2}{(s + 4)^3}\right] = e^{-4t}\mathcal{L}^{-1}\left[\frac{2}{(s - 4 + 4)^3}\right] = e^{-4t}\mathcal{L}^{-1}\left[\frac{2}{s^3}\right]$$

$$= e^{-4t}t^2.$$

Example 5.4.5

Find $\mathscr{L}^{-1}\left[\dfrac{s}{(s-1)^2 + 4}\right]$.

If the denominator had the form $s^2 + 4$, we could handle this using trigonometric functions. Therefore we try the substitution of $s + 1$ for s, which changes $(s - 1)^2$ into s^2. Note that we must also perform this substitution in the numerator:

$$\mathscr{L}^{-1}\left[\frac{s}{(s-1)^2 + 4}\right] = e^t\mathscr{L}^{-1}\left[\frac{s+1}{(s + 1 - 1)^2 + 4}\right]$$

$$= e^t\mathscr{L}^{-1}\left[\frac{s}{s^2 + 4} + \frac{1}{s^2 + 4}\right]$$

$$= e^t\cos 2t + \frac{1}{2} e^t \sin 2t.$$

The last example was presented to us in a form suitable for shifting, but in practice it is more likely to occur in the form $s/(s^2 - 2s + 5)$. We could recover the more useful form by completing the square in the denominator. In general, to deal with terms of the form $(Bs + C)/(s^2 - bs + c)$ we first check to see whether the denominator can be factored. If it can, we use partial fractions; if it can't, we complete the square and shift.

The following example shows how this can come up in the context of solving an initial-value problem.

Example 5.4.6

Solve the initial-value problem

$$(D^2 + 2D + 2)x = 25te^t, \quad x(0) = x'(0) = 0.$$

We transform both sides of the o.d.e. (using the first differentiation formula on the left and the first shift formula on the right):

$$s^2\mathscr{L}[x] + 2s\mathscr{L}[x] + 2\mathscr{L}[x] = \frac{25}{(s-1)^2}.$$

We solve for $\mathscr{L}[x]$:

$$\mathscr{L}[x] = \frac{25}{(s-1)^2(s^2 + 2s + 2)}.$$

Expansion in partial fractions yields

$$\mathscr{L}[x] = \frac{-4}{s-1} + \frac{5}{(s-1)^2} + \frac{4s+7}{s^2+2s+2}$$

so that

$$x = -4\mathscr{L}^{-1}\left[\frac{1}{s-1}\right] + 5\mathscr{L}^{-1}\left[\frac{1}{(s-1)^2}\right] + \mathscr{L}^{-1}\left[\frac{4s+7}{s^2+2x+2}\right].$$

The first two terms in our expression for *x* are easily handled by the shift:

$$-4\mathscr{L}^{-1}\left[\frac{1}{s-1}\right] + 5\mathscr{L}^{-1}\left[\frac{1}{(s-1)^2}\right] = -4e^t\mathscr{L}^{-1}\left[\frac{1}{s}\right] + 5e^t\mathscr{L}^{-1}\left[\frac{1}{s^2}\right]$$

$$= -4e^t + 5te^t.$$

The third term requires us to complete the square in the denominator before shifting:

$$\mathscr{L}^{-1}\left[\frac{4s+7}{s^2+2s+2}\right] = \mathscr{L}^{-1}\left[\frac{4s+7}{(s+1)^2+1}\right]$$

$$= e^{-t}\mathscr{L}^{-1}\left[\frac{4(s-1)+7}{s^2+1}\right] = e^{-t}\mathscr{L}^{-1}\left[\frac{4s+3}{s^2+1}\right]$$

$$= 4e^{-t}\mathscr{L}^{-1}\left[\frac{s}{s^2+1}\right] + 3e^{-t}\mathscr{L}^{-1}\left[\frac{1}{s^2+1}\right]$$

$$= 4e^{-t}\cos t + 3e^{-t}\sin t.$$

Combining these calculations, we obtain the solution of the initial-value problem:

$$x = -4e^t + 5te^t + 4e^{-t}\cos t + 3e^{-t}\sin t.$$

Another important formula comes from differentiating the Laplace transform. By definition,

$$\frac{d}{ds}\mathscr{L}[f(t)] = \frac{d}{ds}\int_0^\infty e^{-st}f(t)\,dt.$$

For the functions we are interested in [$f(t)$ piecewise continuous and of exponential order], the differentiation can be carried out inside the integral sign.

Thus,

$$\frac{d}{ds} \mathcal{L}[f(t)] = \int_0^\infty \left(\frac{\partial}{\partial s} e^{-st} f(t)\right) dt = -\int_0^\infty e^{-st} t f(t) \, dt$$
$$= -\mathcal{L}[tf(t)],$$

or

$$\mathcal{L}[tf(t)] = -\frac{d}{ds} \mathcal{L}[f(t)].$$

Repeated application of this formula gives a more general version.

Fact: Second Differentiation Formula

$$\mathcal{L}[t^n f(t)] = (-1)^n \frac{d^n}{ds^n} \mathcal{L}[f(t)].$$

Example 5.4.7

Find $\mathcal{L}[t^2 \sin 3t]$.
The second differentiation formula says

$$\mathcal{L}[t^2 \sin 3t] = (-1)^2 \frac{d^2}{ds^2} \mathcal{L}[\sin 3t]$$

$$= \frac{d^2}{ds^2}\left(\frac{3}{s^2 + 9}\right) = \frac{d}{ds}\left(\frac{-6s}{(s^2 + 9)^2}\right)$$

$$= \frac{18(s^2 - 3)}{(s^2 + 9)^3}.$$

Example 5.4.8

Find $\mathcal{L}[te^{2t} \cos 3t]$.
We first use the second differentiation formula to find $\mathcal{L}[t \cos 3t]$:

$$\mathcal{L}[t \cos 3t] = -\frac{d}{ds} \mathcal{L}[\cos 3t] = -\frac{d}{ds}\left(\frac{s}{s^2 + 9}\right)$$

$$= \frac{s^2 - 9}{(s^2 + 9)^2}.$$

We can now use the first shift formula to get

$$\mathcal{L}[te^{2t}\cos 3t] = \frac{(s-2)^2 - 9}{((s-2)^2 + 9)^2} = \frac{s^2 - 4s - 5}{(s^2 - 4s + 13)^2}.$$

We note in passing that it is also possible to state the second differentiation formula in terms of the inverse transform (see Note). However, most problems to which this version can be applied can also be handled by the convolution methods we shall see in Section 5.7.

Let's summarize our new manipulation rules.

FURTHER PROPERTIES OF LAPLACE TRANSFORMS

First Shift Formula

Transform version:

$$\mathcal{L}[e^{\alpha t} f(t)] = F(s - \alpha), \quad \text{where } F(s) = \mathcal{L}[f(t)].$$

Inverse transform version:

$$\mathcal{L}^{-1}[F(s)] = e^{\alpha t}\mathcal{L}^{-1}[F(s + \alpha)].$$

Second Differentiation Formula

$$\mathcal{L}[t^n f(t)] = (-1)^n \frac{d^n}{ds^n} \mathcal{L}[f(t)].$$

Note

The second differentiation formula—inverse version

A straightforward restatement of the second differentiation formula in terms of $\mathcal{L}^{-1}$ is the following.

Fact: Second Differentiation Formula—Inverse Version

$$\mathcal{L}^{-1}\left[\frac{d^n}{ds^n} F(s)\right] = (-1)^n t^n \mathcal{L}^{-1}[F(s)].$$

This formula could be used, for example, to calculate $\mathcal{L}^{-1}\left[\dfrac{s}{(s^2 + \beta^2)^2}\right]$, provided we notice that

$$\frac{s}{(s^2 + \beta^2)^2} = -\frac{1}{2}\frac{d}{ds}\left(\frac{1}{s^2 + \beta^2}\right).$$

It follows that

$$\mathcal{L}^{-1}\left[\frac{s}{(s^2 + \beta^2)^2}\right] = -\frac{1}{2}\mathcal{L}^{-1}\left[\frac{d}{ds}\left(\frac{1}{s^2 + \beta^2}\right)\right],$$

which we calculate by the second differentiation formula (inverse version):

$$\mathcal{L}^{-1}\left[\frac{s}{(s^2 + \beta^2)^2}\right] = -\frac{1}{2}(-1)t\mathcal{L}^{-1}\left[\frac{1}{s^2 + \beta^2}\right] = \frac{t}{2\beta}\sin \beta t.$$

EXERCISES

Calculate the Laplace transforms of the functions in Exercises 1 through 13.

1. te^{3t}
2. $t^2 e^{3t}$
3. $3t \sin 2t$
4. $(t + 2)e^{-t}$
5. $t^2 \sin 3t$
6. $t^2 \cos 3t$
7. $t^2 e^{mt}$
8. $t^n e^{mt}$
9. $e^{3t} \sin 2t$
10. $e^{-t} \cos 3t$
11. $te^{2t} \sin 3t$
12. $te^{2t} \cos 3t$
13. $t \sin^2 2t$ (*Hint:* Use trig identities.)

Calculate the inverse Laplace transforms of the functions in Exercises 14 through 24.

14. $\dfrac{1}{(s - 2)^6}$
15. $\dfrac{1}{(s + 2)^4}$
16. $\dfrac{s - 1}{(s - 2)^2(s - 3)}$
17. $\dfrac{1}{s^2 + 6s + 9}$
18. $\dfrac{s + 1}{s^2 + 6s + 9}$
19. $\dfrac{1}{s^3 + 6s^2 + 9s}$
20. $\dfrac{1}{s^2 + 4s + 3}$
21. $\dfrac{1}{s^2 + 4s + 5}$
22. $\dfrac{1}{s^2 + 3s + 3}$
23. $\dfrac{s + 1}{s^2 + 3s + 3}$
24. $\dfrac{1}{s^2 - 1}$

In Exercises 25 through 32, use Laplace transforms to solve the initial-value problems.

25. $(D - 1)x = t^3 e^t$; $\quad x(0) = 0$
26. $(D^2 + 2D - 3)x = 3e^t$; $\quad x(0) = 1, x'(0) = 0$

27. $(D^2 - 2D + 1)x = t^3 e^t$; $x(0) = -1, x'(0) = 2$
28. $(D^2 + 2D + 2)x = 0$; $x(0) = x'(0) = 1$
29. $(D^2 - 1)x = e^t \cos 2t$; $x(0) = x'(0) = 0$
30. $(D^2 - D)x = 4te^{2t}$; $x(0) = x'(0) = 0$
31. $(D^3 - 1)x = 0$; $x(0) = x'(0) = 0, x''(0) = 1$
32. $(D^4 - 2D^2 + 1)x = 0$; $x(0) = x'(0) = x''(0) = 0, x'''(0) = -1$

In Exercise 33 through 37, which refer back to our models, solve the initial-value problem of the indicated exercise.

33. *A Circuit:* Exercise 5, Section 2.1, with $V(t) = 0, R = 2, C = L = 1$, and subject to $Q(0) = 0$ and $I(0) = 1$
34. *Another Circuit:* Exercise 5, Section 2.1, with $V(t) = 10, L = R = 1, C = 2$, and subject to $Q(0) = I(0) = 0$
35. *A Spring:* Exercise 25(b), Section 2.6
36. *A Moving Spring:* Exercise 7(a), Section 5.1
37. *Another Moving Spring:* Exercise 7(b), Section 5.1

More advanced problems:

38. Use the result of Exercise 26, Section 5.2, and the differentiation and shift formulas to find
    a. $\mathscr{L}[t^{1/2}]$ b. $\mathscr{L}[t^{3/2}]$ c. $\mathscr{L}[t^{-1/2}e^{3t}]$

39. a. Use the inverse version of the *first* differentiation formula (Note 3, Section 5.3) to obtain an expression for $\mathscr{L}^{-1}\left[\dfrac{1}{(s^2 + 1)^2}\right] = \mathscr{L}^{-1}\left[\dfrac{1}{s}\dfrac{s}{(s^2 + 1)^2}\right]$ in terms of $\mathscr{L}^{-1}\left[\dfrac{s}{(s^2 + 1)^2}\right]$.

    b. Substitute the value of $\mathscr{L}^{-1}\left[\dfrac{s}{(s^2 + 1)^2}\right]$ found in the note at the end of this section into the result of part (a) and calculate $\mathscr{L}^{-1}\left[\dfrac{1}{(s^2 + 1)^2}\right]$.

    c. Use the inverse version of the *second* differentiation formula to calculate
    $$\mathscr{L}^{-1}\left[\frac{s}{(s^2 + 1)^3}\right] = \mathscr{L}^{-1}\left[-\frac{1}{4}\frac{d}{ds}\left(\frac{1}{(s^2 + 1)^2}\right)\right].$$

    d. Calculate $\mathscr{L}^{-1}\left[\dfrac{1}{(s^2 + 1)^3}\right] = \mathscr{L}^{-1}\left[\dfrac{1}{s}\dfrac{s}{(s^2 + 1)^3}\right]$.

40. a. Use the differentiation formulas to express $\mathscr{L}[tx'(t)]$ in terms of $\mathscr{L}[x(t)]$.
    b. Transform the initial-value problem
    $$\text{(H)}\qquad (D^2 - 2tD + 2)x = 0;\qquad x(0) = 0, x'(0) = 1$$
    to obtain a first-order o.d.e. for $\mathscr{L}[x]$ as a function of s.
    c. Solve the equation obtained in (b) for $\mathscr{L}[x]$; your answer will contain a constant of integration.
    d. Find a value for the constant in (c) so that $\mathscr{L}[x]$ is the transform of a polynomial; check that this polynomial satisfies (H).

41. Follow the procedure described in Exercise 40 to find a polynomial solution of the initial-value problem

$$(D^2 - 2tD + 4)x = 0; \quad x(0) = 1, x'(0) = 0.$$

(*Hint:* $\int s^3 e^{s^2/4} \, ds = \int s^2 s e^{s^2/4} \, ds$ can be calculated using integration by parts.)

42. a. Use the differentiation formulas to express $\mathcal{L}[t^2 x''(t)]$ in terms of $\mathcal{L}[x(t)]$.

 b. Transform the **Bessel equation**

(H) $$t^2 x'' + tx' + (t^2 - n^2)x = 0.$$

Note that the transformed equation is independent of the initial conditions and is no simpler than the original equation.

5.5 FUNCTIONS DEFINED IN PIECES

We saw in Section 5.1 how circuits with switches, and mechanical systems with suddenly changing forces, are most simply and accurately described using **functions defined in pieces**—that is, functions whose values are defined by several formulas, each applied over a different piece of the domain. In this section we develop a quick scheme for computing the transforms of such functions.

Our first step in this scheme is to develop a better notation for functions defined in pieces. This is accomplished through the use of the **unit step function:**

$$u_a(t) = \begin{cases} 0 & \text{if } t < a \\ 1 & \text{if } t \geq a. \end{cases}$$

We sketch the graph of $u_a(t)$ in Figure 5.10. (Recall that we calculated the transform of $u_a(t)$ for $a > 0$ in Example 5.2.7.)

The unit step function $u_a(t)$ has the effect of a mathematical "on" switch at $t = a$. If we multiply a function $f(t)$ by $u_a(t)$, the product will be zero until

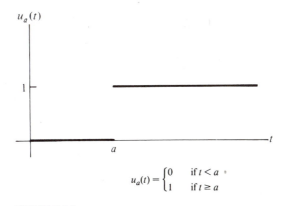

$$u_a(t) = \begin{cases} 0 & \text{if } t < a \\ 1 & \text{if } t \geq a \end{cases}$$

FIGURE 5.10

$t = a$ and will switch to $f(t)$ thereafter:

$$u_a(t)f(t) = \begin{cases} 0 & \text{if } t < a \\ f(t) & \text{if } t \geq a. \end{cases}$$

Thus, for example, the function

$$g(t) = \begin{cases} 0 & \text{if } t < 2 \\ e^{-t} & \text{if } t \geq 2 \end{cases}$$

can be written as

$$g(t) = u_2(t)e^{-t}.$$

Of course, the functions in which we are interested often switch from one nonzero formula to another. For example, we might have

$$g(t) = \begin{cases} t & \text{if } t < 3 \\ e^{-t} & \text{if } t \geq 3. \end{cases}$$

We can think of this as a function that starts out equal to t. At time $t = 3$, a switch does two things: it turns *off* the first formula and turns *on* the second. To turn *off* the formula $g(t) = t$ at $t = 3$, we *subtract* $u_3(t)t$, and to turn *on* $g(t) = e^{-t}$ at $t = 3$, we *add* $u_3(t)e^{-t}$. Thus

$$g(t) = t + u_3(t)(-t + e^{-t}).$$

In general, to express a function defined in pieces by means of step-function notation, we need only remember

1. $u_a(t)$ switches on at $t = a$, and
2. at every interface, we *subtract* a term to switch *off* the previous formula and simultaneously *add* a term to switch *on* the next formula.

We illustrate this procedure with some more examples.

Example 5.5.1

Rewrite $g(t) = |2t - 1|$ in step-function notation.
The definition of absolute value gives

$$g(t) = \begin{cases} -(2t - 1) & \text{if } 2t - 1 < 0 \\ 2t - 1 & \text{if } 2t - 1 \geq 0 \end{cases}$$
$$= \begin{cases} 1 - 2t & \text{if } t < 1/2 \\ 2t - 1 & \text{if } t \geq 1/2. \end{cases}$$

The initial formula is $g(t) = 1 - 2t$. At time $t = 1/2$ we use $u_{1/2}(t)$ to switch *off* $1 - 2t$ and to switch *on* $2t - 1$:

$$g(t) = 1 - 2t + u_{1/2}(t)\{-(1 - 2t) + (2t - 1)\}$$
$$= 1 - 2t + u_{1/2}(t)(4t - 2).$$

Example 5.5.2

Rewrite the following function in step-function notation:

$$g(t) = \begin{cases} 2 & \text{if } t < 1 \\ 3t & \text{if } 1 \le t < 2 \\ 5 & \text{if } 2 \le t. \end{cases}$$

Here, the initial formula is the constant 2, and there are two switching times, $t = 1$ and $t = 2$. At $t = 1$, we use $u_1(t)$ to switch off $g(t) = 2$ and switch on $g(t) = 3t$. At $t = 2$, we use $u_2(t)$ to switch off the *previous* formula, $g(t) = 3t$, and to switch on $g(t) = 5$. Thus

$$g(t) = 2 + u_1(t)(-2 + 3t) + u_2(t)(-3t + 5).$$

Step-function notation lets us rewrite any function defined in pieces as a sum of terms of the form $u_a(t)f(t)$, and reduces the calculation of the transform of such a function to finding the transform of this new kind of term. Keep in mind that $\mathcal{L}$ is an integral over the domain $0 \le t$. Any formula that affects only negative values of t has no effect on the transform. Thus we need only consider transforming terms of the form $u_a(t)f(t)$, with $a \ge 0$.

The definition of $\mathcal{L}$ gives

$$\mathcal{L}[u_a(t)f(t)] = \int_0^\infty e^{-st}u_a(t)f(t)\, dt.$$

We break up this integral at $t = a$ and substitute the value of $u_a(t)$ on each piece of the domain separately:

$$\mathcal{L}[u_a(t)f(t)] = \int_0^a e^{-st}(0)f(t)\, dt + \int_a^\infty e^{-st}(1)f(t)\, dt$$
$$= \int_a^\infty e^{-st}f(t)\, dt.$$

This differs from the Laplace transform of $f(t)$ in that the lower limit of integration is not zero. However, if we set $\tau = t - a$, then $d\tau = dt$, $\tau = 0$ when

$t = a$, and $\tau \to \infty$ as $t \to \infty$. Hence, by rewriting our integral in terms of τ, we have

$$\mathcal{L}[u_a(t)f(t)] = \int_{\tau=0}^{\infty} e^{-s(\tau+a)}f(\tau + a)\, d\tau$$

$$= e^{-as}\int_{0}^{\infty} e^{-s\tau}f(\tau + a)\, d\tau$$

The last integral is equal to the transform of $f(t + a)$. Thus we have the following.

Fact: Second Shift Formula.   *For $a \geq 0$,*

$$\mathcal{L}[u_a(t)f(t)] = e^{-as}\mathcal{L}[f(t + a)].$$

We apply this formula to calculate the transforms of some of the functions we considered earlier in this section.

___

Example 5.5.3

Find $\mathcal{L}[u_2(t)e^{-t}]$.
The second shift formula says

$$\mathcal{L}[u_2(t)e^{-t}] = e^{-2s}\mathcal{L}[e^{-(t+2)}] = e^{-2s}\mathcal{L}[e^{-2}e^{-t}]$$

$$= e^{-2s}e^{-2}\mathcal{L}[e^{-t}]$$

$$= e^{-2(s+1)}\,\frac{1}{s + 1}.$$

___

Example 5.5.4

Find $\mathcal{L}\big[|\,2t - 1\,|\big]$.
We saw in Example 5.5.1 that

$$|2t - 1| = 1 - 2t + u_{1/2}(t)(4t - 2).$$

Thus

$$\mathcal{L}\big[|2t - 1|\big] = \mathcal{L}[1] - \mathcal{L}[2t] + \mathcal{L}[u_{1/2}(t)(4t - 2)]$$

$$= \frac{1}{s} - \frac{2}{s^2} + e^{-s/2}\mathcal{L}\left[4\left(t + \frac{1}{2}\right) - 2\right]$$

$$= \frac{1}{s} - \frac{2}{s^2} + e^{-s/2}\mathcal{L}[4t]$$

$$= \frac{1}{s} - \frac{2}{s^2} + \frac{4e^{-s/2}}{s^2}.$$

Example 5.5.5

Find $\mathcal{L}[g(t)]$, where

$$g(t) = \begin{cases} 2 & \text{if } t < 1 \\ 3t & \text{if } 1 \le t < 2 \\ 5 & \text{if } 2 \le t. \end{cases}$$

Using the result of Example 5.5.2, we have

$$\begin{aligned} \mathcal{L}[g(t)] &= \mathcal{L}[2 + u_1(t)(-2 + 3t) + u_2(t)(-3t + 5)] \\ &= \mathcal{L}[2] + \mathcal{L}[u_1(t)(-2 + 3t)] + \mathcal{L}[u_2(t)(-3t + 5)] \\ &= \frac{2}{s} + e^{-s}\mathcal{L}[-2 + 3(t + 1)] + e^{-2s}\mathcal{L}[-3(t + 2) + 5] \\ &= \frac{2}{s} + e^{-s}\mathcal{L}[1 + 3t] + e^{-2s}\mathcal{L}[-1 - 3t] \\ &= \frac{2}{s} + (e^{-s} - e^{-2s})\left(\frac{1}{s} + \frac{3}{s^2}\right). \end{aligned}$$

If we set $f(t - a) = h(t)$, then the second shift formula says

$$\begin{aligned} \mathcal{L}[u_a(t)f(t - a)] &= \mathcal{L}[u_a(t)h(t)] = e^{-as}\mathcal{L}[h(t + a)] \\ &= e^{-as}\mathcal{L}[f(t)]. \end{aligned}$$

This translates easily into the following inverse transform statement:

Fact: Second Shift Formula—Inverse Version.   *If $\mathcal{L}^{-1}[F(s)] = f(t)$, then*

$$\mathcal{L}^{-1}[e^{-as}F(s)] = u_a(t)f(t - a).$$

Example 5.5.6

Find $\mathcal{L}^{-1}\left[\dfrac{e^{-3s}}{s^2 + 4}\right]$.

We first find

$$f(t) = \mathcal{L}^{-1}\left[\frac{1}{s^2 + 4}\right] = \frac{1}{2}\sin 2t.$$

Then the inverse transform version of the second shift formula says

$$\mathcal{L}^{-1}\left[\frac{e^{-3s}}{s^2 + 4}\right] = u_3(t)f(t - 3) = u_3(t)\frac{\sin 2(t - 3)}{2}$$

$$= \begin{cases} 0 & \text{if } 0 \le t < 3 \\ \dfrac{1}{2}\sin(2t - 6) & \text{if } 3 \le t. \end{cases}$$

Example 5.5.7

Solve the initial-value problem

$$D^2x - x = \begin{cases} t & \text{if } t < 1 \\ 0 & \text{if } t \ge 1 \end{cases} \qquad x(0) = x'(0) = 0.$$

We first rewrite the forcing term using $u_1(t)$:

$$D^2x - x = t + u_1(t)(-t).$$

Now we transform both sides and solve for $\mathcal{L}[x]$:

$$s^2\mathcal{L}[x] - \mathcal{L}[x] = \frac{1}{s^2} + e^{-s}\mathcal{L}[-(t + 1)]$$

$$= \frac{1}{s^2} - e^{-s}\left(\frac{1}{s^2} + \frac{1}{s}\right)$$

$$\mathcal{L}[x] = \frac{1}{s^2(s^2 - 1)} - e^{-s}\left(\frac{1}{s^2(s^2 - 1)} + \frac{1}{s(s^2 - 1)}\right).$$

Thus

$$x = \mathcal{L}^{-1}\left[\frac{1}{s^2(s^2 - 1)}\right] + \mathcal{L}^{-1}\left[e^{-s}\left(-\frac{1}{s^2(s^2 - 1)} - \frac{1}{s(s^2 - 1)}\right)\right].$$

We find and substitute the partial fractions decompositions for $1/s^2(s^2 - 1)$ and $1/s(s^2 - 1)$ to get

$$x = \mathcal{L}^{-1}\left[-\frac{1}{s^2} + \frac{1/2}{s - 1} - \frac{1/2}{s + 1}\right] + \mathcal{L}^{-1}\left[e^{-s}\left(\frac{1}{s} + \frac{1}{s^2} - \frac{1}{s - 1}\right)\right].$$

The first part of our expression is easy to find:

$$\mathcal{L}^{-1}\left[-\frac{1}{s^2} + \frac{1/2}{s - 1} - \frac{1/2}{s + 1}\right] = -t + \frac{1}{2}e^t - \frac{1}{2}e^{-t}.$$

To find the second part, we begin by calculating

$$f(t) = \mathcal{L}^{-1}\left[\frac{1}{s} + \frac{1}{s^2} - \frac{1}{s-1}\right] = 1 + t - e^t$$

and then use the inverse version of the second shift formula:

$$\mathcal{L}^{-1}\left[e^{-s}\left(\frac{1}{s} + \frac{1}{s^2} - \frac{1}{s-1}\right)\right] = u_1(t)f(t-1) = u_1(t)(t - e^{t-1}).$$

Thus

$$x = -t + \frac{1}{2}e^t - \frac{1}{2}e^{-t} + u_1(t)(t - e^{t-1})$$

or, written in pieces,

$$x = \begin{cases} -t + \dfrac{1}{2}e^t - \dfrac{1}{2}e^{-t} & \text{if } t < 1 \\ \left(\dfrac{1}{2} - e^{-1}\right)e^t - \dfrac{1}{2}e^{-t} & \text{if } t \ge 1. \end{cases}$$

We note that even though the forcing function is discontinuous at $t = 1$, our solution x and its derivative x' are both continuous (check this).

The ease with which the Laplace transform lets us handle functions defined in pieces is one of its main advantages.

FUNCTIONS DEFINED IN PIECES

Any function $g(t)$ that is defined in pieces can be written as a sum of terms of the form $u_a(t)f(t)$, where $u_a(t)$ is the **unit step function**

$$u_a(t) = \begin{cases} 0 & \text{if } t < a \\ 1 & \text{if } t \ge a. \end{cases}$$

To obtain this expression for $g(t)$, we note that

1. $u_a(t)$ switches "on" at $t = a$, and
2. at every interface of $g(t)$, we want to switch "off" the immediately preceding formula and switch "on" the next one.

The Laplace transforms of functions defined in pieces are governed by the

Second Shift Formula

Transform version: for $a > 0$,

$$\mathcal{L}[u_a(t)f(t)] = e^{-as}\mathcal{L}[f(t + a)]$$

Inverse transform version: for $a > 0$,

$$\mathcal{L}^{-1}[e^{-as}F(s)] = u_a(t)f(t - a), \quad \text{where } f(t) = \mathcal{L}^{-1}[F(s)].$$

The unit step function $u_a(t)$ is sometimes called the "Heaviside function" after Oliver Heaviside, who pioneered such operational methods in his treatise "Electromagnetic Theory" (1899); another pioneer in this development was George Boole, after whom Boolean algebra is named.

Notes

1. Value at an interface

As we defined it, the unit step function $u_a(t)$ takes on the value 1 at $t = a$. Thus, strictly speaking, the first statement of our summary is true only if, at each interface, the value of $g(t)$ is given by the *later* formula. In general, our procedure for expressing $g(t)$ as a sum of terms of the form $u_a(t)f(t)$ yields a function $h(t)$ that equals $g(t)$ *except possibly* at the interfaces. For example, the function

$$g(t) = \begin{cases} t & \text{if } t \le 2 \\ 1 & \text{if } t > 2 \end{cases}$$

agrees with

$$h(t) = t + u_2(t)(-t + 1)$$

for all values of t except $t = 2$, where $g(2) = 2$, and $h(2) = 1$. Keep in mind, however, that we are interested in step-function notation only as a tool for finding Laplace transforms. Two functions that agree except at a few isolated points have the same integral and hence the same Laplace transform. Thus, we can simply ignore the difference between $g(t)$ and $h(t)$ when taking transforms.

2. A technicality

The interfaces of forcing functions present us with another technical problem, beyond the one discussed in the preceding note. A careful look at the solution we found in Example 5.5.7 will show that

$$D^2x - x = \begin{cases} t & \text{if } t < 1 \\ 0 & \text{if } t > 1 \end{cases}$$

but $D^2x - x$ is not defined at $t = 1$. This illustrates a general problem: if the derivative of a function has a jump discontinuity at a point t_0, then the function cannot be differentiable at t_0. As a result, we must be careful about what we mean by a "solution" to an o.d.e. that has a forcing function with jump discontinuities.

Suppose $Lx = f(t)$ is an nth-order o.d.e. whose forcing function $f(t)$ is piecewise continuous on an interval $a < t < b$. We will say that a function $x = \phi(t)$ is a **solution** of the o.d.e. provided

a. $\phi(t)$ and its first $n - 1$ derivatives are continuous on the entire interval,
b. the nth derivative $\phi^{(n)}(t)$ is defined at all those points of the interval where $f(t)$ is continuous, and
c. substitution of $\phi(t), \phi'(t), \ldots, \phi^{(n)}(t)$ into the o.d.e. yields an identity at all points of the interval where $f(t)$ is continuous.

3. Periodic functions

A function $f(t)$ is **periodic, of period** $p > 0$, provided

$$f(t + p) = f(t)$$

for all t. We have learned to transform two kinds of periodic functions: $\sin \beta t$ and $\cos \beta t$, each with period $2\pi/\beta$. In practice (especially in circuit theory), it is often useful to deal with models involving other kinds of periodic functions. Two common examples are the **square wave**

$$f(t) = \begin{cases} 1 & \text{if } 2n \le t < 2n + 1, \quad n \text{ an integer} \\ -1 & \text{if } 2n + 1 \le t < 2n + 2, \quad n \text{ an integer} \end{cases}$$

of period $p = 2$ (see Figure 5.11), and the **sawtooth function** (also called the fractional part of t)

$$f(t) = t - n \quad \text{when} \quad n \le t < n + 1,$$

of period $p = 1$ (see Figure 5.12).

It seems clear that one ought to be able to determine the transform of a periodic function just from the knowledge of one cycle. In fact, we can obtain such a formula by noting that if $f(t)$ has period p, then

$$f(t) - u_p(t)f(t) = \begin{cases} f(t) & \text{if } 0 \le t < p \\ 0 & \text{if } p \le t. \end{cases}$$

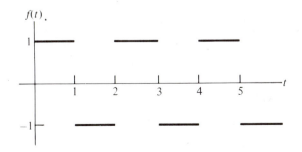

FIGURE 5.11   Square wave

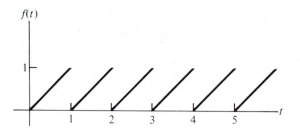

FIGURE 5.12   Sawtooth function

Transforming this expression, we get

$$\mathscr{L}[f(t) - u_p(t)f(t)] = \int_0^\infty e^{-st}\{f(t) - u_p(t)f(t)\}\, dt$$

$$= \int_0^p e^{-st}f(t)\, dt.$$

We can use the second shift formula to rewrite the left side:

$$\mathscr{L}[f(t)] - e^{-ps}\mathscr{L}[f(t + p)] = \int_0^p e^{-st}f(t)\, dt.$$

But if f is periodic with period p, then $f(t + p) = f(t)$ and we have

$$\mathscr{L}[f(t)] - e^{-ps}\mathscr{L}[f(t)] = \int_0^p e^{-st}f(t)\, dt.$$

We can solve this equation for $\mathscr{L}[f(t)]$ to find the formula

$$\mathscr{L}[f(t)] = \frac{1}{1 - e^{-ps}} \int_0^p e^{-st}f(t)\, dt$$

for f periodic of period $p > 0$.

You should use this formula to work out the transforms of the square wave and the sawtooth function (see Exercise 36).

EXERCISES

In Exercises 1 through 12, (a) express $f(t)$ in step-function notation and (b) find $\mathscr{L}[f(t)]$.

1. $f(t) = \begin{cases} 0, & t < 3 \\ t - 3, & t \geq 3 \end{cases}$

2. $f(t) = \begin{cases} t - 3, & t < 3 \\ 0, & t \geq 3 \end{cases}$

3. $f(t) = \begin{cases} 0, & t < \pi \\ \sin t, & t \geq \pi \end{cases}$

4. $f(t) = \begin{cases} 0, & t < 2 \\ e^{-t}, & t \geq 2 \end{cases}$

5. $f(t) = \begin{cases} 0, & t < 1 \\ e^t, & 1 \leq t < 2 \\ 0, & 2 \leq t \end{cases}$

6. $f(t) = \begin{cases} t^2, & t < 1 \\ t^2 - 1, & 1 \leq t < 2 \\ t^2 - 2, & 2 \leq t \end{cases}$

7. $f(t) = |t - 3|$

8. $f(t) = |(t - 2)(t - 1)|$

9. $f(t)$ as in Figure 5.13(a)

10. $f(t)$ as in Figure 5.13(b)

11. $f(t)$ as in Figure 5.13(c)

12. $f(t)$ as in Figure 5.13(d)

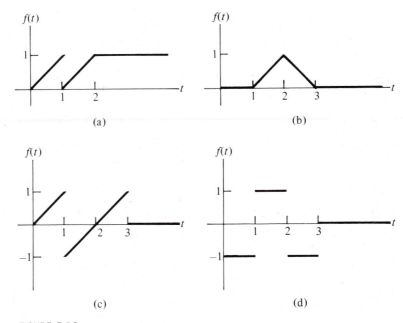

(a)

(b)

(c)

(d)

FIGURE 5.13

Find the inverse Laplace transforms of the functions in Exercises 13 through 20.

13. $\dfrac{e^{-4s}}{s + 4}$

14. $\dfrac{e^{-4s}}{s^2 + 4}$

15. $\dfrac{e^{-4s}}{(s + 4)^2}$

16. $\dfrac{se^{-\pi s}}{s^2 + 4}$

17. $\dfrac{s + e^{-\pi s}}{s^2 + 1}$

18 $\dfrac{e^{-(s+1)}}{s + 1}$

19. $\dfrac{e^{-s}}{s(s + 1)}$

20. $\dfrac{e^{-2s}}{s(s + 1)}$

Solve the initial-value problems in Exercises 21 through 29.

21. $(D - 1)x = \begin{cases} 0, & t < 3 \\ t - 3, & 3 \le t \end{cases}$ $x(0) = -5$

22. $x'' + 2x' + x = \begin{cases} 0, & t < 1 \\ 1, & t \ge 1 \end{cases}$ $x(0) = 0, \, x'(0) = -3$

23. $(D^2 - 4)x = \begin{cases} 0, & t < 2 \\ 2, & t \ge 2 \end{cases}$ $x(0) = 1, \, x'(0) = 0$

24. $(D^2 - 4)x = \begin{cases} 2, & t < 2 \\ 0, & t \ge 2 \end{cases}$ $x(0) = 1, \, x'(0) = 0$

25. $(D^2 + 1)x = \begin{cases} 0, & t < 2 \\ e^{-t}, & t \geq 2 \end{cases}$ $x(0) = x'(0) = 0$

26. $(D^2 + 1)x = \begin{cases} e^{-t}, & t < 2 \\ 0, & t \geq 2 \end{cases}$ $x(0) = x'(0) = 0$

27. $(D^2 + 1)x = \begin{cases} \sin 2t, & t < \pi \\ -\sin 2t, & t \geq \pi \end{cases}$ $x(0) = x'(0) = 0$

28. $(D^3 - D)x = \begin{cases} 1, & t < 2 \\ 0, & t \geq 2 \end{cases}$ $x(0) = x'(0) = x''(0) = 0$

29. $(D + 1)^3 x = \begin{cases} e^{-t}, & t < 3 \\ 0, & t \geq 3 \end{cases}$ $x(0) = x'(0) = x''(0) = 0$

In Exercises 30 through 33, which refer to our models, (a) solve the initial-value problem of the indicated exercise in Section 5.1 and (b) find the value of the solution at each of the indicated times.

30. *A Savings Account:* Exercise 1; $t = 1/2$, $t = 3$
31. *A Tipped Spring:* Exercise 3; $t = 1$, $t = 4$
32. *A Circuit:* Exercise 5, with $L = 1$, $R = 0$, $C = 1/4$; $t = 5$, $t = 20$
33. *More Circuits:* Exercise 6, with $L = 1$, $R = 3$, $C = 1/2$, and
    a. $V_1(t) = 0$, $V_2(t) = 10$; $t = 1$, $t = 3$, $t = 6$
    b. $V_1(t) = 10$, $V_2(t) = 0$; $t = 1$, $t = 3$, $t = 6$
34. *Controlled Immigration:*
    a.   Use Laplace transforms to solve the initial-value problem (N$_3$) of Example 5.1.1.
    b.   Use the result from (a) to estimate the Maniac population in January, 1985.

More advanced problems:

35. Use the result of Exercise 26, Section 5.2, to find $\mathcal{L}[f(t)]$ where

$$f(t) = \begin{cases} 0, & t < k \\ (t - k)^{-1/2}, & t \geq k \end{cases}$$

36. Use the formula for the Laplace transform of a periodic function (Note 3) to find the transform of
    a.   the square wave (see Note 3)
    b.   the sawtooth function (see Note 3)
    c. $|\sin t|$
37. Show that if $k \geq 0$, then

$$\mathcal{L}[f(t + k)] = e^{ks}\mathcal{L}[f(t)] - e^{ks} \int_0^k e^{-st} f(t)\, dt.$$

    [*Hint:* Find $\mathcal{L}[f(t) - u_k(t)g(t - k)]$ where $g(t) = f(t + k)$.]
38. Suppose $f(t + w) = -f(t)$ for all t.
    a.   Show that $f(t)$ is periodic of period $2w$.
    b.   Find a formula for $\mathcal{L}[f(t)]$ in terms of the behavior of $f(t)$ for $0 \leq t \leq w$.
         [*Hint:* Consider $f(t) - u_w(t)f(t)$.]
    c.   The half-wave rectification of $f(t)$ is the function $h(t)$ that is periodic of period

$2w$ and satisfies

$$h(t) = \begin{cases} f(t), & 0 \le t < w \\ 0, & w \le t < 2w. \end{cases}$$

Find $\mathcal{L}[h(t)]$.

5.6 CONVOLUTION

In this section we develop a direct way to inverse-transform products. The method uses an operation on functions, the convolution, which has many uses beyond the specific application we consider here.

Definition: *Given two functions $f(t)$ and $g(t)$, we define a new function, called the **convolution** of f and g and denoted $f * g$, by the rule*

$$(f * g)(t) = \int_0^t f(t - u)g(u)\ du.$$

This formula, like the definition of $\mathcal{L}[f(t)]$, becomes easier to understand if we pause to consider some features of the defining integral.

1. The formula assigns a numerical value, $(f * g)(t)$, to each specific value of t so that as far as the integration is concerned, t *acts like a constant.*
2. The limits of integration refer to u: We integrate from $u = 0$ to $u = t$.
3. Although the second function enters the integrand in a straightforward way, the first has an argument depending on both t and u. In practice, we try to rewrite $f(t - u)$ in terms of functions of t and u alone before integrating.

We calculate a simple example to get a feeling for how this operation works.

Example 5.6.1

Find $(f * g)(t)$ when $f(t) = e^{2t}$ and $g(t) = e^{3t}$.
The definition says

$$(f * g)(t) = \int_0^t f(t - u)g(u)\ du = \int_0^t e^{2(t-u)}e^{3u}\ du$$

$$= e^{2t} \int_0^t e^u\ du = e^{2t}(e^t - 1) = e^{3t} - e^{2t}.$$

The convolution is written so as to resemble a product. In fact, it shares a number of properties with products:

the distributive law

$$f * (c_1 g_1 + c_2 g_2) = c_1(f * g_1) + c_2(f * g_2)$$

the associative law

$$f * (g * h) = (f * g) * h$$

the commutative law

$$f * g = g * f$$

(see Exercises 31 and 32). However, one must be careful not to carry the analogy with ordinary products too far. For example, it is tempting to assume that the convolution of a function $g(t)$ with the constant $f(t) = 1$ leaves $g(t)$ unchanged. But this is *not* so; in fact, for any function $g(t)$,

$$(1 * g)(t) = \int_0^t g(u) \, du.$$

The convolution is useful in calculating inverse Laplace transforms. This is a result of the fact that the transform turns convolutions into products.

Fact: Convolution Formula.   *If $f(t)$ and $g(t)$ both have Laplace transforms, then*

$$\mathcal{L}[(f * g)(t)] = \mathcal{L}[f(t)]\mathcal{L}[g(t)].$$

A proof of this fact is sketched in a note at the end of this section. We make use of this formula primarily in its inverse-transform version:

Fact: Convolution Formula—Inverse Transform Version.   *If $F(s)$ and $G(s)$ have inverse Laplace transforms, then*

$$\mathcal{L}^{-1}[F(s)G(s)] = \mathcal{L}^{-1}[F(s)] * \mathcal{L}^{-1}[G(s)].$$

For practice, we calculate three specific inverse transforms using this formula.

Example 5.6.2

Find $\mathcal{L}^{-1}\left[\dfrac{1}{(s - 2)(s - 3)}\right]$.

Of course, we could do this problem by partial fractions. Alternatively, we can use the convolution formula and the result of Example 5.6.1:

$$\mathcal{L}^{-1}\left[\frac{1}{(s-2)(s-3)}\right] = \mathcal{L}^{-1}\left[\frac{1}{(s-2)}\right] * \mathcal{L}^{-1}\left[\frac{1}{(s-3)}\right]$$

$$= e^{2t} * e^{3t} = e^{3t} - e^{2t}.$$

Example 5.6.3

Find $\mathcal{L}^{-1}\left[\dfrac{s}{(s^2+1)^2}\right]$.

Note that partial fractions can't help us here, since the function $F(s)$ is already written in its partial fractions decomposition. On the other hand, if we regard it as the product of $s/(s^2+1)$ and $1/(s^2+1)$, whose inverse transforms we know, then

$$\mathcal{L}^{-1}\left[\frac{s}{(s^2+1)^2}\right] = \mathcal{L}^{-1}\left[\frac{s}{s^2+1}\cdot\frac{1}{s^2+1}\right] = \mathcal{L}^{-1}\left[\frac{s}{s^2+1}\right] * \mathcal{L}^{-1}\left[\frac{1}{s^2+1}\right]$$

$$= \cos t * \sin t = \int_0^t \cos(t-u)\sin u \, du.$$

We calculate and simplify this integral with the help of several trigonometric identities.

$$\mathcal{L}^{-1}\left[\frac{s}{(s^2+1)^2}\right] = \int_0^t (\cos t \cos u + \sin t \sin u)\sin u \, du$$

$$= \cos t \int_0^t \cos u \sin u \, du + \sin t \int_0^t \sin^2 u \, du$$

$$= \cos t \int_0^t \sin u \, d(\sin u) + \sin t \int_0^t \left(\frac{1}{2} - \frac{1}{2}\cos 2u\right) du$$

$$= \cos t \left(\frac{1}{2}\sin^2 u\right)\bigg]_0^t + \sin t \left(\frac{u}{2} - \frac{1}{4}\sin 2u\right)\bigg]_0^t$$

$$= \frac{1}{2}\cos t \sin^2 t + \frac{t}{2}\sin t - \frac{1}{4}\sin t \sin 2t$$

$$= \frac{1}{2}\cos t \sin^2 t + \frac{t}{2}\sin t - \frac{1}{2}\cos t \sin^2 t$$

$$= \frac{t}{2}\sin t.$$

(You might compare this with the calculation in the note to Section 5.4, which depended on cleverness in recognizing $F(s)$ as a derivative.)

Example 5.6.4

Find $\mathcal{L}^{-1}\left[\dfrac{1}{(s^2 + 1)^2}\right]$.

Again, we regard $F(s)$ as a product and calculate its inverse transform as a convolution, with the aid of trigonometric identities:

$$\mathcal{L}^{-1}\left[\frac{1}{(s^2 + 1)^2}\right] = \mathcal{L}^{-1}\left[\frac{1}{s^2 + 1}\right] * \mathcal{L}^{-1}\left[\frac{1}{s^2 + 1}\right]$$

$$= \sin t * \sin t = \int_0^t \sin(t - u) \sin u\ du$$

$$= \int_0^t (\sin t \cos u - \cos t \sin u) \sin u\ du$$

$$= \sin t \int_0^t \cos u \sin u\ du - \cos t \int_0^t \sin^2 u\ du$$

$$= \sin t \left(\frac{1}{2} \sin^2 t\right) - \cos t \left(\frac{t}{2} - \frac{1}{4} \sin 2t\right)$$

$$= \frac{1}{2} \sin t \sin^2 t - \frac{t}{2} \cos t + \frac{1}{2} \sin t \cos^2 t$$

$$= \frac{1}{2} \sin t - \frac{t}{2} \cos t.$$

(An alternative solution of this problem is outlined in Exercise 39, Section 5.4.)

Inverse Laplace transforms like the two in Examples 5.6.3 and 5.6.4 arise naturally in certain initial-value problems.

Example 5.6.5

Find the solution of the initial-value problem

$$(D^2 + 1)x = \cos t, \quad x(0) = x'(0) = 0.$$

When we transform both sides of the o.d.e., incorporating the initial data via the first differentiation formula, we obtain

$$(s^2 + 1)\mathcal{L}[x] = \mathcal{L}[\cos t] = \frac{s}{s^2 + 1}.$$

Solving for $\mathscr{L}[x]$, we get

$$\mathscr{L}[x] = \frac{s}{(s^2 + 1)^2}.$$

Inverse-transforming to get x, we are led to Example 5.6.3:

$$x = \mathscr{L}^{-1}\left[\frac{s}{(s^2 + 1)^2}\right] = \cos t * \sin t = \frac{t}{2}\sin t.$$

In cases where partial fractions and the convolution are both applicable for finding inverse transforms, the convolution is usually harder. Nevertheless, it can be a useful auxiliary device, as illustrated in the following example.

Example 5.6.6

Solve the initial-value problem

$$(D^4 - 1)x = \begin{cases} 2 & \text{if } t < 3 \\ 0 & \text{if } t \geq 3 \end{cases} \qquad x(0) = x'(0) = x''(0) = 0,\ x'''(0) = 2.$$

In step-function notation, the o.d.e. reads

$$(D^4 - 1)x = 2 - 2u_3(t)$$

so its transform is

$$s^4 \mathscr{L}[x] - 2 - \mathscr{L}[x] = \frac{2}{s} - \frac{2}{s}e^{-3s}.$$

Solving for $\mathscr{L}[x]$, we find

$$\mathscr{L}[x] = \frac{2}{s^4 - 1} + \frac{2}{s(s^4 - 1)} - \frac{2e^{-3s}}{s(s^4 - 1)}.$$

The first term can be decomposed into partial fractions as

$$\frac{2}{s^4 - 1} = \frac{1/2}{s - 1} - \frac{1/2}{s + 1} - \frac{1}{s^2 + 1}$$

so its inverse transform is

$$\mathcal{L}^{-1}\left[\frac{2}{s^4 - 1}\right] = \frac{1}{2}e^t - \frac{1}{2}e^{-t} - \sin t.$$

The second term could also be decomposed into partial fractions. However, note that it is the same as the first term multiplied by $1/s$. We can thus use the convolution formula to obtain its inverse transform from the previous one:

$$\mathcal{L}^{-1}\left[\frac{2}{s(s^4 - 1)}\right] = \mathcal{L}^{-1}\left[\frac{1}{s}\right] * \mathcal{L}^{-1}\left[\frac{2}{s^4 - 1}\right]$$

$$= 1 * \left(\frac{1}{2}e^t - \frac{1}{2}e^{-t} - \sin t\right)$$

$$= \int_0^t \left(\frac{1}{2}e^u - \frac{1}{2}e^{-u} - \sin u\right) du$$

$$= \frac{1}{2}(e^t - 1) + \frac{1}{2}(e^{-t} - 1) + (\cos t - 1)$$

$$= \frac{1}{2}e^t + \frac{1}{2}e^{-t} + \cos t - 2.$$

Finally, the third term is an exponential times the second. We obtain its inverse transform from that of the second term by means of the second shift formula:

$$\mathcal{L}^{-1}\left[\frac{2e^{-3s}}{s(s^4 - 1)}\right] = u_3(t)\left(\frac{1}{2}e^{t-3} + \frac{1}{2}e^{-(t-3)} + \cos(t - 3) - 2\right).$$

Combining all three terms, we have:

$$x = e^t + \cos t - \sin t - 2 - u_3(t)\left(\frac{1}{2}e^{t-3} + \frac{1}{2}e^{-(t-3)} + \cos(t - 3) - 2\right).$$

Let's summarize.

CONVOLUTION

Definition

$$(f * g)(t) = \int_0^t f(t - u)g(u)\, du$$

Algebraic Properties

Distributive: $f * (c_1 g_1 + c_2 g_2) = c_1(f * g_1) + c_2(f * g_2)$

Associative: $f * (g * h) = (f * g) * h$

Commutative: $f * g = g * f$

Convolution Formula

$$\mathcal{L}[(f * g)(t)] = \mathcal{L}[f(t)]\mathcal{L}[g(t)]$$
$$\mathcal{L}^{-1}[F(s)G(s)] = \mathcal{L}^{-1}[F(s)] * \mathcal{L}^{-1}[G(s)]$$

Notes

1. Proof of the convolution formula

Substituting the definition of $f * g$ into the definition of the Laplace transform, we get

$$\mathcal{L}[(f * g)(t) = \int_0^\infty e^{-st}(f * g)(t) \, dt = \int_0^\infty e^{-st} \left(\int_0^t f(t - u)g(u) \, du \right) dt$$
$$= \int_0^\infty \int_0^t e^{-st} f(t - u)g(u) \, du \, dt.$$

The region of integration,

$$0 \le u \le t, \quad 0 \le t < \infty$$

is sketched in Figure 5.14. It can be rewritten as

$$u \le t < \infty, \quad 0 \le u < \infty.$$

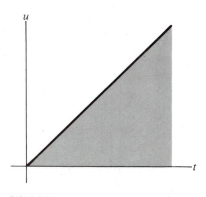

FIGURE 5.14

Thus, reversing the order of integration leads to

$$\mathscr{L}[(f * g)(t)] = \int_0^\infty \int_u^\infty e^{-st} f(t - u) g(u)\, dt\, du.$$

Now, setting $w = t - u$ and $v = u$ and noting the change in the limits ($u \le t < \infty$ becomes $0 \le w < \infty$), we obtain

$$\mathscr{L}[(f * g)(t)] = \int_0^\infty \int_0^\infty e^{-s(w+v)} f(w) g(v)\, dw\, dv,$$

which can be written as a product:

$$\mathscr{L}[(f * g)(t)] = \int_0^\infty e^{-sw} f(w)\, dw \int_0^\infty e^{-sv} g(v)\, dv$$
$$= \mathscr{L}[f(t)] \mathscr{L}[g(t)].$$

2. A special case of the convolution formula
   We have noted already that convolution of a function $g(t)$ with the constant function $f(t) = 1$ leads to the integral of g:

$$(1 * g)(t) = \int_0^t g(u)\, du.$$

Since the constant $f(t) = 1$ is the inverse transform of $F(s) = 1/s$, by setting $G(s) = \mathscr{L}[g(t)]$ we obtain the following integral formula:

$$\mathscr{L}^{-1}\left[\frac{1}{s} G(s)\right] = \mathscr{L}^{-1}\left[\frac{1}{s}\right] * \mathscr{L}^{-1}[G(s)] = \int_0^t \mathscr{L}^{-1}[G(s)]\, du.$$

You might recognize this as the inverse transform version of the first differentiation formula (Note 3, Section 5.3).

EXERCISES

Evaluate the convolutions in Exercises 1 through 7.

1. $t * t^3$ 2. $1 * 1$
3. $t * e^{-4t}$ 4. $t^2 * e^{-4t}$
5. $t * \sin 2t$ 6. $e^t * \cos t$
7. $\cos t * \sin 2t$

Verify the formulas in Exercises 8 through 10.

8. $(\sin \alpha t) * (\cos \alpha t) = \dfrac{t}{2} \sin \alpha t$

9. $(\sin \alpha t) * (\sin \alpha t) = \dfrac{1}{2\alpha} \sin \alpha t - \dfrac{t}{2} \cos \alpha t$

10. $(\cos \alpha t) * (\cos \alpha t) = \dfrac{1}{2\alpha} \sin \alpha t + \dfrac{t}{2} \cos \alpha t$

Use the convolution formula to find the inverse Laplace transforms of the functions in Exercises 11 through 18. You may find the formulas in Exercises 8 through 10 useful in doing some of these problems.

11. $\dfrac{3}{s(s^2 + 4)}$

12. $\dfrac{3}{s(s + 4)}$

13. $\dfrac{3}{s^2(s + 4)}$

14. $\dfrac{1}{(s^2 + 4)^2}$

15. $\dfrac{s}{(s^2 + 4)^2}$

16. $\dfrac{s^2}{(s^2 + 4)^2}$

17. $\dfrac{s}{(s - 1)(s^2 + 1)}$
(*Hint:* Cf. Exercise 6)

18. $\dfrac{s}{(s^2 + 1)(s^2 + 4)}$
(*Hint:* Cf. Exercise 7)

Solve the initial-value problems in Exercises 19 through 25.

19. $(D^2 + 1)^2 x = 0;$ $\quad x(0) = x'(0) = x''(0) = 0, x'''(0) = 1$
20. $(D^2 + 1)^2 x = 0;$ $\quad x(0) = x'(0) = 0, x''(0) = 1, x'''(0) = 0$
21. $(D^2 + 1)^2 x = 0;$ $\quad x(0) = 0, x'(0) = 1, x''(0) = x'''(0) = 0$
22. $(D^2 + 1)^2 x = 0;$ $\quad x(0) = 1, x'(0) = x''(0) = x'''(0) = 0$
    (*Hint:* Partial fractions may help.)
23. $(D^2 + 1)^2 x = \begin{cases} 0, & t < 1 \\ 3, & t \geq 1 \end{cases}$ $\quad x(0) = x'(0) = x''(0) = x'''(0) = 0$
24. $(D^2 + 2D + 2)^2 x = 0;$ $\quad x(0) = x'(0) = x''(0) = 0, x'''(0) = -3$
25. $(D^2 + 2D + 2)^2 x = 0;$ $\quad x(0) = x'(0) = 0, x''(0) = -3, x'''(0) = 0$
26. *Circuits:* Suppose that when $t = 0$, the current in the circuit of Example 5.1.3 is $I_0 = 1$ and the charge is $Q_0 = 0$. Find the charge at times $t = \pi/2$ and $t = 2\pi$ if
    a. $L = 1,$ $R = 0,$ $C = 1,$ $V(t) = \sin t$
    b. $L = 1,$ $R = 0,$ $C = 1,$ $V(t) = \begin{cases} 0, & 0 \leq t < \pi \\ -\sin t, & \pi \leq t \end{cases}$
    c. $L = 1,$ $R = 0,$ $C = \frac{1}{4},$ $V(t) = \sin 2t$
    d. $L = 1,$ $R = 2,$ $C = \frac{1}{2},$ $V(t) = e^{-t} \sin t$

Some more advanced problems:

27. Use Laplace transforms to show that $(\cos t) * \left(\dfrac{1}{\alpha} \sin \alpha t \right) = (\sin t) * (\cos \alpha t)$.

28. a. Use the result of Exercise 26, Section 5.2, and Laplace transforms to show that $(t^{-1/2}) * (t^{-1/2}) = \pi$.
    b. Find $(t^{1/2}) * (t^{1/2})$.
29. Calculate $t * u_a(t)$.
30. a. Evaluate $(\sin t) * (t \sin t)$.

b. Evaluate $(\sin t) * (t \cos t)$.
c. Use (a), (b), and the convolution formula to find

$$\mathcal{L}^{-1}\left[\frac{s}{(s^2 + 1)^3}\right] \quad \text{and} \quad \mathcal{L}^{-1}\left[\frac{1}{(s^2 + 1)^3}\right].$$

d. Solve the initial-value problem

$$(D^2 + 1)^3 x = 0; \qquad x(0) = x'(0) = x''(0) = x'''(0) = 0, \ x^{(iv)}(0) = x^{(v)}(0) = 1.$$

More abstract problems:

31. Use a change of variable in the definition to show that $f * g = g * f$. (*Caution:* Be careful about the limits of integration!)

32. Verify the algebraic properties of the convolution listed in the summary in case the functions f, g, g_1, g_2, and h are continuous and have Laplace transforms.

33. a. Show that if $x = h(t)$ is the solution of the constant-coefficient initial-value problem

(H) $$(D^2 + a_1 D + a_0)x = 0; \qquad x(0) = 0, \ x'(0) = 1$$

then

$$\mathcal{L}[h(t)] = \frac{1}{s^2 + a_1 s + a_0}.$$

b. Show that if $x = x(t)$ is the solution of the initial-value problem

(N) $$(D^2 + a_1 D + a_0)x = E(t); \qquad x(0) = x'(0) = 0$$

then

$$\mathcal{L}[x(t)] = \mathcal{L}[h(t)]\mathcal{L}[E(t)].$$

c. Apply the convolution formula to the equation in (b) to show that

$$x(t) = h(t) * E(t).$$

34. Following the outline of Exercise 33, show that the solution of the constant-coefficient initial-value problem

(N) $$(D^n + a_{n-1}D^{n-1} + \cdots + a_1 D + a_0)x = E(t);$$
$$x(0) = x'(0) = \cdots = x^{(r-1)}(0) = 0$$

is $x(t) = h(t) * E(t)$, where $x = h(t)$ is the solution of the initial-value problem

(H) $$(D^n + a_{n-1}D^{n-1} + \cdots + a_1 D + a_0)x = 0;$$
$$x(0) = x'(0) = \cdots = x^{(n-2)}(0) = 0, x^{(n-1)}(0) = 1.$$

35. Suppose $h_1(t)$ and $h_2(t)$ are linearly independent solutions of the homogeneous o.d.e.

(H) $$D^2 x + a_1(t) Dx + a_0(t)x = 0.$$

a. Use Cramer's rule to show that the specific solution of (H) subject to the conditions at $t = u$

$$x(u) = 0 \quad \text{and} \quad x'(u) = 1$$

is given by the expression

$$H(t, u) = \frac{-h_2(u)}{w(u)} h_1(t) + \frac{h_1(u)}{w(u)} h_2(t)$$

where $w(u)$ is the Wronskian

$$w(u) = \det \begin{bmatrix} h_1(u) & h_2(u) \\ h_1'(u) & h_2'(u) \end{bmatrix}.$$

The function of two variables $H(t, u)$ is called the **Green's function** for (H).

b. Use variation of parameters to show that the general solution of the nonhomogeneous o.d.e.

(N) $$D^2x + a_1(t) Dx + a_0(t)x = E(t)$$

is given by the expression

$$x(t) = \left[-\int_T^t \frac{h_2(u)}{w(u)} E(u)\, du + c_1 \right] h_1(t) + \left[\int_T^t \frac{h_1(u)}{w(u)} E(u)\, du + c_2 \right] h_2[x(t).$$

c. Use (a) and (b) to show that the solution of the initial-value problem

(N) $$D^2x + a_1(t) Dx + a_0(t)x = E(t); \qquad x(T) = x'(T) = 0$$

is given by the expression

$$x(t) = \int_T^t H(t, u)E(u)\, du.$$

d. Find the Green's functions for the homogeneous o.d.e.'s related to the o.d.e.'s given in Exercises 12 and 13 of Section 2.8.

5.7 REVIEW: LAPLACE TRANSFORM SOLUTION OF INITIAL-VALUE PROBLEMS

This section is devoted to working through some initial-value problems that illustrate the role of the methods from the previous sections in the overall solution process. We also make a few comments about some general patterns that come up.

Example 5.7.1

Solve the initial-value problem

$$(D^2 + 3D + 2)x = 2 \sin 2t; \qquad x(0) = x'(0) = 0$$

that models the *LRC* circuit of Example 5.1.3 with inductance $L = 1$, resistance

$R = 3$, and capacitance $C = \frac{1}{2}$, when an alternating voltage $V(t) = 2 \sin 2t$ is plugged in at time $t = 0$.

The transform of the o.d.e. is

$$s^2 \mathcal{L}[x] + 3s\mathcal{L}[x] + 2\mathcal{L}[x] = \frac{4}{s^2 + 4}$$

so

$$\mathcal{L}[x] = \frac{1}{(s^2 + 3s + 2)} \frac{4}{(s^2 + 4)}.$$

The partial fractions decomposition of this expression has the form

$$\frac{4}{(s + 1)(s + 2)(s^2 + 4)} = \frac{A}{s + 1} + \frac{B}{s + 2} + \frac{Cs + E}{s^2 + 4}.$$

The resulting equations for the coefficients are

$$
\begin{aligned}
A + B + C & = 0 \\
2A + B + 3C + E & = 0 \\
4A + 4B + 2C + 3E & = 0 \\
8A + 4B + 2E & = 4
\end{aligned}
$$

with solution

$$A = \frac{4}{5}, \quad B = \frac{-1}{2}, \quad C = \frac{-3}{10}, \quad E = \frac{-2}{10}.$$

Thus

$$
\begin{aligned}
x &= \mathcal{L}^{-1}\left[\frac{4}{(s + 1)(s + 2)(s^2 + 4)} \right] \\
&= \frac{4}{5} \mathcal{L}^{-1}\left[\frac{1}{s + 1} \right] - \frac{1}{2} \mathcal{L}^{-1}\left[\frac{1}{s + 2} \right] - \frac{1}{10} \mathcal{L}^{-1}\left[\frac{3s + 2}{s^2 + 4} \right] \\
&= \frac{4}{5} e^{-t} - \frac{1}{2} e^{-2t} - \frac{3}{10} \cos 2t - \frac{1}{10} \sin 2t.
\end{aligned}
$$

In this example, the transform of the left side of the o.d.e. was $\mathcal{L}[x]$ times $(s^2 + 3s + 2)$, which was the characteristic polynomial of the o.d.e. In general, the initial-value problem

$$P(D)x = E(t); \qquad x(0) = x'(0) = \cdots = x^{(n-1)}(0) = 0$$

will transform (using the first differentiation formula) to

$$P(s)\mathcal{L}[x] = \mathcal{L}[E(t)]$$

so that

$$\mathcal{L}[x] = \frac{1}{P(s)} \mathcal{L}[E(t)].$$

When nonzero initial conditions are imposed, we obtain extra terms, as illustrated in the next example.

Example 5.7.2

Solve the initial-value problem

$$(D^2 + 8D + 15)x = 15; \qquad x(0) = 3, \quad x'(0) = 0,$$

which models a 1-gram mass attached to a horizontal spring with constant $k = 15$, subject to damping $b = 8$ and a constant force $E(t) = 15$. The initial conditions indicate that the mass is pulled 3 centimeters from the equilibrium position and released.

The transform of the o.d.e. is

$$(s^2\mathcal{L}[x] - 3) + 8(s\mathcal{L}[x] - 3) + 15\mathcal{L}[x] = \frac{15}{s}$$

so

$$\mathcal{L}[x] = \frac{15}{s(s^2 + 8s + 15)} + \frac{3s + 24}{s^2 + 8s + 15}.$$

We combine the two terms and look for a single partial fractions decomposition:

$$\frac{15 + 3s^2 + 24s}{s(s^2 + 8s + 15)} = \frac{A}{s} + \frac{B}{s + 3} + \frac{C}{s + 5}.$$

The equations for the coefficients

$$A + B + C = 3$$
$$8A + 5B + 3C = 24$$
$$15A \qquad\qquad = 15$$

have the solution

$$A = 1, \quad B = 5, \quad C = -3.$$

Thus

$$x = \mathcal{L}^{-1}\left[\frac{1}{s} + \frac{5}{s+3} - \frac{3}{s+5}\right] = 1 + 5e^{-3t} - 3e^{-5t}.$$

The expression for $\mathcal{L}[x]$ in this last example is the sum of two terms. The first term has the same form as we obtained in the first example: the transform of the forcing term, divided by the characteristic polynomial. This is the transform of the particular solution of the nonhomogeneous o.d.e. that satisfies initial conditions of absolute rest. The second term, on the other hand, is a polynomial depending on the nonzero initial conditions (and not the forcing term) divided by the characteristic polynomial. This term would be unchanged if we replaced the forcing term with 0 and looked for the solution of the related homogeneous o.d.e. satisfying our given initial condition.

Again, this is a general pattern. The solution of any initial-value problem involving the constant-coefficient o.d.e.

(N) $$P(D)x = E(t)$$

with given initial data at $t = 0$ will lead to an expression of the form

$$\mathcal{L}[x] = \frac{1}{P(s)}\mathcal{L}[E(t)] + \frac{Q(s)}{P(s)},$$

where $Q(s)$ is determined solely by $P(D)$ and the initial data. This is a special case of the fact that the general solution of (N) is the sum of a particular solution of (N) and the general solution of the related homogeneous equation (H): The solution of our initial-value problem breaks into the particular solution of (N) satisfying initial conditions of absolute rest, plus the solution of the related homogeneous equation (H) fitting the given initial data.

Example 5.7.3

Solve the initial-value problem

$$(3D^2 + 6D + 4)x = \frac{11}{2} e^{-t} \sin 2t; \qquad x(0) = 0, \, x'(0) = 1,$$

which models an *LRC* circuit with inductance $L = 3$, resistance $R = 6$, capacitance $C = \frac{1}{4}$, and an alternating voltage of decreasing amplitude $V(t) = (11/2)e^{-t} \sin 2t$. The initial conditions indicate a charge $Q_0 = 0$ and current $I_0 = 1$ at time $t = 0$.

To transform the right-hand side of the o.d.e., we need to apply the first shift formula. First, we find

$$\mathcal{L}[\sin 2t] = \frac{2}{s^2 + 4} = F(s),$$

and then we apply the shift formula:

$$\mathcal{L}\left[\frac{11}{2} e^{-t} \sin 2t\right] = \frac{11}{2} F(s + 1) = \frac{11}{(s + 1)^2 + 4}.$$

We transform the o.d.e. and solve for $\mathcal{L}[x]$:

$$3(s^2 \mathcal{L}[x] - 1) + 6s\mathcal{L}[x] + 4\mathcal{L}[x] = \frac{11}{(s + 1)^2 + 4}.$$

$$\mathcal{L}[x] = \frac{3}{3s^2 + 6s + 4} + \frac{11}{(3s^2 + 6s + 4)([s + 1]^2 + 4)}.$$

Thus

$$x = \mathcal{L}^{-1}\left[\frac{3}{3s^2 + 6s + 4} + \frac{11}{(3s^2 + 6s + 4)([s + 1]^2 + 4)}\right].$$

Upon completion of the square of the first denominator

$$3s^2 + 6s + 4 = 3(s + 1)^2 + 1,$$

we notice that the first shift formula (with $\alpha = -1$) will simplify the entire expression:

$$x = \mathcal{L}^{-1} \left[\frac{3}{3(s + 1)^2 + 1} + \frac{11}{(3[s + 1]^2 + 1)([s + 1]^2 + 4)} \right]$$

$$= e^{-t}\mathcal{L}^{-1} \left[\frac{3}{3s^2 + 1} + \frac{11}{(3s^2 + 1)(s^2 + 4)} \right].$$

Now, we look for a partial fractions decomposition of the second quotient:

$$\frac{11}{(3s^2 + 1)(s^2 + 4)} = \frac{As + B}{3s^2 + 1} + \frac{Cs + E}{s^2 + 4}.$$

The resulting equations

$$
\begin{aligned}
A \quad &+ 3C \quad\quad\quad = 0 \\
B \quad &\quad\quad + 3E = 0 \\
4A \quad &+ C \quad\quad\quad = 0 \\
4B \quad &\quad\quad + E = 11
\end{aligned}
$$

have the solution

$$A = C = 0, \quad B = 3, \quad E = -1.$$

Thus

$$x = e^{-t}\mathcal{L}^{-1} \left[\frac{3}{3s^2 + 1} + \frac{3}{3s^2 + 1} - \frac{1}{s^2 + 4} \right]$$

$$= e^{-t}\mathcal{L}^{-1} \left[\frac{6}{3s^2 + 1} - \frac{1}{s^2 + 4} \right] = e^{-t}\mathcal{L}^{-1} \left[\frac{2}{s^2 + \frac{1}{3}} - \frac{1}{s^2 + 4} \right]$$

$$= 2\sqrt{3}e^{-t} \sin\left(\frac{t}{\sqrt{3}}\right) - \frac{1}{2} e^{-t} \sin 2t.$$

Our final example involves a forcing term defined in pieces.

Example 5.7.4

Solve the initial-value problem

$$(D^2 + 9)x = \begin{cases} 0, & t < 1 \\ 9(t - 1), & 1 < t < 2 \\ 9, & t > 2, \end{cases} \quad x(0) = x'(0) = 0,$$

which models the centrifuge problem in Example 5.1.2 with $m = 1$, $b = 0$, $k = 9$, $t_1 = 1$, $t_2 = 2$, and $F = 9$.

We begin by rewriting the forcing term in step-function notation:

$$(D^2 + 9)x = u_1(t)[9(t - 1)] + u_2(t)[-9(t - 1) + 9]$$
$$= u_1(t)[9(t - 1)] + u_2(t)[-9(t - 2)].$$

We transform the o.d.e., using the second shift formula on the right side:

$$(s^2 + 9)\mathcal{L}[x] = e^{-s}\frac{9}{s^2} - e^{-2s}\frac{9}{s^2}.$$

Solving for $\mathcal{L}[x]$, we have

$$\mathcal{L}[x] = e^{-s}\frac{9}{s^2(s^2 + 9)} - e^{-2s}\frac{9}{s^2(s^2 + 9)}.$$

We will ultimately need to use the second shift formula to inverse-transform both terms above, so we will need to find

$$f(t) = \mathcal{L}^{-1}\left[\frac{9}{s^2(s^2 + 9)}\right].$$

The partial fractions decomposition of the quotient is

$$\frac{9}{s^2(s^2 + 9)} = \frac{1}{s^2} - \frac{1}{s^2 + 9}$$

so

$$f(t) = t - \frac{1}{3}\sin 3t.$$

Now we use the second shift formula to find x:

$$x = \mathcal{L}^{-1}\left[e^{-s}\frac{9}{s^2(s^2 + 9)}\right] - \mathcal{L}^{-1}\left[e^{-2s}\frac{9}{s^2(s^2 + 9)}\right]$$
$$= u_1(t)f(t - 2) - u_2(t)f(t - 2)$$
$$= u_1(t)\left[(t - 1) - \frac{1}{3}\sin 3(t - 1)\right] - u_2(t)\left[(t - 2) - \frac{1}{3}\sin 3(t - 2)\right].$$

Written in pieces, this solution reads

$$x(t) = \begin{cases} 0, & \text{for } t < 1 \\ t - 1 - \dfrac{1}{3}\sin 3(t - 1), & \text{for } 1 < t < 2 \\ 1 - \dfrac{1}{3}\sin 3(t - 1) + \dfrac{1}{3}\sin 3(t - 2), & \text{for } t > 2. \end{cases}$$

We note from this last example that the second shift formula comes into play only when the forcing function is defined in pieces, and the "switch" times for the solution are the same as those of the forcing.

In general, the effective use of the Laplace transform requires familiarity with the basic transforms and the manipulation rules, and comes with practice. We summarize the basic formulas from previous sections and include a few formulas discussed in earlier notes.

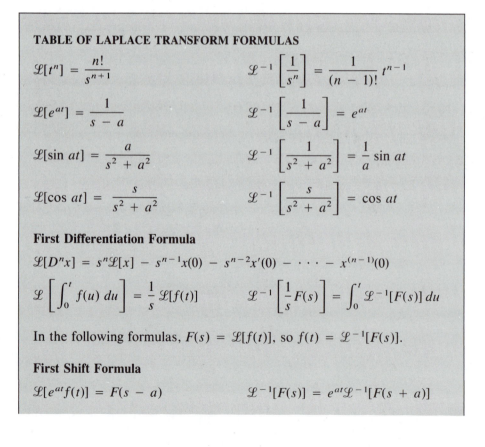

TABLE OF LAPLACE TRANSFORM FORMULAS

$$\mathcal{L}[t^n] = \frac{n!}{s^{n+1}} \qquad \mathcal{L}^{-1}\left[\frac{1}{s^n}\right] = \frac{1}{(n-1)!}\, t^{n-1}$$

$$\mathcal{L}[e^{at}] = \frac{1}{s - a} \qquad \mathcal{L}^{-1}\left[\frac{1}{s - a}\right] = e^{at}$$

$$\mathcal{L}[\sin at] = \frac{a}{s^2 + a^2} \qquad \mathcal{L}^{-1}\left[\frac{1}{s^2 + a^2}\right] = \frac{1}{a}\sin at$$

$$\mathcal{L}[\cos at] = \frac{s}{s^2 + a^2} \qquad \mathcal{L}^{-1}\left[\frac{s}{s^2 + a^2}\right] = \cos at$$

First Differentiation Formula

$$\mathcal{L}[D^n x] = s^n \mathcal{L}[x] - s^{n-1}x(0) - s^{n-2}x'(0) - \cdots - x^{(n-1)}(0)$$

$$\mathcal{L}\left[\int_0^t f(u)\, du\right] = \frac{1}{s}\mathcal{L}[f(t)] \qquad \mathcal{L}^{-1}\left[\frac{1}{s}F(s)\right] = \int_0^t \mathcal{L}^{-1}[F(s)]\, du$$

In the following formulas, $F(s) = \mathcal{L}[f(t)]$, so $f(t) = \mathcal{L}^{-1}[F(s)]$.

First Shift Formula

$$\mathcal{L}[e^{at}f(t)] = F(s - a) \qquad \mathcal{L}^{-1}[F(s)] = e^{at}\mathcal{L}^{-1}[F(s + a)]$$

Second Differentiation Formula

$$\mathcal{L}[t^n f(t)] = (-1)^n \frac{d^n}{ds^n} \mathcal{L}[f(t)] \qquad \mathcal{L}^{-1}\left[\frac{d^n F(s)}{ds^n}\right] = (-1)^n t^n f(t)$$

Second Shift Formula

$$\mathcal{L}[u_a(t)g(t)] = e^{-as}\mathcal{L}[g(t+a)] \qquad \mathcal{L}^{-1}[e^{-as}F(s)] = u_a(t)f(t-a)$$

Convolution

$$\mathcal{L}^{-1}[F(s)G(s)] = \mathcal{L}^{-1}[F(s)] * \mathcal{L}^{-1}[G(s)],$$

where

$$(f * g)(t) = \int_0^t f(t-u)g(u)\,du.$$

Periodic Functions

If $f(t+p) = f(t)$ for all t, then

$$\mathcal{L}[f(t)] = \frac{\int_0^p e^{-st}f(t)\,dt}{1 - e^{-ps}}.$$

EXERCISES (REVIEW PROBLEMS FOR CHAPTER 5)

Solve the initial-value problems in Exercises 1 through 19.

1. $(9D^2 - 4)x = 0$; $x(0) = x'(0) = 3$
2. $(D^2 + 3)x = 0$; $x(0) = x'(0) = -4$
3. $(D^2 + 2D + 5)x = 0$; $x(0) = x'(0) = 2$
4. $(2D^2 + 2D + 1)x = 0$; $x(0) = 0, x'(0) = 2$
5. $x'' + 4x' + 4x = te^{-2t}$; $x(0) = 0, x'(0) = 1$
6. $x'' + 4x' + 4x = te^{-2t}$; $x(0) = 1, x'(0) = 0$
7. $x'' - 2x' + 10x = t^2 e^t$; $x(0) = 0, x'(0) = -6$
8. $(D - 1)x = \begin{cases} 1, & t \le 1 \\ t, & 1 \le t < 2 \\ t^2 - 2, & 2 \le t \end{cases}$ $x(0) = 0$
9. $(D^2 + 3D + 2)x = \begin{cases} e^{2t}, & t < 1 \\ e^{3t}, & t \ge 1 \end{cases}$ $x(0) = x'(0) = 0$

10. $(D^2 - 2D + 1)x = \begin{cases} 0, & t < 2 \\ t - 2, & t \geq 2 \end{cases}$ $x(0) = x'(0) = 1$

11. $(D^2 + 1)x = \begin{cases} 0, & t < 1 \\ t, & t \geq 1 \end{cases}$ $x(0) = 0, x'(0) = 1$

12. $(D^2 + 1)x = \begin{cases} t, & t < 1 \\ 0, & t \geq 1 \end{cases}$ $x(0) = 0, x'(0) = 1$

13. $(D^2 - 2D + 1)x = \begin{cases} t, & t < 2 \\ e^t, & t \geq 2 \end{cases}$ $x(0) = x'(0) = 0$

14. $(D^2 + 1)x = \begin{cases} \sin t, & t \leq \pi \\ 0, & t \geq \pi \end{cases}$ $x(0) = x'(0) = 0$

15. $(D^3 - D)x = 0$; $x(0) = 1, x'(0) = x''(0) = 0$

16. $(D^3 - D)x = 0$; $x(0) = 0, x'(0) = 1, x''(0) = 0$

17. $(D^3 - D)x = 1$; $x(0) = x'(0) = x''(0) = 0$

18. $(D^4 - 1)x = 0$; $x(0) = x'(0) = x''(0) = 0, x'''(0) = 1$

19. $(D^4 - 4D^3 + 6D^2 - 4D + 1)x = 60e^t$; $x(0) = x'(0) = x''(0) = 0, x'''(0) = 1$

Exercises 20 through 22 refer to our models

20. *Ice Water:* How long will it take for the water in Exercise 2, Section 5.1, to reach 32°F?

21. *Fishing:*
     a. Find a formula describing the motion of the box in Exercise 8, Section 5.1.
     b. Show that at least one third of the box will be out of the water at all times.

22. *A Circuit:* Suppose the voltage in the circuit of Example 5.1.3 is the square wave

$$V(t) = \begin{cases} 1 & \text{if } n < t < n + 1, \quad n \text{ even} \\ -1 & \text{if } n < t < n + 1, \quad n \text{ odd} \end{cases}$$

sketched in Figure 5.6. Suppose also that $L = 1$, $R = 0$, $C = \frac{1}{9}$, $Q(0) = 3$, and $I(0) = 0$. Find the charge and current after
     a. $\frac{1}{2}$ sec;   b. $\frac{3}{2}$ sec;   c. $\frac{5}{2}$ sec.

[*Hint:* The values of $V(t)$ for $t > \frac{5}{2}$ do not affect the answer.]

5.8   LAPLACE TRANSFORMS FOR SYSTEMS

The method of Laplace transforms readily extends to initial-value problems for linear differential systems with constant coefficients. We illustrate with several examples.

Example 5.8.1

Solve the system of o.d.e.'s

$$x' = 3x - 2y$$
$$y' = 4x - y,$$

subject to the initial condition

$$x(0) = 1, \quad y(0) = -1.$$

As in the case of a single equation, we transform both sides of each equation, incorporating the initial data by means of the first differentiation formula. This gives

$$s\mathcal{L}[x] - 1 = 3\mathcal{L}[x] - 2\mathcal{L}[y]$$

$$s\mathcal{L}[y] + 1 = 4\mathcal{L}[x] - \mathcal{L}[y].$$

If we move all terms involving $\mathcal{L}[x]$ or $\mathcal{L}[y]$ to one side, we recognize this as a linear algebraic system of equations in the unknowns $\mathcal{L}[x]$ and $\mathcal{L}[y]$:

$$(s - 3)\mathcal{L}[x] + \qquad 2\mathcal{L}[y] = 1$$

$$-4\mathcal{L}[x] + (s + 1)\mathcal{L}[y] \quad = -1.$$

This system can be solved for $\mathcal{L}[x]$ and $\mathcal{L}[y]$ in terms of s (for example, by Cramer's rule):

$$\mathcal{L}[x] = \frac{s + 3}{s^2 - 2s + 5}, \quad \mathcal{L}[y] = \frac{-s + 7}{s^2 - 2s + 5}.$$

We find x and y by applying the inverse transform [this involves completing the square of $s^2 - 2s + 5 = (s - 1)^2 + 4$ and using the first shift formula]:

$$x(t) = \mathcal{L}^{-1}\left[\frac{s + 3}{s^2 - 2s + 5}\right] = e^t \mathcal{L}^{-1}\left[\frac{s + 4}{s^2 + 4}\right]$$

$$= e^t \cos 2t + 2e^t \sin 2t$$

$$y(t) = \mathcal{L}^{-1}\left[\frac{-s + 7}{s^2 - 2s + 5}\right] = e^t \mathcal{L}^{-1}\left[\frac{-s + 6}{s^2 + 4}\right]$$

$$= -e^t \cos 2t + 3e^t \sin 2t.$$

Example 5.8.2

Solve the system of o.d.e.'s

$$DQ = -Q - I_2 \qquad + V(t)$$

$$DI_2 = \quad Q - I_2 + I_3$$

$$DI_3 = \qquad I_2 - I_3 - V(t),$$

where

$$V(t) = \begin{cases} 0, & t < 1 \\ 10, & t > 1 \end{cases}$$

and subject to the initial condition

$$Q(0) = 10, \quad I_2(0) = I_3(0) = 0.$$

This models the circuit of Example 5.1.4 with $R_1 = R_2 = L_1 = L_2 = C = 1$, $V_1(t) = 0$, and $V_2(t) = 10$ (a battery). At $t = 0$ there is a charge of 10 coulombs on the capacitor but no current.

We rewrite $V(t)$ in step-function notation, $V(t) = 10u_1(t)$, and transform the system:

$$s\mathcal{L}[Q] - 10 = -\mathcal{L}[Q] - \mathcal{L}[I_2] \qquad\qquad + 10\,\frac{e^{-s}}{s}$$

$$s\mathcal{L}[I_2] \quad = \quad \mathcal{L}[Q] - \mathcal{L}[I_2] + \mathcal{L}[I_3]$$

$$s\mathcal{L}[I_3] \quad = \qquad\qquad \mathcal{L}[I_2] - \mathcal{L}[I_3] - 10\,\frac{e^{-s}}{s}.$$

Some rearrangement of the terms leads to

$$(s + 1)\mathcal{L}[Q] + \qquad \mathcal{L}[I_2] \qquad\qquad = 10 + 10\,\frac{e^{-s}}{s}$$

$$-\mathcal{L}[Q] + (s + 1)\mathcal{L}[I_2] - \qquad \mathcal{L}[I_3] = 0$$

$$-\mathcal{L}[I_2] + (s + 1)\mathcal{L}[I_3] = -\frac{10e^{-s}}{s}.$$

We solve this system for $\mathcal{L}[Q]$, $\mathcal{L}[I_2]$, and $\mathcal{L}[I_3]$ to get

$$\mathcal{L}[Q] = \frac{10}{s + 1} - \frac{10}{(s + 1)^3} + \frac{10e^{-s}}{s(s + 1)}$$

$$= \frac{10}{s + 1} - \frac{10}{(s + 1)^3} + 10e^{-s}\left(\frac{1}{s} - \frac{1}{s + 1}\right)$$

$$\mathcal{L}[I_2] = \frac{10}{(s + 1)^2}$$

$$\mathcal{L}[I_3] = \frac{10}{(s + 1)^3} - \frac{10e^{-s}}{s(s + 1)}$$

$$= \frac{10}{(s + 1)^3} - 10e^{-s}\left(\frac{1}{s} - \frac{1}{s + 1}\right).$$

We can find the inverse transforms of these expressions using the shift formulas:

$$Q = 10e^{-t} - 5t^2e^{-t} + 10u_1(t)[1 - e^{-(t-1)}]$$

$$I_2 = 10te^{-t}$$

$$I_3 = 5t^2e^{-t} - 10u_1(t)[1 - e^{-(t-1)}].$$

Finally, we note that the Laplace transform can be applied equally well to systems of o.d.e.'s of order higher than 1, as shown by the following example.

Example 5.8.3

Find the solution of the system of o.d.e.'s

$$4D^2x_1 = -8x_1 + 2x_2$$
$$D^2x_2 = 2x_1 - 2x_2 + 24,$$

subject to the initial condition

$$x_1(0) = x_2(0) = x_1'(0) = x_2'(0) = 0.$$

This models the system of Example 5.1.5 with masses $m_1 = 4$ and $m_2 = 1$, spring constants $k_1 = 6$ and $k_2 = 2$, and a constant force $E(t) = 24$, starting from absolute rest.

We transform the system, obtaining

$$4s^2\mathcal{L}[x_1] = -8\mathcal{L}[x_1] + 2\mathcal{L}[x_2]$$

$$s^2\mathcal{L}[x_2] = 2\mathcal{L}[x_1] - 2\mathcal{L}[x_2] + \frac{24}{s}.$$

Rearranging, we obtain

$$4(s^2 + 2)\mathcal{L}[x_1] - 2\mathcal{L}[x_2] = 0$$

$$-2\mathcal{L}[x_1] + (s^2 + 2)\mathcal{L}[x_2] = \frac{24}{s}.$$

Solving for $\mathcal{L}[x_1]$ and $\mathcal{L}[x_2]$, we have

$$\mathcal{L}[x_1] = \frac{12}{s(s^2 + 3)(s^2 + 1)}, \quad \mathcal{L}[x_2] = \frac{24(s^2 + 2)}{s(s^2 + 3)(s^2 + 1)}.$$

Using partial fractions, we obtain

$$x_1 = \mathcal{L}^{-1}\left[\frac{12}{s(s+3)(s^2+1)}\right] = \mathcal{L}^{-1}\left[\frac{4}{s} + \frac{2s}{s^2+3} - \frac{6s}{s^2+1}\right]$$
$$= 4 + 2\cos\sqrt{3}t - 6\cos t$$

$$x_2 = \mathcal{L}^{-1}\left[\frac{24(s^2+2)}{s(s^2+3)(s^2+1)}\right] = \mathcal{L}^{-1}\left[\frac{16}{s} - \frac{4s}{s^2+3} - \frac{12s}{s^2+1}\right]$$
$$= 16 - 4\cos\sqrt{3}t - 12\cos t.$$

Let's summarize.

LAPLACE TRANSFORMS FOR SYSTEMS

To solve an initial-value problem for a system of n linear constant-coefficient o.d.e.'s (of any order) in the unknowns $x_1, \ldots, x_n$:

1. Transform each o.d.e., incorporating the initial data by the first differentiation formula.
2. Gather all terms involving $\mathcal{L}[x_1], \ldots, \mathcal{L}[x_n]$ on one side and all other terms on the other.
3. Solve the resulting linear algebraic system for $\mathcal{L}[x_1], \ldots, \mathcal{L}[x_n]$ in terms of s.
4. Inverse-transform each $\mathcal{L}[x_i]$ to obtain $x_i(t)$.

Note

On spurious solutions

When we apply the Laplace transform to a system of o.d.e.'s with specified initial conditions, we obtain a new system of equations that is satisfied by the transform of a solution to our initial-value problem—*provided such a solution exists*. The existence of a solution is ensured by the theorem in Section 3.3 whenever the system consists of n linear constant-coefficient o.d.e.'s in standard form

$$Dx_1 = a_{11}x_1 + \cdots + a_{1n}x_n + E_1(t)$$

$$\vdots$$

$$Dx_n = a_{n1}x_1 + \cdots + a_{nn}x_n + E_n(t)$$

with the $E_i(t)$'s continuous, and the initial conditions specify a value for each variable at $t = 0$. However, when this theorem does not apply, application of the Laplace transform method can lead to spurious solutions. For example, the system

(S)
$$Dx + y = 0$$
$$D^2x + y + Dy = 1$$

with initial conditions of absolute rest

$$x(0) = x'(0) = 0, \quad y(0) = 0$$

can be formally solved by Laplace transforms [see Exercise 16(a)], but the method yields

$$x = -t, \quad y = 1.$$

This pair of functions solves (S) but fails to satisfy the initial conditions. In fact, no solution of (S) can satisfy the given initial conditions [see Exercise 16(b)].

This example shows that when we use Laplace transforms to solve an initial-value problem for which the system is not in standard form, we should substitute our answer back to check that it actually solves the initial-value problem with which we began.

EXERCISES

In Exercises 1 through 10, find the solution of the given system that satisfies the given initial condition.

1. $Dx = x + 2y$
    $Dy = 2x + y$ $x(0) = 5, y(0) = -5$

2. $Dx = x + 2y + e^t$
    $Dy = 2x + y$ $x(0) = y(0) = 0$

3. $Dx = x + 2y + u_1(t)$
    $Dy = 2x + y$ $x(0) = 5, y(0) = -5$

4. $Dx = x - 2y$
    $Dy = 2x + y$ $x(0) = 5, y(0) = -5$

5. $D^2x = -3x - 4y$
    $Dy = 2x + 3y$ $x(0) = 0, x'(0) = 1, y(0) = 0$

6. $D^2x = 3x + 6y + 6e^{2t}$
    $Dy = x + 2y$ $x(0) = x'(0) = y(0) = y'(0) = 0$

7. $D^2x = 5x - y$
    $D^2y = 3x - 5y$ $x(0) = 8, x'(0) = 16, y(0) = y'(0) = 0$

8. $Dx = x - y$
    $Dy = -y + z$ $x(0) = 0, y(0) = 12, z(0) = 0$
    $Dz = y - z$

9. $Dx = x - y$
    $Dy = -y - z + 12$ $x(0) = y(0) = z(0) = 0$
    $Dz = y - z$

10. $Dx = \quad x + 5y$
    $Dy = -x - \quad y$ $\qquad x(0) = 8, y(0) = z(0) = 0$
    $Dz = \quad x \qquad + 2z + e^{2t}$

Exercises 11 through 15 refer to our models.

11. *Circuits:* Find the solution of the system of equations describing the circuit of Example 3.1.2 under the following assumptions.
    a. $V(t) = 0, R_1 = 2, R_2 = 3, L = 6, C = \frac{1}{6}$, and $I_2(0) = 1, I_1(0) = -3$. [Compare this to Exercise 15(b), Section 3.7.]
    b. $V(t) = 0, R_1 = R_2 = L = C = 1$, and $I_2(0) = 1, I_1(0) = 2$. [Compare this to Exercise 14(b), Section 3.8.]
    c. $V(t) = 0, R_1 = 1, R_2 = 3, C = L = 1$, and $Q(0) = 1, I_2(0) = 2$. [Compare this to Exercise 16(b), Section 3.9.]
    d. $V(t) = 10, R_1 = 1, R_2 = 3, C = L = 1$, and $I_1(0) = I_2(0) = 0$ (see Exercise 9, Section 5.1).

12. *Radioactive Decay:* Solve the initial-value problem of Exercise 19, Section 3.11, describing Professor Kay's experiment.

13. *More Radioactive Decay:* An undergraduate assistant calculates that the reaction described in Exercise 19, Section 3.11 will never run dry, so Professor Kay decides to risk extracting more of substance C. One year after he began the experiment, he starts extracting 125 grams per year of substance C.
    a. Solve the system describing this experiment (see Exercise 10, Section 5.1).
    b. Will the reaction run dry?

14. *A Circuit:* Use Laplace transforms to solve the initial-value problem solved in Example 3.11.3.

15. *Supply and Demand:* Solve the initial-value problem of Exercise 21(b), Section 3.11.

16. *Spurious Solutions:* Suppose x and y satisfy the system

    (S) $\qquad\qquad\qquad\qquad\qquad Dx + y = 0$
    $$D^2x + y + Dy = 1$$

    and initial conditions

    (I) $\qquad\qquad\qquad x(0) = 0, \quad x'(0) = 0, \quad y(0) = 0.$

    a. Show that the Laplace transforms of x and y must be

    $$\mathcal{L}[x] = -\frac{1}{s^2}, \quad \mathcal{L}[y] = \frac{1}{s},$$

    so that $x = -t, y = 1$. Note that these functions don't satisfy (I).
    b. By differentiating the first equation of (S) and substituting in the second, show that the general solution of (S) is

    $$x = c - t, \quad y = 1,$$

    so that the initial-value problem specified by (I) has no solution.

17. *Systems in Matrix Form:* The Laplace transform of a vector valued function

    $$\mathbf{x}(t) = \begin{bmatrix} x_1(t) \\ \cdot \\ \cdot \\ \cdot \\ x_n(t) \end{bmatrix}$$

is defined by the rule

$$\mathscr{L}[\mathbf{x}(t)] = \begin{bmatrix} \mathscr{L}[x_1(t)] \\ \cdot \\ \cdot \\ \cdot \\ \mathscr{L}[x_n(t)] \end{bmatrix}$$

a. Show that $\mathscr{L}[D\mathbf{x}(t)] = s\mathscr{L}[\mathbf{x}(t)] - \mathbf{x}(0)$.
b. If A is an $n \times n$ matrix with constant entries, show that $\mathscr{L}[A\mathbf{x}(t)] = A\mathscr{L}[\mathbf{x}(t)]$.
c. Apply the Laplace transform to both sides of the constant-coefficient system

$$D\mathbf{x} = A\mathbf{x} + \mathbf{E}(t)$$

and regroup the resulting terms to show that

$$(A - sI)\mathscr{L}[\mathbf{x}] = -\mathbf{x}(0) - \mathscr{L}[\mathbf{E}(t)].$$

d. Cramer's rule can now be used to obtain an expression for $\mathscr{L}[x_i(t)]$ as a quotient of determinants, with $\det[A - sI]$ in the denominator (valid for those values of s for which $\det[A - sI] \neq 0$). How does this denominator compare with the characteristic polynomial of A discussed in Section 3.5?
e. When the system is homogeneous ($\mathbf{E}(t) = \mathbf{0}$), the numerator of the expression for $\mathscr{L}[x_i(t)]$ is a polynomial in s of degree less than n. What does this tell us about the relationship between the functions $x_i(t)$ and the roots of the characteristic polynomial? Compare with Sections 3.7 to 3.10.

REVIEW PROBLEMS

See Section 5.7.

6

LINEAR EQUATIONS WITH VARIABLE COEFFICIENTS: POWER SERIES

6.1 TEMPERATURE MODELS: O.D.E.'S FROM P.D.E.'S

Until now we have focused our attention primarily on linear o.d.e.'s with constant coefficients. In this chapter we develop an approach to solving certain o.d.e.'s with variable coefficients. Although the solutions of these equations may not have finite expressions in terms of elementary functions (polynomials, exponentials, trigonometric functions, and so on), we will be able to find expressions for them in terms of power series. You may recall from calculus that the first several terms of a power series expression can be used to obtain a good approximation to the value of a function at a specified point. (Indeed, if our goal is to obtain a decimal expression for such a value, a series expression can be more useful than a closed-form expression involving exponentials and trigonometric functions.) In addition, series expressions can often be used to obtain powerful information about the behavior of solutions of o.d.e.'s.

Although linear o.d.e.'s with variable coefficients can be obtained by varying our earlier models (see Exercises 1 through 5), in practice they usually arise when we look for special solutions to certain partial differential equations (p.d.e.'s). The way such problems lead to o.d.e.'s is illustrated in this section by several versions of the problem of finding a steady-state (ie., time-independent) temperature distribution in a body.

In Section 8.1 we discuss more carefully the derivation of p.d.e.'s modeling temperature distribution. For the present, we note that the steady-state temperature $u(x, y)$ in a two-dimensional plate whose flat surfaces are insulated must satisfy the p.d.e.

$$(\text{L}_2) \qquad \frac{\partial^2 u}{\partial x^2} + \frac{\partial^2 u}{\partial y^2} = 0,$$

and the steady-state temperature $u(x, y, z)$ in a three-dimensional solid satisfies

(L$_3$)
$$\frac{\partial^2 u}{\partial x^2} + \frac{\partial^2 u}{\partial y^2} + \frac{\partial^2 u}{\partial z^2} = 0.$$

These p.d.e.'s are known as the two- and three-dimensional **Laplace equations.**
As a first step toward solving these p.d.e.'s, we look for solutions that can be written as products of functions of one variable. Our hope is that we can replace the p.d.e. by o.d.e.'s for these functions. In the following example we illustrate how the search for such special solutions to the two-dimensional Laplace equation can lead to familiar o.d.e.'s. When we return to this problem in Chapter 8, we will see how the solutions to these o.d.e.'s can be combined to find solutions of the Laplace equation that match given boundary conditions.

▪ Example 6.1.1   Temperature in a Rectangular Plate

Suppose the temperature in a thin rectangular plate, insulated above and below, is controlled by heating (or cooling) elements along its edges (see Figure 6.1). We wish to predict the steady-state temperature, $u(x, y)$, that satisfies the two-dimensional Laplace equation (L$_2$) inside the rectangle.
We look for solutions of (L$_2$) in the special form

(S)
$$u = X(x)Y(y).$$

Substituting (S) into (L$_2$), we obtain

$$\frac{d^2 X}{dx^2} Y + X \frac{d^2 Y}{dy^2} = 0$$

or

$$\frac{1}{X} \frac{d^2 X}{dx^2} = -\frac{1}{Y} \frac{d^2 Y}{dy^2}.$$

Note that the left side of this equation is a function of x alone, but the right is independent of x. The only way this can happen is if both sides have a common *constant* value,

$$\frac{1}{X} \frac{d^2 X}{dx^2} = -\frac{1}{Y} \frac{d^2 Y}{dy^2} = \lambda.$$

When we write these equations separately, we obtain two ordinary differential

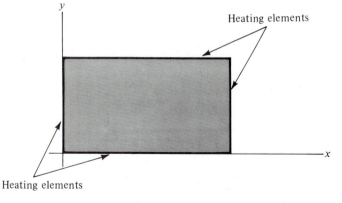

FIGURE 6.1

equations:

$$\frac{1}{X}\frac{d^2X}{dx^2} = \lambda \quad \text{and} \quad -\frac{1}{Y}\frac{d^2Y}{dy^2} = \lambda.$$

These can be rewritten in the more familiar forms

(H$_x$) $[D^2 - \lambda]X = 0$

(H$_y$) $[D^2 + \lambda]Y = 0$

where D stands for d/dx in (H$_x$) and for d/dy in (H$_y$).

Note that the solutions of (H$_x$) and (H$_y$) depend on the specific values of λ; these in turn are determined by the temperature along the edges of the plate (see, for example, Exercise 6). Once we solve (H$_x$) and (H$_y$), we substitute into (S) to obtain solutions to (L$_2$).

In the next examples we will see how o.d.e.'s with variable coefficients arise when we try this procedure on the Laplace equation in polar, cylindrical, or spherical coordinates.

Example 6.1.2   Temperature in a Circular Plate

To investigate the steady-state temperature distribution in a circular plate of radius 1 (insulated above and below) when we know the temperature on the boundary, we are led to the two-dimensional Laplace equation (L$_2$) in the interior of the disc $x^2 + y^2 < 1$. For such problems, it is most convenient to switch to polar coordinates. Rewritten in terms of r and θ, (L$_2$) becomes

(Exercise 9)

(H)
$$\frac{\partial^2 u}{\partial r^2} + \frac{1}{r}\frac{\partial u}{\partial r} + \frac{1}{r^2}\frac{\partial^2 u}{\partial \theta^2} = 0.$$

We look for solutions of (H) in the form

(S)
$$u = R(r)\Theta(\theta).$$

Substituting (S) into (H), we obtain

$$\frac{d^2 R}{dr^2}\,\Theta + \frac{1}{r}\frac{dR}{dr}\,\Theta + \frac{1}{r^2}\,R\,\frac{d^2\Theta}{d\theta^2} = 0.$$

Subtracting the last term from both sides, multiplying by r^2, and dividing by $u = R\Theta$, we come to

$$\frac{r^2}{R}\frac{d^2 R}{dr^2} + \frac{r}{R}\frac{dR}{dr} = -\frac{1}{\Theta}\frac{d^2\Theta}{d\theta^2}.$$

Since the left side of this equation is a function of r alone and the right side depends only on θ, the common value must be a constant, λ. This leads to two o.d.e.'s

(H$_\theta$)
$$-\frac{1}{\Theta}\,D^2\Theta = \lambda$$

(H$_r$)
$$\frac{r^2}{R}\,D^2 R + \frac{r}{R}\,DR = \lambda,$$

where D stands for $d/d\theta$ in (H$_\theta$) and for d/dr in (H$_r$).

The first of these equations can be rewritten as

(H$_\theta$)
$$(D^2 + \lambda)\Theta = 0.$$

Since Θ is a function of the angular coordinate θ, it must satisfy the geometric condition $\Theta(\theta + 2\pi) = \Theta(\theta)$ for all θ. If this is to hold for all solutions of (H$_\theta$), then λ must be a perfect square (Exercise 7):

$$\lambda = n^2, \quad n \geq 0 \text{ an integer.}$$

Of course, the value of λ in (H$_r$) must agree with that in (H$_\theta$). Thus, we are led to the o.d.e. for $R(r)$,

(H$_r$)
$$\frac{r^2}{R}\,D^2 R + \frac{r}{R}\,DR = n^2.$$

Multiplying by R and rearranging terms, we obtain

$$[r^2 D^2 + rD - n^2]R = 0.$$

This is a special case of the **Cauchy-Euler equation**

(CE)
$$p_0 t^2 \frac{d^2 x}{dt^2} + q_0 t \frac{dx}{dt} + r_0 x = 0$$

(with $p_0 = q_0 = 1$ and $r_0 = -n^2$). We will learn to solve (CE) in Section 6.5.

Example 6.1.3   Temperature in a Plug

A metal plug in the shape of a (solid) cylinder is insulated along the surface; the temperature is controlled by two rings on the top and bottom edges (see Figure 6.2). We wish to predict the steady-state temperature in the plug, $u(x, y, z)$, which must satisfy the three-dimensional Laplace equation (L$_3$).

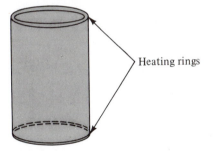

Heating rings

FIGURE 6.2

We start by looking for solutions of (L$_3$) in the form

(S$_1$)
$$u = U(x, y)Z(z).$$

Substituting (S$_1$) into (L$_3$), we get

$$\frac{\partial^2 U}{\partial x^2} Z + \frac{\partial^2 U}{\partial y^2} Z + U \frac{d^2 Z}{dz^2} = 0$$

or

$$\frac{1}{U}\left[\frac{\partial^2 U}{\partial x^2} + \frac{\partial^2 U}{\partial y^2}\right] = -\frac{1}{Z}\frac{d^2 Z}{dz^2}.$$

Since the right side is a function of z alone and the left side is independent of

z, each is constant, and we obtain two equations

(H$_z$)
$$-\frac{1}{Z}\frac{d^2Z}{dz^2} = \lambda$$

(H$_{xy}$)
$$\frac{1}{U}\left[\frac{\partial^2U}{\partial x^2} + \frac{\partial^2U}{\partial y^2}\right] = \lambda.$$

The first equation is again the familiar o.d.e.

$$[D^2 + \lambda]Z = 0,$$

(where $D = d/dz$), whose solutions depend on λ. It turns out that physical considerations lead to the requirement that λ is nonpositive, so

$$\lambda = -\gamma^2 \quad \text{for some } \gamma \geq 0.$$

The shape of the plug leads us to try to rewrite the function U in terms of polar coordinates r and θ. The p.d.e. (H$_{xy}$) now takes the form (Exercise 9)

(H$_{r\theta}$)
$$\frac{1}{U}\left[\frac{\partial^2U}{\partial r^2} + \frac{1}{r}\frac{\partial U}{\partial r} + \frac{1}{r^2}\frac{\partial^2U}{\partial\theta^2}\right] = -\gamma^2.$$

When $\gamma = 0$, this is equivalent to the p.d.e. of the previous example. When $\gamma > 0$, we can still look for solutions to (H$_{r\theta}$) in the separated form

(S$_2$)
$$U = R(r)\Theta(\theta).$$

Substituting (S$_2$) into (H$_{r\theta}$), we obtain

$$\frac{1}{R}\frac{d^2R}{dr^2} + \frac{1}{rR}\frac{dR}{dr} + \frac{1}{r^2\Theta}\frac{d^2\Theta}{d\theta^2} = -\gamma^2$$

or, multiplying by r^2 and rearranging terms,

$$\frac{r^2}{R}\frac{d^2R}{dr^2} + \frac{r}{R}\frac{dR}{dr} + (\gamma r)^2 = -\frac{1}{\Theta}\frac{d\Theta}{d\theta^2}.$$

Once again, the two sides of this equation depend on different variables, so we set each equal to a common constant, κ:

(H$_\theta$)
$$-\frac{1}{\Theta}\frac{d^2\Theta}{d\theta^2} = \kappa$$

(H$_r$)
$$\frac{r^2}{R}\frac{d^2R}{dr^2} + \frac{r}{R}\frac{dR}{dr} + (\gamma r)^2 = \kappa.$$

As in Example 6.1.2, the geometric condition $\Theta(\theta + 2\pi) = \Theta(\theta)$ forces κ to be a perfect square:

$$\kappa = n^2, \quad n \geq 0 \text{ an integer.}$$

The substitution $t = \gamma r$ then puts (H$_r$) in the form of the **Bessel equation**

(B$_\mu$) $$[t^2 D^2 + tD + (t^2 - \mu^2)]x = 0$$

with $D = d/dt$, $x = R$, and $\mu = n$. We will discuss this equation in Sections 6.7 and 6.8.

Example 6.1.4   Temperature in a Planet

Suppose the temperature at the surface of a spherical planet is a function of the latitude alone (see Figure 6.3). We would like to predict the internal temperature distribution u, which we assume to be steady-state.

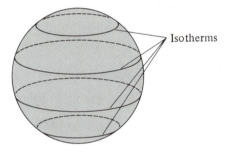

Isotherms

FIGURE 6.3

Again, u must satisfy the three-dimensional Laplace equation (L$_3$), but here we expect it to be most convenient to express u in terms of spherical coordinates ρ, ϕ, and θ (see Figure 6.4). We have assumed that the temperature on the surface depends on the latitude ϕ and not on the longitude θ, so we expect the same to be true inside the planet. When $u = u(\rho, \phi)$ is independent of θ, the three-dimensional Laplace equation becomes (Exercise 10)

(H) $$\rho \frac{\partial^2}{\partial \rho^2} (\rho u) + \frac{1}{\sin \phi} \frac{\partial}{\partial \phi} \left(\sin \phi \frac{\partial u}{\partial \phi} \right) = 0.$$

We look for a solution of (H) in the separated form

(S) $$u = R(\rho)\Phi(\phi).$$

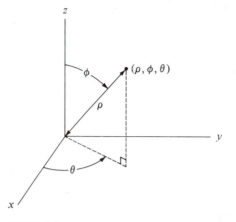

FIGURE 6.4

Substituting (S) into (H), we obtain

$$\rho\Phi \frac{d^2}{d\rho^2}(\rho R) + \frac{R}{\sin\phi} \frac{d}{d\phi}\left(\sin\phi \frac{d\Phi}{d\phi}\right) = 0$$

or

$$\frac{\rho}{R} \frac{d^2}{d\rho^2}(\rho R) = -\frac{1}{\Phi \sin\phi} \frac{d}{d\phi}\left(\sin\phi \frac{d\Phi}{d\phi}\right).$$

As before, the two sides of this equation depend on distinct variables, so the common value is a constant:

(H$_\rho$) $$\frac{\rho}{R} \frac{d^2}{d\rho^2}(\rho R) = \lambda$$

(H$_\phi$) $$-\frac{1}{\Phi \sin\phi} \frac{d}{d\phi}\left(\sin\phi \frac{d\Phi}{d\phi}\right) = \lambda.$$

The first o.d.e. (H$_\rho$) can be written in the form

$$[\rho^2 D^2 + 2\rho D - \lambda]R = 0,$$

which is another case of the Cauchy-Euler equation described in Example 6.1.2 (this time with $p_0 = 1$, $q_0 = 2$, and $r_0 = -\lambda$). The second o.d.e. (H$_\phi$) does not resemble any of our earlier examples. It turns out that if we want solutions of (H$_\phi$) to be twice differentiable at the poles, $\phi = 0$ and $\phi = \pi$, then we are forced to choose $\lambda = n(n + 1)$, where n is a nonnegative integer. If, in addition, we make the substitution $t = \cos \phi$, we obtain (Exercise 8)

$$\frac{1}{\Phi}\frac{d}{dt}\left[(t^2 - 1)\frac{d\Phi}{dt}\right] = n(n + 1).$$

After some manipulation, this takes the form of the **Legendre equation**

$$[(t^2 - 1)D^2 + 2tD - n(n + 1)]\Phi = 0.$$

EXERCISES

1. *A Circuit with Varying Resistance:* The resistance of metals varies with temperature. Suppose the resistor of Exercise 5, Section 2.1 (or Example 3.1.1) is heated so that its resistance at time t is $R(t) = 10 + t$. Find an o.d.e. for the charge Q on the capacitor if $L = C = 1$ and $V(t) = 10$.

2. *A Varying Spring Constant:* When a spring is heated, the spring constant may change. Write an o.d.e. modeling the spring system of Example 2.1.2, with $m = b = 10$, under the assumption that the spring "constant" at time t is $k(t) = 10 - t$.

3. *Varying Damping:* As oil is heated, it becomes slipperier. Write an o.d.e. modeling the spring system of Example 2.1.2, with $m = k = 10$, under the assumption that the spring is attached to a dashpot that is being heated so that the damping coefficient at time t is $b(t) = 10 - t$.

4. *Varying Mass:* Suppose the mass attached to the spring of Example 2.1.2 is a 10 gram block of ice that melts at the rate of 1 gram per second. Write an o.d.e. modeling the motion of the system if $k = 10$ and $b = 11$. [*Note:* When mass varies, Newton's second law takes the form $F = d(mv)/dt$. Use this, *not* $F = ma$, as the basis for this model.]

5. *A Submarine:* Recall the principle of Archimedes (Exercise 7, Section 2.1), which states that water buoys up a submerged object by a force equal to the weight of the water it displaces. Suppose a submarine weighing 50 tons (100,000 lb) is resting on the ocean floor and that the buoyant force exactly counters the force of gravity at this stage. Now, water is pumped out of its tanks at 100 lb/second.
   a. Write a formula for the mass $m(t)$ of the submarine after t seconds.
   b. Write a formula for the net force $F(t)$ acting on the submarine after t seconds.
   c. Use Newton's second law in the form for varying mass ($F = d(mv)/dt$, not $F = ma$) to derive a second-order o.d.e. modeling the position of the submarine, assuming no damping forces.

6. *The Rectangular Plate:* Suppose the temperatures of the bottom edge ($y = 0$) and the top edge ($y = h$) of the plate in Example 6.1.1 are held at $0°$; that is,

   (B) $u(x, 0) = u(x, h) = 0$ for all x.

a. Show that if $u = X(x)Y(y)$ satisfies (B), but is not identically zero, then

(B$_1$) $Y(0) = Y(h) = 0.$

b. Show that if $\lambda < 0$, then the only solution of (H$_y$) that satisfies (B$_1$) is the trivial solution $Y = 0$.

c. Show that if $\lambda = 0$, then (H$_y$) has no nontrivial solutions that satisfy (B$_1$).

d. Show that if $\lambda > 0$, then (H$_y$) has nontrivial solutions that satisfy (B$_1$) only if $\sin h\sqrt{\lambda} = 0$, or $\lambda = n^2\pi^2/h^2$ where n is a nonnegative integer.

e. Show that if $\lambda = n^2\pi^2/h^2$, then the nontrivial solutions of (H$_y$) that satisfy (B$_1$) are constant multiples of

$$y_n = \sin\left(\frac{n\pi y}{h}\right),$$

and (H$_x$) has two linearly independent solutions,

$$x_n = e^{n\pi x/h} \quad \text{and} \quad x_n^* = e^{-n\pi x/h}.$$

f. Corresponding to the solutions of (H$_y$) and (H$_x$) described in (e), obtain solutions to (L$_2$),

$$u_n = e^{-n\pi x/h} \sin\left(\frac{n\pi y}{h}\right) \quad \text{and} \quad u_n^* = e^{-n\pi x/h} \cos\left(\frac{n\pi y}{h}\right)$$

that satisfy (B). Note that any linear combination of these (infinitely many) functions is a solution of (L$_2$) and satisfies (B).

Exercises 7 through 10 fill some of the gaps in the discussion of Examples 6.12 through 6.14.

7. a. Show that if $\lambda < 0$, then the only solution of

(H$_\theta$) $\dfrac{d^2\Theta}{d\theta^2} + \lambda\Theta = 0.$

that satisfies the geometric condition

(G) $\Theta(\theta + 2\pi) = \Theta(\theta) \quad \text{for all } \theta$

is $\Theta = 0$.

b. Show that if $\lambda = 0$, then the only solutions of (H$_\theta$) that satisfy (G) are constant multiples of $\Theta = 1$.

c. Show that if $\lambda > 0$, then (H$_\theta$) has nontrivial solutions that satisfy (G) only if $\lambda = n^2$, where n is a nonnegative integer.

8. a. Show that the substitution $t = \cos\phi$ changes the equation

(H$_\phi$) $-\dfrac{1}{\Phi \sin\phi} \dfrac{d}{d\phi}\left[\sin\phi \dfrac{d\Phi}{d\phi}\right] = n(n+1)$

into

$$\frac{1}{\Phi} \frac{d}{dt}\left[(t^2 - 1)\frac{d\Phi}{dt}\right] = n(n+1).$$

b. Show that this last equation can be rewritten in the form

$$[(t^2 - 1)D^2 + 2tD - n(n + 1)]\Phi = 0.$$

9. *Conversion to Polar Coordinates*
  a. Use the chain rule for partial derivatives to show that if $v = v(x, y)$, where $x = r \cos \theta$ and $y = r \sin \theta$, then

$$\frac{\partial v}{\partial r} = \cos \theta \frac{\partial v}{\partial x} + \sin \theta \frac{\partial v}{\partial y}$$

$$\frac{\partial v}{\partial \theta} = -r \sin \theta \frac{\partial v}{\partial x} + r \cos \theta \frac{\partial v}{\partial y}.$$

  b. Solve the preceding equations for $\partial v/\partial x$ and $\partial v/\partial y$ in terms of $\partial v/\partial r$ and $\partial v/\partial \theta$:

$$\frac{\partial v}{\partial x} = \cos \theta \frac{\partial v}{\partial r} - \frac{1}{r} \sin \theta \frac{\partial v}{\partial \theta}$$

$$\frac{\partial v}{\partial y} = \sin \theta \frac{\partial v}{\partial r} + \frac{1}{r} \cos \theta \frac{\partial v}{\partial \theta}.$$

  c. By applying the equations in (b), first to $v = u(x, y)$ and then to $v = \partial u/\partial x$ and $v = \partial u/\partial y$, show that

$$\frac{\partial^2 u}{\partial x^2} + \frac{\partial^2 u}{\partial y^2} = \frac{\partial^2 u}{\partial r^2} + \frac{1}{r} \frac{\partial u}{\partial r} + \frac{1}{r^2} \frac{\partial^2 u}{\partial \theta^2}.$$

  d. Conclude that the change of variables $x = r \cos \theta$ and $y = r \sin \theta$ changes

$$\frac{\partial^2 u}{\partial x^2} + \frac{\partial^2 u}{\partial y^2} = \beta$$

  into

$$\frac{\partial^2 u}{\partial r^2} + \frac{1}{r} \frac{\partial u}{\partial r} + \frac{1}{r^2} \frac{\partial^2 u}{\partial \theta^2} = \beta.$$

*10. *Conversion to Spherical Coordinates*
  a. Suppose $v = v(x, y, z)$, where $x = \rho \sin \phi \cos \theta$, $y = \rho \sin \phi \sin \theta$, and $z = \rho \cos \phi$. Use the chain rule to express $\partial v/\partial \rho$, $\partial v/\partial \phi$, and $\partial v/\partial \theta$ in terms of $\partial v/\partial x$, $\partial v/\partial y$, and $\partial v/\partial z$.
  b. Solve the equations in (a) for $\partial v/\partial x$, $\partial v/\partial y$, and $\partial v/\partial z$ in terms of $\partial v/\partial \rho$, $\partial v/\partial \phi$, and $\partial v/\partial \theta$.
  c. By applying the equation in (b) to $v = u(x, y, z)$, $v = \partial u/\partial x$, $v = \partial u/\partial y$, and $v = \partial u/\partial z$, show that

$$\frac{\partial^2 u}{\partial x^2} + \frac{\partial^2 u}{\partial y^2} + \frac{\partial^2 u}{\partial z^2} = \frac{1}{\rho} \frac{\partial^2}{\partial \rho^2} (\rho u) + \frac{1}{\rho^2 \sin \phi} \frac{\partial}{\partial \phi} \left(\sin \phi \frac{\partial u}{\partial \phi} \right) + \frac{1}{\rho^2 \sin^2 \phi} \frac{\partial^2 u}{\partial \theta^2}.$$

  d. Conclude that if u is independent of θ, then the change of variables $x = \rho \sin \phi \cos \theta$, $y = \rho \sin \phi \sin \theta$, and $z = \rho \cos \phi$ changes the three-dimensional Laplace equation (L$_3$) into

$$\rho \frac{\partial^2}{\partial \rho^2} (\rho u) + \frac{1}{\sin \phi} \frac{\partial}{\partial \phi} \left(\sin \phi \frac{\partial u}{\partial \phi} \right) = 0.$$

6.2   REVIEW OF POWER SERIES

In this section we recall some features of power series that we will find useful in solving o.d.e.'s. Our review is brief; you can find a more elaborate treatment in most good calculus texts.

A **series** is a formal sum

$$\sum_{k=0}^{\infty} u_k = u_0 + u_1 + \cdots + u_k + \cdots.$$

To evaluate this series, we consider the **partial sums**

$$S_n = \sum_{k=0}^{n} u_k = u_0 + u_1 + \cdots + u_n.$$

If the partial sums approach a limit

$$S = \lim_{n \to \infty} S_n,$$

we call S the **sum** of the series,

$$\sum_{k=0}^{\infty} u_k = S$$

and say the series **converges** to S. If the partial sums do not approach a limit, we say the series **diverges.**

There are many tests for determining whether a series converges. The most useful one for our purposes is the ratio test.

Fact: Ratio Test.   *Given the series $\sum_{k=0}^{\infty} u_k$, suppose that*

$$\rho = \lim_{n \to \infty} \left| \frac{u_{n+1}}{u_n} \right|.$$

Then:

1.   If $\rho < 1$, the series converges.
2.   If $\rho > 1$, the series diverges.
3.   If $\rho = 1$, the test fails. The series may converge or diverge.

We will be dealing with power series. A **power series about $t = t_0$** is a series of the form

$$\sum_{k=0}^{\infty} b_k(t - t_0)^k = b_0 + b_1(t - t_0) + \cdots + 1\, b_k(t - t_0)^k + \cdots,$$

where the b_k's and t_0 are constants, and t is a variable. Note that the partial sums of a power series are polynomials. The basic convergence properties of power series are given by the following statement.

Fact: *Associated to each power series $\Sigma_{k=0}^{\infty} b_k(t - t_0)^k$ is a "number" R, with $0 \leq R \leq \infty$, so that*

1. *if $0 < R < \infty$, the series converges whenever $|t - t_0| < R$ and diverges whenever $|t - t_0| > R$;*
2. *if $R = \infty$, the series converges for all t; and*
3. *if $R = 0$, the series converges only when $t = t_0$.*

The "number" R is called the **radius of convergence** of the power series. In cases 1 and 2, when $R > 0$, we refer to the values of t with $|t - t_0| < R$ (that is, $t_0 - R < t < t_0 + R$) as the **interval of convergence** of the power series. Note that in case 1 nothing was said about the endpoints ($t = t_0 \pm R$) of this interval; the series may converge or diverge at either point.

The radius of convergence of a power series can usually be calculated by means of the ratio test.

Example 6.2.1

Find the interval of convergence of $\Sigma_{k=0}^{\infty} t^k$.
We apply the ratio test. Since

$$\rho = \lim_{n\to\infty} \left| \frac{u_{n+1}}{u_n} \right| = \lim_{n\to\infty} \left| \frac{t^{n+1}}{t^n} \right| = \lim_{n\to\infty} |t| = |t|,$$

the series converges for $|t| < 1$ and diverges for $|t| > 1$. The interval of convergence is $|t| < 1$.

Example 6.2.2

Find the interval of convergence of $\Sigma_{k=0}^{\infty} (t^k/k!)$.
By the ratio test, since

$$\rho = \lim_{n\to\infty} \left| \frac{u_{n+1}}{u_n} \right| = \lim_{n\to\infty} \left| \frac{t^{n+1}/(n+1)!}{t^n/n!} \right|$$

$$= \lim_{n\to\infty} \left| \frac{t^{n+1} n!}{(n+1)! t^n} \right| = \lim_{n\to\infty} \left| \frac{t}{n+1} \right| = 0 < 1,$$

the series always converges. The interval of convergence is the whole real line, $|t| < \infty$.

If a power series has $R > 0$, then it defines a function on its interval of convergence. Conversely, given a function $f(t)$, it may be possible to find a power series about $t = t_0$ so that

$$f(t) = \sum_{k=0}^{\infty} b_k(t - t_0)^k \quad \text{for } |t - t_0| < R,$$

where $R > 0$. If such a series exists, we say that $f(t)$ is **analytic** at $t = t_0$.

As examples, we list some basic functions that are analytic at $t = 0$, together with expressions for these functions as series. (See Examples 6.2.4 and 6.2.6, and Exercise 31.)

Fact: *The following functions are analytic at $t = 0$.*

$$b_0 + b_1 t + \cdots + b_m t^m = b_0 + b_1 t + \cdots + b_m t^m + 0t^{m+1} + \cdots, \quad |t| < \infty$$

$$e^t = \sum_{k=0}^{\infty} \frac{t^k}{k!}, \quad |t| < \infty$$

$$\sin t = \sum_{k=0}^{\infty} \frac{(-1)^k t^{2k+1}}{(2k+1)!}, \quad |t| < \infty$$

$$\cos t = \sum_{k=0}^{\infty} \frac{(-1)^k t^{2k}}{(2k)!}, \quad |t| < \infty$$

$$\frac{1}{1-t} = \sum_{k=0}^{\infty} t^k, \quad |t| < 1.$$

In this chapter, we look for analytic solutions of an o.d.e. by substituting a formal power series $\sum_{k=0}^{\infty} b_k(t - t_0)^k$ into the equation and attempting to determine the constants b_k. In order to carry out this process, we need to know when a power series is identically equal to zero, how to do arithmetic with power series, and how to differentiate power series. Equality of a series with zero is governed by the following (see Exercise 32):

Fact: *If $\sum_{k=0}^{\infty} b_k(t - t_0)^k = 0$ for all t satisfying $|t - t_0| < R$, where $R > 0$, then $b_k = 0$ for every k.*

The basic arithmetic operations for power series at $t = t_0$ are analogous to those for polynomials. In particular, we multiply a series by a number or add two series, termwise.

Fact: *If the series expressions*

$$f(t) = \sum_{k=0}^{\infty} b_k(t - t_0)^k \quad and \quad g(t) = \sum_{k=0}^{\infty} c_k(t - t_0)^k$$

are both valid (at least) for $|t - t_0| < R$, *where* $R > 0$, *then so are the expressions*

$$\alpha f(t) = \sum_{k=0}^{\infty} (\alpha b_k)(t - t_0)^k$$

$$f(t) + g(t) = \sum_{k=0}^{\infty} (b_k + c_k)(t - t_0)^k.$$

The multiplication and division of series is more complicated (see Notes 1 and 2); the multiplication of a power series by a polynomial is illustrated in the following examples.

Example 6.2.3

Find the terms up to t^4 in the expression for

$$e^{2t} - (2 + t)\cos t$$

as a power series about $t = 0$.

The series expression for e^t described above is valid for all values of t. We obtain the expression for e^{2t} by replacing t with $2t$:

$$e^{2t} = 1 + 2t + \frac{(2t)^2}{2} + \frac{(2t)^3}{6} + \frac{(2t)^4}{24} + \cdots$$

$$= 1 + 2t + 2t^2 + \frac{4}{3}t^3 + \frac{2}{3}t^4 + \cdots.$$

Next, we multiply the series for $\cos t$ by $(2 + t)$:

$$(2 + t)\cos t = (2 + t)\left[1 - \frac{t^2}{2} + \frac{t^4}{24} - \cdots\right]$$

$$= (2 + t) - \left(t^2 + \frac{t^3}{2}\right) + \left(\frac{t^4}{12} + \frac{t^5}{24}\right) + \cdots.$$

Finally, we subtract the expression for $(2 + t)\cos t$ from the expression for e^{2t}:

$$e^{2t} = 1 + 2t + 2t^2 + \frac{4}{3}t^3 + \frac{2}{3}t^4 + \cdots$$

$$(2 + t)\cos t = 2 + t - t^2 - \frac{1}{2}t^3 + \frac{1}{12}t^4 + \cdots$$

$$e^{2t} - (2 + t)\cos t = -1 + t + 3t^2 + \frac{11}{6}t^3 + \frac{7}{12}t^4 + \cdots.$$

Example 6.2.4

Verify that

$$\frac{1}{1 - t} = \sum_{k=0}^{\infty} t^k \quad \text{for } |t| < 1.$$

We multiply the series on the right by $(1 - t)$; since this series converges only for $|t| < 1$ (see Example 6.2.1), we must restrict our attention to this interval:

$$(1 - t) \sum_{k=0}^{\infty} t^k = \sum_{k=0}^{\infty} t^k - t \sum_{k=0}^{\infty} t^k = \sum_{k=0}^{\infty} t^k - \sum_{k=0}^{\infty} t^{k+1}, \quad |t| < 1.$$

To perform our subtraction termwise, we must match up the powers of t. We accomplish this by index substitution—substitute $j = k$ in the first series and $j = k + 1$ in the second, remembering that this also affects the limits of summation:

$$(1 - t) \sum_{k=0}^{\infty} t^k = \sum_{j=0}^{\infty} t^j - \sum_{j=1}^{\infty} t^j, \quad |t| < 1.$$

Now we have to match up the limits of summation. We do this by breaking the first sum in two:

$$(1 - t) \sum_{k=0}^{\infty} t^k = \left(1 + \sum_{j=1}^{\infty} t^j\right) - \sum_{j=1}^{\infty} t^j = 1, \quad |t| < 1.$$

Thus, dividing both sides by $1 - t$, we obtain the desired equality:

$$\sum_{k=0}^{\infty} t^k = \frac{1}{1 - t}, \quad |t| < 1.$$

We can differentiate an analytic function by differentiating its power series term by term.

Fact: *If the series expression*

$$f(t) = \sum_{k=0}^{\infty} b_k(t - t_0)^k = b_0 + b_1(t - t_0) + b_2(t - t_0)^2 + \cdots$$

is valid for $|t - t_0| < R$ *(where* $R > 0$*), then so is the expression obtained by termwise differentiation:*

$$f'(t) = \sum_{k=1}^{\infty} kb_k(t - t_0)^{k-1} = b_1 + 2b_2(t - t_0) + \cdots.$$

Repeated application of this differentiation formula yields the following important relation between the derivatives at $t = t_0$ of an analytic function $f(t)$ and the coefficients of its power series expression about $t = t_0$.

Fact: *If* $f(t) = \sum_{k=0}^{\infty} b_k(t - t_0)^k$ *for* $|t - t_0| < R, R > 0$, *then the derivatives of* f *at* $t = t_0$ *are*

$$f^{(k)}(t_0) = k!b_k \quad \text{for } k = 0, 1, \ldots.$$

Example 6.2.5

Find a series expression for $1/(1 - t)^2$ valid for $|t| < 1$.
We know that

$$\frac{1}{1 - t} = \sum_{k=0}^{\infty} t^k, \quad |t| < 1.$$

Differentiation gives

$$\frac{1}{(1 - t)^2} = \sum_{k=1}^{\infty} kt^{k-1}, \quad |t| < 1.$$

The limits of summation can be changed by the index substitution $j = k - 1$ (and $k = j + 1$):

$$\frac{1}{(1 - t)^2} = \sum_{j=0}^{\infty} (j + 1)t^j, \quad |t| < 1.$$

Example 6.2.6

Verify that

$$e^t = \sum_{k=0}^{\infty} \frac{t^k}{k!} \quad \text{for all } t.$$

Denote the series on the right side by $f(t)$, recalling (Example 6.2.2) that it converges for all t:

$$f(t) = \sum_{k=0}^{\infty} \frac{t^k}{k!}, \quad |t| < \infty.$$

Differentiation yields

$$f'(t) = \sum_{k=1}^{\infty} \frac{kt^{k-1}}{k!} = \sum_{k=1}^{\infty} \frac{t^{k-1}}{(k-1)!}$$

and the index substitution $j = k - 1$ (or $k = j + 1$) gives

$$f'(t) = \sum_{j=0}^{\infty} \frac{t^j}{j!} = f(t), \quad |t| < \infty.$$

Thus $x = f(t)$ is a solution of

$$(D - 1)x = 0.$$

Then

$$f(t) = ce^t,$$

and since $f(0) = 1$, we have $c = 1$. It follows that $f(t) = e^t$, or

$$\sum_{k=0}^{\infty} \frac{t^k}{k!} = e^t, \quad |t| < \infty.$$

Our rules for manipulating series enable us to find power series expressions for a wide variety of functions, starting from our expressions for e^t, $\sin t$, $\cos t$, and $1/(1 - t)$. We discuss briefly the problem of finding power series expressions for more complicated functions in the notes following our summary.

POWER SERIES REVIEW

Definitions

A **power series about** $t = t_0$ is a formal expression

$$p(t) = \sum_{k=0}^{\infty} b_k(t - t_0)^k = b_0 + b_1(t - t_0) + \cdots + b_k(t - t_0)^k + \cdots.$$

If for some value $t = t_1$ of t

$$\lim_{n \to \infty} [b_0 + b_1(t_1 - t_0) + \cdots + b_n(t_1 - t_0)^n]$$

exists, then we say the series **converges** at $t = t_1$ and set $p(t_1)$ equal to this limit. If the limit does not exist, we say the series **diverges** at $t = t_1$. Associated with the power series is a **radius of convergence** R, with $0 \leq R \leq \infty$, so that the series converges for all t in the **interval of convergence** $|t - t_0| < R$ and diverges when $|t - t_0| > R$.

A function $f(t)$ is **analytic** at $t = t_0$ if there is a power series about $t = t_0$ with radius of convergence $R > 0$ such that

$$f(t) = \sum_{k=0}^{\infty} b_k(t - t_0)^k, \quad |t - t_0| < R.$$

Properties of Analytic Functions

Suppose that the series expressions

$$f(t) = \sum_{k=0}^{\infty} b_k(t - t_0)^k \quad \text{and} \quad g(t) = \sum_{k=0}^{\infty} c_k(t - t_0)^k$$

are both valid for $|t - t_0| < R, R > 0$.

1. If $f(t) = 0$ for $|t - t_0| < R$ then $b_k = 0$ for $k = 0, 1, 2, \ldots$.

2. $\alpha f(t) = \sum_{k=0}^{\infty} (\alpha b_k)(t - t_0)^k, \quad |t - t_0| < R.$

3. $f(t) + g(t) = \sum_{k=0}^{\infty} (b_k + c_k)(t - t_0)^k, \quad |t - t_0| < R.$

4. $f'(t) = \sum_{k=1}^{\infty} k b_k(t - t_0)^{k-1}, \quad |t - t_0| < R.$

5. $f^{(k)}(t_0) = k! b_k, \quad k = 0, 1, \ldots.$

Notes

1. Products of power series

If the power series expressions

$$f(t) = \sum_{k=0}^{\infty} b_k(t - t_0)^k \quad \text{and} \quad g(t) = \sum_{k=0}^{\infty} c_k(t - t_0)^k$$

are both valid for $|t - t_0| < R$, then so is the expression

$$f(t)g(t) = \sum_{k=0}^{\infty} (b_0c_k + b_1c_{k-1} + \cdots + b_kc_0)(t - t_0)^k.$$

That is, power series can be multiplied like polynomials. For example, we can obtain the first terms of a series about $t = 0$ for $e^t \cos t$ by multiplying the respective series for e^t and $\cos t$:

$$e^t = 1 + t + \frac{1}{2}t^2 + \frac{1}{6}t^3 + \frac{1}{24}t^4 + \cdots$$

$$\cos t = 1 \qquad - \frac{1}{2}t^2 \qquad + \frac{1}{24}t^4 + \cdots$$

$$1 \times e^t = 1 + t + \frac{1}{2}t^2 + \frac{1}{6}t^3 + \frac{1}{24}t^4 + \cdots$$

$$-\frac{1}{2}t^2 \times e^t = \qquad - \frac{1}{2}t^2 - \frac{1}{2}t^3 - \frac{1}{4}t^4 + \cdots$$

$$\frac{1}{24}t^4 \times e^t = \qquad \frac{1}{24}t^4 + \cdots$$

$$e^t \cos t = 1 + t \qquad - \frac{1}{3}t^3 - \frac{1}{6}t^4 + \cdots$$

2. Quotients of analytic functions

If $f(t)$ and $g(t)$ have expressions as power series about $t = t_0$ that are both valid for $|t - t_0| < R$, and if $g(z) \neq 0$ for every complex number z with $|z - t_0| < R$, then $f(t)/g(t)$ has an expression as a power series about $t = t_0$, also valid for $|t - t_0| < R$. Note, however, that *complex* zeros of the denominator affect the convergence of the quotient series even at *real* values of t. For example, the series for $1/(1 + t^2)$ (see Exercise 29) diverges for real t with $|t| > 1$ even though $1 + t^2$ is never zero for real values of t.

One can find any desired number of terms for a quotient series by means of long division. For example, we can find the first terms of the series for $1/(1 - t)$ as follows:

$$
\begin{array}{r}
1 + t + t^2 + \cdots \\
1 - t \overline{\smash{\big)}\ 1 } \\
\underline{1 - t} \\
t \\
\underline{t - t^2} \\
t^2 \\
\underline{t^2 - t^3} \\
t^3 \\
\vdots
\end{array}
$$

3. Taylor series

Suppose we are given a function $f(t)$ for which we want to find a power series expression about $t = t_0$. If there is such an expression

$$
f(t) = \sum_{k=0}^{\infty} b_k(t - t_0)^k
$$

valid for $|t - t_0| < R$, where $R > 0$, then

$$
f^{(k)}(t_0) = k! b_k,
$$

so the coefficients must be given by

$$
b_k = \frac{f^{(k)}(t_0)}{k!}.
$$

The series with these coefficients,

$$
\sum_{k=0}^{\infty} \frac{f^{(k)}(t_0)}{k!} (t - t_0)^k
$$

is known as the **Taylor series for $f(t)$ about $t = t_0$**. It is the only series about $t = t_0$ that could possibly equal $f(t)$.

In practice, it may be easy to decide where the Taylor series converges but difficult to decide whether its limit equals $f(t)$. Indeed, there are examples of functions $f(t)$ whose Taylor series converge, but not to $f(t)$.

Examples 6.2.4 to 6.2.6 illustrated some of the methods used to show that a function is equal to a given series. Another method is to study the "remainder term"

$$
R_n(t, t_0) = f(t) - \sum_{k=0}^{n} \frac{f^{(k)}(t_0)}{k!} (t - t_0)^k.
$$

The function $f(t)$ equals its Taylor series precisely when

$$\lim_{n \to \infty} R_n(t, t_0) = 0.$$

EXERCISES

In Exercises 1 through 9, find the interval of convergence.

1. $\displaystyle\sum_{k=0}^{\infty} \frac{2^k t^k}{(k + 1)!}$

2. $\displaystyle\sum_{k=0}^{\infty} \frac{2^k t^k}{(k + 1)^2}$

3. $\displaystyle\sum_{k=0}^{\infty} \frac{(-1)^k k! t^k}{(k^2 + 1)}$

4. $\displaystyle\sum_{k=0}^{\infty} (-1)^k t^{2k}$

5. $\displaystyle\sum_{k=0}^{\infty} \frac{(k^2 + 1) t^k}{(k + 1) 3^k}$

6. $\displaystyle\sum_{k=0}^{\infty} \frac{k^2 t^k}{(2k)!}$

7. $\displaystyle\sum_{k=0}^{\infty} \frac{(t - 2)^k}{(3k + 1)^2}$

8. $\displaystyle\sum_{k=0}^{\infty} \frac{(2t + 1)^k}{3^k}$

9. $\displaystyle\sum_{k=0}^{\infty} \frac{k!(2t - 1)^k}{(2k + 1)!}$

In Exercises 10 through 17, perform the indicated formal operation.

10. $\displaystyle\sum_{k=0}^{\infty} t^k - \sum_{k=0}^{\infty} (k + 1) t^k$

11. $\displaystyle\sum_{k=0}^{\infty} t^k + \sum_{k=0}^{\infty} (-1)^k t^k$

12. $\displaystyle\sum_{k=0}^{\infty} t^k + \sum_{k=0}^{\infty} t^{k+1}$

13. $\displaystyle (1 + t) \sum_{k=0}^{\infty} (-1)^k t^k$

14. $\displaystyle (t^2 - 1) \sum_{k=0}^{\infty} 3kt^k$

15. $\displaystyle \frac{d}{dt} \left(\sum_{k=0}^{\infty} (3k + 1) t^k \right)$

16. $\displaystyle \frac{d^2}{dt^2} \left(\sum_{k=0}^{\infty} \frac{(-1)^k t^{2k+1}}{(2k + 1)!} \right)$

17. $\displaystyle [t^2 D^2 - 3tD] \sum_{k=0}^{\infty} \frac{3t^k}{k^2}$

In Exercises 18 through 22, find the terms up to t^5 of the given product or quotient (see Notes 1 and 2).

18. $\displaystyle \left(\sum_{k=0}^{\infty} t^k \right) \left(\sum_{k=0}^{\infty} (-1)^k t^k \right)$

19. $\displaystyle \left(\sum_{k=0}^{\infty} \frac{t^k}{k!} \right)^2$

20. $\displaystyle \left(\sum_{k=0}^{\infty} \frac{(-1)^k t^{2k}}{(2k)!} \right) \left(\sum_{k=0}^{\infty} \frac{(-1)^k t^{2k+1}}{(2k + 1)!} \right)$

21. $\displaystyle \left(\sum_{k=0}^{\infty} t^k \right) \Big/ (1 + t^2)$

22. $\displaystyle \left(\sum_{k=0}^{\infty} \frac{2^k t^k}{k!} \right) \Big/ \left(\sum_{k=0}^{\infty} \frac{t^k}{k!} \right)$

Starting with the basic series expressions for $\sin t$, $\cos t$, e^t, and $1/(1 - t)$, find series expressions for the functions in Exercises 23 through 30.

23. $\sin \dfrac{t}{2}$

24. $\cos 2t$

25. e^{-t}

26. $\dfrac{1}{1 + t}$

27. $\dfrac{1}{(1 + t)^2}$

28. $\dfrac{1}{(1 + t)^3}$

29. $\dfrac{1}{1 + t^2}$

30. $\cos^2 t$ (*Hint:* Use a trig identity or see Note 1.)

Some more abstract problems:

31. *Series Expressions for sin t and cos t:*
    a. Show that $\Sigma_{k=0}^{\infty} (-1)^k t^{2k+1}/(2k + 1)!$ converges for all t.
    b. Verify that $\sin t$ and the series in (a) are equal by showing that they are solutions of the same initial value problem:
    $$(D^2 + 1)x = 0, \quad x(0) = 0, x'(0) = 1.$$
    c. Obtain an expression for $\cos t$ by differentiating the expression for $\sin t$.

32. *Equality of a Series with Zero:* Show that if $f(t) = \Sigma_{k=0}^{\infty} b_k(t - t_0)^k = 0$ for all t satisfying $|t - t_0| < R$ $(R > 0)$, then $b_k = 0$, $k = 0, 1, \ldots$ [*Hint:* Express b_k in terms of $f^{(k)}(t_0)$.]

*33. Consider the function
    $$f(t) = \begin{cases} e^{-1/t^2}, & \text{for } t \neq 0 \\ 0, & \text{for } t = 0. \end{cases}$$
    a. Show that $D(e^{-1/t^2}) = (2/t^3)e^{-1/t^2}$ for $t \neq 0$.
    b. Show that in general $D^n(e^{-1/t^2}) = (P_n(t)/t^{3n})e^{-1/t^2}$ for $t \neq 0$, where $P_n(t)$ is a polynomial of degree at most $2(n - 1)$.
    c. Show that $f^{(n)}(0) = 0$ for each n. (*Hint:* Use the fact that for any α, $t^{\alpha}e^{-1/t^2} \to 0$ as $t \to 0$.)
    d. Show that the Taylor series for $f(t)$ expanded about $t = 0$ (see Note 3) converges for all $t \neq 0$, but the sum does not equal $f(t)$ for $t > 0$.

6.3 SOLUTIONS ABOUT ORDINARY POINTS

We will see in this section how to find power series expressions for the solutions to linear o.d.e.'s

(N) $[a_n(t)D^n + a_{n-1}(t)D^{n-1} + \cdots + a_0(t)]x = E(t)$

in case the coefficients $a_n(t), \ldots, a_0(t)$ and the forcing term $E(t)$ are all polynomials. We will find such series by substituting the expression

$$x(t) = \sum_{k=0}^{\infty} b_k(t - t_0)^k$$

into (N) and solving for the coefficients b_k. This straightforward approach always works when t_0 is a point at which the leading coefficient does not vanish. Such points [where $a_n(t_0) \neq 0$] are called the **ordinary points** of (N) to distinguish them from the **singular points** of (N), where $a_n(t_0) = 0$.

The following theorem justifies our search for power series solutions of (N) about an ordinary point. Moreover, it guarantees the validity of the answers we obtain as long as t is closer to t_0 than the nearest complex root of $a_n(t)$.

Theorem: *Suppose that (N) is a linear nth-order o.d.e. with polynomial coefficients and forcing term and that t_0 is an ordinary point of (N). If $a_n(z) \neq 0$ for every complex number z satisfying $|z - t_0| < R$, then any solution of (N) has an expression as a power series about $t = t_0$ that is valid for (at least) $|t - t_0| < R$.*

To see how the procedure works in practice, we start with a familiar o.d.e.

Example 6.3.1

Find a power series expression about $t = 0$ for the general solution of

$$\text{(H)} \qquad\qquad (D^2 - 1)x = 0.$$

Note that (H) has no singular points.

We look for solutions of (H) in the form

$$\text{(S)} \qquad\qquad x(t) = \sum_{k=0}^{\infty} b_k t^k.$$

$$x'(t) = \sum_{k=1}^{\infty} kb_k t^{k-1}, \quad x''(t) = \sum_{k=2}^{\infty} k(k-1)b_k t^{k-2}$$

and substitute into (H):

$$\sum_{k=2}^{\infty} k(k-1)b_k t^{k-2} - \sum_{k=0}^{\infty} b_k t^k = 0.$$

In order to combine these two series into one, we need to match up powers of t. To this end, we perform the index substitutions $j = k - 2$ (or $k = j + 2$) in the first series and $j = k$ in the second; we must remember to adjust the limits of summation as well as the summands. We get

$$\sum_{j=0}^{\infty} (j + 2)(j + 1)b_{j+2}t^j - \sum_{j=0}^{\infty} b_j t^j = 0$$

or

$$\sum_{j=0}^{\infty} [(j + 2)(j + 1)b_{j+2} - b_j]t^j = 0.$$

Since a power series is identically zero only if all its coefficients are zero, we conclude that

$$(j + 2)(j + 1)b_{j+2} - b_j = 0$$

or

(R) $$b_{j+2} = \frac{1}{(j + 2)(j + 1)} b_j \quad \text{for } j = 0, 1, \ldots .$$

An equation such as (R), which relates a given coefficient to earlier ones, is called a **recurrence relation.**

Although (R) tells us nothing about b_0 or b_1, we can use it to find all of the later coefficients in terms of these two. When $j = 0$ and $j = 1$, respectively, (R) gives

$$b_2 = \frac{1}{2 \cdot 1} b_0, \quad b_3 = \frac{1}{3 \cdot 2} b_1.$$

Now these expressions can be substituted back into (R) with $j = 2$ and 3, respectively, to find

$$b_4 = \frac{1}{4 \cdot 3} b_2 = \frac{1}{4 \cdot 3 \cdot 2 \cdot 1} b_0, \quad b_5 = \frac{1}{5 \cdot 4} b_3 = \frac{1}{5 \cdot 4 \cdot 3 \cdot 2} b_1.$$

Continuing in this way, we can express all even-numbered coefficients as multiples of b_0 and all odd-numbered coefficients as multiples of b_1:

$$b_{2m} = \frac{1}{(2m)!} b_0, \quad b_{2m+1} = \frac{1}{(2m + 1)!} b_1.$$

We substitute these values into (S) and separate the terms involving b_0

from those involving b_1 to obtain

$$x(t) = b_0 \sum_{m=0}^{\infty} \frac{1}{(2m)!} t^{2m} + b_1 \sum_{m=0}^{\infty} \frac{1}{(2m + 1)!} t^{2m+1}.$$

Since the leading coefficient of (H) never vanishes, this series expression is valid for all t.

It should not surprise us that our answer involves two arbitrary constants, b_0 and b_1, since we are solving a second-order o.d.e. These constants have a simple relation to initial conditions:

$$x(0) = b_0, \quad x'(0) = b_1.$$

You should verify that substituting $b_0 = b_1 = 1$ (respectively $b_0 = -b_1 = 1$) into (R) and (S) gives a series expression for the solution $x = e^t$ (respectively $x = e^{-t}$), which played an important role in our earlier method for solving (H).

This example illustrates the general pattern. We substitute a formal power series for $x(t)$ into the o.d.e., combine terms, and obtain a recurrence relation. We then use the recurrence relation to determine the coefficients of the series. We next consider a variable-coefficient example.

Example 6.3.2

Solve about $t = 0$

(H) $[(t^2 - 1)D^2 - 2]x = 0.$

Note that $t = 0$ is an ordinary point of (H); the only singular points are $t = \pm 1$.

We look for solutions in the form

(S) $x(t) = \sum_{k=0}^{\infty} b_k t^k.$

Substitution into (H) yields

$$(t^2 - 1) \sum_{k=2}^{\infty} k(k - 1)b_k t^{k-2} - 2 \sum_{k=0}^{\infty} b_k t^k = 0$$

or

$$\sum_{k=2}^{\infty} k(k-1)b_k t^k - \sum_{k=2}^{\infty} k(k-1)b_k t^{k-2} - \sum_{k=0}^{\infty} 2b_k t^k = 0.$$

To match up the powers of t, we substitute $j = k - 2$ (or $k = j + 2$) in the middle series and $j = k$ in the other two:

$$\sum_{j=2}^{\infty} j(j-1)b_j t^j - \sum_{j=0}^{\infty} (j+2)(j+1)b_{j+2} t^j - \sum_{j=0}^{\infty} 2b_j t^j = 0.$$

Now the limits of summation don't match, so we separate the terms for $j = 0$ and $j = 1$ from the last two series.

$$\left[\sum_{j=2}^{\infty} j(j-1)b_j t^j \right] - \left[2b_2 + 6b_3 t + \sum_{j=2}^{\infty} (j+2)(j+1)b_{j+2} t^j \right]$$

$$- \left[2b_0 + 2b_1 t + \sum_{j=2}^{\infty} 2b_j t^j \right] = 0.$$

Combining terms, we have

$$-(2b_2 + 2b_0) - (6b_3 + 2b_1)t$$

$$+ \sum_{j=2}^{\infty} [j(j-1)b_j - (j+2)(j+1)b_{j+2} - 2b_j]t^j = 0.$$

This can happen only if the coefficient of each power of t is zero. Thus

$$b_2 = -b_0, \quad b_3 = -\frac{1}{3}b_1$$

and

(R) $$\qquad b_{j+2} = \frac{j(j-1) - 2}{(j+2)(j+1)} b_j = \frac{j-2}{j+2} b_j, \quad j = 2, 3, \ldots .$$

As in the last example, the even-numbered coefficients are multiples of b_0:

$$b_2 = -b_0, \quad b_4 = \frac{0}{4}b_2 = 0, \quad b_6 = \frac{4}{6}b_4 = 0, \ldots, \quad b_{2m} = 0, \ldots,$$

and the odd-numbered coefficients are multiples of b_1:

$$b_3 = -\frac{1}{3} b_1, \quad b_5 = \frac{1}{5} b_3 = -\frac{1}{3 \cdot 5} b_1,$$

$$b_7 = \frac{3}{7} b_5 = -\frac{1}{5 \cdot 7} b_1, \quad b_9 = \frac{5}{9} b_7 = -\frac{1}{7 \cdot 9} b_1, \dots,$$

$$b_{2m+1} = \frac{2m - 3}{2m + 1} b_{2m-1} = -\frac{1}{(2m + 1)(2m - 1)} b_1, \dots.$$

Note that the general formula for b_{2m} works from $m = 2$ on, whereas the formula for b_{2m+1} works starting from $m = 0$.

Substituting these values back into (S) and separating the terms involving b_0 from those involving b_1, we obtain

$$x(t) = b_0(1 - t^2) + b_1 \sum_{m=0}^{\infty} \left[\frac{-1}{(2m + 1)(2m - 1)} \right] t^{2m+1}.$$

This expression will be valid provided we stay away from the zeros, ± 1, of the leading coefficient of (H)—that is, for $|t| < 1$.

So far we have managed to obtain explicit formulas describing all the coefficients of our solutions. Sometimes the recurrence relations are sufficiently complicated that we cannot find a general pattern. Nonetheless, we can use the recurrence relations to calculate explicitly any specified finite number of coefficients. Since this usually allows us to obtain good approximations to the values of our solutions at specified points (see Section 6.4), we content ourselves with finding the first few terms of the solution.

Example 6.3.3

Solve about $t = 0$:

(H) $$[D^2 - t D + t]x = 0.$$

We look for solutions in the form

$$x(t) = \sum_{k=0}^{\infty} b_k t^k.$$

Substitution into (H) yields

$$\sum_{k=2}^{\infty} k(k-1)b_k t^{k-2} - \sum_{k=1}^{\infty} kb_k t^k + \sum_{k=0}^{\infty} b_k t^{k+1} = 0.$$

We match up powers by substituting $j = k - 2$ (or $k = j + 2$) in the first series, $j = k$ in the second, and $j = k + 1$ (or $k = j - 1$) in the third:

$$\sum_{j=0}^{\infty} (j+2)(j+1)b_{j+2} t^j - \sum_{j=1}^{\infty} jb_j t^j + \sum_{j=1}^{\infty} b_{j-1} t^j = 0.$$

We separate the $j = 0$ term from the first series and combine terms:

$$2b_2 + \sum_{j=1}^{\infty} [(j+2)(j+1)b_{j+2} - jb_j + b_{j-1}]t^j = 0.$$

This can happen only if

$$b_2 = 0$$

and

(R) $$b_{j+2} = \frac{jb_j - b_{j-1}}{(j+2)(j+1)}, \quad j = 1, 2, \ldots .$$

We can use (R) to find the first few coefficients:

$$b_3 = \frac{1}{6}(b_1 - b_0)$$

$$b_4 = \frac{1}{12}(2b_2 - b_1) = -\frac{1}{12}b_1$$

$$b_5 = \frac{1}{20}(3b_3 - b_2) = \frac{1}{40}(b_1 - b_0).$$

Thus, the first few terms of our solution are

$$x(t) = b_0 + b_1 t + 0t^2 + \frac{1}{6}(b_1 - b_0)t^3 - \frac{1}{12}b_1 t^4 + \frac{1}{40}(b_1 - b_0)t^5 + \cdots$$

$$= b_0 \left(1 - \frac{1}{6}t^3 - \frac{1}{40}t^5 + \cdots\right) + b_1 \left(t + \frac{1}{6}t^3 - \frac{1}{12}t^4 + \frac{1}{40}t^5 + \cdots\right).$$

The expression is valid for all t.

If $t = 0$ is a singular point, or if we want to match initial conditions at an ordinary point $t_0 \neq 0$, then we may wish to express our solutions as power series about $t = t_0 \neq 0$. Since most of us find it easier to work with a series about $t = 0$, the usual procedure is to first make the substitution $T = t - t_0$, as shown in the following example.

Example 6.3.4

Find the solution of

(H)
$$[D^2 - (t - 2)]x = 0$$

subject to the condition at $t = 2$

$$x(2) = 1, \quad x'(2) = 0.$$

Since the condition is to be satisfied at $t = 2$, we look for a series about $t = 2$:

(S)
$$x(t) = \sum_{k=0}^{\infty} b_k(t - 2)^k.$$

Our condition tells us that

$$b_0 = 1, \quad b_1 = 0.$$

To simplify our calculations, we make the substitution

$$T = t - 2 \quad (\text{or} \quad t = T + 2),$$

which changes (H) and (S) to

(H')
$$[D^2 - T]x = 0$$

and

(S')
$$x(T + 2) = \sum_{k=0}^{\infty} b_k T^k.$$

Substituting (S') into (H') we obtain

$$\sum_{k=2}^{\infty} k(k - 1)b_k T^{k-2} - \sum_{k=0}^{\infty} b_k T^{k+1} = 0.$$

The substitutions $j = k - 2$ (or $k = j + 2$) in the first series and $j = k + 1$ (or $k = j - 1$) in the second, together with a separation of the $j = 0$ term in the first series, lead to

$$2b_2 + \sum_{j=1}^{\infty} [(j + 2)(j + 1)b_{j+2} - b_{j-1}]T^j = 0.$$

This implies

$$b_2 = 0$$

and

(R) $$b_{j+2} = \frac{1}{(j + 2)(j + 1)} b_{j-1}, \quad j = 1, 2, 3, \ldots.$$

We know specific values for b_0, b_1, and b_2 and have a formula for b_{j+2} in terms of the coefficient *three* steps back, b_{j-1}. Since b_1 and b_2 are both zero, we have

$$0 = b_4 = b_7 = \cdots = b_{3m+1} \cdots$$

$$0 = b_5 = b_8 = \cdots = b_{3m+2} \cdots.$$

The first few terms of the form b_{3m} are

$$b_0 = 1, \quad b_3 = \frac{1}{3 \cdot 2}, b_0 = \frac{1}{3 \cdot 2}, \quad b_6 = \frac{1}{6 \cdot 5} b_3 = \frac{1}{6 \cdot 5 \cdot 3 \cdot 2},$$

$$b_9 = \frac{1}{9 \cdot 8} b_6 = \frac{1}{9 \cdot 8 \cdot 6 \cdot 5 \cdot 3 \cdot 2}.$$

The denominator of b_{3m} looks like $(3m)!$, except that it is missing the factors $1, 4, 7, \ldots, 3m - 2$. Thus

$$b_{3m} = \frac{1 \cdot 4 \cdot 7 \cdots (3m - 2)}{(3m)!}.$$

This general formula starts to work at $m = 1$.

Now putting our expressions back into (S'), we get

$$x(T + 2) = 1 + \sum_{m=1}^{\infty} \frac{1 \cdot 4 \cdot 7 \cdots (3m - 2)}{(3m)!} T^{3m}$$

or, using $T = t - 2$,

$$x(t) = 1 + \sum_{m=1}^{\infty} \frac{1 \cdot 4 \cdot 7 \cdots (3m - 2)}{(3m)!} (t - 2)^{3m}.$$

The expression is valid for all t.

This technique also works nicely for nonhomogeneous equations, as in our final example.

Example 6.3.5

Solve about $t = 0$

(N) $$[D^2 + tD + 1]x = t.$$

We seek solutions in the form

(S) $$x(t) = \sum_{k=0}^{\infty} b_k t^k.$$

Substituting (S) into (N) yields:

$$\sum_{k=2}^{\infty} k(k - 1)b_k t^{k-2} + \sum_{k=1}^{\infty} kb_k t^k + \sum_{k=0}^{\infty} b_k t^k = t.$$

We bring the forcing term to the left-hand side, make index substitutions ($j = k - 2, j = k, j = k$) to match up powers of t, and separate the $j = 0$ and $j = 1$ terms to obtain

$$(2b_2 + b_0) + (6b_3 + 2b_1 - 1)t + \sum_{j=2}^{\infty} [(j + 2)(j + 1)b_{j+2} + (j + 1)b_j]t^j = 0.$$

This leads to

$$b_2 = -\frac{1}{2} b_0, \quad b_3 = \frac{1}{6} - \frac{1}{3} b_1,$$

and

(R) $$b_{j+2} = -\frac{1}{j + 2} b_j, \quad j = 2, 3, \ldots.$$

We use (R) to express the even-numbered coefficients in terms of b_0 and the odd-numbered ones in terms of b_1:

$$b_2 = -\frac{1}{2} b_0, \quad b_4 = -\frac{1}{4} b_2 = \frac{1}{4 \cdot 2} b_0, \quad b_6 = -\frac{1}{6} b_4 = \frac{1}{6 \cdot 4 \cdot 2} b_0, \ldots,$$

$$b_{2m} = \frac{(-1)^m}{(2m) \cdots 6 \cdot 4 \cdot 2} b_0, \ldots$$

$$b_3 = -\frac{1}{3} b_1 + \frac{1}{6}, \quad b_5 = -\frac{1}{5} b_3 = \frac{1}{5 \cdot 3} b_1 - \frac{1}{6 \cdot 5},$$

$$b_7 = -\frac{1}{7} b_5 = \frac{-1}{7 \cdot 5 \cdot 3} b_1 + \frac{1}{7 \cdot 6 \cdot 5}, \ldots,$$

$$b_{2m+1} = \frac{(-1)^m}{(2m + 1) \cdots 5 \cdot 3 \cdot 1} b_1 + \frac{(-1)^{m+1}}{2[(2m + 1) \cdots 5 \cdot 3 \cdot 1]}, \ldots$$

Substitution in (S) gives us

$$x(t) = b_0 \left(1 + \sum_{m=1}^{\infty} \frac{(-1)^m}{(2m) \cdots 4 \cdot 2} t^{2m} \right)$$

$$+ b_1 \left(t + \sum_{m=1}^{\infty} \frac{(-1)^m}{(2m + 1) \cdots 5 \cdot 3 \cdot 1} t^{2m+1} \right)$$

$$+ \frac{1}{2} \sum_{m=1}^{\infty} \frac{(-1)^{m+1}}{(2m + 1) \cdots 5 \cdot 3 \cdot 1} t^{2m+1}.$$

The expression is valid for all t.

Although all our examples were second-order o.d.e.'s, the same procedure works about ordinary points for o.d.e.'s of any order.

SOLUTIONS ABOUT ORDINARY POINTS

Suppose the coefficients $a_i(t)$ and forcing term $E(t)$ in the o.d.e.

(N) $[a_n(t)D^n + \cdots + a_0(t)] x = E(t)$

are all polynomials. We call $t = t_0$ an **ordinary point** of (N) if $a_n(t_0) \neq 0$ and a **singular point** if $a_n(t_0) = 0$.

If $t = 0$ is an ordinary point of (N), we can find a series expression

(S) $$x(t) = \sum_{k=0}^{\infty} b_k t^k$$

for the solutions of (N) as follows:

1. Substitute (S) into (N).

2. Combine terms so as to rewrite the equation in the form of a power series set identically equal to zero. (This requires index substitutions to match up powers of t, and possibly separating out a few extra terms to match up the limits of summation.)

3. Set each (combined) coefficient of the resulting series equal to zero, to obtain a **recurrence relation** (R) between the general coefficient in (S) and earlier ones.

4. If a pattern is evident, write down an explicit formula for the coefficients of (S); otherwise, we can still use (R) to write down any specified finite number of coefficients.

5. Substitute the coefficients back into (S) to obtain a power series expression for the solutions.

The resulting series expression will be valid for $|t| < R$, provided the leading coefficient $a_n(t)$ of (N) has no (complex) zeros with $|z| < R$.

If $t = t_0$ is a nonzero ordinary point of (N), we can find a series expression about $t = t_0$

(S)
$$x(t) = \sum_{k=0}^{\infty} b_k(t - t_0)^k$$

for the solutions of (N). We substitute $T = t - t_0$ (or $t = T + t_0$) and use the procedure described in steps 1 through 5 to find a solution

(S′)
$$x(T + t_0) = \sum_{k=0}^{\infty} b_k T^k$$

to the modified o.d.e.

(N′) $[a_n(T + t_0)D^n + \cdots + a_0(T + t_0)]x = E(T + t_0).$

We then substitute $T = t - t_0$ back into (S′) to get (S).

Note

Concerning our hypotheses

   The assumption in this section that the coefficients and forcing term in (N) are polynomials can be weakened. The general theorem, of which we quoted a special case, is the following.

Theorem: *Suppose that*

1. $a_n(t), \ldots, a_1(t)$, and $E(t)$ are analytic functions whose power series expressions about $t = t_0$ are all valid (at least) for $|t - t_0| < R$, where $R > 0$; and

2. $a_n(z) \neq 0$ *for any complex number z satisfying* $|z - t_0| < R$.

Then every solution of $[a_n(t)D^n + \cdots + a_0(t)]x = E(t)$ *is analytic at* $t = t_0$, *with a power series expression about* $t = t_0$ *valid (at least) for* $|t - t_0| < R$.

The procedure described in the summary actually works under these more general hypotheses. We replace each $a_i(t)$ and $E(t)$ with its power series expression and then proceed as before. However, if one (or more) of the coefficients is not a polynomial, then step 1 will require multiplication of series (see Note 1 in Section 6.2 and Exercises 18 and 19). Also, we cannot in general expect a simple recurrence relation for the b_k's; we can hope to actually calculate only the first few terms.

EXERCISES

In Exercises 1 through 6, (a) find the recurrence relation for the coefficients of series solutions about $t = 0$, and (b) write out the terms to t^5 of the general solution.

1. $[D^2 + tD + 4]x = 0$ 2. $[D^2 + 4t]x = 0$

3. $[D^2 + D + t]x = 0$ 4. $[D^2 + 2tD + t]x = 0$

5. $[D^2 + 3tD - 4]x = t - 2$ 6. $[D^3 + tD]x = 0$

In Exercises 7 through 10, (a) find the recurrence relation for the coefficients of series solutions about $t = 0$, and (b) write out the terms to t^4 of the solution matching the given initial condition. Compare the cited exercise in Section 6.1.

7. $[D^2 + (10 + t)D + 1]x = 10;$ $x(0) = x'(0) = 0$ (Exercise 1)

8. $[10D^2 + 10D + (10 - t)]x = 0;$ $x(0) = 1, x'(0) = 0$ (Exercise 2)

9. $[10D^2 + (10 - t)D + 10]x = 0;$ $x(0) = 1, x'(0) = 0$ (Exercise 3)

10. $[(10 - t)D^2 + 10D + 10]x = 0;$ $x(0) = 1, x'(0) = 0$ (Exercise 4)

In Exercises 11 and 12, (a) find the recurrence relation for the coefficients of series solutions about $t = t_0$, and (b) write out the terms to $(t - t_0)^4$ of the general solution.

11. $[tD^2 - 4]x = 0;$ $t_0 = 1$ 12. $[tD^2 - 4]x = 0;$ $t_0 = 2$

In Exercises 13 and 14, (a) find the recurrence relation for the coefficients of series solutions about $t = 0$, and (b) find the general form for the coefficients.

13. $[D^2 + tD + 1]x = 0$ 14. $[(t^2 + 1)D^2 - 2]x = 0$

Exercises 15 through 17 deal with some important equations from mathematical physics.

15.   a.   Find the recurrence relation for solutions about $t = 0$ of the **Legendre equation** $[(t^2 - 1)D^2 + 2tD - \mu(\mu + 1)]x = 0$, where μ is a constant.

      b.   Show that if μ is a nonnegative integer, then this o.d.e. has a polynomial solution of degree μ.

16.   a.   Find the recurrence relation for solutions of the **Hermite equation** $[D^2 - 2tD + \mu]x = 0$, where μ is a constant.

      b.   Show that if $\mu = 2N$ where N is a nonnegative integer, then this o.d.e. has a polynomial solution of degree N.

17. Find a series expression about $t = 0$ for the general solution of the **Airy equation** $(D^2 + t)x = 0$.

The o.d.e.'s in Exercises 18 through 20 involve analytic functions that are not polynomials (see the note at the end of the section). Find the terms to t^5 of the series solution to the given initial-value problem.

18. $[D^2 + \sin t]x = 0$; $x(0) = 0, x'(0) = 1$
19. $[D^2 + \sin t]x = 0$; $x(0) = 1, x'(0) = 0$
20. $[D^2 + t]x = e^t$; $x(0) = 1, x'(0) = 0$

6.4 POWER SERIES ON COMPUTERS (OPTIONAL)

The recurrence relations that arise in finding series solutions to o.d.e.'s are especially well suited to the use of computers. In this section we discuss general algorithms for finding power series on any programmable device, as well as their implementation in specific BASIC programs.

Our discussion focuses on solutions of the o.d.e.

$$[D^2 + (t + 1)D + 1]x = 0.$$

Substitution of the expression

$$x(t) = \sum_{k=0}^{\infty} b_k t^k$$

into the o.d.e. leads to the recurrence relation

$$b_{j+2} = \frac{-b_j - b_{j+1}}{j + 2}, \quad j = 0, 1, \ldots$$

(check this). Let's begin with an algorithm for calculating coefficients.

Example 6.4.1

Design an algorithm to calculate the coefficients $b_2, b_3, \ldots, b_{10}$ of the series defined by the foregoing recurrence relation for given values of b_0 and b_1.

The heart of our problem is to calculate a given coefficient b_k, assuming we already know $b_0, b_1, \ldots, b_{k-1}$. We can rewrite the recurrence relation as a formula for b_k in terms of earlier data by means of the index substitution

$k = j + 2$ (or $j = k - 2$):

$$b_k = \frac{-b_{k-2} - b_{k-1}}{k}, \quad k = 2, 3, \ldots .$$

Calculation of b_k requires three pieces of previous data: the coefficients two steps and one step back, together with the index k. If these have previously been stored in memory registers A, B, and K.

$$A \leftarrow b_{k-2}, \quad B \leftarrow b_{k-1}, \quad K \leftarrow k$$

then our first step is to calculate b_k and store it for future use:

1. Calculate $\dfrac{-A - B}{K}$ and store it in register C.

Since we want to know the value of b_k, we instruct our computer to print the result of our computation:

2. Print C.

If $k \geq 10$, then we have accomplished our goal. Otherwise, we must proceed to calculate b_{k+1}:

3. If $K \geq 10$, stop. Otherwise, continue.

We would now like to calculate b_{k+1} by repeating step 1, but first we must replace the numbers in registers A, B, and K (b_{k-2}, b_{k-1}, and k, respectively) with the coefficients two steps and one step back from b_{k+1} and the new index (b_{k-1}, b_k, and $k + 1$, respectively):

4. Replace A with B, B with C, K with $K + 1$, and return to step 1.

Note that after we replace A with B, the value of b_{k-2} will no longer be stored in memory. Of course, now that we have calculated b_k, we no longer need b_{k-2}.

Now, if we start by storing the values

$$A \leftarrow b_0, \quad B \leftarrow b_1, \quad K \leftarrow 2$$

and set the machine going at step 1, it will

1. Calculate b_2 and store it in C.
2. Print b_2.
3. Compare 2 to 10. Since $2 < 10$, the machine will continue.

This process will continue until the machine has calculated and printed all the coefficients $b_2, b_3, \ldots, b_{10}$.

The flowchart of Figure 6.5 summarizes this algorithm, and an implementation is provided in the BASIC program of Figure 6.6. Line 10 of the program allows for the specification of the values of b_0 and b_1. The FOR and NEXT statements in lines 20 and 70 instruct the machine to perform a loop, starting with $K = 2$ and ending when $K = 10$. These two steps provide for the specification of the initial value of K, the comparison in step 3 of the algorithm, and the updating of the value of K after each repetition of the loop. Lines 30 through 60 carry out steps 1, 2, and the remaining portions of step 4. When we instruct a computer to run this program, it provides a prompt, asking us to input the data b_0, b_1. If we provide values for these coefficients and press

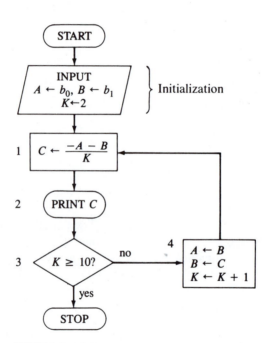

FIGURE 6.5   Calculating coefficients

```
10   INPUT A,B
20   FOR K = 2 TO 10
30   C = (−A − B)/K
40   PRINT "B"K "=" C
50   A = B
60   B = C
70   NEXT K
99   END
```

FIGURE 6.6   BASIC program for coefficients

RETURN, then the computer will calculate and print the values of the later coefficients. Figure 6.7 shows the result of running this program with several choices of b_0 and b_1 on a Digital Equipment Corporation DECsystem-10™ computer.*

	$b_0 = 1, b_1 = 3$	$b_0 = 1, b_1 = -3$	$b_0 = 2, b_1 = 10$
b_2	-2	1	-6
b_3	-0.333333	0.666667	-1.33333
b_4	0.583333	-0.416667	1.83333
b_5	5.00000×10^{-2}	-5.00000×10^{-2}	-0.1
b_6	-8.88889×10^{-2}	7.77778×10^{-2}	-0.288889
b_7	1.98413×10^{-2}	3.96825×10^{-3}	5.55556×10^{-2}
b_8	8.63095×10^{-3}	-9.22619×10^{-3}	2.91667×10^{-2}
b_9	-3.16358×10^{-3}	1.46605×10^{-3}	-9.41358×10^{-3}
b_{10}	-5.46737×10^{-4}	7.76014×10^{-4}	-1.97531×10^{-3}

FIGURE 6.7   Coefficients

In the preceding example we considered the calculation of the coefficients of the series. However, in practice we are usually interested in the numerical value of $x(t)$ for a given choice of t. This value is approximated by the partial sums of the series. One way to calculate these partial sums is to supplement the algorithm of Example 6.4.1 with some extra steps that evaluate t^k, multiply it by b_k, and then sum the result into a memory that stores the partial sum. However, if we are not interested in the coefficients per se, it is more efficient to treat the calculation of the terms $\beta_k = b_k t^k$ of the series directly.

Example 6.4.2

Design an algorithm to calculate the partial sum

$$S_{10} = \sum_{j=0}^{10} b_j t^j = b_0 + b_1 t + \cdots + b_{10} t^{10}$$

of the series for $x(t)$.

The recurrence relation for b_k can be multiplied on both sides by t^k and rewritten to yield an expression for $\beta_k = b_k t^k$ in terms of $\beta_{k-2} = b_{k-2} t^{k-2}$ and $\beta_{k-1} = b_{k-1} t^{k-1}$:

$$\beta_k = b_k t^k = \frac{-b_{k-2} t^k - b_{k-1} t^k}{k} = \frac{-\beta_{k-2} t^2 - \beta_{k-1} t}{k}.$$

---

* DECsystem-10 is a trademark of Digital Equipment Corporation.

This is a recurrence relation analogous to the one in Example 6.4.1. In order to calculate the terms $\beta_2, \ldots, \beta_{10}$, we will store β_{k-2}, β_{k-1}, k, and t in memory. After the calculation of a term β_k, we will store the result in one register and sum it into another register where we will keep track of the partial sum to that point. In all, we will use six memory registers:

$$A \leftarrow \beta_{k-2}, \quad B \leftarrow \beta_{k-1}, \quad K \leftarrow k, \quad C \leftarrow \beta_k$$

$$T \leftarrow t, \quad S \leftarrow \sum_{j=0}^{k-1} \beta_j = \sum_{j=0}^{k-1} b_j t^j.$$

The flowchart in Figure 6.8 describes an algorithm that requires us to provide initial settings for memory registers A, B, K, and T. Note that the initial setting for B is b_1 rather than β_1. The first steps of the algorithm are the calculation of $\beta_1 = b_1 t$ and of $S_1 = b_0 + b_1 t$. The central loop calculates the

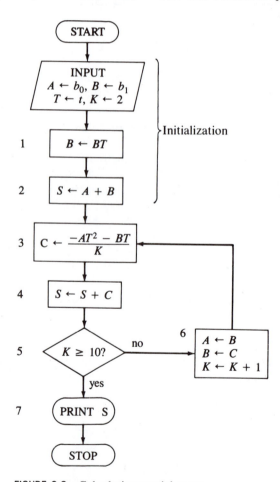

FIGURE 6.8 Calculating partial sums

partial sums $S_2, \ldots, S_{10}$ in turn. Step 7 informs the machine to print the value of S_{10}. If you prefer to see each of the partial sums, you should move the print command to a point between steps 4 and 5.

We have implemented this algorithm in the BASIC program of Figure 6.9. Note that here again, the specification of the initial value of K, the comparison, and the updating of K are handled by FOR and NEXT statements. The table in Figure 6.10 shows the result of running this program with $b_0 = 1$, $b_1 = 3$, and various choices of t.

```
10   INPUT A,B
20   B = B * T
30   S = A + B
40   FOR K = 2 TO 10
50   C = (-A * T * T - B * T)/K
60   S = S + C
70   A = B
80   B = C
90   NEXT K
100  PRINT "S"K "=" S
999  END
```

FIGURE 6.9   BASIC program for partial sums

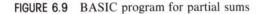

t	.25	.75	1.25
S_{10}	1.622	2.1445	2.02383

FIGURE 6.10   Partial sums ($b_0 = 1$, $b_1 = 3$)

In practical situations, we are usually interested in the numerical value of $x(t)$ to some specified accuracy. Our next example illustrates one approach to deciding whether a given partial sum has achieved a desired degree of accuracy. We compare the partial sum to a fixed number of previously calculated partial sums. If these successive partial sums all agree to within a specified error allowance $\epsilon > 0$, it is hoped that all later partial sums, and hence the sum of the series $x(t)$, will also agree with these to within ϵ. Although this hope is not always justified (see the note at the end of the section), the test does provide a good educated guess for the value of $x(t)$.

Example 6.4.3

Modify the algorithm in Example 6.4.2 so that the calculation terminates the first time that a partial sum differs from each of the previous two by less than $\epsilon = 0.0001$.

The difference between the kth and $(k-1)$st partial sums is

$$\sum_{j=0}^{k} b_j t^j - \sum_{j=0}^{k-1} b_j t^j = b_k t^k = \beta_k,$$

and the difference between the kth and $(k-2)$nd partial sums is

$$\sum_{j=0}^{k} b_j t^j - \sum_{j=0}^{k-2} = b_{k-1} t^{k-1} + b_k t^k = \beta_{k-1} + \beta_k.$$

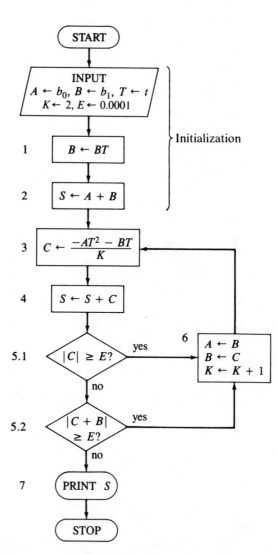

FIGURE 6.11 Calculating partial sums to a desired accuracy

We want the computer to stop as soon as

$$|\beta_k| < 0.0001 \quad \text{and} \quad |\beta_{k-1} + \beta_k| < 0.0001.$$

To incorporate this into our algorithm, we need to have previously stored 0.0001 in memory:

$$E \leftarrow 0.0001.$$

We replace the comparison, step 5, in the algorithm of Example 6.4.2 with two comparisons:

5.1 Calculate $|C|$ and compare it to E. If $|C| \geq E$, go to step 6. Otherwise, go to 5.2.

5.2 Calculate $|C + B|$ and compare it to E. If $|C + B| \geq E$, go to step 6. Otherwise stop.

The resulting algorithm is summarized in Figure 6.11 and implemented in the BASIC program of Figure 6.12. Since we do not know the final value of K at the outset, we cannot use FOR and NEXT statements to specify the loop. Instead, we set the initial data in lines 20 and 30, and provide for the comparisons and updating of K in lines 80 and 90. Figure 6.13 shows the result of running this program with $b_0 = 1$, $b_1 = 3$, and various choices for t. Note that the value of K when the calculation stops increases as t does. This illus-

```
10   INPUT A,B
20   K = 2
30   E = 0.0001
40   B = B * T
50   S = A + B
60   C = (−A * T * T − B * T)/K
70   S = S + C
80   IF ABS(C) >= E THEN 110
90   IF ABS(C + B) >= E THEN 110
100  GO TO 150
110  A = B
120  B = C
130  K = K + 1
140  GO TO 60
150  PRINT "S"K "=" S
999  END
```

FIGURE 6.12   BASIC program for partial sums to a desired accuracy

t	.25	.75	1.25
S_N	$S_6 = 1.622$	$S_{11} = 2.14452$	$S_{15} = 2.02758$

FIGURE 6.13   Partial sums to a desired accuracy ($b_0 = 1$, $b_1 = 3$)

trates the fact that a series $\Sigma\ b_k(t\ -\ t_0)^k$ converges faster for values of t close to t_0.

Our examples illustrate some general principles that apply to calculating the coefficients or numerical value of any power series that is defined by a recurrence relation. We formulate these in the following summary.

POWER SERIES ON COMPUTERS

   To calculate the coefficients b_k or partial sums S_k of a power series defined by a recurrence relation, first design an **algorithm** for the problem:

a.   Express the **recurrence relation** as a formula for the kth coefficient b_k (or term β_k) in terms of earlier ones.

b.   Allocate **memory** for the data needed in this calculation.

c.   Design a **loop** that will calculate b_k (or S_k) from this data, and either terminate the calculation (at the appropriate step) or update the memories and return to the start of the loop with all the data for calculating b_{k+1} (or S_{k+1}) in place.

d.   Check to see whether **additional memory** is needed to retain data not stored in the memories allocated in (b) but needed for repetitions of the loop.

e.   Formulate any necessary **initial steps** that must precede the first repetition of the loop; make sure that at the end of these steps the memories are all set up for this first repetition.

   Now, write a **program** implementing this algorithm. Be careful to allow for a specification of the **initial values** of various memories. Be sure to keep track of the **contents of each memory register** as you run through the program, and don't forget to program your computer to display the result of your calculation in a useful form.

   In order to catch any bugs in your program, try it (or a simple modification) on a calculation that you can verify by hand (such as the calculation of just b_2 and b_3). Now you are ready to carry out more extended calculations.

Note

Accuracy

   Heuristic accuracy tests like the one in Example 6.4.3 are fallible. For example, the series about $t = 0$ for $\sin(t^2)$,

$$\sin(t^2) = 0 + 0 + t^2 + 0 + 0 + 0 - \frac{t^6}{6} + \cdots$$

has as its first few partial sums

$$S_0 = 0 = S_1, \quad S_2 = t^2 = S_3 = S_4 = S_5;$$

a test that compares three (or even four) successive partial sums would conclude that this series sums to t^2. In particular, when $t = \sqrt{\pi}$, the test would suggest that $\sin \pi = \pi$, far from the true value $\sin \pi = 0$.

Nonetheless, we can use a heuristic test to make an educated guess that a specific partial sum $S_N = \sum_{j=0}^{N} b_j t^j$ agrees with the true sum $S = \sum_{j=0}^{\infty} b_j t^j$ within an accuracy specified by $\epsilon > 0$. To be certain of the accuracy, we must then use more rigorous, abstract estimates to show that the "tail" of the series adds up to less than ϵ:

$$|S - S_N| = \left| \sum_{j=N+1}^{\infty} b_j t^j \right| < \epsilon.$$

Even if we have determined that the partial sum S_N is within ϵ of S, there remains the possibility that roundoff error will contribute significantly to our calculation of S_N. Roundoff error results from the fact that every machine uses a finite (decimal or binary) approximation to represent numbers. Such minor errors could in principle accumulate (especially if many calculations are involved) to significantly affect our calculated value of S_N. The analysis of roundoff error is a difficult art, however, and we shall not go into it here.

EXERCISES

In Exercises 1 through 3, we ask for information about the series solution $x(t) = \sum_{k=0}^{\infty} b_k t^k$ to the given initial-value problem. For each initial-value problem,

a. Find the coefficients $b_0, b_1, \ldots, b_{10}$ (as in Example 6.4.1).

b. Approximate $x(t)$, for $t = 0.1, 0.5,$ and 1.0, by calculating the partial sum $S_{10}(t) = \sum_{k=0}^{10} b_k t^k$ (as in Example 6.4.2).

c. Approximate $x(t)$, for $t = 0.1, 0.5,$ and 1.0, by calculating partial sums until a partial sum $S_N(t)$ differs from each of the previous two, $S_{N-1}(t)$ and $S_{N-2}(t)$, by less than $\epsilon = 0.0001$ (as in Example 6.4.3). What is the value of N when the calculation stops?

d. Based on your answers (and ignoring roundoff error), estimate the value of $x(t)$, for $t = 0.1$, $t = 0.5$, and $t = 1.0$, to three decimal places.

1. $[D^2 + (t + 1)D + 1] = 0$ (discussed in the text)
      i. $x(0) = 1, x'(0) = 0$ ii. $x(0) = 0, x'(0) = 1$
     iii. $x(0) = -2, x'(0) = 3$ iv. $x(0) = 3, x'(0) = -1$

2. $[D^2 + tD + 1]x = 0$ (compare Exercise 13, Section 6.3)
      i. $x(0) = 1, x'(0) = 0$ ii. $x(0) = 0, x'(0) = 1$
     iii. $x(0) = -2, x'(0) = 3$ iv. $x(0) = 3, x'(0) = -1$

3. $[10D^2 + (10 - t)D + 10]x = 0$ (compare Exercise 9, Section 6.3)
   i. $x(0) = 1, x'(0) = 0$ ii. $x(0) = 0, x'(0) = 0$
  iii. $x(0) = -2, x'(0) = 3$ iv. $x(0) = 3, x'(0) = -1$

4. $[D^2 + D + t]x = 0$ (compare Exercise 3, Section 6.3)
   i. $x(0) = 1, x'(0) = 0$ ii. $x(0) = 0, x'(0) = 0$
  iii. $x(0) = -2, x'(0) = 3$ iv. $x(0) = 3, x'(0) = -1$

5. $[D^2 + 4t]x = 0$ (compare Exercise 2, Section 6.3)
   i. $x(0) = 1, x'(0) = 0$ ii. $x(0) = 0, x'(0) = 1$
  iii. $x(0) = -2, x'(0) = 3$ iv. $x(0) = 3, x'(0) = -1$

6.5   THE CAUCHY-EULER EQUATION

In this section we investigate the solutions of the second-order **Cauchy-Euler equation**

(H) $$[p_0 t^2 D^2 + q_0 t D + r_0]x = 0,$$

where p_0, q_0, and r_0 are constants. This equation has a singular point at $t = 0$. Although solutions of (H) do not require series, the cases that arise here will help us understand a wider class of equations, which we will solve using series in the next three sections. We will restrict attention to the interval $t > 0$; to find solutions for $t < 0$, we can substitute $T = -t$ and assume $T > 0$.

We begin by noting that if we substitute $x = t^m$ into the left-hand side of (H), we obtain a multiple of t^m:

$$[p_0 t^2 D^2 + q_0 t D + r_0]t^m = (p_0 m(m - 1) + q_0 m + r_0)t^m.$$

Thus, we see immediately that *if m satisfies the equation*

(I) $$p_0 m(m - 1) + q_0 m + r_0 = 0,$$

then $x = t^m$ is a solution of (H). This quadratic equation is called the **indicial equation** of (H).

Since distinct powers of t are linearly independent (see Exercise 16), we see that *if the indicial equation has distinct real roots m_1 and m_2, then the general solution of (H) is*

$$x = c_1 t^{m_1} + c_2 t^{m_2}.$$

Example 6.5.1

Solve

(H)
$$[2t^2D^2 + tD - 1]x = 0, \quad t > 0.$$

Substituting $x = t^m$ into the left-hand side gives

$$[2t^2D^2 + tD - 1]t^m = (2m(m - 1) + m - 1)t^m,$$

so the indicial equation of (H) is

(I)
$$2m(m - 1) + m - 1 = 0$$

or

$$(2m + 1)(m - 1) = 0.$$

The roots of (I) are

$$m = -\frac{1}{2}, \quad m = 1.$$

Corresponding to these roots we obtain the solutions $x_1 = t^{-1/2}$ and $x_2 = t$, which generate the general solution of (H):

$$x = c_1 t^{-1/2} + c_2 t.$$

Example 6.5.2

Solve

(H)
$$[t^2D^2 - 2tD + 2]x = 0, \quad t > 0.$$

Substituting $x = t^m$ into the left-hand side of (H), we get

$$[t^2D^2 - 2tD + 2]t^m = (m(m - 1) - 2m + 2)t^m$$

and the indicial equation of (H) is

(I)
$$m(m - 1) - 2m + 2 = 0$$

or

$$(m - 1)(m - 2) = 0.$$

We obtain the general solution of (H) from the roots, $m = 1$ and $m = 2$, of (I):

$$x = c_1 t + c_2 t^2.$$

If the indicial equation has only one root m_1, then $x = t^{m_1}$ is still a solution of (H), but we need another to generate the general solution of (H). In the following example, we use a method similar to variation of parameters to find this second solution.

Example 6.5.3

Solve

(H) $$[t^2 D^2 + 5tD + 4]x = 0, \quad t > 0.$$

Substitution of $x = t^m$ into (H) gives us

$$[t^2 D^2 + 5tD + 4]t^m = (m(m - 1) + 5m + 4)t^m = 0,$$

so the indicial equation is

(I) $$m(m - 1) + 5m + 4 = 0$$

or

$$m^2 + 4m + 4 = 0.$$

This equation has only one root, $m = -2$. Corresponding to this root we have one solution of (H),

$$x_1 = t^{-2}.$$

We look for a second solution of the form

(S$_2$) $$x_2 = x_1 k(t) = t^{-2}k(t).$$

The first two derivatives of this function are

$$x_2' = t^{-2}k'(t) - 2t^{-3}k(t)$$

$$x_2'' = t^{-2}k''(t) - 4t^{-3}k'(t) + 6t^{-4}k(t).$$

Substitution into (H) gives

$$[k''(t) - 4t^{-1}k'(t) + 6t^{-2}k(t)] + [5t^{-1}k'(t) - 10t^{-2}k(t)] + 4t^{-2}k(t) = 0,$$

or

$$k''(t) + t^{-1}k'(t) = 0.$$

We rewrite this in the form

$$\frac{1}{k'(t)} \frac{dk'(t)}{dt} = \frac{-1}{t}$$

and integrate both sides to get

$$\ln |k'(t)| = -\ln t + a.$$

Thus

$$k'(t) = \frac{b}{t}$$

and

$$k(t) = b \ln t + c.$$

Taking $b = 1$ and $c = 0$, we substitute back into (S$_2$) to obtain our second solution to (H):

$$x_2 = t^{-2} \ln t.$$

The solutions x_1 and x_2 are linearly independent on $t > 0$ (see Exercise 15), so the general solution of (H) is

$$x = c_1 x_1 + c_2 x_2 = c_1 t^{-2} + c_2 t^{-2} \ln t.$$

The pattern of Example 6.5.3 is a general one: *When the indicial equation has only one root $m = m_1$, the general solution of (H) on $t > 0$ is*

$$x = c_1 t^{m_1} + c_2 t^{m_1} \ln t.$$

We leave it to the reader to verify this (see Exercise 19).

There is one final possibility: The indicial equation may have complex roots.

■ Example 6.5.4

Solve

(H) $$[t^2 D^2 + 3tD + 2]x = 0, \quad t > 0.$$

Substitution of $x = t^m$ in the left-hand side gives

$$[t^2 D^2 + 3tD + 2]t^m = (m(m - 1) + 3m + 2)t^m,$$

so the indicial equation is

(I) $$m(m - 1) + 3m + 2 = 0$$

or

$$m^2 + 2m + 2 = 0.$$

The roots of (I) are

$$m = -1 \pm i.$$

As in Sections 2.6 and 3.8, we reason that (H) should have two complex solutions

$$x_1(t) = t^{-1+i}, \quad x_2 = t^{-1-i}.$$

To make use of these expressions, we recall the definition of t^m for general m,

$$t^m = e^{m \ln t},$$

and Euler's formula,

$$e^{u+iv} = e^u(\cos v + i \sin v).$$

Using these formulas, we rewrite x_1:

$$x_1(t) = t^{-1+i} = t^{-1}t^i = t^{-1}e^{i \ln t}$$
$$= t^{-1}[\cos(\ln t) + i \sin(\ln t)].$$

Similarly

$$x_2(t) = t^{-1}[\cos(\ln t) - i \sin(\ln t)].$$

The functions

$$\frac{1}{2}[x_1(t) + x_2(t)] = t^{-1} \cos(\ln t)$$

and

$$\frac{1}{2i}[x_1(t) - x_2(t)] = t^{-1} \sin(\ln t)$$

are real-valued solutions of (H), as can be verified by substitution. Since they are linearly independent on $t > 0$ (Exercise 18), the general solution of (H) is

$$x = c_1 t^{-1} \cos(\ln t) + c_2 t^{-1} \sin(\ln t).$$

This example is also typical (see Exercise 20). *When the indicial equation has complex roots $\alpha \pm \beta i$, the general solution of (H) is*

$$x = c_1 t^\alpha \cos(\beta \ln t) + c_2 t^\alpha \sin(\beta \ln t).$$

Notice that

$$t^{\alpha + \beta i} = t^\alpha t^{\beta i} = t^\alpha e^{i\beta \ln t}$$
$$= t^\alpha \cos(\beta \ln t) + i t^\alpha \sin(\beta \ln t),$$

so that the functions generating the general solution are the real and imaginary parts of the complex solution $t^{\alpha + \beta i}$.

We summarize our observations.

THE CAUCHY-EULER EQUATION

To solve the Cauchy-Euler equation

(H) $$[p_0 t^2 D^2 + q_0 t D + r_0]x = 0, \quad t > 0:$$

1. Calculate the effect of substituting $x = t^m$ into the left-hand side:

$$[t^2 p_0 D^2 + t q_0 D + r_0]t^m = (p_0 m(m - 1) + q_0 m + r_0)t^m.$$

2. Find the roots $m = m_1$ and $m = m_2$ of the **indicial equation**

(I) $\qquad\qquad p_0 m(m - 1) + q_0 m + r_0 = 0.$

3. If the roots are real and distinct, the general solution of (H) for $t > 0$ is

$$x = c_1 t^{m_1} + c_2 t^{m_2}.$$

4. If there is only one root, $m = m_1 = m_2$, the general solution of (H) for $t > 0$ is

$$x = c_1 t^{m_1} + c_2 t^{m_1} \ln t.$$

5. If the roots are complex, $m = \alpha \pm \beta i$, the general solution of (H) for $t > 0$ is generated by the real and imaginary parts of $t^{\alpha + \beta i}$:

$$x = c_1 t^\alpha \cos(\beta \ln t) + c_2 t^\alpha \sin(\beta \ln t).$$

EXERCISES

In Exercises 1 through 14, find the general solution for $t > 0$.

1. $[t^2 D^2 + tD - 1]x = 0$
2. $[4t^2 D^2 - 3]x = 0$
3. $[t^2 D^2 - tD + 1]x = 0$
4. $[t^2 D^2 + tD + 1]x = 0$
5. $[9t^2 D^2 + 15tD + 1]x = 0$
6. $[t^2 D^2 - 2tD - 18]x = 0$
7. $[t^2 D^2 + 3tD + 5]x = 0$
8. $[t^2 D^2 - 6tD + 12]x = 0$
9. $[t^2 D^2 + tD - 3]x = 0$
10. $[t^2 D^2 + 7tD + 9]x = 0$
11. $[t^2 D^2 - 3tD + 5]x = 0$
12. $[3t^2 D^2 - 3tD + 1]x = 0$
13. $[t^2 D^2 + tD - 1]x = t^2$
14. $[t^2 D^2 - tD + 1]x = t^2$

15. *Temperature in a Circular Plate:* Recall, from Example 6.1.2, that if $u = R(r)\Theta(\theta)$ is a solution of the two-dimensional Laplace equation (in polar coordinates), then $\Theta(\theta)$ satisfies

(H$_\theta$) $\qquad\qquad \dfrac{d^2 \Theta}{d\theta^2} + n^2 \Theta = 0,$

and $R(r)$ satisfies

(H$_r$)
$$r^2 \frac{d^2R}{dr^2} + r \frac{dR}{dr} - n^2R = 0.$$

a. Find the general solution of (H$_r$).
b. Use the fact that the temperature in the plate must be bounded as r approaches 0 to conclude that the physically meaningful solutions of (H$_r$) are constant multiples of $R_n(r) = r^n$.
c. Corresponding to each nonnegative integer n, obtain two solutions of the two-dimensional Laplace equation by multiplying R_n with generators for the general solution of (H$_\theta$).
d. Find a linear combination of the functions described in (c) (possibly with different values of n) that satisfies the boundary condition

$$u(1, \theta) = \cos \theta - 2 \sin 3\theta, \quad 0 \le \theta \le 2\pi.$$

More abstract problems:

16. If m_1 and m_2 are distinct real numbers (not necessarily positive integers), show that the functions t^{m_1} and t^{m_2} are linearly independent on the interval $0 < t < \infty$.

17. Show that the functions t^{m_1} and $t^{m_1} \ln t$ are linearly independent on $0 < t < \infty$.

18. Show that if $\beta \ne 0$, the functions $t^\alpha \cos(\beta \ln t)$ and $t^\alpha \sin(\beta \ln t)$ are linearly independent on $0 < t < \infty$.

19. Suppose the indicial equation has a double root $m = m_1$.
    a. Show that $q_0 = p_0 - 2p_0 m_1$ and $r_0 = p_0 m_1^2$.
    b. Show that t^{m_1} and $t^{m_1} \ln t$ are solutions of the Cauchy-Euler equation.

20. Suppose the indicial equation has roots $\alpha \pm \beta i$ with $\beta \ne 0$.
    a. Show that $q_0 = p_0 - 2\alpha p_0$ and $r_0 = p_0(\alpha^2 + \beta^2)$.
    b. Show that $t^\alpha \cos(\beta \ln t)$ and $t^\alpha \sin(\beta \ln t)$ are solutions of the Cauchy-Euler equation.

21. a. Show that upon substituting $T = -t$, the Cauchy-Euler equation

$$p_0 t^2 \frac{d^2x}{dt^2} + q_0 t \frac{dx}{dt} + r_0 x = 0$$

becomes

$$p_0 T^2 \frac{d^2x}{dT^2} + q_0 T \frac{dx}{dT} + r_0 x = 0.$$

    b. Show that if $x = t^m$ is a solution of the Cauchy-Euler equation for $t > 0$, then $x = |t|^m$ is a solution for $t < 0$.

22. Use the result of Exercise 21(b) to find the general solution for $t < 0$ of
    a. $t^2 x'' - 6x = 0$
    b. $t^2 x'' + tx' - 4x = 0$
    c. $t^2 x'' - 3tx' + 4x = 0$
    d. $t^2 x'' + tx' + 4x = 0$

23. a. Show that for $t > 0$, the substitution $s = \ln t$ (so that $t = e^s$) transforms the Cauchy-Euler equation into a homogeneous equation with constant coefficients (cf. Exercise 22, Section 2.7).
    b. Relate the solutions of this transformed equation to the three cases treated in the summary.

6.6   REGULAR SINGULAR POINTS: FROBENIUS SERIES

In this and the following two sections, we use series to handle a class of second-order equations with relatively well-behaved singular points. The method, developed by the German mathematicians L. Fuchs (1866) and G. Frobenius (1873) [building on techniques used by Euler (1764)], is known as the **method of Frobenius**.

Recall that $t = t_0$ is a singular point of an o.d.e. with polynomial coefficients if it is a root of the leading coefficient. This means that $t - t_0$ is a factor of the leading coefficient; we will deal with singular points at which the powers of $t - t_0$ dividing the coefficients are subject to certain restrictions:

Definition: *We say that $t = t_0$ is a **regular singular point** of an o.d.e. with polynomial coefficients provided the equation can be written in the form*

(H) $$[(t - t_0)^2 p(t)D^2 + (t - t_0)q(t)D + r(t)]x = 0,$$

where $p(t)$, $q(t)$, and $r(t)$ are polynomials and $p(t_0) \neq 0$.

The Cauchy-Euler equation discussed in the previous section,

$$[t^2 p_0 D^2 + t q_0 D + r_0]x = 0,$$

has a regular singular point at $t = 0$. So does **Bessel's equation:**

$$[t^2 D^2 + tD + (t^2 - \mu^2)]x = 0$$

Another o.d.e. that comes up frequently in mathematical physics is **Laguerre's equation**

$$[tD^2 + (1 - t)D + \mu]x = 0.$$

This o.d.e. has a regular singular point at $t = 0$; to see this, multiply through by t to get

$$[t^2 D^2 + t(1 - t)D + \mu t]x = 0.$$

Another example is the **Legendre equation**

$$[(1 - t^2)D^2 - 2tD + \mu(\mu + 1)]x = 0,$$

with regular singular points at $t = \pm 1$ (multiply through by $1 \mp t$ to check this).

We shall concentrate on the case $t_0 = 0$. As we saw in Section 6.3, other cases can be reduced to this one by substituting $T = t - t_0$. We shall also concentrate on finding solutions for $t > 0$; solutions with $t < 0$ can be found by substituting $T = -t$ with $T > 0$.

We know from the Cauchy-Euler equation that *integer* powers do not suffice for solving o.d.e.'s with regular singular points. Instead, we look for solutions of the form

$$x(t) = t^m \sum_{k=0}^{\infty} b_k t^k = \sum_{k=0}^{\infty} b_k t^{m+k},$$

where $b_0 \neq 0$, and m need not be an integer. We refer to an expression of this form as a **Frobenius series**.

Let's see how substituting a Frobenius series into an o.d.e. can lead to the general solution.

Example 6.6.1

Solve

(H) $$[2t^2 D^2 + (t^2 + t)D - 1]x = 0, \quad t > 0.$$

Note that (H) has a regular singular point at $t = 0$, with $p(t) = 2$, $q(t) = t + 1$, and $r(t) = -1$. We look for solutions of the form

(S) $$x = t^m \sum_{k=0}^{\infty} b_k t^k = \sum_{k=0}^{\infty} b_k t^{k+m}, \quad b_0 \neq 0.$$

In order to substitute (S) into (H), we first calculate the effect of the operator $L = 2t^2 D^2 + (t^2 + t)D - 1$ on an arbitrary power of t:

$$\begin{aligned}
L[t^s] &= 2s(s - 1)t^s + st^{s+1} + st^s - t^s \\
&= (2s + 1)(s - 1)t^s + st^{s+1}.
\end{aligned}$$

Using this, we calculate the effect of L on the Frobenius series (S):

$$Lx = \sum_{k=0}^{\infty} b_k L[t^{m+k}]$$

$$= \sum_{k=0}^{\infty} b_k[(2m + 2k + 1)(m + k - 1)t^{m+k} + (m + k)t^{m+k+1}]$$

$$= \sum_{j=0}^{\infty} b_j(2m + 2j + 1)(m + j - 1)t^{m+j} + \sum_{j=1}^{\infty} b_{j-1}(m + j - 1)t^{m+j}$$

$$= b_0(2m + 1)(m - 1)t^m$$

$$\quad + \sum_{j=1}^{\infty} [b_j(2m + 2j + 1)(m + j - 1) + b_{j-1}(m + j - 1)]t^{m+j}.$$

If this is to be identically zero, the coefficient of each power of t must vanish. Thus

$$b_0(2m + 1)(m - 1) = 0$$

and

(R) $b_j(2m + 2j + 1)(m + j - 1) + b_{j-1}(m + j - 1) = 0, \quad j = 1, 2, \ldots .$

Since $b_0 \neq 0$, the first of these equations gives

(I) $$(2m + 1)(m - 1) = 0,$$

which has two roots,

$$m = -\frac{1}{2} \quad \text{and} \quad m = 1.$$

If we choose the second root, $m = 1$, then (R) reads

(R$_1$) $$b_j(2j + 3)(j) + b_{j-1}(j) = 0$$

or

$$b_j = -\frac{1}{2j + 3} b_{j-1}.$$

This recurrence relation lets us find the coefficients in terms of b_0:

$$b_1 = -\frac{1}{5} b_0, \quad b_2 = \frac{(-1)^2}{5 \cdot 7} b_0, \ldots, \quad b_k = \frac{(-1)^k}{5 \cdot 7 \cdots (2k + 3)} b_0, \ldots .$$

Setting $b_0 = 1$, this gives us a solution to (H):

(S$_1$) $$x_1(t) = t \left(1 + \sum_{k=1}^{\infty} \frac{(-1)^k}{5 \cdot 7 \cdots (2k + 3)} t^k \right) .$$

On the other hand, if we take $m = -1/2$, then (R) reads

(R$_2$) $$b_j(2j) \left(j - \frac{3}{2} \right) + b_{j-1} \left(j - \frac{3}{2} \right) = 0$$

or

$$b_j = -\frac{1}{2j} b_{j-1}.$$

Thus

$$b_1 = -\frac{1}{2} b_0, \quad b_2 = \frac{(-1)^2}{2 \cdot 2^2} b_0, \quad b_3 = \frac{(-1)^3}{2 \cdot 3 \cdot 2^3} b_0,$$

$$b_4 = \frac{(-1)^4}{2 \cdot 3 \cdot 4 \cdot 2^4} b_0, \ldots, \quad b_k = \frac{(-1)^k}{k! 2^k} b_0, \ldots.$$

Setting $b_0 = 1$, we obtain a second solution of (H):

(S₂)
$$x_2(t) = t^{-1/2} \sum_{k=0}^{\infty} \frac{(-1)^k}{k! 2^k} t^k.$$

We note that $x_1(t)$ and $x_2(t)$ define linearly independent solutions of (H) for $t > 0$ (see Exercise 22), so the general solution to (H) for $t > 0$ is

$$x = c_1 x_1(t) + c_2 x_2(t)$$

$$= c_1 t \left(1 + \sum_{k=1}^{\infty} \frac{(-1)^k}{5 \cdot 7 \cdots (2k+3)} t^k \right) + c_2 t^{-1/2} \left(\sum_{k=0}^{\infty} \frac{(-1)^k}{k! 2^k} t^k \right).$$

This example illustrates a general pattern. If

$$L = t^2 p(t) D^2 + t q(t) D + r(t),$$

then

$$L[t^s] = p(t) s(s-1) t^s + q(t) s t^s + r(t) t^s.$$

Since $p(t)$, $q(t)$, and $r(t)$ are polynomials, this can be expanded in powers of t (with coefficients depending on s):

$$L[t^s] = f_0(s) t^s + f_1(s) t^{s+1} + \cdots + f_n(s) t^{s+n}.$$

Note that the coefficient $f_0(s)$ of t^s can be described in terms of the constant coefficients of $p(t)$, $q(t)$, and $r(t)$:

$$f_0(s) = p_0 s(s-1) + q_0 s + r_0.$$

To solve the homogeneous o.d.e.

(H)
$$Lx = [t^2 p(t) D^2 + t q(t) D + r(t)] x = 0$$

we substitute the Frobenius series

$$x = t^m \sum_{k=0}^{\infty} b_k t^k = \sum_{k=0}^{\infty} b_k t^{m+k}, \quad b_0 \neq 0$$

and obtain

$$0 = Lx = \sum_{k=0}^{\infty} b_k L[t^{m+k}]$$

$$= \sum_{k=0}^{\infty} b_k f_0(m + k)t^{m+k} + \sum_{k=0}^{\infty} b_k f_1(m + k)t^{m+k+1}$$

$$+ \cdots + \sum_{k=0}^{\infty} b_k f_n(m + k)t^{m+k+n}$$

$$= \sum_{j=0}^{\infty} b_j f_0(m + j)t^{m+j} + \sum_{j=1}^{\infty} b_{j-1} f_1(m + j - 1)t^{m+j}$$

$$+ \cdots + \sum_{j=n}^{\infty} b_{j-n} f_n(m + j - n)t^{m+j}.$$

For this last expression to be zero, the combined coefficient of each power of t should be zero. The coefficient of t^m is

$$b_0 f_0(m) = 0,$$

and the coefficients of higher powers give the recurrence relation

(R) $b_j f_0(m + j) + b_{j-1} f_1(m + j - 1) + \cdots + b_0 f_j(m) = 0,$
        $j = 1, 2, 3, \ldots ,$

where we take the functions $f_k(s)$ to be identically zero if $k > n$. Since $b_0 \neq 0$, the first equation can be rewritten $f_0(m) = 0$, or

(I) $p_0 m(m - 1) + q_0 m + r_0 = 0.$

This is called the **indicial equation** of (H) (compare the previous section, in which (H) is the Cauchy-Euler equation). *Once we find the roots, $m = m_1$ and $m = m_2$, of the indicial equation (I), we try to use the recurrence relation (R) to determine the coefficients of the corresponding solution to (H).*

Of course, if $m_1 = m_2$ we cannot expect to obtain two independent solutions in this way. In fact, even when $m_1 \neq m_2$, we may find it impossible to obtain two independent solutions as Frobenius series; if the roots differ by an integer, say $m_1 = m_2 + j(j > 0)$, then $f_0(m_2 + j) = f_0(m_1) = 0$, and we cannot divide (R) by $f_0(m_2 + j)$ to obtain an expression for b_j. We consider two examples with roots differing by an integer. In the first, we will still manage to find two independent Frobenius series solutions, but in the second we will be able to find only one.

Example 6.6.2

Solve

(H) $$Lx = [t^2D^2 + (t^2 + t)D - 1]x = 0, \quad t > 0.$$

Again we look for Frobenius series solutions

(S) $$x = t^m \sum_{k=0}^{\infty} b_k t^k = \sum_{k=0}^{\infty} b_k t^{m+k}, \quad b_0 \neq 0.$$

Since

$$L[t^s] = s(s - 1)t^s + (t^2 + t)st^{s-1} - t^s = (s + 1)(s - 1)t^s + st^{s+1},$$

substituting (S) into (H) gives

$$
\begin{aligned}
0 = Lx &= \sum_{k=0}^{\infty} b_k L[t^{m+k}] \\
&= \sum_{k=0}^{\infty} b_k(m + k + 1)(m + k - 1)t^{m+k} + \sum_{k=0}^{\infty} b_k(m + k)t^{m+k+1} \\
&= \sum_{j=0}^{\infty} b_j(m + j + 1)(m + j - 1)t^{m+j} + \sum_{j=1}^{\infty} b_{j-1}(m + j - 1)t^{m+j} \\
&= b_0(m + 1)(m - 1)t^m \\
&\quad + \sum_{j=1}^{\infty} [b_j(m + j + 1)(m + j - 1) + b_{j-1}(m + j - 1)]t^{m+j}.
\end{aligned}
$$

We need

(I) $$(m + 1)(m - 1) = 0$$

and

(R) $$b_j(m + j + 1)(m + j - 1) + b_{j-1}(m + j - 1) = 0, \quad j = 1, 2, 3, \ldots.$$

The roots of the indicial equation (I) are

$$m = \pm 1.$$

Substituting $m = 1$ into (R) gives

(R$_1$) $$b_j(j + 2)j + b_{j-1}j = 0$$

or

$$b_j = -\frac{1}{j + 2} b_{j-1}.$$

Thus

$$b_1 = -\frac{1}{3} b_0, \quad b_2 = \frac{(-1)^2}{4 \cdot 3} b_0, \ldots,$$

$$b_k = \frac{(-1)^k}{(k + 2) \cdots 4 \cdot 3} b_0 = \frac{(-1)^k \cdot 2}{(k + 2)!} b_0, \ldots.$$

If we set $b_0 = 1$, the resulting solution of (H) is

(S$_1$) $$x_1(t) = t \left(1 + \sum_{k=1}^{\infty} \frac{(-1)^k 2}{(k + 2)!} t^k \right).$$

Substituting the other root, $m = -1$, into (R) leads to

(R$_2$) $$b_j j(j - 2) + b_{j-1}(j - 2) = 0.$$

When $j = 1$, this gives

$$b_1 = -b_0.$$

When $j = 2$, (R$_2$) reads $0 = 0$. This means that there is no restriction on b_2; in particular, we can pick

$$b_2 = 0.$$

Now (R$_2$) gives

$$b_3 = 0, \quad b_4 = 0, \ldots, \quad b_k = 0, \ldots.$$

Setting $b_0 = 1$ yields the solution of (H):

(S$_2$) $$x_2(t) = t^{-1}(1 - t).$$

The functions $x_1(t)$ and $x_2(t)$ are linearly independent solutions of (H) for

$t > 0$, so the general solution (for $t > 0$) is

$$x = c_1 x_1(t) + c_2 x_2(t) = c_1 t \left(1 + \sum_{k=1}^{\infty} \frac{(-1)^k 2}{(k+2)!} t^k \right) + c_2 t^{-1}(1 - t).$$

Example 6.6.3

Solve

(H) $$Lx = [t^2 D^2 + (t^3 + t)D - 1]x = 0, \quad t > 0.$$

You can check that substituting the Frobenius series

(S) $$x = t^m \sum_{k=0}^{\infty} b_k t^k = \sum_{k=0}^{\infty} b_k t^{m+k}, \quad b_0 \neq 0$$

into (H) leads to

$$0 = Lx = b_0(m + 1)(m - 1)t^m + b_1(m + 2)mt^{m+1}$$

$$+ \sum_{j=2}^{\infty} [b_j(m + j + 1)(m + j - 1) + b_{j-2}(m + j - 2)]t^{m+j}.$$

This implies

(I) $$(m + 1)(m - 1) = 0$$
$$b_1(m + 2)m = 0$$

and

(R) $b_j(m + j + 1)(m + j - 1) + b_{j-2}(m + j - 2) = 0, \quad j = 2, 3, \ldots$

The roots of the indicial equation (I) are

$$m = \pm 1.$$

Using $m = +1$, we get

$$3b_1 = 0$$

and

(R$_1$) $$b_j(j + 2)j + b_{j-2}(j - 1) = 0, \quad j = 2, 3, \ldots$$

Thus

$$b_2 = \frac{-1}{2^2 \cdot 2} b_0, \quad b_4 = \frac{(-1)^2 \cdot 3}{2^4(3 \cdot 2)(2 \cdot 1)} b_0, \ldots$$

and

$$b_1 = 0, \quad b_3 = 0, \ldots.$$

The general pattern here is difficult to discover, but it can be verified that if $b_0 = 1$, we have, for $k = 1, 2, \ldots,$

$$b_{2k} = \frac{(-1)^k(2k - 1)!}{2^{3k-1}(k + 1)!k!(k - 1)!}, \quad b_{2k-1} = 0.$$

Thus, one solution of (H) is

$$(S_1) \qquad x_1(t) = t \left(1 + \sum_{k=1}^{\infty} \frac{(-1)^k(2k - 1)!}{2^{3k-1}(k + 1)!k!(k - 1)!} t^{2k} \right).$$

On the other hand, substitution of $m = -1$ leads to

$$-b_1 = 0$$

and

$$(R_2) \qquad b_j j(j - 2) + b_{j-2}(j - 3) = 0, \quad j = 2, 3, \ldots.$$

But for $j = 2$, (R_2) reads

$$-b_0 = 0.$$

This means in particular that *we cannot find a Frobenius series solution (with $b_0 \neq 0$) corresponding to $m = -1$.* We will return to this example in Section 6.8.

Another potential source of trouble is the case when *the roots of the indicial equation are complex.* Fortunately, the device that worked for the Cauchy-Euler equation saves us here: *The general solution to (H) is generated by the real and imaginary parts of the complex solution that corresponds to one of the complex roots.*

Example 6.6.4

Solve

(H) $$Lx = [t^2D^2 + tD + (1 - t)]x = 0, \quad t > 0.$$

You can verify that the indicial equation and recurrence relation for (H) are, respectively,

(I) $$m^2 + 1 = 0$$

and

(R) $$b_j[(m + j)^2 + 1] - b_{j-1} = 0, \quad j = 1, 2, \ldots.$$

The roots of the indicial equation (I) are

$$m = \pm i.$$

Using $m = i$, the recurrence relation becomes

(R$_1$) $$b_j[(j + i)^2 + 1] - b_{j-1} = 0$$

or

$$b_j = \frac{b_{j-1}}{(j + 2i)j}.$$

We rationalize the denominator by multiplying top and bottom by $j - 2i$:

$$b_j = \frac{j - 2i}{(j^2 + 4)j} b_{j-1}.$$

If we set $b_0 = 1$, we find

$$b_1 = \frac{1 - 2i}{5 \cdot 1}, \quad b_2 = \frac{2 - 2i}{8 \cdot 2} b_1 = \frac{-(1 + 3i)}{40}, \ldots.$$

We won't attempt a general pattern here, but we content ourselves with these coefficients. The corresponding complex Frobenius series is

$$x = t^i \left[1 + \left(\frac{1 - 2i}{5} \right) t - \left(\frac{1 + 3i}{40} \right) t^2 + \cdots \right]$$

$$= [\cos(\ln t) + i \sin(\ln t)] \left[1 + \left(\frac{1 - 2i}{5} \right) t - \left(\frac{1 + 3i}{40} \right) t^2 + \cdots \right]$$

$$= \left\{ \cos(\ln t) \left[1 + \frac{t}{5} - \frac{t^2}{40} + \cdots \right] + \sin(\ln t) \left[\frac{2t}{5} + \frac{3t^2}{40} + \cdots \right] \right\}$$

$$+ i \left\{ \cos(\ln t) \left[-\frac{2}{5}t - \frac{3}{40}t^2 + \cdots \right] + \sin(\ln t) \left[1 + \frac{t}{5} - \frac{t^2}{40} + \cdots \right] \right\}.$$

The general solution to (H) is generated by the real and imaginary parts of this function:

$$x = c_1 \left\{ \cos(\ln t) \left[1 + \frac{t}{5} - \frac{t^2}{40} + \cdots \right] + \sin(\ln t) \left[\frac{2t}{5} + \frac{3t^2}{40} + \cdots \right] \right\}$$

$$+ c_2 \left\{ \cos(\ln t) \left[-\frac{2t}{5} - \frac{3t^2}{40} + \cdots \right] + \sin(\ln t) \left[1 + \frac{t}{5} - \frac{t^2}{40} + \cdots \right] \right\}.$$

We now summarize the method, noting that (as in Section 6.3) the solutions it yields will be valid for $0 < t < R$, provided the leading coefficient of (H), $p(t)$, has no complex roots z with $|z| < R$.

REGULAR SINGULAR POINTS: FROBENIUS SERIES

An o.d.e. has a **regular singular point** at $t = 0$ if it can be written in the form

(H) $$Lx = [t^2 p(t)D^2 + tq(t)D + r(t)]x = 0,$$

where $p(t)$, $q(t)$, and $r(t)$ are polynomials and $p(0) \neq 0$.

In this situation we look for solutions of (H) in the form of **Frobenius series**,

(S) $$x = t^m \sum_{k=0}^{\infty} b_k t^k = \sum_{k=0}^{\infty} b_k t^{m+k}, \quad b_0 \neq 0$$

as follows:

1. Calculate $L[t^s]$ and use the result to find $Lx = \sum_{k=0}^{\infty} b_k L[t^{m+k}]$.

2. Rewrite Lx as a Frobenius series by combining like powers of t (using index substitutions).

3. Set each combined coefficient equal to zero to obtain the **indicial equation**

(I) $$p_0 m(m-1) + q_0 m + r_0 = 0$$

and a general **recurrence relation** (R).

4. If the indicial equation (I) has distinct real roots, then substituting these values for m in (R) leads to solutions of (H) that generate the general solution, except possibly when the roots differ by an integer.

5. If the indicial equation (I) has complex roots, then substituting either root into (R) leads to a complex solution of (H) whose real and imaginary parts generate the general solution.

6. If the indicial equation (I) has a double root, we can still find one solution of (H), using (R), but we need a second solution to generate the general solution of (H).

The Frobenius series solutions obtained this way will be valid for $0 < t < R$, provided $p(z) \neq 0$ for every complex number z with $|z| < R$.

Note

Extending the method

Our discussion of Frobenius series in this section focused on o.d.e.'s

(H) $$Lx = [t^2 p(t)D^2 + tq(t)D + r(t)]x = 0$$

with $p(t)$, $q(t)$, and $r(t)$ polynomials. The method just summarized also applies when $p(t)$, $q(t)$, and $r(t)$ are given by power series

$$p(t) = \sum_{n=0}^{\infty} p_n t^n, \quad q(t) = \sum_{k=0}^{\infty} q_n t^n, \quad r(t) = \sum_{n=0}^{\infty} r_n t^n.$$

The indicial equation maintains the same form

(I) $$p_0 m(m-1) + q_0 m + r_0 = 0,$$

but now the recurrence relation, which arises from a multiplication of series, can be extremely complicated. The Frobenius series expressions for the solutions are valid for $0 < t < R$ provided all three coefficient series converge for $|t| < R$ and $p(z) \neq 0$ for all complex z with $|z| < R$.

EXERCISES

In Exercises 1 through 12, find an expression for the general solution in terms of Frobenius series. In Exercises 1 through 8, you need only find the first four coefficients of each of the two generating solutions. In Exercises 9 through 12, find the general pattern for the coefficients.

1. $[9t^2D^2 + 9tD + (6t - 1)]x = 0, \quad 0 < t$
2. $[3t^2D^2 + t(3t + 4)D + (t - 2)]x = 0, \quad 0 < t$
3. $[4t^2D^2 + 4tD + (4t^2 - 1)]x = 0, \quad 0 < t$
4. $[9t^2D^2 + 9tD + (9t^2 - 4)]x = 0, \quad 0 < t$
5. $[4t^2D^2 + 4tD + (4t^2 - 9)]x = 0, \quad 0 < t$
6. $[t^2D^2 + t(1 - t)D + 1]x = 0, \quad 0 < t$
7. $[t^2(t + 2)D^2 + t(t + 1)D + (t^2 - 1)]x = 0, \quad 0 < t < 2$
8. $[t^2D^2 - tD + 5(1 - t)]x = 0, \quad 0 < t$
9. $[2t^2D^2 - t(t - 1)D - 1]x = 0, \quad 0 < t$
10. $[4t^2(t + 1)D^2 + (t - 3)]x = 0, \quad 0 < t < 1$
11. $[2tD^2 + D + t]x = 0, \quad 0 < t$
12. $[3t^2(t - 1)D^2 + 4tD - 2]x = 0, \quad 0 < t < 1$

In Exercises 13 through 16, find a Frobenius series solution. Check that any other Frobenius series solution is a constant multiple of the one you find.

13. $[t^2D^2 + tD + t^2]x = 0, \quad 0 < t$
14. $[t^2D^2 - tD + (t + 1)]x = 0, \quad 0 < t$
15. $[t^2D^2 + tD + (t - 1)]x = 0, \quad 0 < t$
16. $[t^2D^2 + tD + (t^2 - 4)]x = 0, \quad 0 < t$

In Exercises 17 through 21, find the general solution by making an appropriate change of variable and solving the resulting equation.

17. $[2t^2D^2 - t(t - 1)D - 1]x = 0, \quad t < 0$ (compare Exercise 9)
18. $[2tD^2 + D + t]x = 0, \quad t < 0$ (compare Exercise 11)
19. $[2(t - 1)^2D^2 + t(t - 1)D - 1]x = 0, \quad t > 1$
20. $[2(t - 1)^2D^2 + t(t - 1)D - 1]x = 0, \quad t < 1$
21. $[(t^2 + 2t + 1)D^2 + (t^2 + 3t + 2)D - 1]x = 0, \quad t > -1$
22. *Linear Independence of Frobenius Series:* Suppose that $f(t) = t^{m_1} \sum_{k=0}^{\infty} b_k t^k$ and $g(t) = t^{m_2} \sum_{k=0}^{\infty} c_k t^k$, where $b_0 \neq 0$, $c_0 \neq 0$, and $m_1 < m_2$. Show that $f(t)$ and $g(t)$ are linearly independent on any interval of the form $0 < t < a$. [*Hint:* Multiply the equation $Af(t) + Bg(t) = 0$ by t^{-m_1} and see what happens as $t \to 0$.]

6.7 A CASE STUDY: BESSEL FUNCTIONS OF THE FIRST KIND

In this section we use Frobenius series to find nontrivial solutions for Bessel's equation,

(B_μ) $[t^2D^2 + tD + (t^2 - \mu^2)]x = 0, \quad t > 0$

where $\mu \geq 0$. The "Bessel functions" that arise in solving (B_μ) play an important role in many problems of mathematical physics.

The indicial equation of (B_μ),

$$m^2 - \mu^2 = 0$$

has roots $m = \pm\mu$. We saw in the last section that we can always find a nontrivial Frobenius series solution corresponding to the larger root, $m = \mu$. We begin by finding such solutions to Bessel's equation with $\mu = 0$ and $\mu = 1$.

Example 6.7.1

Find a nontrivial Frobenius series solution to Bessel's equation with $\mu = 0$,

(B_0) $$[t^2 D^2 + t D + t^2]x = 0, \quad t > 0.$$

Substituting

(S) $$x = t^m \sum_{k=0}^{\infty} b_k t^k = \sum_{k=0}^{\infty} b_k t^{m+k}, \quad b_0 \neq 0$$

into (B_0) yields the equation

$$\sum_{k=0}^{\infty} (m+k)(m+k-1)b_k t^{m+k} + \sum_{k=0}^{\infty} (m+k)b_k t^{m+k} + \sum_{k=0}^{\infty} b_k t^{m+k+2} = 0$$

or

$$m^2 b_0 t^m + (m+1)^2 b_1 t^{m+1} + \sum_{j=2}^{\infty} [(m+j)^2 b_j + b_{j-2}]t^{m+j} = 0.$$

Thus

$$m^2 b_0 = 0, \quad (m+1)^2 b_1 = 0$$

and

$$(m+j)^2 b_j + b_{j-2} = 0, \quad j = 2, 3, \ldots.$$

Since $b_0 \neq 0$, we must take $m = 0$. Then

$$b_1 = 0$$

and

$$b_j = -\frac{1}{j^2} b_{j-2}, \quad j = 2, 3, \ldots.$$

Since $b_1 = 0$, every odd-numbered coefficient vanishes. On the other hand, the even-numbered coefficients satisfy

$$b_2 = \frac{-b_0}{2^2}, \quad b_4 = \frac{-b_2}{4^2} = \frac{(-1)^2 b_0}{4^2 \cdot 2^2}, \ldots,$$

$$b_{2n} = \frac{(-1)^n b_0}{(2n)^2 (2n-2)^2 \cdots 2^2} = \frac{(-1)^n b_0}{2^{2n} (n!)^2}, \ldots.$$

The solution of (B_0) obtained by taking $b_0 = 1$ is called the **Bessel function of the first kind of order 0** and is denoted $J_0(t)$:

$$J_0(t) = \sum_{n=0}^{\infty} \frac{(-1)^n t^{2n}}{2^{2n}(n!)^2} = \sum_{n=0}^{\infty} \frac{(-1)^n}{(n!)^2} \left(\frac{t}{2}\right)^{2n}.$$

Since 0 is the only root of the indicial equation of (B_0), any Frobenius series solution of (B_0) is a constant multiple of $J_0(t)$.

Example 6.7.2

Find a nontrivial Frobenius series solution of Bessel's equation with $\mu = 1$,

(B$_1$) $[t^2 D^2 + tD + (t^2 - 1)]x = 0, \quad t > 0.$

Substituting

(S) $x = t^m \sum_{k=0}^{\infty} b_k t^k = \sum_{k=0}^{\infty} b_k t^{m+k}, \quad b_0 \neq 0$

into (B$_1$) yields the equation

$$\sum_{k=0}^{\infty} (m+k)(m+k-1)b_k t^{m+k}$$

$$+ \sum_{k=0}^{\infty} (m+k)b_k t^{m+k} + \sum_{k=0}^{\infty} b_k t^{m+k+2} - \sum_{k=0}^{\infty} b_k t^{m+k} = 0$$

or

$$(m^2 - 1)b_0 t^m + (m^2 + 2m)b_1 t^{m+1} + \sum_{j=2}^{\infty} \{[(m+j)^2 - 1]b_j + b_{j-2}\} t^{m+j} = 0.$$

Thus

$$(m^2 - 1)b_0 = 0$$

$$(m^2 + 2m)b_1 = 0$$

and

(R) $$[(m+j)^2 - 1]b_j + b_{j-2} = 0, \quad j = 2, 3, \ldots .$$

Since $b_0 \neq 0$, the first two equations give

$$m = \pm 1$$

and

$$b_1 = 0.$$

If we take $m = +1$, the recurrence relation (R) gives

$$b_j = -\frac{1}{(j+2)j} b_{j-2}, \quad j = 2, 3, \ldots .$$

Since $b_1 = 0$, the odd-numbered coefficients vanish. The even-numbered coefficients satisfy

$$b_2 = -\frac{b_0}{4 \cdot 2} = -\frac{b_0}{2^2(2 \cdot 1)}, \quad b_4 = -\frac{b_2}{6 \cdot 4} = \frac{b_0}{2^4(3 \cdot 2)(2 \cdot 1)}, \ldots ,$$

$$b_{2n} = \frac{(-1)^n b_0}{2^{2n}(n+1)!n!}, \ldots .$$

The solution obtained when $b_0 = 1/2$ is called the **Bessel function of the first kind of order 1** and is denoted $J_1(t)$:

$$J_1(t) = \frac{1}{2} \sum_{n=0}^{\infty} \frac{(-1)^n t^{2n+1}}{2^{2n}(n+1)!n!} = \sum_{n=0}^{\infty} \frac{(-1)^n}{(n+1)!n!} \left(\frac{t}{2}\right)^{2n+1}.$$

If we take $m = -1$, the recurrence relation reads

$$(-2j + j^2)b_j + b_{j-2} = 0, \quad j = 2, 3, \ldots .$$

Taking $j = 2$, we see that $b_0 = 0$. Thus there are no nontrivial Frobenius series solutions corresponding to $m = -1$; any Frobenius series solution of (B$_1$) is a constant multiple of $J_1(t)$.

In general, substituting

(S) $$x = t^m \sum_{k=0}^{\infty} b_k t^k = \sum_{k=0}^{\infty} b_k t^{m+k}, \quad b_0 \neq 0$$

into Bessel's equation

(B$_\mu$) $$[t^2 D^2 + tD + (t^2 - \mu^2)]x = 0$$

leads to the equation

$$(m^2 - \mu^2)b_0 t^m + [(m + 1)^2 - \mu^2]b_1 t^{m+1}$$

$$+ \sum_{j=2}^{\infty} \{[(m + j)^2 - \mu^2]b_j + b_{j-2}\}t^{m+j} = 0.$$

Hence

(I) $$m^2 = \mu^2$$

(R$_1$) $$[(m + 1)^2 - \mu^2]b_1 = 0$$

and

(R$_j$) $$[(m + j)^2 - \mu^2]b_j + b_{j-2} = 0, \quad j = 2, 3, \ldots .$$

If we take $m = \mu$, we get

$$(2\mu + 1)b_1 = 0$$

and

$$b_j = - \frac{1}{(2\mu + j)j} b_{j-2}, \quad j = 2, 3, \ldots .$$

Thus b_1 and the other odd-numbered coefficients vanish; the even-numbered coefficients satisfy

$$b_{2n} = \frac{(-1)^n b_0}{2^{2n}(n!)[(\mu + n) \cdots (\mu + 2)(\mu + 1)]}.$$

When μ is an integer, the choice

$$b_0 = \frac{1}{2^\mu(\mu!)}$$

leads to a solution of (B_μ) that can be written in a neat form:

$$J_\mu(t) = \sum_{n=0}^{\infty} \frac{(-1)^n t^{2n+\mu}}{2^{2n+\mu} n!(\mu + n)!} = \sum_{n=0}^{\infty} \frac{(-1)^n}{n!(\mu + n)!} \left(\frac{t}{2}\right)^{2n+\mu}.$$

When μ is not an integer, an appropriate choice of b_0 still yields a solution that has a neat form. To describe this choice, we need to extend the notion of factorial to nonintegers. This is done by the **gamma function**, defined for $s > 0$ by the integral formula

$$\Gamma(s) = \int_0^\infty e^{-t} t^{s-1} \, dt.$$

This improper integral converges for $s > 0$ and defines a function with the following properties (see Exercise 1):

1. $\Gamma(s + 1) = s\Gamma(s)$ for all $s > 0$.
2. $\Gamma(n + 1) = n!$ for all integers $n \geq 0$.

Our choices for b_0 when μ is an integer can be written as

$$b_0 = \frac{1}{2^\mu \Gamma(\mu + 1)}.$$

Even when μ is not an integer, we can make this same choice of b_0. Using property 1, we then see that

$$b_{2n} = \frac{(-1)^n}{2^{2n+\mu} n![(\mu + n) \cdots (\mu + 2)(\mu + 1)]\Gamma(\mu + 1)}$$

$$= \frac{(-1)^n}{2^{2n+\mu} n!\Gamma(\mu + n + 1)}.$$

The resulting solution of (B_μ) is called the **Bessel function of the first kind of order** μ and is denoted $J_\mu(t)$:

$$J_\mu(t) = \sum_{n=0}^{\infty} \frac{(-1)^n}{n!\Gamma(\mu + n + 1)} \left(\frac{t}{2}\right)^{2n+\mu}.$$

We have, then, the following:

Fact: *For any* $\mu \geq 0$, $x = J_\mu(t)$ *is a nontrivial solution of* (B_μ).

In Example 6.7.1 and 6.7.2, the Frobenius series solutions of (B_μ) were all constant multiples of $J_\mu(t)$. The following two examples illustrate cases in which the smaller root $m = -\mu$ of the indicial equation leads to a solution of (B_μ) independent of $J_\mu(t)$.

Example 6.7.3

Find the general solution to Bessel's equation with $\mu = 1/3$,

($B_{1/3}$) $$\left[t^2 D^2 + tD + \left(t^2 - \frac{1}{9} \right) \right] x = 0, \quad t > 0.$$

We already know a Frobenius series solution corresponding to $m = 1/3$, namely, the Bessel function of the first kind

$$J_{1/3}(t) = \sum_{n=0}^{\infty} \frac{(-1)^n}{n! \Gamma(n + \frac{4}{3})} \left(\frac{t}{2} \right)^{2n + 1/3}.$$

In addition, since the roots $m = \pm 1/3$ of the indicial equation differ by 2/3, which is not an integer, we know that ($B_{1/3}$) also has a Frobenius series solution corresponding to $m = -1/3$.

If we substitute $m = -1/3$ into equations (R_1) and (R_j), we get

$$\left[\left(\frac{2}{3} \right)^2 - \frac{1}{9} \right] b_1 = 0$$

and

$$\left[\left(-\frac{1}{3} + j \right)^2 - \frac{1}{9} \right] b_j + b_{j-2} = 0, \quad j = 2, 3, \ldots.$$

It follows that the odd-numbered coefficients vanish, and

$$b_{2n} = \frac{(-1)^n b_0}{2^{2n}(n!)[(-\frac{1}{3} + n) \ldots (-\frac{1}{3} + 2)(-\frac{1}{3} + 1)]}, \quad n = 1, 2, \ldots.$$

The choice

$$b_0 = \frac{1}{2^{-1/3} \Gamma(-\frac{1}{3} + 1)}$$

gives us the **Bessel function of the first kind of order** $-1/3$:

$$J_{-1/3}(t) = \sum_{n=0}^{\infty} \frac{(-1)^n}{n! \Gamma(-\frac{1}{3} + n + 1)} \left(\frac{t}{2} \right)^{2n - 1/3}.$$

This function is independent of $J_{1/3}(t)$ (Exercise 7), so the general solution of $(B_{1/3})$ is

$$x = c_1 J_{1/3}(t) + c_2 J_{-1/3}(t).$$

Example 6.7.4

Find the general solution of Bessel's equation with $\mu = 1/2$,

$(B_{1/2})$
$$\left[t^2 D^2 + tD + \left(t^2 - \frac{1}{4} \right) \right] x = 0, \quad t > 0.$$

The Bessel function of the first kind

$$J_{1/2}(t) = \sum_{n=0}^{\infty} \frac{(-1)^n}{n!\Gamma(n + \frac{3}{2})} \left(\frac{t}{2} \right)^{2n + 1/2}$$

is a solution corresponding to the larger root $m = 1/2$ of the indicial equation. Since the difference between the roots is 1, which is an integer, we are not guaranteed a solution corresponding to $m = -1/2$. Nonetheless, we can try to find one.

Substitution of $m = -1/2$ into (R_1) and (R_j) gives

$$\left[\left(1 - \frac{1}{2} \right)^2 - \left(\frac{1}{2} \right)^2 \right] b_1 = 0$$

and

$$\left[\left(-\frac{1}{2} + j \right)^2 - \left(\frac{1}{2} \right)^2 \right] b_j + b_{j-2} = 0, \quad j = 2, 3, \ldots.$$

Since the left side of the first of these equations is 0, there is no restriction on b_1. We pick $b_1 = 0$, so that all odd-numbered coefficients vanish. The even-numbered coefficients satisfy

$$b_{2n} = \frac{(-1)^n b_0}{2^{2n} n! (-\frac{1}{2} + n) \cdots (-\frac{1}{2} + 2)(-\frac{1}{2} + 1)}.$$

Following our earlier pattern, we set

$$b_0 = \frac{1}{2^{-1/2} \Gamma(-\frac{1}{2} + 1)}$$

to obtain the **Bessel function of the first kind of order** $-1/2$:

$$J_{-1/2}(t) = \sum_{n=0}^{\infty} \frac{(-1)^n}{n!\Gamma(-\frac{1}{2} + n + 1)} \left(\frac{t}{2}\right)^{2n - 1/2}.$$

This is independent of $J_{1/2}(t)$, so the general solution of ($B_{1/2}$) is

$$x = c_1 J_{1/2}(t) + c_2 J_{-1/2}(t).$$

In general, when the difference 2μ between the roots $\pm\mu$ is not an integer, we are guaranteed a solution corresponding to $-\mu$. Substitution of $m = -\mu$ into (R_1) and (R_j) gives

$$(-2\mu + 1)b_1 = 0$$

and

$$(-2\mu + j)jb_j + b_{j-2} = 0, \quad j = 2, 3, \ldots.$$

It follows that the odd-numbered coefficients are 0, and the even-numbered ones are given by

$$b_{2n} = \frac{(-1)^n b_0}{2^{2n}(n!)(-\mu + n)(-\mu + n - 1) \cdots (-\mu + 1)}.$$

Thus we are led to the solution

$$x = b_0 t^{-\mu} + \sum_{n=1}^{\infty} \frac{(-1)^n b_0}{2^{2n}(n!)(-\mu + n)(-\mu + n - 1) \cdots (-\mu + 1)} t^{2n - \mu}.$$

Notice that this formula makes sense as long as μ is not an integer (even if 2μ is an integer). Indeed, substitution into (B_μ) shows that it defines a solution of (B_μ) even when μ is half an odd integer.

We would like to follow the pattern already set by choosing $b_0 = 1/2^{-\mu}\Gamma(-\mu + 1)$. However, when $-\mu + 1 < 0$, this requires us to define $\Gamma(s)$ for negative values of s. We start by noting that if $-1 < s < 0$, then $s + 1$ is positive, so $\Gamma(s + 1)$ makes sense. To preserve property 1 of the gamma function, we define

$$\Gamma(s) = \frac{\Gamma(s + 1)}{s}, \quad -1 < s < 0.$$

Now, if $-2 < s < -1$, then $-1 < s + 1 < 0$, so $\Gamma(s + 1)$ makes sense. In order to preserve property 1, we define

$$\Gamma(s) = \frac{\Gamma(s + 1)}{s} = \frac{\Gamma(s + 2)}{(s + 1)s}, \quad -2 < s < 0.$$

Continuing this way, we can define $\Gamma(s)$ for all $s < 0$ except $-1, -2, \ldots$, in such a way that property 1 remain true.

If μ is not an integer, the choice

$$b_0 = \frac{1}{2^{-\mu}\Gamma(-\mu + 1)}$$

leads to the **Bessel function of the first kind of order $-\mu$,**

$$J_{-\mu}(t) = \sum_{n=0}^{\infty} \frac{(-1)^n}{n!\Gamma(-\mu + n + 1)} \left(\frac{t}{2}\right)^{2n - \mu}.$$

This function is a solution of (B_μ), independent of $J_\mu(t)$. Thus we have the following.

Fact: *If μ is not an integer, then the general solution of (B_μ) is*

$$x = c_1 J_\mu(t) + c_2 J_{-\mu}(t).$$

When the order μ is an integer, it is not possible to find a second Frobenius series solution that is independent of $J_\mu(t)$. This case requires the techniques of the next section and leads to "Bessel functions of the second kind."

Bessel functions were studied in special cases by Euler, the Bernoullis, Lagrange, Fourier, and Poisson. They were first systematically studied by the German mathematician F. W. Bessel (1824) and are now standard tools of applied mathematicians. Summaries of their properties, as well as tables of values, can be found in numerous handbooks of mathematical physics. We will develop a few of these properties in the exercises that follow our summary.

BESSEL FUNCTIONS OF THE FIRST KIND

The **gamma function** is defined for all real numbers $s \neq 0, -1, -2, \ldots$ by

$$\Gamma(s) = \begin{cases} \int_0^\infty e^{-t}t^{s-1}\, dt & \text{if } s > 0; \\ \dfrac{\Gamma(s + j)}{(s + j - 1) \cdots (s + 1)s} & \text{if } -j < s < -j + 1, \\ & \quad j \text{ a positive integer.} \end{cases}$$

It has the following properties:

1. $\Gamma(s + 1) = s\Gamma(s)$ whenever both sides are defined.
2. $\Gamma(n + 1) = n!$ for any integer $n \geq 0$.

For any real number $\rho \neq -1, -2, \ldots$, the **Bessel function of the first kind of order ρ** is defined for $t > 0$ by

$$J_\rho(t) = \sum_{n=0}^{\infty} \frac{(-1)^n}{n!\,\Gamma(\rho + n + 1)} \left(\frac{t}{2}\right)^{2n+\rho}.$$

The Bessel equation

$$(B_\mu) \qquad\qquad [t^2 D^2 + tD + (t^2 - \mu^2)]x = 0,$$

where $\mu \geq 0$, always has $x = J_\mu(t)$ as a solution. If μ is not an integer, then the general solution of (B_μ) is

$$x = c_1 J_\mu(t) + c_2 J_{-\mu}(t).$$

If μ is an integer, then every Frobenius series solution of (B_μ) is a constant multiple of $J_\mu(t)$.

EXERCISES

Exercises 1 through 4 deal with the gamma functions.

1. a. Show that $\Gamma(s) > 0$ for $0 < s \leq 1$.
    b. Use integration by parts and L'Hôpital's rule to show that $\Gamma(s + 1) = s\Gamma(s)$ for all $s > 0$.
    c. Show that $\Gamma(1) = 1$.
    d. Use the results of parts (b) and (c) to show that $\Gamma(n + 1) = n!$ for all integers $n \geq 0$.

2. a. By substituting $x = t^{1/2}$ in the formula defining $\Gamma(1/2)$, show that $\Gamma(1/2) = 2 \int_0^\infty e^{-x^2}\, dx$ so that

    $$\left[\Gamma\left(\frac{1}{2}\right)\right]^2 = 4 \int_0^\infty e^{-x^2}\, dx \int_0^\infty e^{-y^2}\, dy = 4 \int_0^\infty \int_0^\infty e^{-x^2 - y^2}\, dx\, dy.$$

    b. By converting the double integral in (a) to polar coordinates, show that

    $$\left[\Gamma\left(\frac{1}{2}\right)\right]^2 = 4 \int_0^{\pi/2} \int_0^\infty e^{-r^2} r\, dr\, d\theta = \pi,$$

    or $\Gamma(1/2) = \sqrt{\pi}$.

3. a. Use the fact that $\Gamma(1/2) = \sqrt{\pi}$ (see Exercise 2) to show that $\Gamma(3/2) = \sqrt{\pi}/2$.

b. More generally, show that if k is a positive integer, then

$$\Gamma\left(\frac{2k + 1}{2}\right) = \frac{1 \cdot 3 \cdot 5 \cdots (2k - 1)}{2^k}\sqrt{\pi}.$$

4. a. Use the fact that $\Gamma(1/2) = \sqrt{\pi}$ (see Exercise 2) to show that $\Gamma(-1/2) = -2\sqrt{\pi}$.

b. Find $\Gamma\left(-\dfrac{2k + 1}{2}\right)$, where k is a positive integer.

In Exercises 5 through 17 we develop some properties of Bessel functions of the first kind, $J_\rho(t)$, where ρ is any real number other than one of the negative integers: $\rho \neq -1, -2, \ldots$.

5. Use the ratio test to show that the series $\sum_{n=0}^{\infty} (-1)^n (t/2)^{2n}/n!\Gamma(\rho + n + 1)$ converges for all t, so $J_\rho(t)$ is defined (at least) for $0 < t < \infty$.

6. a. Check that $J_0(0) = 1$, and $J_\rho(0) = 0$ for $\rho > 0$.
   b. Show that if $0 > \rho \neq -1, -2, \ldots$, then $J_\rho(t)$ becomes unbounded as t approaches 0. [*Hint:* What happens to $t^{-\rho}J_\rho(t)$?]

7. Use the results of Exercise 6 to show that if $\mu > 0$ is not an integer, then $J_\mu(t)$ and $J_{-\mu}(t)$ are linearly independent on $0 < t < \infty$.

8. Differentiate the series expression for $J_0(t)$ to show that $J_0'(t) = -J_1(t)$.

9. a. Multiply the series expression for $J_\rho(t)$ by t^ρ and differentiate to show that

$$\frac{d}{dt}[t^\rho J_\rho(t)] = t^\rho J_{\rho-1}(t).$$

   [When $\rho = 0$, interpret $J_{\rho-1}(t) = J_{-1}(t)$ to mean $-J_1(t)$.]
   b. Expand the left side of the equation in (a) and cancel $t^{\rho-1}$ to obtain

$$tJ_\rho'(t) + \rho J_\rho(t) = tJ_{\rho-1}(t).$$

10. By an argument similar to the one used in Exercise 9, show that

   a. $\dfrac{d}{dt}[t^{-\rho}J_\rho(t)] = -t^{-\rho}J_{\rho+1}(t)$

   b. $tJ_\rho'(t) - \rho J_\rho(t) = -tJ_{\rho+1}(t)$

11. By subtracting the formula in Exercise 10(b) from the one in Exercise 9(b), show that

$$J_{\rho+1}(t) = \frac{2\rho J_\rho(t) - tJ_{\rho-1}(t)}{t}.$$

12. Use the formula in Exercise 11 to express $J_2(t)$, $J_3(t)$, and $J_4(t)$ in terms of $J_0(t)$ and $J_1(t)$.

13. a. Using the result of Exercise 3(b), show that

$$J_{-1/2}(t) = \sqrt{\frac{2}{\pi t}} \sum_{n=0}^{\infty} \frac{(-1)^n t^{2n}}{(2n)!} = \sqrt{\frac{2}{\pi t}} \cos t.$$

   b. Using the formula in Exercise 10(b), show that $J_{1/2}(t) = \sqrt{2/\pi t}\sin t$.
   c. Using the formula in Exercise 11, first with $\rho = 1/2$ and then with $\rho = -1/2$, obtain expressions for $J_{3/2}(t)$ and $J_{-3/2}(t)$ in terms of $\sin t$, $\cos t$,

powers of t, and constants. [*Note:* By continuing in this way, we can obtain expressions for any Bessel function of the form $J_{n/2}(t)$, where n is an odd integer. It turns out that these are the only Bessel functions that have such (finite) expressions in terms of elementary functions.]

It can be shown that each function $J_\rho(t)$ has infinitely many zeros. In Exercises 14 through 16, we outline a proof that the zeros of $J_\rho(t)$ and $J_{\rho+1}(t)$ on $0 < t < \infty$ are distinct and alternate.

14. a. Use the formula in Exercise 10(b) to show that if $J_\rho(t) = J_{\rho+1}(T) = 0$, where $T > 0$, then $J'_\rho(T) = 0$ as well.
   b. Use the uniqueness of solutions of initial-value problems to show that if $J_\rho(T) = J_{\rho+1}(T) = 0$, where $T > 0$, then $J_\rho(t) = 0$ for all $t > 0$.
   c. Conclude that $J_\rho(T)$ and $J_{\rho+1}(T)$ cannot have a common zero $T > 0$.

15. Suppose T_1 and T_2 are zeros of $J_\rho(t)$, where $0 < T_1 < T_2$. Show that $J_{\rho+1}(t)$ has a zero between T_1 and T_2. [*Hint:* Apply Rolle's theorem to $f(t) = t^{-\rho}J_\rho(t)$ on the interval $T_1 \le t \le T_2$, and use the formula in Exercise 10(a).]

16. Suppose T_1 and T_2 are zeros of $J_{\rho+1}(t)$, where $0 < T_1 < T_2$. Show that $J_\rho(t)$ has a zero between T_1 and T_2. [*Hint:* Apply Rolle's theorem to $g(t) = t^{\rho+1}J_{\rho+1}(t)$, and use the formula in Exercise 9(a) with ρ replaced by $\rho + 1$.]

17. *The Graphs of $J_0(t)$ and $J_1(t)$:*
   a. Write a program for your computer that takes the initial data t, μ, and $b_0 = 1/2^\mu \mu!$ and then uses the recurrence relation

$$b_{2k}t^{\mu+2k} = \frac{-t^2}{4(\mu + k)k} b_{2k-2}t^{\mu+2k-2}$$

to calculate the partial sums

$$S_{2n} = \sum_{k=0}^{n} b_{2k}t^{\mu+2k}$$

until $|b_{2n}t^{\mu+2n}|$ is less than $\epsilon = 0.001$. Use your program to obtain approximate values for $J_0(t)$ and $J_1(t)$ with $t = 0.5, 1, 1.5, \ldots, 9$. [*Hint:* It might be useful to write your program in such a way that once $J_\mu(t)$ has been calculated, all registers are set at appropriate values for calculating $J_\mu(t + 0.5)$.]
   b. Using the results from (a), sketch the graphs of $J_0(t)$ and $J_1(t)$ for $0 \le t \le 9$. [*Hint:* It helps to use different scales on the horizontal and vertical axes. The values of $J_0(t)$ and $J_1(t)$ all lie in the interval $-0.5 \le x \le 1.0$ when $t > 0$.] Note that your graphs should oscillate with decreasing magnitude as t increases. This behavior is exhibited on $0 < t < \infty$ by each of the functions $J_\rho(t)$.

6.8   REGULAR SINGULAR POINTS: EXCEPTIONAL CASES

Out discussion in Section 6.6 of the equation

(H) $$[t^2p(t)D^2 + tq(t)D + r(t)]x = 0, \quad t > 0$$

with a regular singular point at $t = 0$, is incomplete in two cases. When the roots m_1 and m_2 of the indicial equation

(I) $$p_0 m(m - 1) + q_0 m + r_1 = 0$$

are equal ($m_1 - m_2 = 0$), or *sometimes* when they differ by a nonzero integer ($m_1 - m_2 = n > 0$), the method discussed in Section 6.6 yields a Frobenius series solution $x_1(t)$ corresponding to m_1, but not the general solution. In this section we complete our discussion of these cases by finding a solution to (H) independent of $x_1(t)$.

We treat first the case of a double root $m_1 = m_2$ of (I). Recall that when the indicial equation of the Cauchy-Euler equation had a double root, then one solution $x_1(t)$ was a power of t, and a second was obtained by multiplying $x_1(t)$ by $\ln t$. In the present case, the first solution is multiplied by $\ln t$ and then adjusted by adding a second Frobenius series.

Fact: *If the indicial equation (I) has a double root $m_1 = m_2 = m$, then the general solution of (H) is $x = c_1 x_1(t) + c_2 x_2(t)$, where the first solution is a Frobenius series*

$$x_1(t) = t^m \sum_{k=0}^{\infty} b_k t^k, \quad b_0 \neq 0$$

and the second has the form

$$x_2(t) = x_1(t) \ln t + t^m \sum_{k=0}^{\infty} b_k^* t^k.$$

We note that the second solution can be replaced by any function of the form $x_2(t) + c x_1(t)$. By adjusting c, we can obtain a second solution with any value we wish for b_0^*. It is customary in this case to take $b_0^* = 0$.

Example 6.8.1

Find the general solution of

(H) $$Lx = [t^2 D^2 + t(t + 3)D + 1]x = 0.$$

We start by looking for Frobenius series solutions. Since

$$L[t^s] = t^2 s(s - 1)t^{s-2} + t(t + 3)st^{s-1} + t^s = (s + 1)^2 t^s + st^{s+1}$$

we have

$$L\left[\sum_{k=0}^{\infty} b_k t^{m+k}\right] = \sum_{k=0}^{\infty} (m + k + 1)^2 b_k t^{m+k} + \sum_{k=0}^{\infty} (m + k) b_k t^{m+k+1}$$

$$= (m + 1)^2 b_0 t^m + \sum_{j=1}^{\infty} [(m + j + 1)^2 b_j + (m + j - 1)b_{j-1}]t^m$$

For this to be identically zero, we need

(I) $$(m + 1)^2 = 0$$

and

(R) $$(m + j + 1)^2 b_j + (m + j - 1)b_{j-1} = 0, \quad j = 1, 2, \ldots.$$

The indicial equation (I) has a double root $m = -1$. Substituting this root into (R) gives the recurrence relation

$$b_j = \frac{-(j - 2)}{j^2} b_{j-1}.$$

Setting $b_0 = 1$, we get

$$b_0 = 1, \quad b_1 = \frac{-(1 - 2)}{1} = 1, \quad b_2 = 0, \quad b_3 = 0, \ldots, \quad b_n = 0, \ldots.$$

Thus, one solution of (H) is

(S$_1$) $$x_1(t) = t^{-1}(1 + t) = t^{-1} + 1.$$

We next look for a second solution of the form

(S$_2$)
$$x_2(t) = x_1(t) \ln t + t^{-1} \sum_{k=0}^{\infty} b_k^* t^k$$

$$= (t^{-1} + 1) \ln t + \sum_{k=0}^{\infty} b_k^* t^{-1+k}.$$

We can use our earlier calculation of the effect of L on an arbitrary Frobenius series to find its effect on the second summand of $x_2(t)$:

$$L\left[\sum_{k=0}^{\infty} b_k^* t^{-1+k}\right] = \sum_{j=1}^{\infty} [j^2 b_j^* + (j - 2)b_{j-1}^*]t^{-1+j}.$$

The effect of L on the first summand is

$$
\begin{aligned}
L[(t^{-1} + 1) \ln t] &= t^2 D^2[(t^{-1} + 1) \ln t] \\
&\quad + (t^2 + 3t)D[(t^{-1} + 1) \ln t] + (t^{-1} + 1) \ln t \\
&= t^2(-3t^{-3} - t^{-2} + 2t^{-3} \ln t) \\
&\quad + (t^2 + 3t)(t^{-2} + t^{-1} - t^{-2} \ln t) + (t^{-1} + 1) \ln t \\
&= 3 + t.
\end{aligned}
$$

Combining these, we see that

$$Lx_2(t) = 3 + t + \sum_{j=1}^{\infty} [j^2 b_j^* + (j - 2)b_{j-1}^*]t^{-1+j}$$

$$= (3 + b_1^* - b_0^*) + (1 + 4b_2^*)t + \sum_{j=3}^{\infty} [j^2 b_j^* + (j - 2)b_{j-1}^*]t^{-1+j}.$$

If this is to be identically zero, we must have

$$b_1^* = b_0^* - 3$$

$$b_2^* = -\frac{1}{4}$$

and

(R*) $$b_j^* = -\frac{j - 2}{j^2} b_{j-1}^*, \quad j = 3, 4, \ldots .$$

There is no restriction on b_0^*. If we choose

$$b_0^* = 0$$

then

$$b_1^* = -3, \quad b_2^* = -\frac{1}{4} = -\frac{1}{2^2}$$

and

$$b_3^* = -\frac{1}{3^2} b_2^* = \frac{1}{3^2 2^2},$$

$$b_4^* = -\frac{2}{4^2} b_3^* = -\frac{2}{4^2 3^2 2^2},$$

$$b_5^* = -\frac{3}{5^2} b_4^* = \frac{3 \cdot 2}{5^2 4^2 3^2 2^2}, \cdots,$$

$$b_k^* = \frac{(-1)^{k+1}}{(k!)^2} (k - 2)! = \frac{(-1)^{k-1}}{k!k(k - 1)}, \cdots.$$

Note that the general formula for b_k^* is valid only from $k = 2$ on. Our second solution of (H) is

(S2)
$$x_2(t) = (t^{-t} + 1) \ln t - 3 + \sum_{k=2}^{\infty} \frac{(-1)^{k-1}}{k!k(k - 1)} t^{-1+k}$$

$$= (t^{-1} + 1) \ln t - 3 + \sum_{n=1}^{\infty} \frac{(-1)^n}{(n + 1)!(n + 1)n} t^n.$$

The general solution of (H) is

$$x = c_1 x_1(t) + c_2 x_2(t)$$

$$= c_1(t^{-1} + 1) + c_2 \left[(t^{-1} + 1) \ln t - 3 + \sum_{n=1}^{\infty} \frac{(-1)^n}{(n + 1)!(n + 1)n} t^n \right].$$

In Example 6.8.1, substitution of $(t^{-1} + 1) \ln t$ into L eliminated the terms involving $\ln t$. It is easy to verify (Exercise 16) that this is a general pattern. Indeed, *if $x_1(t)$ is a solution of (H), then*

(E)
$$L[x_1(t) \ln t] = 2tp(t)x_1'(t) + [q(t) - p(t)]x_1(t).$$

We will use this fact to simplify our work in the next example.

Example 6.8.2

Find the general solution of Bessel's equation with $\mu = 0$,

(B0)
$$[t^2 D^2 + tD + t^2]x = 0.$$

We have already seen in Example 6.7.1 that the indicial equation of (B0)

has a double root $m = 0$, and that a corresponding Frobenius series solution is the Bessel function of the first kind

(S₁)
$$J_0(t) = \sum_{k=0}^{\infty} \frac{(-1)^k}{(k!)^2} \left(\frac{t}{2}\right)^{2k} = \sum_{k=0}^{\infty} \frac{(-1)^k t^{2k}}{(k!)^2 2^{2k}}.$$

We look for a second solution in the form

(S₂)
$$K_0(t) = J_0(t) \ln t + t^0 \sum_{k=0}^{\infty} b_k^* t^k.$$

Substitution of (S₂) into (B₀) gives

$$2t J_0'(t) + b_1^* t + \sum_{j=2}^{\infty} [j^2 b_j^* + b_{j-2}^*] t^j = 0$$

or

$$b_1^* t + \sum_{j=2}^{\infty} [j^2 b_j^* + b_{j-2}^*] t^j = -2t J_0'(t)$$

$$= \sum_{k=0}^{\infty} \frac{(-1)^{k+1} 4k}{(k!)^2 2^{2k}} t^{2k}.$$

Since the right side of the last equation has no odd powers of t, we conclude that

$$b_1^* = 0$$

and

$$j^2 b_j^* + b_{j-2}^* = 0, \quad j = 3, 5, \ldots.$$

It follows that

$$b_j^* = 0 \quad \text{for odd } j.$$

On the other hand, comparing even powers of t, we have (taking $j = 2k$)

$$(2k)^2 b_{2k}^* + b_{2k-2}^* = \frac{(-1)^{k+1} 4k}{(k!)^2 2^{2k}}$$

or

$$b_{2k}^* = -\frac{1}{4k^2} b_{2k-2}^* + \frac{(-1)^{k+1}}{4^k (k!)^2 k}, \quad k = 1, 2, \ldots.$$

Once again, there is no restriction on b_0^*. Taking

$$b_0^* = 0$$

we see that

$$b_2^* = -\frac{1}{4} b_0^* \qquad + \frac{(-1)^2}{4(1!)^2} \quad = \frac{1}{4}$$

$$b_4^* = -\frac{1}{4 \cdot 2^2} b_2^* + \frac{(-1)^3}{4^2(2!)^2 2} = -\frac{1}{4^2 2^2} \left[1 + \frac{1}{2} \right]$$

$$b_6^* = -\frac{1}{4 \cdot 3^2} b_4^* + \frac{(-1)^4}{4^3(3!)^2 3} = \frac{1}{4^3(3!)^2} \left[1 + \frac{1}{2} + \frac{1}{3} \right]$$

and, in general,

$$b_{2k}^* = \frac{(-1)^{k+1}}{4^k(k!)^2} \left[1 + \frac{1}{2} + \frac{1}{3} + \cdots + \frac{1}{k} \right].$$

Thus, our second solution to (B_0) is

$$(S_2) \qquad K_0(t) = J_0(t) \ln t + \sum_{k=1}^{\infty} \frac{(-1)^{k+1} H_k}{(k!)^2} \left(\frac{t}{2} \right)^{2k},$$

where H_k is the kth partial sum of the harmonic series,

$$H_k = 1 + \frac{1}{2} + \frac{1}{3} + \cdots + \frac{1}{k}.$$

The general solution of (B_0) is

$$x = c_1 J_0(t) + c_2 K_0(t).$$

Any function that satisfies (B_0) and is not a multiple of $J_0(t)$ [such as $K_0(t)$] is called a **Bessel function of the second kind of order 0.** It is customary when working with Bessel functions to use as the second solution a specially chosen linear combination of $J_0(t)$ and $K_0(t)$, which is usually denoted $Y_0(t)$.

Besides the case of a double root, we often need to go beyond Frobenius series to handle the case in which the roots, although distinct, differ by an integer, say $m_1 - m_2 = n > 0$. In this case, we can always find a Frobenius series solution $x_1(t)$ corresponding to the larger root m_1. We can also try to find a second Frobenius series solution corresponding to the smaller root m_2.

If this works, as in Examples 6.6.2 and 6.7.4, we are done. If not, we look for a second solution in a form similar to that in the double root case.

Fact: *If the indicial equation (I) has roots m_1 and m_2 that differ by a positive integer, $m_1 - m_2 = n > 0$, then the general solution of (H) is of the form $x = c_1 x_1(t) + c_2 x_2(t)$, where the first solution is a Frobenius series corresponding to m_1:*

$$x_1(t) = t^{m_1} \sum_{k=0}^{\infty} b_k t^k, \quad b_0 \neq 0$$

and the second is either a Frobenius series corresponding to m_2 or else is such a series plus $x_1(t) \ln t$:

$$x_2(t) = t^{m_2} \sum_{k=0}^{\infty} b_k^* t^k, \quad b_0^* \neq 0$$

or

$$x_2(t) = x_1(t) \ln t + t^{m_2} \sum_{k=0}^{\infty} b_k^* t^k.$$

When (I) had a double root, we were free to choose any value we wished for b_0^*, including 0. When the roots differ by an integer, $m_1 - m_2 = n$, we are free to choose any value we wish for b_n^*.

Example 6.8.3

Find the first terms of the general solution to

(H) $$Lx = [t^2 D^2 + (t^3 + t)D - 1]x = 0, \quad t > 0.$$

Recall from Example 6.6.3 that the roots of the indicial equation are $m_1 = 1$ and $m_2 = -1$, that a solution corresponding to $m = 1$ is

(S₁) $$x_1(t) = t - \frac{1}{8} t^3 + \frac{1}{64} t^5 + \cdots$$

and that there is no Frobenius series solution corresponding to $m = -1$. Accordingly, we look for a second solution in the form

(S₂)
$$x_2(t) = x_1(t) \ln t + t^{-1} \sum_{k=0}^{\infty} b_k^* t^k.$$

The effect of L on an arbitrary Frobenius series was calculated in Example 6.6.3; using that calculation, we see that

$$L\left[t^{-1} \sum_{k=0}^{\infty} b_k^* t^k \right] = -b_1^* - b_0^* t + 3b_3^* t^2 + (8b_4^* + b_2^*)t^3$$
$$+ (15b_5^* + 2b_3^*)t^4 + \cdots.$$

The effect of L on the first summand of $x_2(t)$ can be worked out using the formula (E) preceding Example 6.8.2:

$$L[x_1(t) \ln t] = 2t x_1'(t) + (t^2 + 1 - 1)x_1(t)$$
$$= \left[2t - \frac{3}{4} t^3 + \frac{5}{32} t^5 + \cdots \right] + \left[t^3 - \frac{1}{8} t^5 + \cdots \right]$$
$$= 2t + \frac{1}{4} t^3 + \frac{1}{32} t^5 + \cdots.$$

Combining these, we see that

$$L[x_2(t)] = -b_1^* + (2 - b_0^*)t + 2b_3^* t^2 + \left(8b_4^* + b_2^* + \frac{1}{4} \right) t^3$$
$$+ (15b_5^* + 2b_3^*)t^4 + \cdots.$$

Setting coefficients of t^k equal to zero, we find

$$b_1^* = 0, \quad b_0^* = 2, \quad b_3^* = 0, \quad b_4^* = -\frac{1}{8} b_2^* - \frac{1}{32}, \quad b_5^* = 0, \ldots.$$

There is no restriction on b_2^*. If we choose

$$b_2^* = 0$$

then

$$b_4^* = -\frac{1}{32}$$

and our second solution is

$$(S_2) \quad x_2(t) = x_1(t) \ln t + \sum_{k=0}^{\infty} b_k^* t^{k-1}$$

$$= \left(t - \frac{1}{8} t^3 + \frac{1}{64} t^5 + \cdots \right) \ln t + \left(2t^{-1} - \frac{1}{32} t^3 + \cdots \right).$$

The general solution of (H) is

$$x = c_1 x_1(t) + c_2 x_2(t)$$

$$= c_1 \left(t - \frac{1}{8} t^3 + \frac{1}{64} t^5 + \cdots \right) + c_2 \left(2t^{-1} - \frac{1}{32} t^3 + \cdots \right)$$

$$+ c_2 \left(t - \frac{1}{8} t^3 + \frac{1}{64} t^5 + \cdots \right) \ln t.$$

Example 6.8.4

Find the general solution to Bessel's equation with $\mu = 1$,

$$(B_1) \qquad Lx = [t^2 D^2 + tD + (t^2 - 1)]x = 0, \quad t > 0.$$

We know from Example 6.7.2 that the roots of the indicial equation are $m_1 = 1$ and $m_2 = -1$, that a solution corresponding to m_1 is

$$J_1(t) = \sum_{k=0}^{\infty} \frac{(-1)^k}{k!(k+1)!} \left(\frac{t}{2} \right)^{2k+1} = \sum_{k=0}^{\infty} \frac{(-1)^k t^{2k+1}}{k!(k+1)!2^{2k+1}}$$

and that there is no Frobenius series solution corresponding to -1. We look for a second solution of the form

$$K_1(t) = J_1(t) \ln t + t^{-1} \sum_{k=0}^{\infty} b_k^* t^k.$$

The effect of L on the first summand of $K_1(t)$ is

$$L[J_1(t) \ln t] = 2tJ_1'(t) = \sum_{k=0}^{\infty} \frac{(-1)^k (2k+1)}{k!(k+1)!2^{2k}} t^{2k+1}$$

and its effect on the second summand is

$$L\left[\sum_{k=0}^{\infty} b_k^* t^{-1+k}\right] = -b_1^* + \sum_{j=2}^{\infty} [j(j-2)b_j^* + b_{j-2}^*]t^{j-1}.$$

These two series must cancel, or

$$-b_1^* + \sum_{j=2}^{\infty} [j(j-2)b_j^* + b_{j-2}^*]t^{j-1} = -\sum_{k=0}^{\infty} \frac{(-1)^k(2k+1)}{k!(k+1)!2^{2k}} t^{2k+1}.$$

Since the right-hand side contains no even powers of t,

$$-b_1^* = 0$$

and

$$j(j-2)b_j^* + b_{j-2}^* = 0, \quad j = 3, 5, 7, \ldots.$$

It follows that

$$b_j^* = 0 \quad \text{for odd } j.$$

Equating coefficients of t^{2n-1} (take $j = 2n$ on the left and $k = n - 1$ on the right), we see that

$$2n(2n-2)b_{2n}^* + b_{2n-2}^* = \frac{(-1)^n(2n-1)}{(n-1)!n!2^{2n-2}}, \quad n = 1, 2, \ldots.$$

Thus

$$b_0^* = -1$$

and

$$b_{2n}^* = \frac{1}{4n(n-1)}\left[\frac{(-1)^n(2n-1)}{4^{n-1}n!(n-1)!} - b_{2n-2}^*\right], \quad n = 2, 3, \ldots.$$

These relations place no restrictions on b_2^*. The recurrence relation is difficult to translate into a pattern, but you can verify that the choice

$$b_2^* = 1$$

leads to

$$b_{2n}^* = \frac{(-1)^n}{4^n n!(n-1)!}\,[H_n + H_{n-1}], \quad n = 2, 3, \ldots,$$

where, as before, H_n is the nth partial sum of the harmonic series. Thus, our second solution is

(S₂) $$K_1(t) = J_1(t)\ln t - t^{-1} + t + \sum_{j=2}^{\infty} \frac{(-1)^j[H_j + H_{j-1}]}{4^j j!(j-1)!}\,t^{2j-1}.$$

The general solution of (B₁) is

$$x = c_1 J_1(t) + c_2 K_1(t).$$

Again, any function [such as $K_1(t)$] that satisfies (B₁) and is not a multiple of $J_1(t)$ is a **Bessel function of the second kind of order 1.** The solutions of (B₁) are usually described in terms of $J_1(t)$ and a specially chosen linear combination of $J_1(t)$ and $K_1(t)$.

These exceptional cases exhaust the possibilities for regular singular points.

REGULAR SINGULAR POINTS: EXCEPTIONAL CASES

Suppose the equation

(H) $$L[x] = [t^2 p(t)D^2 + tq(t)D + r(t)]x = 0, \quad t > 0$$

with a regular singular point at $t = 0$, has an indicial equation

(I) $$p_0 m(m-1) + q_0 m + r_0 = 0$$

with roots m_1 and m_2 that differ by an integer, $m_1 - m_2 = n \geq 0$. Then (H) always has a Frobenius series solution

$$x_1(t) = t^{m_1} \sum_{k=0}^{\infty} b_k t^k, \quad b_0 \neq 0.$$

If every Frobenius series solution of (H) is a constant multiple of $x_1(t)$, then (H) has a second solution of the form

(S$_2$) $x_2(t) = x_1(t) \ln t + t^{m_2} \displaystyle\sum_{k=0}^{\infty} b_k^* t^k.$

To find $x_2(t)$, we substitute into (H), make a free choice of b_n^*, and determine the remaining coefficients b_k^*. The substitution can be facilitated by the formula

(E) $L[x_1(t) \ln t] = 2tp(t)x_1'(t) + [q(t) - p(t)]x_1(t).$

EXERCISES

In Exercises 1 through 10, find the general solution. In Exercises 1 through 6, you need only find the first four coefficients of each of the two generating solutions. In Exercises 7 through 10, find the general pattern for the coefficients.

1. $[t^2D^2 - tD + (t + 1)]x = 0,$ $0 < t$ (see Exercise 14, Section 6.6)
2. $[t^2D^2 + tD + (t - 1)]x = 0,$ $0 < t$ (see Exercise 15, Section 6.6)
3. $[t^2D^2 + tD + (t^2 - 4)]x = 0,$ $0 < t$ (see Exercise 16, Section 6.6)
4. $[8t^2D^2 + 2t(t + 4)D + (t - 2)]x = 0,$ $0 < t$
5. $[9t^2D^2 + 15tD + (t + 1)]x = 0,$ $0 < t$
6. $[t^2(t + 1)D^2 + t(t + 1)D - (4t + 1)]x = 0,$ $0 < t < 1$
7. $[t^2D^2 + t(1 - t)D - 1]x = 0,$ $0 < t$
8. $[tD^2 + (1 - t)D + 1]x = 0,$ $0 < t$
9. $[t^2(t - 2)D^2 - 2t(1 - t)D - 2t]x = 0,$ $0 < t < 2$
10. $[t^2D^2 + t(t + 4)D + (t + 2)]x = 0,$ $0 < t$
11. Show that if $\mu = 0$ or 1, then the only solutions of the Bessel equation that remain bounded as t approaches 0 are the constant multiples of $J_\mu(t)$. [*Hint:* What happens to $K_\mu(t)$ as t approaches 0?]
12. a. Show that if μ is a nonnegative integer, then the Laguerre equation

    (H) $[tD^2 + (1 - t)D + \mu]x = 0$

    has a nonzero solution $x_\mu(t)$ that is a polynomial.
    b. Show that the only solutions of (H) that remain bounded as t approaches 0 are the constant multiples of $x_\mu(t)$.
13. Consider the Legendre equation

    (H) $[(t^2 - 1)D_2 + 2tD - \mu(\mu + 1)]x = 0$

    with μ a nonnegative integer.
    a. Show that the general solution of (H) is generated by a polynomial $x_\mu(1 - t)$

and a solution that becomes unbounded as t approaches 1. (*Hint:* Substitute $T = 1 - t$. What form does the solution of the resulting equation take?)

b. Show that the only solutions of (H) that remain bounded as t approaches 1 are the constant multiples of $x_\mu(1 - t)$.

c. Show that the only solutions of (H) that remain bounded as t approaches -1 are the constant multiples of a polynomial solution $x_\mu^*(1 + t)$. (*Hint:* Substitute $T = 1 + t$.)

d. Show that $x_\mu(1 - t)$ and $x_\mu^*(1 + t)$ are constant multiples of each other.

14. *Legendre Polynomials:* For each nonnegative integer n, define a polynomial by

$$P_n(t) = \frac{1}{2^n n!} D^n (t^2 - 1)^n.$$

a. Find $P_0(t)$, $P_1(t)$, and $P_2(t)$.

b. Show that $(t^2 - 1)D(t^2 - 1)^n - 2nt(t^2 - 1)^n = 0$.

c. Differentiate this equation to show that

$$(t^2 - 1)D^2(t^2 - 1)^n - 2(n - 1)t\, D(t^2 - 1)^n - 2n(t^2 - 1)^n = 0.$$

d. Show that repeated differentiation of this equation leads to

$$(t^2 - 1)D^{k+2}(t^2 - 1)^n - 2(n - k - 1)t\, D^{k+1}(t^2 - 1)^n$$
$$- 2[n + (n - 1) + \cdots + (n - k)]D^k(t^2 - 1)^n = 0.$$

In particular, $k = n$ gives

$$(t^2 - 1)D^{n+2}(t^2 - 1)^n + 2t\, D^{n+1}(t^2 - 1)^n$$
$$- 2[n + (n - 1) + \cdots + 1]D^n(t^2 - 1)^n = 0.$$

e. Use the fact that $n + (n - 1) + \cdots + 1 = n(n + 1)/2$ to show that $x = P_n(t)$ solves the Legendre equation with $\mu = n$,

(H) $$[(t^2 - 1)D^2 + 2tD - n(n + 1)]x = 0.$$

f. Use the result of Exercise 13 to show that any solution of (H) that remains bounded as t approaches 1 or -1 is a constant multiple of $P_n(t)$.

15. *Temperature in a Planet:* The temperature distribution in the planet of Example 6.1.4 is modeled by the equation

(H) $$\rho \frac{\partial^2}{\partial \rho^2}(\rho u) + \frac{1}{\sin \phi} \frac{\partial}{\partial \phi}\left(\sin \phi \frac{\partial u}{\partial \phi}\right) = 0.$$

Recall that if $u = R(\rho)\Phi(\phi)$ solves (H), then R must solve the Cauchy-Euler equation

(H$_\rho$) $$\rho^2 \frac{d^2 R}{d\rho^2} + 2\rho \frac{dr}{d\rho} - n(n + 1)R = 0$$

and Φ must solve the Legendre equation

(H$_t$) $$(t^2 - 1)\frac{d^2\Phi}{dt^2} + 2t\frac{d\Phi}{dt} - n(n + 1)\Phi = 0,$$

where $t = \cos \phi$ and n is a nonnegative integer.

a. Use the fact that the temperature in the planet has to remain bounded near the origin to conclude that the physically meaningful solutions of (H$_\rho$) are constant multiples of $R_n = \rho^n$.

b. Use the boundedness of the temperature near the poles ($\phi = 0$ and $\phi = \pi$), together with the result of Exercise 14, to conclude that the physically meaningful solutions of (H$_t$) are constant multiples of $\Phi_n = P_n(t)$.

c. Check that any linear combination of the functions

$$u_n = R_n \Phi_n = \rho^n P_n(t) = \rho^n P_n(\cos \phi)$$

satisfies (H).

d. Find a linear combination of the functions u_n that matches the boundary condition $u(1, \phi) = \sin^2 \phi, \ 0 < \phi < \pi$. [*Hint:* $u(1, \phi) = 1 - t^2 = \frac{2}{3} P_0(t) - \frac{2}{3} P_2(t)$.]

16. Show that if $x_1(t)$ is a solution of

(H) $Lx = [t^2 p(t) D^2 + t q(t) D + r(t)]x = 0, \quad t > 0,$

then $L[x_1(t) \ln t] = 2tp(t)x_1'(t) + [q(t) - p(t)]x_1(t).$

REVIEW PROBLEMS

In Exercises 1 through 18, use the methods of this chapter to find an expression for the general solution on the given interval. When you cannot find a pattern for the terms of a series, find the first four coefficients.

1. $[(t^2 - 1)D^2 + tD - 1]x = 0, \quad 0 < t < 1$
2. $[t^2 D^2 + 4tD + 2]x = 0, \quad 0 < t$
3. $[16t^2 D^2 + 16tD + (16t^2 - 1)]x = 0, \quad 0 < t$
4. $[t^2 D^2 + t(4 - t)D + 2(1 - t)]x = 0, \quad 0 < t$
5. $[9t^2 D^2 - 3tD + 4]x = 0, \quad 0 < t$
6. $[t^2(t - 2)D^2 - 2t(1 - t)D - 6t]x = 0, \quad 0 < t < 2$
7. $[D^2 - 4tD]x = 0, \quad 0 < t$
8. $[t^2(2 + t)D^2 + t(1 + t)D - 1]x = 0, \quad 0 < t < 2$
9. $[3t^2(t + 1)D^2 - 4tD + 2]x = 0, \quad 0 < t < 1$
10. $[D^2 + t]x = 0, \quad 0 < t$
11. $[t^2 D^2 + tD - (t + 1)]x = 0, \quad 0 < t$
12. $[8t^2 D^2 + 2t(4 - t)D - (t + 2)]x = 0, \quad 0 < t$
13. $[t^2(t + 1)D^2 + 3t(1 - t)D + 1]x = 0, \quad 0 < t < 1$
14. $[2t^2 D^2 + 6tD + 12]x = 0, \quad 0 < t$
15. $[(t^2 + 1)D^2 + (t + 1)D + 1]x = 0, \quad 0 < t < 1$
16. $[t^2(t + 1)D^2 + t(t + 1)D + 1]x = 0, \quad 0 < t < 1$
17. $[t^2 D^2 + t(4 - t)D + (2 - t)]x = 0, \quad 0 < t$
18. $[3t^2 D^2 + t(4 - 3t)D - (t + 2)]x = 0, \quad 0 < t$
19. Find a series expression for the solution of the equation in Exercise 1 that satisfies the initial conditions $x(0) = 1, \ x'(0) = 2$.
20. Find the terms up to t^5 in a power series expression for the solution of the o.d.e. $[(t^2 - 1)D^2 - 1]x = t^2 + 3$ satisfying $x(0) = -3, \ x'(0) = -5$.

21. *A Submarine:* The submarine of Exercise 5, Section 6.1, starts from rest at the bottom of the ocean.

   a. Find a series equation for the distance from the submarine to the ocean bottom at time t.

   b. Use the first three nonzero terms of this series to estimate the position of the submarine after 10 seconds.

22. Find the first five terms for a power series about $t_0 = -1$ expressing the general solution to $[tD^2 - 4]x = 0, -2 < t < 0$.

23. Find the first five terms of a power series about $t_0 = 1$ expressing the general solution to $[tD^2 + tD + 1]x = 0, 0 < t < 2$.

24. Show that if μ is a nonnegative integer, then the **Chebyshev equation** $[(1 - t^2)D^2 - tD + \mu^2]x = 0$ has a polynomial solution of degree at most μ.

25. The **Airy equation** is

   (H) $$[D^2 + t]x = 0.$$

   a. Show that the substitution $x = t^{1/2}y$ changes (H) into

   (H$_1$) $$\left[t^2D^2 + tD + \left(t^3 - \frac{1}{4} \right) \right] y = 0.$$

   b. Show that the substitution $t = \left(\frac{3}{2} s \right)^{2/3}$ leads to

   $$\frac{dy}{dt} = \left(\frac{3}{2} s \right)^{1/3} \frac{dy}{ds}$$

   $$\frac{d^2y}{dt^2} = \left(\frac{3}{2} s \right)^{1/3} \left[\left(\frac{3}{2} s \right)^{1/3} \frac{d^2y}{ds^2} + \frac{1}{2} \left(\frac{3}{2} s \right)^{-2/3} \frac{dy}{ds} \right]$$

   and converts (H$_1$) into Bessel's equation with $\mu = 1/3$,

   (H$_2$) $$\left[s^2 \frac{d^2y}{ds^2} + s \frac{dy}{ds} + \left(s^2 - \frac{1}{9} \right) \right] y = 0.$$

   c. Use (a) and (b) to show that the general solution to the Airy equation on $0 < t$ can be written in the form

   $$x(t) = c_1 \sqrt{t} \, J_{1/3} \left(\frac{2}{3} t^{3/2} \right) + c_2 \sqrt{t} \, J_{-1/3} \left(\frac{2}{3} t^{3/2} \right),$$

   where $J_{1/3}$ and $J_{-1/3}$ are the Bessel functions of the first kind of orders $1/3$ and $-1/3$, respectively.

7

NUMERICAL APPROXIMATION OF SOLUTIONS

7.1 SOME NONLINEAR MODELS

Although the techniques we have seen so far let us handle many basic physical models, other models can lead to o.d.e.'s that cannot be solved by these methods. For such problems, numerical methods are often used to give *approximate* predictions. In this chapter we investigate several numerical schemes that yield approximations to the value, at a specified point $t = \tau$, of the solution $x(t)$ to an initial-value problem. Approximations for various values of τ can be combined to yield a table of values or graph of $x(t)$. A reasonably accurate table or graph is sometimes more useful for predicting the behavior of a given model than an exact but complicated formula giving x as a function of t.

Numerical approximations are usually calculated on computers. We shall discuss the ideas of each scheme, formulate it in terms of a general algorithm, and implement this algorithm in a BASIC program. Implementation of the algorithms in other languages is fairly straightforward, but we shall not discuss the specifics of such implementations. We shall look instead at the application of these schemes to specific o.d.e.'s and try to interpret the answers we obtain.

In this section we look briefly at some models leading to nonlinear o.d.e.'s for which numerical methods are appropriate. Some of these models will already be familiar to readers who have worked through the qualitative theory in Chapter 4.

Example 7.1.1   The Simple Pendulum (Revisited)

The o.d.e. modeling the swinging of a pendulum (see Example 2.10.2) is

(D)
$$\frac{d^2\theta}{dt^2} = -\frac{g}{L} \sin \theta,$$

where θ is the angular displacement from the bottom position, L is the length of the pendulum, and g is a constant ($g \approx 981$ cm/sec^2). When $|\theta|$ is small, (D) can be approximated by a linear o.d.e. However, in general (and even to check the accuracy of the linear approximation) we need to study (D) directly.

A trick due to Newton allows us to change the second-order o.d.e. for θ into a first-order o.d.e. for the angular velocity, which is defined as

$$\omega = \frac{d\theta}{dt}.$$

Note that

$$\frac{d^2\theta}{dt^2} = \frac{d\omega}{dt} = \frac{d\theta}{dt}\frac{d\omega}{d\theta} = \omega\frac{d\omega}{d\theta} = \frac{d}{d\theta}\left(\frac{\omega^2}{2}\right).$$

Therefore, we can write (D) as

$$\frac{d}{d\theta}\left(\frac{\omega^2}{2}\right) = -\frac{g}{L}\sin\theta,$$

which can be integrated to give

$$\omega^2 = \frac{2g}{L}(\cos\theta + k).$$

Since $\omega = d\theta/dt$, this gives

$$\frac{d\theta}{dt} = \pm\sqrt{\frac{2g}{L}}\sqrt{\cos\theta + k}.$$

Separation of variables leads to the implicit solution

$$t = \pm\sqrt{\frac{L}{2g}}\int_0^{\theta(t)}\frac{d\theta}{\sqrt{\cos\theta + k}}.$$

This formula is certainly not suitable for predicting the value of θ at a given time t. In theory, it could be used to predict when we shall reach a predetermined position θ, but even here we are troubled by the fact that the integral cannot be expressed in terms of elementary functions. Thus, we need to resort to numerical approximations.

We note that the second-order equation (D) in the preceding example can be rewritten as a system of two first-order equations:

$$\frac{d\theta}{dt} = \omega$$

$$\frac{d\omega}{dt} = -\frac{g}{L}\sin\theta.$$

This is, in fact, illustrative of a general pattern (which some readers may have seen in Section 3.2). Given a second-order o.d.e.

$$x'' = F(t, x, x'),$$

we can introduce a new variable $y = x'$, and write

$$x' = y$$

$$y' = F(t, x, y).$$

Our discussion in this chapter focuses on first-order o.d.e.'s, and in the last section we will see how numerical schemes can be used to handle the systems that arise in this way from o.d.e.'s of order 2 or more.

Example 7.1.2   Nonlinear Damping

We noted at the beginning of Example 2.1.2 that the motion of a mass might be subject to significant resistance from the surrounding medium. In that example, we assumed that the resisting force was proportional to velocity, $F_{\text{damp}} = -bv$. This is physically realistic when the medium is a highly viscous fluid (such as oil) moving relatively slowly. However, when the fluid is turbulent, the resisting force has a power of the speed as an extra factor, $F_{\text{damp}} = -bv|v|^{\beta}$. Thus the equation for a mass-spring system suspended in a turbulent fluid could have the form

(D) $$mx'' + bx'|x'| + kx = E(t).$$

As a result of the presence of the absolute value in the middle term, we cannot find the general solution by the methods of Chapters 1 through 6. Instead, we will apply a numerical method to the system of o.d.e.'s equivalent to (D):

$$x' = y$$

$$y' = \frac{1}{m}[E(t) - kx - by|y|].$$

Example 7.1.3   Van der Pol's Equation

In 1920, the Dutch electrical engineer B. Van der Pol noted that the experimental data concerning the reaction of a vacuum-tube receiver to an incoming signal could be reproduced qualitatively by a modified *LRC* circuit (see Section 3.1) in which the resistor, instead of obeying Ohm's law (resistance is constant), has a resistance varying with the charge $Q = \int I \, dt$ as a quadratic function. After a change of variable of the form $x = aQ + b$ (for a suitable choice of a and b), this model leads to the o.d.e.

(D) $$x'' + \epsilon(x^2 - 1)x' + kx = A \sin \omega t,$$

which is known as the (forced) **Van der Pol equation.**

The analytic methods of Chapters 2, 3, and 5 do not apply to this equation, and a search for series solutions would be cumbersome. The qualitative discussion in Example 4.6.1 relied on sophisticated computer techniques. In this chapter we apply a numerical method to obtain quantitative information about the system equivalent to (D):

$$x' = y$$
$$y' = \epsilon(1 - x^2)y - kx + A \sin \omega t.$$

EXERCISES

The following models lead to o.d.e.'s and systems for which numerical methods are appropriate. These include linear o.d.e.'s whose solutions involve integrals that cannot be expressed in closed form in terms of elementary functions, nonlinear o.d.e.'s and systems, and a system with nonconstant coefficients.

1. *Controlled Immigration:* Suppose the government of the country in Example 1.1.4 decides to allow immigration at the decreasing rate of $1/(t + 1)$ million per year.
   a. Find an o.d.e. for the population t years from now.
   b. Use variation of parameters to find an integral expression for the solution of this o.d.e.

2. *Ice Water:* A pail of water at 60°F is placed outdoors and loses heat according to Newton's law of cooling (Section 1.4) with constant of proportionality $\gamma = 1/10$. Suppose the temperature outdoors is dropping so that at time t it is $40/(1 + t^2)$.
   a. Find an o.d.e. for the temperature of the water at time t.
   b. Use variation of parameters to find an integral expression for the solution of this o.d.e.

3. *A Population in an Improving Environment:* Modify Example 1.1.6 by assuming that the population limit increases so that at time t it is $10 + t/10$, whereas α

remains fixed at 1/10. Find an o.d.e. for the population. (Note that this o.d.e., unlike the one in Example 1.1.6, is not separable.)

4. *A Moving Spring:* Modify Exercise 3, Section 2.1, by assuming that at time t the elevator is $h(t) = 80t/(t + 1)$ ft off the ground.
   a. Find a second-order o.d.e. for the distance $z(t)$ from the weight to the ground if $b = 3$ and $k = L = 4$.
   b. Use variation of parameters to find an expression for $z(t)$.
   c. Find a system of two first-order o.d.e.'s for $z(t)$ and $y(t) = z'(t)$.

5. *A Damped Pendulum:* Suppose the pendulum of Example 2.10.2 encounters resistance with moment $M_{\text{damp}} = -d\theta/dt$.
   a. Find an o.d.e. for θ if the arm of the pendulum is 1 meter long and the bob has mass $m = 1$ gram.
   b. Replace this o.d.e. by two first-order o.d.e.'s for $x = \theta$ and $y = \theta'$.

6. *Diffusion:* Modify Exercise 8(a), Section 3.1, by assuming that the permeability of the membrane varies so that at time t it is $P(t) = 1/(10 + t)$. Find a system of two first-order equations for the concentrations of the two solutions, if the volume of liquid on each side of the membrane is $V = 5$ liters.

7. *Interacting Populations (The Volterra-Lotka Model):* Suppose that a population of rabbits protected from foxes would have a constant per capita growth rate g_R, and a population of foxes deprived of rabbits would decrease at the constant per capita rate d_F. In a common environment, foxes kill and eat rabbits. The killing of rabbits slows the growth of the rabbit population, and at the same time, eating rabbits retards the decrease in the fox population. Assume that each of these effects is expressed by a term proportional to the product of the two population sizes. Find a system of two first-order equations for the sizes of the rabbit and fox populations.

7.2 EULER'S METHOD

In this section we investigate a simple scheme for approximating the value, at a specified point $t = \tau$, of the solution $x(t)$ to a given first-order initial-value problem:

(D) $$x' = f(t, x)$$

(I) $$x(t_0) = x_0.$$

This scheme, known as **Euler's method,** is related to the graphing technique of Section 1.7.

We divide the interval $t_0 \leq t \leq \tau$ into n shorter intervals $t_{i-1} \leq t \leq t_i$, $i = 1, 2, \ldots, n$, of equal lengths $t_i - t_{i-1} = h$. The **number of steps** n and the **step size** h are then related by the formula

$$h = \frac{\tau - t_0}{n},$$

and the endpoints of the shorter intervals satisfy

$$t_i = t_{i-1} + h.$$

You should check that this means

$$t_i = t_0 + ih.$$

We proceed by estimating the value of $x(t)$ at each of the successive check-points $t = t_i$.

The value of $x(t)$ at $t = t_0$ is specified by the initial condition (I):

$$x(t_0) = x_0.$$

To estimate the value of $x(t)$ at $t = t_1$, we note that on a small interval the derivative $x'(t)$ is nearly constant. The derivative at $t = t_0$ is given by the o.d.e. (D):

$$x'(t_0) = f(t_0, x_0).$$

If the derivative were actually constant for $t_0 \le t \le t_1 = t_0 + h$, then the value of $x(t)$ at $t = t_1$ would be given by

(E₁) $$x_1 = x_0 + hf(t_0, x_0).$$

We adopt this as our estimate for the value of $x(t_1)$ and use it to predict $x(t_2)$. Note that x_1 is the value we would get if we moved from (t_0, x_0) along the line tangent to the curve $x = x(t)$ instead of moving along the curve itself (see Figure 7.1).

Similar reasoning over the interval $t_1 \le t \le t_2 = t_1 + h$ leads to an

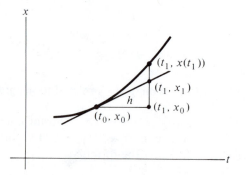

FIGURE 7.1

estimate for the value of $x(t)$ at $t = t_2$:

(E$_2$) $$x_2 = x_1 + hf(t_1, x_1).$$

Indeed, continuing in this way, we obtain estimates for the values of $x(t)$ at the successive checkpoints $t = t_i$ via the recurrence relation

(E$_i$)
$$x_i = x_{i-1} + hf(t_{i-1}, x_{i-1})$$
$$= x_{i-1} + hf(t_0 + (i - 1)h, x_{i-1}).$$

The final estimate x_n corresponds to $t = t_n = \tau$, and this is our approximation for $x(\tau)$.

Recurrence relations like (E$_i$) are easily handled on computers or hand calculators. When programming such devices to implement Euler's method for a specific initial-value problem, it is useful to first map out strategy, as described in the example below.

Example 7.2.1   Algorithm for Euler's Method

The basic calculation of x_i using (E$_i$) requires the values of t_{i-1}, x_{i-1}, and h. We calculate t_{i-1} from the starting time t_0, the index i, and the step size h. Thus, we need as previous data the values of t_0, i, h, and x_{i-1}. We assume these have been stored in memory registers labeled $T0$, I, H, and X, respectively:

$$T0 \leftarrow t_0, \quad I \leftarrow i, \quad H \leftarrow h, \quad X \leftarrow x_{i-1}.$$

Using these values, we wish to calculate x_i according to (E$_i$), and store the result. Note that, once the calculation has been performed, we no longer need the values of t_{i-1} and x_{i-1}. In particular, we can use X to store x_i:

1. Calculate $x_i = X + H \cdot f(T0 + (I - 1) \cdot H, X)$ and store it in X.

To decide when to stop calculating values of x_i, we assume that the total number of steps n is stored in memory register N:

$$N \leftarrow n.$$

We make our decision via a *conditional branching*:

2. Is $I < N$?
    a.   If so, replace I by $I + 1$, then return to step 1 and start over.
    b.   If not, print X and stop.

To run Euler's method, we need to enter initial values for i, t, and x in registers, I, $T0$, and X, the number of steps n in N, and the step size h in H. Often we really want to feed in the ending time $\tau = I_n$ and let the machine compute h; we can use a register named TN to store the value of τ. We enter t_0, τ, x_0, and n as initial data,

$$T0 \leftarrow t_0, \quad TN \leftarrow \tau, \quad X \leftarrow x_0, \quad N \leftarrow n$$

and instruct the machine to calculate h and store it in H:

0. Calculate $h = (TN - T0)/N$ and store it in H.

Then we enter $i = 1$ in I:

00. Store $i = 1$ in I.

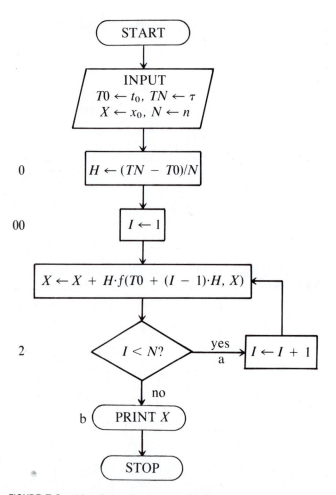

FIGURE 7.2 Algorithm for Euler's Method

We follow this with steps 1 and 2, which will automatically repeat until $I \geq N$. When the calculation stops, x_n will be in memory X and will be printed.

An outline of program steps, like 0 to 3 above, is called an **algorithm.** The algorithm we have described is sketched in the flowchart in Figure 7.2.

We will apply each of our numerical schemes for first-order o.d.e.'s to the same three sample initial-value problems. For two of these, exact solutions can be found by methods from Chapter 1. This will allow us to compare the approximations we obtain to the exact value.

Example 7.2.2

Our first problem is to approximate $x(1)$, where $x(t)$ is the solution of the initial-value problem

(D$_1$) $$x' = 6t - 6x + 1$$

(I$_1$) $$x(0) = 0.1.$$

You can check that the exact solution of (D$_1$) satisfying (I$_1$) is $x(t) = t + 0.1e^{-6t}$, so $x(1) = 1 + 0.1e^{-6}$; to five decimals this is

$$x(1) = 1.00025.$$

Let's see how Euler's method works on this problem.

We give a BASIC program implementing the algorithm of Example 7.2.1 for the o.d.e. (D$_1$) in Figure 7.3. Readers familiar with other programming languages are invited to write their own programs implementing this algorithm. Note that to adapt our program to a different o.d.e., we need only change the definition of $f(t, x)$ in line 10. Line 20 allows for the specification of t_0, τ, x_0,

```
10   DEF FNF(T, X)=6*T−6*X+1
20   INPUT T0, TN, X, N
30   H=(TN−T0)/N
40   FOR I=1 TO N
50   X=X+H*FNF(T0+(I−1)*H, X)
60   NEXT I
70   PRINT X
99   END
```

Initial data: t_0, τ, x_0, n

FIGURE 7.3   BASIC Program for Euler's Method for $x' = 6t - 6x + 1$

and n. Line 30 carries out step 0 of the algorithm. The FOR and NEXT statements in lines 40 and 60 instruct the machine to perform a loop, starting with $I = 1$ and ending when $I = N$. These two statements specify the initial value of I (step 00 in the algorithm), carry out the comparison in step 2, and update the value of I after each repetition of the loop. When the loop is completed (at $I = N$), the computer continues to line 70, printing the current value in X—that is, x_n—and stopping. When we instruct the computer to carry out this program, it prompts us to input the data t_0, τ, x_0, and n (line 20). If we enter these values (*in the specified order* and separated by commas) and press RETURN, the computer will carry out the program, print x_n, and stop.

The table in Figure 7.4 shows the result of running this program with $t_0 = 0$, $\tau = 1$, $x_0 = 0.1$, and various choices of n on a Digital Equipment Corporation DECsystem-10™ computer. The last column shows the difference between x_n and the true value of $x(1) = 1 + (0.1)e^{-6}$, as calculated by the computer. This difference is known as the **error**. Note that increasing the number of steps decreases the error—in other words, it improves the accuracy. Keep in mind that various computers or versions of BASIC perform arithmetic in slightly different ways. This can result in values that differ in one or two digits from those in our table.

n	x_n	Error $[x_n - x(1)]$
10	1.00001	-2.37400×10^{-4}
20	1.00008	-1.68079×10^{-4}
50	1.00017	-8.03266×10^{-5}
100	1.00021	-4.23882×10^{-5}
200	1.00023	-2.17501×10^{-5}
500	1.00024	-8.86059×10^{-6}
1000	1.00024	-4.44985×10^{-6}
2000	1.00025	-2.21467×10^{-6}

FIGURE 7.4   Euler's Method Predictions for $x(1)$ where $x' = 6t - 6x + 1$ and $x(0) = 0.1$
(Exact solution: $x(1) = 1 + (0.1)e^{-6} \approx 1.0002479$)

Example 7.2.3

Our second problem is to approximate $x(1)$, where $x(t)$ is the solution of the initial-value problem

(D_1) $$x' = \frac{6x}{t + 1}$$

(I_2) $$x(0) = 1.$$

The solution of (D$_2$) satisfying (I$_2$) is $x(t) = (t + 1)^6$, so

$$x(1) = 64.$$

Let's use Euler's method to approximate this value.

To adapt the program of Figure 7.3 for (D$_2$), we change line 10 to read

10   DEF FNF(T, X) = 6 * X/(T + 1)

Figure 7.5 shows the result of running the BASIC program with various numbers of steps, n. Again, increasing the number of steps improves the accuracy. Note, however, that our estimates in this example are much less accurate than in Example 7.2.2. With 10 steps in Figure 7.4, the error was already only 0.02% of the exact value, whereas with 2000 steps in Figure 7.5, the error is still 0.37% of the exact value.

n	x_n	Error $[x_n - x(1)]$
10	35.3846	-28.6154
20	45.9913	-18.0087
50	55.5144	-8.48561
100	59.4944	-4.5056
200	61.6757	-2.32426
500	63.0523	-0.947665
1000	63.5231	-0.476897
2000	63.7608	-0.239217

FIGURE 7.5   Euler's Method Predictions for $x(1)$ where $x' = 6x/(t + 1)$ and $x(0) = 1$
(Exact solution: $x(1) = 64$)

We noted in the preceding examples that increasing the number of steps n (equivalently, reducing the step size h) improved the accuracy of our predictions. This is intuitively reasonable, since the basic assumption of our method—that $x'(t)$ does not change very much between checkpoints—comes closer to the truth as we reduce h. However, the very different accuracies of our predictions in the two sample problems point out our need for a better idea of the accuracy of our prediction in any given problem. This need becomes especially acute when we realize that, *in practice, we have no exact solution* for comparison with our numerical prediction. Indeed, if we had an exact solution, numerical schemes would be redundant.

Example 7.2.4

Our third problem is one for which we have no exact solution. We will approximate $x(1)$, where $x(t)$ is the solution of the initial-value problem

(D$_3$) $x' = x^2 - t^2$

(I$_3$) $x(0) = 0.5.$

Since we cannot find a closed-form solution for this problem, we must rely on numerical predictions.

As in Example 7.2.3, we can implement Euler's method for (D$_3$) by changing the program step that calculates $f(t, x)$. The table in Figure 7.6 shows the predictions of $x(1)$ for various choices of n.

What can we say about $x(1)$ from this table? We expect the predictions lower in the table (corresponding to higher n) to be more accurate. Suppose, for example, that we need to know $x(1)$ to three decimal places. The usual heuristic test is to compare the prediction using n steps with the prediction using $2n$ steps and to accept the answer as correct to three places if these predictions agree to three places. This test, applied to the last two lines of the table, suggests that 0.506 is correct. However, without a careful analysis of the error, we cannot be absolutely certain of this answer.

n	x_n
10	0.565218
20	0.537231
50	0.518632
100	0.512113
200	0.508793
500	0.50678
1000	0.506106
2000	0.505769

FIGURE 7.6   Euler's Method Predictions for $x(1)$ where $x' = x^2 - t^2$
and $x(0) = 0.5$
(Exact solution unavailable)

In the next section, we will analyze the accuracy of Euler's method. The outcome of this analysis is that for any initial-value problem, the prediction x_n for $x(\tau)$ satisfies an error estimate of the form

$$|x_n - x(\tau)| < \frac{C}{n},$$

where the constant C is determined from the problem. Thus, in theory, we could obtain as accurate a prediction as we like by applying Euler's method with n sufficiently large.

In practice, however, there are two difficulties with this. One is the fact that each operation on a computer may contribute a roundoff error. Although this error for any individual calculation is quite small, the accumulated roundoff error can become a major problem when the number of steps becomes large. The second problem is cost, in both time and money. The simple examples we consider here take very little time to carry out on a computer, but many practical problems lead to much more complicated o.d.e.'s, for which a large number of steps may be noticeably slow or expensive to calculate. For these reasons, it is desirable to use numerical schemes that are more efficient than Euler's method. We will consider two such schemes in Sections 7.4 and 7.5. Although they are more complicated to implement, those schemes usually give equal or better accuracy in fewer steps than does Euler's method.

Numerical methods for solving o.d.e.'s are as old as o.d.e.'s themselves: Newton considered such methods in his *Principia*, and Euler proposed his method in 1768. Of course, the implementation of such schemes was a very different proposition before the computer revolution!

EULER'S METHOD

To predict $x(\tau)$ via Euler's method, where $x(t)$ satisfies

(D) $$x' = f(t, x)$$

(I) $$x(t_0) = x_0$$

pick the number of steps n, define the step size $h = (\tau - t_0)/h$, and find x_n using the initial condition (I) together with the recurrence relation

$$x_i = x_{i-1} + hf(t_0 + (i - 1)h, x_{i-1}).$$

Note

Hypotheses

It is important, before we expend time or money on numerical schemes for predicting the value $x(\tau)$ of a solution to an initial-value problem, that we know the problem has a unique solution. Otherwise, we are likely to obtain nonsensical answers from numerical schemes. You should recall the existence and uniqueness theorem in Section 1.6 (a proof of which is sketched in Appendix B at the end of this chapter). In each example in Sections 7.3 to 7.5, the o.d.e. will satisfy the hypotheses of this theorem, and τ will lie in an open interval containing t_0, on which the initial-value problem has a unique solution. We will see in Section 7.6 what can happen if this is not the case.

EXERCISES

The calculations in Exercise 1 should be performed by hand or on an unprogrammed calculator. In Exercises 2 through 9, use a computer.

1. Use Euler's method with $n = 2$ to approximate $x(0.2)$ where $x(t)$ is the solution of $x' = 6t - 6x + 1$ that satisfies $x(0) = 0.1$.

In Exercises 2 through 7, use Euler's method with $n = $ 10, 20, 50, 100, 200, and 500 to approximate $x(\tau)$, where $x(t)$ is the solution of the given initial-value problem. In Exercises 2 through 4, find the exact value of $x(\tau)$ and the error for each approximation.

2. $x' = -x^2(1 + t)$, $x(0) = 2$

    a. $\tau = 1$ b. $\tau = 2$ c. $\tau = 5$

3. $x' = \dfrac{2(t + 1)^3}{x}$, $x(0) = 1$

    a. $\tau = 1$ b. $\tau = 2$ c. $\tau = 5$

4. $x' = x - e^t$, $x(1) = 0$

    a. $\tau = 0$ b. $\tau = 2$ c. $\tau = 5$

5. $x' = 0.05x + \dfrac{1}{t + 1}$, $x(0) = 10$ (compare Exercise 1, Section 7.1)

    a. $\tau = 1$ b. $\tau = 5$ c. $\tau = 10$

6. $x' = -0.1x + \dfrac{4}{1 + t^2}$, $x(0) = 60$ (compare Exercise 2, Section 7.1)

    a. $\tau = 1$ b. $\tau = 7$ c. $\tau = 8$

7. $x' = 0.005\left(10 + \dfrac{t}{10} - x\right)x$, $x(0) = 10$ (compare Exercise 3, Section 7.1)

    a. $\tau = 1$ b. $\tau = 5$ c. $\tau = 10$

8. *Calculating* ln 2: Use Euler's method with $n = 50$ and 100 to approximate $x(2)$, where $x(t)$ is the solution of $x' = 1/t$ that satisfies $x(1) = 0$. Note that the exact value of $x(2)$ is ln 2.

9. *Calculating* π: Use Euler's method with $n = 50$ and 100 to approximate $x(\tau)$, where $x(t)$ is the solution of the given initial-value problem.

    a. $x' = \dfrac{6}{\sqrt{1 - t^2}}$, $x(0) = 0$; $\tau = 0.5$

    b. $x' = \dfrac{4}{1 + t^2}$, $x(0) = 0$; $\tau = 1$

    Note that in either case, the exact value of $x(\tau)$ is π.

10. *Accuracy:* For a given value of n, the Euler approximations in Exercise 3 to $x(1)$, $x(2)$, and $x(5)$ use step sizes $1/n$, $2/n$, and $5/n$, respectively. We expect approximations using larger step sizes to be less accurate. Yet the error in this problem goes down as τ increases from 1 to 5. How can we explain this? (*Hint:* What percentage of the exact values does each error represent?)

7.3 ERROR ESTIMATES FOR EULER'S METHOD (OPTIONAL)

In this section we analyze the difference between the exact solution $x(\tau)$ to an initial-value problem.

(D) $$x' = f(t, x)$$

(I) $$x(t_0) = x_0$$

and the Euler's method prediction x_n, obtained from the recurrence relation

(E$_i$) $$x_i = x_{i-1} + hf(t_{i-1}, x_{i-1})$$

with $h = (\tau - t_0)/n$ and $t_i = t_{i-1} + h$, $i = 1, 2, \ldots, n$. We will estimate the absolute value of the error,

$$e_n = x_n - x(t_n) = x_n - x(\tau)$$

by obtaining a recurrence relation for the **error at step** i:

$$e_i = x_i - x(t_i), \quad i = 0, 1, \ldots, n.$$

We will then relate this recurrence relation to one we can solve exactly. Note that we assume the sequence x_i is given exactly by (E$_i$), so we are ignoring roundoff error.

Let's start by analyzing the error in the first of our sample problems.

Example 7.3.1

Analyze e_n for the Euler approximations to $x(1)$, where $x(t)$ is the solution of the initial-value problem

(D$_1$) $$x' = 6t - 6x + 1$$

(I$_1$) $$x(0) = 0.1.$$

Here $x_0 = 0.1$, and the recurrence relation defining the Euler approximation is

(E$_i$) $$x_i = x_{i-1} + h(6t_{i-1} - 6x_{i-1} + 1), \quad i = 1, 2, \ldots, n.$$

On the other hand, we can relate the value of the exact solution at $t = t_i$ to its value at $t = t_{i-1}$ by means of the Taylor series for $x(t)$ expanded about $t = t_{i-1}$. Using the Lagrange form of the remainder, we have

$$x(t_i) = x(t_{i-1} + h) = x(t_{i-1}) + hx'(t_{i-1}) + \frac{h^2}{2} x''(s_i),$$

where $t_{i-1} < s_i < t_i$. Substituting the expression (D$_1$) for x', we can write this as

(T$_i$) $$x(t_i) = x(t_{i-1}) + h[6t_{i-1} - 6x(t_{i-1}) + 1] + \frac{h^2}{2} x''(s_i).$$

If we subtract (T$_i$) from (E$_i$), we get

$$x_i - x(t_i) = (1 - 6h)[x_{i-1} - x(t_{i-1})] - \frac{h^2}{2} x''(s_i).$$

Using the definitions of e_i and e_{i-1}, we see that this is a recurrence relation for e_i:

$$e_i = (1 - 6h)e_{i-1} - \frac{h^2}{2} x''(s_i).$$

Our real interest is in estimating $|e_i|$. If we assume $h < 1/6$ (equivalently, $n > 6$), then $|1 - 6h| = 1 - 6h$. Using the recurrence relation for e_i, we obtain the inequality

$$|e_i| \le (1 - 6h)|e_{i-1}| + \frac{h^2}{2} |x''(s_i)|.$$

In order to estimate the last term, we differentiate both sides of the o.d.e. (D$_1$) with respect to t to obtain the expression

$$x'' = 6 - 6x' = 6 - 6(6t - 6x + 1) = 36(x - t).$$

Since the exact solution of the initial-value problem is $x = t + 0.1e^{-6t}$, and we are interested only in values of t with $0 \le t \le 1$, we have

$$|x''(t)| = 36|x - t| = 36|0.1e^{-6t}| \le 3.6.$$

(When we don't know the exact solution, we need indirect estimates here.) Substituting this inequality into our inequality for e_i, we obtain

$$|e_i| \le (1 - 6h)|e_{i-1}| + 1.8h^2.$$

To see the implications of this last inequality for $|e_n|$, we consider the sequence r_i starting from

$$r_0 = e_0 = 0$$

and satisfying the recurrence relation

$$r_i = (1 - 6h)r_{i-1} + 1.8h^2.$$

Two facts about r_i are easily established (Exercises 5 and 6): first (remember, we are assuming $h < 1/6$),

$$|e_i| \le r_i;$$

second, the solution to the last recurrence relation is

$$r_i = \frac{1 - (1 - 6h)^i}{1 - (1 - 6h)}(1.8h^2) = [1 - (1 - 6h)^i](0.3h).$$

Noting that $0 < (1 - 6h)^i < 1$, we have

$$|e_i| \le r_i < 0.3h.$$

In particular, for $n > 6$,

$$|x_n - x(1)| = |e_n| < 0.3h = \frac{0.3}{n}.$$

This guarantees that, for example, the approximation x_n obtained by taking $n = 100$ will differ from the exact value by less than 0.003. Note that our actual comparison in Figure 7.3 indicates that the approximation is, in fact, much better than that.

The error analysis of the Euler prediction for any initial-value problem

(D) $$x' = f(t, x)$$

(I) $$x(t_0) = x_0$$

follows the pattern of the preceding example. However, some of the details require more work, and some of our estimates will be rougher.

Suppose we use n steps to predict $x(\tau)$, where $\tau > t_0$. Our interest is then limited to the values of t satisfying

$$t_0 \leq t \leq \tau.$$

We assume that we also have some a priori estimates concerning the solution $x(t)$ and the function $f(t, x)$. (We will see how to determine these in the following examples.) In particular, we assume that the values of $x(t)$ and the Euler values x_i all stay within known bounds:

$$a \leq x \leq A, \qquad x = x_i \text{ or } x(t), \quad t_0 \leq t \leq \tau.$$

If f and its first partial derivatives $f_t = \partial f / \partial t$ and $f_x = \partial f / \partial x$ are all continuous, it is then possible to find constants M and K so that for all values of t and x satisfying these bounds,

$$|f_x(t, x)| \leq M$$

$$|f_t(t, x) + f(t, x)f_x(t, x)| \leq K.$$

The Euler prediction is determined by $x_0 = x(t_0)$ and the recurrence relation

(E$_i$) $$x_i = x_{i-1} + hf(t_{i-1}, x_{i-1}).$$

Using the Taylor series with remainder, we can write

$$x(t_i) = x(t_{i-1}) + hx'(t_{i-1}) + \frac{h^2}{2} x''(s_i)$$

where $t_{i-1} \leq s_i \leq t_i$. We can use (D) to rewrite $x'(t_i)$, so

(T$_i$) $$x(t_i) = x(t_{i-1}) + hf(t_{i-1}, x(t_{i-1})) + \frac{h^2}{2} x''(s_i).$$

Subtracting (T$_i$) from (E$_i$), we obtain a recurrence relation for e_i:

$$e_i = e_{i-1} + h[f(t_{i-1}, x_{i-1}) - f(t_{i-1}, x(t_{i-1}))] - \frac{h^2}{2} x''(s_i).$$

The mean-value theorem [applied to $g(x) = f(t_{i-1}, x)$] tells us that

$$f(t_{i-1}, x_{i-1}) - f(t_{i-1}, x(t_{i-1})) = e_{i-1}f_x(t_{i-1}, u_{i-1}),$$

where u_{i-1} is between x_{i-1} and $x(t_{i-1})$. Thus,

$$e_i = e_{i-1}(1 + hf_x(t_{i-1}, u_{i-1})) - \frac{h^2}{2} x''(s_i).$$

Taking absolute values, we have

$$|e_i| \leq |e_{i-1}|(1 + h|f_x(t_{i-1}, u_{i-1})|) + \frac{h^2}{2} |x''(s_i)|.$$

Together with our bound on $|f_x(t, x)|$, this gives

$$|e_i| \leq |e_{i-1}|(1 + hM) + \frac{h^2}{2} |x''(s_i)|.$$

To estimate $x''(s_i)$, note that differentiating both sides of (D) with respect to t gives

$$x''(t) = f_t(t, x) + f_x(t, x)x'(t) = f_t(t, x) + f_x(t, x)f(t, x).$$

The absolute value of the right-hand side of this expression is bounded by K. Thus $|x''(s_i)| \leq K$, and

$$|e_i| \leq |e_{i-1}|(1 + hM) + \frac{h^2 K}{2}.$$

We now set $r_0 = e_0 = 0$, and define

$$r_i = (1 + hM)r_{i-1} + \frac{h^2 K}{2}, \quad i = 1, 2, \ldots, n.$$

As in Example 7.3.1,

$$|e_i| \leq r_i = \left[\frac{(1 + hM)^i - 1}{(1 + hM) - 1} \right] \frac{h^2 K}{2}.$$

Since $1 + hM \leq e^{hM}$ (see Exercise 4),

$$|e_i| \leq r_i \leq \frac{(e^{ihM} - 1)K}{2M} h.$$

Taking $i = n$, and using the fact that $nh = \tau - t_0$, we have

$$|e_n| \le \frac{(e^{M(\tau - t_0)} - 1)K}{2M}\, h = \frac{(e^{M(\tau - t_0)} - 1)K(\tau - t_0)}{2M}\, \frac{1}{n}.$$

Let's use these observations to carry out error estimates for the second of our sample problems.

Example 7.3.2

Estimate the error when Euler's method with $n = 2000$ is used to approximate $x(1)$, where $x(t)$ is the solution of the initial-value problem

$$\text{(D}_2\text{)}\qquad\qquad\qquad x' = \frac{6x}{t + 1}$$

$$\text{(I}_2\text{)}\qquad\qquad\qquad x(0) = 1.$$

The exact solution of the initial-value problem is $x(t) = (t + 1)^6$. Thus, for

$$0 \le t \le 1$$

we have

$$0 \le x(t) \le 64.$$

Starting with $x_0 = 1$, the Euler values x_i are given by

$$\text{(E}_i\text{)}\qquad\qquad\qquad x_i = x_{i-1} + h\left(\frac{6x_{i-1}}{t_{i-1} + 1}\right).$$

Since the quantity in parentheses is nonnegative, this is an increasing sequence. Thus, each x_i is at most x_n, which we know is less than 64 (see Figure 7.5).

$$0 \le x_i \le 64.$$

Here $f(t, x) = 6x/(t + 1)$, so

$$f_x = \frac{6}{t + 1} \quad \text{and} \quad f_t = \frac{-6x}{(t + 1)^2}.$$

For $0 \le t \le 1$ and $0 \le x \le 64$,

$$|f_x| \le 6 = M$$

and

$$|f_t + f_x f| = \left| \frac{-6x}{(t+1)^2} + \frac{36x}{(t+1)^2} \right| = \frac{30|x|}{(t+1)^2} \le 1920 = K.$$

Also, we have $\tau - t_0 = 1$. Thus, we know that

$$|e_n| \le \frac{(e^{M(\tau - t_0)} - 1)K(\tau - t_0)}{2M} \frac{1}{n} = \frac{(e^6 - 1)1920}{12} \frac{1}{n}$$

$$\le \frac{64,640}{2000} = 32.32.$$

In this light, our observed error of 0.24 doesn't look so bad.

———————————————

In Examples 7.3.1 and 7.3.2, we used the exact solutions of the initial-value problems to obtain bounds on x. When we don't know the exact solution, we can often use the following theorem about **differential inequalities** to obtain such bounds.

Theorem: *Suppose both $f(t, x)$ and $\partial f / \partial x$ are continuous and that*

$$g(t, x) \le f(t, x) \quad \text{for} \quad t_0 \le t \le \tau.$$

If $y(t)$ and $z(t)$ are solutions of the initial-value problems

$$y' = g(t, y), \quad y(t_0) = \alpha$$

and

$$z' = f(t, z), \quad z(t_0) = \alpha$$

then

$$y(t) \le z(t) \quad \text{for} \quad t_0 \le t \le \tau.$$

Let's use this theorem to help in our analysis of the error in our third sample problem.

Example 7.3.3

Estimate the error when Euler's method with $n = 2000$ is used to approximate $x(1)$, where $x(t)$ solves the initial-value problem

(D$_3$) $$x' = x^2 - t^2$$

(I$_3$) $$x(0) = 0.5.$$

As in Examples 7.3.1. and 7.3.2, we are interested only in values of t satisfying

$$0 \le t \le 1.$$

For these values of t we have

$$-t^2 \le x^2 - t^2 \le x^2.$$

Let $y(t)$ and $z(t)$ be the solutions of

$$y' = -t^2 \quad \text{and} \quad z' = z^2$$

that satisfy

$$y(0) = z(0) = x(0) = 0.5.$$

Then the theorem tells us that

$$y(t) \le x(t) \le z(t) \quad \text{for} \quad 0 \le t \le 1.$$

The o.d.e.'s for y and z can be solved by separating variables:

$$y(t) = \frac{3 - 2t^3}{6}, \quad z(t) = \frac{1}{2 - t}.$$

We see that $y(t)$ is a decreasing function, whereas $z(t)$ is increasing. Thus,

$$y(t) \ge y(1) = \frac{1}{6} \quad \text{and} \quad z(t) \le z(1) = 1.$$

We conclude that

$$\frac{1}{6} \le x(t) \le 1.$$

It can be shown (Exercise 11) that the values x_i also lie within these bounds.

Now $f(t, x) = x^2 - t^2$, so for $0 \leq t \leq 1$ and $\frac{1}{6} \leq x \leq 1$,

$$|f_x| = |2x| \leq 2 = M$$

and

$$|f_t + f_x f| = |-2t + 2x(x^2 - t^2)| \leq |2t| + |2x^3| + |2xt^2| \leq 6 = K.$$

Since $\tau - t_0 = 1$, we have

$$|e_n| \leq \frac{(e^2 - 1)6}{2 \cdot 2} \frac{1}{n} = \frac{3}{2} \frac{(e^2 - 1)}{2000} \leq 0.0048.$$

Using the value for x_{2000} from Figure 7.6 on page 588, we see that

$$0.5009 < x(1) < 0.5106.$$

Thus we cannot be certain that the estimated value $x(1) = 0.506$ is correct to three places (compare Examples 7.4.4 and 7.5.3).

We now summarize our error estimates.

ERROR ESTIMATES FOR EULER'S METHOD

   To estimate the accuracy of the Euler prediction x_n for $x(\tau)$, where $x(t)$ satisfies

(D) $$x' = f(t, x)$$

(I) $$x(t_0) = x_0:$$

1. Find numbers a and A so that for

$$t_0 \leq t \leq \tau$$

the exact solution $x(t)$ and the predictions x_i all satisfy

$$a \leq x \leq A.$$

2. Find numbers M and K so that whenever t and x lie within these bounds,

$$|f_x| \le M \quad \text{and} \quad |f_t + ff_x| \le K.$$

Then

$$|x_n - x(\tau)| \le \frac{C}{n},$$

where

$$C = \frac{K}{2M}(\tau - t_0)(e^{(\tau - t_0)M} - 1).$$

This estimate ignores roundoff error.

EXERCISES

In Exercises 1 through 3, we use differential inequalities to obtain bounds on the solutions to the initial-value problems from Exercises 5(a), 6(a), and 7(a) of Section 7.2.

1. Let $x(t)$ be the solution of the initial-value problem

$$x' = 0.05x + \frac{1}{t + 1}, \quad x(0) = 10.$$

   a. Check that for $0 \le t \le 1$ we have

   $$0.05x + 0.5 \le 0.05x + \frac{1}{t + 1} \le 0.05x + 1.$$

   b. Solve the initial-value problems $y' = 0.05y + 0.5$, $y(0) = 10$ and $z' = 0.05z + 1$, $z(0) = 10$ to obtain functions satisfying $y(t) \le x(t) \le z(t)$ for $0 \le t \le 1$.

   c. By determining the minimum of $y(t)$ and the maximum of $z(t)$, show that for $0 \le t \le 1$ we have

   $$10 \le x(t) \le 30e^{0.05} - 20 < 11.6.$$

2. Let $x(t)$ be the solution of the initial-value problem

$$x' = -0.1x + \frac{4}{1 + t^2}, \quad x(0) = 60.$$

   a. Check that for $0 \le t \le 1$ we have

   $$-0.1x + 2 \le -0.1x + \frac{4}{1 + t^2} \le -0.1x + 4.$$

  b. Solve the initial-value problems $y' = -0.1y + 2$, $y(0) = 60$ and $z' = -0.1z + 4$, $z(0) = 60$ to obtain functions satisfying $y(t) \le x(t) \le z(t)$ for $0 \le t \le 1$.

  c. By determining the minimum of $y(t)$ and the maximum of $z(t)$, show that for $0 \le t \le 1$ we have

$$56 < 40e^{-0.1} + 20 \le x(t) \le 60.$$

3. Let $x(t)$ be the solution of the initial-value problem

$$x' = 0.005\left(10 + \frac{t}{10} - x\right)x, \quad x(0) = 10.$$

  a. Check that for $0 \le t \le 1$ we have

$$0.005(10 - x)x \le 0.005\left(10 + \frac{t}{10} - x\right)x.$$

  b. Use the inequality in (a) to show that for $0 \le t \le 1$ we have

$$10 \le x(t).$$

  c. Now check that for $0 \le t \le 1$ we have

$$0.005\left(10 + \frac{t}{10} - x\right)x \le 0.005\left(10 + \frac{t}{10} - 10\right)x \le 0.0005x.$$

  d. Solve the initial-value problem $z' = 0.0005z$, $z(0) = 10$, to obtain a function satisfying $x(t) \le z(t)$ for $0 \le t \le 1$.

  e. Show that for $0 \le t \le 1$ we have

$$10 \le x(t) \le 10.01.$$

Exercises 4 through 7 provide information that is useful when estimating the sizes of elements of sequences given by recurrence relations.

4. Show that $1 + t \le e^t$ for all t. [*Hint:* Find the minimum value of the function $f(t) = e^t - t - 1$.]

5. Show that the solution $\{r_i\}$ to a recurrence relation of the form

$$r_i = Ar_{i-1} + B, \quad i = 1, 2, \ldots,$$

  can be written as

$$r_i = A^i r_0 + (A^{i-1} + A^{i-2} + \cdots + 1)B = A^i r_0 + \frac{1 - A^i}{1 - A}B.$$

6. Assume that the sequences $\{y_i\}$ and $\{z_i\}$ start from the same value, $y_0 = z_0$, and that the terms z_i are given by a recurrence relation $z_i = F(z_{i-1})$, where $F(x)$ is an increasing function [i.e., $a \le b$ implies $F(a) \le F(b)$]. Assume further that $y_i \le F(y_{i-1})$ for $i \ge 1$. Show that $y_i \le z_i$ for $i \ge 0$. (*Hint:* Use induction on i.)

7. a. Assume that the sequences $\{y_i\}$ and $\{z_i\}$ start from the same value, $y_0 = z_0$, and are defined by recurrence relations $y_i = G(i - 1, y_{i-1})$ and $z_i = F(i - 1, z_{i-1})$, where $F(i - 1, x)$ is an increasing function of x. Assume further that $G(i - 1, x) \le F(i - 1, x)$ for $1 \le i \le n$. Show that $y_i \le z_i$ for $0 \le i \le n$.

  b. Check that a similar argument works if we assume that $G(i - 1, x)$ is increasing, rather than $F(i - 1, x)$.

In Exercises 8 through 11, we use the results of Exercises 4 through 7 to obtain bounds on the Euler values x_i from Exercises 5(a), 6(a), and 7(a) of Section 7.2 and from Examples 7.3.3.

8. Let $\{x_i\}$ be defined by

$$x_0 = 10 \quad \text{and} \quad x_i = x_{i-1} + h\left[0.05x_{i-1} + \frac{1}{t_{i-1} + 1}\right] \quad \text{for } t \geq 1,$$

where $h = 1/n$ and $t_{i-1} = (i - 1)h$ for $i \geq 1$.

a. Check that for $1 \leq i \leq n$ we have $0 \leq t_{i-1} \leq 1$, and hence

$$(1 + 0.05h)x_{i-1} + 0.5h \leq x_{i-1} + h\left(0.05x_{i-1} + \frac{1}{t_{i-1} + 1}\right)$$

$$\leq (1 + 0.05h)x_{i-1} + h.$$

b. Solve the two recurrence relations $y_i = (1 + 0.05h)y_{i-1} + 0.5h$, $y_0 = 10$ and $z_i = (1 + 0.05h)z_{i-1} + h$, $z_0 = 10$ to obtain sequences satisfying $y_i \leq x_i \leq z_i$ for $0 \leq i \leq n$.

c. Check that the sequences $\{y_i\}$ and $\{z_i\}$ are both increasing, and that for $0 \leq i \leq n$ we have

$$10 = y_0 \leq x_i \leq z_n \leq 30e^{0.5} - 20 < 11.6.$$

9. Let $\{x_i\}$ be defined by

$$x_0 = 60 \quad \text{and} \quad x_i = x_{i-1} + h\left(-0.1x_{i-1} + \frac{4}{1 + t_{i-1}^2}\right) \quad \text{for } i \geq 1,$$

where $h = 1/n \leq 1/10$ and $t_{i-1} = (i - 1)h$ for $i \geq 1$.

a. Check that for $1 \leq i \leq n$ we have $0 \leq t_{i-1} \leq 1$, and hence

$$(1 - 0.1h)x_{i-1} + 2h \leq x_{i-1} + h\left(-0.1x_{i-1} + \frac{4}{1 + t_{i-1}^2}\right)$$

$$\leq (1 - 0.1h)x_{i-1} + 4h.$$

b. Solve the two recurrence relations $y_i = (1 - 0.1h)y_{i-1} + 2h$, $y_0 = 60$ and $z_i = (1 - 0.1h)z_{i-1} + 4h$, $z_0 = 60$ to obtain sequences satisfying $y_i \leq x_i \leq z_i$ for $0 \leq i \leq n$.

c. Check that the sequences $\{y_i\}$ and $\{z_i\}$ are both decreasing, and that for $0 \leq i \leq n$ we have

$$20 \leq y_n \leq x_i \leq z_0 = 60.$$

10. Let $\{x_i\}$ be defined by

$$x_0 = 10 \quad \text{and} \quad x_i = x_{i-1} + 0.005\left(10 + \frac{t_{i-1}}{10} - x_{i-1}\right)x_{i-1} \quad \text{for } i \geq 1,$$

where $h = 1/n$ and $t_{i-1} = (i - 1)h$ for $i \geq 1$.

a. Show that for $1 \leq i \leq n$ we have $0 \leq t_{i-1} \leq 1$, and hence that

$$x_{i-1} + 0.005(10 - x_{i-1})x_{i-1} \leq x_{i-1} + 0.005\left(10 + \frac{t_{i-1}}{10} - x_{i-1}\right)x_{i-1}.$$

b. Use this inequality to show that for $1 \le i \le n$ we have

$$10 \le x_i.$$

c. Now check that for $1 \le i \le n$ we have

$$x_{i-1} + 0.005\left(10 + \frac{t_{i-1}}{10} - x_{i-1}\right) \le 1.0005x_{i-1}.$$

d. Solve the recurrence relation $z_i = 1.0005z_{i-1}$, $z_0 = 10$ to obtain a sequence satisfying $x_i \le z_i$ for $0 \le i \le n$.

e. Check that the sequence $\{z_i\}$ is increasing, and that for $0 \le i \le n$ we have

$$10 \le x_i \le z_n \le 10e^{0.0005} < 10.01.$$

11. Let $\{x_i\}$ be defined by

$$x_0 = 0.5 \quad \text{and} \quad x_i = x_{i-1} + h(x_{i-1}^2 - t_{i-1}^2) \quad \text{for } i \ge 1,$$

where $h = 1/n$ and $t_{i-1} = (i - 1)h$ for $i \ge 1$.

a. Check that for $1 \le i \le n$ we have $0 \le t_{i-1} \le 1$, and hence that

$$x_{i-1} - ht_{i-1}^2 \le x_{i-1} + h(x_{i-1}^2 - t_{i-1}^2) \le x_{i-1} + hx_{i-1}^2.$$

b. Let $\{y_i\}$ be defined by $y_0 = 0.5$ and $y_i = y_{i-1} - ht_{i-1}^2$. Use the fact that

$$1^2 + 2^2 + \cdots + (i - 1)^2 = \frac{(i - 1)i(2i - 1)}{6}$$

to show that

$$y_i = \frac{1}{2} - h^3\left[\frac{(i - 1)i(2i - 1)}{6}\right] \ge \frac{1}{6} \quad \text{for } 0 \le i \le n.$$

c. Let $\{z_i\}$ be defined by $z_0 = 0.5$ and $z_i = z_{i-1} + hz_{i-1}^2$ for $(i \ge 1)$ and let $z(t) = 1/(2 - t)$. Show that $z_i \le z(t_i) \le 1$ for $0 \le i \le n$. [*Hint:* Assume $z_{i-1} \le z(t_{i-1})$ and find the minimum for $t \ge t_{i-1}$ of the function $s(t) = 1/(2 - t) - [z_{i-1} + z_{i-1}^2(t - t_{i-1})]$, which gives the vertical separation between $z(t)$ and the line used to obtain the Euler value z_i.]

d. Use the result of Exercise 7 to conclude that $\frac{1}{6} \le x_i \le 1$ for $0 \le i \le n$.

In Exercises 12 through 14, we use the bounds on $x(t)$ and x_i found in Exercises 1 through 3 and 8 through 10 to estimate the error when Euler's method is used for the initial-value problems of Exercises 5(a), 6(a), and 7(a) of Section 7.2.

12. Let $f(t, x) = 0.05x + \dfrac{1}{t + 1}$.

a. Show that if $0 \le t \le 1$ and $10 \le x \le 11.6$, then $|f_x| \le 0.05$ and $|f_t + ff_x| \le 1.08$.

b. Estimate the error when Euler's method with $n = 1000$ is used to approximate $x(1)$, where $x(t)$ is the solution of $x' = f(t, x)$ satisfying $x(0) = 1$.

13. Let $f(t, x) = -0.1x + \dfrac{4}{1 + t^2}$.

a. Show that if $0 \le t \le 1$ and $20 \le x \le 60$, then $|f_x| \le 0.1$ and $|f_t + ff_x| \le 9$.

b. Estimate the error when Euler's method with $n = 1000$ is used to approximate $x(1)$, where $x(t)$ is the solution of $x' = f(t, x)$ satisfying $x(0) = 60$.

14. Let $f(t, x) = 0.005\left(10 + \dfrac{t}{10} - x\right) x$.

   a. Show that if $0 \le t \le 1$ and $10 \le x \le 10.01$, then $|f_x| \le 0.16$ and $|f_t + f f_x| \le 0.16$.

   b. Estimate the error when Euler's method with $n = 1000$ is used to approximate $x(1)$, where $x(t)$ is the solution of $x' = f(t, x)$ satisfying $x(0) = 10$.

7.4   A MODIFIED EULER METHOD

To obtain a reasonably accurate prediction of $x(\tau)$ by Euler's method, we may be required to perform an impracticably large number of calculations. In this section and the next we discuss schemes that generally yield equally accurate predictions in fewer steps.

   We begin by noting that an initial-value problem

(D) $$x'(t) = f(t, x(t))$$

(I) $$x(t_0) = x_0$$

can be changed formally into an integration problem by writing $F(t)$ in place of $f(t, x(t))$, integrating both sides of (D), and substituting (I):

$$x(t) = x_0 + \int_{t_0}^{t} F(t)\, dt, \quad \text{where } F(t) = f(t, x(t)).$$

In particular,

$$x(\tau) = x_0 + \int_{t_0}^{\tau} F(t)\, dt.$$

Of course, we cannot normally integrate $F(t)$ because its definition involves the unknown function $x(t)$. Instead, we estimate this integral formula numerically. To do this, we first break the interval $t_0 \le t \le \tau$ into n equal parts of length $h = (\tau - t_0)/n$. We then look for approximations x_i to the values of $x(t)$ at the checkpoints $t_i = t_0 + ih$, $i = 0, 1, \ldots, n$. Note that the exact values satisfy the recurrence relation

(S$_i$) $$x(t_i) = x(t_{i-1}) + \int_{t_{i-1}}^{t_i} F(t)\, dt.$$

If we find reasonable approximations for the integrals in (S$_i$),

$$\int_{t_{i-1}}^{t_i} F(t)\, dt \approx A_i,$$

then the recurrence relation

(R_i) $x_i = x_{i-1} + A_i, \quad i = 1, \ldots, n$

will yield reasonable approximations for the values $x(t_i)$. Our different numerical schemes for approximating $x(\tau)$, including Euler's method, result from taking different estimates for these integrals.

The simplest way to estimate an integral is to replace the integrand by a constant. If we replace $F(t)$ for $t_{i-1} \le t \le t_i$ by the constant value $F(t_{i-1})$, the integral in (S_i) is replaced by the area of a rectangle of width $t_i - t_{i-1} = h$ and height $F(t_{i-1})$ (Figure 7.7). This gives us the estimate

$$\int_{t_{i-1}}^{t_i} F(t) \, dt \approx hF(t_{i-1}).$$

If we have already chosen an approximation x_{i-1} for $x(t_{i-1})$, then

$$F(t_{i-1}) = f(t_{i-1}, x(t_{i-1})) \approx f(t_{i-1}, x_{i-1}),$$

so

$$\int_{t_{i-1}}^{t_i} F(t) \, dt \approx hf(t_{i-1}, x_{i-1}).$$

Corresponding to this approximation for the integral, we obtain the recurrence relation defining Euler's method:

$$x_i = x_{i-1} + hf(t_{i-1}, x_{i-1}).$$

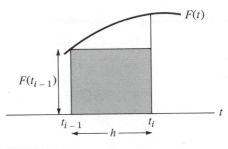

FIGURE 7.7

A more accurate estimate for the integral is obtained by replacing the integrand $F(t)$ with a first-degree polynomial that agrees with $F(t)$ at the endpoints $t = t_{i-1}$ and $t = t_i$. The integral is then replaced by the area of a trapezoid

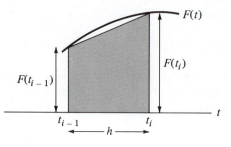

FIGURE 7.8

(Figure 7.8), yielding the **trapezoid rule:**

$$\int_{t_{i-1}}^{t_i} F(t)\, dt \approx \frac{h}{2}\,[F(t_{i-1}) + F(t_i)].$$

As before,

$$F(t_{i-1}) \approx f(t_{i-1}, x_{i-1}).$$

On the other hand,

$$F(t_i) = f(t_i, x(t_i)) \approx f(t_i, x_i^*),$$

where we take x_i^* to be the estimate of $x(t_i)$ given by Euler's method:

$$x_i^* = x_{i-1} + hf(t_{i-1}, x_{i-1}).$$

Combining these considerations, we obtain a recurrence relation defining a new scheme for predicting $x(\tau)$:

$$(\mathbf{M}_i) \qquad x_i = x_{i-1} + \frac{h}{2}\,[f(t_{i-1}, x_{i-1}) + f(t_i, x_i^*)],$$

where

$$x_i^* = x_{i-1} + hf(t_{i-1}, x_{i-1}).$$

This scheme is known as the **modified Euler method** (or average-slope method).

Whereas Euler's method with n steps leads to predictions for $x(\tau)$ with reliable accuracy of the form C/n, it turns out (see the note at the end of the section) that the reliable accuracy of the modified Euler method has the form C/n^2. This means, for example, that doubling the number of steps in the new scheme cuts the estimated error by one fourth, as opposed to a cut of only one half for Euler's method. On the other hand, the modified Euler method is somewhat harder to implement, as seen in the following algorithm.

Example 7.4.1   Modified Euler Algorithm

We map out an algorithm for the recurrence relation (M_i) along lines similar to Example 7.2.1.

In order to calculate x_i, we must first calculate x_i^*. This is precisely the same calculation as for Euler's method, so we again allocate memories I, $T0$, X, H, and N for i, t_0, x_{i-1}, h, and n:

$$I \leftarrow i, \quad T0 \leftarrow t_0, \quad X \leftarrow x_{i-1}, \quad H \leftarrow h, \quad N \leftarrow n.$$

Looking ahead, we note that after calculating x_i^* we will still need the value of x_{i-1} in order to calculate x_i. Thus we must store x_i^* in a memory register other than X, say XS:

1. Calculate $x_i^* = X + H \cdot f(T0 + (I - 1) \cdot H, X)$ and store it in XS.

Now, we calculate x_i and store it in X:

2. Calculate $x_i = X + (H/2) \cdot [f(T0 + (I - 1) \cdot H, X) + f(T0 + I \cdot H, XS)]$ and store it in X.

To decide whether to continue or to stop, we compare I with N:

3. Is $I < N$?
    a. If so, replace I with $I + 1$, return to step 1, and start over.
    b. If not, print X and stop.

As in Example 7.2.1, it is convenient to enter t_0, τ, x_0, and n:

$$T0 \leftarrow t_0, \quad TN \leftarrow \tau, \quad X \leftarrow x_0, \quad N \leftarrow n.$$

We then calculate h and store it in H:

0. Calculate $h = (TN - T0)/N$ and store it in H.

Finally, we set the initial value $i = 1$ in I:

00. Store $i = 1$ in I.

Steps 1 through 3 constitute a loop calculating x_i for $i = 1, 2, \ldots, n$. This algorithm is sketched in a flowchart in Figure 7.9. When the program is run, the calculation stops with x_n in X, which is then printed.

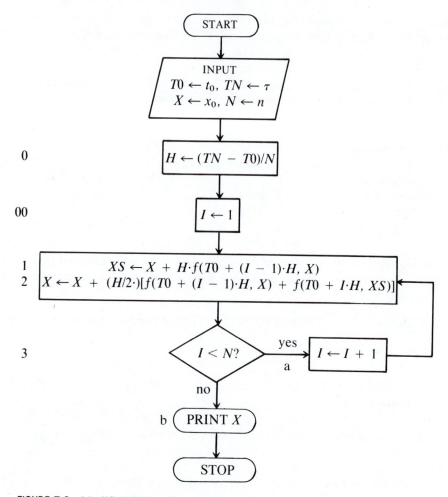

FIGURE 7.9   Modified Euler Algorithm

Let's investigate the performance of the modified Euler scheme in our sample problems.

Example 7.4.2

Use the modified Euler method to approximate $x(1)$, where

(D_1) $$x' = 6t - 6x + 1$$

(I_1) $$x(0) = 0.1.$$

We have given a BASIC program implementing the preceding algorithm in Figure 7.10. The table in Figure 7.11 gives the outcome of trials using the BASIC program with $t_0 = 0$, $x_0 = 0.1$, $\tau = 1$, and various choices of n. Note the improved accuracy compared with Figure 7.4: The modified Euler prediction with $n = 50$ is not much further off than the Euler prediction with $n = 1000$.

```
10   DEF FNF(T, X) = 6*T − 6*X + 1
20   INPUT T0, TN, X, N
30   H = (TN − T0)/N
40   FOR I = 1 TO N
50   XS = X + H*FNF(T0 + (I − 1)*H, X)
60   X = X + (H/2)*(FNF(T0 + (I − 1)*H, X) + FNF(T0 + I*H, XS))
70   NEXT I
80   PRINT X
99   END
```

FIGURE 7.10   BASIC Program for the Modified Euler Method for $x' = 6t − 6x + 1$

n	x_n	Error
10	1.00043	1.82932×10^{-4}
20	1.00028	2.95397×10^{-5}
50	1.00025	3.93951×10^{-6}
100	1.00025	9.21471×10^{-7}
200	1.00025	2.29116×10^{-7}
500	1.00025	5.03023×10^{-8}
1000	1.00025	-9.30231×10^{-9}
2000	1.00025	5.59885×10^{-9}

FIGURE 7.11   Modified Euler Predictions for $x(1)$ where $x' = 6t − 6x + 1$ and $x(0) = 0.1$
(Exact solution: $x(1) = 1 + 0.1e^{-6} \approx 1.0002479$)

Example 7.4.3

Use the modified Euler scheme to predict $x(1)$, where

$$(D_2) \qquad\qquad x' = \frac{6x}{t + 1}$$

$$(I_2) \qquad\qquad x(0) = 1.$$

The table in Figure 7.12 shows the result of running a BASIC program for (D$_2$) with $t_0 = 0$, $x_0 = 1$, $\tau = 1$, and various choices of n. The improvement in accuracy here, compared with Figure 7.5, is even more dramatic. The mod-

n	x_n	Error
10	58.2831	-5.71689
20	62.2573	-1.74266
50	63.6874	-0.3126
100	63.9189	-8.10647×10^{-2}
200	63.9794	-2.0627×10^{-2}
500	63.9967	-3.32975×10^{-3}
1000	63.9992	-8.41618×10^{-4}
2000	63.9998	-2.18868×10^{-4}

FIGURE 7.12 Modified Euler Predictions for $x(1)$ where $x' = 6x/(t + 1)$ and $x(0) = 1$ (Exact solution: $x(1) = 64$)

ified Euler prediction with $n = 10$ is almost as accurate as the Euler prediction with $n = 100$, and modified Euler with $n = 50$ (a calculation taking 3 minutes on a calculator) gives a more accurate prediction than Euler with $n = 1000$ (which takes nearly an hour on a calculator). The modified Euler prediction with $n = 1000$ is accurate to 0.0013%.

Example 7.4.4

Use the modified Euler scheme to predict $x(1)$, where

$$(D_3) \qquad\qquad x' = x^2 - t^2$$

$$(I_3) \qquad\qquad x(0) = 0.5.$$

The table in Figure 7.13 lists the outcome of trials with various values of n. Here we cannot compare the results with an exact solution. However, com-

n	x_n
10	0.504902
20	0.505329
50	0.505418
100	0.505428
200	0.50543
500	0.505431
1000	0.505431
2000	0.505431

FIGURE 7.13. Modified Euler Predictions for $x(1)$ where $x' = x^2 - t^2$ and $x(0) = 0.5$
(Exact solution unavailable)

parison with the Euler predictions in Figure 7.6 shows consistently lower estimates in the present trials. In particular, the numbers in Figure 7.13 strongly suggest that the third digit of $x(1)$ is 5, not 6. This shows that the heuristic test, that we accept an answer when the predictions using n and $2n$ agree up to significant figures, can be misleading.

Let's summarize.

MODIFIED EULER METHOD

To predict $x(\tau)$ by the modified Euler method, where $x(t)$ satisfies

(D) $$x' = f(t, x)$$

(I) $$x(t_0) = x_0$$

pick the number of steps n, define the step size $h = (\tau - t_0)/h$, and find x_n using the initial condition (I) together with the recurrence relation

$$x_i = x_{i-1} + \frac{h}{2}[f(t_{i-1}, x_{i-1}) + f(t_i, x_i^*)]$$

where

$$x_i^* = x_{i-1} + hf(t_{i-1}, x_{i-1}).$$

Note

Error estimates for the modified Euler scheme

We outline here a partial analysis of the errors

$$e_i = x_i - x(t_i)$$

in using the modified Euler scheme

(M$_i$)
$$\begin{cases} x_i^* = x_{i-1} + hf(t_{i-1}, x_{i-1}), \\ x_i = x_{i-1} + \dfrac{h}{2}[f(t_{i-1}, x_{i-1}) + f(t_i, x_i^*)] \end{cases}$$

to predict the value at $t = \tau$ of the solution $x(t)$ to

(I) $$x(t_0) = x_0$$

(D) $$x' = f(t, x).$$

A careful analysis of the global error e_n is quite fussy, but the central idea is similar to that in Section 7.3.

Using Taylor series with remainder, we express the values of the exact solution at the checkpoints in the recurrence relation

(T$_i$) $$x(t_i) = x(t_{i-1}) + hf(t_{i-1}, x(t_{i-1})) + \frac{h^2}{2} \mathscr{F}(t_{i-1}, x(t_{i-1})) + h^3 R(t_{i-1}, x(t_{i-1})),$$

where

$$\mathscr{F}(t, x) = f_t(t, x) + f(t, x)f_x(t, x)$$

is the expression for $x''(t)$ obtained from (D) and R is a remainder term that can be expressed in terms of f and its derivatives to order 3. Similarly, using Taylor series, we can express $f(t_i, x_i^*)$ in terms of $f(t_{i-1}, x_{i-1})$ by

$$f(t_i, x_i^*) = f(t_{i-1}, x_{i-1}) + h\mathscr{F}(t_{i-1}, x_{i-1}) + \frac{h^2}{2} R_i,$$

where $\mathscr{F}$ is as above, and R_i is a remainder term that can also be written in terms of f and its partial derivatives to order 2. Thus, the recurrence relation (M$_i$) can be written

(M$_i$)
$$x_i = x_{i-1} + \frac{h}{2}\left[f(t_{i-1}, x_{i-1}) + f(t_{i-1}, x_{i-1}) + h\mathscr{F}(t_{i-1}, x_{i-1}) + \frac{h^2}{2} R_i \right]$$
$$= x_{i-1} + hf(t_{i-1}, x_{i-1}) + \frac{h^2}{2} \mathscr{F}(t_{i-1}, x_{i-1}) + \frac{h^3}{4} R_i.$$

Subtracting (T$_i$) from (M$_i$), we have a recurrence relation for the error:

$$e_i = e_{i-1} + h[f(t_{i-1}, x_{i-1}) - f(t_{i-1}, x(t_{i-1}))]$$
$$+ \frac{h^2}{2} [\mathscr{F}(t_{i-1}, x_{i-1}) - \mathscr{F}(t_{i-1}, x(t_{i-1}))] + h^3 \left[\frac{1}{4} R_i - R(t_{i-1}, x(t_{i-1}))\right].$$

Now, using the properties of f, we can obtain estimates of the following sort: For all $t_0 \le t \le \tau$ and $a \le x, y \le b$,

$$|f(t, x) - f(t, y)| < M|x - y|$$
$$|\mathscr{F}(t, x) - \mathscr{F}(t, y)| < 2M|x - y|$$
$$|R_i| < \frac{K}{8}, \quad |R(t, x)| < \frac{K}{2}.$$

Then we have

$$|e_i| \leq |e_{i-1}||1 + Mh + Mh^2| + h^3 K,$$

and so, setting $r_0 = 0$ and

$$r_i = r_{i-1}(1 + Mh + Mh^2) + h^3 K$$

we have

$$|e_i| \leq r_i = \frac{(1 + Mh + Mh^2)^i - 1}{(1 + Mh + Mh^2) - 1} h^3 K \leq \frac{e^{i(Mh + Mh^2)} - 1}{M + Mh} h^2 K.$$

Hence, using $nh = \tau - t_0$, we have the estimate at $i = n$,

$$|e_n| \leq h^2 \frac{K}{M(1 + h)} (e^{(\tau - t_0)M(1 + h)} - 1),$$

or, assuming $0 < h < 1$,

$$|e_n| \leq h^2 \frac{K}{M} (e^{2M(\tau - t_0)} - 1).$$

This should be compared with the estimate of the error for Euler's method in Section 7.3. The salient feature is that here the error estimate is proportional to h^2, which for small values of h is much less than h.

EXERCISES

To facilitate comparison, Exercises 1 through 9 here repeat Exercises 1 through 9 in Section 7.2, with Euler's method replaced by the modified Euler method. The calculations in Exercise 1 should be performed by hand or on an unprogrammed calculator. In Exercises 2 through 9, use a computer.

1. Use the modified Euler method with $n = 2$ to approximate $x(0.2)$, where $x(t)$ is the solution of $x' = 6t - 6x + 1$ that satisfies $x(0) = 0.1$.

In Exercises 2 through 7, use the modified Euler method with $n = 10, 20, 50, 100, 200,$ and 500 to approximate $x(\tau)$, where $x(t)$ is the solution of the given initial-value problem. In Exercises 2 and 3, find the exact value of $x(\tau)$ and the error for each approximation.

2. $x' = -x^2(1 + t),$ $x(0) = 2$
   a. $\tau = 1$ b. $\tau = 2$ c. $\tau = 5$

3. $x' = \dfrac{2(t + 1)^3}{x},$ $x(0) = 1$
   a. $\tau = 1$ b. $\tau = 2$ c. $\tau = 5$

4. $x' = x - e^t$, $x(1) = 0$
   a. $\tau = 0$ b. $\tau = 2$ c. $\tau = 5$

5. $x' = 0.05x + \dfrac{1}{t + 1}$, $x(0) = 10$
   a. $\tau = 1$ b. $\tau = 5$ c. $\tau = 10$

6. $x' = -0.1x + \dfrac{4}{1 + t^2}$, $x(0) = 60$
   a. $\tau = 1$ b. $\tau = 7$ c. $\tau = 8$

7. $x' = 0.005\left(10 + \dfrac{t}{10} - x\right)x$, $x(0) = 10$
   a. $\tau = 1$ b. $\tau = 5$ c. $\tau = 10$

8. *Calculating* ln 2: Use the modified Euler method with $n = 50$ and 100 to approximate $x(2)$, where $x(t)$ is the solution of $x' = 1/t$ that satisfies $x(1) = 0$.

9. *Calculating* π: Use the modified Euler method with $n = 50$ and 100 to approximate $x(\tau)$, where $x(t)$ is the solution of the given initial-value problem.

   a. $x' = \dfrac{6}{\sqrt{1 - t^2}}$, $x(0) = 0$; $\tau = 0.5$.

   b. $x' = \dfrac{4}{1 + t^2}$, $x(0) = 0$; $\tau = 1$

10. *Graphing* e^{-t}:
    a. Modify your computer program for the modified Euler method so that it prints (or pauses to display) each x_i in turn. Use this modified program to find approximations to $x(0.1)$, $x(0.2)$, . . . , $x(0.9)$, and $x(1)$, where $x(t)$ is the solution of $x' = -x$ that satisfies $x(0) = 1$.
    b. Using the values obtained in part (a), sketch the graph of $x(t) = e^{-t}$ for $0 \le t \le 1$ by computer (if possible) or by hand.

7.5 A RUNGE-KUTTA SCHEME

The efficiency of the modified Euler scheme is greatly surpassed by a different approximation scheme, first suggested in the work of the German mathematicians C. Runge (1895), K. Heun (1900), and W. Kutta (1901). This scheme, known as the **Runge-Kutta method,** satisfies an error estimate of the form

$$|x_n - x(\tau)| < \frac{C}{n^4}.$$

It is the basis of a great many packaged calculator and computer programs for solving o.d.e.'s.

The Runge-Kutta method is based on replacing the trapezoid rule for integrals with **Simpson's rule.** In the trapezoid rule, the integrand $F(t)$ is approximated by a first-order polynomial that agrees with F at the endpoints. In

Simpson's rule, we use the quadratic polynomial that agrees with F at the ends $t = t_{i-1}$ and $t = t_i$, and at the midpoint $t = \bar{t}_i$; upon integrating it, we obtain (Exercise 11)

$$\int_{t_{i-1}}^{t_i} F(t)\, dt \approx \frac{h}{6} [F(t_{i-1}) + 4F(\bar{t}_i) + F(t_i)],$$

where $h = t_i - t_{i-1}$ is the length of the interval.

Just as in the previous section, we turn the recurrence relation

$$x(t_i) = x(t_{i-1}) + \int_{t_{i-1}}^{t_i} F(t)\, dt \qquad [\text{where } F(t) = f(t, x(t))]$$

into a scheme for predicting $x(\tau)$ by estimating each of the F-values in Simpson's rule [using appropriate guesses for $x(t)$] and replacing the integral with Simpson's approximation. The natural guess for $x(t_{i-1})$ comes from our previous step, $x(t_{i-1}) \approx x_{i-1}$. This gives an estimate for the value of $F(t_{i-1})$, which we denote by k_1:

$$F(t_{i-1}) \approx k_1 = f(t_{i-1}, x_{i-1}).$$

For the midpoint value, we average two estimates obtained in a way similar to the modified Euler scheme. A first guess for $x(\bar{t}_i)$ (where $\bar{t}_i = t_{i-1} + h/2$) is obtained from an Euler step using k_1:

$$\bar{x}_i = x_{i-1} + \frac{hk_1}{2}.$$

This gives our first estimate for $F(\bar{t}_i)$, which we call k_2:

$$k_2 = f(\bar{t}_i, \bar{x}_i) = f\left(t_{i-1} + \frac{h}{2}, x_{i-1} + \frac{k_1 h}{2}\right).$$

Now, we use k_2 to make a second guess for $x(\bar{t}_i)$:

$$\bar{\bar{x}}_i = x_{i-1} + \frac{hk_2}{2}.$$

Substituting this into f gives k_3, a second estimate for $F(\bar{t}_i)$:

$$k_3 = f(\bar{t}_i, \bar{\bar{x}}_i) = f\left(t_{i-1} + \frac{h}{2}, x_{i-1} + \frac{k_2 h}{2}\right).$$

We will use the average of k_2 and k_3 in place of $F(\bar{t}_i)$, or

$$4F(\bar{t}_i) \approx 2k_2 + 2k_3.$$

Finally, using k_3, we make a preliminary Euler guess at $x(t_i)$:

$$x_i^* = x_{i-1} + hk_3.$$

This is used to estimate $F(t_i)$, with k_4:

$$F(t_i) \approx k_4 = f(t_i, x_i^*) = f(t_{i-1} + h, x_{i-1} + hk_3).$$

Combining these in Simpson's estimate of the integral, we obtain the Runge-Kutta scheme for predicting $x(\tau)$:

(RK$_i$) $$x_i = x_{i-1} + \frac{h}{6}[k_1 + 2k_2 + 2k_3 + k_4],$$

where, we recall,

$$k_1 = f(t_{i-1}, x_{i-1})$$

$$k_2 = f\left(t_{i-1} + \frac{h}{2}, x_{i-1} + \frac{hk_1}{2}\right)$$

$$k_3 = f\left(t_{i-1} + \frac{h}{2}, x_{i-1} + \frac{hk_2}{2}\right)$$

$$k_4 = f(t_{i-1} + h, x_{i-1} + hk_3).$$

[Of course, $x_0 = x(0)$.]

Let's see how the Runge-Kutta scheme compares with our other two schemes when applied to our sample problems.

Example 7.5.1

Figure 7.14 shows a BASIC program implementing the Runge-Kutta scheme for

(D$_1$) $$x' = 6t - 6x + 1$$

(I$_1$) $$x(0) = 0.1.$$

We calculate the four quantities, k_1, k_2, k_3, and k_4 in lines 50 through 80, then implement formula (RK$_i$) in line 90.

```
10   FNF(T, X)=6*T-6*X+1
20   INPUT T0, TN, X, N
30   H=(TN-T0)/N
40   FOR I=1 to N
50   K1=FNF(T0+(I-1)*H, X)
60   K2=FNF(T0+(I-1/2)*H, X+H*K1/2)
70   K3=FNF(T0+(I-1/2)*H, X+H*K2/2)
80   K4=FNF(T0+I*H, X+H*K3)
90   X=X+(H/6)*(K1+2*K2+2*K3+K4)
100  NEXT I
110  PRINT X
999  END
```

FIGURE 7.14   BASIC Program for the Runge-Kutta Method for $x' = 6t - 6x + 1$

Figure 7.15 shows the predictions for $x(1)$ of the Runge-Kutta scheme with various numbers of steps. We have also listed the error of the Runge-Kutta, Euler (Figure 7.4), and modified Euler (Figure 7.11) predictions. The improvement in accuracy is dramatic. The Runge-Kutta predictions printed by the computer all agree with the exact value to five decimals. When the extra places kept by the computer are taken into account (as in the error columns), we see that 10 Runge-Kutta steps give a prediction more accurate (by a factor of about 10) than 200 Euler steps, and about twice as accurate as 50 modified Euler steps.

When we interpret the "error" columns in Figure 7.15, we must keep in mind the inherent limitations of computer arithmetic. A number is represented in a computer as a finite-digit approximation. The number of digits is determined by several factors, most importantly the particular computer language we use and the computer on which we implement it. The calculations used to create the tables in this book carry eight significant digits. Whenever we see a figure

n	x_n	Error for Runge-Kutta	Error for Modified Euler	Error for Euler
10	1.00025	2.67291×10^{-6}	1.82932×10^{-4}	-2.37400×10^{-4}
20	1.00025	1.24808×10^{-7}	2.95397×10^{-5}	-1.68079×10^{-4}
50	1.00025	5.59885×10^{-9}	3.93951×10^{-6}	-8.03266×10^{-5}
100	1.00025	-9.30231×10^{-9}	9.29471×10^{-7}	-4.23882×10^{-5}
200	1.00025	-9.30231×10^{-9}	2.29116×10^{-7}	-2.17501×10^{-5}
500	1.00025	5.59885×10^{-9}	5.03023×10^{-8}	-8.86059×10^{-6}
1000	1.00025	-2.42035×10^{-8}	-9.30231×10^{-9}	-4.44985×10^{-6}
2000	1.00025	-9.30231×10^{-9}	5.59885×10^{-9}	-2.21467×10^{-6}

FIGURE 7.15   Runge-Kutta Predictions of $x(1)$ where $x' = 6t - 6x + 1$, $x(0) = 0.1$ (Exact solution: $x(1) = 1 + 0.1e^{-6} \approx 1.0002479$)

for the "error" that begins after the eighth digit of our answer, as in the Runge-Kutta column from $n = 50$ on, we must realize that this error is at the level of "machine noise," so variations in the error within this range are meaningless. The somewhat greater error for the Runge-Kutta prediction with $n = 1000$ is probably due to accumulated roundoff error, which we discuss further in Section 7.6.

Example 7.5.2

Figure 7.16 lists the data corresponding to Figure 7.15 for the predictions of $x(1)$, where

$$(D_2) \qquad\qquad x' = \frac{6x}{t + 1}$$

$$(I_2) \qquad\qquad x(0) = 1.$$

Here, the known value of the exact solution is $x(1) = 64$. The increased accuracy of the Runge-Kutta scheme is even more dramatic than in the previous problem. Runge-Kutta with $n = 10$ is more accurate than Euler with $n = 1000$ or modified Euler with $n = 200$.

Note that the computer printed "0" error for the Runge-Kutta prediction with $n = 500$. This means that the predicted value agrees with the exact value in all digits carried by the computer. A calculation carrying more digits would be likely to show *some* error, although it would, of course, be very small.

n	x_n	Error for Runge-Kutta	Error for Modified Euler	Error for Euler
10	63.9355	-6.45447×10^{-2}	-5.71689	$-2.86154 \times 10^{+1}$
20	63.9951	-4.94766×10^{-3}	-1.74266	$-1.80087 \times 10^{+1}$
50	63.9999	-1.42097×10^{-4}	-3.12600	-8.48561
100	64.	-4.76837×10^{-6}	-8.10647×10^{-2}	-4.5056
200	64.	-4.29153×10^{-6}	-2.06270×10^{-2}	-2.32426
500	64.	0	-3.32975×10^{-3}	-9.47665×10^{-1}
1000	64.	9.53674×10^{-6}	-8.41618×10^{-4}	-0.476897
2000	64.	-4.76837×10^{-7}	-2.18868×10^{-4}	-0.239217

FIGURE 7.16   Runge-Kutta Predictions of $x(1)$ where $x' = 6x/(t + 1)$, $x(0) = 1$ (Exact solution: $x(1) = 64$)

■   **Example 7.5.3**

Figure 7.17 compares the predictions of the three schemes for $x(1)$, where

(D$_3$)
$$x' = x^2 - t^2$$

(I$_3$)
$$x(0) = 0.5.$$

Here we have no exact solution available, but we can compare the extent to which the numerical predictions stabilize with increasing n. Notice that with the Euler scheme, there is still some uncertainty about the third digit even when $n = 2000$ (recall Example 7.3.3), whereas all the Runge-Kutta predictions agree on the first five digits. The stability of these predictions, and their agreement with the high-n modified Euler predictions, suggests that we can be confident of a value of $x(1) = 0.50543$.

n	x_n (Euler)	x_n (Modified Euler)	x_n (Runge-Kutta)
10	0.565218	0.504902	0.505434
20	0.537231	0.505329	0.505431
50	0.518632	0.505418	0.505431
100	0.512113	0.505428	0.505431
200	0.508793	0.505430	0.505431
500	0.506780	0.505431	0.505431
1000	0.506106	0.505431	0.505431
2000	0.505769	0.505431	0.505431

FIGURE 7.17   Predictions of $x(1)$ where $x' = x' - t^2$, $x(0) = 0.5$ (Exact solution unavailable)

The scheme we have described here is one of a large class of "Runge-Kutta schemes" with varying degrees of accuracy. This one, which is properly called a fourth-order Runge-Kutta scheme (because the error is proportional to $1/n^4$), is the one most often used in actual practice.

RUNGE-KUTTA SCHEME

To predict $x(\tau)$ via the Runge-Kutta method, where

(D)
$$x' = f(t, x)$$

(I)
$$x(t_0) = x_0$$

pick the number of steps n, define the step size $h = (\tau - t_0)/n$, and find x_n using the initial condition (I) together with the recurrence relation

$$x_i = x_{i-1} + \frac{h}{6}[k_1 + 2k_2 + 2k_3 + k_4],$$

where

$$k_1 = f(t_{i-1}, x_{i-1})$$

$$k_2 = f\left(t_{i-1} + \frac{h}{2}, x_{i-1} + \frac{hk_1}{2}\right)$$

$$k_3 = f\left(t_{i-1} + \frac{h}{2}, x_{i-1} + \frac{hk_2}{2}\right)$$

$$k_4 = f(t_{i-1} + h, x_{i-1} + hk_3).$$

Note

Error estimates for Runge-Kutta

Precise error estimates for the Runge-Kutta and other sophisticated schemes are very difficult to obtain. Usually, one attempts only to estimate the local error, which is the maximum difference between x_i and $\bar{x}(t_i)$, where $\bar{x}(t)$ is the exact solution of (D) whose value at t_{i-1} is x_{i-1}. To obtain an estimate of just the order of this error, one uses Taylor series without trying to estimate the size of the remainder. The extrapolation of the global error can then be found from the following:

Fact: *Suppose a numerical scheme of the form*

(H$_i$) $x_i = x_{i-1} + hg(t_{i-1}, x_{i-1}, h)$

satisfies a local error estimate

(L) $|x_i - \bar{x}(t_i)| < \lambda(h).$

Suppose further that the function g satisfies

$$|g(t, x, h) - g(t, y, h)| < M|x - y|$$

in the region described by the inequalities

$$t_0 \le t \le \tau, \qquad a \le x, y \le A, \qquad \text{and} \qquad h < H.$$

If $x(t)$ and x_i satisfy the foregoing estimates on x in this region, then the global error satisfies

$$|x_n - x(\tau)| \le \frac{\lambda(h)}{M} e^{ML},$$

where $L = \tau - t_0$ is the length of the interval over which we predict the solution.

EXERCISES

To facilitate comparison, Exercises 1 through 9 here repeat Exercises 1 through 9 in Sections 7.2 and 7.4, with Euler's method and the modified Euler method replaced by the Runge-Kutta method. The calculations in Exercise 1 should be performed by hand or on an unprogrammed calculator. In Exercises 2 through 9, use a computer.

1. Use the Runge-Kutta method with $n = 2$ to approximate $x(0.2)$, where $x(t)$ is the solution of $x' = 6t - 6x + 1$ that satisfies $x(0) = 0.1$.

In Exercises 2 through 7, use the Runge-Kutta method with $n = 10, 20, 50, 100, 200,$ and 500, to approximate $x(\tau)$, where $x(t)$ is the solution of the given initial-value problem. In Exercises 2 and 3, find the exact value of $x(\tau)$ and the error for each approximation.

2. $x' = -x^2(1 + t), \quad x(0) = 2$
   a. $\tau = 1$ b. $\tau = 2$ c. $\tau = 5$

3. $x' = \dfrac{2(t + 1)^3}{x}, \quad x(0) = 1$
   a. $\tau = 1$ b. $\tau = 2$ c. $\tau = 5$

4. $x' = x - e^t, \quad x(1) = 0$
   a. $\tau = 0$ b. $\tau = 2$ c. $\tau = 5$

5. $x' = 0.05x + \dfrac{1}{t + 1}, \quad x(0) = 10$
   a. $\tau = 1$ b. $\tau = 5$ c. $\tau = 10$

6. $x' = -0.1x + \dfrac{4}{1 + t^2}, \quad x(0) = 60$
   a. $\tau = 1$ b. $\tau = 7$ c. $\tau = 8$

7. $x' = 0.005 \left(10 + \dfrac{t}{10} - x \right) x, \quad x(0) = 10$
   a. $\tau = 1$ b. $\tau = 5$ c. $\tau = 10$

8. *Calculating ln 2:* Use the Runge-Kutta method with $n = 50$ and 100 to approximate $x(2)$, where $x(t)$ is the solution of $x' = 1/t$ that satisfies $x(1) = 0$.

9. *Calculating π:* Use the Runge-Kutta method with $n = 50$ and 100 to approximate $x(\tau)$, where $x(t)$ is the solution of the given initial-value problem.

a. $x' = \dfrac{6}{\sqrt{1 - t^2}}, \quad x(0) = 0; \qquad \tau = 0.5$

b. $x' = \dfrac{4}{1 + t^2}, \quad x(0) = 0; \qquad \tau = 1$

10. *Graphing* $f(t) = \int_0^t e^{s^2} \, ds$:
   a. Modify your computer program for the Runge-Kutta method so that it prints each x_i in turn. Use this modified program to find approximations to $x(0.1)$, $x(0.2), \ldots, x(0.9)$, and $x(1)$, where $x(t)$ is the solution of $x' = e^{t^2}$ that satisfies $x(0) = 0$.
   b. Using the values obtained in (a), sketch the graph of $x(t)$ for $0 \le t \le 1$.

11. *Simpson's Rule:* Show that if $Q(t) = at^2 + bt + c$, then

$$\int_{t_{i-1}}^{t_i} Q(t) \, dt = \frac{h}{6} [Q(t_{i-1}) + 4Q(\bar{t}_i) + Q(t_i)],$$

where $h = t_i - t_{i-1}$, and $\bar{t}_i = t_{i-1} + h/2$.

7.6 SOME WORDS OF WARNING

In this section we consider some examples that illustrate phenomena that can seriously affect the reliability of numerical predictions. The intelligent user of numerical schemes should be aware of such limitations when interpreting the results of numerical simulation of o.d.e.'s. We deal throughout with relatively simple examples, but keep in mind that the problems that come up in practice can be very complex, and detecting some of the phenomena we describe can be a very subtle problem in these cases.

Example 7.6.1   Roundoff Error

Based on our experience so far, we might guess (especially if we read the error analysis in Sections 7.3 and 7.4) that by taking n sufficiently large we can attain arbitrarily good accuracy with any numerical scheme. However, this reasoning assumes that the actual calculations in the recurrence relation for a given scheme are carried out exactly. This is rarely the case on real computers. Every time a computer performs an arithmetic operation on real numbers, it employs only a finite-digit approximation for each number entering the operation, and it stores only an equal number of digits of the answer, rounding or truncating the rest. This results in **roundoff error.**

The effect of roundoff error has been negligible in most of the examples we have seen so far in this chapter. However, when a large number of calculations is to be performed (because of small step size, or because we are predicting the solution over a long time interval, or because the function appearing on the right side of the o.d.e. is complicated), the cumulative effect of roundoff error can become visible. For example, Figure 7.18 shows the Runge-Kutta predictions for the solution at $t = 1$ to the initial-value problem

$$(D_4) \qquad\qquad x' = \frac{6x^3}{(t + 1)^{13}}$$

$$(I_4) \qquad\qquad x(0) = 1$$

for several values of n between 100 and 5000. You should check that the exact solution (obtained via separation of variables) satisfies $x(1) = 64$. The accuracy of our predictions is reasonably good and improves up to $n = 1000$. But then the error *quadruples* going from $n = 1000$ to $n = 1500$, and does not improve significantly after that. A fivefold increase in the number of steps (from $n = 1000$ to $n = 5000$) not only buys us no improvement, but leads to a deterioration in the accuracy because of accumulated roundoff error. This deterioration even affects the printed values for the prediction.

n	x_n	Error
100	63.9983	-1.65606×10^{-3}
500	64.0001	9.82285×10^{-5}
1000	64.0001	5.72205×10^{-5}
1500	64.0002	2.01225×10^{-4}
2000	64.0001	1.43051×10^{-4}
2500	64.0004	3.81470×10^{-4}
5000	63.9997	-2.77514×10^{-4}

FIGURE 7.18   Runge-Kutta Predictions for $x(1)$ where $x' = 6x^3/(t + 1)^{13}$, $x(0) = 1$ (Exact solution: $x(1) = 64$)

Roundoff error may vary from one calculating device to another. (You may already have noticed the effect of such differences when comparing your own calculations with values in the tables.) In particular, results obtained on devices that use decimal arithmetic (such as most calculators) may differ from those obtained on devices that use binary arithmetic (such as most computers). Minor differences can be observed resulting from seemingly innocent differences in the way we implement formulas. For example, in our algorithms we calculated the time t_i from the closed formula $t_i = t_0 + ih$ instead of using the recursive definition $t_i = t_{i-1} + h$, because the latter calculation is more prone to accumulated roundoff error.

Example 7.6.2   Singularities

Figure 7.19 shows predictions by the Runge-Kutta method for $x(1)$, where $x(t)$ satisfies our third sample o.d.e.

(D₃) $x' = x^2 - t^2$

with various initial conditions.

n	x_n with $x_0 = 0.5$	x_n with $x_0 = 1.0$	x_n with $x_0 = 1.5$
10	0.505434	24.3590	overflow
20	0.505431	26.2407	overflow
50	0.505431	26.6640	overflow
100	0.505431	26.6854	overflow
200	0.505430	26.6870	overflow
500	0.505431	26.6871	overflow

FIGURE 7.19   Runge-Kutta Predictions for $x(1)$ where $x' = x^2 - t^2$

The predictions for $x_0 = 0.5$ (investigated in Example 7.5.3) and for $x_0 = 1.0$ stabilize to several digits. Our computer printed overflow messages when we took $x_0 = 1.5$.

To understand the overflow, we try to estimate the growth of the exact solution $x(t)$ of (D₃) with initial condition

$$x(0) = 1.5$$

by means of differential inequalities. From the fact that

$$x'(t) = x^2 - t^2 \geq x^2 - 1, \quad 0 \leq t \leq 1$$

we can conclude (see Section 7.3) that the function $y(t)$ that satisfies

$$y'(t) = y^2 - 1 \quad \text{and} \quad y(0) = 1.5$$

is a lower bound for x:

$$x(t) \geq y(t), \quad 0 \leq t \leq 1.$$

Using separation of variables, we find that

$$y(t) = \frac{5 + e^{2t}}{5 - e^{2t}}.$$

Note that $y(t)$ tends to infinity as t approaches $(\ln 5)/2$, and that $(\ln 5)/2$ lies in the interval $0 \leq t \leq 1$. Since $x(t)$ is even larger than $y(t)$, it also blows up by the time t approaches $(\ln 5)/2$. Thus, the domain of this solution to (D_3) is at most $t < (\ln 5)/2$. Since $x(t)$ is not defined at $t = 1$, the prediction called for in the last column of Figure 7.19 is meaningless.

A similar analysis applied to the function $z(t)$ satisfying $z' = z^2$ and $z(0) = x(0) = x_0$ shows that $x(t) \leq z(t)$, and $z(t)$ is well defined and bounded for $0 \leq t \leq 1$, provided $x_0 \leq 1$. Thus, the solution $x(t)$ whose values are predicted by the first two columns of Figure 7.19 can be guaranteed to exist.

▪ Example 7.6.3   Unstable Problems

Figure 7.20 shows several predictions, from the modified Euler and Runge-Kutta schemes, for the value at $t = 1$ of the solution to the initial-value problem

(D_4) $$x' = 100x + 100t^2 - 100 - 2t$$

(I_4) $$x(0) = 1.$$

The values in the table suggest a phenomenon similar to the previous example. It is easy to check, however, that the exact solution to the problem is

$$x(t) = 1 - t^2,$$

which has value $x(1) = 0$. Thus, a different phenomenon is occurring here.

n	x_n (Modified Euler)	x_n (Runge-Kutta)
10	5.94453×10^{14}	3.99410×10^{24}
20	7.87547×10^{21}	2.05647×10^{32}
50	8.88183×10^{30}	overflow
100	2.07426×10^{35}	overflow

FIGURE 7.20 Predictions for $x(1)$ where $x' = 100(x + t^2 - 1) - 2t$, $x(0) = 1$ (Exact solution: $x(1) = 0$)

The presence (and persistence) of large numbers in our predictions is easier to understand if we note that the general solution to (D_4) is

$$x(t) = 1 - t^2 + Ce^{100t}.$$

This shows that solutions starting very near ours (i.e., with small but nonzero C) quickly spread apart. This sensitivity to initial conditions means that the error made in the first step of our scheme, which leads us to start predicting with $x(t_1) = x_1$ instead of the true value, becomes magnified by the equation itself. Such problems, in which slight changes in the initial data yield large changes in the exact solution, are often referred to as **unstable** or **ill-conditioned** problems.

Example 7.6.4   Stiff Problems

Figure 7.21 shows Euler, modified Euler, and Runge-Kutta predictions for the solution to

$$(D_5) \qquad\qquad\qquad x' = -75x$$

$$(I_5) \qquad\qquad\qquad x(0) = 1$$

for n varying from 10 to 1000. The exact solution is $x(t) = e^{-75t}$, whose value at $t = 1$ is $e^{-75} \approx 2.67864 \times 10^{-33}$. Unlike the previous example, this solution is stable; in fact, all exact solutions approach zero very quickly. Nonetheless, the predictions of both the Euler and modified Euler methods are erratic. The Runge-Kutta predictions eventually reach the right ballpark, but this takes far more steps than might be expected from the behavior of the exact solutions. Such equations are called **stiff**.

n	x_n (Euler)	x_n (Modified Euler)	x_n (Runge-Kutta)
10	1.34627×10^8	2.23646×10^{13}	1.57958×10^{19}
20	6.11863×10^8	4.27968×10^{12}	2.74586×10^{11}
50	8.88178×10^{-16}	6.22302×10^{-11}	6.96472×10^{-29}
100	underflow	3.38762×10^{-28}	3.88086×10^{-33}
200	underflow	2.72902×10^{-32}	2.72433×10^{-33}
500	5.12227×10^{-36}	3.66959×10^{-33}	2.67960×10^{-33}
1000	1.38590×10^{-34}	2.88549×10^{-33}	2.67869×10^{-33}

FIGURE 7.21   Euler, Modified Euler, and Runge-Kutta Predictions for $x(1)$ where $x' = -75x$, $x(0) = 1$
(Exact solution: $x(1) = e^{-75} \approx 2.67864 \times 10^{-33}$)

The phenomenon of stiffness is best understood by referring to Figure 7.22(a), where we have sketched the graphs of several solutions to an equation of this type and indicated how an Euler scheme would behave. The tangent line to a solution near $t = 0$ has large negative slope and greatly overshoots the appropriate solution (and even the t-axis). This means that the value of x_1 (for moderate h) is very far from correct, and again the tangent line to this solution has large slope and overshoots the appropriate value, and so on. You can see that for very small step sizes, this problem does not occur [Figure 7.22(b)].

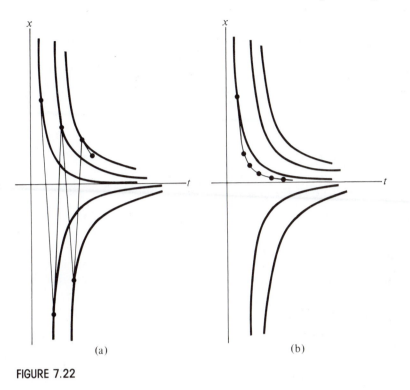

(a)   (b)

FIGURE 7.22

Note that the improved methods, which average the slope at several points, may cut down the magnitude of the problem. Nonetheless, to get reasonable answers, we have to push our scheme much harder than might be expected in normal problems.

Even more sophisticated schemes of various kinds have been developed for dealing with some of the difficulties noted in these examples. Some of these schemes use a simple basic recurrence relation, such as the ones defining the modified Euler and Runge-Kutta schemes, but use more sophisticated means to extrapolate an answer from this formula. Some programs modify the step size when stiffness (or instability) seems to be present. There are also multistep methods, such as the Adams-Bashforth scheme, in which preliminary guesses are made with a Runge-Kutta or other sophisticated scheme (which is repeated with decreasing h until the prediction stabilizes); then these guesses are used to make an averaged prediction of the next x value. The development and use of such schemes is beyond the scope of our discussion. We note that as an all-around method to attain moderate accuracy (say, to 10^{-4}) on many problems, the Runge-Kutta scheme of Section 7.5 is a reasonable choice.

WARNING

When interpreting the predictions of a numerical scheme for solving o.d.e.'s, watch out for

1.   roundoff error
2.   singularities
3.   instability
4.   stiffness

When any of these seems to be a serious problem, consult a numerical analyst or other expert.

EXERCISES

1.   Investigate the errors calculated in Exercises 2 through 4 of Sections 7.4 and 7.5 to determine whether the effect of roundoff is observable. In particular, see if there are any cases where roundoff has affected the printed values.

2.   The solution of the initial-value problem $x' = 1/\sqrt{1 - t^2}, x(0) = 0$, is $x = \arcsin t$, whose value at $t = 1$ is $x(1) = \arcsin(1) = \pi/2$. Although we can use the modified Euler and Runge-Kutta methods to approximate $x(\tau)$ for values of τ just short of 1, we cannot take $\tau = 1$. Why not?

3.   Let $x_1(t) = e^{9t}$ and $x_2(t) = e^{9t} + e^{-9t}$. Use your calculator to find $x_1(\tau)$, $x_2(\tau)$, and $x_1(\tau) - x_2(\tau)$ for $\tau = 1$ and 2. (Note that it will be difficult for any numerical scheme for second-order o.d.e.'s to distinguish between these two linearly independent solutions of $(D^2 - 81)x = 0$ on $t \geq 1$.)

4.   Let $x(t)$ be the solution of $x' = 9x - 18e^{-9t}$ that satisfies $x(0) = 1$.
     a.   Use the modified Euler and Runge-Kutta methods with $n = 10, 100, 500$, and 1000 to approximate $x(1)$.
     b.   Check that the exact value is $x(1) = e^{-9}$ and that in each case the error $x_n - x(1)$ is of the same order of magnitude as x_n itself.
     c.   How do you explain this?

5.   Let $x(t)$ be the solution of the initial-value problem $x' = 5x^4/t^{16}$, $x(1) = 1$.
     a.   Check that $x(t) = t^5$, so that $x(2) = 32$.
     b.   Use the modified Euler and Runge-Kutta methods with $n = 50, 100, 200, 500$, 1000, 2000, 3000, 4000, and 5000 to approximate $x(2)$, and calculate the error $x_n - x(2)$ for each of these approximations. (Note that when $n = 1000$, the modified Euler prediction differs from the true value by more than 10%, and even when $n = 5000$ it differs from the true value in the third digit. The Runge-Kutta prediction improves to about 0.5% when $n = 500$, but fails to improve with larger n, and in fact the error when $n = 5000$ is more than twice as large as when $n = 500$.)

6.   Let $x(t)$ be the solution of $x' = 1 + x^2$ that satisfies $x(0) = 0$.

a. Use the modified Euler and Runge-Kutta methods, with $n = 50$ and 100, to look for an approximation to $x(2)$.
b. Find the exact value of $x(t)$ and explain why the attempt to find $x(2)$ was doomed to failure.

7. Let $x(t)$ be the solution of $x' = 1 + x^2 + t$ that satisfies $x(0) = 0$.
    a. Check that attempts to find $x(2)$ by the modified Euler and Runge-Kutta methods, with $n = 50$ and 100, result in very large numbers or overflow messages.
    b. Use the fact that for $0 \leq t \leq 2$ we have $1 + x^2 \leq 1 + x^2 + t$ to show that $x(t)$ becomes unbounded before t reaches 2.

8. The o.d.e. of Example 7.6.2 is of the form $x' = f(t, x)$ with both f and $\partial f / \partial x$ continuous for all values of t and x. Yet the solution that satisfies $x(0) = 0$ isn't defined for all t. Why doesn't this contradict the existence and uniqueness theorem discussed in Section 1.6?

7.7 A RUNGE-KUTTA SCHEME FOR DIFFERENTIAL SYSTEMS

In this section, we consider briefly how the Runge-Kutta scheme for a single first-order o.d.e. can be extended to handle a system of two o.d.e.'s

$$(S) \qquad \begin{aligned} x' &= f(t, x, y) \\ y' &= g(t, x, y) \end{aligned}$$

such as those obtained from the models in Section 7.1. The reader familiar with Chapter 3 may note that the extension we discuss here can be carried out for systems of any order by simply rewriting our earlier formulas in vector notation.

A solution of the system (S) is given by a pair of functions $x = x(t)$ and $y = y(t)$ satisfying the two equations simultaneously. The initial conditions that determine a specific solution have the form

$$(I) \qquad x(t_0) = x_0, \quad y(t_0) = y_0.$$

Numerical schemes for solving a single first-order o.d.e. adapt to solving (S) by replacing each x-value with a pair of values (one for x and one for y) and each function evaluation with a pair of evaluations [one for $f(t, x, y)$ and one for $g(t, x, y)$].

A quick review of the discussion of the Runge-Kutta method in Section 7.5, viewed with an eye toward (S), leads to the following recurrence relation for an approximate solution to (S):

$$(RK_i) \qquad \begin{cases} x_i = x_{i-1} + \dfrac{h}{6} (k_1 + 2k_2 + 2k_3 + k_4) \\[2mm] y_i = y_{i-1} + \dfrac{h}{6} (\ell_1 + 2\ell_2 + 2\ell_3 + \ell_4), \end{cases}$$

where

$$
\begin{cases}
k_1 = f(t_{i-1}, x_{i-1}, y_{i-1}) \\
\ell_1 = g(t_{i-1}, x_{i-1}, y_{i-1})
\end{cases}
$$

$$
\begin{cases}
k_2 = f\left(t_{i-1} + \dfrac{h}{2}, x_{i-1} + \dfrac{hk_1}{2}, y_{i-1} + \dfrac{h\ell_1}{2}\right) \\[2mm]
\ell_2 = g\left(t_{i-1} + \dfrac{h}{2}, x_{i-1} + \dfrac{hk_1}{2}, y_{i-1} + \dfrac{h\ell_1}{2}\right) \\[2mm]
k_3 = f\left(t_{i-1} + \dfrac{h}{2}, x_{i-1} + \dfrac{hk_2}{2}, y_{i-1} + \dfrac{h\ell_2}{2}\right) \\[2mm]
\ell_3 = g\left(t_{i-1} + \dfrac{h}{2}, x_{i-1} + \dfrac{hk_2}{2}, y_{i-1} + \dfrac{h\ell_2}{2}\right)
\end{cases}
$$

$$
\begin{cases}
k_4 = f(t_{i-1} + h, x_{i-1} + hk_3, y_{i-1} + h\ell_3) \\
\ell_4 = g(t_{i-1} + h, x_{i-1} + hk_3, y_{i-1} + h\ell_3).
\end{cases}
$$

Here, $h = (\tau - t_0)/n$, and our approximations to $x(\tau)$ and $y(\tau)$ are given by x_n and y_n, respectively.

Let's use this Runge-Kutta scheme on several initial-value problems related to our models in Section 7.1.

Example 7.7.1

Use the Runge-Kutta method to predict $x(1)$ and $y(1)$, where

$$(S_1) \qquad \begin{aligned} x' &= y \\ y' &= -9.81 \sin x \end{aligned}$$

and

$$(I_1) \qquad x(0) = \frac{\pi}{4}, \quad y(0) = 0.$$

This models the pendulum of Example 7.1.1, with an arm 1 meter long, starting at rest from a 45° angle with the vertical.

In Figure 7.23 we give a BASIC program for the Runge-Kutta scheme, as applied to (S_1). To adapt this program to a different system, we need only change the definitions of $f(t, x, y)$ and $g(t, x, y)$ in lines 10 and 15. If we input the initial data t_0, τ, x_0, y_0, and n the computer will calculate and print x_n and y_n.

```
10   DEF FNF(T, X, Y) = Y
15   DEF FNG(T, X, Y) = -9.81*SIN(X)
20   INPUT T0, TN, X, Y, N
30   H = (TN - T0)/N
40   FOR I = 1 TO N
50   K1 = FNF(T0 + (I - 1)*H, X, Y)
55   L1 = FNG(T0 + (I - 1)*H, X, Y)
60   K2 = FNF(T0 + (I - 1/2)*H, X + H*K1/2, Y + H*L1/2)
65   L2 = FNG(T0 + (I - 1/2)*H, X + H*K1/2, Y + H*L1/2)
70   K3 = FNF(T0 + (I - 1/2)*H, X + H*K2/2, Y + H*L2/2)
75   L3 = FNG(T0 + (I - 1/2)*H, X + H*K2/2, Y + H*L2/2)
80   K4 = FNF(T0 + I*H, X + H*K3, Y + H*L3)
85   L4 = FNG(T0 + I*H, X + H*K3, Y + H*L3)
90   X = X + (H/6)*(K1 + 2*K2 + 2*K3 + K4)
95   Y = Y + (H/6)*(L1 + 2*L2 + 2*L3 + L4)
100  NEXT I
110  PRINT X, Y
999  END
```

Initial data: t_0, τ, x_0, y_0, n

FIGURE 7.23   BASIC Program for the Runge-Kutta Method for $\begin{cases} x' = y \\ y' = -9.81 \sin x \end{cases}$

Figure 7.24 shows the outcome of trials with various choices for n, using the value 0.785398 for $x_0 = \pi/4$. The stability of these predictions suggests that we can be reasonably confident of the values $x(1) = -0.778954$ and $y(1) = -0.298522$.

n	x_n	y_n
10	-0.778893	-0.298911
50	-0.778954	-0.298523
100	-0.778954	-0.298522
500	-0.778954	-0.298522

FIGURE 7.24   Predictions for $x(1)$ and $y(1)$
where $\begin{cases} x' = y \\ y' = -9.81 \sin x \end{cases}$ and $x(0) = \pi/4$, $y(0) = 0$

In Example 2.10.2 we modeled the same pendulum by the linear o.d.e.

(H) $$[D^2 + 9.81]x = 0.$$

The exact solution of this equation satisfying

(I) $$x(0) = \frac{\pi}{4}, \quad x'(0) = 0$$

is $x(t) = (\pi/8)(e^{t\sqrt{9.81}} + e^{-t\sqrt{9.81}})$, with value at $t = 1$

$$x(1) = \frac{\pi}{8}(e^{\sqrt{9.81}} + e^{-\sqrt{9.81}}) \approx 9.018534.$$

We see that for relatively large displacement, the behavior of the linear model differs strikingly from that of the nonlinear one. Note that the approximation we used in deriving (H), that $\sin x \approx x$, is far from true when $x = \pi/4$.

Example 7.7.2

Use the Runge-Kutta method to predict $x(1)$ and $y(1)$, where

$$(S_2) \qquad \begin{aligned} x' &= y \\ y' &= 1 - x - y|y| \end{aligned}$$

and

$$(I_2) \qquad x(0) = 1, \quad y(0) = 1.$$

This is the system of Example 7.1.2, with $m = b = k = E(t) = 1$.
    Figure 7.25 shows the predictions for various numbers of steps. These predictions are even more stable than the ones in Example 7.7.1; they all agree to six significant figures.

n	x_n	y_n
10	1.59432	0.240829
50	1.59432	0.240829
100	1.59432	0.240829
500	1.59432	0.240829

FIGURE 7.25   Predictions for $x(1)$ and $y(1)$
where $\begin{cases} x' = y \\ y' = 1 - x - y|y| \end{cases}$ and $x(0) = y(0) = 1$

Example 7.7.3

Use the Runge-Kutta method to predict $x(1)$ and $y(1)$, where

$$(S_3) \qquad \begin{aligned} x' &= y \\ y' &= y - x^2y - x + 0.01 \sin t \end{aligned}$$

and

(I_3) $x(0) = 1, \quad y(0) = 0.$

This system represents the Van der Pol equation with $\epsilon = K = \omega = 1$ and $A = 0.01$.

The predictions for this problem are given in Figure 7.26. Here again, the predictions are quite stable.

n	x_n	y_n
10	0.499347	-1.03876
50	0.499346	-1.03876
100	0.499346	-1.03876
500	0.499346	-1.03876

FIGURE 7.26 Predictions for $x(1)$ and $y(1)$

where $\begin{cases} x' = y \\ y' = y - x^2y - x + 0.01 \sin t \end{cases}$ and $x(0) = 1, y(0) = 0$

Let's summarize.

A RUNGE-KUTTA SCHEME FOR DIFFERENTIAL SYSTEMS

To adapt a numerical scheme designed for solving a single first-order o.d.e. to solving the system

(S) $\begin{aligned} x' &= f(t, x, y) \\ y' &= g(t, x, y) \end{aligned}$

we replace each x-value with a pair of values (x and y) and each function evaluation with a pair (f and g). For the Runge-Kutta method, the resulting recurrence relation is given by the equations

(RK_i) $\begin{cases} x_i = x_{i-1} + \dfrac{h}{6}(k_1 + 2k_2 + 2k_3 + k_4) \\[2mm] y_i = y_{i-1} + \dfrac{h}{6}(\ell_1 + 2\ell_2 + 2\ell_3 + \ell_4), \end{cases}$

where

$$\begin{cases} k_1 = f(t_{i-1}, x_{i-1}, y_{i-1}) \\ \ell_1 = g(t_{i-1}, x_{i-1}, y_{i-1}) \end{cases}$$

$$\begin{cases} k_2 = f\left(t_{i-1} + \dfrac{h}{2}, x_{i-1} + \dfrac{hk_1}{2}, y_{i-1} + \dfrac{h\ell_1}{2}\right) \\ \ell_2 = g\left(t_{i-1} + \dfrac{h}{2}, x_{i-1} + \dfrac{hk_1}{2}, y_{i-1} + \dfrac{h\ell_1}{2}\right) \end{cases}$$

$$\begin{cases} k_3 = f\left(t_{i-1} - \dfrac{h}{2}, x_{i-1} + \dfrac{hk_2}{2}, y_{i-1} + \dfrac{h\ell_2}{2}\right) \\ \ell_3 = g\left(t_{i-1} + \dfrac{h}{2}, x_{i-1} + \dfrac{hk_2}{2}, y_{i-1} + \dfrac{h\ell_2}{2}\right) \end{cases}$$

$$\begin{cases} k_4 = f(t_{i-1} + h, x_{i-1} + hk_3, y_{i-1} + h\ell_3) \\ \ell_4 = g(t_{i-1} + h, x_{i-1} + h\ell_3, y_{i-1} + h\ell_3). \end{cases}$$

EXERCISES

The calculations in Exercise 1 should be performed by hand or on an unprogrammed calculator. Those in Exercises 2 through 8 are best done on a computer.

1. Use the Runge-Kutta method with $n = 2$ to approximate $x(0.2)$ and $y(0.2)$, where $x' = y$, $y' = x + t$, and $x(0) = y(0) = 1$.

In Exercises 2 through 8, use the Runge-Kutta method with $n = 10, 50, 100,$ and 500 to approximate $x(\tau)$ and $y(\tau)$, where the pair of functions $x(t)$ and $y(t)$ is the solution of the given initial-value problem.

2. $x' = y$, $y' = -9.81 \sin x$, $x(0) = 0.1$, $y(0) = 0$
    a. $\tau = 1$ b. $\tau = 5$ c. $\tau = 10$

    (Note that this models the pendulum of Example 7.1.1, with an arm 1 meter long, starting at rest from an angle of 0.1 radian with the vertical.)

3. $x' = y$, $y' = -9.81x$, $x(0) = 0.1$, $y(0) = 0$
    a. $\tau = 1$ b. $\tau = 5$ c. $\tau = 10$

    (Note that this is a linear approximation to the system of the preceding exercise.)

4. $x' = y$, $y' = -y - 9.81 \sin x$, $x(0) = 0.1$, $y(0) = 0$
    a. $\tau = 1$ b. $\tau = 5$ c. $\tau = 10$

    (Note that this models the damped pendulum of Exercise 5, Section 7.1.)

5. $x' = y$, $y' = -8x - 6y + 64\left[9 - \dfrac{10}{t + 1}\right]$, $x(0) = y(0) = 0$.

   a. $\tau = 1$ b. $\tau = 5$ c. $\tau = 10$

   (compare Exercise 4, Section 7.1)

6. $x' = \dfrac{y - x}{50 + 5t}$, $y' = \dfrac{x - y}{50 + 5t}$, $x(0) = 0$, $y(0) = 0.1$

   a. $\tau = 1$ b. $\tau = 5$ c. $\tau = 10$

   (compare Exercise 6, Section 7.1)

7. $x' = 0.1x - 0.00005xy$, $y' = -0.08y + 0.00001xy$, $x(0) = 10,000$, $y(0) = 1000$

   a. $\tau = 10$ b. $\tau = 25$
   c. $\tau = 50$ d. $\tau = 75$

   (compare Exercise 7, Section 7.1)

8. *Graphing* sin *t and* cos *t:*
   a. Modify your program for the Runge-Kutta method so that it prints each pair x_i and y_i in turn. Use this modified program to find approximations to $x(\tau)$ and $y(\tau)$ at $\tau = 0.5, 1.0, 1.5, \ldots, 5.0$, where $x(t)$ and $y(t)$ satisfy $x' = y$, $y' = -x$ and the initial conditions $x(0) = 0$, $y(0) = 1$.
   b. Using the values obtained in part (a), sketch the graphs of $x(t) = \sin t$ and $y(t) = \cos t$ for $0 \le t \le 5$.

REVIEW PROBLEMS

In Problems 1 through 8:

i. Use the Euler, modified Euler, and Runge-Kutta methods, with $n = 10, 20, 50, 100, 200$, and 500 to approximate $x(\tau)$, where $x(t)$ is the solution of the given initial-value problem.

ii. Interpret your data. Have the predictions stabilized enough for you to be reasonably confident of the value of $x(\tau)$ to four significant figures? If not, is it likely that further trials with larger choices for n will provide enough data to allow a more confident estimate of this value?

1. $x' = e^{t^2}$, $x(0) = 1$
    a. $\tau = 0.1$ b. $\tau = 0.2$ c. $\tau = 1$
2. $x' = 2tx + 1$, $x(0) = 1$
    a. $\tau = 1$ b. $\tau = 2$ c. $\tau = 5$
3. $x' = 2tx + \cos x$, $x(0) = 1$
    a. $\tau = 1$ b. $\tau = 2$ c. $\tau = 3$
4. $x' = 2tx + \sin x$, $x(0) = 1$
    a. $\tau = 1$ b. $\tau = 2$ c. $\tau = 3$
5. $x' = \sqrt{1 + xt}$, $x(0) = 1$
    a. $\tau = 1$ b. $\tau = 5$ c. $\tau = 10$

6. $x' = x^2 - x \cos t, \quad x(0) = 1$
    a. $\tau = 1$ b. $\tau = 1.5$ c. $\tau = 2$
7. $x' = x^2 - x \sin t, \quad x(0) = 1$
    a. $\tau = 1$ b. $\tau = 1.1$ c. $\tau = 1.25$
8. $x' = x^2 - xe^t, \quad x(0) = 1$
    a. $\tau = 1$ b. $\tau = 2$ c. $\tau = 3$

In Problems 9 through 11:

(i)   Use the Runge-Kutta method for systems, with $n = 10, 50,$ and 100, to approximate $x(1)$ and $y(1)$, where the pair of functions $x(t)$ and $y(t)$ is the solution of the given initial-value problem.

(ii) Interpret your data, as in Problems 1 through 8.

9. $x' = y, \quad y' = x^2 + y$
    a. $x(0) = 1, \quad y(0) = 0$ b. $x(0) = 0, \quad y(0) = 1$
10. $x' = y, \quad y' = x^2 + y^2$
    a. $x(0) = 1, \quad y(0) = 0$ b. $x(0) = 0, \quad y(0) = 1$
11. $x' = xy, \quad y' = x + y + 1$
    a. $x(0) = 1, \quad y(0) = 0$ b. $x(0) = 0, \quad y(0) = 1$

In Problems 12 through 15, use the Runge-Kutta method, with $n = 50$ and 100, to approximate the desired values. Compare these approximations to the exact values, which can be obtained by the methods of Chapters 1 through 3.

12. *Compound Interest:*
    a. Suppose \$10,000 is deposited in a bank account accruing interest at 8% per year, compounded continuously (see Exercise 18, Section 1.1). Find the amount of money in the account after
        (i) 1 year   (ii) 5 years   (iii) 10 years
    b. Repeat (a) under the assumption that the interest rate is 8.5% per year.

13. *A Mixing Problem:* How much oil is left in the tanker of Exercise 24, Section 1.1, after 240 hours? (Compare Exercise 22, Section 1.2.)

14. *A Damped Spring:*
    a. Replace the o.d.e. of Example 2.1.2 with a system of o.d.e.'s for x and $y = x'$.
    b. Find the position of the mass after 10 seconds if $m = b = k = 1, E(t) = 0$, and
        (i) $x(0) = 0, \quad x'(0) = 5$ (ii) $x(0) = x'(0) = -1$
        (iii) $x(0) = 1, \quad x'(0) = 0$

    (Compare Exercise 5(a), Section 2.8.)

15. *An LRC Circuit:* Suppose the circuit of Example 3.1.1 (or Exercise 5, Section 2.1), with $V = 0, R = 5, L = 1/2$, and $C = 1/8$, starts with $I(0) = 0$ and $Q(0) = 1$. Find the charge on the capacitor after 1 second. (Compare Exercise 14, Section 3.7.)

APPENDIX

B

EXISTENCE AND UNIQUENESS OF SOLUTIONS

We observed in the note at the end of Section 7.2 the practical importance of knowing in advance that a given initial-value problem has a unique solution. In some cases we can exhibit a solution, so existence is not an issue. But even in these cases, it is important to know whether this is the *only* solution. Furthermore, the cases in which numerical schemes are most useful are generally those in which solutions cannot be exhibited, so existence is also an issue.

The existence and uniqueness theorem quoted in Section 1.6, gives fairly general hypotheses that ensure both existence and uniqueness for the solution of an initial-value problem involving a first-order o.d.e.:

Theorem: *Suppose that a first-order o.d.e. can be written in the form* $dx/dt = f(t, x)$, *with both* $f(t, x)$ *and* $\partial f/\partial x$ *continuous in some rectangular region of the tx-plane. Then for any real numbers* t_0 *and* α, *with* (t_0, α) *in the region, there is an open interval containing* t_0 *on which there exists precisely one solution of the o.d.e. that satisfies the initial condition* $x(t_0) = \alpha$.

In this appendix, we shall see how ideas from this chapter can be adapted to prove this theorem. Although a thoroughly rigorous proof requires sophisticated ideas from real analysis, we can sketch the basic ideas of the proof here. The proof we sketch can be extended to apply to an o.d.e. of any order by converting the o.d.e. into an equivalent system of first-order o.d.e.'s as in Section 3.2 or Section 7.1.

We begin as at the start of Section 7.4 by transforming the initial-value o.d.e. problem

(A) $$x' = f(t, x), \quad x(t_0) = \alpha$$

into an integral equation. The solutions of (A) are precisely the same as the

637

continuous functions $x(t)$ that satisfy

(B)
$$x(t) = \alpha + \int_{t_0}^{t} f(s, x(s)) \, ds.$$

Thus we can prove the theorem by showing the existence and uniqueness of a solution to (B) on an open interval I of the form $t_0 - T < t < t_0 + T$.

Our arguments are simplified if we assume that f and $\partial f / \partial x$ are continuous for all t and x, and that two inequalities hold for all points (t, x):

(i)
$$|f(t, x) - f(t, \bar{x})| \le M|x - \bar{x}|$$

(ii)
$$MT = L, \quad \text{where } 0 < L < 1.$$

We will discuss the relation of these simplifying assumptions to the hypothesis of the theorem at the end of the argument.

We first tackle the uniqueness question, under the assumption that (i) and (ii) hold. Given two functions $\phi_1(t)$ and $\phi_2(t)$ that satisfy (B) for all t in the interval I,

$$\phi_i(t) = \alpha + \int_{t_0}^{t} f(s, \phi_i(s)) \, ds, \quad i = 1, 2$$

how do we know that $\phi_1(t) = \phi_2(t)$ for all t in I? We estimate $|\phi_1(t) - \phi_2(t)|$ using basic properties of integrals together with (i):

$$|\phi_1(t) - \phi_2(t)| = \left| \int_{t_0}^{t} [f(s, \phi_1(s)) - f(s, \phi_2(s))] \, ds \right|$$

$$\le \left| \int_{t_0}^{t} |f(s, \phi_1(s)) - f(s, \phi_2(s))| \, ds \right|$$

$$\le \left| \int_{t_0}^{t} M|\phi_1(s) - \phi_2(s)| \, ds \right|$$

If we denote the maximum of $|\phi_1(t) - \phi_2(t)|$ on the interval I by $\| \phi_1 - \phi_2 \|$, we see that for t in I,

$$|\phi_1(t) - \phi_2(t)| \le \left| \int_{t_0}^{t} M \| \phi_1 - \phi_2 \| \, ds \right| = M |t - t_0| \| \phi_1 - \phi_2 \|.$$

Note, in particular, that for t in I, $|t - t_0| \le T$, so

$$|\phi_1(t) - \phi_2(t)| \le MT \| \phi_1 - \phi_2 \|.$$

Now, the right side is independent of t, so taking the maximum on I of the left side, we have

$$\| \phi_1 - \phi_2 \| \leq MT \| \phi_1 - \phi_2 \|.$$

Since $MT = L$ and $0 < L < 1$, this implies $\| \phi_1 - \phi_2 \| = 0$, so

$$\phi_1(t) = \phi_2(t) \quad \text{for all } t \text{ in } I.$$

This establishes the uniqueness of solutions to (B) [and hence also (A)], assuming (i) and (ii).

To show that (B) actually has a solution, we use an algorithm known as **Picard's method** to construct a sequence of approximate solutions. Starting with the constant function

$$x_0(t) = \alpha$$

we construct a sequence of functions, defined for t in I, via a recurrence relation obtained by modifying (B):

$$x_i(t) = \alpha + \int_{t_0}^{t} f[s, x_{i-1}(s)] \, ds.$$

Our object is to show the existence at each fixed t in I of the pointwise limit

$$x(t) = \lim_{n \to \infty} x_n(t)$$

and to show that the function $x(t)$ so defined is continuous and satisfies (B).

We first note that $x_n(t)$ is the nth partial sum of the series

(S)
$$x_0(t) + \sum_{i=1}^{\infty} [x_i(t) - x_{i-1}(t)].$$

Thus, to show that the sequence $\{x_n(t)\}$ converges, it suffices to show that this series converges. By an argument similar to the one used in the uniqueness proof, we can show that for $i \geq 2$,

$$\| x_i - x_{i-1} \| \leq L \| x_{i-1} - x_{i-2} \|$$

(where again we denote the maximum over I of a function $|g(t)|$ by $\| g \|$). It follows from this inequality that

$$\| x_i - x_{i-1} \| \leq L^{i-1} \| x_1 - x_0 \|, \quad i = 1, 2, \ldots .$$

We now see that the absolute value of each term of (S) is less than or equal to the corresponding term of the series

$$|x_0(t)| + \sum_{i=1}^{\infty} L^{i-1} \| x_1 - x_0 \|.$$

Since $L < 1$, the ratio test (see Section 5.2) can be used to show that this series converges. By comparison, our series (S) is absolutely convergent, so its partial sums $\{x_n(t)\}$ converge. The reader acquainted with real analysis may note that this argument actually shows that the sequence $\{x_n(t)\}$ is uniformly convergent. Since each $x_n(t)$ is continuous, it follows that the function $x(t)$ is also continuous.

To see that $x(t)$ satisfies (B), we note that for each $i \geq 1$,

$$\left| x(t) - \left[\alpha + \int_{t_0}^{t} f(s, x(s)) \, ds \right] \right|$$
$$\leq |x(t) - x_i(t)| + \left| x_i(t) - \left[\alpha + \int_{t_0}^{t} f(s, x(s)) \, ds \right] \right|.$$

Using the definition of $x_i(t)$ and an argument similar to the one used in the uniqueness proof, we see that

$$\left| x(t) - \left[\alpha + \int_{t_0}^{t} f(s, x(s)) \, ds \right] \right|$$
$$\leq |x(t) - x_i(t)| + \left| \int_{t_0}^{t} [f(s, x_{i-1}(s)) - f(s, x(s))] \, ds \right|$$
$$\leq |x(t) - x_i(t)| + MT \| x_{i-1} - x \|.$$

As $i \to \infty$, the right side of this inequality goes to zero. Hence the left side, which is independent of i, must equal zero. Thus

$$x(t) = \alpha + \int_{t_0}^{t} f(s, x(s)) \, ds$$

as desired.

Recall that our argument so far made use of several simplifying assumptions: that f and $f_x = \partial f / \partial x$ are continuous on the whole tx-plane and that the inequalities (i) and (ii) hold universally. Now, if we assume continuity of f and f_x only in some rectangular region of the plane,

$$t_1 < t < t_2, \quad x_1 < x < x_2,$$

then given a point (t_0, α) in this region, it is possible to find a smaller (bounded) rectangular region R of the form

$$t_0 - T \le t \le t_0 + T, \quad \alpha - X \le x \le \alpha + X$$

that is contained in the original region. Continuity guarantees that each of f_x and f has a finite maximum on R:

$$M = \max |f_x(t, x)| \quad \text{and} \quad N = \max |f(t, x)|.$$

It follows from the mean-value theorem that (i) holds for any pair of points $(t, \bar{x})$ in R. Furthermore, if we relace T with a smaller value, then (i) will continue to hold with the same value of M. By taking T small enough, we will have (ii) as well. Finally, we note that we may have to further restrict T to make sure that the points (t, x) that arise in the argument do not lie either above or below the region R. For the uniqueness argument, we note that any solution of (B) has $\phi(t_0) = \alpha$ and has slope bounded by N. If we take T small enough so that $NT < X$, then the points $(t, \phi(t))$ all lie in R, as long as t is in I. Similar but somewhat more involved arguments show that by taking T small enough, we can also ensure that for our sequence $\{x_i(t)\}$ of functions, the points $(t, x_i(t))$ lie in R whenever t is in I.

PARTIAL DIFFERENTIAL EQUATIONS: FOURIER SERIES

8.1 MODELS OF HEAT FLOW

We turn our attention in this chapter to situations modeled by functions of several variables, solving *partial* differential equations (p.d.e.'s). In this section, we look at some models of heat flow. We considered some steady-state temperature distributions in Section 6.1. Here, our temperature distributions are allowed to vary with time.

Example 8.1.1 One-Dimensional Heat Flow

We consider first the temperature in a wire of length ℓ that is insulated so that heat cannot flow through its sides. We assume that the wire is of uniform density, that it has a constant cross section, and that the temperature at any point depends only on the distance x from the left end of the wire [Figure 8.1(a)] and the time t. We denote the temperature by $u = u(x, t)$.

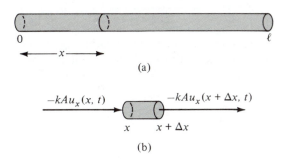

FIGURE 8.1

Our analysis proceeds from two principles of heat flow, which refine Newton's law of cooling (see Section 1.4):

1. The heat flow across a small surface is proportional to the rate of decrease of temperature across the surface times its area.
2. The temperature of a body increases in proportion to the flow of heat into the body divided by its mass.

Let's start by applying the first principle to a small piece of the wire with left endpoint x and right endpoint $x + \Delta x$ [Figure 8.1(b)]. Since the sides of the wire are insulated, heat can flow in or out of the piece only through the ends. The first principle tells us that the heat flow *into* the left end is proportional to $-u_x(x, t)$ times the cross-sectional area A (here $u_x = \partial u/\partial x$), and the heat flow *out* from the right end is proportional to $-u_x(x + \Delta x, t)A$. The net flow into our piece of wire is given by

$$\Delta Q = kA[u_x(x + \Delta x, t) - u_x(x, t)].$$

Using the mean value theorem, we can replace the quantity in brackets by Δx times the derivative of $\partial u/\partial x$ at an intermediate point $\bar{x}$:

$$\Delta Q = kA(\Delta x)u_{xx}(\bar{x}, t), \qquad x < \bar{x} < x + \Delta x.$$

The second principle relates ΔQ to the mass of the piece and to the rate of change of the temperature of the piece. The mass is $\delta A \Delta x$, where δ is the density of the wire. The rate of change of the temperature will generally be different at different points of the wire, but if Δx is small we expect it to be approximately the same throughout the piece. Thus, the second principle gives

$$u_t(x, t) = \frac{s \, \Delta Q}{\delta A \, \Delta x}.$$

Combining our expressions, we have

$$u_t(x, t) = \frac{skA(\Delta x)u_{xx}(\bar{x}, t)}{\delta A \, \Delta x} = \frac{sk}{\delta} u_{xx}(\bar{x}, t).$$

To obtain successively better approximations to the behavior of the temperature, we apply this analysis to successively shorter pieces of wire. In the limit as $\Delta x \to 0$, the points x and $\bar{x}$ merge and we obtain the **one-dimensional heat equation**

$$\frac{\partial^2 u}{\partial x^2} = \frac{1}{\alpha} \frac{\partial u}{\partial t},$$

where $\alpha = sk/\delta$.

In specific physical situations, we usually know the temperature distribution at some particular time $t = 0$. This is an *initial value* for u:

(I) $$u(x, 0) = f(x), \quad 0 < x < \ell.$$

In addition, we usually have some *boundary conditions* on the behavior of u for all time. For example, if each end is maintained at $0°$, we have boundary conditions of the form

(B₁) $$u(0, t) = u(\ell, t) = 0 \quad \text{for all } t.$$

On the other hand, if the ends are simply insulated, then there is no heat flow across the ends; by the first principle, this amounts to boundary conditions of the form

(B₂) $$u_x(0, t) = u_x(\ell, t) = 0 \quad \text{for all } t.$$

A third kind of boundary condition arises if we assume a slightly different physical configuration. We bend the wire around and join the ends so that we are looking at heat flow in a ring (Figure 8.2). In this situation, it proves convenient to denote the length by 2ℓ and parametrize location by $-\ell \leq x \leq \ell$. The *periodic boundary conditions* imposed by this configuration demand that the values of u and u_x at $x = \ell$ agree with those at $x = -\ell$:

(B₃) $$u(\ell, t) - u(-\ell, t) = u_x(\ell, t) - u_x(-\ell, t) = 0 \quad \text{for all } t.$$

We shall study the solutions of the one-dimensional heat equation with various initial conditions (I) and boundary conditions (B₁), or (B₂), or (B₃) in the next few sections. Other boundary conditions are considered in Exercise 1.

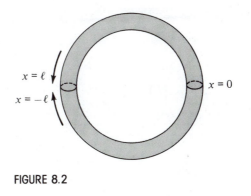

FIGURE 8.2

Example 8.1.2   Two-Dimensional Heat Flow

We consider next the problem of heat flow in a thin plate, insulated on its surfaces. A point in the plate is located by Cartesian coordinates (x, y), and the temperature is a function of three variables, $u = u(x, y, t)$.

We start by applying the first principle of heat flow to a small rectangular plate, with one corner at (x, y) and dimensions Δx by Δy (Figure 8.3). To do so, we need to analyze the heat flow to the right across the vertical edges and upward across the horizontal edges of our piece. The rate of decrease of temperature across the vertical edges is given at each point by $-u_x$. Although this may vary with the point, when Δy is small we expect it to be approximately the same as $-u_x(x, y, t)$ along the left edge and $-u_x(x + \Delta x, y, t)$ along the right. Similarly, when Δx is small, we expect the rate of decrease of temperature across the bottom and top edges to be approximately the same as $-u_y(x, y, t)$ and $-u_y(x, y + \Delta y, t)$, respectively. If the plate has (constant) thickness T, the area of each vertical edge is $T\Delta y$, and that of each horizontal edge is $T\Delta x$. Thus, the first principle gives us the heat flow into the piece as

$$\Delta Q = k\{T(\Delta y)[u_x(x + \Delta x, y, t) - u_x(x, y, t)]$$
$$+ T(\Delta x)[u_y(x, y + \Delta y, t) - u_y(x, y, t)]\}.$$

Applying the mean value theorem to each of the quantities in brackets, we can rewrite this as

$$\Delta Q = kT(\Delta x)(\Delta y)\{u_{xx}(\bar{x}, y, t) + u_{yy}(x, \bar{y}, t)\},$$

where $x < \bar{x} < x + \Delta x$ and $y < \bar{y} < y + \Delta y$.

The rest of our analysis proceeds as in Example 8.1.1. The mass of the

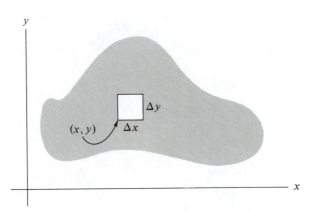

FIGURE 8.3

piece is $m = \delta T \, \Delta x \, \Delta y$, where δ is the (uniform) density, and the rate of change of the temperature is approximately the same throughout the piece. Thus, the second principle of heat flow tells us that

$$u_t(x, y, t) = \frac{s \, \Delta Q}{\delta T \, \Delta x \, \Delta y}.$$

Combining this with our expression for ΔQ, we have

$$u_t(x, y, t) = \frac{sk}{\delta} \{u_{xx}(\bar{x}, y, t) + u_{yy}(x, \bar{y}, t)\}.$$

Passing to the limit as $\Delta x \to 0$ and $\Delta y \to 0$, we obtain the **two-dimensional heat equation** in Cartesian coordinates

(H₂)
$$\frac{\partial^2 u}{\partial x^2} + \frac{\partial^2 u}{\partial y^2} = \frac{1}{\alpha} \frac{\partial u}{\partial t}.$$

The data for a two-dimensional heat flow problem are analogous to those in the one-dimensional case. We usually have an initial condition

(I)
$$u(x, y, 0) = f(x, y)$$

together with a boundary condition. For example, in a rectangular plate defined by $0 < x < \ell$, $0 < y < h$, we have a boundary consisting of four edges. If the vertical edges are maintained at given temperature distributions and the horizontal ones are insulated, we obtain boundary conditions of the form

(B)
$$u(0, y, t) = T_0(y), \quad u(\ell, y, t) = T_\ell(y), \qquad 0 < y < h$$
$$u_y(x, 0, t) = u_y(x, h, t) = 0, \qquad\qquad 0 < x < \ell.$$

You can imagine other possible combinations of boundary conditions (Exercise 2).

It is worth noting that, as in Example 6.1.2, the shape of the plate may lead us to use a different coordinate system. For example, the evolution of temperature in a circular plate of radius 1 is best expressed in terms of polar coordinates, $u = u(r, \theta, t)$. We are then led to the two-dimensional heat equation in polar coordinates (Exercise 6)

(H₂)
$$\frac{\partial^2 u}{\partial r^2} + \frac{1}{r} \frac{\partial u}{\partial r} + \frac{1}{r^2} \frac{\partial^2 u}{\partial \theta^2} = \frac{1}{\alpha} \frac{\partial u}{\partial t}.$$

An initial condition for this problem takes the form

(I)
$$u(r, \theta, 0) = f(r, \theta); \qquad 0 \le r \le 1, \quad 0 \le \theta < 2\pi.$$

If the plate is insulated along the edge, we have the boundary condition

(B) $u_r(1, \theta, t) = 0, \qquad 0 \le \theta < 2\pi$ and all t.

Of course, the geometry of the situation imposes conditions on the θ-dependence of u, which can be expressed as $u(r, \theta + 2\pi, t) = u(r, \theta, t)$ or as an additional boundary condition,

(B') $u(r, 0, t) - u(r, 2\pi, t) = u_\theta(r, 0, t) - u_\theta(r, 2\pi, t) = 0.$

We will solve some two-dimensional heat flow problems in Sections 8.6 and 8.7.

Example 8.1.3 Three-Dimensional Heat Flow

The evolution of temperature in a solid body is governed by a p.d.e. for the temperature as a function of three spatial coordinates and time, $u = u(x, y, z, t)$. The analysis leading to this p.d.e. is similar to that in the preceding examples.

We apply the principles of heat flow to a small rectangular solid piece of the body, with one corner at (x, y, z) and dimensions Δx by Δy by Δz (Figure 8.4). This piece has six faces across which heat flow can occur. The net flow

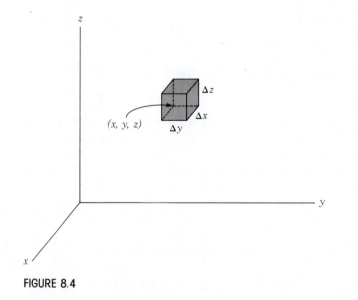

FIGURE 8.4

into the piece is approximated by

$$\Delta Q = k\{(\Delta y)(\Delta z)[u_x(x + \Delta x, y, z, t) - u_x(x, y, z, t)]$$
$$+ (\Delta x)(\Delta z)[u_y(x, y + \Delta y, z, t) - u_y(x, y, z, t)]$$
$$+ (\Delta x)(\Delta y)[u_z(x, y, z + \Delta z, t) - u_z(x, y, z, t)]\}.$$

or, using the mean value theorem, by

$$\Delta Q = k(\Delta x)(\Delta y)(\Delta z)\{u_{xx}(\bar{x}, y, z, t) + u_{yy}(x, \bar{y}, z, t) + u_{zz}(x, y, \bar{z}, t)\},$$

where

$$x < \bar{x} < x + \Delta x, \quad y < \bar{y} < y + \Delta y, \quad \text{and} \quad z < \bar{z} < x + \Delta z.$$

The second principle gives

$$u_t(x, y, z, t) = \frac{s \, \Delta Q}{\delta \, \Delta x \, \Delta y \, \Delta z},$$

where δ is the density of the body. Combining these equations, and taking limits as $\Delta x \to 0$, $\Delta y \to 0$, and $\Delta z \to 0$, we obtain the **three-dimensional heat equation** in Cartesian coordinates

(H₃)
$$\frac{\partial^2 u}{\partial x^2} + \frac{\partial^2 u}{\partial y^2} + \frac{\partial^2 u}{\partial z^2} = \frac{1}{\alpha} \frac{\partial u}{\partial t}.$$

Once again, our natural data includes an initial condition

(I)
$$u(x, y, z, 0) = f(x, y, z)$$

and boundary conditions. For a rectangular solid, $0 < x < \ell$, $0 < y < w$, $0 < z < h$, insulated on all six faces, the boundary conditions would take the form

(B)
$$u_x(0, y, z, t) = u_x(\ell, y, z, t) = 0$$
$$u_y(x, 0, z, t) = u_y(x, w, z, t) = 0$$
$$u_z(x, y, 0, t) = u_z(x, y, h, t) = 0.$$

The two most important ways of rewriting the three-dimensional heat equation are in cylindrical coordinates (Exercise 6),

$$\frac{\partial^2 u}{\partial r^2} + \frac{1}{r} \frac{\partial u}{\partial r} + \frac{1}{r^2} \frac{\partial^2 u}{\partial \theta^2} + \frac{\partial^2 u}{\partial z^2} = \frac{1}{\alpha} \frac{\partial u}{\partial t},$$

which are especially useful when the body has rotational symmetry about the z-axis, and spherical coordinates (Exercise 7),

$$\frac{1}{\rho^2}\left[\rho\frac{\partial^2}{\partial\rho^2}(\rho u) + \frac{1}{\sin\phi}\frac{\partial}{\partial\phi}\left(\sin\phi\frac{\partial u}{\partial\phi}\right) + \frac{1}{\sin^2\phi}\frac{\partial^2 u}{\partial\theta^2}\right] = \frac{1}{\alpha}\frac{\partial u}{\partial t}$$

useful in case of rotational symmetry about the origin.

Note

The Laplacian

You may have noted the similarity between the left side of the various forms of the heat equation in this section and the left side of the Laplace equation in Section 6.1. In fact, a steady-state temperature distribution can be thought of as a solution of the heat equation for which $\partial u/\partial t = 0$. Substituting this into the right side of the heat equation immediately reduces it to the corresponding Laplace equation.

The various expressions that appear on the left side of the heat and Laplace equations appear so often that they are known collectively as the **Laplacian** of u and are denoted by $\nabla^2 u$, or Δu. When a quantity u depends on the Cartesian coordinates of a point, possibly together with other variables, then $\nabla^2 u$ is defined as the sum of the second partials of u with respect to the Cartesian coordinates. Thus, in one dimension

$$\nabla^2 u = \frac{\partial^2 u}{\partial x^2},$$

in two dimensions

$$\nabla^2 u = \frac{\partial^2 u}{\partial x^2} + \frac{\partial^2 u}{\partial y^2},$$

and in three dimensions

$$\nabla^2 u = \frac{\partial^2 u}{\partial x^2} + \frac{\partial^2 u}{\partial y^2} + \frac{\partial^2 u}{\partial z^2}.$$

With this notation, the heat equation takes the unified form

(H) $$\nabla^2 u = \frac{1}{\alpha}\frac{\partial u}{\partial t},$$

and the Laplace equation takes the form

(L) $$\nabla^2 u = 0.$$

The versions of the heat and Laplace equations that occur in other coordinate systems come from expressing $\nabla^2 u$ in terms of these coordinates.

EXERCISES

1. Determine the boundary conditions for the temperature $u(x, t)$ in the wire of Example 8.1.1 if the left end is kept at $0°$ and
    a. the right end is insulated
    b. the right end is kept at T_ℓ degrees.

2. Determine the boundary conditions for the temperature $u(x, y, t)$ in the rectangular plate of Example 8.1.2 if
    a. the temperature at each point (x, h) on the top edge is $f(x)$ and the remaining edges are kept at $0°$.
    b. the temperature distributions along the top and right edges are given by the functions $f(x)$ and $g(y)$, respectively, and the left and bottom edges are kept at $0°$.

3. Determine the boundary conditions for the temperature $u(r, \theta, t)$ in a semicircular plate, defined by $0 \le \theta \le \pi$ and $0 \le r \le 1$, if the temperature along the straight edge is kept at $0°$ and the temperature at each point of the curved edge depends only on the angle θ and is given by $f(\theta)$.

4. Suppose an electric current in the wire of Example 8.1.1 adds heat to the wire at a constant rate per unit length, G.
    a. Check that the heat flow into a piece of wire extending from x to $x + \Delta x$ is $\Delta Q = (kAu_{xx}(\bar{x}, t) + G)\Delta x$, where $x < \bar{x} < x + \Delta x$.
    b. Equate this expression with $\Delta Q = (\delta A/s)u_t(x, t)\Delta x$, cancel Δx, and take limits as $\Delta x \to 0$, to obtain a p.d.e. for $u(x, t)$.

5. Suppose that, instead of being constant, the values k, A, s, and δ in Example 8.1.1 vary with x—say, $kA = p(x)$ and $\delta A/s = r(x)$.
    a. Check that the heat flow into a piece of wire extending from x to $x + \Delta x$ is

    $$\Delta Q = p(x + \Delta x)u_x(x + \Delta x, t) - p(x)u_x(x, t)$$
    $$= [p(\bar{x})u_{xx}(\bar{x}, t) + p'(\bar{x})u_x(\bar{x}, t)]\Delta x, \quad \text{where } x < \bar{x} < x + \Delta x.$$

    b. By comparing this with $\Delta Q = r(x)u_t(x, t)\Delta x$, obtain a p.d.e. for $u(x, t)$.

6. *a. Show that if $x = r \cos \theta$ and $y = r \sin \theta$, then

    $$\frac{\partial^2 u}{\partial x^2} + \frac{\partial^2 u}{\partial y^2} = \frac{\partial^2 u}{\partial r^2} + \frac{1}{r}\frac{\partial u}{\partial r} + \frac{1}{r^2}\frac{\partial^2 u}{\partial \theta^2}.$$

    (*Hint:* Follow the outline of Exercise 9, Section 6.1.)
    b. Use the result of (a) to obtain the polar-coordinate and cylindrical-coordinate versions of the two- and three-dimensional heat equations.

7. *a. Show that if $x = \rho \sin \phi \cos \theta$, $y = \rho \sin \phi \sin \theta$, and $z = \rho \cos \phi$, then

    $$\frac{\partial^2 u}{\partial x^2} + \frac{\partial^2 u}{\partial y^2} + \frac{\partial^2 u}{\partial z^2} = \frac{1}{\rho}\frac{\partial^2}{\partial \rho^2}(\rho u) + \frac{1}{\rho^2 \sin \phi}\frac{\partial}{\partial \phi}\left(\sin \phi \frac{\partial u}{\partial \phi}\right) + \frac{1}{\rho^2 \sin^2 \phi}\frac{\partial^2 u}{\partial \theta^2}.$$

    (*Hint:* follow the outline of Exercise 10, Section 5.1.)
    b. Obtain the spherical-coordinate version of the three-dimensional heat equation.

8.2   THE ONE-DIMENSIONAL HEAT EQUATION

In this section we look for solutions of the one-dimensional heat equation

(H)
$$\frac{\partial^2 u}{\partial x^2} = \frac{1}{\alpha} \frac{\partial u}{\partial t}$$

that satisfy one of the boundary conditions of Example 8.1.1:

(B$_1$)
$$u(0, t) = u(\ell, t) = 0$$

or

(B$_2$)
$$u_x(0, t) = u_x(\ell, t) = 0$$

or

(B$_3$)
$$u(\ell, t) - u(-\ell, t) = u_x(\ell, t) - u_x(-\ell, t) = 0$$

(here $u_x = \partial u / \partial x$). Note that if u_1 and u_2 satisfy (H) and any one of the boundary conditions (B), then so does any linear combination $c_1 u_1 + c_2 u_2$. When we solved linear homogeneous o.d.e.'s, which also had this property, we found finite sets of independent functions whose linear combinations constituted complete collections of solutions. In the present case, we will find infinitely many independent functions $u_j = u_j(x, t), j = 1, 2, \ldots$, each of which is a solution of (H) and satisfies our boundary conditions (B). The solutions matching some initial conditions,

(I)
$$u(x, 0) = f(x)$$

can be written as linear combinations of finitely many of the u_j's, but we will see that other initial conditions require "infinite combinations"

$$u = \sum_{j=1}^{\infty} c_j u_j.$$

The method we use to find the u_j's is known as **separation of variables.** We look for solutions to (H) of the form $u = XT$, where $X = X(x)$ is a function of x alone, $T = T(t)$ is a function of t alone, and neither is identically zero. (We touched on this method in Section 6.1.)

Example 8.2.1

Let's first investigate solutions of the heat equation

(H)
$$\frac{\partial^2 u}{\partial x^2} = \frac{1}{\alpha} \frac{\partial u}{\partial t}, \qquad 0 < x < \ell, \quad 0 < t$$

that satisfy the boundary conditions

(B₁)
$$u(0, t) = u(\ell, t) = 0.$$

We look for nontrivial solutions of (H) that satisfy (B₁) and are of the form

$$u = XT$$

with $X = X(x)$ and $T = T(t)$. Substitution into (H) gives

$$X''T = \frac{1}{\alpha} XT'$$

or, separating the terms involving x from those involving t,

$$\frac{X''}{X} = \frac{1}{\alpha} \frac{T'}{T}.$$

Since the left-hand side of this equation does not depend on t, whereas the right-hand side does not depend on x, the common value must be a constant:

$$\frac{X''}{X} = \frac{1}{\alpha} \frac{T'}{T} = -\lambda.$$

Thus, we need

(H_x)
$$X'' + \lambda X = 0$$

(H_t)
$$T' + \alpha \lambda T = 0.$$

Note that these are *ordinary* differential equations, which we know how to solve.

Since we want $u = XT$ to satisfy (B₁), we also need

$$X(0)T(t) = X(\ell)T(t) = 0.$$

$T(t)$ is not identically zero, so

(B$'_1$) $$X(0) = X(\ell) = 0.$$

If $\lambda < 0$, the general solution of (H$_x$) is

$$X = c_1 e^{x\sqrt{-\lambda}} + c_2 e^{-x\sqrt{-\lambda}}.$$

Substitution into (B$'_1$) leads to the algebraic equations

$$
\begin{aligned}
c_1 \qquad\quad + c_2 \qquad\qquad &= 0 \\
c_1 e^{\ell\sqrt{-\lambda}} + c_2 e^{-\ell\sqrt{-\lambda}} &= 0,
\end{aligned}
$$

which have only the trivial solution

$$c_1 = c_2 = 0.$$

Thus, there are no nontrivial solutions corresponding to negative values of λ.
If $\lambda = 0$, the general solution of (H$_x$) is

$$X = c_1 + c_2 x.$$

Substitution into (B$'_1$) leads to

$$c_1 = c_2 = 0,$$

so again there are no nontrivial solutions corresponding to $\lambda = 0$.
Now if $\lambda > 0$, the general solution of (H$_x$) is

$$X = c_1 \sin x\sqrt{\lambda} + c_2 \cos x\sqrt{\lambda},$$

and substitution into (B$'_1$) gives

$$
\begin{aligned}
c_2 \qquad\qquad\qquad &= 0 \\
c_1 \sin \ell\sqrt{\lambda} + c_2 \cos \ell\sqrt{\lambda} &= 0.
\end{aligned}
$$

In order to get a nontrivial solution, we need $\sin \ell\sqrt{\lambda} = 0$. This can happen only if $\ell\sqrt{\lambda} = n\pi$ or

$$\lambda = \frac{n^2 \pi^2}{\ell^2}, \qquad \text{where } n \text{ is a positive integer.}$$

Corresponding to each such choice of λ, equation (H$_x$) has nontrivial solution that satisfies (B$_1'$),

$$X_n = \sin \frac{n\pi x}{\ell},$$

and equation (H$_t$) has a nontrivial solution

$$T_n = e^{-n^2\pi^2\alpha t/\ell^2}.$$

The product of these,

$$u_n(x, t) = X_n T_n = e^{-n^2\pi^2\alpha t/\ell^2} \sin \frac{n\pi x}{\ell},$$

is a solution of (H) that satisfies (B$_1$).

Suppose now that we wish to find a solution of (H) that satisfies (B$_1$) and also matches the initial condition

(I$_1$) $$u(x, 0) = 5 \sin \frac{\pi x}{\ell} - 2 \sin \frac{3\pi x}{\ell}, \qquad 0 < x < \ell.$$

Since

$$u_n(x, 0) = \sin \frac{n\pi x}{\ell},$$

none of the solutions $u_n(x, t)$ matches (I$_1$). Note, however, that

$$u(x, 0) = 5u_1(x, 0) - 2u_3(x, 0).$$

The function

$$u(x, t) = 5u_1(x, t) - 2u_3(x, t)$$
$$= 5e^{-\pi^2\alpha t/\ell} \sin \frac{\pi x}{\ell} - 2e^{-9\pi^2\alpha t/\ell} \sin \frac{3\pi x}{\ell}$$

is a solution of (H), satisfies (B$_1$), and matches (I$_1$).

Note that to match the initial condition

$$u(x, 0) = \sum_{j=1}^{m} b_j \sin \frac{n\pi x}{\ell} = \sum_{j=1}^{m} b_j u_j(x, 0), \qquad 0 < x < \ell,$$

we can choose the solution

$$u(x, t) = \sum_{j=1}^{m} b_j u_j(x, t) = \sum_{j=1}^{m} b_j e^{-j^2 \pi^2 \alpha t / \ell^2} \sin \frac{j\pi x}{\ell}.$$

In order to handle more general initial conditions,

(I) $u(x, 0) = f(x), \qquad 0 < x < \ell$

we have to consider solutions obtained by allowing series expressions of the form

$$u(x, t) = \sum_{j=1}^{\infty} b_j u_j(x, t) = \sum_{j=1}^{\infty} b_j e^{-j^2 \pi^2 \alpha t / \ell^2} \sin \frac{j\pi x}{\ell}.$$

Such a series satisfies

$$u(x, 0) = \sum_{j=1}^{\infty} b_j \sin \frac{j\pi x}{\ell}.$$

Since we want this to match (I), the problem reduces to determining constants b_j for which

$$f(x) = \sum_{j=1}^{\infty} b_j \sin \frac{j\pi x}{\ell}, \qquad 0 < x < \ell.$$

We will learn to express general functions as trigonometric series in Sections 8.3 and 8.4.

Example 8.2.2

Find solutions of the heat equation

(H) $$\frac{\partial^2 u}{\partial x^2} = \frac{1}{\alpha} \frac{\partial u}{\partial t}, \qquad 0 < x < \ell, \ \ 0 < t$$

that satisfy the boundary conditions

(B$_2$) $u_x(0, t) = u_x(\ell, t) = 0.$

Substitution of $u = XT$ into (H) leads to the same equations (H$_x$) and

(H$_t$) as in Example 8.2.1, and the boundary conditions (B$_2$) lead to the conditions

(B$_2'$) $$X'(0) = X'(\ell) = 0.$$

As in Example 8.2.1, when $\lambda < 0$ there are no nontrivial solutions of (H$_x$) that satisfy (B$_2'$).

If $\lambda = 0$, the general solution of (H$_x$) is

$$X = c_1 + c_2 x.$$

Substitution into (B$_2'$) gives

$$c_2 = 0$$

but no restriction on c_1. Thus, corresponding to $\lambda = 0$, we obtain a nontrivial solution of (H$_x$) satisfying (B$_2'$), $X_0 = 1$. If we multiply X_0 by the solution of (H$_t$) corresponding to $\lambda = 0$, $T_0 = 1$, we obtain a function

$$u_0^*(x, t) = X_0 T_0 = 1$$

that is a solution of (H) and satisfies (B$_2$).

If $\lambda > 0$, the general solution of (H$_x$) is

$$X = c_1 \sin x\sqrt{\lambda} + c_2 \cos x\sqrt{\lambda},$$

and our boundary conditions (B$_2'$) give

$$c_1 = 0$$
$$c_1 \cos \ell\sqrt{\lambda} - c_2 \sin \ell\sqrt{\lambda} = 0.$$

Since X is to be nontrivial, $\ell\sqrt{\lambda} = n\pi$ or

$$\lambda = \frac{n^2\pi^2}{\ell^2}, \qquad \text{where } n \text{ is a positive integer.}$$

Corresponding to each positive integer n, equation (H$_x$) has a nontrivial solution that satisfies (B$_2'$),

$$X_n^* = \cos \frac{n\pi x}{\ell},$$

and equation (H_t) has a nontrivial solution

$$T_n = e^{-n^2\pi^2\alpha t/\ell^2},$$

so (H) has a nontrivial solution that satisfies (B_2)

$$u_n^*(x, t) = X_n^* T_n = e^{-n^2\pi^2\alpha t/\ell^2} \cos \frac{n\pi x}{\ell}.$$

Let's now look for a solution of (H) that satisfies (B_1) and also matches the initial condition

$$(I_2) \qquad u(x, 0) = 7 - \cos \frac{\pi x}{\ell} + 3 \cos \frac{2\pi x}{\ell}, \qquad 0 < x < \ell.$$

Since $u_n^*(x, 0) = \cos n\pi x/\ell$, this initial condition can be written as

$$u(x, 0) = 7u_0^*(x, 0) - u_1^*(x, 0) + 3u_2^*(x, 0).$$

The function

$$u(x, t) = 7u_0^*(x, t) - u_1^*(x, t) + 3u_2^*(x, t)$$

$$= 7 - e^{-\pi^2\alpha t/\ell^2} \cos \frac{\pi x}{\ell} + 3e^{-4\pi^2\alpha t/\ell^2} \cos \frac{2\pi x}{\ell}$$

is a solution of (H), satisfies (B_2), and matches (I_2).

In this case, more general initial conditions

$$(I) \qquad\qquad u(x, 0) = f(x), \qquad 0 < x < \ell$$

require us to consider solutions of the form

$$u(x, t) = \frac{a_0}{2} u_0^*(x, t) + \sum_{j=1}^{\infty} a_j u_j^*(x, t) = \frac{a_0}{2} + \sum_{j=1}^{\infty} a_j e^{-j^2\pi^2\alpha t/\ell^2} \cos \frac{j\pi x}{\ell}.$$

Substitution of $t = 0$ gives

$$u(x, 0) = \frac{a_0}{2} + \sum_{j=1}^{\infty} a_j \cos \frac{j\pi x}{\ell}.$$

Since we want to match (I), the problem reduces to determining constant a_j for which

$$f(x) = \frac{a_0}{2} + \sum_{j=1}^{\infty} a_j \cos \frac{j\pi x}{\ell}, \qquad 0 < x < \ell.$$

Example 8.2.3

Find solutions of the heat equation

(H)
$$\frac{\partial^2 u}{\partial x^2} = \frac{1}{\alpha}\frac{\partial u}{\partial t}, \qquad -\ell < x < \ell, \quad 0 < t$$

that satisfy the boundary conditions

(B_3)
$$u(\ell, t) - u(-\ell, t) = u_x(\ell, t) - u_x(-\ell, t) = 0.$$

Substitution of $u = XT$ into (H) leads to the equations (H_x) and (H_t) of Example 8.2.1, and (B_3) leads to

(B_3')
$$X(\ell) - X(-\ell) = X'(\ell) - X'(-\ell) = 0.$$

As in Example 8.2.2, there are no nontrivial solutions corresponding to negative values of λ, and $\lambda = 0$ leads to a solution

$$u_0^*(x, t) = 1.$$

If $\lambda > 0$, the general solution of (H_x) is

$$X = c_1 \sin x\sqrt{\lambda} + c_2 \cos x\sqrt{\lambda}.$$

Since $\sin(-\ell\sqrt{\lambda}) = -\sin \ell\sqrt{\lambda}$ and $\cos(-\ell\sqrt{\lambda}) = \cos \ell\sqrt{\lambda}$, our boundary conditions (B_3') give

$$2c_1 \sin \ell\sqrt{\lambda} \qquad\qquad = 0$$
$$2\sqrt{\lambda}c_2 \sin \ell\sqrt{\lambda} = 0.$$

In order to get nontrivial solutions, we need $\sin \ell\sqrt{\lambda} = 0$, or

$$\lambda = \frac{n^2\pi^2}{\ell^2}, \quad \text{where } n \text{ is a positive integer;}$$

there are no restrictions on c_1 and c_2. Corresponding to each positive integer n, we obtain two linearly independent solutions of (H_x) that satisfy (B_3'),

$$X_n = \sin \frac{n\pi x}{\ell} \qquad \text{and} \qquad X_n^* = \cos \frac{n\pi x}{\ell}$$

and a nontrivial solution of (H_t)

$$T_n = e^{-n^2\pi^2\alpha t/\ell^2}.$$

These yield two independent functions

$$u_n(x, t) = X_n T_n = e^{-n^2 \pi^2 \alpha t / \ell^2} \sin \frac{n \pi x}{\ell}$$

$$u_n^*(x, t) = X_n^* T_n = e^{-n^2 \pi^2 \alpha t / \ell^2} \cos \frac{n \pi x}{\ell},$$

each of which is a solution of (H) and satisfies (B$_3'$).

Let's now find a solution of (H) that satisfies (B$_3$) and also matches the initial condition

(I$_3$) $$u(x, 0) = 3 - \sin \frac{3 \pi x}{\ell} + \frac{1}{2} \cos \frac{7 \pi x}{\ell}, \qquad -\ell < x < \ell.$$

Since

$$u(x, 0) = 3 u_0^*(x, 0) - u_3(x, 0) + \frac{1}{2} u_7^*(x, 0),$$

the function

$$u(x, t) = 3 u_0^*(x, t) - u_3(x, t) + \frac{1}{2} u_7^*(x, t)$$

$$= 3 - e^{-9 \pi^2 \alpha t / \ell^2} \sin \frac{3 \pi x}{\ell} + \frac{1}{2} e^{-49 \pi^2 \alpha t / \ell^2} \cos \frac{7 \pi x}{\ell}$$

is a solution of (H), satisfies (B$_3$), and matches (I$_3$).

To handle more general initial conditions

(I) $$u(x, 0) = f(x), \qquad -\ell < x < \ell$$

we must allow solutions of the form

$$u(x, t) = \frac{a_0}{2} u_0^*(x, t) + \sum_{j=1}^{\infty} (a_j u_j^*(x, t) + b_j u_j(x, t))$$

$$= \frac{a_0}{2} + \sum_{j=1}^{\infty} e^{-j^2 \pi^2 \alpha t / \ell^2} \left(a_j \cos \frac{j \pi x}{\ell} + b_j \sin \frac{j \pi x}{\ell} \right).$$

Since we want to match (I), we must find constants a_j and b_j for which

$$f(x) = \frac{a_0}{2} + \sum_{j=1}^{\infty} \left(a_j \cos \frac{j \pi x}{\ell} + b_j \sin \frac{j \pi x}{\ell} \right), \qquad -\ell < x < \ell.$$

Our summary describes the information we have obtained about solutions of the one-dimensional heat equation that satisfy one of the three boundary conditions (B$_1$), (B$_2$), or (B$_3$), and that match given initial conditions. Exercises 9 through 16 deal with different boundary conditions as well as different p.d.e.'s.

THE ONE-DIMENSIONAL HEAT EQUATION

Boundary Condition 1

We look for the solution of

(H)
$$\frac{\partial^2 u}{\partial x^2} = \frac{1}{\alpha} \frac{\partial u}{\partial t}, \qquad 0 < x < \ell, \ \ 0 < t$$

(B$_1$)
$$u(0, t) = u(\ell, t) = 0, \qquad 0 < t$$

(I)
$$u(x, 0) = f(x), \qquad 0 < x < \ell$$

in the form

$$u(x, t) = \sum_{j=1}^{\infty} b_j u_j(x, t) = \sum_{j=1}^{\infty} b_j e^{-j^2 \pi^2 \alpha t / \ell^2} \sin \frac{j \pi x}{\ell},$$

where the b_j's are determined by expressing $f(x)$ in the form

$$f(x) = \sum_{j=1}^{\infty} b_j \sin \frac{j \pi x}{\ell}, \qquad 0 < x < \ell.$$

Boundary Condition 2

We look for the solution of

(H)
$$\frac{\partial^2 u}{\partial x^2} = \frac{1}{\alpha} \frac{\partial u}{\partial t}, \qquad 0 < x < \ell, \ \ 0 < t$$

(B$_2$)
$$u_x(0, t) = u_x(\ell, t) = 0, \qquad 0 < t$$

(I)
$$u(x, 0) = f(x), \qquad 0 < x < \ell$$

in the form

$$u(x, t) = \frac{a_0}{2} + \sum_{j=1}^{\infty} a_j e^{-j^2 \pi^2 \alpha t / \ell^2} \cos \frac{j \pi x}{\ell},$$

where the a_js are determined by expressing $f(x)$ in the form

$$f(x) = \frac{a_0}{2} + \sum_{j=1}^{\infty} a_j \cos\frac{j\pi x}{\ell}, \qquad 0 < x < \ell.$$

Boundary Condition 3

We look for the solution of

(H) $\qquad \dfrac{\partial^2 u}{\partial x^2} = \dfrac{1}{\alpha}\dfrac{\partial u}{\partial t}, \qquad -\ell < x < \ell, \quad 0 < t$

(B₃) $\quad u(-\ell, t) - u(\ell, t) = u_x(-\ell, t) - u_x(\ell, t) = 0, \qquad 0 < t$

(I) $\qquad\qquad u(x, 0) = f(x), \qquad -\ell < x < \ell$

in the form

$$u(x, t) = \frac{a_0}{2} + \sum_{j=1}^{\infty} e^{-j^2\pi^2\alpha t/\ell^2}\left(a_j \cos\frac{j\pi x}{\ell} + b_j \sin\frac{j\pi x}{\ell}\right),$$

where the a_j's and b_j's are determined by expressing $f(x)$ in the form

$$f(x) = \frac{a_0}{2} + \sum_{j=1}^{\infty}\left(a_j \cos\frac{j\pi x}{\ell} + b_j \sin\frac{j\pi x}{\ell}\right), \qquad -\ell < x < \ell.$$

Notes

1. A technicality

You may have noticed that in each of the examples, we asked for solutions valid in a specified region of the tx-plane. In Examples 8.2.1 and 8.2.2 the region was the half-infinite strip $0 < x < \ell$, $0 < t$, which we sketch in Figure 8.5. Note that the inequalities are strict so that the region *excludes the boundaries* $x = 0$, $x = \ell$, and $t = 0$. This is typical. When dealing with a p.d.e., we look for continuous functions that satisfy the p.d.e. *inside* a region. Strictly speaking, the function need not even be defined on the boundary of the region. We do, however, require that the limit of the function as one approaches the boundary must exist and must match the given boundary and/or initial conditions. Thus, technically speaking, the boundary and initial conditions of Example 8.2.1 are

$$\lim_{x\to 0} u(x, t) = \lim_{x\to\ell} u(x, t) = 0 \qquad \text{for all } t, \quad 0 < t$$

$$\lim_{t\to 0} u(x, t) = 5\sin\frac{\pi x}{\ell} - 2\sin\frac{3\pi x}{\ell} \qquad \text{for all } x, \quad 0 < x < \ell.$$

FIGURE 8.5

The conditions of Example 8.2.2 are

$$\lim_{x \to 0} u_x(x, t) = \lim_{x \to \ell} u_x(x, t) = 0 \qquad \text{for all } t, \quad 0 < t$$

$$\lim_{t \to 0} u(x, t) = 7 - \cos \frac{\pi x}{\ell} + 3 \cos \frac{2\pi x}{\ell} \qquad \text{for all } x, \quad 0 < x < \ell.$$

2.   On the existence and uniqueness of solutions to p.d.e.'s
    The theorems governing the existence and uniqueness of solutions to second-order p.d.e.'s are much more complicated than the corresponding result for nth-order o.d.e.'s. We will not attempt to state them here. However, the boundary–initial-value problems we will deal with do have unique solutions.

3.   On nonhomogeneous boundary conditions
    If u_1 and u_2 both satisfy one of the boundary conditions (B$_1$) through (B$_3$), then so does any linear combination $c_1 u_1 + c_2 u_2$. Boundary conditions with this property are called **homogeneous.**
    Suppose we have a **nonhomogeneous** boundary condition like

(B$_N$) $u(0, t) = T_0, \quad u(\ell, t) = T_\ell$

(where T_0 and T_ℓ are given constants). We can still use separation of variables to look for some function $u_P(x, t)$ that is a solution of (H) and satisfies (B$_N$). However, the function we find this way may not match the initial condition

(I) $u(x, 0) = f(x).$

To match (I), we must adjust our particular solution $u_P(x, t)$ by an appropriate function $u_H(x, t)$, which is a solution of (H) and satisfies the related homogeneous boundary conditions

(B$_H$) $u(0, t) = u(\ell, t) = 0.$

If we choose $u_H(x, t)$ so that

$$u_H(x, 0) = f(x) - u_P(x, 0),$$

then

$$u(x, t) = u_H(x, t) + u_P(x, t)$$

will be a solution of (H) that satisfies (B_N) and matches (I). You are asked to work examples of this kind in Exercises 14 through 16.

4. Eigenvalues and eigenfunctions

In each of Examples 8.2.1 through 8.2.3, the bulk of the work consists of finding the solutions of (H_x) that satisfy the boundary conditions (B_i'). If we restrict our attention to those functions $X(x)$ that satisfy the boundary conditions (B_i') and if we use A to denote the operator that assigns to each such function the value $AX = -X''$, then we can describe this step as follows:

i.   Find the values of λ for which $AX = \lambda X$ has nontrivial solutions.

ii. Corresponding to each such value of λ, find those solutions.

In this form, we can see the striking similarity between this problem and the problem we dealt with in Section 3.5. By analogy with the terminology we used there, it is customary to refer to the values of λ for which $AX = \lambda X$ has nontrivial solutions as *eigenvalues* of the boundary-value problem, and to the corresponding nontrivial solutions as *eigenfunctions*. Thus the boundary-value problem (H_x) and (B_i') has eigenvalues $n^2\pi^2/\ell^2$ and eigenfunctions $X_n = \sin(n\pi x/\ell)$, $n = 1, 2, \ldots$. The boundary-value problem (H_x) and (B_3') has eigenvalues 0 and $n^2\pi^2/\ell^2$ and eigenfunctions $X_0^* = 1$, $X_n = \sin(n\pi x/\ell)$, and $X_n^* = \cos(n\pi x/\ell)$, $n = 1, 2, \ldots$.

EXERCISES

In Exercises 1 through 8, find the solution of the one-dimensional heat equation that matches the given boundary and initial conditions. Note that the boundary conditions are the same as (B_1) through (B_3), with specific values for ℓ. In particular, we have already determined (in Examples 8.2.1 through 8.2.3) the form we expect such solutions to take.

1. $u(0, t) = u(1, t) = 0$; $u(x, 0) = \sin 3\pi x + \sin 5\pi x$, $0 < x < 1$.

2. $u(0, t) = u(1, t) = 0$; $u(x, 0) = \sin 2\pi x - 3 \sin 4\pi x$, $0 < x < 1$.

3. $u(0, t) = u(2, t) = 0$; $u(x, 0) = \sin 2\pi x - 3 \sin 4\pi x$, $0 < x < 2$.

4. $u(0, t) = u(2\pi, t) = 0$; $u(x, 0) = 2 \sin x - \sin 2x$, $0 < x < 2\pi$.

5. $u_x(0, t) = u_x(1, t) = 0$; $u(x, 0) = 7 - \cos 3\pi x$, $0 < x < 1$.

6. $u_x(0, t) = u_x(\pi, t) = 0$; $u(x, 0) = \sin^2 x$, $0 < x < \pi$. (*Hint:* Think!)

7. $u(1, t) - u(-1, t) = u_x(1, t) - u_x(-1, t) = 0$;
    $u(x, 0) = \sin 3\pi x + \sin 5\pi x$, $-1 < x < 1$.

8. $u(1, t) - u(-1, t) = u_x(1, t) - u_x(-1, t) = 0$;
    $u(x, 0) = 7 - \cos 3\pi x + \sin 5\pi x$, $-1 < x < 1$.

Some more advanced problems:

9. a. Show that the only solutions of the one-dimensional heat equation that are of
        the form $u = X(x)T(t)$ and satisfy the boundary conditions of Exercise 1(a),
        Section 8.1,

    (B$_4$) $u(0, t) = u_x(\ell, t) = 0,$

    are the multiples of

    $$u_n(x, t) = e^{-\alpha(2n+1)^2\pi^2 t/4\ell^2} \sin \frac{(2n+1)\pi x}{2\ell},$$

    where n is a nonnegative integer.

    b. Find the solution of the one-dimensional heat equation that satisfies (B$_4$) and
        matches the initial condition

    $$u(x, 0) = \sin \frac{5\pi x}{2\ell} - \sin \frac{\pi x}{2\ell}, \quad 0 < x < \ell.$$

10. a. Show that substituting $u = X(x)T(t)$ into the *one-dimensional wave equation*

    (W) $\dfrac{\partial^2 u}{\partial x^2} = \dfrac{1}{\alpha^2} \dfrac{\partial^2 u}{\partial t^2}, \quad 0 < x < \ell, \quad 0 < t$

    leads to o.d.e.'s for X and T.

    b. Show that the only solutions of (W) that are of the form $u = XT$ and satisfy
        the boundary conditions

    (B) $u(0, t) = u(\ell, t) = 0$

    are the functions $u = c_n u_n + c_n^* u_n^*$, where n is a positive integer,

    $$u_n = \sin \frac{n\pi x}{\ell} \cos \frac{\alpha n\pi t}{\ell} \quad \text{and} \quad u_n^* = \sin \frac{n\pi x}{\ell} \sin \frac{\alpha n\pi t}{\ell}.$$

    c. Find the solution of (W) that satisfies (B) and matches the initial conditions

    $$u(x, 0) = \sin \frac{3\pi x}{\ell} \quad \text{and} \quad u_t(x, 0) = 2 \sin \frac{5\pi x}{\ell}, \quad 0 < x < \ell.$$

11. a. Show that substituting $u = X(x)Y(y)$ into the *two-dimensional Laplace equa-
        tion*

    (L) $\dfrac{\partial^2 u}{\partial x^2} + \dfrac{\partial^2 u}{\partial y^2} = 0, \quad 0 < x < \ell, \quad 0 < y < h$

    leads to o.d.e.'s for X and Y.

    b. Show that the only solutions of (L) that are of the form $u = XY$ and satisfy
        the boundary conditions

    (B) $u(0, y) = u(\ell, y) = 0, \quad 0 < y < h \quad$ and $\quad u(x, 0) = 0, \quad 0 < x < \ell$

    are the multiples of

    $$u_n(x, y) = \sin \frac{n\pi x}{\ell} (e^{n\pi y/\ell} - e^{-n\pi y/\ell}),$$

    where n is a positive integer.

c. Find the solution of (L) that satisfies (B) as well as the additional boundary condition

$$u(x, h) = \sin \frac{\pi x}{\ell} - 2 \sin \frac{5\pi x}{\ell}, \qquad 0 < x < \ell.$$

12. a. Show that substituting $u = X(x)T(t)$ into the p.d.e. of Exercise 5(b), Section 8.1,

$$p(x) \frac{\partial^2 u}{\partial x^2} + p'(x) \frac{\partial u}{\partial x} = r(x) \frac{\partial u}{\partial t}$$

leads to o.d.e.'s for X and T. (Note that we dealt with o.d.e.'s like the one for X in Chapter 6.)

b. Check that after substituting $u = X(x)T(t)$ into the p.d.e.

$$\frac{\partial^2 u}{\partial x^2} + t \frac{\partial u}{\partial x} = \frac{\partial u}{\partial t}$$

it is impossible to separate the terms depending on x from those depending on t. (This means that separation of variables doesn't always work.)

13. a. Show that if $u = u_P(x, t)$ is a solution of the p.d.e. of Exercise 4, Section 8.1,

(N) $$\frac{\partial^2 u}{\partial x^2} + \frac{G}{\beta} = \frac{1}{\alpha} \frac{\partial u}{\partial t}, \qquad 0 < x < \ell, \; 0 < t$$

and if $u = u_H(x, t)$ is a solution of the one-dimensional heat equation, then $u = u_H(x, t) + u_P(x, t)$ is also a solution of (N).

b. Find a solution to (N) that is of the form $u = u_P(x)$ and satisfies the conditions $u_P(0) = u_P(\ell) = 0$.

c. By adding an appropriate function $u_H(x, t)$ to $u_P(x)$, obtain the solution of (N) that satisfies the boundary conditions $u(0, t) = u(\ell, t) = 0$ and matches the initial condition $u(x, 0) = u_P(x) + \sin(\pi x/\ell)$, $0 < x < \ell$.

In Exercises 14 through 16, find the solution of the one-dimensional heat equation that satisfies the given nonhomogeneous boundary condition (see Note 3) and matches the given initial condition.

14. $u(0, t) = u(\ell, t) = 10$; $u(x, 0) = 10 - 2 \sin \dfrac{3\pi x}{\ell}$, $0 < x < \ell$

15. $u(0, t) = 0$, $u(\ell, t) = 10$; $u(x, 0) = \dfrac{10x}{\ell} - 2 \sin \dfrac{3\pi x}{\ell}$, $0 < x < \ell$

(compare Exercise 1(b), Section 8.1)

16. $u_x(0, t) = u_x(\ell, t) = 10$; $u(x, 0) = 10x + \cos \dfrac{3\pi x}{\ell}$, $0 < x < \ell$

8.3 FOURIER SERIES

In order to solve problems like the ones in the previous section, we need to express the initial temperature $f(x)$ as a trigonometric series. In this section we will learn how to find expressions of the form

$$f(x) = \frac{a_0}{2} + \sum_{j=1}^{\infty} \left(a_j \cos\frac{j\pi x}{\ell} + b_j \sin\frac{j\pi x}{\ell} \right), \qquad -\ell < x < \ell.$$

Series of this form are called **Fourier series** after Jean Baptiste Fourier, who introduced them in his study of heat flow (1822).

Our discussion will make use of some simple facts about functions with symmetric graphs. Recall that if $g(-x) = -g(x)$ for all x, then the graph of $y = g(x)$ is symmetric about the origin; in this case, we say that $g(x)$ is **odd.** The functions whose graphs are sketched in Figure 8.6 are examples of odd functions. Using the interpretation of the definite integral as a sum of signed areas, we see that *if $g(x)$ is odd and integrable, then $\int_{-\ell}^{\ell} g(x)\, dx = 0$.* For example, since $\sin(j\pi x/\ell)$ and $\sin(j\pi x/\ell)\cos(k\pi x/\ell)$ are odd, we have

$$\int_{-\ell}^{\ell} \sin\frac{j\pi x}{\ell}\, dx = \int_{-\ell}^{\ell} \sin\frac{j\pi x}{\ell} \cos\frac{k\pi x}{\ell}\, dx = 0.$$

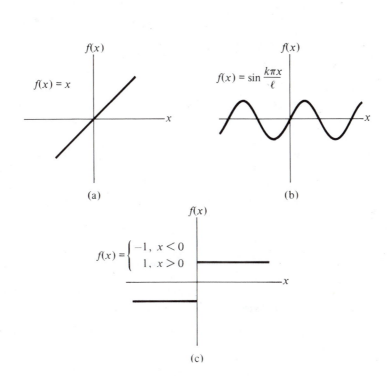

FIGURE 8.6 Some odd functions

If $g(-x) = g(x)$ for all x, then the graph of $y = g(x)$ is symmetric about the y-axis, and we say that $g(x)$ is **even.** The functions whose graphs are sketched in Figure 8.7 are even functions. *If $g(x)$ is even and integrable, then*

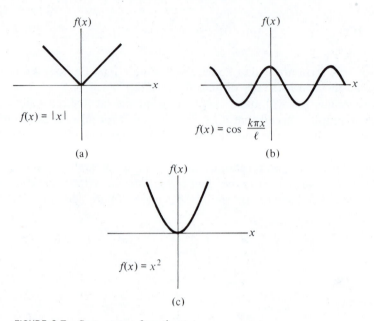

$f(x) = |x|$

(a)

$f(x) = \cos \dfrac{k\pi x}{\ell}$

(b)

$f(x) = x^2$

(c)

FIGURE 8.7   Some even functions

$\int_{-\ell}^{\ell} g(x)\, dx = 2 \int_{0}^{\ell} g(x)\, dx$. For example, $\cos(j\pi x/\ell)$ is even, so

$$\int_{-\ell}^{\ell} \cos \frac{j\pi x}{\ell}\, dx = 2 \int_{0}^{\ell} \cos \frac{j\pi x}{\ell}\, dx = \frac{2\ell}{j\pi} \sin \frac{j\pi x}{\ell} \Bigg]_{0}^{\ell} = 0.$$

We leave it to you to check (Exercises 13 and 14) that $\cos(j\pi x/\ell)\cos(k\pi x/\ell)$ and $\sin(j\pi x/\ell)\sin(k\pi x/\ell)$ are even, and that

$$\int_{-\ell}^{\ell} \cos \frac{j\pi x}{\ell} \cos \frac{k\pi x}{\ell}\, dx = \int_{-\ell}^{\ell} \sin \frac{j\pi x}{\ell} \sin \frac{k\pi x}{\ell}\, dx = \begin{cases} 0 & \text{if } j \neq k \\ \ell & \text{if } j = k \end{cases}.$$

We can use these observations about the integrals of products of trigonometric functions to determine the connection between the coefficients of a Fourier series and the function the series defines. If

$$f(x) = \frac{a_0}{2} + \sum_{j=1}^{\infty} \left(a_j \cos \frac{j\pi x}{\ell} + b_j \sin \frac{j\pi x}{\ell} \right),$$

then (assuming Fourier series can be integrated term by term)

$$\int_{-\ell}^{\ell} f(x)\, dx = \int_{-\ell}^{\ell} \frac{a_0}{2}\, dx + \sum_{j=1}^{\infty} \left(a_j \int_{-\ell}^{\ell} \cos \frac{j\pi x}{\ell}\, dx + b_j \int_{-\ell}^{\ell} \sin \frac{j\pi x}{\ell}\, dx \right)$$

$$= \int_{-\ell}^{\ell} \frac{a_0}{2}\, dx = a_0 \ell.$$

If we multiply $f(x)$ by $\cos(k\pi x/\ell)$ and integrate, we see that

$$\int_{-\ell}^{\ell} f(x) \cos \frac{k\pi x}{\ell}\, dx = \frac{a_0}{2} \int_{-\ell}^{\ell} \cos \frac{k\pi x}{\ell}\, dx$$

$$+ \sum_{j=1}^{\infty} \left(a_j \int_{-\ell}^{\ell} \cos \frac{j\pi x}{\ell} \cos \frac{k\pi x}{\ell}\, dx \right.$$

$$\left. + b_j \int_{-\ell}^{\ell} \sin \frac{j\pi x}{\ell} \cos \frac{k\pi x}{\ell}\, dx \right)$$

$$= a_k \int_{-\ell}^{\ell} \cos \frac{k\pi x}{\ell} \cos \frac{k\pi x}{\ell}\, dx = a_k \ell.$$

Similarly (Exercise 15),

$$\int_{-\ell}^{\ell} f(x) \sin \frac{k\pi x}{\ell}\, dx = b_k \ell.$$

Thus we have

(FS$_0$) $$a_0 = \frac{1}{\ell} \int_{-\ell}^{\ell} f(x)\, dx$$

(FS$_1$) $$a_k = \frac{1}{\ell} \int_{-\ell}^{\ell} f(x) \cos \frac{k\pi x}{\ell}\, dx, \qquad k = 1, 2, \ldots$$

(FS$_2$) $$b_k = \frac{1}{\ell} \int_{-\ell}^{\ell} f(x) \sin \frac{k\pi x}{\ell}\, dx, \qquad k = 1, 2, \ldots.$$

We use these formulas to associate a Fourier series to a given function $f(x)$.

Definition: *The **Fourier series of f(x)** on the interval $-\ell < x < \ell$ is the series*

$$\frac{a_0}{2} + \sum_{j=1}^{\infty} \left(a_j \cos \frac{j\pi x}{\ell} + b_j \sin \frac{j\pi x}{\ell} \right),$$

where the constants a_j and b_j are given by the integral formulas (FS$_0$) through (FS$_2$).

Example 8.3.1

Find the Fourier series on $-2 < x < 2$ of

$$f(x) = \begin{cases} -1, & -2 < x < 0 \\ 1, & 0 < x < 2 \end{cases}.$$

Note that $f(x)$ and $f(x)\cos(k\pi x/2)$ are odd functions. Thus, (FS$_0$) and (FS$_1$) give

$$a_0 = \frac{1}{2} \int_{-2}^{2} f(x)\, dx = 0$$

$$a_k = \frac{1}{2} \int_{-2}^{2} f(x) \cos \frac{k\pi x}{2} \, dx = 0, \qquad k = 1, 2, \ldots.$$

Since $f(x)\sin(k\pi x/2)$ is an even function, (FS$_2$) gives

$$b_k = \frac{1}{2} \int_{-2}^{2} f(x) \sin \frac{k\pi x}{2} \, dx = \int_{0}^{2} f(x) \sin \frac{k\pi x}{2} \, dx = \int_{0}^{2} \sin \frac{k\pi x}{2} \, dx$$

$$= -\frac{2}{k\pi} \cos \frac{k\pi x}{2} \Big]_{0}^{2} = -\frac{2}{k\pi} (\cos k\pi - 1) = \begin{cases} 0 & \text{if } k \text{ is even} \\ \dfrac{4}{k\pi} & \text{if } k \text{ is odd.} \end{cases}$$

Thus, the Fourier series of $f(x)$ on $-2 < x < 2$ is

$$\frac{4}{\pi} \sin \frac{\pi x}{2} + \frac{4}{3\pi} \sin \frac{3\pi x}{2} + \cdots = \frac{4}{\pi} \sum_{n=1}^{\infty} \frac{1}{(2n+1)} \sin \frac{(2n+1)\pi x}{2}.$$

Example 8.3.2

Find the Fourier series on $-\pi < x < \pi$ of $f(x) = |x|$.
Since $|x|$ is even, (FS$_0$) gives

$$a_0 = \frac{1}{\pi} \int_{-\pi}^{\pi} |x|\, dx = \frac{2}{\pi} \int_{0}^{\pi} |x|\, dx = \frac{2}{\pi} \int_{0}^{\pi} x\, dx = \pi.$$

We use the fact that $|x|\cos kx$ is even, and integration by parts, to find

$$a_k = \frac{1}{\pi} \int_{-\pi}^{\pi} |x|\cos kx\, dx = \frac{2}{\pi} \int_{0}^{\pi} |x|\cos kx\, dx = \frac{2}{\pi} \int_{0}^{\pi} x \cos kx\, dx$$

$$= \frac{2}{k\pi} \left\{ x \sin kx \Big]_{0}^{\pi} - \int_{0}^{\pi} \sin kx\, dx \right\} = \frac{2}{k^2 \pi} \cos kx \Big]_{0}^{\pi}$$

$$= \frac{2}{k^2 \pi} (\cos k\pi - 1) = \begin{cases} 0 & \text{if } k \text{ is even} \\ -\dfrac{4}{k^2 \pi} & \text{if } k \text{ is odd.} \end{cases}$$

Since $|x|\sin kx$ is odd,

$$b_k = \frac{1}{\pi} \int_{-\pi}^{\pi} |x|\sin kx\, dx = 0, \quad k = 1, 2, \ldots.$$

Thus, the Fourier series of $|x|$ on $-\pi < x < \pi$ is

$$\frac{\pi}{2} - \frac{4}{\pi} \cos \pi x - \frac{4}{9\pi} \cos 3\pi x + \cdots = \frac{\pi}{2} - \frac{4}{\pi} \sum_{n=0}^{\infty} \frac{1}{(2n + 1)^2} \cos(2n + 1)x.$$

Example 8.3.3

Find the Fourier series on $-1 < x < 1$ of $f(x) = x - x^2$.
Substitution into (FS$_0$) gives

$$a_0 = \int_{-1}^{1} (x - x^2)\, dx = \frac{x^2}{2} - \frac{x^3}{3}\Bigg]_{-1}^{1} = -\frac{2}{3}.$$

Substitution into (FS$_1$) leads to an integral that we can simplify by using the facts that $x \cos k\pi x$ is odd and $x^2 \cos k\pi x$ is even; the resulting integral requires integration by parts twice:

$$a_k = \int_{-1}^{1} x \cos k\pi x\, dx - \int_{-1}^{1} x^2 \cos k\pi x\, dx = -2 \int_{0}^{1} x^2 \cos k\pi x\, dx$$

$$= -\frac{2}{k\pi}\left\{ x^2 \sin k\pi x\Big]_0^1 - \int_0^1 2x \sin k\pi x\, dx \right\} = \frac{4}{k\pi} \int_0^1 x \sin k\pi x\, dx$$

$$= -\frac{4}{k^2\pi^2}\left\{ x \cos k\pi x\Big]_0^1 - \int_0^1 \cos k\pi x\, dx \right\}$$

$$= -\frac{4}{k^2\pi^2}\left\{ \cos k\pi - \frac{1}{k\pi}\Big[\sin k\pi x\Big]_0^1 \right\}$$

$$= (-1)^{k+1} \frac{4}{k^2\pi^2}, \qquad k = 1, 2, \ldots.$$

We can use the facts that $x \sin k\pi x$ is even and $x^2 \sin k\pi x$ is odd to reduce the calculation of b_k to an integral that requires only one integration by parts:

$$b_k = \int_{-1}^{1} x \sin k\pi x\, dx - \int_{-1}^{1} x^2 \sin k\pi x\, dx = 2 \int_0^1 x \sin k\pi x\, dx$$

$$= -\frac{2}{k\pi}\left\{ x \cos k\pi x\Big]_0^1 - \int_0^1 \cos k\pi x\, dx \right\} = -\frac{2}{k\pi}\left\{ \cos k\pi - \frac{1}{k\pi}\Big[\sin k\pi x\Big]_0^1 \right\}$$

$$= (-1)^{k+1} \frac{2}{k\pi}, \qquad k = 1, 2, \ldots.$$

The Fourier series of $x - x^2$ on $-1 < x < 1$ is

$$-\frac{1}{3} + \sum_{k=1}^{\infty} \left[(-1)^{k+1} \frac{4}{k^2\pi^2} \cos k\pi x + (-1)^{k+1} \frac{2}{k\pi} \sin k\pi x \right].$$

The Fourier series of a function may not always converge; even if it does converge, it may not converge to the value of the function. We can, however, describe simple conditions that guarantee convergence to $f(x)$. Recall that a function $g(x)$ is **piecewise continuous** on an interval $-\ell < x < \ell$ provided there are points $-\ell = x_0 < x_1 < \cdots < x_n = \ell$ so that (i) $g(x)$ is continuous on each subinterval $x_j < x < x_{j+1}$, and (ii) the one-sided limits

$$g(x_j + 0) = \lim_{x \to x_j^+} g(x) \quad \text{and} \quad g(x_{j+1} - 0) = \lim_{x \to x_{j+1}^-} g(x)$$

exist, for $j = 0, 1, \ldots, n - 1$. Our convergence criterion requires $f(x)$ and $f'(x)$ to be piecewise continuous.

Theorem: *Suppose $f(x)$ and $f'(x)$ are both piecewise continuous on the interval $-\ell < x < \ell$. Then the Fourier series of $f(x)$ converges*

1. *to $f(x)$ at each point x, $-\ell < x < \ell$, at which $f(x)$ is continuous, and*
2. *to the average of the one-sided limits, $[f(x - 0) + f(x + 0)]/2$, at each point x, $-\ell < x < \ell$, at which $f(x)$ is not continuous.*

This theorem guarantees that the Fourier series of the function $f(x)$ in Example 8.3.1 converges to $f(x)$ if $-2 < x < 0$ or if $0 < x < 2$; it converges to 0 if $x = 0$. The Fourier series of $|x|$ converges to $|x|$ on $-\pi < x < \pi$. The Fourier series of $x - x^2$ converges to $x - x^2$ on $-1 < x < 1$.

In our next example, we use Fourier series to solve a heat flow problem.

Example 8.3.4

Find the solution of the one-dimensional heat equation

(H)
$$\frac{\partial^2 u}{\partial x^2} = \frac{1}{\alpha} \frac{\partial u}{\partial t}, \quad -1 < x < 1, \quad 0 < t$$

that satisfies the boundary condition

(B₃)
$$u(1, t) - u(-1, t) = u_x(1, t) - u_x(-1, t) = 0$$

and the initial condition

(I)
$$u(x, 0) = x^3 - x, \quad -1 < x < 1.$$

Since $x^3 - x$ and its derivative are continuous, our convergence theorem tells us that $x^3 - x$ is equal to its Fourier series on $-1 < x < 1$. Since

$x^3 - x$ and $(x^3 - x) \cos k\pi x$ are odd, substitution into (FS$_0$) and (FS$_1$) gives

$$a_k = 0, \qquad k = 0, 1, 2, \ldots.$$

Substitution into (FS$_2$) leads to an integral that requires integration by parts three times:

$$
\begin{aligned}
b_k &= 2 \int_0^1 (x^3 - x) \sin k\pi x \, dx \\
&= -\frac{2}{k\pi} \left\{ (x^3 - x) \cos k\pi x \Big]_0^1 - \int_0^1 (3x^2 - 1) \cos k\pi x \, dx \right\} \\
&= \frac{2}{k^2\pi^2} \left\{ (3x^2 - 1) \sin k\pi x \Big]_0^1 - \int_0^1 6x \sin k\pi x \, dx \right\} \\
&= \frac{12}{k^3\pi^3} \left\{ x \cos k\pi x \Big]_0^1 - \int_0^1 \cos k\pi x \, dx \right\} \\
&= \frac{12}{k^3\pi^3} \left\{ \cos k\pi = \left[\frac{1}{k\pi} \sin k\pi x \right]_0^1 \right\} = (-1)^k \frac{12}{k^3\pi^3}, \qquad k = 1, 2, \ldots.
\end{aligned}
$$

Thus,

$$x^3 - x = \sum_{k=1}^{\infty} (-1)^k \frac{12}{k^3\pi^3} \sin k\pi x, \qquad -1 < x < 1.$$

In the notation of Example 8.2.3, we have

$$u(x, 0) = x^3 - x = \sum_{k=1}^{\infty} (-1)^k \frac{12}{k^3\pi^3} u_k(x, 0).$$

The function

$$u(x, t) = \sum_{k=1}^{\infty} (-1)^k \frac{12}{k^3\pi^3} u_k(x, t) = \sum_{k=1}^{\infty} (-1)^k \frac{12}{k^3\pi^3} e^{-k^2\pi^2\alpha t} \sin k\pi x$$

is a solution of (H), satisfies (B$_3$), and matches (I).

Let's summarize our information and Fourier series.

FOURIER SERIES

The **Fourier series** of $f(x)$ on the interval $-\ell < x < \ell$ is the series

$$\frac{a_0}{2} + \sum_{j=1}^{\infty} \left(a_j \cos \frac{j\pi x}{\ell} + b_j \sin \frac{j\pi x}{\ell} \right),$$

where

(FS$_0$) $a_0 = \dfrac{1}{\ell} \displaystyle\int_{-\ell}^{\ell} f(x)\,dx$

(FS$_1$) $a_k = \dfrac{1}{\ell} \displaystyle\int_{-\ell}^{\ell} f(x)\cos\dfrac{k\pi x}{\ell}\,dx,$ $k = 1, 2, \ldots.$

(FS$_2$) $b_k = \dfrac{1}{\ell} \displaystyle\int_{-\ell}^{\ell} f(x)\sin\dfrac{k\pi x}{\ell}\,dx,$ $k = 1, 2, \ldots.$

If $f(x)$ and $f'(x)$ are piecewise continuous on $-\ell < x < \ell$, then the Fourier series of $f(x)$ converges

1.   to $f(x)$ at each point x, $-\ell < x < \ell$, at which $f(x)$ is continuous, and

2.   to $[f(x - 0) + f(x + 0)]/2$ at each point x, $-\ell < x < \ell$, at which $f(x)$ is not continuous.

When calculating Fourier series, it is worth checking to see whether the integrands have symmetric graphs. If an integrand $g(x)$ is **odd** [that is, if $g(-x) = -g(x)$ for all x], then $\int_{-\ell}^{\ell} g(x)\,dx = 0$. If $g(x)$ is **even** [$g(-x) = g(x)$ for all x], then $\int_{-\ell}^{\ell} g(x)\,dx = 2\int_0^{\ell} g(x)\,dx$.

Notes

1.   A technicality

Although we have enough information to conclude that *formally* the function $u(x, t)$ in Example 8.3.4 solves the given boundary–initial-value problem, we have not actually *proved* that it does. To do so, we would have to show that the series defining $u(x, t)$ converges to a continuous function in the region $-\ell < x < \ell$, $0 < t$, that $u(x, t)$ has partial derivatives $\partial^2 u/\partial x^2$ and $\partial u/\partial t$ that yield an identity when substituted into (H), and that $u(x, t)$ matches the given boundary and initial conditions (see Note 1, Section 8.2). We will not attempt to do this here. However, the formal solutions to the specific boundary–initial-problems we will be dealing with are indeed solutions to these problems.

2.   On the convergence of Fourier series

If $f(x)$ is defined on $-\ell < x < \ell$, then the *periodic extension* of $f(x)$ (of period 2ℓ) is the function $F(x)$ that agrees with $f(x)$ on $-\ell < x < \ell$ and repeats every 2ℓ units:

$$F(x) = f(x), \qquad -\ell < x < \ell$$

$$F(x + 2\ell) = f(x) \qquad \text{for all } x.$$

If $f(x)$ and $f'(x)$ are piecewise continuous on $-\ell < x < \ell$, then the Fourier series of $f(x)$ converges, not only on $-\ell < x < \ell$ but *for all* x; it converges to $F(x)$ at each point x at which $F(x)$ is continuous and to $[F(x - 0) + F(x + 0)]/2$ at each point x at which $F(x)$ is not continuous. For example, the Fourier series of the functions $f(x)$ in Examples 8.3.1 and 8.3.2 converge to the functions whose graphs are sketched in Figures 8.8 and 8.9.

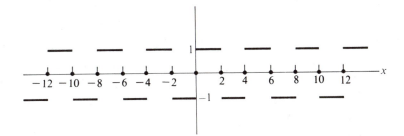

FIGURE 8.8 The Fourier series of $f(x) = \begin{cases} -1, & -2 < x < 0 \\ 1, & 0 < x < 2 \end{cases}$ on $-2 < x < 2$

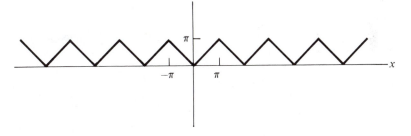

FIGURE 8.9 The Fourier series of $f(x) = |x|$ on $-\pi < x < \pi$

3. On the uniqueness of Fourier series expressions

We saw that if

(A) $$f(x) = \frac{a_0}{2} + \sum_{j=1}^{\infty} \left(a_j \cos\frac{j\pi x}{\ell} + b_j \sin\frac{j\pi x}{\ell} \right), \qquad -\ell < x < \ell,$$

then each of the coefficients a_k and b_k could be expressed as an integral involving $f(x)$ and the trigonometric functions $\cos(k\pi x/\ell)$ and $\sin(k\pi x/\ell)$. This means that the coefficients in such an expression are uniquely determined. After all, if $f(x)$ could be expressed in terms of the same trigonometric functions, on the same interval, but using other coefficients,

$$f(x) = \frac{a_0^*}{2} + \sum_{j=1}^{\infty} \left(a_j^* \cos\frac{j\pi x}{\ell} + b_j^* \sin\frac{j\pi x}{\ell} \right), \qquad -\ell < x < \ell,$$

then corresponding coefficients a_k and a_k^* (respectively, b_k and b_k^*) would be given by the same integral formula. This observation can often be used to find the coefficients in a Fourier series without calculating integrals (see Exercise 8). Keep in mind, however, that we have *not* said that (A) is the only trigonometric series expression for $f(x)$. We will see, in the exercises and examples in this chapter, that $f(x)$ may have expressions in terms of different trigonometric functions, or on different intervals, that bear little resemblance to (A).

EXERCISES

1. Decide whether $f(x)$ is odd, even, or neither.
    a. $f(x) = x - 5x^7$
    b. $f(x) = 1 + x^3$
    c. $f(x) = 1 - 3x^6$
    d. $f(x) = |1 - x|$
    e. $f(x) = 1 - |x|$
    f. $f(x) = x^2 \sin 3\pi x$
    g. $f(x) = e^x$
    h. $f(x) = e^x + e^{-x}$
    i. $f(x) = e^x - e^{-x}$

In Exercises 2 through 9,

a. find the Fourier series of $f(x)$ on the given interval, and
b. use the convergence theorem of this section to determine the function to which the Fourier series converges on that interval.

2. $f(x) = \begin{cases} 0, & -\ell < x < 0 \\ 1, & 0 \le x < \ell \end{cases}; \quad -\ell < x < \ell$
3. $f(x) = x; \quad -\ell < x < \ell$
4. $f(x) = 1 - |x|; \quad -1 < x < 1$
5. $f(x) = \begin{cases} 2 + x, & -2 < x < 0 \\ x, & 0 \le x < 2 \end{cases}; \quad -2 < x < 2$
6. $f(x) = x^2; \quad -1 < x < 1$
7. $f(x) = x^3; \quad -\pi < x < \pi$
8. $f(x) = \sin^2 x; \quad -\pi < x < \pi \quad$ (*Hint:* Think.)
9. $f(x) = e^x; \quad -1 < x < 1$

In Exercises 10 through 12, find the solution of the one-dimensional heat equation that satisfies the boundary conditions

(B$_3$) $u(1, t) - u(-1, t) = u_x(1, t) - u_x(-1, t) = 0$

and matches the given initial condition. This models a wire (of length 2) with the given initial temperature distribution, which is suddenly bent to form a circle.

10. $u(x, 0) = \begin{cases} 0, & -1 < x < 0 \\ 1/2, & x = 0 \\ 1, & 0 < x < 1 \end{cases}$
11. $u(x, 0) = 1 - x^2; \quad -1 < x < 1$
12. $u(x, 0) = 2x + x^2; \quad -1 < x < 1$

Some more abstract problems:

13. a. Show that if $f(x)$ and $g(x)$ are both even, then $f(x)g(x)$ is even.

b.   Show that if $f(x)$ and $g(x)$ are both odd, then $f(x)g(x)$ is even.

c.   Show that if $f(x)$ is even and $g(x)$ is odd, then $f(x)g(x)$ is odd.

14. Show that

$$\int_{-\ell}^{\ell} \cos\frac{j\pi x}{\ell} \cos\frac{k\pi x}{\ell}\, dx = \int_{-\ell}^{\ell} \sin\frac{j\pi x}{\ell} \sin\frac{k\pi x}{\ell}\, dx = \begin{cases} 0 & \text{if } j \neq k \\ \ell & \text{if } j = k. \end{cases}$$

(*Hint:* Use the trigonometric identities $\cos A \cos B = \frac{1}{2}[\cos(A - B) + \cos(A + B)]$ and $\sin A \sin B = \frac{1}{2}[\cos(A - B) - \cos(A + B)]$.)

15. Show that if

$$f(x) = \frac{a_0}{2} + \sum_{j=1}^{\infty} \left(a_j \cos\frac{j\pi x}{\ell} + b_j \sin\frac{j\pi x}{\ell} \right),$$

then

$$\int_{-\ell}^{\ell} f(x) \sin\frac{k\pi x}{\ell}\, dx = b_k \ell.$$

16. a.   Show that if $f(x)$ is even, then its Fourier series on $-\ell < x < \ell$ is of the form

$$\frac{a_0}{2} + \sum_{j=1}^{\infty} a_j \cos\frac{j\pi x}{\ell},$$

where

$$a_k = \frac{2}{\ell} \int_{0}^{\ell} f(x) \cos\frac{k\pi x}{\ell}\, dx, \qquad k = 0, 1, 2, \ldots.$$

b.   Show that if $f(x)$ is odd, then its Fourier series on $-\ell < x < \ell$ is of the form

$$\sum_{j=1}^{\infty} b_j \sin\frac{j\pi x}{\ell},$$

where

$$b_k = \frac{2}{\ell} \int_{0}^{\ell} f(x) \sin\frac{k\pi x}{\ell}\, dx, \qquad k = 1, 2, \ldots.$$

17. a.   Show that the coefficients of the Fourier series of $cf(x)$ on $-\ell < x < \ell$ can be gotten by multiplying the coefficients of the Fourier series of $f(x)$ on $-\ell < x < \ell$ by c.

b.   Show that the coefficients of the Fourier series of $[f(x) + g(x)]$ on $-\ell < x < \ell$ can be gotten by adding the Fourier coefficients of $f(x)$ on $-\ell < x < \ell$ to the corresponding Fourier coefficients of $g(x)$ on $-\ell < x < \ell$. (*Note:* The intervals have to be the same.)

18. Use the observations in Exercise 17 to obtain each of the following Fourier series from the Fourier series calculated in the preceding examples.

a. $h(x) = \pi - |x|;\qquad -\pi < x < \pi$

b. $h(x) = \begin{cases} 0, & -2 < x < 0 \\ 1, & 0 < x < 2 \end{cases};\qquad -2 < x < 2$

c. $h(x) = x^3 - x^2;\qquad -1 < x < 1$

19. Suppose $f(\ell y) = \alpha(\ell)f(y)$ for all y. Show that the coefficients of the Fourier

series of $f(x)$ on $-\ell < x < \ell$ can be obtained by multiplying the coefficients of the Fourier series of $f(x)$ on $-1 < x < 1$ by $\alpha(\ell)$.

20. Use the observation in Exercise 19 to obtain the following Fourier series from the Fourier series calculated in the preceding exercises and examples.

   a. $f(x) = x^2$; $-\ell < x < \ell$
   b. $f(x) = x^3$; $-1 < x < 1$
   c. $f(x) = |x|$; $-\ell < x < \ell$
   d. $f(x) = \begin{cases} -1, & -\ell < x < 0, \\ 1, & 0 < x < \ell; \end{cases}$ $-\ell < x < \ell$

8.4   SINE SERIES AND COSINE SERIES

In order to solve problems like the ones in Examples 8.2.1 and 8.2.2, we need to express the initial temperature distribution $f(x)$, which is defined for $0 < x < \ell$, in the form

$$f(x) = \sum_{j=1}^{\infty} b_j \sin \frac{j\pi x}{\ell}, \qquad 0 < x < \ell,$$

or

$$f(x) = \frac{a_0}{2} + \sum_{j=1}^{\infty} a_j \cos \frac{j\pi x}{\ell}, \qquad 0 < x < \ell.$$

Such series are called **sine series** and **cosine series,** respectively. In this section we find such expressions by calculating the Fourier series on $-\ell < x < \ell$ of the **odd extension** of $f(x)$,

$$O_f(x) = \begin{cases} -f(-x), & -\ell < x < 0 \\ f(x), & 0 < x < \ell \end{cases}$$

and the **even extension** of $f(x)$

$$E_f(x) = \begin{cases} f(-x), & -\ell < x < 0 \\ f(x), & 0 < x < \ell. \end{cases}$$

Note that the formulas defining these functions are precisely what is needed to guarantee that $O_f(x)$ and $E_f(x)$ *agree with* $f(x)$ *for* $0 < x < \ell$, *that* $O_f(x)$ *is an odd function, and that* $E_f(x)$ *is an even function.* We have sketched the **graphs** of a function $f(x)$ and its odd and even extensions in Figure 8.10. Since $O_f(x)$ and $O_f(x)\cos(k\pi x/\ell)$ are odd, we see that

$$\int_{-\ell}^{\ell} O_f(x)\, dx = \int_{-\ell}^{\ell} O_f(x) \cos \frac{k\pi x}{\ell} = 0.$$

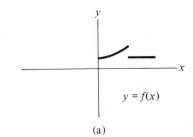

(a)

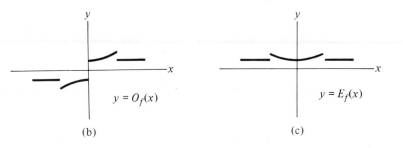

(b)   (c)

FIGURE 8.10

Thus, *the Fourier series of $O_f(x)$ is a sine series*

$$\sum_{j=1}^{\infty} b_j \sin \frac{j\pi x}{\ell},$$

where [using the fact that $O_f(x)\sin (k\pi x/\ell)$ is even]

$$b_k = \frac{1}{\ell} \int_{-\ell}^{\ell} O_f(x) \sin \frac{k\pi x}{\ell}\, dx = \frac{2}{\ell} \int_{0}^{\ell} O_f(x) \sin \frac{k\pi x}{\ell}\, dx$$

$$= \frac{2}{\ell} \int_{0}^{\ell} f(x) \sin \frac{k\pi x}{\ell}\, dx, \quad k = 1, 2, \ldots.$$

Similarly [since $E_f(x)$ and $E_f(x)\cos(k\pi x/\ell)$ are even, and $E_f(x)\sin(k\pi x/\ell)$ is odd], *the Fourier series of $E_f(x)$ is a cosine series*

$$\frac{a_0}{2} + \sum_{j=1}^{\infty} a_j \cos \frac{j\pi x}{\ell},$$

where

$$a_0 = \frac{2}{\ell} \int_{0}^{\ell} f(x)\, dx$$

and

$$a_k = \frac{2}{\ell} \int_0^\ell f(x) \cos \frac{k\pi x}{\ell} \, dx, \qquad k = 1, 2, \ldots .$$

If $f(x)$ and $f'(x)$ are piecewise continuous on $0 < x < \ell$, then $O_f(x)$, $O_f'(x)$, $E_f(x)$, and $E_f'(x)$ will be piecewise continuous on $-\ell < x < \ell$. The convergence theorem of Section 8.3 tells us that the Fourier series of $O_f(x)$ converges on $-\ell < x < \ell$ to $O_f(x)$ or to $[O_f(x - 0) + O_f(x + 0)]/2$, and the Fourier series of $E_f(x)$ converges to $E_f(x)$ or to $[E_f(x - 0) + E_f(x + 0)]/2$. Since $O_f(x) = E_f(x) = f(x)$ for $0 < x < \ell$, we see that *the Fourier series of $O_f(x)$ and $E_f(x)$ both converge to $f(x)$ at each point x, $0 < x < \ell$, at which $f(x)$ is continuous, and to $[f(x - 0) + f(x + 0)]/2$ at each point x, $0 < x < \ell$, at which $f(x)$ is not continuous.*

Example 8.4.1

Find a sine series expression for $f(x) = 1$ on $0 < x < 2$.
The odd extension of $f(x)$ is

$$O_f(x) = \begin{cases} -1, & -2 < x < 0 \\ 1, & 0 < x < 2. \end{cases}$$

The Fourier series of $O_f(x)$, which we calculated in Example 8.3.1, provides a sine series expression for $f(x)$:

$$1 = \frac{4}{\pi} \sum_{n=0}^{\infty} \frac{1}{2n + 1} \sin \frac{(2n + 1)\pi x}{2}, \qquad 0 < x < 2.$$

Note that the cosine series expression for $f(x)$ is just $f(x) = 1$.

Example 8.4.2

Find a cosine series expression for $f(x) = x$ on $0 < x < \pi$.
The even extension of $f(x)$ is

$$E_f(x) = \begin{cases} -x, & -\pi < x < 0 \\ x, & 0 < x < \pi. \end{cases}$$

That is, $E_f(x) = |x|$. The Fourier series of $|x|$, which we calculated in Example 8.3.2, provides a cosine series expression for $f(x) = x$:

$$x = \frac{\pi}{2} - \frac{4}{\pi} \sum_{n=0}^{\infty} \frac{1}{(2n + 1)^2} \cos(2n + 1)x, \qquad 0 < x < \pi.$$

Example 8.4.3

Find sine series and cosine series expressions for $f(x) = x - x^2$ on $0 < x < 1$.

The Fourier series of $O_f(x)$ is a sine series with coefficients

$$
\begin{aligned}
b_k &= 2 \int_0^1 f(x) \sin k\pi x \, dx = 2 \int_0^1 (x - x^2) \sin k\pi x \, dx \\
&= -\frac{2}{k\pi} \left\{ (x - x^2) \cos k\pi x \Big|_0^1 - \int_0^1 (1 - 2x) \cos k\pi x \, dx \right\} \\
&= \frac{2}{k^2 \pi^2} \left\{ (1 - 2x) \sin k\pi x \Big|_0^1 + 2 \int_0^1 \sin k\pi x \, dx \right\} \\
&= \frac{4}{k^3 \pi^3} \cos k\pi x \Big|_0^1 = -\frac{4}{k^3 \pi^3} (\cos k\pi - 1) \\
&= \begin{cases} 0 & \text{if } k \text{ is even} \\ \dfrac{8}{k^3 \pi^3} & \text{if } k \text{ is odd.} \end{cases}
\end{aligned}
$$

Thus

$$
x - x^2 = \frac{8}{\pi^3} \sum_{n=0}^{\infty} \frac{1}{(2n + 1)^3} \sin(2n + 1)\pi x, \qquad 0 < x < 1.
$$

The Fourier series of $E_f(x)$ is a cosine series with coefficients

$$
a_0 = 2 \int_0^1 f(x) \, dx = 2 \int_0^1 (x - x^2) \, dx = \frac{1}{3}
$$

and

$$
\begin{aligned}
a_k &= 2 \int_0^1 f(x) \cos k\pi x \, dx = 2 \int_0^1 (x - x^2) \cos k\pi x \, dx \\
&= \frac{2}{k\pi} \left\{ (x - x^2) \sin k\pi x \Big|_0^1 - \int_0^1 (1 - 2x) \sin k\pi x \, dx \right\} \\
&= \frac{2}{k^2 \pi^2} \left\{ (1 - 2x) \cos k\pi x \Big|_0^1 + 2 \int_0^1 \cos k\pi x \, dx \right\} \\
&= \frac{2}{k^2 \pi^2} \left\{ (- \cos k\pi - 1) + \frac{2}{k\pi} \left[\sin k\pi x \Big|_0^1 \right] \right\} \\
&= \begin{cases} -\dfrac{4}{k^2 \pi^2} & \text{if } k \text{ is even} \\ 0 & \text{if } k \text{ is odd.} \end{cases}
\end{aligned}
$$

Thus

$$x - x^2 = \frac{1}{6} - \frac{1}{\pi^2} \sum_{n=1}^{\infty} \frac{1}{n^2} \cos 2n\pi x, \qquad 0 < x < 1.$$

Note that the Fourier series of $x - x^2$ on $-1 < x < 1$ (Example 8.3.3), the sine series of $x - x^2$ on $0 < x < 1$, and the cosine series of $x - x^2$ on $0 < x < 1$ are three different series.

In our next two examples, we use sine and cosine series to solve heat flow problems.

Example 8.4.4

Find the solution of the heat equation

(H)
$$\frac{\partial^2 u}{\partial x^2} = \frac{1}{\alpha} \frac{\partial u}{\partial t}, \qquad 0 < x < 1, \quad 0 < t$$

that satisfies the boundary conditions

(B$_1$)
$$u(0, t) = u(1, t) = 0$$

and matches the initial condition

(I)
$$u(x, 0) = x - x^2, \qquad 0 < x < 1.$$

Recall from Example 8.2.1 that in order to solve this problem we must express $u(x, 0) = x - x^2$ in terms of the functions $u_n(x, 0) = \sin n\pi x$. We found in Example 8.4.3 that

$$x - x^2 = \frac{8}{\pi^3} \sum_{n=0}^{\infty} \frac{1}{(2n + 1)^3} \sin(2n + 1)\pi x, \qquad 0 < x < 1.$$

The function

$$u(x, t) = \frac{8}{\pi^3} \sum_{n=0}^{\infty} \frac{1}{(2n + 1)^3} e^{-(2n + 1)^2 \pi^2 \alpha t} \sin(2n + 1)\pi x$$

is a solution of (H), satisfies (B$_1$), and matches (I).

Note that this boundary–initial-value problem models a rod with both ends kept at $0°$. For any x, $\lim_{t\to\infty} u(x, t) = 0$. Thus the temperature at each point tends toward zero over time.

Example 8.4.5

Find the solution of the heat equation

(H)
$$\frac{\partial^2 u}{\partial x^2} = \frac{1}{\alpha} \frac{\partial u}{\partial t}, \qquad 0 < x < 2, \quad 0 < t$$

that satisfies the boundary conditions

(B₂)
$$u_x(0, t) = u_x(2, t) = 0$$

and matches the initial condition

(I)
$$u(x, 0) = \begin{cases} 3, & 0 < x < 1 \\ 3/2, & x = 1 \\ 0, & 1 < x < 2. \end{cases}$$

Recall from Example 8.2.2 that in order to solve this problem we need a cosine series expression for $u(x, 0)$. The Fourier series of the even extension of $u(x, 0)$ is such a series; it has coefficients

$$a_0 = \frac{2}{2} \int_0^2 u(x, 0) \, dx = \int_0^1 3 \, dx = 3$$

and

$$a_k = \int_0^2 u(x, 0) \cos \frac{k\pi x}{2} \, dx = \int_0^1 3 \cos \frac{k\pi x}{2} \, dx = \frac{6}{k\pi} \sin \frac{k\pi x}{2} \Big]_0^1$$

$$= \frac{6}{k\pi} \sin \frac{k\pi}{2} = \begin{cases} 0 & \text{if } k \text{ is even} \\ (-1)^n \dfrac{6}{(2n + 1)\pi} & \text{if } k = 2n + 1. \end{cases}$$

Thus

$$u(x, 0) = \frac{3}{2} + \frac{6}{\pi} \sum_{n=0}^{\infty} \frac{(-1)^n}{2n + 1} \cos \frac{(2n + 1)\pi x}{2}, \qquad 0 < x < 2.$$

The function

$$u(x, t) = \frac{3}{2} + \frac{6}{\pi} \sum_{n=0}^{\infty} \frac{(-1)^n}{2n + 1} e^{-(2n + 1)^2 \pi^2 \alpha t / 4} \cos \frac{(2n + 1)\pi x}{2}$$

is a solution of (H), satisfies (B₂), and matches (I).

Note that this boundary–initial-value problem models a rod with insulated ends, which starts with half of it at 3° and the other half at 0°. Even though the initial temperature distribution is not continuous, $u(x, t)$ is continuous throughout the region $0 < x < 2$, $0 < t$. Thus the distribution of heat smooths out immediately. Note also that for any x, $\lim_{t \to \infty} u(x, t) = 3/2$. Thus, over time the temperature at each point tends toward the average value of the initial temperature distribution.

Let's summarize.

SINE SERIES AND COSINE SERIES

If $f(x)$ is defined on $0 < x < \ell$, then the **odd extension** of $f(x)$, $O_f(x)$, and the **even extension,** $E_f(x)$, are, respectively, the odd and even functions that agree with $f(x)$ on $0 < x < \ell$. The **sine series** of $f(x)$ on $0 < x < \ell$ is obtained by calculating the Fourier series of $O_f(x)$:

$$\sum_{j=1}^{\infty} b_j \sin \frac{j\pi x}{\ell},$$

where

$$b_k = \frac{2}{\ell} \int_0^{\ell} f(x) \sin \frac{k\pi x}{\ell}\, dx, \qquad k = 1, 2, \dots .$$

The **cosine series** of $f(x)$ on $0 < x < \ell$ is obtained by calculating the Fourier series of $E_f(x)$:

$$\frac{a_0}{2} + \sum_{j=1}^{\infty} a_j \cos \frac{j\pi x}{\ell},$$

where

$$a_k = \frac{2}{\ell} \int_0^{\ell} f(x) \cos \frac{k\pi x}{\ell}\, dx, \qquad k = 0, 1, 2, \dots .$$

If $f(x)$ and $f'(x)$ are piecewise continuous on $0 < x < \ell$, then both of these series converge to $f(x)$ at each point x, $0 < x < \ell$, at which $f(x)$ is continuous, and to $[f(x - 0) + f(x + 0)]/2$ at each point x, $0 < x < \ell$, at which $f(x)$ is not continuous.

EXERCISES

In Exercises 1 through 7,

a. find the cosine series of $f(x)$ on the given interval,
b. find the sine series of $f(x)$ on the given interval, and
c. determine the function to which these series converge on the given interval

1. $f(x) = 1; \quad 0 < x < \ell$ (compare Example 8.4.1)

2. $f(x) = \begin{cases} 1, & 0 < x < 1 \\ -1, & 1 \le x < 2 \end{cases}; \quad 0 < x < 2$

3. $f(x) = x; \quad 0 < x < \ell$ (compare Exercise 3 of Section 8.3, and Example 8.4.2)

4. $f(x) = 1 - x; \quad 0 < x < 1$ (compare Exercise 4, Section 8.3)

5. $f(x) = \begin{cases} x, & 0 < x < 1 \\ 2 - x, & 1 < x < 2 \end{cases}; \quad 0 < x < 2$

6. $f(x) = x^2; \quad 0 < x < 1$ (compare Exercise 6, Section 8.3)

7. $f(x) = \cos x; \quad 0 < x < \pi$
   (*Hint:* $\sin A \cos B = \frac{1}{2}[\sin(A + B) + \sin(A - B)]$)

In Exercises 8 through 11, find the solution of the one-dimensional heat equation that matches the given initial condition and

a. satisfies the boundary conditions

    (B$_2$) $u_x(0, t) = u_x(2, t) = 0,$

b. satisfies the boundary conditions

    (B$_1$) $u(0, t) = u(2, t) = 0.$

8. $u(x, 0) = 1; \quad 0 < x < 2$

9. $u(x, 0) = x; \quad 0 < x < 2$

10. $u(x, 0) = \begin{cases} x, & 0 < x < 1 \\ 2 - x, & 1 < x < 2 \end{cases}$

11. $u(x, 0) = \begin{cases} 0, & 0 < x < 1 \\ 1/2, & x = 1 \\ 1, & 1 < x < 2 \end{cases}$

Some more abstract problems:

12. a. The average value of a function $f(x)$ on an interval $a < x < b$ is defined to be $f_{avg} = (\int_a^b f(x)\, dx)/(b - a)$. Show that as $t \to \infty$, the temperature at each point in a wire insulated at both ends approaches the average value of the initial temperature distribution.

    b. How does the temperature distribution in a wire whose ends are kept at $0°$ behave as $t \to \infty$?

    c. How does the temperature distribution in a wire bent into a circle behave as $t \to \infty$?

13. Suppose $F(x)$ and $F'(x)$ are piecewise continuous on $0 < x < 2\ell$, and the graph

of $F(x)$ is symmetric about the line $x = \ell$. Show that the sine series of $F(x)$ on $0 < x < 2\ell$ is of the form

$$\sum_{n=0}^{\infty} c_n \sin \frac{(2n + 1)\pi x}{2\ell},$$

where

$$c_n = \frac{4}{2\ell} \int_0^{\ell} F(x) \sin \frac{(2n + 1)\pi x}{2\ell} \, dx.$$

[*Hint:* The function $g(u) = F(u + \ell)$ is symmetric about $u = 0$. What about $\sin\{k\pi(u + \ell)/2\ell\}$?]

14. a. By applying the result of Exercise 13 to the function $F(x)$ that agrees with $f(x)$ on $0 < x < 1$ and is symmetric about the line $x = 1$, find an expression for each of the following in terms of the functions $\sin\{(2n + 1)\pi x/2\}$:
        (i) $f(x) = 1, \quad 0 < x < 1$ (ii) $f(x) = x, \quad 0 < x < 1$
    b. For each of these choices of $f(x)$, find the solution of the one-dimensional heat equation that satisfies the boundary conditions $u(0, t) = u_x(1, t) = 0$ and matches the initial condition $u(x, 0) = f(x)$. (*Hint:* See Exercise 9, Section 8.2.)

8.5   THE ONE-DIMENSIONAL WAVE EQUATION

In the preceding sections we focused on solutions of the one-dimensional heat equation. In this section, we shall see how similar techniques can be adapted to boundary-value problems involving the **one-dimensional wave equation**

(W)
$$\frac{\partial^2 u}{\partial x^2} = \frac{1}{\alpha^2} \frac{\partial^2 u}{\partial t^2},$$

which has somewhat different characteristics. We start with a model that shows how this p.d.e. is derived from a physical situation.

Example 8.5.1   The Vibrating String

A string is stretched between two fixed ends. We wish to model its vibrations in response to being plucked, struck, or otherwise displaced.

We think of the string as stretched horizontally so that at equilibrium it is a straight line segment, say $0 < x < \ell$, $y = 0$ (Figure 8.11). We assume that all horizontal forces are balanced, so the net force is vertical and each point on the string undergoes only vertical displacement. We denote the displacement from equilibrium of the point x at time t by the function of two variables $u(x, t)$. If t_0 is fixed, then $u(x, t_0)$ is a function of x whose graph gives the

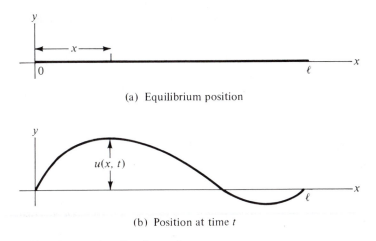

(a) Equilibrium position

(b) Position at time t

FIGURE 8.11   The vibrating string

shape of the string at time t_0. If x_0 is fixed, then $u(x_0, t)$ is a function of t describing the motion of a single point on the string.

Our analysis is based on a physical principle, reminiscent of Hooke's law, which states that the vertical component of the force of tension at any point is proportional to the slope u_x of the string there.

A small piece of string with left endpoint x and right endpoint $x + \Delta x$ is subject to external forces from the tension at each end (Figure 8.12). At the right end, it is pulled up by a tensile force $F_+(t)$ proportional to the slope $u_x(x + \Delta x, t)$, and at the left end it is pulled down by a force $F_-(t)$ proportional to $u_x(x, t)$. The net upward force is then

$$F(t) = F_+(t) - F_-(t) = k[u_x(x + \Delta x, t) - u_x(x, t)]$$

Applying the mean value theorem to u_x, we can rewrite this as

$$F(t) = k(\Delta x)u_{xx}(\bar{x}, t), \quad \text{where } x < \bar{x} < x + \Delta x.$$

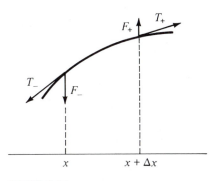

FIGURE 8.12

On the other hand, Newton's second law of motion, $F = ma$, relates this force to the mass $m = \delta \Delta x$ of the piece (δ is the density of the string) and the acceleration. Although the acceleration generally varies along the string, when Δx is small each piece of the string behaves approximately as if the acceleration were constant throughout the piece: $a = u_{tt}(x, t)$. Thus

$$k(\Delta x)u_{xx}(\bar{x}, t) = \delta(\Delta x)u_{tt}(x, t)$$

or, setting $\alpha = \sqrt{k/\delta}$,

$$u_{xx}(\bar{x}, t) = \frac{1}{\alpha^2} u_{tt}(x, t).$$

In the limit as $\Delta x \to 0$, the points x and $\bar{x}$ merge, and we obtain the one-dimensional wave equation

(W)
$$\frac{\partial^2 u}{\partial x^2} = \frac{1}{\alpha^2} \frac{\partial^2 u}{\partial t^2}.$$

The most natural boundary condition for this problem is that the ends stay fixed:

(B)
$$u(0, t) = u(\ell, t) = 0.$$

The wave equation differs from the heat equation in that the right side contains the second derivative with respect to time instead of the first. Correspondingly, our initial conditions should specify both the initial configuration and the initial velocity at each point of the string:

(I)
$$u(x, 0) = f(x) \quad \text{and} \quad u_t(x, 0) = g(x).$$

The boundary-value problem derived in the preceding example can be solved by methods similar to those we used for the heat equation. Ignoring for the moment the initial conditions, we look for nontrivial solutions of (W) that satisfy (B) and are of the form

(S)
$$u = X(x)T(t).$$

Substitution into (W) gives $X''T = XT''/\alpha^2$, or

$$\frac{X''}{X} = \frac{T''}{\alpha^2 T}.$$

Since each side of this equation depends on a different variable, both must have a common constant value. If we call this constant $-\lambda$, we are led to two o.d.e.'s:

(W_x) $$X'' + \lambda X = 0$$

(W_t) $$T'' + \lambda \alpha^2 T = 0.$$

Substitution of (S) into (B) leads to

(B') $$X(0) = X(\ell) = 0.$$

We found in Example 8.2.1 that (W_x) has nontrivial solutions satisfying (B') only if

$$\lambda = \frac{n^2 \pi^2}{\ell^2}, \qquad n > 0 \text{ an integer}$$

and that corresponding to each such choice of λ the nontrivial solutions of (W_x) are constant multiples of

$$X_n = \sin \frac{n\pi x}{\ell}.$$

Using the same value of λ in (H_t), we find two independent solutions

$$T_n = \cos \frac{\alpha n \pi t}{\ell} \qquad \text{and} \qquad T_n^* = \sin \frac{\alpha n \pi t}{\ell}.$$

Substituting these forms for X and T into (S), we find that for each integer $n > 0$ there are two independent solutions to (W) that satisfy (B):

$$u_n = \sin \frac{n\pi x}{\ell} \cos \frac{\alpha n \pi t}{\ell} \qquad \text{and} \qquad u_n^* = \sin \frac{n\pi x}{\ell} \sin \frac{\alpha n \pi t}{\ell}.$$

Of course, our problem also had initial conditions. Based on our experience with the heat equation, we expect to find a solution in the form of an infinite combination of the functions u_n and u_n^*,

$$u(x, t) = \sum_{n=1}^{\infty} (b_n u_n + b_n^* u_n^*)$$

$$= \sum_{n=1}^{\infty} \sin \frac{n\pi x}{\ell} \left(b_n \cos \frac{\alpha n \pi t}{\ell} + b_n^* \sin \frac{\alpha n \pi t}{\ell} \right).$$

Note that when $t = 0$,

$$u(x, 0) = \sum_{n=1}^{\infty} b_n \sin \frac{n\pi x}{\ell}.$$

Thus, *to match the initial condition $u(x, 0) = f(x)$, we must take b_n to be the nth coefficient in the sine series for $f(x)$ on $0 < x < \ell$.* On the other hand (assuming we can differentiate term by term),

$$u_t(x, 0) = \sum_{n=1}^{\infty} \frac{\alpha n \pi}{\ell} b_n^* \sin \frac{n\pi x}{\ell}.$$

To match the initial condition $u_t(x, 0) = g(x)$, we must take $(\alpha n\pi/\ell)b_n^*$ to be the nth coefficient of the sine series for $g(x)$ on $0 < x < \ell$.

▪ Example 8.5.2

Find the solution to (W), subject to (B) with $\ell = 3$, and satisfying the initial conditions

$$u(x, 0) = 5 \sin \frac{\pi x}{3} - \sin \frac{2\pi x}{3} \quad \text{and} \quad u_t(x, 0) = 2 \sin \frac{4\pi x}{3}.$$

To match our initial conditions with the series

$$u(x, t) = \sum_{n=1}^{\infty} \sin \frac{n\pi x}{3} \left(b_n \cos \frac{\alpha n \pi t}{3} + b_n^* \sin \frac{\alpha n \pi t}{3} \right)$$

we need to choose the coefficients so that

$$5 \sin \frac{\pi x}{3} - \sin \frac{2\pi x}{3} = u(x, 0) = \sum_{n=1}^{\infty} b_n \sin \frac{n\pi x}{3}$$

and

$$2 \sin \frac{4\pi x}{3} = u_t(x, 0) = \sum_{n=1}^{\infty} \frac{\alpha n \pi}{3} b_n^* \sin \frac{n\pi x}{3}.$$

Thus, we need to take

$$b_1 = 5, \quad b_2 = -1, \quad \frac{4\alpha\pi}{3} b_4^* = 2$$

and all the other coefficients to be zero. The solution is

$$u(x, t) = 5 \sin \frac{\pi x}{3} \cos \frac{\alpha \pi t}{3} - \sin \frac{2\pi x}{3} \cos \frac{2\alpha \pi t}{3} + \frac{3}{2\alpha \pi} \sin \frac{4\pi x}{3} \sin \frac{4\alpha \pi t}{3}.$$

Example 8.5.3

Find the solution of (W), subject to (B) with $\ell = 1$, and satisfying the initial conditions

$$u(x, 0) = x(1 - x) \quad \text{and} \quad u_t(x, 0) = 0.$$

This corresponds to a string of length 1, which is deformed into the shape of a parabola and then released.

We expect our solution to be of the form

$$u(x, t) = \sum_{n=1}^{\infty} \sin n\pi x \, (b_n \cos \alpha n\pi t + b_n^* \sin \alpha n\pi t),$$

where b_n is the nth sine series coefficient of $x(1 - x)$ (see Example 8.4.3),

$$b_n = \begin{cases} 0 & \text{if } n \text{ is even} \\ \dfrac{8}{n^3 \pi^3} & \text{if } n \text{ is odd} \end{cases}$$

and $\alpha n\pi b_n^*$ is the nth sine series coefficient of 0, so

$$b_n^* = 0.$$

Thus, setting $n = 2k + 1$, we have (formally) the solution to our problem:

$$u(x, t) = \sum_{k=0}^{\infty} \frac{8}{(2k + 1)^3 \pi^3} \sin(2k + 1)\pi x \cos \alpha(2k + 1)\pi t.$$

Example 8.5.4

Find the solution of (W), subject to (B) with $\ell = \pi$, and satisfying the initial conditions

$$u(x, 0) = \sin x \quad \text{and} \quad u_t(x, 0) = x.$$

The solution is of the form

$$u(x, t) = \sum_{n=1}^{\infty} \sin nx \, (b_n \cos \alpha nt + b_n^* \sin \alpha nt),$$

where $b_1 = 1$, $b_n = 0$ for $n \neq 1$, and $\alpha n b_n^*$ is the nth sine series coefficient of x. We can use the integral formula of Section 8.4 to calculate this coefficient:

$$\alpha n b_n^* = \frac{2}{\pi} \int_0^{\pi} x \sin nx \, dx = (-1)^{n+1} \left(\frac{2}{n} \right).$$

Thus,

$$u(x, t) = \sin x \cos \alpha t + \sum_{n=1}^{\infty} (-1)^{n+1} \frac{2}{\alpha n^2} \sin nx \sin \alpha nt.$$

Our method gives the motion of the string in a form with a useful physical interpretation, which is particularly neat when the initial velocity is $u_t(x, 0) = 0$. In this case, the motion is a sum

$$u(x, t) = \sum_{n=1}^{\infty} b_n \sin \frac{n\pi x}{\ell} \cos \frac{\alpha n \pi t}{\ell}$$

of terms

$$w_n(x, t) = b_n \sin \frac{n\pi x}{\ell} \cos \frac{\alpha n \pi t}{\ell}.$$

The motion described by a single term $w_n(x, t)$ is called a **standing wave**. For each point x, w_n represents a regular oscillation of period t_n given by $\alpha n \pi t_n / \ell = 2\pi$, or $t_n = 2\ell/\alpha n$. This period is independent of x, so that under w_n the whole string goes through a periodic deformation, returning to its original shape at intervals of time t_n. The shape at time $t = 0$ is the graph of $y = b_n \sin(n\pi x/\ell)$ (Figure 8.13). As t increases, the y-coordinate is shrunk by the factor $\cos(\alpha n \pi t/\ell)$; the shape is gradually squashed until it is flat when $\cos(\alpha n \pi t/\ell) = 0$. The y-coordinate then stretches again, on the other side of the x-axis, until the string takes on the mirror image of the original position when $\cos(\alpha n \pi t/\ell) = -1$. This process is reversed over the second half of the period. We have sketched various stages of the process for w_n when $n = 1$ in Figure 8.14 on page 694.

Our solution $u(x, t)$ is a superposition of standing waves. The frequency of the first standing wave, $1/t_1 = \alpha/2\ell$, is called the **fundamental frequency;** the remaining frequencies, $1/t_n = n\alpha/2\ell$, are called **overtones** or **harmonics.**

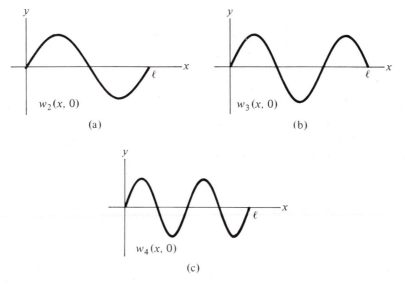

FIGURE 8.13

The amplitudes b_n of the various standing waves are determined by the sine series coefficients of the initial configuration $u(x, 0)$ and in turn determine the musical sound of the string.

The following summary describes our method for solving (W), subject to the boundary conditions (B) and initial conditions (I_1) and (I_2). Similar techniques apply to problems with other homogeneous boundary conditions; the exercises include such problems.

THE ONE-DIMENSIONAL WAVE EQUATION

We look for the solution of the **one-dimensional wave equation**

(W)
$$\frac{\partial^2 u}{\partial x^2} = \frac{1}{\alpha} \frac{\partial^2 u}{\partial t^2}$$

with boundary conditions

(B) $u(0, t) = u(\ell, t) = 0$

and initial data

(I) $u(x, 0) = f(x)$ and $u_t(x, 0) = g(x)$

in the form of a superposition of the special solutions of (W) obtained from separation of variables:

$$u(x, t) = \sum_{n=1}^{\infty} \sin \frac{n\pi x}{\ell} \left(b_n \cos \frac{\alpha n \pi t}{\ell} + b_n^* \sin \frac{\alpha n \pi t}{\ell} \right).$$

To match the initial conditions, we must take b_n and $\alpha n \pi b_n^*/\ell$ to be the nth coefficients in the sine series for $f(x)$ and $g(x)$, respectively.

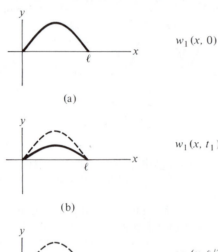

$w_1(x, 0)$

(a)

$w_1(x, t_1), 0 < t_1 < \ell/2\alpha$

(b)

$w_1(x, \ell/2\alpha)$

(c)

$w_1(x, t_2), \ell/2\alpha < t_2 < \ell/\alpha$

(d)

$w_1(x, \ell/\alpha)$

(e)

FIGURE 8.14

Note

D'Alembert's solution

Another approach to solving the one-dimensional wave equation, devised by d'Alembert (1746) and expanded by Euler (1748), gives the solution in a form that lends itself to a different physical interpretation. By an argument known as the method of characteristics (see Exercise 11) it can be shown that every solution of the one-dimensional wave equation can be written in the form

$$u(x, t) = \phi_L(x + \alpha t) + \phi_R(x - \alpha t),$$

where ϕ_L and ϕ_R are functions of one variable. For each fixed value of t, the graph of $\phi_L(x + \alpha t)$ can be obtained by translating the graph of $y = \phi_L(x)$ to the left by an amount αt. Thus we can view $\phi_L(x + \alpha t)$ as a rigid wave, having the shape of the graph of $y = \phi_L(x)$ and moving to the left at the constant speed α. Similarly, $\phi_R(x - \alpha t)$ represents a rigid wave moving to the right at the same speed. This picture of $u(x, t)$ as a sum of two *traveling waves* is the one generally used to analyze wave patterns resulting from nonhomogeneous boundary data.

EXERCISES

In Exercises 1 through 6, find the solution of the one-dimensional wave equation that satisfies the boundary conditions $u(0, t) = u(\ell, t) = 0$ for the given value of ℓ and matches the given initial conditions on $0 < x < \ell$.

1. $\ell = 1$; $u(x, 0) = 3 \sin \pi x$, $u_t(x, 0) = 7 \sin \pi x + 5 \sin 3\pi x$
2. $\ell = 1$; $u(x, 0) = 3 \sin \pi x$, $u_t(x, 0) = 1$
3. $\ell = 2$; $u(x, 0) + \begin{cases} x, & 0 < x < 1 \\ 2 - x, & 1 \le x < 2 \end{cases}$, $u_t(x, 0) = 0$
   (*Hint:* See Exercise 5, Section 8.4.)
4. $\ell = 1$; $u(x, 0) = 0$, $u_t(x, 0) = x$
   (*Hint:* See Exercise 3, Section 8.4.)
5. $\ell = 1$; $u(x, 0) = x(1 - x)$, $u_t(x, 0) = 1 - x$
   (*Hint:* See Example 8.4.3 and Exercise 4, Section 8.4.)
6. $\ell = \pi$; $u(x, 0) = \sin x$, $u_t(x, 0) = \cos x$
   (*Hint:* See Exercise 7, Section 8.4.)

Some more advanced problems:

7. Suppose the left side of a vibrating string of length 1 is kept fixed at 0, and the right end is free to move vertically but is constrained so that the slope at this end is always 0. Then the displacement $u(x, t)$ must satisfy the boundary conditions

(B′) $u(0, t) = u_x(1, t) = 0.$

   a. Show that the only solutions of the one-dimensional wave equation that are of the form $u = X(x)T(t)$ and satisfy (B′) are the functions $c_n u_n + c_n^* u_n^*$, where n is a nonnegative integer,

$$u_n = \sin \frac{(2n+1)\pi x}{2} \cos \frac{\alpha(2n+1)\pi t}{2}$$

$$u_n^* = \sin \frac{(2n+1)\pi x}{2} \sin \frac{\alpha(2n+1)\pi t}{2}.$$

b. Find the solution of the one-dimensional wave equation that satisfies (B') and matches the initial conditions

(i) $u(x, 0) = \sin \dfrac{\pi x}{2} - \sin \dfrac{3\pi x}{2};$ $u_t(x, 0) = \sin \dfrac{3\pi x}{2}$

(ii) $u(x, 0) = 0;$ $u_t(x, 0) = 1$ (See Exercise 14(a), Section 8.4.)
(Compare Exercise 9, Section 8.2.)

8. *A Forced Vibrating String:* When a vibrating string is subject to a constant external force, the displacement satisfies an equation of the form

(N) $$\frac{\partial^2 u}{\partial x^2} = \frac{1}{\alpha^2} \frac{\partial^2 u}{\partial t^2} + G.$$

a. Show that if $u = u_P(x, t)$ satisfies (N) and $u = u_W(x, t)$ satisfies the one-dimensional wave equation, then $u = u_P(x, t) + u_W(x, t)$ also satisfies (N).
b. Find a time-independent solution $u = u_P(x)$ to (N) that satisfies the boundary conditions $u_P(0) = u_P(1) = 0$.
c. By adding an appropriate function $u_W(x, t)$ to $u_P(x)$, obtain the solution of (N) that satisfies the boundary conditions $u(0, t) = u(1, t) = 0$ and matches the initial conditions $u(x, 0) = u_t(x, 0) = 0, 0 < x < 1$.
(Compare Exercise 13, Section 8.2.)

9. *A Nonhomogeneous Boundary Condition:* Adapt the method described in Note 3, Section 8.2, to the one-dimensional wave equation to find the solution that satisfies the boundary conditions $u(0, t) = 0$ and $u(1, t) = 1$ and matches the initial conditions $u(x, 0) = x$ and $u_t(x, 0) = \sin \pi x$ for $0 < x < 1$.

10. *A Damped Vibrating String:* A string with fixed ends and vibrating subject to damping is modeled by an equation of the form

(W') $$\frac{\partial^2 u}{\partial x^2} = \frac{1}{\alpha^2} \frac{\partial^2 u}{\partial t^2} + \beta \frac{\partial u}{\partial t}$$

and the boundary conditions

(B) $$u(0, t) = u(\ell, t) = 0.$$

a. Show that if $0 < \beta < 2\pi/\ell\alpha$, then the method of separation of variables leads to formal solutions of the form

$$u = \sum_{n=1}^{\infty} \sin \frac{n\pi x}{\ell} e^{-\alpha^2 \beta t/2} (b_n \cos \mu_n t + b_n^* \sin \mu_n t),$$

where $\mu_n = (\sqrt{4\alpha^2 n^2 \pi^2 - \beta^2 \ell^2 \alpha^4})/2\ell$.

b. How do these solutions behave as $t \to \infty$?
c. Show that in order to match the initial conditions $u(x, 0) = f(x)$ and $u_t(x, 0) = g(x)$, we should take b_n to be the nth coefficient of the sine series for $f(x)$ and b_n^* so that $(-\alpha^2 \beta/2)b_n + \mu_n b_n^*$ is the nth coefficient of the sine series for $g(x)$.

11. *D'Alembert's Solution:* Let $\xi = x - \alpha t$ and $\eta = x + \alpha t$.
   a. Use the chain rule to show that

$$\frac{\partial^2 u}{\partial x^2} = \frac{\partial^2 u}{\partial \xi^2} + 2 \frac{\partial^2 u}{\partial \xi \, \partial \eta} + \frac{\partial^2 u}{\partial \eta^2}$$

and

$$\frac{1}{\alpha^2} \frac{\partial^2 u}{\partial t^2} = \frac{\partial^2 u}{\partial \xi^2} - 2 \frac{\partial^2 u}{\partial \xi \, \partial \eta} + \frac{\partial^2 u}{\partial \eta^2}$$

so that the one-dimensional wave equation becomes

(W*)
$$\frac{\partial^2 u}{\partial \xi \, \partial \eta} = 0.$$

   b. Show that any solution of (W*) is of the form

$$u = \phi_L(\eta) + \phi_R(\xi) = \phi_L(x + \alpha t) + \phi_R(x - \alpha t),$$

where ϕ_L and ϕ_R are functions of one variable.

12. *Standing Waves:* If the initial velocity of the vibrating string in Example 8.5.1 is not zero, then its displacement is of the form $u = \sum_{n=1}^{\infty} w_n(x, t)$, where

$$w_n(x, t) = \sin \frac{n \pi x}{\ell} \left(b_n \cos \frac{\alpha n \pi t}{\ell} + b_n^* \sin \frac{\alpha n \pi t}{\ell} \right).$$

Show that if $B_n = \sqrt{b_n^2 + (b_n^*)^2}$, and if γ_n satisfies $\cos(\alpha n \pi \gamma_n / \ell) = b_n / B_n$ and $\sin(\alpha n \pi \gamma_n / \ell) = b_n^* / B_n$, then

$$w_n(x, t) = B_n \sin \frac{n \pi x}{\ell} \cos \frac{\alpha n \pi (t - \gamma_n)}{\ell}.$$

[*Hint:* $\cos(A - B) = \cos A \cos B + \sin A \sin B$.]
Note that starting when $t = \gamma_n$, a string undergoing the motion described by the single term $w_n(x, t)$ goes through the same kind of periodic deformation we described just before the summary. Thus, even in the case when $u_t(x, 0) \neq 0$, each $w_n(x, t)$ represents a standing wave.

8.6 THE TWO-DIMENSIONAL LAPLACE EQUATION: THE DIRICHLET PROBLEM

The one-dimensional heat and wave equations are examples of two basic types of p.d.e.'s that come up frequently in simple physical problems. In this section, we look at an example of the third basic type, the **Laplace equation** in two dimensions,

(L)
$$\frac{\partial^2 u}{\partial x^2} + \frac{\partial^2 u}{\partial y^2} = 0,$$

which we encountered in Section 6.1. We look for solutions to (L) in a specified region of the plane, with given boundary values:

(B) $u(x, y) = f(x, y)$ on the boundary.

This problem is known as the **Dirichlet problem,** after the nineteenth-century German mathematician who made fundamental contributions to its solution.

■ Example 8.6.1

A thin rectangular plate, defined by $0 < x < \ell$ and $0 < y < h$, is insulated on its faces. The bottom and side edges are kept at temperature zero, and the top is maintained at the temperature distribution $f(x)$ (Figure 8.15). We wish to find a steady-state (i.e., time-independent) temperature distribution in the plate.

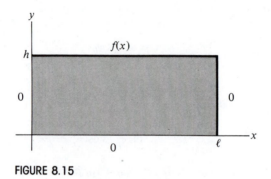

FIGURE 8.15

In general, the temperature distribution in the plate is modeled by a function of three variables $u(x, y, t)$ satisfying the two-dimensional heat equation

(H) $$\frac{\partial^2 u}{\partial x^2} + \frac{\partial^2 u}{\partial y^2} = \frac{1}{\alpha}\frac{\partial u}{\partial t}.$$

For a time-independent distribution $u = u(x, y)$, we have $\partial u/\partial t = 0$. Substituting this in (H) gives the Laplace equation

(L) $$\frac{\partial^2 u}{\partial x^2} + \frac{\partial^2 u}{\partial y^2} = 0.$$

Our boundary conditions are

(B) $u(0, y) = u(\ell, y) = 0,$ $0 < y < h$
             $u(x, 0) = 0, \quad u(x, h) = f(x),$ $0 < x < \ell.$

To solve this problem, we start by looking for nontrivial solutions of (L) in the form

(S) $$u = X(x)Y(y).$$

Substitution into (L) leads to the o.d.e.'s

(L$_x$) $$X'' + \lambda X = 0$$

(L$_y$) $$Y'' - \lambda Y = 0.$$

The boundary conditions $u = 0$ along the side and bottom edges of the rectangle, rewritten in terms of X and Y, give

(B$_1$) $$X(0) = X(\ell) = 0$$

(B$_2$) $$Y(0) = 0.$$

For a nontrivial solution of (L$_x$) to satisfy (B$_1$), we must have

$$\lambda = \frac{n^2\pi^2}{\ell^2},$$

where n is a positive integer. The corresponding nontrivial solutions of (L$_x$) are constant multiples of

$$X_n = \sin\frac{n\pi x}{\ell}.$$

Furthermore, with this choice of λ, the general solution of (L$_y$) is

$$Y = c_1 e^{n\pi y/\ell} + c_2 e^{-n\pi y/\ell}.$$

The condition (B$_2$) gives $c_1 + c_2 = 0$, so the solutions of (L$_y$) that satisfy (B$_2$) are constant multiples of

$$Y_n = e^{n\pi y/\ell} - e^{-n\pi y/\ell}.$$

Substitution into (S) gives solutions of (L) that are 0 along the side and bottom edges,

$$u_n(x, y) = \sin\frac{n\pi x}{\ell}(e^{n\pi y/\ell} - e^{-n\pi y/\ell}).$$

To fit the boundary condition along the top edge, we need in general to look for a solution in the form of a series

$$u(x, y) = \sum_{n=1}^{\infty} a_n u_n(x, y) = \sum_{n=1}^{\infty} a_n \sin \frac{n\pi x}{\ell} (e^{n\pi y/\ell} - e^{-n\pi y/\ell}).$$

When $y = h$, the factor in $u_n(x, y)$ involving y becomes

$$C_n = e^{n\pi h/\ell} - e^{-n\pi h/\ell}$$

so that

$$u(x, h) = \sum_{n=1}^{\infty} a_n C_n \sin \frac{n\pi x}{\ell}.$$

To match the condition $u(x, h) = f(x)$, we need to take $a_n C_n$ to be the nth coefficient of the sine series for $f(x)$. Using the integral formula for these coefficients (see Section 8.4), we obtain a formula for a_n:

$$a_n = \frac{2}{C_n \ell} \int_0^{\ell} f(x) \sin \frac{n\pi x}{\ell}\, dx.$$

The solution of the Dirichlet problem on a rectangle with more complicated boundary values can be found by superposition of functions like those found in Example 8.6.1, as illustrated by the following example.

Example 8.6.2

Find the steady-state temperature distribution in the plate of Example 8.6.1 if the top edge has temperature $f(x)$, the right edge has temperature $g(y)$, and the other two edges have temperature zero (Figure 8.16).

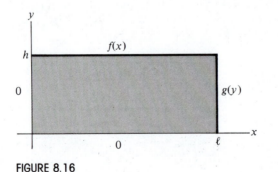

FIGURE 8.16

Again, we have the Dirichlet problem in a rectangle, this time with boundary values

(B) $$u(0, y) = 0, \quad u(\ell, y) = g(y), \quad\quad 0 < y < h$$
$$u(x, 0) = 0, \quad u(x, h) = f(x), \quad\quad 0 < x < \ell.$$

We will solve this problem by finding two solutions $u_T(x, y)$ and $u_R(x, y)$ of (L), the first satisfying the boundary data

(B$_T$) $$u_T(0, y) = u_T(\ell, y) = 0, \quad\quad\quad 0 < y < h$$
$$u_T(x, 0) = 0, \quad u_T(x, h) = f(x), \quad\quad 0 < x < \ell$$

and the second satisfying

(B$_R$) $$u_R(0, y) = 0, \quad u_R(\ell, y) = g(y), \quad\quad 0 < y < h$$
$$u_R(x, 0) = u_R(x, h) = 0, \quad\quad\quad\quad 0 < x < \ell.$$

The sum of these functions will solve (L) and satisfy the required boundary conditions (B).

Now, the Dirichlet problem with boundary values (B$_T$) was solved in Example 8.6.1; $u_T(x, y)$ has the form

$$u_T(x, y) = \sum_{n=1}^{\infty} a_n \sin \frac{n\pi x}{\ell} (e^{n\pi y/\ell} - e^{-n\pi y/\ell}),$$

where the a_n's come from the sine series for $f(x)$:

$$a_n = \frac{2}{C_n \ell} \int_0^{\ell} f(x) \sin \frac{n\pi x}{\ell} \, dx, \quad\quad C_n = e^{n\pi h/\ell} - e^{-n\pi h/\ell}.$$

The Dirichlet problem with boundary values (B$_R$) is strictly analogous to that with (B$_T$), with the roles of x and y interchanged and $f(x)$ replaced by $g(y)$. You can check that the solution is

$$u_R(x, y) = \sum_{n=1}^{\infty} b_n \sin \frac{n\pi y}{h} (e^{n\pi x/h} - e^{-n\pi x/h}),$$

where the b_n's are related to the sine series coefficients of $g(y)$:

$$b_n = \frac{2}{E_n h} \int_0^{h} g(y) \sin \frac{n\pi y}{h} \, dy, \quad\quad E_n = e^{n\pi \ell/h} - e^{-n\pi \ell/h}.$$

The solution to our original problem is

$$u(x, y) = u_T(x, y) + u_R(x, y).$$

As our final example, we turn to the Dirichlet problem on the unit disc (compare Example 6.12).

Example 8.6.3

A circular plate, defined by $x^2 + y^2 < 1$, is insulated on its faces; the temperature on the boundary is defined in terms of the angular variable θ by

(B) $u = f(\theta)$ when $x = \cos \theta, \quad y = \sin \theta.$

This problem is best handled in polar coordinates. We recall from Example 6.1.2 (see also Exercise 6, Section 8.1) that the Laplace equation in polar coordinates is

(L) $$\frac{\partial^2 u}{\partial r^2} + \frac{1}{r}\frac{\partial u}{\partial r} + \frac{1}{r}\frac{\partial^2 u}{\partial \theta^2} = 0.$$

We look for solutions of (L) in the form

(S) $u = R(r)\Theta(\theta).$

This leads to the o.d.e.'s

(L$_\theta$) $\Theta'' + \lambda\Theta = 0$

(B$_r$) $r^2 R'' + rR' - \lambda R = 0.$

The solutions of (L$_\theta$) must satisfy the geometric condition

$$\Theta(\theta + 2\pi) = \Theta(\theta),$$

which forces λ to be a perfect square:

$$\lambda = n^2, \quad n \text{ a nonnegative integer.}$$

The corresponding solutions of (L$_\theta$) are the constant multiples of

$$\Theta_0^* = 1 \quad (n = 0)$$

and the linear combinations of

$$\Theta_n^* = \cos n\theta \quad \text{and} \quad \Theta_n = \sin n\theta \qquad (n > 0).$$

The equation (L_r), with $\lambda = n^2$, is a Cauchy-Euler equation (see Section 6.5); the indicial equation

$$m(m - 1) + m - n^2 = 0$$

has roots $m = \pm n$.

If $n = 0$, the general solution of (L_r) is

$$R = c_1 + c_2 \ln r.$$

However, since $\ln r$ is unbounded as $r \to 0$, the physically meaningful solutions in this case are constant multiples of

$$R_0 = 1.$$

Thus the choice $n = 0$ leads to a constant solution of (L),

$$u_0^*(r, \theta) = R_0 \Theta_0^* = 1.$$

If $n > 0$, the general solution of (L_r) is

$$R = c_1 r^n + c_2 r^{-n}.$$

Again, r^{-n} is unbounded as $r \to 0$, so the physically meaningful solutions are constant multiples of

$$R_n = r^n.$$

Substitution into (S) leads to the solutions of (L)

$$u_n(r, \theta) = R_n \Theta_n = r^n \sin n\theta$$
$$u_n^*(r, \theta) = R_n \Theta_n^* = r^n \cos n\theta.$$

To fit our boundary values, we need in general to look for a series solution to (L) in the form

$$u(r, \theta) = \frac{a_0}{2} + \sum_{n=1}^{\infty} r^n (a_n \cos n\theta + b_n \sin n\theta).$$

Substituting $r = 1$ and matching (B), we have

$$f(\theta) = u(1, \theta) = \frac{a_0}{2} + \sum_{n=1}^{\infty} (a_n \cos n\theta + b_n \sin n\theta).$$

We recognize this as the Fourier series expression for $f(\theta)$. The coefficients are given (see Section 8.3) by

$$a_0 = \frac{1}{\pi} \int_{-\pi}^{\pi} f(\theta)\, d(\theta)$$

$$a_n = \frac{1}{\pi} \int_{-\pi}^{\pi} f(\theta) \cos n\theta\, d\theta, \quad n > 0$$

$$b_n = \frac{1}{\pi} \int_{-\pi}^{\pi} f(\theta) \sin n\theta\, d\theta, \quad n > 0.$$

Our summary describes the solution of the Dirichlet problem on the rectangle and the disc.

THE TWO-DIMENSIONAL LAPLACE EQUATION: THE DIRICHLET PROBLEM

The solution of the **Laplace equation**

$$\frac{\partial^2 u}{\partial x^2} + \frac{\partial^2 u}{\partial y^2} = 0$$

on the rectangle $0 < x < \ell, 0 < y < h$, with given boundary values

$$u(0, y) = q(y), \quad u(\ell, y) = g(y), \qquad 0 < y < h$$

$$u(x, 0) = p(x), \quad u(x, h) = f(x), \qquad 0 < x < \ell$$

is obtained by adding the solutions to four boundary-value problems, in each of which u is zero along three edges and u is a function of a single variable along the fourth edge. Each of the simpler boundary-value problems is solved by a series, in terms of functions obtained by separation of variables, with coefficients related to those of the sine series for the value of u on the fourth edge.

To solve the Laplace equation in polar coordinates,

$$\frac{\partial^2 u}{\partial r^2} + \frac{1}{r}\frac{\partial u}{\partial r} + \frac{1}{r^2}\frac{\partial^2 u}{\partial \theta^2} = 0$$

on the unit disc $r < 1$, with given boundary values

$$u(1, \theta) = f(\theta)$$

we use the geometric condition $u(r, \theta + 2\pi) = u(r, \theta)$ and the observation that physically meaningful solutions are bounded as $r \to 0$. The solution is a series, in terms of functions obtained by separation of variables, whose coefficients are the same as in the Fourier series of $f(\theta)$.

Note

On the classification of p.d.e.'s

The simplest physical problems leading to p.d.e.'s involve equations of the general form

(P) $$A\frac{\partial^2 u}{\partial \xi^2} + B\frac{\partial^2 u}{\partial \eta\,\partial \xi} + C\frac{\partial^2 u}{\partial \eta^2} = E\frac{\partial u}{\partial \xi} + F\frac{\partial u}{\partial \eta} + Gu.$$

A linear change of variables can transform this into one of three standard forms, depending on the number $AC - B^2$.

If $AC - B^2 > 0$, then (P) is called an *elliptic* p.d.e., and it can be written in the form

$$\frac{\partial^2 u}{\partial x^2} + \frac{\partial^2 u}{\partial y^2} = f\left(\frac{\partial u}{\partial x}, \frac{\partial u}{\partial y}, u, x, y\right).$$

The two-dimensional Laplace equation is elliptic.

If $AC - B^2 = 0$, then (P) is called a *parabolic* p.d.e., and it can be written in the form

$$\frac{\partial^2 u}{\partial x^2} = f\left(\frac{\partial u}{\partial x}, \frac{\partial u}{\partial t}, u, x, t\right).$$

The one-dimensional heat equation is parabolic.

If $AC - B^2 < 0$, then (P) is called a *hyperbolic* p.d.e., and it can be written in the form

$$\frac{\partial^2 u}{\partial y\,\partial x} = f\left(\frac{\partial u}{\partial x}, \frac{\partial u}{\partial y}, u, x, y\right).$$

The one-dimensional wave equation is hyperbolic; the form shown is related to the d'Alembert solution for traveling waves, described in the note at the end of Section 8.5.

The different characteristics of these three kinds of p.d.e.'s are reflected in the different kinds of data we have used to specify a solution, as well as in the different smoothness properties of solutions. For a detailed discussion of these differences, consult any good introductory book on p.d.e.'s.

EXERCISES

In Exercises 1 through 6, find the steady-state temperature distribution in the plate of Example 8.6.1, with $\ell = h = 1$, if the temperature along the edges is as given.

1. $u(0, y) = u(1, y) = u(x, 0) = 0, \quad u(x, 1) = 1; \qquad 0 < x, y < 1$
2. $u(0, y) = 0, \quad u(1, y) = y, \quad u(x, 0) = u(x, 1) = 0; \qquad 0 < x, y < 1$
3. $u(0, y) = u(1, y) = 0, \quad u(x, 0) = 1 - x, \quad u(x, 1) = 0; \qquad 0 < x, y < 1$
4. $u(0, y) = 1, \quad u(1, y) = 0, \quad u(x, 0) = u(x, 1) = 0; \qquad 0 < x, y < 1$
5. $u(0, y) = 0, \quad u(1, y) = y, \quad u(x, 0) = 0, \quad u(x, 1) = 1; \qquad 0 < x, y < 1$
6. $u(0, y) = 1, \quad u(1, y) = y, \quad u(x, 0) = 1 - x, \quad u(x, 1) = 1; \qquad 0 < x, y < 1$

In Exercises 7 and 8, find the steady-state temperature in the disc of Example 8.6.3 if the temperature along the boundary is as given.

7. $u(1, \theta) = \sin^2 \theta, \quad -\pi < \theta < \pi$

8. $u(1, \theta) = \begin{cases} 0, & -\pi < \theta < 0 \\ 1/2, & \theta = 0, \pi \\ 1, & 0 < \theta < \pi \end{cases}$

9. A semicircular plate, defined by $0 \le \theta \le \pi$ and $0 \le r \le 1$, is insulated on its faces. Find the steady-state temperature in the plate if the temperature along the bottom edge is kept at $0°$ $[u(r, 0) = u(r, \pi) = 0$ for $0 \le r \le 1]$, and the temperature along the upper part of the boundary is given by $u(1, \theta) = f(\theta), 0 < \theta < \pi$. (Compare Exercise 3, Section 8.1.)

10. Find the steady-state temperature distribution in the rectangular plate of Example 8.6.1 if the horizontal edges are insulated $[u_y(x, 0) = u_y(x, h) = 0$ for $0 < x < \ell]$ and the temperature distributions along the vertical edges are
    a. $u(0, y) = 0, \quad u(\ell, y) = g(y); \qquad 0 < y < h.$
    b. $u(0, y) = a(y), \quad u(\ell, y) = 0; \qquad 0 < y < h.$
    c. $u(0, y) = a(y), \quad u(\ell, y) = g(y); \qquad 0 < y < h.$
    (Compare Example 8.1.2.) Note that these are *not* examples of the Dirichlet problem, since the boundary conditions specify the values of $u_y(x, 0)$ and $u_y(x, h)$ rather than the values of $u(x, 0)$ and $u(x, h)$.

8.7   HIGHER-DIMENSIONAL EQUATIONS

We saw that the solutions to certain boundary-value problems for the one-dimensional heat and wave equations and the two-dimensional Laplace equations could be found as series combinations of the special solutions we obtained by separation of variables. The coefficients of these series were found by determining the Fourier series (or sine or cosine series) of a given function. In this section we use three examples to illustrate how the method changes when more spatial dimensions are involved. In particular, we will see that such problems may require other kinds of series expressions for given functions.

Our first example illustrates the occurrence of multiple (Fourier) series.

Example 8.7.1   Heat Flow in a Rectangular Plate

A rectangular plate defined by $0 < x < 1$ and $0 < y < 2$ is insulated on its faces and along the top and bottom edges and maintained at temperature zero along the right and left edges. The model for the heat flow in this situation is the two-dimensional heat equation (we take $\alpha = 1$)

(H)
$$\frac{\partial^2 u}{\partial x^2} + \frac{\partial^2 u}{\partial y^2} = \frac{\partial u}{\partial t}$$

with boundary conditions

(B)
$$u(0, y, t) = u(1, y, t) = 0, \qquad 0 < y < 2, \quad \text{all } t$$
$$u_y(x, 0, t) = u_y(x, 2, t) = 0, \qquad 0 < x < 1, \quad \text{all } t.$$

Let's investigate the evolution of the initial temperature distribution

(I)
$$u(x, y, 0) = x(1 - x)y^2(3 - y), \qquad 0 < x < 1, \quad 0 < y < 2.$$

We begin by looking for nontrivial solutions in the separated form

(S)
$$u = X(x)Y(y)T(t).$$

Substitution into (H) gives

$$X''YT + Y''XT = XYT'$$

or, dividing by $u = XYT$ and rearranging terms,

$$\frac{X''}{X} = \frac{T'}{T} - \frac{Y''}{Y}.$$

The common value of the two sides of this equation must be a constant, which we call $-\lambda$. Thus, we have

(H$_x$)
$$X'' + \lambda X = 0$$

(H$_{yt}$)
$$\frac{Y''}{Y} = \frac{T'}{T} + \lambda.$$

Again, the common value of the two sides of (H_{yt}) must be a constant, which we call $-\mu$. This leads to

(H$_y$) $Y'' + \mu Y = 0$

(H$_t$) $T' + (\lambda + \mu)T = 0.$

Substitution of (S) into (B) leads to the boundary conditions

(B$_x$) $X(0) = X(1) = 0$

(B$_y$) $Y'(0) = Y'(2) = 0.$

We saw in Example 8.2.1 that we can find nontrivial solutions to (H_x) that satisfy (B_x) only if we take

$$\lambda = n^2\pi^2, \quad n \text{ a positive integer;}$$

the nontrivial solutions are then constant multiples of

$$X_n = \sin n\pi x.$$

We dealt with an o.d.e. boundary-value problem equivalent to the one determined by (H_y) and (B_y) in Example 8.2.2. We need

$$\mu = \frac{m^2\pi^2}{4}, \quad m \text{ a nonnegative integer.}$$

The corresponding nontrivial solutions of (H_y) are constant multiples of

$$Y_m = \cos\frac{m\pi y}{2}$$

(of course, $Y_0 = \cos 0 = 1$).

If we choose $\lambda = n^2\pi^2$ and $\mu = m^2/4$, equation (H_t) becomes

$$T' + c(n, m)T = 0 \quad \text{where} \quad c(n, m) = n^2\pi^2 + \frac{m^2\pi^2}{4}.$$

The solutions of this equation are constant multiples of

$$T_{n,m} = e^{-c(n,m)t}.$$

Substitution into (S) yields solutions of (H) that satisfy (B):

$$u_{n,m}(x, y, t) = X_n Y_m T_{n,m} = e^{-c(n,m)t} \sin n\pi x \cos\frac{m\pi y}{2}.$$

We expect to find solutions of (H) that satisfy (B) in the form of series combinations of the functions $u_{n,m}(x, y, t)$:

$$u(x, y, t) = \sum_{n=1}^{\infty} \sum_{m=0}^{\infty} a_{n,m} e^{-c(n,m)t} \sin n\pi x \cos \frac{m\pi y}{2}.$$

If we substitute $t = 0$ and try to fit the initial condition (I), we find that we need to express the initial temperature distributions as a **double trigonometric series**

$$x(1 - x)y^2(3 - y) = \sum_{n=1}^{\infty} \sum_{m=0}^{\infty} a_{n,m} \sin n\pi x \cos \frac{m\pi y}{2}.$$

We can compute the coefficients $a_{n,m}$ in the following way. For each fixed value of y, we express the temperature distribution as a sine series in x, with coefficients depending on y:

$$x(1 - x)y^2(3 - y) = \sum_{n=1}^{\infty} A_n(y) \sin n\pi x.$$

If we treat y as a constant and use the method of Section 8.4, we are led to the integral

$$A_n(y) = 2y^2(3 - y) \int_0^1 x(1 - x) \sin n\pi x \, dx,$$

which we evaluate (see Example 8.4.3) to find

$$A_n(y) = \begin{cases} 0 & \text{if } n \text{ is even} \\ \dfrac{8y^2(3 - y)}{n^3 \pi^3} & \text{if } n \text{ is odd.} \end{cases}$$

Now we expand each of the functions $A_n(y)$ in a cosine series of the form

$$A_n(y) = \sum_{m=0}^{\infty} a_{n,m} \cos \frac{m\pi y}{2}.$$

If n is even, $A_n(y) = 0$, so $a_{n,m} = 0$ for all m. If n is odd, we use the method of Section 8.4 to find

$$2a_{n,0} = \int_0^2 \frac{8y^2(3 - y)}{n^3 \pi^3} \, dy = \frac{32}{n^3 \pi^3},$$

and, when $m > 0$,

$$a_{n,m} = \int_0^2 \frac{8y^2(3 - y)}{n^3 \pi^3} \cos \frac{m\pi y}{2} \, dy = \begin{cases} 0 & \text{if } m \text{ is even} \\ -\dfrac{1536}{n^3 m^4 \pi^7} & \text{if } m \text{ is odd.} \end{cases}$$

Thus, we obtain the solution to (H), (B), and (I),

$$u(x, y, t) = \sum_{n=1}^{\infty} \sum_{m=0}^{\infty} a_{n,m} e^{-c(n,m)t} \sin n\pi x \cos \frac{m\pi y}{2},$$

where

$$a_{n,m} = \begin{cases} 0 & \text{if } n \text{ is even or if } m > 0 \text{ is even} \\ \dfrac{16}{n^3 \pi^3} & \text{if } n \text{ is odd and } m = 0 \\ -\dfrac{1536}{n^3 m^4 \pi^7} & \text{if } n \text{ and } m \text{ are both odd.} \end{cases}$$

Our next two examples indicate that boundary-value problems can require expansions of a given function as a series of special functions other than sines and cosines.

Example 8.7.2   Internal Temperature of a Planet

Assuming that the surface temperature of a spherical planet depends only on latitude, an attempt to predict the steady-state internal temperature distribution leads to the three-dimensional Laplace equation in spherical coordinates, with no θ-dependence:

(L) $$\rho \frac{\partial^2}{\partial \rho^2} (\rho u) + \frac{1}{\sin \phi} \frac{\partial}{\partial \phi} \left(\sin \phi \frac{\partial u}{\partial \phi} \right) = 0.$$

Let's find the solution of (L) that satisfies the boundary condition

(B) $$u(1, \phi) = \sin^2 \phi, \qquad 0 < \phi < \pi.$$

(This is zero at the poles and has a maximum value of 1 at the equator. We have picked units so that the planet has radius 1.)

This is a special case of Example 6.1.4, where we saw that separation of variables

(S) $$u(\rho, \phi) = R(\rho)\Phi(\phi)$$

and the substitution $s = \cos \phi$ led to the o.d.e.'s

(L$_r$) $$\rho^2 \frac{d^2R}{dr^2} + 2\rho \frac{dR}{dr} - n(n + 1)R = 0$$

(L$_s$) $$(s^2 - 1) \frac{d^2\Phi}{ds} + 2s \frac{d\Phi}{ds} - n(n + 1)\Phi = 0,$$

where n is a nonnegative integer.

The o.d.e. (L$_r$) is a Cauchy-Euler equation whose general solution (see Section 6.5) is

$$R(\rho) = \alpha\rho^n + bp^{-(n+1)}.$$

In order for our solution to be bounded at the origin, we need $b = 0$. Thus, a given choice of n in (L$_r$) leads (up to constant multiples) to the solution

$$R_n(\rho) = \rho^n.$$

The o.d.e. (L$_s$) is a Legendre equation. As we saw in Exercises 13 and 14 of Section 6.8, the general solution of (L$_s$) is generated by two independent functions—a **Legendre polynomial**, which can be written

$$P_n(s) = \frac{1}{2^n n!} \frac{d^n}{ds^n} (s^2 - 1)^n,$$

and a function that is unbounded near $s = \pm 1$ ($\phi = 0, \pi$). The physically meaningful solutions of (L$_s$) are constant multiples of

$$\Phi_n = P_n(s).$$

We expect to find a solution to our boundary-value problem in the form of a combination of the functions $u_n = R_n\Phi_n$:

$$u(\rho, s) = \sum_{n=0}^{\infty} a_n\rho^n P_n(s).$$

In terms of $s = \cos \phi$, our boundary condition (B) reads

$$u(1, s) = 1 - s^2.$$

In order to match this condition, we need to choose the constants a_n so that

$$\sum_{n=0}^{\infty} a_n P_n(s) = 1 - s^2.$$

The first three Legendre polynomials are

$$P_0(s) = 1, \quad P_1(s) = s, \quad P_2(s) = \frac{1}{2}(3s^2 - 1),$$

and we see that

$$1 - s^2 = \frac{2}{3}P_0(s) - \frac{2}{3}P_2(s).$$

The solution to our problem is

$$u = \frac{2}{3}P_0(s) - \frac{2}{3}\rho^2 P_2(s) = \frac{2}{3} - \frac{1}{3}\rho^2(3x^2 - 1)$$

$$= \frac{2}{3} - \frac{1}{3}\rho^2(3\cos^2\phi - 1).$$

In this case, the function $1 - s^2 = u(1, \phi)$ is a finite linear combination of Legendre polynomials. Techniques expressing more complicated functions as infinite series in Legendre polynomials closely resemble the Fourier coefficient techniques of Sections 8.3 and 8.4. We will briefly discuss these techniques in the note at the end of the section.

■ **Example 8.7.3   The Vibrating Drumhead**

The vibrations of a thin membrane, stretched tightly and fixed to a circular frame of radius 1, can be modeled by the two-dimensional wave equation in polar coordinates,

(W) $$\frac{\partial^2 u}{\partial r^2} + \frac{1}{r}\frac{\partial u}{\partial r} + \frac{1}{r}\frac{\partial^2 u}{\partial \theta^2} = \frac{1}{\alpha^2}\frac{\partial^2 u}{\partial t^2}$$

with boundary condition

(B) $$u(1, \theta, t) = 0.$$

If we start with rotationally symmetric initial data

(I) $$u(r, \theta, 0) = f(r) \quad \text{and} \quad u_t(r, \theta, 0) = g(r),$$

then we expect a solution with no θ-dependence, $u = u(r, t)$.
Substitution of the separated form

(S) $$u = R(r)T(t)$$

into (W) leads to the equation

$$\frac{R''}{R} + \frac{R'}{rR} = \frac{T''}{\alpha T}.$$

It turns out that, for physical reasons, the common (constant) value of the two sides of this equation has to be negative. If we call this value $-\beta^2$, with $\beta > 0$, we are led to the o.d.e.'s

(W$_r$) $\qquad\qquad\qquad r^2 R'' + rR' + (r\beta)^2 R = 0$

(W$_t$) $\qquad\qquad\qquad T'' + (\alpha\beta)^2 T = 0.$

Substitution of (S) into (B) leads to

(B$_r$) $\qquad\qquad\qquad\qquad R(1) = 0.$

The o.d.e. (W$_r$) becomes, upon substituting $s = r\beta$, the Bessel equation (with $\mu = 0$),

(B$_0$) $\qquad\qquad s^2 \dfrac{d^2R}{ds^2} + s\dfrac{dR}{ds} + s^2 R = 0.$

We saw in Sections 6.7 and 6.8 that the general solution of (B$_0$) is

$$R = a J_0(s) + b Y_0(s) = a J_0(r\beta) + b Y_0(r\beta),$$

where $J_0(s)$ and $Y_0(s)$ are Bessel functions of the first and second kinds, respectively. However, $Y_0(s)$ is unbounded near $s = 0$, so the physically meaningful solutions are constant multiples of

$$R_\beta(r) = J_0(r\beta).$$

Substitution into (B$_r$) gives

$$J_0(\beta) = 0.$$

This equation turns out to have a sequence of positive roots

$$0 < \beta_1 < \beta_2 < \beta_3 < \cdots,$$

and these are the possible choices for β. (These roots are specific real numbers, whose approximate values can be found in tables of Bessel functions.)

Corresponding to each choice $\beta = \beta_n$, we obtain two independent solutions of (W$_t$),

$$T_n(t) = \cos \alpha\beta_n t \qquad \text{and} \qquad T_n^*(t) = \sin \alpha\beta_n t.$$

Substitution into (S) gives two independent solutions of (W) that satisfy (B),

$$u_n(r, t) = J_0(r\beta_n) \cos \alpha\beta_n t \quad \text{and} \quad u_n^*(r, t) = J_0(r\beta_n) \sin \alpha\beta_n t.$$

We expect a solution of our problem to be of the form

$$u(r, t) = \sum_{n=1}^{\infty} J_0(r\beta_n)(a_n \cos \alpha\beta_n t + b_n \sin \alpha\beta_n t).$$

In order to match our initial condition (I), we need to represent $f(r)$ and $g(r)$ as **Bessel series:**

$$f(r) = u(r, 0) = \sum_{n=1}^{\infty} a_n J_0(r\beta_n)$$

$$g(r) = u_t(r, 0) = \sum_{n=1}^{\infty} \alpha\beta_n b_n J_0(r\beta_n).$$

Techniques for finding such representations also resemble those of Sections 8.3 and 8.4, and are discussed in the note at the end of the section.

Let's summarize.

HIGHER-DIMENSIONAL EQUATIONS

   Boundary-value problems involving the two- or three-dimensional heat or wave equation, or the three-dimensional Laplace equation, can require series expressions for a given function in terms of trigonometric functions, Bessel functions, or Legendre polynomials.

Note

Orthogonal functions

   A sequence of functions $g_n(x)$ is said to be **orthogonal** on $a < x < b$ with respect to $w(x)$ provided that whenever $m \neq n$,

$$\int_a^b g_m(x)g_n(x)w(x) \, dx = 0.$$

We saw in Section 8.3 that the trigonometric functions $\sin(n\pi x/\ell)$ and $\cos(n\pi x/\ell)$ are orthogonal on $-\ell < x < \ell$ with respect to $w(x) = 1$. It turns out that the Legendre polynomials $P_n(x/\ell)$ are also orthogonal on $-\ell < x < \ell$ with respect to $w(x) = 1$, and the Bessel functions $J_0(\beta_n x)$ are orthogonal on $0 < x < \ell$ with respect to $w(x) = x$.

If a function $f(x)$ has a series expression in terms of an orthogonal sequence of functions

$$f(x) = \sum_{n=0}^{\infty} a_n g_n(x),$$

then (assuming we can integrate term by term),

$$\int_a^b g_k(x)f(x)w(x)\, dx = \sum_{n=0}^{\infty} \left[a_n \int_a^b g_k(x)g_n(x)w(x)\, dx \right]$$

$$= a_k \int_a^b g_k(x)g_k(x)w(x)\, dx$$

or

$$a_k = \frac{1}{\int_b^a [g_k(x)]^2 w(x)\, dx} \int_a^b g_m(x)f(x)w(x)\, dx.$$

We can use this formula to associate a series to any given function $f(x)$. In general, however, the resulting series may not converge to the given function. In the case of the Legendre polynomials $P_n(x/\ell)$ on $-\ell < x < \ell$ and the Bessel functions $J_0(\beta_n x)$ on $0 < x < \ell$, if $f(x)$ and $f'(x)$ are piecewise continuous, then the resulting series converges to $f(x)$ at each point in the interval at which $f(x)$ is continuous and to $[f(x - 0) + f(x + 0]/2$ at each point at which $f(x)$ is not continuous.

EXERCISES

In Exercises 1 through 3, find the temperature $u(x, y, t)$ in the plate of Example 8.7.1 if the initial temperature distribution there is replaced with the given function.

1. $u(x, y, 0) = (1 - \cos \pi y)(\sin \pi x + 2 \sin 3\pi x)$
2. $u(x, y, 0) = (1 - \cos \pi y)x$
3. $u(x, y, 0) = xy$.

In Exercises 4 through 6, find the steady-state internal temperature in the planet of Example 8.7.2 if the boundary condition there is replaced with the given function.

4. $u(1, \phi) = 1 + 3 \cos \phi$
5. $u(1, \phi) = 1 \cos^2 \phi$
6. $u(1, \phi) = 2 \sin^2 \dfrac{\phi}{2}$ (*Hint:* Use a trig identity.)

7. Determine the displacement $u(r, t)$ of the membrane of Example 8.7.3 if we start with $u(r, 0) = 3J_0(r\beta_1) + J_0(r\beta_3)$ and $u_t(r, 0) = J_0(r\beta_2)$.

8. Show that as $t \to \infty$, the temperature at each point of the plate of Example 8.7.1 approaches 0.

9. a. By substituting $u = X(x)Y(y)Z(z)T(t)$ and separating variables one at a time (as in Example 8.7.1), show that the only solutions of the three-dimensional heat equation,

    (H₃) $$\frac{\partial^2 u}{\partial x^2} + \frac{\partial^2 u}{\partial y^2} + \frac{\partial^2 u}{\partial z^2} = \frac{1}{\alpha} \frac{\partial u}{\partial t}, \qquad 0 < x, y, z < \pi$$

    that have this form and satisfy the boundary conditions

    $$u_x(0, y, z, t) = u_x(\pi, y, z, t) = 0$$

    (B) $$u_y(x, 0, z, t) = u_y(x, \pi, z, t) = 0$$

    $$u_z(x, y, 0, t) = u_z(x, y, \pi, t) = 0$$

    are the multiples of

    $$u_{n,m,p} = e^{-(n^2 + m^2 + p^2)\alpha t} \cos nx \cos my \cos pz,$$

    where n, m, and p are nonnegative integers.

   b. Show that to match the initial condition $u(x, y, z, 0) = f(x, y, z)$ with a solution of the form $u = \sum_{p=0}^{\infty} \sum_{m=0}^{\infty} \sum_{n=0}^{\infty} a_{n,m,p} u_{n,m,p}$, we need to find an expression for $f(x, y, z)$ as a triple trigonometric series

    $$f(x, y, z) = \sum_{p=0}^{\infty} \sum_{m=0}^{\infty} \sum_{n=0}^{\infty} a_{n,m,p} \cos nx \cos my \cos pz.$$

   c. Find the solution of (H₃) that satisfies (B) and matches the initial condition $u(x, y, z, 0) = (1 - \cos 3x) \cos 2y \cos z$.

*10. If $u = u(r, t)$ is a θ-independent temperature distribution for the unit disc, $0 \le r \le 1$ and $0 \le \theta \le 2\pi$, then u must satisfy

    (H) $$\frac{\partial^2 u}{\partial r^2} + \frac{1}{r} \frac{\partial u}{\partial r} = \frac{1}{\alpha} \frac{\partial u}{\partial t}.$$

    Suppose $u = R(r)T(t)$ is a nontrivial solution of (H) that satisfies

    (B) $$u_r(1, t) = 0.$$

   a. Obtain o.d.e.'s for R and T.
   b. Use the fact that the temperature must remain bounded as $t \to \infty$ to show that T is a multiple of $T_\lambda = e^{-\lambda \alpha t}$, where $\lambda \ge 0$.
   c. Suppose $\lambda = 0$. Use the fact that the temperature must remain bounded as $r \to 0$ to conclude that R is a constant so that u is a multiple of $u_0 = 1$.
   d. Suppose $\lambda > 0$. Show that the substitution $s = r\sqrt{\lambda}$ changes the o.d.e. for R into the Bessel equation

    (H₃) $$s^2 \frac{d^2 R}{ds^2} + s \frac{dR}{ds} + s^2 R = 0.$$

    Use the facts that the solutions of (Hₛ) that remain bounded as $s \to 0$ are the multiples of $J_0(s) = J_0(r\sqrt{\lambda})$ (see Exercise 11, Section 6.8), that $dJ_0(r\sqrt{\lambda})/dr = -\sqrt{\lambda} J_1(r\sqrt{\lambda})$ (see Exercise 8, Section 6.7), and that u satisfies (B) to conclude that u is a multiple of

$$u_n = J_0(r\gamma_n)e^{-\alpha\gamma_n^2 t},$$

where $\gamma_1, \gamma_2, \ldots$ are the positive roots of J_1.

e. What form do you expect for an arbitrary solution of (H) that satisfies (B)?

REVIEW PROBLEMS

In Problems 1 through 4, find

a. the Fourier series of $f(x)$ on $-\ell < x < \ell$,
b. the cosine series of $f(x)$ on $0 < x < \ell$, and
c. the sine series of $f(x)$ on $0 < x < \ell$,

1. $f(x) = \ell - x$
2. $f(x) = \ell - |x|$
3. $f(x) = \ell^2 - x^2$
4. $f(x) = \ell x - x^2$

In Problems 5 through 10, find the solution of the one-dimensional heat equation

(H) $$\frac{\partial^2 u}{\partial x^2} = \frac{1}{\alpha}\frac{\partial u}{\partial t}$$

that satisfies the specified boundary conditions

(B$_1$) $u(0, t) = u(\ell, t) = 0$
(B$_2$) $u_x(0, t) = u_x(\ell, t) = 0$

or

(B$_3$) $u(\ell, t) - u(-\ell, t) = u_x(\ell, t) - u_x(-\ell, t) = 0$

and matches the given initial condition.

5. (B$_1$); $u(x, 0) = \ell - x$, $0 < x < \ell$
6. (B$_1$); $u(x, 0) = \ell x - x^2$, $0 < x < \ell$
7. (B$_2$); $u(x, 0) = \ell - x$, $0 < x < \ell$
8. (B$_2$); $u(x, 0) = \ell^2 - x^2$, $0 < x < \ell$
9. (B$_3$); $u(x, 0) = \ell - x$, $-\ell < x < \ell$
10. (B$_3$); $u(x, 0) = \ell - |x|$, $-\ell < x < \ell$

In Problems 11 and 12, find the solution of the one-dimensional wave equation

(W) $$\frac{\partial^2 u}{\partial x^2} = \frac{1}{\alpha^2}\frac{\partial^2 u}{\partial t^2}$$

that satisfies the boundary conditions

(B) $u(0, t) = u(\ell, t) = 0$

and matches the given initial conditions.

11. $u(x, 0) = \ell x - x^2$, $\quad u_t(x, 0) = \sin \dfrac{2\pi x}{\ell}$; $\quad 0 < x < \ell$

12. $u(x, 0) = \ell x - x^2$, $\quad u_t(x, 0) = \ell - x$; $\quad 0 < x < \ell$

In Problems 13 through 15, find the steady-state temperature in a square plate, $0 < x < \ell, 0 < y < \ell$, if the flat surfaces are insulated, and the temperature along the edges is as given.

13. $u(0, y) = u(\ell, y) = u(x, 0) = 0$, $\quad u(x, \ell) = \ell - x$; $\quad 0 < x, y < \ell$
14. $u(0, y) = 0$, $\quad u(\ell, y) = \ell - y$, $\quad u(x, 0) = u(x, \ell) = 0$; $\quad 0 < x, y < \ell$
15. $u(0, y) = 0$, $\quad u(\ell, y) = \ell - y$, $\quad u(x, 0) = 0$, $\quad u(x, \ell) = \ell - x$; $\quad 0 < x, y < \ell$
16. *An O.D.E. Boundary-Value Problem:* Consider the o.d.e.

(H$_x$)
$$\frac{d^2 X}{dx^2} + \lambda X = 0, \qquad 0 \le x \le 1$$

together with the boundary conditions

(B′)
$$X(0) = X'(1) + \gamma X(1) = 0,$$

where γ is a positive constant.

a. Show that if $\lambda \le 0$, then (H$_x$) has no nontrivial solutions that satisfy (B′).
b. Show that if $\lambda > 0$ and (H$_x$) has a nontrivial solution that satisfies (B′), then $\tan \sqrt{\lambda} = -\sqrt{\lambda}/\gamma$.
c. By graphing the two functions $\tan \mu$ and $-\mu/\gamma$, convince yourself that the equation $\tan \mu = -\mu/\gamma$ has infinitely many positive solutions for μ, $0 < \mu_1 < \mu_2 < \cdots$.
d. Conclude that the nontrivial solutions of (H$_x$) that satisfy (B′) are the constant multiples of $X_n = \sin \mu_n x$.

17. *Another Wire:* Suppose the temperature at the left end of the wire in Example 8.1.1 is kept at 0, and the right end loses heat at a rate proportional to the temperature there. Then, taking $\ell = 1$, we have

(B)
$$u(0, t) = u_x(1, t) + \gamma u(1, t) = 0,$$

where γ is a positive constant.

a. Using the result of the preceding problem, determine the functions of the form $u = X(x)T(t)$ that satisfy the one-dimensional heat equation and the boundary conditions (B).
b. What form do you expect for an arbitrary solution of the one-dimensional heat equation that satisfies (B)?

18. *Another Vibrating String:* Suppose the left end of the string in Example 8.5.1 is kept fixed, and the slope of the right end is always a negative constant multiple of its displacement. Then, taking $\ell = 1$, we have

(B)
$$u(0, t) = u_x(1, t) + \gamma u(1, t) = 0,$$

where γ is a positive constant.

a. Using the result of Problem 16, determine the functions of the form $u = X(x)T(t)$ that satisfy the one-dimensional wave equation and the boundary conditions (B).

b. What form do you expect for an arbitrary solution of the one-dimensional wave equation that satisfies (B)?

SUPPLEMENTARY READING LIST

GENERAL REFERENCES ABOUT DIFFERENTIAL EQUATIONS

Boyce, W. E., and DiPrima, R. C., *Elementary Differential Equations and Boundary Value Problems*, 4th ed., Wiley, New York, 1986.

Braun, M., *Differential Equations and Their Applications*, 2nd ed., Springer-Verlag, New York, 1978.

Hildebrand, F. B., *Advanced Calculus for Applications*, 2nd ed., Prentice-Hall, Englewood Cliffs, NJ, 1976.

Kaplan, W., *Ordinary Differential Equations*, Addison-Wesley, Reading, MA, 1958.

Koçak, H., *Differential and Difference Equations through Computer Experiments*, 2nd ed., Springer-Verlag, New York, 1989.

Kreider, D. L., Kuller, R. G., Ostberg, D. R., and Perkins, F. W., *An Introduction to Linear Analysis*, Addison-Wesley, Reading, MA, 1966.

Rainville, E. D., and Bedient, P. E., *Elementary Differential Equations*, 6th ed., Macmillan, New York, 1981.

Simmons, G. F., *Differential Equations with Applications and Historical Notes*, McGraw-Hill, New York, 1972.

APPLICATIONS

Burghes, D. N., and Barrie, M. S., *Modelling with Differential Equations*, Ellis Horwood, Chichester, 1981.

Cannon, R. H., *Dynamics of Physical Systems*, McGraw-Hill, New York, 1967.

720

Papoulis, A., *Circuits and Systems: A Modern Approach*, Holt, Rinehart and Winston, New York, 1980.

Pielou, E., *Mathematical Ecology*, Wiley, New York, 1977.

Pollard, H., *Applied Mathematics: An Introduction*, Addison-Wesley, Reading, MA, 1972.

Thomson, W. T., *Theory of Vibrations with Applications*, 2nd ed., Prentice-Hall, Englewood Cliffs, NJ, 1981.

MORE THEORETICAL TREATMENTS OF DIFFERENTIAL EQUATIONS

Arnold, V. I., *Ordinary Differential Equations*, translated by R. Silverman, MIT Press, Cambridge, MA, 1973.

Birkhoff, G., and Rota, G. C., *Ordinary Differential Equations*, 3rd ed., Wiley, New York, 1978.

Coddington, E. A., and Levinson, N., *Theory of Ordinary Differential Equations*, McGraw-Hill, New York, 1955.

Hirsch, W. H., and Smale, S., *Differential Equations, Dynamical Systems and Linear Algebra*, Academic Press, New York, 1974.

Hurewicz, W., *Lectures on Ordinary Differential Equations*, MIT Press, Cambridge, MA, 1958.

HISTORY

Boyer, C. B., *A History of Mathematics*, Wiley, New York, 1968.

Eves, H., *An Introduction to the History of Mathematics*, 5th ed., Saunders College Publishing, Philadelphia, 1983.

Kline, M., *Mathematical Thought from Ancient to Modern Times*, Oxford University Press, New York, 1972.

LINEAR ALGEBRA

Anton, H., *Elementary Linear Algebra*, 3rd ed., Wiley, New York, 1981.

Hoffman, K., and Kunze, R., *Linear Algebra*, 2nd ed., Prentice-Hall, Englewood Cliffs, NJ, 1971.

Kumpel, P. G., and Thorpe, J. A., *Linear Algebra with Applications to Differential Equations*, Saunders College Publishing, Philadelphia, 1983.

O'Nan, *Linear Algebra*, 2nd ed., Harcourt Brace Jovanovich, New York, 1976.

NUMERICAL METHODS

Atkinson, K. E., *An Introduction to Numerical Analysis*, Wiley, New York, 1978.
Dahlquist, G., and Björck, A., *Numerical Methods*, translated by N. Anderson, Prentice-Hall, Englewood Cliffs, NJ, 1974.
Henrici, P., *Discrete Variable Methods in Ordinary Equations*, Wiley, New York, 1962.

PARTIAL DIFFERENTIAL EQUATIONS

Powers, D. L., *Boundary Value Problems*, 2nd ed., Academic Press, New York, 1979.
Sagan, H., *Boundary and Eigenvalue Problems in Mathematical Physics*, Wiley, New York, 1961.
Sommerfeld, A., *Partial Differential Equations in Physics*, translated by E. Straus, Academic Press, New York, 1949.

ANSWERS TO ODD-NUMBERED EXERCISES

CHAPTER ONE

Section 1.1

1. a. 3   b. 4   c. 7   d. 3
3. No
5. Yes   7. $k = -2, 3$
9. $k = -7$ 11. $k = 1/2$
13. a. $x = \dfrac{t^3}{2} + \dfrac{t^2}{2} + c_1 t + c_2$ b. $x = \dfrac{t^3}{2} + \dfrac{t^2}{2} + 3t + 2$

15. a. $x = t^3 + c_1 \dfrac{t^2}{2} + c_2 t + c_3 = t^3 + k_1 t^2 + c_2 t + c_3$

    b. $x = t^3 - 3t^2 + 3t - 1$
17. a. $x = te^t - 2e^t + c_1 t + c_2$ b. $x = te^t - 2e^t + t + 2$
19. $\dfrac{dx}{dt} = .08x + 400$ 21. $\dfrac{dx}{dt} = -\dfrac{x}{2} + 15$

23. a. $\dfrac{dx}{dt} = -\dfrac{x}{4}$ b. $\dfrac{dx}{dt} = \dfrac{x(25 - x)}{100}$

25. $\dfrac{dx}{dt} = \gamma(M - x)$

27. a. $AB = \sqrt{400 - y^2}$ b. $\tan\theta = -\dfrac{y}{\sqrt{400 - y^2}}$ c. $\dfrac{dy}{dx} = -\dfrac{y}{\sqrt{400 - y^2}}$

Section 1.2

1. $x = ke^{2t/3}$ 3. $x = -\dfrac{e^{-2t}}{2} + e^{-t} + c$

5. $x = kte^{t^2/2} - 1$ 7. $t = 2 \ln |x^5 + 3x + 2| + c$

723

9. $x = \dfrac{-1}{2 \tan 2t + C}$

11. $x = C\left[\dfrac{t - 2}{t + 1}\right]$

13. $-(x + 1)e^{-x} = (2t - 1)e^{2t} + C$

15. $x = 3e^{-5t/3}$

17. $x = \dfrac{2t}{2 - t}$

19. $x = \tan\left(1 - \dfrac{1}{t}\right)$

21. $x = \dfrac{1}{9}(t^{3/2} - 21)^2$

23. a. $x = x_0 e^{-\gamma t}$ b. $\gamma = \dfrac{\ln 2}{T}$ c. $x = x_0 2^{-t/22}$

25. $20 \ln 2 \approx 13.9$ years

27. a. $x(t) = \dfrac{10e^{.05t}}{C + e^{.05t}}$

29. a. $x = 50e^{-t/4} \to 0$ as $t \to \infty$

   b. $x = \dfrac{50}{2 - e^{-t/4}} = \dfrac{50e^{t/4}}{2e^{t/4} - 1} \to 25$ as $t \to \infty$

Section 1.3

1. a. linear, nonhomogeneous   b. linear, homogeneous
   c. linear, homogeneous   d. nonlinear
   e. linear, nonhomogeneous   f. nonlinear
   g. nonlinear   h. linear, nonhomogeneous

3. a. $-\infty < t < \infty$ b. $-\infty < t < 1$ or $1 < t < \infty$
   c. $-1 < t < \infty$ d. $-\infty < t < -2, -2 < t < 2$ or $2 < t < \infty$

5. $x = ce^{-2t/3} + \frac{1}{2}$

7. $x = ce^{-3t} + \dfrac{t}{3} - \dfrac{1}{9}$

9. $x = c \cos t + t \cos t$

11. $x = ce^{-3t} - \dfrac{t^2}{3} + \dfrac{2t}{9} - \dfrac{11}{27}$

13. $x = c\sqrt{\dfrac{t + 1}{t - 1}} - \dfrac{2}{\sqrt{t - 1}}$

15. $x = -\dfrac{2}{5}\sin t - \dfrac{1}{5}\cos t + Ce^{2t}$

17. $x = \dfrac{1}{2}e^{3t} \ln\left|\dfrac{t - 1}{t + 1}\right| + Ce^{3t}$

19. $x = 5e^{2t} - 4$

21. $x = 3\sqrt{t^2 + 1}$

23. $x = \frac{3}{2}e^{t^2/2} - 1$

25. $x = 6000e^{.08t} - 5000$

27. a. $x(1) = -28.5 + 29e^{1/2} \approx 19.31$
    b. $x(6) = -26 + 29e^3 \approx 556.48$
    c. $x(12) = -23 + 29e^6 \approx 11676.44$

31. b. It is an exponential

33. b. The o.d.e. is not normal on any interval containing $t = 0$

35. $a_1(t)x' + a_0(t)x = ba_0(t)$, b constant

Section 1.4

1. a. $\dfrac{dx}{dt} = 8 - 0.04x, \quad x(0) = 20, \quad$ so $\quad x(t) = 200 - 180e^{-0.04t}$ grams

    b. $\lim\limits_{t \to \infty} x(t) = 200$

3. $\dfrac{dx}{dt} = \begin{cases} 8 - .04x & \text{for } t < 10 \\ \quad - .04x & \text{for } t > 10 \end{cases}$ so $x(10) = 200 - 180e^{.0.4} \approx 79.34$

    $x(10 + t) = x(10)e^{-.04t}, \quad$ so $x(20) = (200 - 180e^{-0.4}) \approx 53.18$

5. a. $\dfrac{dx}{dt} = 25 - \dfrac{x}{100}, \quad x(0) = 100, \quad$ so $\quad x(t) = 2500 - 2400e^{-t/100}$

    b. $\lim\limits_{t \to \infty} x(t) = 2500$

7. $10\,\dfrac{\ln(3/4)}{\ln(11/12)} \approx 33.06$ minutes

9. $I = 6 - 3e^{-t/2}; \quad \lim\limits_{t \to \infty} I(t) = 6$

11. $I = I_0 e^{-2t} + 10te^{-2t}; \quad \lim\limits_{t \to \infty} I(t) = 0$

13. a. $v(t) = 16 - 6e^{-2t}, \quad x(t) = 16t - 3 + 3e^{2t}$ b. $\lim\limits_{t \to \infty} v(t) = 16$

15. a. $x(t) = 8t^2 + v_0 t + x_0$ b. $\sqrt{5/2} \approx 1.58$ seconds

    c. $\dfrac{3\sqrt{65} - 3}{16} \approx 1.32$ seconds

17. a. $8 + 4e^{-2} \approx 8.54$ feet   b. $16 + 4e^{-4} \approx 16.07$ feet

    c. The solution of $8t + 4e^{-2t} = 20; \quad t \approx 2.4966081$ seconds

19. $x(t) = \dfrac{2t}{t + 9}; \quad 9$ seconds

21. a. $\dfrac{dx}{dt} = \ln\left(\dfrac{28}{27}\right)(3 - x)(4 - x)$

    b. $x(t) = \dfrac{9(28^t) - 8(27^t)}{3(28^t) - 2(27^t)}$

    c. $\lim\limits_{t \to \infty} x(t) = 3$

23. a. $x(t) = \dfrac{ab(e^{(a-b)kt} - 1)}{ae^{(a-b)kt} - b}; \quad \lim\limits_{t \to \infty} x(t) = b$

    b. $x(t) = \dfrac{a^2 kt}{1 + akt}; \quad \lim\limits_{t \to \infty} x(t) = a$

25. $y = cx$

Section 1.5

1. $x(t) = \dfrac{-t}{c + \ln t}$

3. not homogeneous

5. $x(t) = \dfrac{3}{1 + ce^{3t}}$

7. $x(t) = [ce^{3t^2/2} - 2]^{1/3}$

9. $x(t) = c_1 e^t + c_2 - t$

11. $x(t) = c_1 t^4 + c_2 t^2 + c_3 t + c_4$

13. a. $x = t - \dfrac{t}{\ln t + c}$ b. $x(t) = t + \dfrac{3t}{1 - 3 \ln t}$

15. a. $x(t) = \dfrac{t^4}{2} + c_1 \ln t + c_2$ b. $x(t) = \dfrac{t^4}{2} - 2 \ln t - \frac{1}{2}$

17. a. $x(t) = \dfrac{t^3}{3} + \dfrac{t^2}{2} + \dfrac{c_1}{t} + c_2 t + c_3$

     b. $x(t) = \dfrac{t^3}{6} + \dfrac{t^2}{2} + \dfrac{6}{t} + \dfrac{3t}{2} - \dfrac{1}{6}$

19. a. $x(t) = \dfrac{1}{ce^{-t} - 4}$ b. $x(t) = \dfrac{1}{5e^{-t} - 4}$

Section 1.6

1. $(0, \alpha)$ 3. $(t_0, 0)$
5. $(t_0, 1)$ 7. no exceptions
9. a. $-\infty < t < \infty,\ -\infty < x < \infty$ b. $-\infty < t < \infty$
11. a. $-\infty < t < \infty,\ -\infty < x < \infty$ b. $-\infty < t < \frac{1}{3}$
13. a. $0 < t < \infty,\ -\infty < x < \infty$ b. $0 < t < \infty$
15. a. $0 < t < \infty,\ 0 < x < \infty$ b. $\frac{1}{3} < t < \infty$
17. infinitely many solutions   19. unique solution
21. no solutions
25. $\partial f / \partial x = 2 x^{-1/3}$ is undefined at $x = 0$, so the region to which the theorem applies is $-\infty < t < \infty, 0 < x < \infty$. The two functions in question agree for $0 < t$, and both leave this region at $t = 0$.

Section 1.7

1.

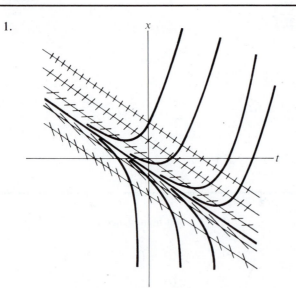

3.

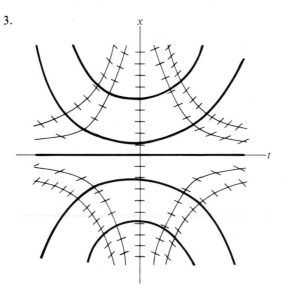

5.

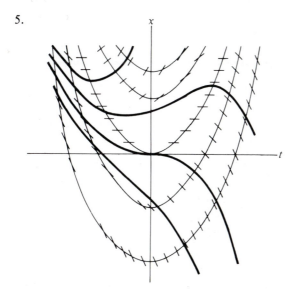

7.

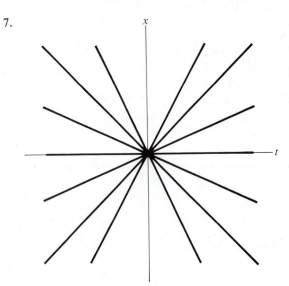

9.

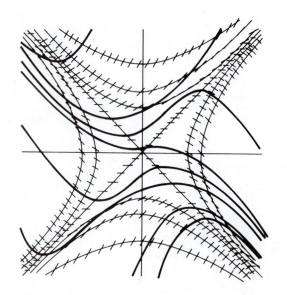

11.

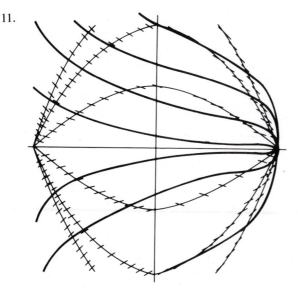

Section 1.8

1. a.　$\rightarrow\!\!-\!\!(-2)\!\!-\!\!\leftarrow\!\!-\!\!(1)\!\!-\!\!\rightarrow\!\!-$
    b.　-2 attractor,　1 repeller
    c.　If $x(0) < -2$, $x(t)$ increases, with $\lim_{t\to\infty} x(t) = -2$; if $-2 < x(0) < 1$, $x(t)$ decreases, with $\lim_{t\to\infty} x(t) = -2$; if $1 < x(0)$, $x(t)$ increases without bound.

3. a.　$-\!\!-\!\!\leftarrow\!\!-\!\!(-3)\!\!-\!\!\rightarrow\!\!-\!\!(-2)\!\!-\!\!\leftarrow\!\!-\!\!(1)\!\!-\!\!\rightarrow\!\!-$
    b.　-3, 1 repellers,　-2 attractor
    c.　If $x(0) < -3$, $x(t)$ decreases without bound; if $-3 < x(0) < -2$, $x(t)$ increases, with $\lim_{t\to\infty} x(t) = -2$; if $-2 < x(0) < 1$, $x(t)$ decreases, with $\lim_{t\to\infty} x(t) = -2$; if $1 < x(0)$, $x(t)$ increases without bound.

5. a.　$-\!\!-\!\!\leftarrow\!\!-\!\!(-2)\!\!-\!\!\rightarrow\!\!-\!\!(1)\!\!-\!\!\rightarrow\!\!-$
    b.　-2 repeller,　1 neither
    c.　If $x(0) < -2$, $x(t)$ decreases without bound; if $-2 < x(0) < 1$, $x(t)$ increases, with $\lim_{t\to\infty} x(t) = 1$; if $1 < x(0)$, $x(t)$ increases without bound.

7. a.　$-\!\!-\!\!\leftarrow\!\!-\!\!(-1)\!\!-\!\!\rightarrow\!\!-\!\!(3)\!\!-\!\!\leftarrow\!\!-$
    b.　-1 repeller,　3 attractor
    c.　If $x(0) < -1$, $x(t)$ decreases without bound; if $-1 < x(0) < 3$, $x(t)$ increases, with $\lim_{t\to\infty} x(t) = 3$; if $3 < x(0)$, $x(t)$ decreases, with $\lim_{t\to\infty} x(t) = 3$.

9. a.　$\rightarrow\!\!-\!\!(-1)\!\!-\!\!\leftarrow\!\!-\!\!(0)\!\!-\!\!\leftarrow\!\!-\!\!(2)\!\!-\!\!\rightarrow\!\!-$
    b.　-1 attractor,　0 neither,　2 repeller
    c.　If $x(0) < -1$, $x(t)$ increases, with $\lim_{t\to\infty} x(t) = -1$; if $-1 < x(0) < 0$, $x(t)$ decreases, with $\lim_{t\to\infty} x(t) = -1$; if $0 < x(0) < 2$, $x(t)$ decreases, with $\lim_{t\to\infty} x(t) = 0$; if $2 < x(0)$, $x(t)$ increases without bound.

11. a.　$\rightarrow\!\!-\!\!(-5)\!\!-\!\!\leftarrow\!\!-\!\!(1)\!\!-\!\!\leftarrow\!\!-\!\!(3)\!\!-\!\!\rightarrow\!\!-\!\!(4)\!\!-\!\!\rightarrow\!\!-$
    b.　-5 attractor,　3 repeller,　14 neither

   c. If $x(0) < -5$, $x(t)$ increases, with $\lim_{t \to \infty} x(t) = -5$; if $-5 < x(0) < 1$, $x(t)$ decreases, with $\lim_{t \to \infty} x(t) = -5$; if $1 < x(0) < 3$, $x(t)$ decreases, with $\lim_{t \to \infty} x(t) = 1$; if $3 < x(0) < 4$, $x(t)$ increases, with $\lim_{t \to \infty} x(t) = 4$; if $4 < x(0)$, $x(t)$ increases without bound.

13. a. $\longleftarrow\!\!\!\leftarrow\!(-3)\!\rightarrow\!\!\!\rightarrow\!\!-\!(3)\!-\!\leftarrow$
     b. -3 repeller, 3 attractor
     c. If $x(0) < -3$, $x(t)$ decreases without bound; if $-3 < x(0) < 3$, $x(t)$ increases, with $\lim_{t \to \infty} x(t) = 3$; if $3 < x(0)$, $x(t)$ decreases, with $\lim_{t \to \infty} x(t) = 3$.

15. Attractor   17. Repeller

19. Repeller   21. Tends to 2500.

23. x tends to a

25. b. Equilibrium at $x = q/p$, with $f' = -p$ in the derivative test

27. $x' = x^n$ has repeller at $x = 0$ if n odd, neither if n even; $x' = -x^n$ has attractor at $x = 0$ if n odd.

Section 1.9

1. Exact, $\dfrac{t^2}{2} + \dfrac{x^2}{2} = c$, so $x = \pm\sqrt{k - t^2}$

3. Not exact   5. Not exact

7. Not exact   9. Not exact

11. Exact, $x = \dfrac{c(t^2 + 1)}{t}$ 13. $x = \dfrac{(1 - t)(t + 1)^2}{t^3 - t + 1}$

15. $x = (1 + te^t)^{-1}$ 17. $(x^3 + x^2 t)e^t = C$.

19. a. $2\rho = t\,\dfrac{d\rho}{dt}$ b. $\rho = t^2$ c. $x = ct^{-3} - \dfrac{5t}{8}$

21. a. $t\,\dfrac{t\rho}{dt} = -\dfrac{\rho}{2}$ b. $\rho = t^{-1/2}$ c. $xt^{1/2} - \cos x = C$

23. a. $x\,\dfrac{d\rho}{dt} = \rho$ b. $\rho = x$ c. $x^2 t + x^3 \sin t = C$

25. a. $\dfrac{d\rho}{dt} = \rho$ b. $\rho = e^x$ c. $e^x(\sin t + 2x - 1) = C$

Review Problems

1. a. $x = k\sqrt{(t + 1)/(t - 1)}$ b. $x = \sqrt{(t + 1)/3(t - 1)}$

3. a. $\dfrac{x}{1 + x} = ke^t$, so $x = \dfrac{ke^t}{1 - ke^t}$ b. $x = \dfrac{e^t}{2 - e^t}$

5. a. $x = k/t$ b. $x = 2/t$

7. a. $3x - 3 \ln |x + 1| = \ln t - \dfrac{1}{t} + c, \quad x = -1$

    b. $3x - 3 \ln |x + 1| = \ln t - \dfrac{1}{t} + \dfrac{7}{2} - 4 \ln 2$

9. a. $x = \cos t(\ln |\sec t + \tan t| + c)$
    b. $x = \cos t(\ln |\sec t + \tan t| + 1)$

11. a. $x^2 + 2tx - t^2 = c, \quad \text{so } x = -t \pm \sqrt{2t^2 + c}$
    b. $x = -t + \sqrt{2t^2 + 2}$

13. a. $x = cte^{-1/t} + t$ b. $x = t$

15. a. $(x - t)e^{x/t} = k$ b. $(x - t)e^{x/t} = -e^{1/2}$

17. a. $x = 4[2t \cos 2t - \sin 2t - c]^{-1}$ b. $x = 4[2t \cos 2t - \sin 2t + 4]^{-1}$

19. a. $x = \dfrac{1}{3} t^{3/2} e^{t/2} + ce^{t/2}$ b. $x = \dfrac{1}{3} (t^{3/2} - 1)e^{t/2}$

21.

23. a. $x(t)$ decreases without bound   b. $x(t) = 1$
    c. $x(t)$ decreases, with $\lim_{t \to \infty} x(t) = 1$ d. $x(t)$ increases without bound.

25. a. $\dfrac{dx}{dt} = \alpha x,$ where we want $x(1) = 3x(0),$ so $\alpha = \ln 3$

    b. $x(7) = 3^7 = 2187$

    c. $\dfrac{dx}{dt} = (\ln 3) \dfrac{x(6000 - x)}{6000}$ d. $x(7) = \dfrac{6000(3^7)}{5999 + 3^7} \approx 1603$

27. $x(t) = \dfrac{4 - 4(9/8)^t}{1 - 2(9/8)^t}$

29. a. $\dfrac{d^2x}{dt^2} = \dfrac{-gR^2}{x^2}$ b. $v \dfrac{dv}{dx} = \dfrac{-gR^2}{x^2}$

    c. $v^2 = \dfrac{2gR^2}{x} + v_0^2 - 2gR$ e. $x = \left[\dfrac{3}{2} tR\sqrt{2g} + R^{3/2} \right]^{2/3}$

CHAPTER TWO

Section 2.1

1. $\dfrac{w}{32}\dfrac{d^2x}{dt^2} + b\dfrac{dx}{dt} + kx = w$

3. a. $\dfrac{d^2z}{dt^2} = -2b\dfrac{dz}{dt} + 2k(y - z - L) - 32$

    b. $\dfrac{d^3z}{dt^3} + 2b\dfrac{d^2z}{dt^2} + 2k\dfrac{dz}{dt} = 4k$

5. $L\dfrac{d^2Q}{dt^2} + R\dfrac{dQ}{dt} + \dfrac{1}{C}Q = V(t)$

7. a. $\dfrac{d^2x}{dt^2} + 125x = 32$ b. $\dfrac{d^2x}{dt^2} + 2b\dfrac{dx}{dt} + 125x = 32$

9. a. $y = \dfrac{1}{44}\dfrac{d^2x}{dt^2} + \dfrac{62.5x}{22} - \dfrac{16}{22}$ b. $\dfrac{d^2y}{dt^2} = -44y + 7$

    c. $\dfrac{d^4x}{dt^4} + 169\dfrac{d^2x}{dt^2} + 5500x = 1716$

Section 2.2

1. Linear, nonhomogeneous; $(D^2 - tD - 5t)x = -25$
3. Not linear   5. Not linear
7. Linear, nonhomogeneous; $(D^4 + 5t^3D)x = \sqrt{t^2 - 1}$
9. a. No   b. Yes   c. Yes   d. No
11. $L[e^t] = 0,\ \ L[3e^{2t}] = 3e^{2t}$
13. $L[e^{2t}] = 7e^{2t},\ \ L[t^3] = 6t + 9t^2 - 3t^3,$
     $L[5e^{2t} - 2t^3] = 35e^{2t} - 12t - 18t^2 + 6t^3$
15. $L[t] = 1,\ \ L[e^t] = e^t,\ \ L[te^t] = (t + 1)e^t$
17. $x = -\frac{1}{4}\sin 2t + \dfrac{3t}{2} - \dfrac{3\pi}{2}$
19. $x = (t - 3)e^t + t^2 + 2t + 3$
21. $x = \dfrac{t^5}{40} - \dfrac{t^4}{24}$
23. a. $A = \frac{1}{2}, B = -\frac{9}{4}$ b. $x = c_1e^{2t} + c_2e^{-t} + \frac{1}{2}t - \frac{9}{4}$
25. a. $A = -2$ b. $x = c_1 + c_2e^{2t} + c_3e^{-t} - 2t$
27. a. $A = -2,\ B = 0,\ C = -1$ b. $x = c_1e^t + c_2e^{-t} - 2 - t^2$

Section 2.3

1. Yes
3. Yes
5. Yes
7. Yes
9. Yes
11. Yes
13. No
15. a. e^t, e^{-t} b. Yes
17. a. $1, t^{-1}$ b. Yes
19. a. e^t, e^{2t}, e^{3t} b. Yes
21. a. e^{-2t}, e^{2t} b. No
23. Any condition with $x''(0) \neq 2x(0) - 2x'(0)$: e.g., $x(0) = x'(0) = 0, x''(0) = 1$
25. Any condition with $x''(0) \neq x(0)$: e.g., $x(0) = x'(0) = 0, x''(0) = 1$
27. Any condition with $x(0) - 2x'(0) + x''(0) \neq -4$: e.g., $x(0) = x'(0) = 0, x''(0) = 1$
29. Any condition with $x'''(0) \neq 8x(0) - 4x'(0) + 2x'''(0)$:
    e.g., $x(0) = x'(0) = x''(0) = 0, x'''(0) = 1$

Section 2.4

1. Dependent
3. Independent
5. Dependent
7. Dependent
9. Independent
11. Independent
19. b. $f(t) = t, \quad g(t) = t + 1$

Section 2.5

1. $x = c_1 e^{t\sqrt{5}} + c_2 e^{-t\sqrt{5}}$
3. $x = c_1 + c_2 e^{5t/2}$
5. $x = c_1 e^{2t/3} + c_2 t e^{2t/3}$
7. $x = c_1 + c_2 t + c_3 t^2 + c_4 e^t + c_5 t e^t$
9. $x = c_1 + c_2 t + c_3 e^{-t} + c_4 t e^{-t} + c_5 t^2 e^{-t} + c_6 e^{2t} + c_7 e^{-5t/3} + c_8 e^{3t/2}$
11. $x = c_1 e^{t\sqrt{2}} + c_2 e^{-t\sqrt{2}} + c_3 e^{-(1+\sqrt{5})t/2} + c_4 e^{-(1-\sqrt{5})t/2}$
13. $x = c_1 e^t + c_2 t e^t + c_3 e^{-t} + c_4 t e^{-t}$
15. $x = 8 + 2e^{5t}$
17. $x = -e^t + 3t e^t + e^{-2t}$
19. $Q = t e^{-t}$
21. a. $r = \dfrac{-b \pm \sqrt{b^2 + 2mk(-3 \pm \sqrt{5})}}{2m}$

   b. $x = (c_1 + c_2 t + c_3 e^{t\sqrt[4]{5}} + c_4 e^{-t\sqrt[4]{5}}) e^{-(1+\sqrt{5})t/2}$
23. $L[e^t] = 4e^t, \quad L[e^t \sin t] = e^t(3 \sin t + 4 \cos t), \quad L[e^{-t} \sin t] = -e^{-t} \sin t$
25. $L[e^{\alpha t} \cos \beta t] = L[e^{\alpha t} \sin \beta t] = 0$
27. $(D - \lambda)^{k-2}$

Section 2.6

1. $x = c_1 \cos \dfrac{t}{3} + c_2 \sin \dfrac{t}{3}$
3. $x = c_1 e^{2t} + c_2 e^{-2t} + c_3 \cos 2t + c_4 \sin 2t$

5. $x = c_1 \cos t + c_2 \sin t + c_3 e^{-t/2} \cos \dfrac{t\sqrt{3}}{2} + c_4 e^{-t/2} \sin \dfrac{t\sqrt{3}}{2}$

7. $x = c_1 e^t + c_2 \cos 3t + c_3 \sin 3t$

9. $x = c_1 e^t + c_2 e^{-t} + c_3 \cos t + c_4 \sin t$

11. $x = (c_1 + c_2 t + c_3 t^2 + c_4 t^3) \cos t + (c_5 + c_6 t + c_7 t^2 + c_8 t^3) \sin t$

13. $x = 2 \cos \dfrac{t}{4} + 36 \sin \dfrac{t}{4}$ 15. $x = 0$

17. $x = \dfrac{5}{2} e^{-t/5} \sin \dfrac{2t}{5}$ 19. $(D - 3)(D - 1)$

21. $(D - 2)^3 (D - 1)^2$ 23. $(D^2 + 4)^2$

25. a. $x = \left(c_1 \cos \dfrac{t}{2} + c_2 \sin \dfrac{t}{2}\right) e^{-3t/2}$

 b. $x = \left(-2 \cos \dfrac{t}{2} - 4 \sin \dfrac{t}{2}\right) e^{-3t/2}$

27. $x = c_1 e^{-t/5} \cos \dfrac{2t}{5} + c_2 e^{-t/5} \sin \dfrac{2t}{5}$

29. $Q = Q_0 \cos (t/\sqrt{LC})$

Section 2.7

1. $x = c_1 e^{-t} + c_2 t e^{-t} + 3 + t$

3. $x = c_1 + c_2 t + c_3 e^{-2t} + c_4 t e^{-2t} + \dfrac{1}{4} t^2$

5. $x = c_1 \cos t + c_2 \sin t + \tfrac{1}{2} t \sin t$

7. $x = c_1 e^{-3t} \cos t + c_2 e^{-3t} \sin t - 2e^t \cos t + 4e^t \sin t$

9. $x = c_1 e^{t/3} + c_2 e^{-t/3} - \tfrac{9}{50} \cos t - \tfrac{1}{10} t \sin t$

11. $x = \tfrac{3}{8} \sin t - \tfrac{1}{8} \sin 3t$

13. $x = 9e^{-t/5} \cos \dfrac{2t}{5} - 8e^{-t/5} \sin \dfrac{2t}{5} - 10 + 5t$

15. $p(t) = k_1 + k_2 t + k_3 t^2 + (k_4 + k_5 t)e^{-t/2} \cos t\sqrt{5} + (k_6 + k_7 t)e^{-t/2} \sin t\sqrt{5}$
$\quad + k_8 t e^{3t}$

17. a. $x = c_1 \cos 5t\sqrt{5} + c_2 \sin 5t\sqrt{5} + \tfrac{32}{125}$
 b. $x = c_1 e^{-10t} \cos 5t + c_2 e^{-10t} \sin 5t + \tfrac{32}{125}$
 c. $x = c_1 e^{-25t} + c_2 e^{-5t} + \tfrac{32}{125}$

21. $x = c_1 \cos 5t\sqrt{5} + c_2 \sin 5t\sqrt{5} + c_3 \cos 2t\sqrt{11} + c_4 \sin 2t\sqrt{11} + \tfrac{39}{125}$

23. $x = c_1 t^2 + c_2 t + 4 - t \ln t$

25. $x = c_1 t^{-2} + c_2 t^{-2} \ln t + \dfrac{1}{16} t^2 - \dfrac{1}{3} t$

Section 2.8

1. $x = c_1 e^t + c_2 e^{2t} + 3(t - 1)e^{2t}$
3. $x = c_1 \cos t + c_2 \sin t - \cos t \ln |\sec t + \tan t|$
5. $x = c_1 e^{t/3} + c_2 t e^{t/3} + \frac{1}{28} t^{7/3} e^{t/3}$
7. $x = (c_1 + c_2 t + c_3 t^2 + \frac{1}{2} \ln t + \frac{3}{4}) e^t$
9. $x = (t + 2) \sin t + (1 + \ln |\cos t|) \cos t$

11. $x = \dfrac{-4}{(\ln 2)^3} e^t + \dfrac{3}{(\ln 2)^4} t e^t + t^{-3} e^t$

13. $x = c_1 t^{-1} + c_2 e^t - \dfrac{t}{2} - 1$

15. $x = c_1 t + c_2 t^2 + c_3 t^{-1} + \frac{5}{36} t^{-1} + \frac{1}{6} t^{-1} \ln t$
17. $x = c_1 t^{-2} + c_2 t^{-1} + \frac{1}{42} t^5$ 19. $x = c_1 + c_2 t^{-1} + \frac{4}{15} t^{3/2}$

21. $x = (-c_1 e^{1/t} + c_2) t$ 23. $x = -c_1(t + 1) + c_2 e^t + 4$
25.   a. $x = c_1 e^{-3t} + c_2 e^{-t} - \frac{1}{4} e^{-t} + \frac{1}{2} t e^{-t}$
     b. $x = c_1 \cos 2t + c_2 \sin 2t + \cos^2 2t - \frac{1}{3} \cos^4 2t + \frac{1}{3} \sin^4 2t$
        $= c_1 \cos 2t + c_2 \sin 2t + \frac{2}{3} - \frac{1}{3} \sin^2 2t$

27.   a. $(D^2 + 6D + 8)x = 8(h(t) - L - 4)$
     b. $x = c_1 e^{-2t} + c_2 e^{-4t} + 4e^{-2t} \int h e^{2t} \, dt - 4e^{-4t} \int h e^{4t} \, dt - L - 4$

Section 2.9

3.   a. $x = \frac{1}{10} \sin 10t$

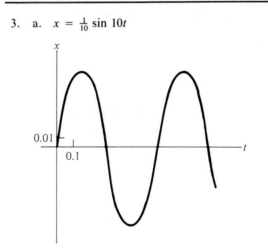

b. $x = \frac{1}{8}e^{-6t} \sin 8t$

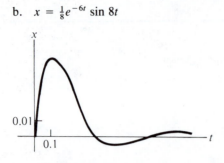

c. $x = \frac{1}{6}e^{-8t} \sin 6t$

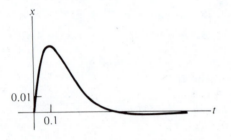

d. $x = te^{-10t}$

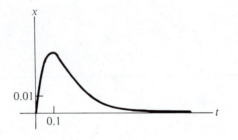

e. $x = \frac{1}{48}(-e^{-50t} + e^{-2t})$

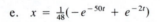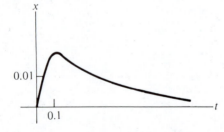

5. a. (i)   10.0269   (ii) 9.9968   (iii) 9.9978
    b. (i)   10.1859   (ii) 9.5949   (iii) 10.3679
    c. (i)   10.0023   (ii) 9.9990   (iii) 10.0005

7. Here, the forcing term E depends on the mass.

9. a. $e^{-t}[2 \cos t + 2 \sin t]$

    b. $\sqrt{5} \cos (t - \alpha)$,   where $\alpha = \arccos(-2/\sqrt{5})$

    c.

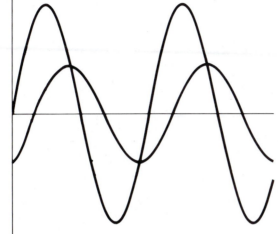

11. $R^2 - \dfrac{4L}{C} < 0$ (underdamped): $Q = Ae^{-\sigma t} \cos(\omega t - \alpha)$,

    $I = Be^{-\sigma t} \cos(\omega t - \beta)$

    $R^2 - \dfrac{4L}{C} > 0$ (overdamped): $Q = c_1 e^{-\sigma_1 t} + c_2 e^{-\sigma_2 t}$,

    $I = -\sigma_1 c_1 e^{-\sigma_1 t} - \sigma_2 c_2 e^{-\sigma_2 t}$

    $R^2 - \dfrac{4L}{C} = 0$ (critically damped): $Q = c_1 e^{-\sigma t} + c_2 t e^{-\sigma t}$,

    $I = (-c_1 \sigma + c_2)e^{-\sigma t} - \sigma c_2 t e^{-\sigma t}$.

13. a. $A \cos \left(\dfrac{20t\sqrt{5}}{\sqrt{w}} - \alpha \right) + \dfrac{2w}{125}$ b. $w = \dfrac{500}{\pi^2 f^2}$ (where f = frequency).

Section 2.10

1. In case a, the shaft turns counterclockwise, then reverses direction, gradually approaching equilibrium. In case b, the shaft turns clockwise, overshooting equilibrium, then turns back and approaches equilibrium. In case c, the shaft turns clockwise toward equilibrium.

3. $s = \frac{1}{6}x^4 - 2x^3 + 36x;$ lowest point $x = 3,$ $s = \frac{135}{62}$

5. a. $\left(D^2 - \frac{g}{L}\right)\theta = 0$ b. No

7. a. $F_{\text{tan}} = w \sin \theta + (5 - 5\sqrt{2} \sin \theta) \cos \theta$

   c. $w \frac{d^2\theta}{dt^2} = 16w\sqrt{2} \sin \theta + 80\sqrt{2} \cos \theta - 160 \sin \theta \cos \theta$

   d. $w \frac{d^2\theta}{dt^2} + 16(5 - w)\theta = 4\pi(5 - w) + 16w$

   e. $\theta = -\frac{1}{4} \cos 8t + \frac{1}{4}(\pi + 1)$

Section 2.11

1. $x_1 = ce^{t/5} - 2,$ $x_2 = \frac{1}{4} ce^{t/5}$

3. $x_1 = c_1 e^{-t} + c_2 e^{11t},$ $x_2 = -\frac{12}{7} e^{11t}$

5. $x_1 = c_1 e^{-2t} \cos t + c_2 e^{-2t} \sin t + 8,$
   $x_2 = (-2c_1 + c_2)e^{-2t} \cos t - (c_1 + 2c_2)e^{-2t} \sin t - 10$

7. $x_1 = c_1 e^{-t/2} + c_2 e^t - 4 + 3e^{2t},$ $x_2 = -3c_1 e^{-t/2} - \frac{3}{2} c_2 e^t + 8 - 3e^{2t}$

9. $x_1 = c_1 e^{7t/2} + c_2 e^{-t} + 1,$ $x_2 = -c_1 e^{7t/2} + \frac{2}{7} c_2 e^{-t}$

11. $x_1 = c_1 e^t + c_2 e^{-t} + c_3 t e^{-t} - 3,$ $x_2 = (2c_2 + c_3)e^{-t} + 2c_3 t e^{-t} - 3$

13. $x_1 = c_1 e^{2t} + c_2 e^{-2t} + c_3 \cos 2t + c_4 \sin 2t - 8,$
    $x_2 = -c_1 e^{2t} + 3c_2 e^{-2t} + (c_3 + 2c_4) \cos 2t - (2c_3 - c_4) \sin 2t$

15. $x_1 = 3e^{-t} - \frac{8}{3}e^{-2t} - \frac{1}{3}e^t,$ $x_2 = -3e^{-t} + \frac{4}{3}e^{-2t} + \frac{2}{3}e^t + 1$

17. $x_1 = 6e^{-t} + 4te^{-t} + 2e^t + 3,$ $x_2 = -2e^{-t} - 4te^{-t} + 2e^t$

19. a. $(D - \gamma a)p + bs = \gamma b,$ $-\beta p + Ds = 0$

    b. $p = c_1 e^{(\gamma a + \mu)t/2} + c_2 e^{(\gamma a - \mu)t/2},$
    $s = k_1 e^{(\gamma a + \mu)t/2} + k_2 e^{(\gamma a - \mu)t/2} + b,$ where $\mu = \sqrt{\gamma^2 a^2 - 4\beta\gamma}$

    c. $p = c_1 e^{\gamma a t/2} \cos \frac{vt}{2} + c_2 e^{\gamma a t/2} \sin \frac{vt}{2}$

    $s = k_1 e^{\gamma a t/2} \cos \frac{vt}{2} + k_2 e^{\gamma a t/2} \sin \frac{vt}{2} + b,$ where $v = \sqrt{4\beta\gamma - \gamma^2 a^2}$

21. $x_2 = f(t),$ $x_1 = \frac{1}{2}(e^t - f(t)) + c$

Review Problems

1. a. $L[1] = L[t] = L\left[\frac{1}{t}\right] = 0,$ $L[t^3] = 24t$ b. $x = c_1 + c_2 t + c_3 \frac{1}{t} + t^3$

3. $x = c_1 e^{t/3} + c_2 t e^{t/3} + \dfrac{1}{t} e^{t/3}$

5. $x = (c_1 + c_2 t + c_3 t^2 + t^2 \ln t - \frac{3}{2}t^2)e^{-t}$

7. $x = c_1 + c_2 t + c_3 t^2 + c_4 \cos t + c_5 \sin t + \frac{1}{2}t^3 - e^{-t}$

9. $x = 2e^{-t/2} \cos \dfrac{3t}{2} - \dfrac{2}{3} e^{-t/2} \sin \dfrac{3t}{2}$ 11. $x = -e^t + 4te^t + 2e^{-t}$

13. $x = c_1 \cos(\ln t) + c_2 \sin(\ln t) + \dfrac{3}{2t}$

15. $x = c_1 \cos t + c_2 \sin t - \frac{1}{2}t \cos t$ 17. $x = c_1 t + c_2 t \ln t + \dfrac{1}{4t}$

19. $x = c_1 e^{t/2} + c_2 e^{-t/2} + \frac{16}{3}e^t - 2te^t + 2e^{t/2}$

21. $x_1 = c_1 e^{2t} + c_2 e^{-t} - 3e^t + 5, \quad x_2 = c_1 e^{2t} + \frac{1}{2}c_2 e^{-t} - \frac{5}{2}e^t + 3$

23. $x_1 = c_1 e^{-t/2} \cos \dfrac{t\sqrt{3}}{2} + c_2 e^{-t/2} \sin \dfrac{t\sqrt{3}}{2} + \dfrac{2}{3} e^t,$

$x_2 = \left(\dfrac{1}{2} c_1 - \dfrac{\sqrt{3}}{2} c_2 \right) e^{-t/2} \cos \dfrac{t\sqrt{3}}{2} + \left(\dfrac{\sqrt{3}}{2} c_1 + \dfrac{1}{2} c_2 \right) e^{-t/2} \sin \dfrac{t\sqrt{3}}{2} + \dfrac{1}{3} e^t$

Appendix A

1. -2	3. 0	5. 20	7. 8
9. $6e^{2t}$	11. 120	13. 0	15. 120
17. No	19. Yes	21. Yes	23. Yes
25. None	27. Unique	29. $x = \dfrac{8}{7}$	31. $x = 1$

CHAPTER THREE

Section 3.1

1. $\dfrac{dI_1}{dt} = -2I_1 + 2I_2 + 4e^{-4t}$

   $\dfrac{dI_2}{dt} = 2I_1 - 2I_2$

3. $\dfrac{dQ}{dt} = I_2$

   $\dfrac{dI_2}{dt} = -\dfrac{1}{2} Q - I_2 + e^{-t/2}$

5. $\dfrac{dQ_1}{dt} = I_1 = \dfrac{-2}{3} Q_1 - \dfrac{1}{2} Q_2 + 10$

   $\dfrac{dQ_2}{dt} = I_2 = -\dfrac{1}{3} Q_1 - \dfrac{1}{2} Q_2 + 5$

7. a. $\dfrac{dA}{dt} = -k_A A, \qquad \dfrac{dB}{dt} = k_A A - k_B B, \qquad \dfrac{dC}{dt} = k_B B$

   b. $\dfrac{dA}{dt} = -k_A A + \alpha, \qquad \dfrac{dB}{dt} = k_A A - k_B B, \qquad \dfrac{dC}{dt} = k_B B - \gamma$

9. $\dfrac{dx}{dt} = \dfrac{1}{v}[P_1(G - x) + P_2(y - x)] = \dfrac{(P_1 + P_2)}{v} x + \dfrac{P_2}{v} y + \dfrac{P_1 G}{v},$

   $\dfrac{dy}{dt} = \dfrac{1}{V} P_2(x - y) \qquad\qquad = \qquad \dfrac{P_2}{V} x - \dfrac{P_2}{V} y$

Section 3.2

1. a. $\begin{bmatrix} 0 \\ 2 \end{bmatrix}$ b. $\begin{bmatrix} 0 \\ 9 \end{bmatrix}$ c. $\begin{bmatrix} 7 \\ 7 \end{bmatrix}$ d. $\begin{bmatrix} -1 \\ 8 \end{bmatrix}$ e. $\begin{bmatrix} -1 \\ -2 \end{bmatrix}$ f. $\begin{bmatrix} -2 \\ 26 \end{bmatrix}$

   g. $\begin{bmatrix} -3 & 3 \\ -6 & -9 \end{bmatrix}$ h. $\begin{bmatrix} -1 & -5 \\ 2 & 1 \end{bmatrix}$ i. $\begin{bmatrix} 4 \\ 10 \end{bmatrix}$

In Exercises 3 through 9, $\mathbf{x} = \begin{bmatrix} x \\ y \end{bmatrix}$ or $\mathbf{x} = \begin{bmatrix} x \\ y \\ z \end{bmatrix}$.

3. Linear, nonhomogeneous, order 2; $\quad D\mathbf{x} = \begin{bmatrix} -1 & 1 \\ 2 & 0 \end{bmatrix} \mathbf{x} + \begin{bmatrix} t \\ t \end{bmatrix}$

5. Linear, homogeneous, order 2; $\quad D\mathbf{x} = \begin{bmatrix} 5 & -6 \\ 2 & 1 \end{bmatrix} \mathbf{x}$

7. Linear, nonhomogeneous, order 3; $\quad D\mathbf{x} = \begin{bmatrix} 0 & -t & -1 \\ -1/t & 0 & -1/t \\ 1 & -t & 0 \end{bmatrix} \mathbf{x} + \begin{bmatrix} t \\ 1 \\ 0 \end{bmatrix}$

9. Linear, nonhomogeneous, order 3; $\quad D\mathbf{x} = \begin{bmatrix} 1 & 3 & 0 \\ 2 & 1 & 0 \\ 1 & 3t & 0 \end{bmatrix} \mathbf{x} = \begin{bmatrix} t^2 \\ t \\ -t^2 \end{bmatrix}$

In Exercises 11 to 17, $\mathbf{x} = \begin{bmatrix} x_1 \\ x_2 \end{bmatrix} = \begin{bmatrix} x \\ x' \end{bmatrix}$ or $\mathbf{x} = \begin{bmatrix} x_1 \\ x_2 \\ x_3 \end{bmatrix} = \begin{bmatrix} x \\ x' \\ x'' \end{bmatrix}$.

11. a. $\begin{aligned} x_1' &= \quad x_2 \\ x_2' &= x_1 \ + t \end{aligned}$ b. $\begin{aligned} x_1 &= x = c_1 e^t + c_2 e^{-t} - t \\ x_2 &= x' = c_1 e^t - c_2 e^{-t} - 1 \end{aligned}$

    c. $D\mathbf{x} = \begin{bmatrix} 0 & 1 \\ 1 & 0 \end{bmatrix} \mathbf{x} + \begin{bmatrix} 0 \\ t \end{bmatrix}$ d. $\mathbf{x} = c_1 \begin{bmatrix} e^t \\ e^t \end{bmatrix} + c_2 \begin{bmatrix} e^{-t} \\ -e^{-t} \end{bmatrix} + \begin{bmatrix} -t \\ -1 \end{bmatrix}$

13. a. $\begin{aligned} x_1' &= \quad x_2 \\ x_2' &= -x_1 \ + 1 \end{aligned}$ b. $\begin{aligned} x_1 &= x = \quad c_1 \cos t + c_2 \sin t + 1 \\ x_2 &= x' = -c_1 \sin t + c_2 \cos t \end{aligned}$

    c. $D\mathbf{x} = \begin{bmatrix} 0 & 1 \\ -1 & 0 \end{bmatrix} \mathbf{x} + \begin{bmatrix} 0 \\ 1 \end{bmatrix}$ d. $\mathbf{x} = c_1 \begin{bmatrix} \cos t \\ -\sin t \end{bmatrix} + c_2 \begin{bmatrix} \sin t \\ \cos t \end{bmatrix} + \begin{bmatrix} 1 \\ 0 \end{bmatrix}$

15. a. $x_1' = x_2$
$x_2' = x_3$
$x_3' = -x_1 + x_2 + x_3 + 4$

b. $x_1 = x = c_1e^t + c_2te^t + c_3e^{-t} + 4$
$x_2 = x' = c_1e^t + c_2(t+1)e^t - c_3e^{-t}$
$x_3 = x'' = c_1e^t + c_2(t+2)e^t + c_3e^{-t}$

c. $D\mathbf{x} = \begin{bmatrix} 0 & 1 & 0 \\ 0 & 0 & 1 \\ -1 & 1 & 1 \end{bmatrix} \mathbf{x} + \begin{bmatrix} 0 \\ 0 \\ 4 \end{bmatrix}$

d. $\mathbf{x} = c_1\begin{bmatrix} e^t \\ e^t \\ e^t \end{bmatrix} + c_2\begin{bmatrix} te^t \\ (t+1)e^t \\ (t+2)e^t \end{bmatrix} + c_3\begin{bmatrix} e^{-t} \\ -e^{-t} \\ e^{-t} \end{bmatrix} + \begin{bmatrix} 4 \\ 0 \\ 0 \end{bmatrix}$

17. a. $x_1' = x_2$
$x_2' = x_3$
$x_3' = x_2 + 1$

b. $x_1 = x = c_1 + c_2e^t + c_3e^{-t} - t$
$x_2 = x' = c_2e^t - c_3e^{-t} - 1$
$x_3 = x'' = c_2e^t + c_3e^{-t}$

c. $D\mathbf{x} = \begin{bmatrix} 0 & 1 & 0 \\ 0 & 0 & 1 \\ 0 & 1 & 0 \end{bmatrix} \mathbf{x} + \begin{bmatrix} 0 \\ 0 \\ 1 \end{bmatrix}$

d. $\mathbf{x} = c_1\begin{bmatrix} 1 \\ 0 \\ 0 \end{bmatrix} + c_2\begin{bmatrix} e^t \\ e^t \\ e^t \end{bmatrix} + c_3\begin{bmatrix} e^{-t} \\ -e^{-t} \\ e^{-t} \end{bmatrix} + \begin{bmatrix} -t \\ -1 \\ 0 \end{bmatrix}$

19. a. $\begin{bmatrix} e^t \\ 3e^{3t} \end{bmatrix}, \begin{bmatrix} 3e^{3t} \\ 3e^{3t} \end{bmatrix}$

b. $\begin{bmatrix} e^t + 2e^{3t} \\ 3e^{3t} \end{bmatrix}, \begin{bmatrix} 3e^{3t} \\ 3e^{3t} \end{bmatrix}$

c. $\mathbf{x}_1$ is not a solution; $\mathbf{x}_2$ is

21. a. $\begin{bmatrix} 1 \\ 0 \end{bmatrix}, \begin{bmatrix} 0 \\ 1 \end{bmatrix}$ b. $\begin{bmatrix} 1 \\ 0 \end{bmatrix}, \begin{bmatrix} 0 \\ 1 \end{bmatrix}$ c. Both are solutions

23. a. $\begin{bmatrix} e^t \\ 0 \\ 0 \end{bmatrix}, \begin{bmatrix} 2e^t \\ 0 \\ 0 \end{bmatrix}$ b. $\begin{bmatrix} e^t \\ 0 \\ 0 \end{bmatrix}, \begin{bmatrix} 2e^t + 1 \\ -2 \\ 4 \end{bmatrix}$

c. $\mathbf{x}_1$ is a solution; $\mathbf{x}_2$ is not

25.

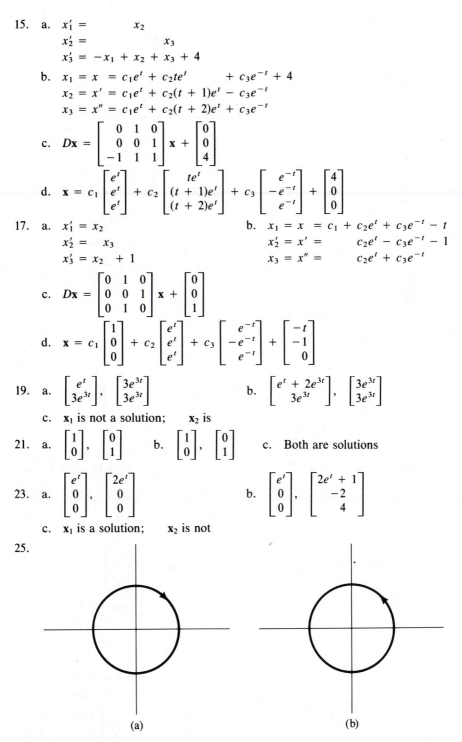

(a)   (b)

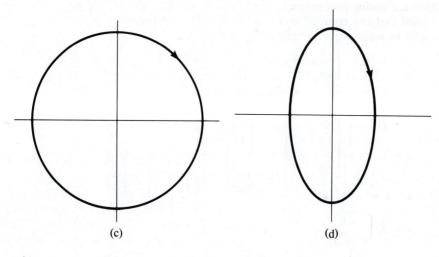

(c)   (d)

27. $x' = \qquad\qquad v$
     $v' = -2kx - 2bv + 2ky - 2kL - 32$
     $y' = \qquad\qquad\qquad\qquad\quad 2$

33. b.   No   c.   No

Section 3.3

1. a.   Both are solutions   b.   Yes
3. a. h_1 is a solution, h_2 is not   b.   No
5. a. h_1 is a solution, h_2 is not   b.   No
7. a.   All are solutions   b.   No   9.   a.   All are solutions   b.   Yes
11. Yes   13.   No
15. Yes

Section 3.4

1. Independent   3.   Dependent   5.   Independent
7. Independent   9.   Independent

Section 3.5

1. a. $\lambda^2 - 3\lambda + 2$ b. $\lambda = 1, 2$
3. a. $\lambda^2 - 2\lambda - 2$ b. $\lambda = 1 \pm \sqrt{3}$
5. a. $(2 - \lambda)(\lambda - 3)(\lambda - 1)$ b. $\lambda = 1, 2, 3$

When comparing your answers for Exercises 7 through 13 with the answers here, keep in mind that any nonzero linear combination of eigenvectors corresponding to λ will also be an eigenvector corresponding to λ.

7. $\lambda = 2, \quad \mathbf{v} = \begin{bmatrix} 1 \\ 1 \end{bmatrix}; \quad \lambda = -2, \quad \mathbf{w} = \begin{bmatrix} -1 \\ 3 \end{bmatrix}$

9. $\lambda = 1, \quad \mathbf{v} = \begin{bmatrix} 1 \\ 0 \\ 0 \end{bmatrix}; \quad \lambda = 2, \quad \mathbf{w} = \begin{bmatrix} 1 \\ 1 \\ 0 \end{bmatrix}; \quad \lambda = -3, \quad \mathbf{u} = \begin{bmatrix} -3 \\ 2 \\ 10 \end{bmatrix}$

11. $\lambda = 1, \quad \mathbf{v} = \begin{bmatrix} 1 \\ 0 \\ 0 \end{bmatrix}; \quad \lambda = -1, \quad \mathbf{w} = \begin{bmatrix} 1 \\ 2 \\ 0 \end{bmatrix}$

13.   a. $\lambda = -2$ b. $\mathbf{x} = e^{-2t}\mathbf{v} = \begin{bmatrix} e^{-2t} \\ e^{-2t} \end{bmatrix}$

15. $\mathbf{x} = c_1 \begin{bmatrix} e^{4t} \\ -e^{4t} \end{bmatrix} + c_2 \begin{bmatrix} e^{-2t} \\ e^{-2t} \end{bmatrix}$

17. $\mathbf{x} = c_1 \begin{bmatrix} e^{t} \\ 0 \\ 0 \end{bmatrix} + c_2 \begin{bmatrix} e^{2t} \\ e^{2t} \\ 0 \end{bmatrix} + c_3 \begin{bmatrix} -3e^{-3t} \\ 2e^{-3t} \\ 10e^{-3t} \end{bmatrix}$

19.   a. $\mathbf{x} = c_1 \begin{bmatrix} e^{-2t} \\ e^{-2t} \end{bmatrix}$ b. $\mathbf{x}(0) = \begin{bmatrix} 1 \\ 0 \end{bmatrix}$

Section 3.6

1. $\begin{bmatrix} 1 & 0 & 0 & 1 \\ 0 & 1 & 0 & 0 \\ 0 & 0 & 1 & 0 \\ 0 & 0 & 0 & 0 \end{bmatrix}$

3. $\begin{bmatrix} 1 & 0 & 2 & 0 \\ 0 & 1 & 0 & 0 \\ 0 & 0 & 0 & 1 \\ 0 & 0 & 0 & 0 \end{bmatrix}$

5.   a. $x = y = z = w = u = 0$ b. $\mathbf{x} = 0$

7.   a. $x = -a + 2, \quad y = 1, \quad z = a$ b. $\mathbf{x} = a \begin{bmatrix} -1 \\ 0 \\ 1 \end{bmatrix} + \begin{bmatrix} 2 \\ 1 \\ 0 \end{bmatrix}$

9.   a. $x_1 = 1/2, \quad x_2 = 0, \quad x_3 = -1/2, \quad x_4 = 0$ b. $\mathbf{x} = \begin{bmatrix} 1/2 \\ 0 \\ -1/2 \\ 0 \end{bmatrix}$

11.   No solutions   13. $\mathbf{u} = \begin{bmatrix} (-7a + 5)/3 \\ -(a + 1)/3 \\ a \end{bmatrix}$

In Exercises 14 through 19, the eigenvectors are the nonzero vectors of the given forms.

15. $\lambda = 0, \quad \mathbf{v} = a \begin{bmatrix} -1 \\ 1 \end{bmatrix}; \quad \lambda = 2, \quad \mathbf{w} = b \begin{bmatrix} 1 \\ 1 \end{bmatrix}$

17. $\lambda = 2$, $\mathbf{v} = a \begin{bmatrix} -1 \\ 1 \end{bmatrix}$

19. $\lambda = 1$, $\mathbf{v} = a \begin{bmatrix} 2 \\ 0 \\ 1 \end{bmatrix} + b \begin{bmatrix} -2 \\ 1 \\ 0 \end{bmatrix}$; $\lambda = -2$, $\mathbf{u} = c \begin{bmatrix} -1 \\ 1 \\ 2 \end{bmatrix}$

21. a. $x = c_1 e^t + c_2 e^{-t} + c_3 \cos t + c_4 \sin t$; $x = \frac{3}{2} e^t - \frac{1}{2} \cos t - \frac{1}{2} \sin t$
    b. $x = c_1 e^t + c_2 t e^t + c_3 e^{-t} + c_4 t e^{-t}$; $x = \frac{5}{4} e^t - \frac{1}{2} t e^t - \frac{1}{4} e^{-t}$

25. $x + y + z = 1$
    $x + y + z = 2$

Section 3.7

1. $\mathbf{x} = c_1 \begin{bmatrix} 2e^t \\ e^t \end{bmatrix} + c_2 \begin{bmatrix} e^{2t} \\ e^{2t} \end{bmatrix}$

3. $\mathbf{x} = c_1 \begin{bmatrix} e^{(1+\sqrt{3})t} \\ \sqrt{3} e^{(1+\sqrt{3})t} \end{bmatrix} + c_2 \begin{bmatrix} e^{(1-\sqrt{3})t} \\ -\sqrt{3} e^{(1-\sqrt{3})t} \end{bmatrix}$

5. $\mathbf{x} = c_1 \begin{bmatrix} 2e^t \\ 0 \\ e^t \end{bmatrix} + c_2 \begin{bmatrix} -2e^t \\ e^t \\ 0 \end{bmatrix} + c_3 \begin{bmatrix} -e^{-2t} \\ e^{-2t} \\ 2e^{-2t} \end{bmatrix}$

7. a. $x_1 = c_1 e^{3t} + c_2 e^{-t}$ b. $x_1 = 2e^{3t} - e^{-t}$
    $x_2 = c_1 e^{3t} - c_2 e^{-t}$ $x_2 = 2e^{3t} + e^{-t}$

9. a. $\mathbf{x} = c_1 \begin{bmatrix} e^{2t} \\ e^{2t} \\ 0 \end{bmatrix} + c_2 \begin{bmatrix} -1 \\ 1 \\ 0 \end{bmatrix} + c_3 \begin{bmatrix} 0 \\ 0 \\ e^{-t} \end{bmatrix}$ b. $\mathbf{x} = \begin{bmatrix} 3e^{2t} - 1 \\ 3e^{2t} + 1 \\ 2e^{-t} \end{bmatrix}$

11. a. $x_1 = c_1 e^{2t} + c_2 e^t$ b. $x_1 = e^t$
    $x_2 = 3c_1 e^{2t} + 2c_2 e^t$ $x_2 = 2e^t$
    $x_3 = c_1 e^{2t} + c_2 e^t + c_3 e^{-t}$ $x_3 = e^t + 2e^{-t}$

13. a. $x_1 = 4c_2 e^{2t} + 6c_3 e^t$ b. $x_1 = 25e^{2t} - 24e^t$
    $x_2 = c_1 e^{2t}$ $x_2 = e^{2t}$
    $x_3 = 3c_2 e^{2t} + 5c_3 e^t + c_4 e^{-2t}$ $x_3 = \frac{75}{4} e^{2t} - 20e^t + \frac{9}{4} e^{-2t}$
    $x_4 = 4c_2 e^{2t} + 5c_3 e^t$ $x_4 = 25e^{2t} - 20e^t$

15. a. $Q = -c_1 e^{-t} - 2c_2 e^{-5t/2}$ b. $Q = -e^{-t} + 2e^{-5t/2}$
    $I_2 = 2c_1 e^{-t} + c_2 e^{-5t/2}$ $I_2 = 2e^{-t} - e^{-5t/2}$

17. a. $x_1 = c_1 - c_2 e^{-t/25}$ b. $x_1 = c_1 - 2c_2 e^{-3t/100}$
    $x_2 = c_1 + c_2 e^{-t/25}$ $x_2 = c_1 + c_2 e^{-3t/100}$

Section 3.8

1. $\mathbf{x} = \begin{bmatrix} \sin t \\ \cos t \end{bmatrix} + c_2 \begin{bmatrix} -\cos t \\ \sin t \end{bmatrix}$

3. $\mathbf{x} = c_1 \begin{bmatrix} 0 \\ e^{2t} \\ 0 \end{bmatrix} + c_2 \begin{bmatrix} -e^t \sin t \\ 0 \\ e^t \cos t \end{bmatrix} + c_3 \begin{bmatrix} e^t \cos t \\ 0 \\ e^t \sin t \end{bmatrix}$

5. $\mathbf{x} = c_1 \begin{bmatrix} 2e^{-t} \\ 0 \\ e^{-t} \end{bmatrix} + c_2 \begin{bmatrix} e^{-t/2}\left(-\cos\dfrac{t\sqrt{3}}{2} + \sqrt{3}\sin\dfrac{t\sqrt{3}}{2}\right) \\ 2e^{-t/2}\cos\dfrac{t\sqrt{3}}{2} \\ 0 \end{bmatrix}$

$+ c_3 \begin{bmatrix} e^{-t/2}\left(-\sqrt{3}\cos\dfrac{t\sqrt{3}}{2} - \sin\dfrac{t\sqrt{3}}{2}\right) \\ 2e^{-t/2}\sin\dfrac{t\sqrt{3}}{2} \\ 0 \end{bmatrix}$

7. $\mathbf{x} = c_1 \begin{bmatrix} -e^{-t}\sin 2t \\ e^{-t}\sin 2t \\ 0 \\ 2e^{-t}\cos 2t \end{bmatrix} + c_2 \begin{bmatrix} e^{-t}\cos 2t \\ -e^{-t}\cos 2t \\ 0 \\ 2e^{-t}\sin 2t \end{bmatrix} + c_3 \begin{bmatrix} e^{-t}(-\cos 2t + 2\sin 2t) \\ 0 \\ 5e^{-t}\cos 2t \\ 0 \end{bmatrix}$

$+ c_4 \begin{bmatrix} e^{-t}(-2\cos 2t - \sin 2t) \\ 0 \\ 5e^{-t}\sin 2t \\ 0 \end{bmatrix}$

9. $\mathbf{x} = c_1 \begin{bmatrix} -e^{t}\sin t \\ e^{t}\cos t \end{bmatrix} + c_2 \begin{bmatrix} e^{t}\cos t \\ e^{t}\sin t \end{bmatrix}$ b. $\mathbf{x} = \begin{bmatrix} -3e^{t}\sin t + 2e^{t}\cos t \\ 3e^{t}\cos t + 2e^{t}\sin t \end{bmatrix}$

11. a. $x_1 = c_1 e^{t}(-\cos t\sqrt{2} - \sqrt{2}\sin t\sqrt{2}) + c_2 e^{t}(\sqrt{2}\cos t\sqrt{2} - \sin t\sqrt{2})$
    $x_2 = 3c_1 e^{t}\cos t\sqrt{2} \qquad\qquad + 3c_2 e^{t}\sin t\sqrt{2}$

b. $x_1 = e^{t}\left(2\cos t\sqrt{2} - \dfrac{3\sqrt{2}}{2}\sin t\sqrt{2}\right), \quad x_2 = e^{t}\left(\cos t\sqrt{2} + \dfrac{7\sqrt{2}}{2}\sin t\sqrt{2}\right)$

13. $Q = c_1 e^{-t/2}\left(-\cos\dfrac{t\sqrt{3}}{2} + \sqrt{3}\sin\dfrac{t\sqrt{3}}{2}\right) + c_2 e^{-t/2}\left(-\sqrt{3}\cos\dfrac{t\sqrt{3}}{2} - \sin\dfrac{t\sqrt{3}}{2}\right)$

$I = 2c_1 e^{-t/2}\cos\dfrac{t\sqrt{3}}{2} \qquad\qquad + 2c_2 e^{-t/2}\sin\dfrac{t\sqrt{3}}{2}$

15. $x_1 = c_1\sin t - c_2\cos t - 2\sqrt{6}c_3\sin t\sqrt{6} + 2\sqrt{6}c_4\cos t\sqrt{6}$
    $v_1 = c_1\cos t + c_2\sin t - 12c_3\cos t\sqrt{6} - 12c_4\sin t\sqrt{6}$
    $x_2 = 2c_1\sin t - 2c_2\cos t + \sqrt{6}c_3\sin t\sqrt{6} - \sqrt{6}c_4\cos t\sqrt{6}$
    $v_2 = 2c_1\cos t + 2c_2\sin t + 6c_3\cos t\sqrt{6} + 6c_4\sin t\sqrt{6}$

Section 3.9

1. a. $\begin{bmatrix} 3 & 2 & 1 \\ 2 & 1 & 1 \\ -1 & -1 & 0 \end{bmatrix}$ b. $\begin{bmatrix} 6 \\ 3 \\ -3 \end{bmatrix}$ c. $\begin{bmatrix} 6 \\ 3 \\ -3 \end{bmatrix}$

d. $\begin{bmatrix} 4 & -1 & 1 \\ 2 & 0 & 1 \\ 2 & -1 & 0 \end{bmatrix}$ e. $\begin{bmatrix} 8 & -3 & -1 \\ 2 & -1 & 0 \\ -3 & 1 & 0 \end{bmatrix}$ f. $\begin{bmatrix} 22 & -8 & -3 \\ 5 & -2 & -1 \\ -8 & 3 & 1 \end{bmatrix}$

g. $\begin{bmatrix} 12 & 7 & 5 \\ 7 & 4 & 3 \\ -5 & -3 & -2 \end{bmatrix}$ h. $\begin{bmatrix} 9 & 12 & -3 \\ 3 & 3 & 0 \\ -4 & -5 & 1 \end{bmatrix}$ i. $\begin{bmatrix} 16 & 0 & 0 \\ 5 & 1 & 0 \\ 0 & 0 & 1 \end{bmatrix}$

3. $\mathbf{h}(t) = \begin{bmatrix} (1-t)e^t \\ -te^t \end{bmatrix}$

5. $\mathbf{x} = c_1 \begin{bmatrix} (1-t)e^{2t} \\ te^{2t} \end{bmatrix} + c_2 \begin{bmatrix} -te^{2t} \\ (1+t)e^{2t} \end{bmatrix}$

7. $\mathbf{x} = c_1 \begin{bmatrix} e^{-t} \\ 2e^{-t} \\ 0 \end{bmatrix} + c_2 \begin{bmatrix} e^t \\ 0 \\ 0 \end{bmatrix} + c_3 \begin{bmatrix} -te^t \\ -e^t \\ 2e^t \end{bmatrix}$

9. $\mathbf{x} = c_1 \begin{bmatrix} -e^t \sin t \\ 2e^t \sin t \\ e^t(\sin t + \cos t) \\ 2e^t \cos t \end{bmatrix} + c_2 \begin{bmatrix} e^t \cos t \\ -2e^t \cos t \\ -e^t(\sin t + \cos t) \\ 2e^t \sin t \end{bmatrix} + c_3 \begin{bmatrix} e^{2t} \\ 0 \\ 0 \\ 0 \end{bmatrix} + c_4 \begin{bmatrix} te^{2t} \\ 0 \\ e^{2t} \\ 0 \end{bmatrix}$

11. a. $\mathbf{x} = c_1 \begin{bmatrix} (1-2t)e^{2t} \\ -4te^{2t} \end{bmatrix} + c_2 \begin{bmatrix} te^{2t} \\ (1-2t)e^{2t} \end{bmatrix}$ b. $\mathbf{x} = \begin{bmatrix} (3-2t)e^{2t} \\ (4-4t)e^{2t} \end{bmatrix}$

13. a. $x_1 = c_1(1-t)e^{3t} - c_2 te^{3t}$, $x_2 = c_1 te^{2t} + c_2(1+t)e^{3t}$
    b. $x_1 = (3-7t)e^{3t}$, $x_2 = (4-7t)e^{3t}$

15. $Q = c_1(1+t)e^{-t} + c_2 te^{-t}$, $I = -c_1 te^{-t} + c_2(1-t)e^{-t}$

19. b. Yes. For example, $\begin{bmatrix} 1 & 1 \\ -1 & -1 \end{bmatrix}$.

Section 3.10

1. $\mathbf{h}(t) = \begin{bmatrix} (1 - \frac{1}{2}t^2)e^{-t} \\ te^{-t} \\ \frac{1}{2}t^2 \end{bmatrix}$

3. $\mathbf{x} = c_1 \begin{bmatrix} (1-t+\frac{1}{2}t^2)e^t \\ \frac{1}{2}t^2 e^t \\ (t+\frac{1}{2}t^2)e^t \end{bmatrix} + c_2 \begin{bmatrix} (t-t^2)e^t \\ (1-t-t^2)e^t \\ (-3t-t^2)e^t \end{bmatrix} + c_3 \begin{bmatrix} \frac{1}{2}t^2 e^t \\ (t+\frac{1}{2}t^2)e^t \\ (1+2t+\frac{1}{2}t^2)e^t \end{bmatrix}$

5. $\mathbf{x} = c_1 \begin{bmatrix} 0 \\ e^{-t} \\ e^{-t} \\ e^{-t} \end{bmatrix} + c_2 \begin{bmatrix} (1+t)e^t \\ 0 \\ 0 \\ -te^t \end{bmatrix} + c_3 \begin{bmatrix} -te^t \\ -e^t \\ e^t \\ te^t \end{bmatrix} + c_4 \begin{bmatrix} te^t \\ 0 \\ 0 \\ (1-t)e^t \end{bmatrix}$

7. $\mathbf{x} = c_1 \begin{bmatrix} -2t \cos t + \sin 2t \\ 2t \sin 2t \\ 4 \sin 2t \\ 4 \cos 2t \end{bmatrix} + c_2 \begin{bmatrix} -\cos 2t - 2t \sin 2t \\ -2t \cos 2t \\ -4 \cos 2t \\ 4 \sin 2t \end{bmatrix}$

$+ c_3 \begin{bmatrix} \sin 2t \\ \cos 2t \\ 0 \\ 0 \end{bmatrix} + c_4 \begin{bmatrix} -\cos 2t \\ \sin 2t \\ 0 \\ 0 \end{bmatrix}$

9. $\mathbf{x} = c_1 \begin{bmatrix} e^{2t} \\ 0 \\ te^{2t} \\ te^{2t} \end{bmatrix} + c_2 \begin{bmatrix} 0 \\ e^{2t} \\ 0 \\ 0 \end{bmatrix} + c_3 \begin{bmatrix} 0 \\ 0 \\ (1-t)e^{2t} \\ -te^{2t} \end{bmatrix} + c_4 \begin{bmatrix} 0 \\ 0 \\ te^{2t} \\ (1+t)e^{2t} \end{bmatrix}$

11. $\mathbf{x} = c_1 \begin{bmatrix} 0 \\ e^{-t} \\ 0 \\ 0 \\ 0 \end{bmatrix} + c_2 \begin{bmatrix} 0 \\ 0 \\ e^{-t} \\ 0 \\ 0 \end{bmatrix} + c_3 \begin{bmatrix} (1+\frac{1}{2}t^2) \\ t \\ 1 \\ \frac{1}{2}t^2 \\ t \end{bmatrix} + c_4 \begin{bmatrix} 1 \\ 0 \\ 0 \\ 1 \\ 0 \end{bmatrix} + c_5 \begin{bmatrix} t \\ 1 \\ 0 \\ t \\ 1 \end{bmatrix}$

13. a. $x_1 = c_1e^{-t} + c_2te^{-t}$, $\quad x_2 = c_2e^{-t}$, $\quad x_3 = c_3e^{-t} + c_4te^{-t} + c_4te^{-t}$, $\quad x_4 = c_4e^{-t}$

   b. $x_1 = (1+t)e^{-t}$, $\quad x_2 = e^{-t}$, $\quad x_3 = -e^{-t}$, $\quad x_4 = 0$

Section 3.11

1. $\mathbf{x} = c_1 \begin{bmatrix} 2e^t \\ e^t \end{bmatrix} + c_2 \begin{bmatrix} e^{2t} \\ e^{2t} \end{bmatrix} + \begin{bmatrix} -e^t \\ -e^t \end{bmatrix}$

3. $\mathbf{x} = c_1 \begin{bmatrix} e^{(1+\sqrt{3})t} \\ \sqrt{3}e^{(1+\sqrt{3})t} \end{bmatrix} + c_2 \begin{bmatrix} e^{(1-\sqrt{3})t} \\ -\sqrt{3}e^{(1-\sqrt{3})t} \end{bmatrix} + \begin{bmatrix} 0 \\ -1 \end{bmatrix}$

5. $\mathbf{x} = c_1 \begin{bmatrix} e^t \\ e^t \\ e^t \end{bmatrix} + c_2 \begin{bmatrix} 0 \\ 2e^{2t} \\ e^{2t} \end{bmatrix} + c_3 \begin{bmatrix} -e^{3t} \\ 5e^{3t} \\ 3e^{3t} \end{bmatrix} + \begin{bmatrix} te^t \\ te^t \\ te^t \end{bmatrix}$

7. $\mathbf{x} = c_1 \begin{bmatrix} 0 \\ e^{2t} \\ e^{2t} \end{bmatrix} + c_2 \begin{bmatrix} 2e^{2t}\cos t \\ e^{2t}(\cos t + \sin t) \\ 2e^{2t}\cos t \end{bmatrix} + c_3 \begin{bmatrix} 2e^{2t}\sin t \\ e^{2t}(\sin t - \cos t) \\ 2e^{2t}\sin t \end{bmatrix} + \begin{bmatrix} -2e^{2t} \\ -e^{2t} \\ -2e^{2t} \end{bmatrix}$

9. $\mathbf{x} = c_1 \begin{bmatrix} (1-t)e^{-2t} \\ -te^{-2t} \end{bmatrix} + c_2 \begin{bmatrix} te^{-2t} \\ (t+1)e^{-2t} \end{bmatrix} + \begin{bmatrix} (t^2 - \frac{1}{6}t^3)e^{-2t} \\ (\frac{1}{2}t^2 - \frac{1}{6}t^3)e^{-2t} \end{bmatrix}$

11. $\mathbf{x} = c_1 \begin{bmatrix} e^t \\ te^t \\ te^t \end{bmatrix} + c_2 \begin{bmatrix} 0 \\ (1-t)e^t \\ -te^t \end{bmatrix} + c_3 \begin{bmatrix} 0 \\ te^t \\ (1+t)e^t \end{bmatrix} + \begin{bmatrix} 0 \\ 1 \\ 0 \end{bmatrix}$

13. a. $x_1 = c_1 + c_2e^{2t} - \dfrac{1}{3}e^{3t}$ b. $x_1 = e^{2t} - \dfrac{1}{3}e^{3t}$

       $x_2 = c_1 + 2c_2e^{2t} - \dfrac{4}{3}e^{3t}$ $x_2 = 2e^{2t} - \dfrac{4}{3}e^{3t}$

15. a. $x_1 = c_1e^{2t} - c_2\sin 3t + c_3\cos 3t + te^{2t} - 1$

     $x_2 = \qquad\qquad c_2\cos 3t + c_3\sin 3t$

     $x_3 = c_1e^{2t} \qquad\qquad\qquad\quad + te^{2t}$

   b. $x_1 = 3e^{2t} - 2\sin 3t - \cos 3t + te^{2t} - 1$

     $x_2 = \qquad\quad 2\cos 3t - \sin 3t$

     $x_3 = 3e^{2t} \qquad\qquad\qquad + te^{2t}$

17. $Q = -c_1e^{-t}\sin t + c_2e^{-t}\cos t$

    $I_2 = c_1e^{-t} + c_2e^{-t}\sin t + e^{-t}$

19. a. $A = 500$, $B = -2000e^{-t/10} + 2500$, $C = 2000e^{-t/10} - 1500 + 150t$

   b. Yes

21. a. $p = -c_1e^{-t/10} - c_2e^{-t/2}$
     $s = 500c_1e^{-t/10} + 100c_2e^{-t/2} + 5000$

     b. $p = 5e^{-5.2} - e^{-26} \approx .03$
     $s = -2500e^{-5.2} + 100e^{-26} + 5000 \approx 4986$
     $w = -3000e^{-5.2} + 600e^{-26} + 5000 \approx 4983$

Review Problems

1. a. $x_1 = c_1e^{3t}(\cos 2t - \sin 2t) + c_1e^{3t}(\cos 2t + \sin 2t)$
     $x_2 = 2c_1e^{3t}\cos 2t + 2c_2e^{3t}\sin 2t + 2e^{3t}$
     b. $x_1 = -e^{3t}(\cos 2t + 5\sin 2t),\quad x_2 = e^{3t}(4\cos 2t - 6\sin 2t + 2)$

3. a. $x_1 = c_1(1 + 4t)e^{-t} - 4c_2te^{-t} + 1 + (t + 2t^2)e^{-t}$,
     $x_2 = 4c_1te^{-t} + c_2(1 - 4t)e^{-t} + 1 + 2t^2e^{-t}$
     b. $x_1 = 1 + (1 + 5t + 2t^2)e^{-t},\qquad x_2 = 1 + (4t + 2t^2)e^{-t}$

5. a. $x_1 = 6c_1e^{4t}$
     $x_2 = 4c_1e^{4t} + c_2e^{2t} + c_3e^{-2t} + e^{3t}$
     $x_3 = c_1e^{4t} + c_3e^{-2t} + e^{3t}$
     b. $x_1 = 6e^{4t}$
     $x_2 = 4e^{4t} - 4e^{2t} - e^{-2t} + e^{3t}$
     $x_3 = e^{4t} - e^{-2t} + e^{3t}$

7. a. $x_1 = -2c_1e^{2t} + 2te^{2t}$ b. $x_1 = (2t - 2)e^{2t}$
     $x_2 = c_2e^{2t} + c_3e^{4t} + e^{4t}$ $x_2 = 8e^{2t} - e^{4t}$
     $x_3 = c_1e^{2t} + c_3e^{4t} - te^{2t}$ $x_3 = (-t + 1)e^{2t} - 2e^{4t}$

9. a. $x_1 = c_1e^{2t} + c_3te^{2t} - \frac{1}{4}te^{-2t}$ b. $x_1 = \frac{5}{4}te^{2t} - \frac{1}{4}te^{-2t}$
     $x_2 = c_2e^{2t} + c_3te^{2t} + \frac{1}{16}e^{-2t}$ $x_2 = \frac{5}{16}e^{2t} + \frac{5}{4}te^{2t} + \frac{1}{16}e^{-2t}$
     $x_3 = c_3e^{2t} - \frac{1}{4}e^{-2t}$ $x_3 = \frac{5}{4}e^{2t} - \frac{1}{4}e^{-2t}$

11. a. $x_1 = c_1e^t - c_2te^t + 4e^{2t},\quad x_2 = -c_3e^{-t} + c_4e^{-t} + \frac{2}{3}e^{2t}$,
     $x_3 = c_2e^t + c_4e^{-t} + \frac{4}{3}e^{2t},\quad x_4 = c_3e^{-t} + \frac{2}{3}e^{2t}$
     b. $x_1 = 4e^{2t},\quad x_2 = \frac{13}{3}e^{-t} + \frac{2}{3}e^{2t},\quad x_3 = \frac{14}{3}e^{-t} + \frac{4}{3}e^{2t},\quad x_4 = \frac{1}{3}e^{-t} + \frac{2}{3}e^{2t}$

13. a. $I_1 = c_1 - c_2e^{-4t} + (2t - \frac{1}{2})e^{-4t}$
     $I_2 = c_1 + c_2e^{-4t} - (2t + \frac{1}{2})e^{-4t}$
     b. $Q = c_1(1 + \frac{1}{2}t)e^{-t/2} + c_2te^{-t/2} + \frac{1}{2}t^2e^{-t/2}$,
     $I_2 = -\frac{1}{4}c_1te^{-1/2} + c_2(1 - \frac{1}{2}t)e^{-t/2} + (t - \frac{1}{4}t^2)e^{-t/2}$
     c. $Q = c_1e^{-t/2}\left(\sin\frac{t}{2} - \cos\frac{t}{2}\right) - c_2e^{-t/2}\left(\sin\frac{t}{2} + \cos\frac{t}{2}\right) + 4e^{-t/2}$

     $I_2 = c_1e^{-t/2}\cos\frac{t}{2} + c_2e^{-t/2}\sin\frac{t}{2} - 2e^{-t/2}$

     d. $Q = c_1e^{-t/2}\left(\sin\frac{t}{2} + \cos\frac{t}{2}\right) + c_2e^{-t/2}\left(\sin\frac{t}{2} - \cos\frac{t}{2}\right) + 8e^{-t/2}$

     $I_3 = c_1e^{-t/2}\cos\frac{t}{2} + c_2e^{-t/2}\sin\frac{t}{2} + 4e^{-t/2}$

     e. $Q_1 = 3c_1e^{-t} - c_2e^{-t/6} + 15$
     $Q_2 = 12c_1e^{-t} + c_2e^{-t/6}$

CHAPTER FOUR

Section 4.1

1.

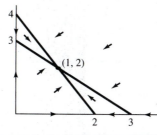

Solutions appear to approach $(1, 2)$, $(0, 3)$, or $(2, 0)$ as $t \to \infty$.

3.

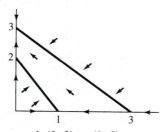

Solutions appear to approach $(0, 3)$ or $(1, 0)$ as $t \to \infty$.

5.

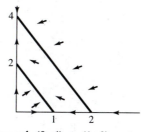

Solutions appear to approach $(0, 4)$ or $(1, 0)$ as $t \to \infty$.

7.

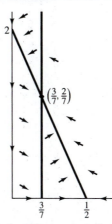

Long-term behavior of solutions is not clear from signs alone.

9.

Solutions appear to approach $(0, 0)$ or $(1, 0)$ or to have $x \to \infty$ as $t \to \infty$.

Section 4.2

1. $\lambda = \pm i\sqrt{2}$

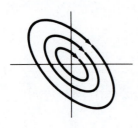

3. $\lambda = \dfrac{1}{2} \pm \dfrac{i\sqrt{3}}{2}$

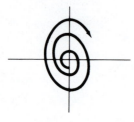

5. $\lambda = 1, 2; \quad \mathbf{v}_1 = \begin{bmatrix} 3 \\ 2 \end{bmatrix} \quad \mathbf{v}_2 = \begin{bmatrix} 2 \\ 1 \end{bmatrix}$

7. $\lambda = -\sqrt{5}, \sqrt{5}$; $\mathbf{v}_1 = \begin{bmatrix} 1 \\ -2 + \sqrt{5} \end{bmatrix}$ $\mathbf{v}_2 = \begin{bmatrix} 1 \\ -2 - \sqrt{5} \end{bmatrix}$

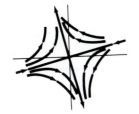

9. $\lambda = 1, 3$; $\mathbf{v}_1 = \begin{bmatrix} 1 \\ -1 \end{bmatrix}$ $\mathbf{v}_2 = \begin{bmatrix} 2 \\ -1 \end{bmatrix}$

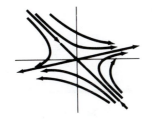

11. $\lambda = -4, 2$; $\mathbf{v}_1 = \begin{bmatrix} 1 \\ -1 \end{bmatrix}$ $\mathbf{v}_2 = \begin{bmatrix} 5 \\ 1 \end{bmatrix}$

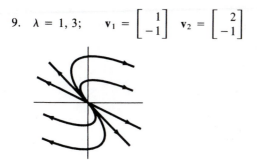

13. $\lambda = -2, -1$; $\mathbf{v}_1 = \begin{bmatrix} 1 \\ 1 \end{bmatrix}$ $\mathbf{v}_2 = \begin{bmatrix} 2 \\ 3 \end{bmatrix}$

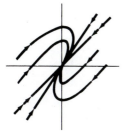

15. $\lambda = 3;$ all $\mathbf{v} \neq \mathbf{0}$

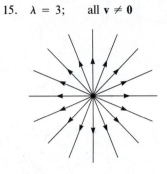

17.

19.

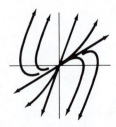

21.

23.

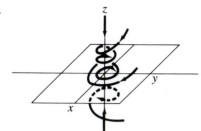

27. Both approach 0 as $t \to \infty$.

29. b. $x_2(t) = x_1(t + a)$ for all t

Section 4.3

1. Unstable   3. Stable

5. Unstable   7. Unstable

9. (0, 4), (3, 0) stable, attractors; (0, 0) unstable, repeller; (1, 2) unstable, neither.

11. (3, 0), $(-1, 4)$ stable, attractors; (0, 0) unstable, repeller; (0, 2) unstable, neither.

13. (2, 0) stable, attractor; (0, 0), $(3, -2)$ unstable, neither.

15. (1, 2) stable, attractor; (0, 0), (2, 0) unstable, neither.

17. $(-5, 0)$ stable, attractor; (0, 0) unstable, repeller; $(0, -1)$, (3, 8), $(-2, 3)$ unstable, neither.

19. (0, 0) unstable, neither; (2, 2) unstable, repeller

21. (0, 0) unstable, repeller; $(0, \frac{1}{2})$, (1, 0), $(-\frac{3}{2}, -1)$ unstable, neither.

23. (0, 0, 0) unstable, repeller; $(8, -4, -8)$ unstable, neither.

Section 4.4

1. b.   (0, 0) minimum   c.   (0, 0) stable

3. b.   (0, 0), $(-1, 0)$, (0, 1) saddles; $(-\frac{1}{3}, \frac{1}{3})$ minimum

     c.   (0, 0), $(-1, 0)$, (0, 1) unstable; $(-\frac{1}{3}, \frac{1}{3})$ stable

5. b.   (1, 2) minimum;   c.   (1, 2) stable.

7. b.   all points $(m\pi, n\pi)$ or $((m + \frac{1}{2})\pi, (n + \frac{1}{2})\pi)$ for m, $n = 0, \pm 1, \pm 2, \ldots$; $(m\pi, n\pi)$ all saddles; $((m + \frac{1}{2})\pi, (n + \frac{1}{2})\pi)$ maximum if $m + n$ odd, minimum if $m + n$ even.

     c. $(m\pi, n\pi)$ unstable, $((m + \frac{1}{2})\pi, (n + \frac{1}{2})\pi)$ stable.

9. d.   If $ac - b^2 > 0$, then (0, 0) is a stable equilibrium. If $ac - b^2 < 0$, then (0, 0) is unstable, but not a repeller.

11. a.   (0, 0) stable   b.   (0, 0) unstable   c.   (1, 0) stable, $(-1, 0)$ unstable
     d.   (0, 0) unstable   e. $(\pm 1, 0)$ stable, (0, 0) unstable.

13. a. $\dfrac{dx}{dt} = y$

$\dfrac{dy}{dt} = \sin x$ $\qquad$ $(n\pi, 0)$ stable if n odd, unstable if n even

b. $\dfrac{dx}{dt} = x$

$\dfrac{dy}{dt} = -y$ $\qquad$ $(0, 0)$ unstable

c. $\dfrac{dx}{dt} = y$

$\dfrac{dy}{dt} = -x$ $\qquad$ $(0, 0$ stable$)$

d. $\dfrac{dx}{dt} = y$

$\dfrac{dy}{dt} = 4(x - x^3)$ $\qquad$ $(0, 0)$ unstable, $(\pm 1, 0)$ stable.

Section 4.5

1. $(0, 0)$ attractor

3. $(-\tfrac{1}{3}, \tfrac{1}{3})$ attractor; $(0, 0)$, $(-1, 0)$, $(0, 1)$ neither.

5. $(1, 2)$ attractor

7. $((m + \tfrac{1}{2})\pi, (n + \tfrac{1}{2})\pi)$ attractor if $m + n$ even, repeller if $m + n$ odd; $(m\pi, n\pi)$ neither.

Section 4.6

1. The only equilibrium not on an axis is $(1, 2)$. The linearization here has eigenvalues of opposite sign.

3. All equilibria lie on the axes, which cannot be crossed by an integral curve.

5. The only equilibrium not on an axis is $(\tfrac{3}{7}, \tfrac{2}{7})$, and

$$\frac{\partial f}{\partial x} + \frac{\partial g}{\partial y} = -1 - x - y,$$

which is always negative in the first quadrant.

7. No integral curve can cross the x-axis, and the only equilibrium is at the origin. Also,

$$\frac{\partial f}{\partial x} + \frac{\partial g}{\partial y} = e^x > 0 \qquad \text{everywhere.}$$

9. On the circle of radius 1 about the origin ($x^2 + y^2 = 1$),

$$\frac{dr}{dt} = 2 - y^2 \geq 2 - 1 > 0,$$

and on the circle of radius 2 about the origin ($x^2 + y^2 = 4$),

$$\frac{dr}{dt} = \frac{1}{4}(-4 - y^2) < 0.$$

11. On the circle of radius 1 about the origin ($x^2 + y^2 = 1$),

$$\frac{dr}{dt} = y^2(3 - e) > 0,$$

and on the circle of radius $\sqrt{2}$ about the origin ($x^2 + y^2 = 2$),

$$\frac{dr}{dt} = y^2(3 - e^2) \leq 0.$$

Review Problems

1.

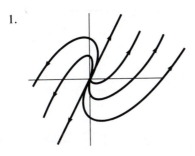

repeller

3.

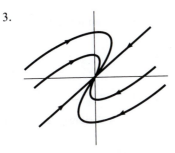

attractor

5.

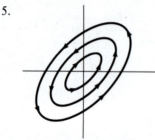

stable, but not attractor

7.

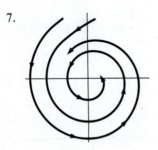

attractor

9. $(n\pi, 0)$: attractor if n even, neither if n odd.

11. $(n\pi, 0)$: attractor if n even, neither if n odd.

13. $(0, 0)$: repeller

15. $(2, -4)$ repeller; $(0, 0)$ neither.

21. a.

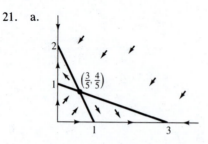

b. $(0, 2)$, $(3, 0)$ attractors, $(\frac{3}{5}, \frac{4}{5})$ neither
c. Only equilibrium in first quadrant is $(\frac{3}{5}, \frac{4}{5})$, and the linearization here has eigenvalues of opposite sign.
d. Most solutions appear to approach $(0, 2)$ or $(3, 0)$ as $t \to \infty$.

23. a.

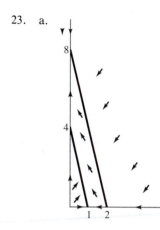

b. (0, 8) attractor; (0, 0) repeller; (1, 0) neither
c. no equilibria in the first quadrant
d. Most solutions appear to approach (0, 8) as $t \to \infty$.

25. $(-2, -1)$ is a saddle point for E, hence an unstable equilibrium.

27. $(-2, -1)$ is a saddle point for E, hence an unstable equilibrium.

29. $\dfrac{\partial f}{\partial x} + \dfrac{\partial g}{\partial y} = x^2 + 3y^2 + 1 > 0$

31. $\dfrac{dr}{dt} = 4 - (x^2 + y^2) - e^{-x}$

This is positive for $x^2 + y^2 = 3$ and negative for $x^2 + y^2 = 4$. Furthermore, the only equilibrium is at the origin, which is not in the region between these circles.

CHAPTER FIVE

Section 5.1

1. $(D - 0.08)x = \begin{cases} 400, & 0 \le t < 2 \\ 500, & t \ge 2 \end{cases};$ $x(0) = 1000$

3. $(D^2 + 4)x = \begin{cases} 0, & 0 \le t < 2 \\ 32, & t \ge 2 \end{cases};$ $x(0) = -\dfrac{1}{2}, \quad x'(0) = 0$

5. $\left(LD^2 + RD + \dfrac{1}{C} \right) Q = \begin{cases} t, & 0 \le t < 10 \\ 10, & t \ge 10 \end{cases};$ $Q(0) = Q'(0) = 0$

7. a. $(D^3 + 4D^2 + 4D)z = 8;$ $z(0) = z'(0) = 0, \quad z''(0) = -8$
    b. $(D^3 + 4D^2 + 4D)z = 8;$ $z(0) = z'(0) = 0, \quad z''(0) = 8$

9. $DQ = -Q - I_2 + 10, \quad DI_2 = Q - 3I_2;$ $Q(0) = 10, \quad I_2(0) = 0$

Section 5.2

1. $F(s) = \dfrac{1}{s-4}, \quad s > 4$

3. $F(s) = \dfrac{2}{s^3}, \quad s > 0$

5. $F(s) = \dfrac{2}{s^2+4}, \quad s > 0$

7. $F(s) = \begin{cases} 3, & s = 0 \\ (1 - e^{-3s})/s, & s \neq 0 \end{cases}$

9. $F(s) = \dfrac{24}{s^5}$

11. $F(s) = \dfrac{2}{s^3} - \dfrac{7}{s} + \dfrac{s}{s^2+4}$

13. $f(s) = \dfrac{3}{s} + \dfrac{12}{s^2} + \dfrac{242}{s^4} - \dfrac{3}{s-2}$

15. $F(s) = \dfrac{2}{s^3} + \dfrac{3}{s^2} + \dfrac{2}{s}$

17. $f(t) = e^{2t}$

19. $f(t) = \dfrac{1}{2} e^{t/2}$

21. $f(t) = \dfrac{1}{12} t^4$

23. $f(t) = \dfrac{1}{\sqrt{3}} \sin t\sqrt{3}$

25. No

29. a. All except ii are of exponential order
     b. All except i, v, vi are piecewise continuous on $[0, h]$

Section 5.3

1. $\mathcal{L}[x] = \dfrac{-3}{s-1}$

3. $\mathcal{L}[x] = \dfrac{1}{(s-2)(s^2-1)}$

5. $\mathcal{L}[x] = \dfrac{s}{(s^2+9)(s^2+1)}$

7. $\mathcal{L}[f(t)] = \dfrac{s}{s^2 + \beta^2}$

9. $\mathcal{L}[f(t)] = \dfrac{2s+3}{(s+1)^2}$

11. $-\dfrac{1}{3} + \dfrac{1}{4} e^{-t} + \dfrac{1}{12} e^{3t}$

13. $-\dfrac{1}{2} \sin t + \dfrac{1}{4} e^{t} - \dfrac{1}{4} e^{-t}$

15. $-\dfrac{1}{4} \cos 2t + \dfrac{1}{4} t$

17. $x = -\dfrac{1}{3} + \dfrac{9}{4} e^{-t} + \dfrac{1}{12} e^{3t}$

19. $x = -\dfrac{1}{5} \cos 2t - \dfrac{1}{10} \sin 2t + \dfrac{1}{5} e^{t}$

21. $x = e^t + e^{2t}$

23. $x = 3 - \dfrac{1}{3} e^{t} + \dfrac{7}{2} e^{2t} - \dfrac{7}{6} e^{-2t}$

25. $x = -1 + \dfrac{1}{2} e^{t} + \dfrac{1}{2} e^{-t} + \cos t$

27. $x = 5e^{t-1} - 3t$

29. $x = \dfrac{1}{4} e^{\pi - t} - \dfrac{1}{4} e^{t-\pi} - \dfrac{1}{2} \sin t$

31. $x = 70 - 3t$

33. $x = 4 - \dfrac{1}{4} e^{-4t} + \dfrac{3}{4} e^{-2t}$

37. a. $A_1 = \dfrac{p(m_1)}{(m_1 - m_2) \cdots (m_1 - m_k)}$

Section 5.4

1. $\dfrac{1}{(s-3)^2}$

3. $\dfrac{12s}{(s^2+4)^2}$

5. $\dfrac{18(s^2-3)}{(s^2+9)^3}$

7. $\dfrac{2}{(s-m)^3}$

9. $\dfrac{2}{s^2-6s+13}$

11. $\dfrac{6(s-2)}{(s^2-4s+13)^2}$

13. $\dfrac{1}{2s^2}+\dfrac{16-s^2}{2(s^2+16)^2}$

15. $\dfrac{t^3 e^{-2t}}{6}$

17. te^{-3t}

19. $\dfrac{1}{9}-\dfrac{1}{9}e^{-3t}-\dfrac{1}{3}te^{-3t}$

21. $e^{-2t}\sin t$

23. $e^{-3t/2}\left(\cos\dfrac{t\sqrt{3}}{2}-\dfrac{\sqrt{3}}{3}\sin\dfrac{t\sqrt{3}}{2}\right)$

25. $x=\dfrac{t^4 e^t}{4}$

27. $x=-e^t+3te^t+\dfrac{1}{20}t^5 e^t$

29. $x=\dfrac{1}{8}e^{-t}-\dfrac{1}{8}e^t\cos 2t+\dfrac{1}{8}e^t\sin 2t$

31. $x=\dfrac{1}{3}e^t-\dfrac{1}{3}e^{-t/2}\left(\cos\dfrac{t\sqrt{3}}{2}+\sqrt{3}\sin\dfrac{t\sqrt{3}}{2}\right)$

33. $Q=te^{-t}$

35. $x=-2e^{-3t/2}\cos\dfrac{t}{2}-4e^{-3t/2}\sin\dfrac{t}{2}$

37. $x=2t-2te^{-2t}$

39. a. $\displaystyle\int_0^t \mathcal{L}^{-1}\left[\dfrac{s}{(s^2+1)^2}\right]ds$ b. $-\dfrac{1}{2}t\cos t+\dfrac{1}{2}\sin t$

    c. $\dfrac{1}{8}t\sin t-\dfrac{1}{8}t^2\cos t$ d. $\dfrac{3}{8}\sin t-\dfrac{3}{8}t\cos t-\dfrac{1}{8}t^2\sin t$

41. $x=1-2t^2$

Section 5.5

1. $f(t)=u_3(t)(t-3),\quad \mathcal{L}[f(t)]=\dfrac{e^{-3s}}{s^2}$

3. $f(t)=u_\pi(t)\sin t,\quad \mathcal{L}[f(t)]=\dfrac{-e^{-\pi s}}{s^2+1}$

5. $f(t)=u_1(t)e^t-u_2(t)e^t,\quad \mathcal{L}[f(t)]=\dfrac{e^{-(s-1)}}{s-1}-\dfrac{e^{-2(s-1)}}{s-1}$

7. $f(t)=3-t+2u_3(t)(t-3),\quad \mathcal{L}[f(t)]=\dfrac{3}{s}-\dfrac{1}{s^2}+\dfrac{2e^{-3s}}{s^2}$

9. $f(t) = t - u_1(t) + u_2(t)(2 - t)$, $\mathscr{L}[f(t)] = \dfrac{1}{s^2} - \dfrac{e^{-s}}{s} - \dfrac{e^{-2s}}{s^2}$

11. $f(t) = t - 2u_1(t) + u_3(t)(2 - t)$, $\mathscr{L}[f(t)] = \dfrac{1}{s^2} - \dfrac{2e^{-s}}{s} - e^{-3s}\left(\dfrac{1}{s} + \dfrac{1}{s^2}\right)$

13. $u_4(t)e^{-4(t-4)}$ 15. $u_4(t)(t - 4)e^{-4(t-4)}$

17. $\cos t + u_\pi(t)\sin(t - \pi) = \cos t - u_\pi(t)\sin t$

19. $u_1(t)[1 - e^{-(t-1)}]$ 21. $x = -5e^t + u_3(t)(2 - t + e^{t-3})$

23. $x = \dfrac{1}{2}e^{2t} + \dfrac{1}{2}e^{-2t} + u_2(t)\left[-\dfrac{1}{2} + \dfrac{1}{4}e^{2(t-2)} + \dfrac{1}{4}e^{-2(t-2)}\right]$

25. $x = u_2(t)\left[\dfrac{1}{2}e^{-t} - \dfrac{e^{-2}}{2}\cos(t - 2) + \dfrac{e^{-2}}{2}\sin(t - 2)\right]$

27. $x = \dfrac{2}{3}\sin t - \dfrac{1}{3}\sin 2t + u_\pi(t)\left(\dfrac{4}{3}\sin t + \dfrac{2}{3}\sin t\right)$

29. $x = \dfrac{1}{6}t^3e^{-t} - \dfrac{1}{6}u_3(t)(t - 3)^3e^{-t}$

31. a. $x = -\dfrac{1}{2}\cos 2t + u_2(t)[8 - 8\cos 2(t - 2)]$

   b. $x(1) = -\dfrac{1}{2}\cos 2 \approx 0.2081$, $x(4) = -\dfrac{1}{2}\cos 8 - 8\cos 4 + 8 \approx 13.3019$

33.   (i)   a. $Q = u_2(t)[5 + 5e^{-2(t-2)} - 10e^{-(t-2)}] - u_5(t)[5 + 5e^{-2(t-5)} - 10e^{-(t-5)}]$

      b. $Q(1) = 0$, $Q(3) = 5 + 5e^{-2} - 10e^{-1} \approx 1.9979$,
            $Q(6) = 5e^{-8} - 10e^{-4} - 5e^{-2} + 10e^{-1} \approx 2.8206$

      (ii)   a. $Q = 5 + 5e^{-2t} - 10e^{-t} - u_2(t)[5 + 5e^{-2(t-2)} - 10e^{-(t-2)}]$
                  $+ u_5(t)[5 + 5e^{-2(t-5)} - 10e^{-(t-5)}]$

      b. $Q(1) = 5 + 5e^{-2} - 10e^{-1} \approx 1.9979$
            $Q(3) = 5e^{-6} - 10e^{-3} - 5e^{-2} + 10e^{-1} \approx 2.5166$
            $Q(6) = 5 + 5e^{-12} - 10e^{-6} - 5e^{-8} + 10e^{-4} + 5e^{-2} - 10e^{-1} \approx 2.1546$

35. $\mathscr{L}[f(t)] = e^{-ks}\sqrt{\pi/s}$

Section 5.6

1. $\dfrac{1}{20}t^5$

3. $\dfrac{1}{4}t + \dfrac{1}{16}e^{-4t} - \dfrac{1}{16}$

5. $\dfrac{1}{2}t - \dfrac{1}{4}\sin 2t$

7. $\dfrac{2}{3}\cos t - \dfrac{2}{3}\cos 2t$

11. $\dfrac{3}{4} - \dfrac{3}{4}\cos 2t$

13. $\dfrac{3}{4}t - \dfrac{3}{16} + \dfrac{3}{16}e^{-4t}$

15. $\dfrac{1}{4}t \sin 2t$

17. $\dfrac{1}{2}\sin t - \dfrac{1}{2}\cos t + \dfrac{1}{2}e^t$

19. $x = \dfrac{1}{2}\sin t - \dfrac{1}{2}t \cos t$

21. $x = \dfrac{3}{2}\sin t - \dfrac{1}{2}t \cos t$

23. $x = 3u_1(t)[1 - \cos(t - 1) - \dfrac{1}{2}(t - 1)\sin(t - 1)]$

25. $x = -3e^{-t}\left(\dfrac{3}{2}\sin t + \dfrac{1}{2}t\sin t - \dfrac{3}{2}t\cos t\right)$

29. $\dfrac{1}{2}u_a(t)(t-a)^2$

35. d. (Exercise 12, Section 2.8) $H(t, u) = \dfrac{t}{1-u} - \dfrac{ue^{t-u}}{1-u}$

  (Exercise 13, Section 2.8) $H(t, u) = \dfrac{-u^2t^{-1}}{u+1} + \dfrac{ue^{t-u}}{u+1}$

Section 5.7

1. $x = \dfrac{15}{4}e^{2t/3} - \dfrac{3}{4}e^{-2t/3}$ 3. $x = 2e^{-t}(\cos 2t + \sin 2t)$

5. $x = e^{-2t}\left(t + \dfrac{1}{6}t^3\right)$

7. $x = e^t\left(-2\sin 3t + \dfrac{2}{81}\cos 3t - \dfrac{2}{81} + \dfrac{1}{9}t^2\right)$

9. $x = \dfrac{1}{12}e^{2t} + \dfrac{1}{4}e^{-2t} - \dfrac{1}{3}e^{-t}$

  $+ u_1(t)\left(\dfrac{1}{20}e^{3t} + \dfrac{1}{5}e^{5-2t} - \dfrac{1}{4}e^{4-t} - \dfrac{1}{12}e^{2t} - \dfrac{1}{4}e^{4-2t} + \dfrac{1}{3}e^{3-t}\right)$

11. $x = \sin t + u_1(t)[t - \sin(t-1) - \cos(t-1)]$

13. $x = 2 + t + (t-2)e^t + u_2(t)\left[\dfrac{1}{2}(t-2)^2e^t + (10 - 3t)e^{2-t} - t - 2\right]$

15. $x = 1$ 17. $x = -t + \dfrac{1}{2}e^t - \dfrac{1}{2}e^{-t}$

19. $x = \dfrac{1}{6}t^3e^t + \dfrac{5}{2}t^4e^t$

21. a. $x = \dfrac{32}{125} - \dfrac{3}{500}\cos 5t\sqrt{5} + \dfrac{4}{125}u_3(t)[1 - \cos 5(t-3)\sqrt{5}]$

Section 5.8

1. $x = 5e^{-t}, \quad y = -5e^{-t}$

3. $x = 5e^{-t} + u_1(t)\left[\dfrac{1}{3} + \dfrac{1}{6}e^{3(t-1)} - \dfrac{1}{2}e^{-(t-1)}\right],$

  $y = -5e^{-t} + u_1(t)\left[-\dfrac{2}{3} + \dfrac{1}{6}e^{3(t-1)} + \dfrac{1}{2}e^{-(t-1)}\right]$

5. $x = e^t(t - t^2), \quad y = t^2e^t$

7. $x = 9e^{2t} - \cos 2t - \sin 2t, \quad y = 3e^{2t} - 3\cos 2t - 3\sin 2t$

9. $x = 9 + 6t - 8e^t - e^{-2t}, \quad y = 3 + 6t - 3e^{-2t}, \quad z = -3 + 6t + 3e^{-2t}$

11. a. $Q = -e^{-t} + 2e^{-5t/2}$, $I_2 = 2e^{-t} - e^{-5t/2}$

   b. $Q = -2e^{-t}(\cos t + \sin t)$, $I_2 = 2e^{-t}(\cos t - \sin t)$

   c. $Q = e^{-2t}(1 - t)$, $I_2 = e^{-2t}(2 - t)$

   d. $Q = \dfrac{15}{2} + \dfrac{5}{2}e^{-2t} + 5te^{-2t}$, $I_2 = \dfrac{5}{2} - \dfrac{5}{2}e^{-2t} + 5te^{-2t}$

13. a. $A = 500$, $B = 2500 - 2000e^{-t/10}$

   $C = 150t - 1500 + 2000e^{-t/10} - 25u_1(t)(t - 1)$

   b. No

15. $p(52) = -e^{-26} + 5e^{-5.2} \approx .03$,

   $s(52) = 5000 + 100e^{-26} - 2500e^{-5.2} \approx 4986$,

   $w(52) = 5000 + 600e^{-26} - 3000e^{-5.2} \approx 4983$

17. d. The same, with λ replaced by s.

   e. Partial fractions lead to expressions for the functions $x_i(t)$ as linear combinations of functions of the form $t^k e^{\lambda t}$, $t^j e^{\alpha t} \cos \beta t$, and $t^j e^{\alpha t} \sin \beta t$, where λ is a real root of multiplicity m_λ and $0 \le k < m_\lambda$, and $\alpha \pm \beta i$ are complex roots of multiplicity $m_{\alpha + \beta i}$ and $0 \le j < m_{\alpha + \beta i}$.

CHAPTER SIX

Section 6.1

1. $[D^2 + (10 + t)D + 1]Q = 10$

5. a. $m = \dfrac{(100{,}000 - 100t)}{32}$

   c. $[(1000 - t)D^2 - D]x = 32t$

3. $[10D^2 + (10 - t)D + 10]x = 0$

   b. $F(t) = 100t$

Section 6.2

1. $|t| < \infty$

5. $|t| < 3$

9. $|2t - 1| < \infty$

13. 1

15. $\displaystyle\sum_{k=1}^{\infty} k(3k + 1)t^{k-1} = \sum_{j=0}^{\infty} (j + 1)(3j + 4)t^j$

17. $\displaystyle\sum_{k=1}^{\infty} \dfrac{(3k - 12)}{k} t^k$

19. $1 + 2t + 2t^2 + \dfrac{4}{3}t^3 + \dfrac{2}{3}t^4 + \dfrac{4}{15}t^5 + \cdots$

21. $1 + t + t^4 + t^5 + \cdots$

3. Converges only when $t = 0$

7. $|t - 2| < 1$

11. $\displaystyle\sum_{n=0}^{\infty} 2t^{2n}$

23. $\sin \dfrac{t}{2} = \displaystyle\sum_{k=0}^{\infty} \dfrac{(-1)^k t^{2k+1}}{(2k+1)!2^{2k+1}}, \quad |t| < \infty$

25. $e^{-t} \displaystyle\sum_{k=0}^{\infty} \dfrac{(-1)^k}{k!} t^k, \quad |t| < \infty$

27. $\dfrac{1}{(1+t)^2} = \displaystyle\sum_{j=0}^{\infty} (-1)^j (j+1) t^j, \quad |t| < 1$

29. $\dfrac{1}{1+t^2} = \displaystyle\sum_{k=0}^{\infty} (-1)^k t^{2k}, \quad |t| < 1$

Section 6.3

1. a. $b_{j+2} = \dfrac{-(j+4)b_j}{(j+2)(j+1)}, \quad j = 0, 1, 2, \ldots$

   b. $x = b_0(1 - 2t^2 + t^4 - \cdots) + b_1 \left(t - \dfrac{5}{6} t^3 + \dfrac{7}{24} t^5 - \cdots \right)$

3. a. $b_2 = -\dfrac{1}{2} b_1$, and $b_{j+2} = \dfrac{-(j+1)b_{j+1} - b_{j-1}}{(j+2)(j+1)}$ for $j = 1, 2, \ldots$

   b. $x = b_0 \left(1 - \dfrac{1}{6} t^3 + \dfrac{1}{24} t^4 - \dfrac{1}{120} t^5 + \cdots \right)$

   $\qquad + b_1 \left(t - \dfrac{1}{2} t^2 + \dfrac{1}{6} t^3 - \dfrac{1}{8} t^4 + \dfrac{1}{20} t^5 - \cdots \right)$

5. a. $b_2 = 2b_0 - 1, \quad b_3 = \dfrac{b_1 + 1}{6}$, and $b_{j+2} = \dfrac{(4 - 3j)b_j}{(j+2)(j+1)}$ for $j = 2, 3, \ldots$

   b. $x = b_0 \left(1 + 2t^2 - \dfrac{1}{3} t^4 + \cdots \right) + b_1 \left(t + \dfrac{1}{6} t^3 - \dfrac{1}{24} t^5 + \cdots \right)$

   $\qquad + \left(-t^2 + \dfrac{1}{6} t^3 + \dfrac{1}{6} t^4 - \dfrac{1}{24} t^5 + \cdots \right)$

7. a. $b_2 = \dfrac{-b_0 - 10b_1 + 10}{2}$, and $b_{j+2} = \dfrac{-10b_{j+1} - b_j}{j+2}$ for $j = 1, 2, \ldots$

   b. $x = 5t^2 - \dfrac{50}{3} t^3 + \dfrac{485}{12} t^4 - \cdots$

9. a. $b_{j+2} = \dfrac{-10(j+1)b_{j+1} + (j-10)b_j}{10(j+2)(j+1)}, \quad j = 0, 1, 2, \ldots$

   b. $x = 1 - \dfrac{1}{2} t^2 + \dfrac{1}{6} t^3 - \dfrac{1}{120} t^4 + \cdots$

11. a. $b_{j+2} = \dfrac{-j(j+1)b_{j+1} + 4b_j}{(j+2)(j+1)}, \quad j = 0, 1, 2, \ldots$

   b. $x = b_0 \left[1 + 2(t-1)^2 - \dfrac{2}{3}(t-1)^3 + (t-1)^4 + \cdots \right]$

   $\qquad + b_1 \left[(t-1) + \dfrac{2}{3}(t-1)^3 - \dfrac{1}{3}(t-1)^4 + \cdots \right]$

13. a. $b_{j+2} = \dfrac{-b_j}{j+2}, \quad j = 0, 1, 2, \ldots$

b. $x = b_0 \displaystyle\sum_{m=0}^{\infty} \dfrac{(-1)^m t^{2m}}{2^m m!} + b_1 \sum_{m=0}^{\infty} \dfrac{(-1)^m t^{2m+1}}{1 \cdot 3 \cdot 5 \cdots (2m+1)}$

15. a. $b_{j+2} = \dfrac{(j + \mu + 1)(j - \mu)}{(j+2)(j+1)} b_j, \quad j = 0, 1, 2, \ldots$

17. $x = b_0 \displaystyle\sum_{m=0}^{\infty} \dfrac{(-1)^m 1 \cdot 4 \cdot 7 \cdots (3m-2)}{(3m)!} t^{3m}$

$\qquad + b_1 \left[t + \displaystyle\sum_{m=1}^{\infty} \dfrac{(-1)^m \cdot 2 \cdot 5 \cdots (3m-1)}{(3m+1)!} t^{3m+1} \right]$

19. $x = 1 - \dfrac{1}{6} t^3 + \dfrac{1}{20} t^5 + \cdots$

Section 6.4

1. (i) a. $1, \ 0, \ -0.5, \ 0.166667, \ 8.33333 \times 10^{-2}, \ -0.05,$
   $-5.55556 \times 10^{-3}, \ 7.93651 \times 10^{-3}, \ -2.97619 \times 10^{-4},$
   $-8.48765 \times 10^{-4}, \ 1.14638 \times 10^{-4}$

   b. $x(0.1) \approx 0.995174; \qquad x(0.5) \approx 0.899452; \qquad x(1) \approx 0.701349$

   c. $x(0.1) \approx 0.995175; \ N = 5; \qquad x(0.5) \approx 0.899454, \ N = 7;$
   $x(1) \approx 0.701401, \ N = 12$

   d. $x(0.1) \approx 0.995; \qquad x(0.5) \approx 0.899; \qquad x(1) \approx 0.701$

   (ii) a. $0, \ 1, \ -0.5, \ -0.166667, \ 0.166667, \ 0, \ -2.77778 \times 10^{-2},$
   $3.96825 \times 10^{-3}, \ 2.97619 \times 10^{-3}, \ -7.71605 \times 10^{-4},$
   -2.20459×10^{-4}

   b. $x(0.1) \approx 9.48500 \times 10^{-2}; \qquad x(0.5) \approx 0.36419; \qquad x(1) \approx 0.478175$

   c. $x(0.1) \approx 0.9485, \ N = 5; \qquad x(0.5) \approx 0.364192, \ N = 8;$
   $x(1) \approx 0.478268, \ N = 13$

   d. $x(0.1) \approx 0.095; \qquad x(0.5) \approx 0.364; \qquad x(1) \approx 0.478$

   (iii) a. $-2, \ 3, \ -0.5, \ -0.833333, \ 0.333333, \ 0.1, \ -7.22222 \times 10^{-2},$
   $-3.96825 \times 10^{-3}, \ 9.52381 \times 10^{-3}, \ -6.17284 \times 10^{-4},$
   -8.90653×10^{-4}

   b. $x(0.1) \approx -1.7058; \qquad x(0.5) \approx -0.706333; \quad x(1) \approx 3.18254 \times 10^{-2}$

   c. $x(0.1) \approx 1.7058, \ N = 5; \qquad x(0.5) \approx -0.706331, \ N = 6;$
   $x(1) \approx 3.20099 \times 10^{-2}, \ N = 13$

   d. $x(0.1) \approx -1.706; \qquad x(0.5) \approx -0.706; \qquad x(1) \approx 0.003$

   (iv) a. $3, \ -1, \ -1, \ 0.666667, \ 8.33333 \times 10^{-2}, \ -0.15,$
   $1.11111 \times 10^{-2}, \ 1.98413 \times 10^{-2}, \ -3.86905 \times 10^{-3},$
   $-1.77469 \times 10^{-3}, \ 5.64374 \times 10^{-4}$

   b. $x(0.1) \approx 2.89067; \qquad x(0.5) \approx 2.33416; \qquad x(1) \approx 1.62587$

   c. $x(0.1) \approx 2.89067, \ N = 5; \qquad x(0.5) \approx 2.33416, \ N = 9;$
   $x(1) \approx 1.62593, \ N = 12$

   d. $x(0.1) \approx 2.891; \qquad x(0.5) \approx 2.334; \qquad x(1) \approx 1.626$

3. (i) a. $1, \ 0, \ -0.5, \ 0.166667, \ -8.33333 \times 10^{-3}, \ -4.16667 \times 10^{-3},$

8.61111×10^{-4}, -7.34127×10^{-5}, 3.02579×10^{-6}, -3.03131×10^{-8}, -3.69268×10^{-9}

    b. $x(0.1) \approx 0.995166$; $x(0.5) \approx 0.895195$; $x(1) \approx 0.654957$

    c. $x(0.1) \approx 0.995166$, $N = 5$; $x(0.5) \approx 0.895195$, $N = 7$; $x(1) \approx 0.654957$, $N = 8$

    d. $x(0.1) \approx 0.995$; $x(0.5) \approx 0.895$; $x(1) \approx 0.655$

  (ii)   a.   0,   1, -0.5, 1.66667×10^{-2}, 2.91667×10^{-2}, -6.41667×10^{-3}, 4.86111×10^{-4}, 6.94444×10^{-6}, -4.34028×10^{-6}, 4.53318×10^{-7}, -3.56867×10^{-8}

    b. $x(0.1) \approx 9.50195 \times 10^{-2}$; $x(0.5) \approx 0.378713$; $x(1) \approx 0.539906$

    c. $x(0.1) \approx 9.5019 \times 10^{-2}$, $N = 4$; $x(0.5) \approx 0.378713$, $N = 7$; $x(1) \approx 0.539905$, $N = 8$

    d. $x(0.1) \approx 0.010$; $x(0.5) \approx 0.379$; $x(1) \approx 0.540$

 (iii)   a. -2,   3, -0.5, -0.283333, 0.104167, -1.09167×10^{-2}, -2.63889×10^{-4}, 1.67659×10^{-4}, -1.90724×10^{-5}, 1.42058×10^{-6}, -9.96748×10^{-8}

    b. $x(0.1) \approx -1.70527$; $x(0.5) \approx -0.65425$; $x(1) \approx 0.309803$

    c. $x(0.1) \approx -1.70527$, $N = 5$; $x(0.5) \approx -0.65425$, $N = 7$; $x(1) \approx 0.309803$, $N = 9$

    d. $x(0.1) \approx -1.705$; $x(0.5) \approx -0.654$; $x(1) \approx 0.310$

 (iv)   a.   3, -1, -1, 0.483333, -5.41667×10^{-2}, -6.08333×10^{-3}, 2.09722×10^{-3}, -2.27183×10^{-4}, 1.34177×10^{-5}, -5.44257×10^{-7}, 2.46087×10^{-8}

    b. $x(0.1) \approx 2.89048$; $x(0.5) \approx 2.30687$; $x(1) \approx 1.42497$

    c. $x(0.1) \approx 2.89048$, $N = 5$; $x(0.5) \approx 2.30687$, $N = 7$; $x(1) \approx 1.42497$, $N = 9$

    d. $x(0.1) \approx 2.890$; $x(0.5) \approx 2.307$; $x(1) \approx 1.425$

5.   (i)   a.   1,   0,   0, -0.666667,   0,   0, 8.88889×10^{-2},   0,   0, -4.93827×10^{-3},   0

    b. $x(0.1) \approx 0.999333$; $x(0.5) \approx 0.918046$; $x(1) \approx 0.417284$

    c. $x(0.1) \approx 1$, $N = 2$; $x(0.5) \approx 1$, $N = 2$; $x(1) \approx 1$, $N = 2$

    d. $x(0.1) \approx 0.999$; $x(0.5) \approx 0.918$; $x(1) \approx 0.417$

  (ii)   a.   0,   1,   0,   0, -0.333333,   0,   0, 3.17460×10^{-2}, 0,   0, -1.41093×10^{-3}

    b. $x(0.1) \approx 9.99667 \times 10^{-2}$; $x(0.5) \approx 0.479413$; $x(1) \approx 0.697002$

    c. $x(0.1) \approx 0.1$, $N = 2$; $x(0.5) \approx 0.5$, $N = 2$; $x(1) \approx 1$, $N = 2$

    d. $x(0.1) \approx 0.010$; $x(0.5) \approx 0.479$; $x(1) \approx 0.697$

 (iii)   a. -2,   3,   0, 1.33333, -1,   0, -0.177778, 9.52381×10^{-2}, 0, 9.87654×10^{-3}, -4.23280×10^{-3}

    b. $x(0.1) \approx -1.679877$; $x(0.5) \approx -0.397852$; $x(1) \approx 1.25644$

    c. $x(0.1) \approx -1.69877$, $N = 6$; $x(0.5) \approx -0.397848$, $N = 9$; $x(1) \approx 1.25625$, $N = 15$

    d. $x(0.1) \approx -1.699$; $x(0.5) \approx -0.398$; $x(1) \approx 1.256$

 (iv)   a.   3, -1,   0,   2, 0.333333,   0, 0.266667, -3.17460×10^{-2}, 0, -1.48148×10^{-2}, 1.41093×10^{-3}

    b. $x(0.1) \approx 2.89803$; $x(0.5) \approx 2.27472$; $x(1) \approx 0.55485$

    c. $x(0.1) \approx 2.89803$, $N = 5$; $x(0.5) \approx 2.27472$, $N = 9$; $x(1) \approx 0.555263$, $N = 14$

    d. $x(0.1) \approx 2.898$; $x(0.5) \approx 2.275$; $x(1) \approx 0.555$

Section 6.5

1. $x = c_1 t + c_2 t^{-1}$ 3. $x = c_1 t + c_2 t \ln t$

5. $x = c_1 t^{-1/3} + c_2 t^{-1/3} \ln t$

7. $x = c_1 t^{-1} \cos(2 \ln t) + c_2 t^{-1} \sin(2 \ln t)$

9. $x = c_1 t^{\sqrt{3}} + c_2 t^{-\sqrt{3}}$ 11. $x = c_1 t^2 \cos(\ln t) + c_2 t^2 \sin(\ln t)$

13. $x = \dfrac{t^2}{3} + c_1 t + c_2 t^{-1}$

15. a. $R(r) = c_1 r^n + c_2 r^{-n}$ c. $u_n = r^n \cos n\theta, \quad u_n^* = r^n \sin n\theta$
    d. $u(r, \theta) = r \cos \theta - 2r^3 \sin 3\theta$

23. a. $p_0 \dfrac{d^2 x}{ds^2} + (q_0 - p_0) \dfrac{dx}{ds} + r_0 x = 0$

    b. If the indicial equation has distinct real roots, m_1 and m_2, then $x = c_1 e^{m_1 s} + c_2 e^{m_2 s}$. If there is only one root m_1, then $x = c_1 e^{m_1 s} + c_2 s e^{m_1 s}$. If the roots are complex, $\alpha \pm \beta i$, then $x = c_1 e^{\alpha s} \cos \beta s + c_2 e^{\alpha s} \sin \beta s$.

Section 6.6

1. $x = c_1 t^{1/3} \left(1 - \dfrac{2}{5} t + \dfrac{1}{20} t^2 - \dfrac{1}{330} t^3 + \cdots \right)$

   $+ c_2 t^{-1/3} \left(1 - 2t + \dfrac{1}{2} t^2 - \dfrac{1}{21} t^3 + \cdots \right)$

3. $x = c_1 t^{1/2} \left(1 - \dfrac{1}{6} t^2 + \cdots \right) + c_2 t^{-1/2} \left(1 - \dfrac{1}{2} t^2 + \cdots \right)$

5. $x = c_1 t^{3/2} \left(1 - \dfrac{1}{10} t^2 + \cdots \right) + c_2 t^{-3/2} \left(1 + \dfrac{1}{2} t^2 + \cdots \right)$

7. $x = c_1 t \left(1 - \dfrac{1}{5} t - \dfrac{1}{70} t^2 + \dfrac{23}{1890} t^3 + \cdots \right)$

   $+ c_2 t^{-1/2} \left(1 + \dfrac{1}{4} t - \dfrac{17}{32} t^2 + \dfrac{121}{1152} t^3 + \cdots \right)$

9. $x = c_1 t \displaystyle\sum_{k=1}^{\infty} \dfrac{t^k}{5 \cdot 7 \cdots (3 + 2k)} + c_2 t^{-1/2} \displaystyle\sum_{k=0}^{\infty} \dfrac{t^k}{2^k k!}$

11. $x = c_1 t^{1/2} \displaystyle\sum_{n=0}^{\infty} \dfrac{(-1)^n t^{2n}}{2^n (n!) 1 \cdot 5 \cdots (4n + 1)}$

    $+ c_2 \left[1 + \displaystyle\sum_{n=1}^{\infty} \dfrac{(-1)^n t^{2n}}{2^n (n!) 3 \cdot 7 \cdots (4n - 1)} \right]$

13. $x = \displaystyle\sum_{n=0}^{\infty} \dfrac{(-1)^n t^{2n}}{2^{2n} (n!)^2}$ 15. $x = t \displaystyle\sum_{k=0}^{\infty} \dfrac{(-1)^k 2 t^k}{k!(k + 2)!}$

17. $x = -c_1 t \left[1 + \displaystyle\sum_{k=1}^{\infty} \dfrac{t^k}{5 \cdot 7 \cdots (2k + 3)} \right] + c_2 |t|^{-1/2} \displaystyle\sum_{k=0}^{\infty} \dfrac{t^k}{k! 2^k}$

19. $x = c_1(t - 1)\left[1 + \displaystyle\sum_{k=1}^{\infty} \frac{(-1)^k(t - 1)^k}{5 \cdot 7 \cdots (2k + 3)}\right] + c_2(t - 1)^{-1/2} \displaystyle\sum_{k=0}^{\infty} \frac{(-1)^k(t - 1)^k}{2^k k!}$

21. $x = c_1(t + 1) \displaystyle\sum_{k=0}^{\infty} \frac{(-1)^k 2(t + 1)^k}{(k + 2)!} - c_2(t + 1)^{-1}t$

Section 6.7

13. c. $J_{3/2}(t) = \sqrt{\dfrac{2}{\pi t}} \left[\dfrac{\sin t - t \cos t}{t}\right], \quad J_{-3/2}(t) = -\sqrt{\dfrac{2}{\pi t}} \left[\dfrac{\cos t + t \sin t}{t}\right]$

17. a.

t	$J_0(t)$	$J_1(t)$	t	$J_0(t)$	$J_1(t)$
0.5	0.9384766	0.2422689	5.0	−0.1776033	−0.3275530
1.0	0.7651910	0.4400499	5.5	−0.0068876	−0.3414492
1.5	0.5118315	0.5579109	6.0	0.1506638	−0.2767514
2.0	0.2238889	0.5767361	6.5	0.2600867	−0.1538123
2.5	−0.0484110	0.4970892	7.0	0.3000329	−0.0046954
3.0	−0.2600409	0.3390610	7.5	0.2663599	0.1351787
3.5	−0.3801323	0.1373984	8.0	0.1716420	0.2346672
4.0	−0.3971882	−0.0660519	8.5	0.0418914	0.2731083
4.5	−0.3205267	−0.2311232	9.0	−0.0903122	0.2452409

17. b.

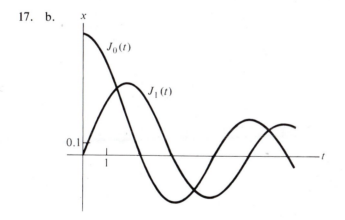

Section 6.8

1. $x = c_1 x_1(t) + c_2 \left[x_1(t) \ln t + 2t^2 - \dfrac{3}{4} t^3 + \dfrac{11}{108} t^4 - \cdots\right)\right], \quad$ where

$\quad\quad x_1(t) = t - t^2 + \dfrac{1}{4} t^3 - \dfrac{1}{36} t^4 + \cdots$

3. $x = c_1 x_1(t) + c_2[x_1(t) \ln t - 16t^{-2} - 4 + \cdots]$,   where

$$x_1(t) = t^2 - \frac{1}{12} t^4 + \cdots$$

5. $x = c_1 x_1(t) + c_2 \left[x_1(t) \ln t + \frac{2}{9} t^{2/3} - \frac{1}{108} t^{5/3} + \frac{11}{78,732} t^{8/3} + \cdots \right]$,   where

$$x_1(t) = t^{-1/3} - \frac{1}{9} t^{2/3} + \frac{1}{324} t^{5/3} - \frac{1}{26,244} t^{8/3} + \cdots$$

7. $x = c_1 t \sum\limits_{k=0}^{\infty} \dfrac{2t^k}{(k+2)!} + c_2 t^{-1}(1+t)$

9. $x = c_1(1-t) + c_2 \left[(1-t) \ln t + \dfrac{5}{2} t - \sum\limits_{k=2}^{\infty} \dfrac{(k+1)t^k}{2^k k(k-1)} \right]$

15. d. $u(\rho, \phi) = \dfrac{2}{3} + \rho^2 \left(\dfrac{1}{3} - \cos^2 \phi \right)$

Review Problems

1. $x = b_0 \left[1 + \sum\limits_{n=1}^{\infty} \dfrac{1 \cdot 3 \cdot 5 \cdots (2n-3)}{2^n n!} t^{2n} \right] + b_1 t$

3. $x = c_1 t^{1/4} \sum\limits_{n=0}^{\infty} \dfrac{(-1)^n t^{2n}}{(n!)1 \cdot 5 \cdot 9 \cdots (4n+1)}$

$+ c_2 t^{-1/4} \left[1 + \sum\limits_{n=1}^{\infty} \dfrac{(-1)^n t^{2n}}{(n!)3 \cdot 7 \cdot 11 \cdots (4n-1)} \right]$

$= k_1 J_{1/4}(t) + k_2 J_{-1/4}(t)$

5. $x = c_1 t^{2/3} + c_2 t^{2/3} \ln t$ 7. $x = b_0 + b_1 \sum\limits_{n=0}^{\infty} \dfrac{2^n t^{2n+1}}{n!(2n+1)}$

9. $x = c_1 t^2 \left[1 + \sum\limits_{k=1}^{\infty} \dfrac{(-1)^k 3^k (k+1)! t^k}{8 \cdot 11 \cdot 14 \cdots (3k+5)} \right]$

$+ c_2 t^{1/3} \left[1 + \sum\limits_{k=1}^{\infty} \dfrac{1 \cdot 4 \cdot 7 \cdots (3k-2)t^k}{3^k k!} \right]$

11. $x = c_1 x_1(t) + c_2 \left[x_1(t) \ln t - 2t^{-1} + 2 - \dfrac{4}{9} t^2 + \cdots \right]$,

where $x_1(t) = t \sum\limits_{k=0}^{\infty} \dfrac{2t^k}{(k+2)!k!}$

13. $x = c_1(t^{-1} - 5) + c_2 \left[(t^{-1} - 5) \ln t + 16 - 5t - \dfrac{5}{3} t^2 - \dfrac{5}{12} t^3 - \dfrac{1}{20} t^4 \right]$

15. $x = b_0 \left(1 - \dfrac{1}{2} t^2 + \dfrac{1}{6} t^3 + \cdots \right) + b_1 \left(t - \dfrac{1}{2} t^2 - \dfrac{1}{6} t^3 + \cdots \right)$

17. $x = c_1 t^{-t} + c_2 \left[t^{-1} \ln t - t^{-2} + t^{-2} \sum\limits_{k=2}^{\infty} \dfrac{t^k}{k!(k-1)} \right]$

19. $x = 1 + 2t + \sum\limits_{n=1}^{\infty} \dfrac{1 \cdot 3 \cdot 5 \cdots (2n-3)}{2^n n!} t^{2n}$

21. a. $x = \sum\limits_{j=3}^{\infty} \dfrac{2t^j}{125j(1000)^{j-3}}$ b. 5.3737

23. $x = b_0 \left[1 - \dfrac{1}{2}(t-1)^2 + \dfrac{1}{3}(t-1)^3 - \dfrac{1}{8}(t-1)^4 + \cdots \right]$

$$+ b_1 \left[(t-1) - \dfrac{1}{2}(t-1)^2 + \dfrac{1}{8}(t-1)^4 + \cdots \right]$$

CHAPTER SEVEN

Section 7.1

1. a. $x' = .05x + \dfrac{1}{t+1}$ b. $x = \left(c + \displaystyle\int \dfrac{1}{t+1} e^{-.05t}\, dt \right) e^{.05t}$

3. $x' = .005 \left(10 + \dfrac{t}{10} - x \right) x$

5. a. $\theta'' + \theta' + 9.81 \sin\theta = 0$ b. $x' = y,\ \ y' = -y - 9.81x$

7. $R' = g_R R - \alpha RF,\ \ F' = -d_F F + \beta RF$

Section 7.2

1. $x(0.2) \approx .216$

The computer calculations for this chapter were done on a DECsystem-10™ computer,*
using programs written in BASIC. When comparing your answers to ours, keep in mind
that different devices may round differently.

3. a. $x(1) = 4$ b. $x(2) = 9$

n	X_n	Error	X_n	Error
10	3.94079	-5.92130×10^{-2}	8.80757	-1.92431×10^{-1}
20	3.97062	-2.93820×10^{-2}	8.97062	-2.93820×10^{-2}
50	3.9883	-1.17006×10^{-2}	8.9883	-1.17006×10^{-2}
100	3.99416	-5.84179×10^{-3}	8.99416	-5.84179×10^{-3}
200	3.99708	-2.91890×10^{-3}	8.99708	-2.91890×10^{-3}
500	3.99883	-1.16724×10^{-3}	8.99883	-1.16724×10^{-3}

* DECsystem-10 is a trademark of Digital Equipment Corporation.

c. $x(5) = 36$

n	x_n	Error
10	35.0264	-0.973562
20	35.5074	-0.492609
50	35.8017	-0.198294
100	35.9007	-9.93443×10^{-2}
200	35.9503	-4.97203×10^{-2}
500	35.9801	-1.98994×10^{-2}

5.

	a.	b.	c.
n	$x_n \approx x(1)$	$x_n \approx x(5)$	$x_n \approx x(10)$
10	11.2487	15.1515	20.3608
20	11.237	15.0454	20.1049
50	11.2301	14.9863	19.9711
100	11.2278	14.9675	19.9306
200	11.2267	14.9582	19.9111
500	11.226	14.9527	19.8997

7.

	a.	b.	c.
n	$x_n \approx x(1)$	$x_n \approx x(5)$	$x_n \approx x(10)$
10	10.0022	10.0528	10.199
20	10.0023	10.0553	10.2072
50	10.0024	10.0568	10.212
100	10.0024	10.0573	10.2136
200	10.0024	10.0575	10.2143
500	10.0024	10.0577	10.2148

9. a. 3.13699, 3.13928   b. 3.16153, 3.15158

Section 7.3

1. b. $y = 20e^{.05t} - 10$, $z = 30e^{-.05t} - 20$
3. d. $x = 10e^{.0005t}$
9. b. $y_i = 40(1 - .1h)^i + 20$, $z_i = 20(1 - .1h)^i + 40$
13. $|x_n - x(\tau)| < 5 \times 10^{-3}$

Section 7.4

1. $x(.2) \approx .23364$
3. a. $x(1) = 4$ b. $x(2) = 9$

n	x_n	Error	x_n	Error
10	4.00389	3.89129×10^{-3}	9.01886	1.88621×10^{-2}
20	4.00095	9.54092×10^{-4}	9.00457	4.56727×10^{-3}
50	4.00015	1.50979×10^{-4}	9.00072	7.18474×10^{-4}
100	4.00004	3.76701×10^{-5}	9.00018	1.78695×10^{-4}
200	4.00001	9.41753×10^{-6}	9.00004	4.44651×10^{-5}
500	4.	1.54972×10^{-6}	9.00001	7.27177×10^{-6}

c. $x(5) = 36$

n	x_n	Error
10	36.1339	0.13386
20	36.0317	3.16701×10^{-2}
50	36.0049	4.93574×10^{-3}
100	36.0012	1.22452×10^{-3}
200	36.0003	3.05176×10^{-4}
500	36.	4.91142×10^{-5}

5.

n	a. $x_n \approx x(1)$	b. $x_n \approx x(5)$	c. $x_n \approx x(10)$
10	11.2262	14.975	20.0184
20	11.2257	14.9556	19.9257
50	11.2255	14.9501	19.8977
100	11.2255	14.9493	19.8936
200	11.2255	14.9491	19.8926
500	11.2255	14.949	19.8923

7.

n	a. $x_n \approx x(1)$	b. $x_n \approx x(5)$	c. $x_n \approx x(10)$
10	10.0025	10.0578	10.2154
20	10.0025	10.0578	10.2152
50	10.0025	10.0578	10.2151
100	10.0025	10.0578	10.2151
200	10.0025	10.0578	10.2151
500	10.0025	10.0578	10.2151

9. a. 3.14163, 3.14160   b. 3.14153, 3.14158

Section 7.5

1. $x(.02) \approx .230184036$

3. a. $x(1) = 4$ b. $x(2) = 9$

n	x_n	Error		x_n	Error
10	4.	3.03984×10^{-6}		9.00004	3.58820×10^{-5}
20	4.	1.78814×10^{-7}		9.	2.14577×10^{-6}
50	4.	0		9.	0
100	4.	-2.98023×10^{-8}		9.	1.19209×10^{-7}
200	4.	0		9.	0
500	4.	0		9.	0

c. $x(5) = 36$

n	x_n	Error
10	36.0006	6.35624×10^{-4}
20	36.	3.57628×10^{-5}
50	36.	9.53674×10^{-7}
100	36.	0
200	36.	0
500	36.	0

5.

	a.	b.	c.
n	$x_n \approx x(1)$	$x_n \approx x(5)$	$x_n \approx x(10)$
10	11.2255	14.9492	19.8947
20	11.2255	14.949	19.8924
50	11.2255	14.949	19.8923
100	11.2255	14.949	19.8922
200	11.2255	14.949	19.8922
500	11.2255	14.949	19.8922

7.

	a.	b.	c.
n	$x_n \approx x(1)$	$x_n \approx x(5)$	$x_n \approx x(10)$
10	10.0025	10.0578	10.2151
20	10.0025	10.0578	10.2151
50	10.0025	10.0578	10.2151
100	10.0025	10.0578	10.2151
200	10.0025	10.0578	10.2151
500	10.0025	10.0578	10.2151

9. a. 3.14159, 3.13159   b.   3.14159, 3.14159

Section 7.6

1.   In Example 2 of Section 7.5, the absolute value of the error increases between $n = 200$ and $n = 500$; in Exercises 4(b) and 4(c), it goes up between $n = 100$ [$n = 50$ for 4(b)] and $n = 500$. In all of these cases, the error affects only the last significant digit kept by the machine, and the printed values are not affected.

3.

τ	$x_1(τ)$	$x_2(τ)$	$x_1(τ) - x_2(τ)$
1	8103.08	8103.08	-1.22070×10^{-4}
2	65659969	65659969	0

5. b.

n	x_n (Mod. Euler)	Error	x_n (Runge-Kutta)	Error
50	5.81551	-26.1845	25.3321	-6.66791
100	8.73132	-23.2687	31.29	-0.71003
200	13.3001	-18.6999	31.9748	-2.51863×10^{-2}
500	21.8387	-10.1613	32.0156	1.56212×10^{-2}
1000	27.716	-4.284	32.0254	2.54459×10^{-2}

Section 7.7

1. $x(.2) \approx 1.22273841, \quad y(.2) \approx 1.24146931$

3. a.

n	$x_n \approx x(1)$	$y_n \approx y(1)$
10	-9.99888×10^{-2}	-3.05139×10^{-3}
50	-9.99955×10^{-2}	-2.97578×10^{-3}
100	-9.99955×10^{-2}	-2.97566×10^{-3}
500	-9.99955×10^{-2}	-2.97566×10^{-3}

b.

n	$x_n \approx x(5)$	$y_n \approx y(5)$
10	-4.48801×10^{-2}	-3.78583×10^{-2}
50	-9.98490×10^{-2}	-1.52472×10^{-2}
100	-9.98858×10^{-2}	-1.48971×10^{-2}
500	-9.98872×10^{-2}	-1.48730×10^{-2}

c.

n	$x_n \approx x(10)$	$y_n \approx y(10)$
10	-84.8936	154.591
50	9.72004×10^{-2}	3.97149×10^{-2}
100	9.94613×10^{-2}	3.04484×10^{-2}
500	9.95490×10^{-2}	2.97135×10^{-2}

5.

a.

n	$x_n \approx x(1)$	$y_n \approx y(1)$
10	12.9775	25.0739
50	12.9772	25.075
100	12.9772	25.075
500	12.9772	25.075

b.

$x_n \approx x(5)$	$y_n \approx y(5)$
56.5590	2.91839
56.5328	3.05001
56.5328	3.05006
56.5328	3.05006

c.

n	$x_n \approx x(10)$	$y_n \approx y(10)$
10	-1.98272×10^8	7.93089×10^8
50	64.1692	0.769223
100	64.1692	0.769380
500	64.1692	0.769386

7. a.

n	$x_n \approx x(10)$	$y_n \approx y(10)$
10	14807.2	1574.57
50	14807.2	1574.57
100	14807.2	1574.57
500	14807.2	1574.57

b.

$x_n \approx x(25)$	$y_n \approx y(25)$
9091.56	3580.01
9091.21	3580.09
9091.21	3580.09
9091.21	3580.09

c.

n	$x_n \approx x(50)$	$y_n \approx y(50)$
10	3670.93	1591.86
50	3670.76	1590.71
100	3670.76	1590.71
500	3670.76	1590.70

d.

$x_n \approx x(75)$	$y_n \approx y(75)$
10926.0	1050.24
10971.7	1048.22
10971.8	1048.22
10971.8	1048.22

Review Problems

1. a. (i)

n	x_n (Euler)	x_n (Mod. Euler)	x_n (Runge-Kutta)
10	1.10029	1.10034	1.10033
20	1.10031	1.10033	1.10033
50	1.10032	1.10033	1.10033
100	1.10033	1.10033	1.10033
200	1.10033	1.10033	1.10033
500	1.10033	1.10033	1.10033

(ii) Yes, 1.100

b. (i)

n	x_n (Euler)	x_n (Mod. Euler)	x_n (Runge-Kutta)
10	1.2023	1.20271	1.2027
20	1.2025	1.2027	1.2027
50	1.20262	1.2027	1.2027
100	1.20266	1.2027	1.2027
200	1.20268	1.2027	1.2027
500	1.20269	1.2027	1.2027

(ii) Yes, 1.203

c. (i)

n	x_n (Euler)	x_n (Mod. Euler)	x_n (Runge-Kutta)
10	2.38126	2.46717	2.46265
20	2.42083	2.46378	2.46265
50	2.44565	2.46283	2.46265
100	2.45411	2.4627	2.46265
200	2.45837	2.46266	2.46265
500	2.46094	2.46265	2.46265

(ii) Yes, 2.463

3. a. (i)

n	x_n (Euler)	x_n (Mod. Euler)	x_n (Runge-Kutta)
10	2.65461	2.88993	2.88591
20	2.76252	2.88709	2.88591
50	2.83435	2.88612	2.88591
100	2.85973	2.88596	2.88591
200	2.87271	2.88592	2.88591
500	2.88061	2.88591	2.88591

(ii)   Yes, 2.886

b. (i)

n	x_n (Euler)	x_n (Mod. Euler)	x_n (Runge-Kutta)
10	20.9795	49.7586	55.6417
20	31.3412	53.9254	55.6581
50	43.1804	55.1278	55.7131
100	48.7608	55.5673	55.6656
200	52.0324	55.6399	55.6669
500	54.1641	55.6625	55.6669

(ii)   Yes, 55.67

c. (i)

n	x_n (Euler)	x_n (Mod. Euler)	x_n (Runge-Kutta)
10	279.774	3103.57	7656.1
20	945.723	5702.32	8203.79
50	2902.87	7669.43	8256.53
100	4683.39	8088.72	8261.86
200	6149.61	8213.07	8260.1
500	7317.93	8254.98	8262.49

(ii)   No, but further trials of Runge-Kutta might stabilize.

5. a. (i)

n	x_n (Euler)	x_n (Mod. Euler)	x_n (Runge-Kutta)
10	2.31909	2.36398	2.36383
20	2.34115	2.36387	2.36383
50	2.35468	2.36383	2.36383
100	2.35924	2.36383	2.36383
200	2.36153	2.36383	2.36383
500	2.36291	2.36383	2.36383

(ii)   Yes, 2.364

b. (i)

n	x_n (Euler)	x_n (Mod. Euler)	x_n (Runge-Kutta)
10	21.2781	25.1752	25.3188
20	23.1948	25.2817	25.3191
50	24.4432	25.313	25.3191
100	24.8766	25.3175	25.3191
200	25.0967	25.3187	25.3191
500	25.2299	25.319	25.3191

(ii)   Yes, 25.32

c. (i)

n	x_n (Euler)	x_n (Mod. Euler)	x_n (Runge-Kutta)
10	105.321	138.217	140.516
20	121.74	139.908	140.528
50	132.714	140.425	140.529
100	136.571	140.503	140.529
200	138.537	140.522	140.529
500	139.729	140.528	140.529

(ii)   Yes, 140.5

7. a. (i)

n	x_n (Euler)	x_n (Mod. Euler)	x_n (Runge-Kutta)
10	3.22745	4.34479	4.55509
20	3.66392	4.49184	4.55655
50	4.10164	4.54512	4.55666
100	4.30535	4.55368	4.55666
200	4.42393	4.55591	4.55666
500	4.50169	4.55654	4.55666

(ii)   Yes, 4.557

b. (i)

n	x_n (Euler)	x_n (Mod. Euler)	x_n (Runge-Kutta)
10	3.88258	6.45187	7.39469
20	4.75192	7.06855	7.41487
50	5.85097	7.34813	7.41669
100	6.48072	7.39855	7.41674
200	6.8964	7.41208	7.41674
500	7.19332	7.41599	7.41675

(ii)   Yes, 7.417

c. (i)

n	x_n (Euler)	x_n (Mod. Euler)	x_n (Runge-Kutta)
10	5.43671	19.2218	84.5204
20	8.27689	39.8123	224.102
50	15.7995	124.142	1690.13
100	27.5959	404.406	75723.5
200	51.9942	3711.95	4.96903×10^{12}
500	149.843	3.33856×10^{12}	overflow

(ii)   No, No!

9. a. (i)

n	x_n	y_n
10	1.87445	2.48707
50	1.87448	2.48714
100	1.87449	2.48714

(ii) Yes, $x \approx 1.874$, $y \approx 2.487$

b. (i)

	x_n	y_n
	1.92625	3.78429
	1.92628	3.78432
	1.92628	3.78432

(ii) Yes, $x \approx 1.926$, $y \approx 3.784$

11. a. (i)

n	x_n	y_n
10	5.03573	4.41120
50	5.03645	4.41165
100	5.03645	4.41165

b. (i)

x_n	y_n
0	4.43656
0	4.43656
0	4.43656

(ii) Yes, $x \approx 5.036$, $y \approx 4.412$

(ii) Yes, $x \approx 0$,
$y \approx 4.437$

13.

n	x_n	Error
50	9071.80	7.32422×10^{-4}
100	9071.79	-3.66211×10^{-4}

15.

n	x_n	Error
50	0.180335	3.72529×10^{-9}
100	0.180335	1.67638×10^{-8}

CHAPTER EIGHT

Section 8.1

1. a. $u(0, t) = u_x(\ell, t) = 0$ b. $u(0, t) = 0$, $u(\ell, t) = T_\ell$
3. $u(1, \theta, t) = f(\theta)$, $0 < \theta < \pi$
$u(r, 0, t) = u(r, \pi, t) = 0$, $0 \le r < 1$
5. b. $p(x) \dfrac{\partial^2 u}{\partial x^2} + p'(x) \dfrac{\partial u}{\partial x} = r(x) \dfrac{\partial u}{\partial t}$

Section 8.2

1. $u(x, t) = e^{-9\pi^2 \alpha t} \sin 3\pi x + e^{-25\pi^2 \alpha t} \sin 5\pi x$
3. $u(x, t) = e^{-4\pi^2 \alpha t} \sin 2\pi x - 3e^{-16\pi^2 \alpha t} \sin 4\pi x$
5. $u(x, t) = 7 - e^{-9\pi^2 \alpha t} \cos 3\pi x$
7. $u(x, t) = e^{-9\pi^2 \alpha t} \sin 3\pi x + e^{-25\pi^2 \alpha t} \sin 5\pi x$
9. b. $u(x, t) = e^{-25\pi^2 \alpha t / 4\ell^2} \sin \dfrac{5\pi x}{2\ell} - e^{-\pi^2 \alpha t / 4\ell^2} \sin \dfrac{\pi x}{2\ell}$
11. a. $X'' + \lambda X = 0$, $Y'' - \lambda Y = 0$
c. $u(x, y) = \left(\sin \dfrac{\pi x}{\ell} \right) \left(\dfrac{e^{\pi y/\ell} - e^{-\pi y/\ell}}{e^{\pi h/\ell} - e^{-\pi h/\ell}} \right) - 2 \left(\sin \dfrac{5\pi x}{\ell} \right) \left(\dfrac{e^{5\pi y/\ell} - e^{-5\pi y/\ell}}{e^{5\pi h/\ell} - e^{-5\pi h/\ell}} \right)$
13. b. $u_P(x) = \dfrac{G}{2\beta} x(\ell - x)$ c. $u(x,t) = \dfrac{G}{2\beta} x(\ell - x) + e^{-\pi^2 \alpha t / \ell^2} \sin \dfrac{\pi x}{\ell}$
15. $u(x, t) = \dfrac{10x}{\ell} - 2e^{-9\pi^2 \alpha t / \ell^2} \sin \dfrac{3\pi x}{\ell}$

Section 8.3

1. a. Odd   b. Neither   c. Even   d. Neither   e. Even
   f. Odd   g. Neither   h. Even   i. Odd

3. a. $\dfrac{2\ell}{\pi} \displaystyle\sum_{k=1}^{\infty} \dfrac{(-1)^{k+1}}{k} \sin \dfrac{k\pi x}{\ell}$ b. $f(x)$

5. a. $1 - \dfrac{2}{\pi} \displaystyle\sum_{n=1}^{\infty} \dfrac{1}{n} \sin n\pi x$ b. $f(x)$ if $x \neq 0$,   1 if $x = 0$

7. a. $2 \displaystyle\sum_{k=1}^{\infty} \dfrac{(-1)^{k+1}(k^2\pi^2 - 6)}{k^3} \sin kx$ b. $f(x)$

9. a. $(e + e^{-1}) \left[\dfrac{1}{2} + \displaystyle\sum_{k=1}^{\infty} \dfrac{(-1)^k}{(1 + k^2\pi^2)} (\cos k\pi x - k\pi \sin k\pi x) \right]$ b. $f(x)$

11. $u(x, t) = \dfrac{2}{3} - \dfrac{4}{\pi^2} \displaystyle\sum_{k=1}^{\infty} \dfrac{(-1)^k}{k^2} e^{-k^2\pi^2\alpha t} \cos k\pi x$

Section 8.4

1. a. 1   b. $\dfrac{4}{\pi} \displaystyle\sum_{n=0}^{\infty} \dfrac{1}{(2n + 1)} \sin \dfrac{(2n + 1)\pi x}{\ell}$ c. $f(x)$

3. a. $\dfrac{\ell}{2} - \dfrac{4\ell}{\pi^2} \displaystyle\sum_{n=0}^{\infty} \dfrac{1}{(2n + 1)^2} \cos \dfrac{(2n + 1)\pi x}{\ell}$

   b. $\dfrac{2\ell}{\pi} \displaystyle\sum_{k=1}^{\infty} \dfrac{(-1)^{k+1}}{k} \sin \dfrac{k\pi x}{\ell}$ c. $f(x)$

5. a. $\dfrac{1}{2} - \dfrac{4}{\pi^2} \displaystyle\sum_{m=0}^{\infty} \dfrac{1}{(2m + 1)^2} \cos(2m + 1)\pi x$

   b. $\dfrac{8}{\pi^2} \displaystyle\sum_{n=0}^{\infty} \dfrac{(-1)^n}{(2n + 1)^2} \sin \dfrac{(2n + 1)\pi x}{2}$ c. $f(x)$

7. a. $\cos x$ b. $\dfrac{8}{\pi} \displaystyle\sum_{n=1}^{\infty} \dfrac{n}{(4n^2 - 1)} \sin 2nx$ c. $f(x)$

9. a. $u(x, t) = 1 - \dfrac{8}{\pi^2} \displaystyle\sum_{n=0}^{\infty} \dfrac{1}{(2n + 1)^2} e^{-(2n+1)^2\pi^2\alpha t/4} \cos \dfrac{(2n + 1)\pi x}{2}$

   b. $u(x, t) = \dfrac{4}{\pi} \displaystyle\sum_{k=1}^{\infty} \dfrac{(-1)^{k+1}}{k} e^{-k^2\pi^2\alpha t/4} \sin \dfrac{k\pi x}{2}$

11. a. $u(x, t) = \dfrac{1}{2} + \dfrac{2}{\pi} \displaystyle\sum_{n=0}^{\infty} \dfrac{(-1)^{n+1}}{(2n + 1)} e^{-(2n+1)^2\pi^2\alpha t/4} \cos \dfrac{(2n + 1)\pi x}{2}$

   b. $u(x, t) = \dfrac{2}{\pi} \displaystyle\sum_{n=0}^{\infty} \dfrac{1}{(2n + 1)} e^{-(2n+1)^2\pi^2\alpha t/4} \sin \dfrac{(2n + 1)\pi x}{2}$

   $\qquad - \dfrac{2}{\pi} \displaystyle\sum_{m=0}^{\infty} \dfrac{1}{(2m + 1)} e^{-(2m+1)^2\pi^2\alpha t} \sin(2m + 1)\pi x$

Section 8.5

1. $u(x, t) = (\sin \pi x)\left(3 \cos \alpha \pi t + \dfrac{7}{\alpha \pi} \sin \alpha \pi t\right) + \dfrac{5}{3\alpha \pi} \sin 3\pi x \sin 3\alpha \pi t$

3. $u(x, t) = \dfrac{8}{\pi^2} \displaystyle\sum_{n=0}^{\infty} \dfrac{(-1)^n}{(2n + 1)^2} \sin \dfrac{(2n + 1)\pi x}{2} \cos \dfrac{\alpha(2n + 1)\pi t}{2}$

5. $u(x, t) = \dfrac{8}{\pi^3} \displaystyle\sum_{n=0}^{\infty} \dfrac{1}{(2n + 1)^3} \sin(2n + 1)\pi x \cos \alpha(2n + 1)\pi t$

    $\qquad + \dfrac{2}{\alpha \pi^2} \displaystyle\sum_{k=1}^{\infty} \dfrac{1}{k^2} \sin k\pi x \sin \alpha k\pi t$

7. b.   (i) $u(x, t) = \sin \dfrac{\pi x}{2} \cos \dfrac{\alpha \pi t}{2} - \sin \dfrac{3\pi t}{2} \cos \dfrac{3\alpha \pi t}{2} + \dfrac{2}{3\alpha \pi} \sin \dfrac{3\pi x}{2} \sin \dfrac{3\alpha \pi t}{2}$

    (ii) $u(x, t) = \dfrac{8}{\alpha \pi^2} \displaystyle\sum_{n=0}^{\infty} \dfrac{1}{(2n + 1)^2} \sin \dfrac{(2n + 1)\pi x}{2} \sin \dfrac{\alpha(2n + 1)\pi t}{2}$

9. $u(x, t) = x + \dfrac{1}{\alpha \pi} \sin \pi x \sin \alpha \pi t$

Section 8.6

1. $u(x, y) = \dfrac{4}{\pi} \displaystyle\sum_{n=0}^{\infty} \dfrac{1}{(2n + 1)} [\sin(2n + 1)\pi x] \left[\dfrac{e^{(2n+1)\pi y} - e^{-(2n+1)\pi y}}{e^{(2n+1)\pi} - e^{-(2n+1)\pi}}\right]$

3. $u(x, y) = \dfrac{2}{\pi} \displaystyle\sum_{k=1}^{\infty} \dfrac{1}{k} [\sin k\pi x] \left[\dfrac{e^{k\pi y} - e^{k\pi(2 - y)}}{1 - e^{2k\pi}}\right]$

5. $u(x, y) = u_1 + u_2$, where u_i is the solution to problem i, $i = 1, 2$

7. $u(r, \theta) = \dfrac{1}{2} - \dfrac{1}{2} r^2 \cos 2\theta$

9. $u(r, \theta) = \displaystyle\sum_{k=1}^{\infty} c_k r^k \sin k\theta$, where $c_k = \dfrac{2}{\pi} \displaystyle\int_0^{\pi} f(\theta) \sin k(\theta) \, d\theta$

Section 8.7

1. $u(x, y, t) = (e^{-\pi^2 t} + e^{-2\pi^2 t} \cos \pi y) \sin \pi x + 2(e^{-9\pi^2 t} - e^{-10\pi^2 t} \cos \pi y) \sin 3\pi x$

3. $u(x, y, t) = \displaystyle\sum_{k=1}^{\infty} \left\{\left[\dfrac{2}{\pi} - \dfrac{16}{\pi^3} \displaystyle\sum_{j=0}^{\infty} \dfrac{1}{(2j + 1)^2} e^{(2j+1)^2 \pi^2 t/4} \cos \dfrac{(2j + 1)\pi y}{2}\right]\right.$

    $\qquad\qquad\qquad\qquad\qquad \times \left.\left[\dfrac{(-1)^{k+1}}{k} \sin k\pi x\right]\right\}$

5. $u(\rho, \phi) = \dfrac{4}{3} + \rho^2\left(\cos^2 \phi - \dfrac{1}{3}\right)$

7. $u(r, t) = 3J_0(r\beta_1) \cos \alpha \beta_1 t + \dfrac{1}{\alpha \beta_2} J_0(r\beta_2) \sin \alpha \beta_2 t + J_0(r\beta_3) \cos \alpha \beta_3 t$

9. c. $u(x, y, z, t) = (e^{-5\alpha t} - e^{-14\alpha t} \cos 3x) \cos 2y \cos z_3$

Review Problems

1. a. $\ell + \dfrac{2\ell}{\pi} \displaystyle\sum_{k=1}^{\infty} \dfrac{(-1)^k}{k} \sin \dfrac{k\pi x}{\ell}$

   b. $\dfrac{\ell}{2} + \dfrac{4\ell}{\pi^2} \displaystyle\sum_{n=0}^{\infty} \dfrac{1}{(2n+1)^2} \cos \dfrac{(2n+1)\pi x}{\ell}$

   c. $\dfrac{2\ell}{\pi} \displaystyle\sum_{k=1}^{\infty} \dfrac{1}{k} \sin \dfrac{k\pi x}{\ell}$

3. a. and b. $\dfrac{2\ell^2}{3} + \dfrac{4\ell^2}{\pi^2} \displaystyle\sum_{k=1}^{\infty} \dfrac{(-1)^{k+1}}{k^2} \cos \dfrac{k\pi x}{\ell}$

   c. $\ell^2 \displaystyle\sum_{k=1}^{\infty} \left[\dfrac{2}{k\pi} - \dfrac{4\{(-1)^k - 1\}}{k^3\pi^3} \right] \sin \dfrac{k\pi x}{\ell}$

5. $u(x, t) = \dfrac{2\ell}{\pi} \displaystyle\sum_{k=1}^{\infty} \dfrac{1}{k} e^{-k^2\pi^2\alpha t/\ell^2} \sin \dfrac{k\pi x}{\ell}$

7. $u(x, t) = \dfrac{\ell}{2} + \dfrac{4\ell}{\pi^2} \displaystyle\sum_{n=0}^{\infty} \dfrac{1}{(2n+1)^2} e^{-(2n+1)^2\pi^2\alpha t/\ell^2} \cos \dfrac{(2n+1)\pi x}{\ell}$

9. $u(x, t) = \ell + \dfrac{2}{\ell} \displaystyle\sum_{k=1}^{\infty} \dfrac{(-1)^k}{k} e^{-k^2\pi^2\alpha t/\ell^2} \sin \dfrac{k\pi x}{\ell}$

11. $u(x, t) = \dfrac{\ell}{2\alpha\pi} \sin \dfrac{2\alpha\pi}{\ell} \sin \dfrac{2\alpha\pi t}{\ell} + \dfrac{8\ell^2}{\pi^3} \displaystyle\sum_{n=0}^{\infty} \dfrac{1}{(2n+1)^3} \sin \dfrac{(2n+1)\pi x}{\ell} \cos \dfrac{\alpha(2n+1)\pi t}{\ell}$

13. $u(x, y) = \dfrac{2\ell}{\pi} \displaystyle\sum_{k=1}^{\infty} \dfrac{1}{k} \left[\sin \dfrac{k\pi x}{\ell} \right] \left[\dfrac{e^{k\pi y/\ell} - e^{-k\pi y/\ell}}{e^{k\pi} - e^{-k\pi}} \right]$

15. $u(x, y) = u_{13} + u_{14}$, where u_i is the solution to problem i, $i = 13, 14$

17. a. $u(x, t) = c_n e^{-\mu_n^2\alpha t} \sin \pi_n x$ b. $u(x, t) = \displaystyle\sum_{n=1}^{\infty} c_n e^{-\pi_n^2\alpha t} \sin \pi_n x$

INDEX